LABORATORY MANUAL

Human Anatomy

Fifth Edition

ERIC WISE

Santa Barbara City College

HUMAN ANATOMY LABORATORY MANUAL, FIFTH EDITION

Published by McGraw-Hill Education, 2 Penn Plaza, New York, NY 10121.

Some ancillaries, including electronic and print components, may not be available to customers outside the United States.

This book is printed on acid-free paper.

2 3 4 5 LMN 19 18 17 16

ISBN 978-1-259-68383-1
MHID 1-259-68383-4

Senior Vice President, Products & Markets: *Kurt L. Strand*
Vice President, General Manager, Products & Markets: *Marty Lange*
Vice President, Content Design & Delivery: *Kimberly Meriwether David*
Managing Director: *Michael S. Hackett*
Brand Manager: *Chloe Bouxsein*
Director, Product Development: *Rose Koos*
Product Developer: *Donna Nemmers*
Marketing Managers: *Jessica Cannavo / James F. Connely*
Director of Digital Content: *Michael G. Koot, PhD*
Digital Product Analyst: *John J. Theobald*
Director, Content Design & Delivery: *Linda Avenarius*
Program Manager: *Angela R. FitzPatrick*
Content Project Managers: *Vicki Krug / Brent dela Cruz*
Buyer: *Laura M. Fuller*
Design: *David Hash*
Content Licensing Specialists: *Lori Hancock / Lorraine Buczek*
Cover Image: *Human head, x-ray, © Pasieka/Science Photo Library/Getty Images RF; White matter fibers of the human brain, © Pasieka/Science Photo Library/Getty Images; Pyramidal neurons of the cerebral cortex impregnated with the Golgi method, © Jose Luis Calvo/Shutterstock.*
Compositor: *MPS Limited*
Printer: *LSC Communications*

Some of the laboratory experiments included in this manual may be hazardous if materials are handled improperly or if procedures are conducted incorrectly. Safety precautions are necessary when you are working with chemicals, glass test tubes, hot water baths, sharp instruments, and the like, or for any procedures that generally require caution. Consult page v of the front matter for Laboratory Safety Guidelines. Your school may have set regulations regarding safety procedures that your instructor will explain to you. Should you have any problems with materials or procedures, please ask your instructor for help.

The Internet addresses listed in the text were accurate at the time of publication. The inclusion of a website does not indicate an endorsement by the authors or McGraw-Hill Education, and McGraw-Hill Education does not guarantee the accuracy of the information presented at these sites.

mheducation.com/highered

Contents

Correlation Guide

of Exercises to Saladin: *Human Anatomy*, Fifth Edition

This lab manual can be used independently or with Saladin's *Human Anatomy* text. Below is a correlation guide listing the chapters in Saladin's *Human Anatomy*, 5th edition text that correspond to the exercises in this lab manual.

Wise Exercises	Saladin Chapters
1. Organs, Systems, and Organization of the Body	1. The Study of Human Anatomy
2. Microscopy	2. Cytology—The Study of Cells
3. Cell Structure	2. Cytology—The Study of Cells
4. Tissues	3. Histology—The Study of Tissues
5. The Integumentary System	5. The Integumentary System
6. Introduction to the Skeletal System	6. The Skeletal System I: Bone Tissue
7. Axial Skeleton 1: Skull	7. The Skeletal System II: Axial Skeleton
8. Axial Skeleton 2: Vertebrae, Ribs, Sternum	7. The Skeletal System II: Axial Skeleton
9. Appendicular Skeleton	8. The Skeletal System III: Appendicular Skeleton
10. Joints	9. The Skeletal System IV: Joints
11. Axial Muscles 1: Muscles of the Head and Neck	10. The Muscular System I: Introduction 11. The Muscular System II: Axial Musculature
12. Axial Muscles 2: Muscles of the Trunk	11. The Muscular System II: Axial Musculature
13. Appendicular Muscles 1: Muscles of the Shoulder and Upper Limb	12. The Muscular System III: Appendicular Musculature
14. Appendicular Muscles 2: Muscles of the Hip, Thigh, Leg, and Foot	12. The Muscular System III: Appendicular Musculature
15. Introduction to the Nervous System	13. The Nervous System I: Nervous Tissue
16. Spinal Cord and Spinal Nerves	14. The Nervous System II: Spinal Cord and Spinal Nerves 16. The Nervous System IV: Autonomic Nervous System and Visceral Reflexes
17. Brain and Cranial Nerves	15. The Nervous System III: Brain and Cranial Nerves
18. Sensory Receptors	17. The Nervous System V: Sense Organs
19. The Endocrine System	18. The Endocrine System
20. Blood Cells	19. The Circulatory System I: Blood
21. The Heart	20. The Circulatory System II: The Heart
22. Introduction to Blood Vessels and Blood Vessels 1: Blood Vessels of the Axial Region	21. The Circulatory System III: Blood Vessels
23. Blood Vessels 2: Blood Vessels of the Appendicular Region	21. The Circulatory System III: Blood Vessels
24. The Lymphatic System	22. The Lymphatic System and Immunity
25. The Respiratory System	23. The Respiratory System
26. The Digestive System	24. The Digestive System
27. The Urinary System	25. The Urinary System
28. The Male Reproductive System	26. The Reproductive System
29. The Female Reproductive System and Development	26. The Reproductive System

Laboratory Safety Guidelines

The following is a brief list of safety guidelines for you to follow in the anatomy laboratory. More complete descriptions of safety procedures are found throughout the manual.

1. Read all of the laboratory material prior to coming to class. This is a safety issue. Failure to read or understand the laboratory material can result in hazards. Unauthorized experiments are not allowed in the laboratory.
2. Locate the first-aid kit, eyewash station, shower station, fire blanket, fire extinguisher, and other safety areas in the laboratory prior to beginning the first laboratory class. Be familiar with how to use the equipment in the event of an emergency.
3. Clean up spills. Inform your instructor of any spill in the laboratory. Be careful if the material is toxic or caustic. If you are not sure if the material is hazardous, ask your instructor for the proper procedure for cleanup.
4. Assume all bodily fluids in the laboratory are infectious. Follow precautions when handling bodily fluids, such as wearing protective gloves, laboratory coats, and protective eye wear. Never use any instrument twice that comes into contact with bodily fluid. Once the instrument is used, either dispose of it in a biohazard bag or clean it in a container of 10% bleach or other disinfecting solution. Clean all laboratory surfaces with a bleach or other disinfecting solution at the end of a laboratory involving bodily fluids, even if you think no fluid has come in contact with the table surface.
5. Keep the laboratory clean and free of clutter. Place all backpacks, purses, and umbrellas in safe areas and not on laboratory tables.
6. Do not eat, smoke, or chew gum in the laboratory. Many reagents in the laboratory are toxic, so do not drink them. Never pipette anything by mouth. Use a pipette bulb or pipette pump when pipetting.
7. Keep your hair secured so it does not catch fire or dip into beakers containing solutions. Never heat volatile material over an open flame—an explosion might occur.
8. Do not wear contact lenses in the laboratory. Notify your instructor if you wear contact lenses.
9. Do not throw sharp material such as glass or cutting blades in the normal trash containers in the laboratory. They are to be disposed of in an appropriate container such as a "sharps" container. Report any glassware breakage to your instructor, and dispose of it in the appropriate container.
10. Never point a test tube that is heating over a Bunsen burner in the direction of someone else. Never walk away from anything that is being heated. Pay attention to material on hot plates and remove material with appropriate mitts or tongs. Heat material only in appropriate heat-resistant containers. Turn off and unplug hot plates immediately after use.
11. Dissect with the blade cutting away from you and your laboratory partners. If you do cut yourself, make sure you wash the wound well with soap and water and notify your instructor.
12. If you have an allergic reaction to the preserving fluid (usually restricted breathing, a flushed feeling, or a skin rash), notify your instructor immediately. Notify your instructor if you are pregnant or have any medical condition.
13. Do not apply cosmetics in the laboratory.
14. Wear closed-toed shoes in laboratory, not sandals.
15. Wash your hands after laboratory classes, especially before eating or going to the restroom.

Instructor Preface

Anatomy is usually the first of several classes in the allied health field; as such, it represents the course that frequently determines whether a student will enter the field or look for another career. As an instructor of anatomy, I decided to write a student-friendly laboratory manual for the undergraduate student of anatomy, consisting of 29 exercises designed to help students learn basic human anatomy.

The diversity of interests in today's anatomy students is due, in part, to the number of majors that either require or recommend the subject. This lab manual provides a framework for understanding anatomy for students interested in nursing, radiology, physical or occupational therapy, physical education, dental hygiene, or other allied health careers.

This lab manual can be used with Saladin's *Human Anatomy* text, or it can be used independently. The illustrations are labeled; therefore, students do not need to bring their lecture texts to the lab. The lab manual can be used in either a one-term or a full-year course. The illustrations are outstanding, and the balanced combination of line art and photographs provides effective coverage of the material. The amount of lecture material in the manual is limited, so there is little material included that is not part of the laboratory experience.

Practical lab experience is an invaluable opportunity to reinforce lecture concepts, enrich students' understanding of anatomy, and allow them to explore new dimensions in the subject area. The educational benefit of reinforcing lecture material with hands-on experiences and acquiring knowledge with a learn-by-doing philosophy makes the anatomy laboratory a very important educational environment. Many of us use laboratory experiences to provide those students who have different learning styles another avenue for learning.

The 29 exercises in this lab manual provide a comprehensive overview of the human body. Each exercise presents the core elements of the subject matter. This manual may be tailored to match your own vision of the course, or it may be used in its entirety. The labs generally take between 2 and 3 hours to complete.

This lab manual was written for three types of anatomy courses. For courses that use the cat as the primary dissection animal, cat dissections appear at the end of the lab manual, and mammalian organ dissections follow the material on human anatomy. For courses that use models or charts, numerous cadaver photographs are included, allowing students to see the representative structures as they exist in the cadaver. Finally, for courses that use cadavers, this manual can be used by studying the human material and omitting the cat dissection sections.

New to the Fifth Edition

The nomenclature for the 5th edition has been updated to match the terminology in the 5th edition of Saladin's *Human Anatomy* text. The organization of the lab manual has also been revised to match the sequence in Saladin's *Human Anatomy* text. Numerous photos and illustrations in the lab manual have been improved and updated. In the muscle section, individual muscles have been grouped according to the part of the body they act on, following the same organization used in the Saladin: *Human Anatomy* text. The list below outlines the most impactful changes in the fifth edition of this lab manual.

- Exercise 1—Cardiovascular System was renamed to Circulatory System. The body cavities section has been rewritten.
- Exercise 2—The sections on how to use a microscope are reorganized for better flow. The description of depth of field is expanded.
- Exercise 3—The functions of some of the organelles have been updated to reflect our current understanding.
- Exercise 6—The discussion on bone shapes has been removed.
- Exercise 7—The Skull is moved to follow Exercise 6, to match the sequence of Saladin: *Human Anatomy*. Some of the illustrations are upgraded, and the hyoid is now included in this exercise.
- Exercise 9—The Appendicular Skeleton is moved to follow Exercise 8, to match the sequence of Saladin: *Human Anatomy*. The *pelvic girdle* was renamed to the *hip bones*, and the section was revised.
- Exercise 10—Articulations has been renamed Joints, and several figures are updated.
- Exercise 11—Axial musculature (head and neck) is moved to follow Joints, and the ranking of the muscles presented is changed to match the sequence in Saladin: *Human Anatomy*. Attachment points have been used synonymously with origins and insertions, to align with this same change in Saladin: *Human Anatomy*.
- Exercise 12—Muscles of the Trunk is moved to match the sequence of Saladin: *Human Anatomy*. Origins and

Insertions have been renamed as attachment points. Muscles in this exercise have been grouped to follow the format of Saladin: *Human Anatomy*.

- Exercise 13—The first part of the appendicular muscles (muscles of the upper limb) is reordered to coordinate with the sequence of Saladin: *Human Anatomy*. Attachment points have been used synonymously with origins and insertions. The presentation of the order of the muscles also aligns with the sequence of presentation in Saladin: *Human Anatomy*.
- Exercise 14—The second part of the appendicular muscles (muscles of the hip and lower limb) aligns with the topic sequence in Saladin: *Human Anatomy*.
- Exercise 16—Spinal Cord and Spinal Nerves is moved to match the sequence of Saladin: *Human Anatomy*.
- Exercise 17—Brain and Cranial Nerves follows the Spinal Cord and Spinal Nerves exercise to coordinate with the topic sequence of Saladin: *Human Anatomy*, and there is minor rewriting of the material in the exercise.
- Exercise 22—Introduction to Blood Vessels and Blood Vessels 1: Blood Vessels of the Axial Region now follows Exercise 21: The Heart, to correspond to the topic order in Saladin: *Human Anatomy*. The pulmonary circuit is added to this exercise, as well as the blood vessels of the head, neck and trunk.
- Exercise 23—Appendicular blood vessels are now included in this exercise.
- Exercise 24—This exercise now exclusively covers the lymphatic system.
- Exercise 29—The ovarian and menstrual cycles, and contraception material, are removed from this exercise.

Key Features

1. **Dynamic art program.** All of the illustrations are state-of-the-art, extremely accurate, use bold and appealing colors, and offer a unique, three-dimensional view.

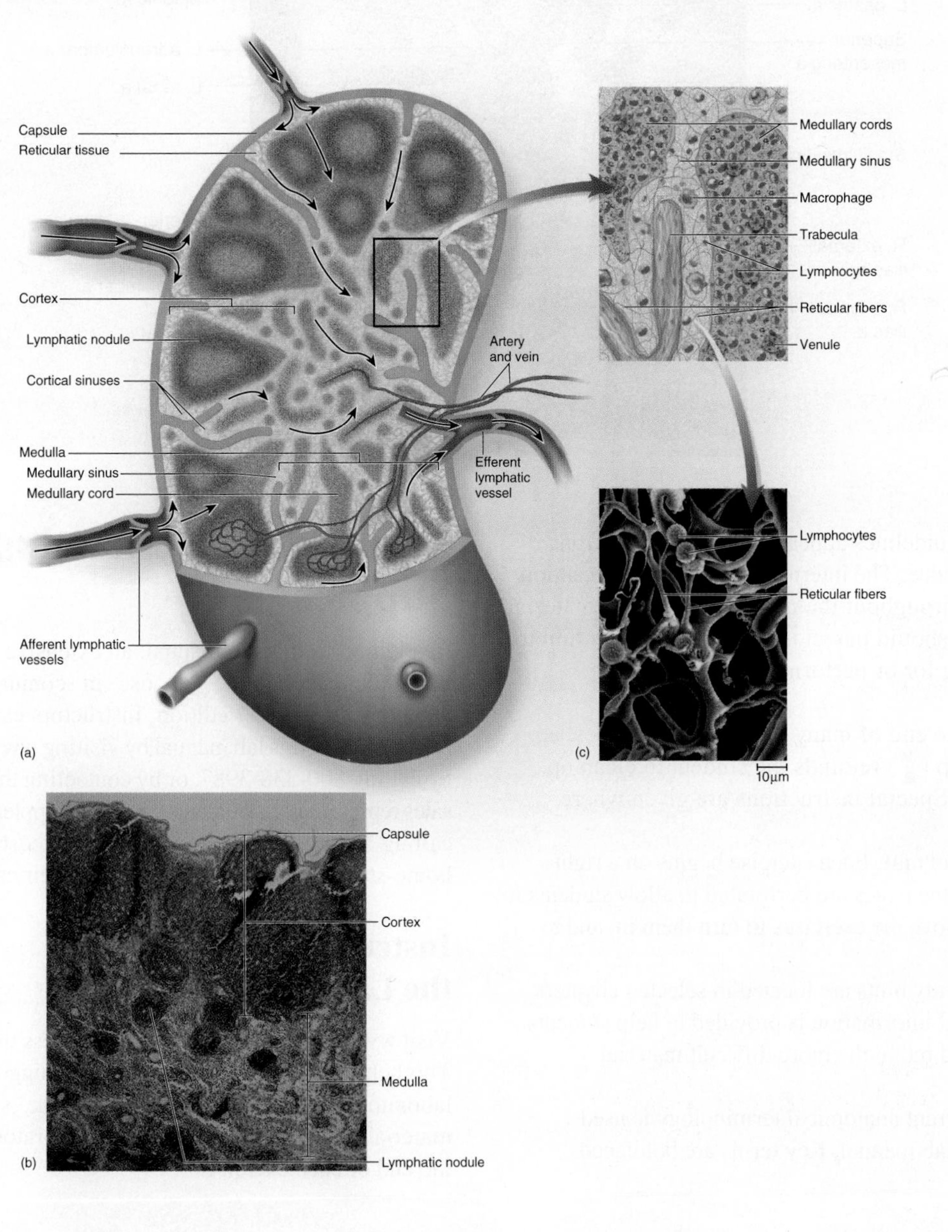

2. **Instructional photographs.** Numerous color photographs, including detailed histological light micrographs and cadaver and cat dissections, show detailed structures.
3. **Labels.** Illustrations are labeled for students to learn the names and terminology by looking at real-life examples or models, and by referring to the illustrations in the manual.
4. **Focus on the laboratory.** This manual focuses primarily on the material necessary for the laboratory, and does not repeat the material presented in the lecture text, with the expectation that students will look up material in the lecture text when necessary.
5. **Use of the cat for a dissection specimen.** Cat dissections are located in a separate section at the end of the manual, so that laboratories focusing on the cat can view all of the cat anatomy photos in one location.

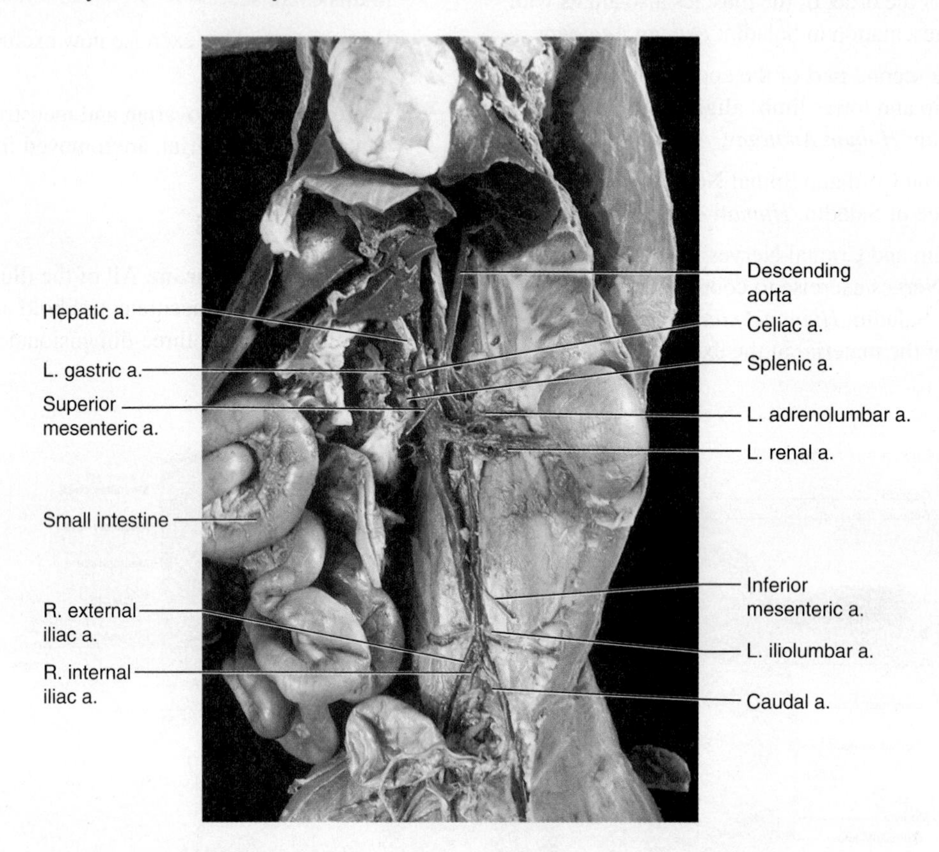

6. **Safety.** Safety guidelines appear on page v of the front matter for reference. The international symbol for caution ( ) is used throughout the manual to identify material that the reader should pay close and special attention to when preparing for or performing the laboratory exercise.
7. **Cleanup.** At the end of many laboratory exercises, an icon for cleanup ( ) reminds the student to clean up the laboratory. Special instructions are given where appropriate.
8. **User-friendly format.** Each exercise begins on a right-hand page, and the pages are perforated to allow students to more easily remove the exercises to turn them in, and to store them later.
9. **Study hints.** Study hints are located in selected chapters where additional information is provided to help students comprehend and retain the more difficult material presented.
10. **Key terms.** Current anatomical terminology is used throughout the lab manual. Key terms are boldfaced.

Teaching and Learning Supplements

In addition to this lab manual, an extensive array of supplemental materials is available for use in conjunction with Saladin's *Human Anatomy*, 5th edition. Instructors can obtain teaching aids to accompany this lab manual by visiting www.mhhe.com/catalogs, by calling 800-338-3987, or by contacting their local McGraw-Hill sales representative. Students can order supplement study materials by calling 800-262-4729, by visiting http://shop.mheducation.com/home-student.html, or by contacting their campus bookstore.

Instructor's Manual for the Laboratory Manual

Visit www.mhhe.com/labcentral to access the Instructor's Manual. This helpful preparation guide includes suggestions for coordinating laboratory exercises with the textbook, setup instructions and materials lists, and answers to the laboratory review questions at the end of each exercise.

Integrated and Adaptive Learning Systems

LearnSmart Labs is a superadaptive simulated lab experience that brings meaningful scientific exploration to students. Through a series of adaptive questions, LearnSmart Labs identifies a student's knowledge gaps and provides resources to quickly and efficiently close those gaps. Once students have mastered the necessary basic skills and concepts, they engage in a highly realistic simulated lab experience that allows for mistakes and the execution of the scientific method.

LEARNSMART
Mc Graw Hill Education | LABS®

The primary goal of **LearnSmart Prep** is to help students who are unprepared to take college-level courses. Using superadaptive technology, the program identifies what a student doesn't know and then provides "teachable moments" designed to mimic the office hour experience. When combined with a personalized learning plan, an unprepared or struggling student has all the tools needed to quickly and effectively learn the foundational knowledge and skills necessary to be successful in a college-level course.

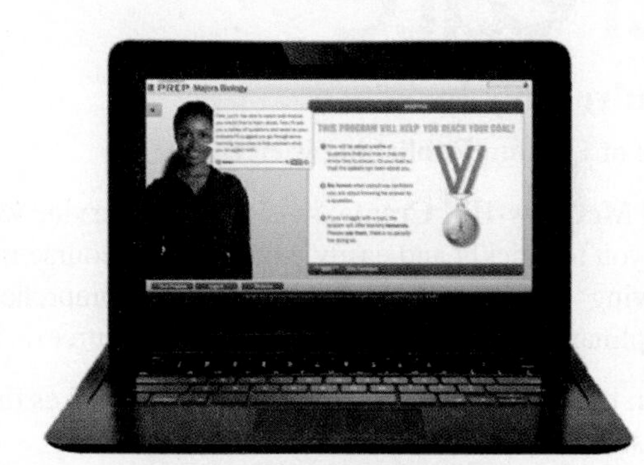

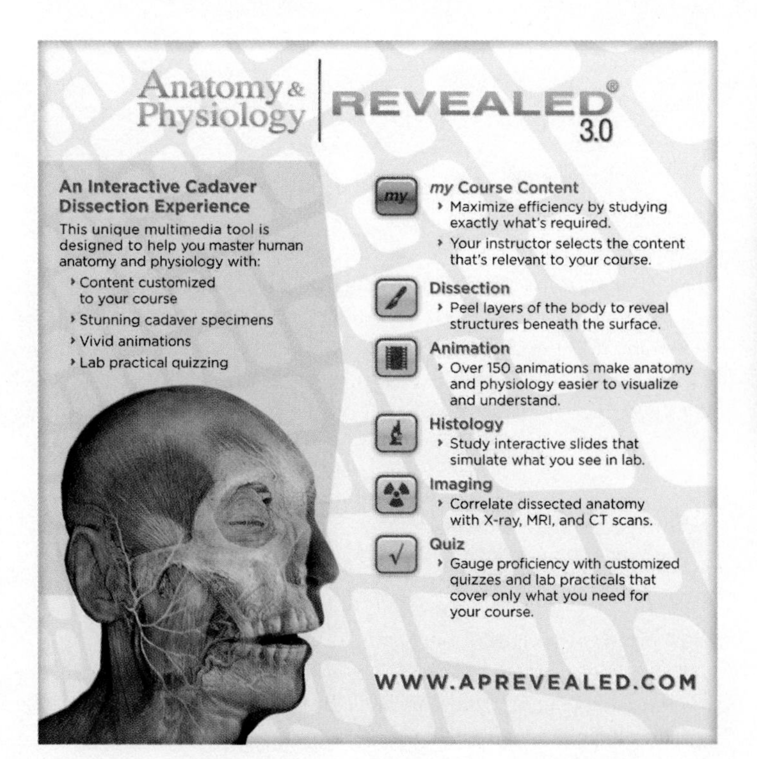

Anatomy & Physiology|REVEALED® is now mobile! It is also available in Cat and Fetal Pig versions.

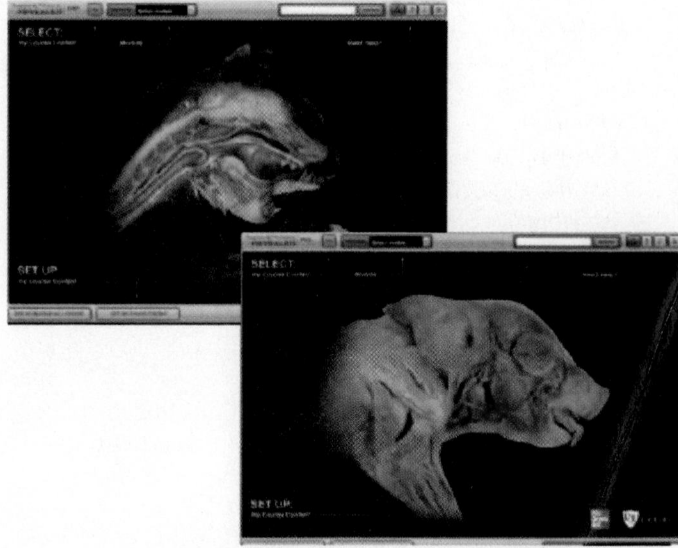

Mc Graw Hill Education connect®

Connect is a teaching and learning platform that is proven to deliver better results for students and instructors. Connect empowers students by continually adapting to deliver precisely what they need, when they need it and how they need it, so your class time is more engaging and effective. With Connect *Human Anatomy*, instructors can deliver assignments, quizzes, and tests easily online. Students can practice important skills at their own pace and on their own schedule.

What You've Only Imagined

The Future of Custom Publishing is Here.

Introducing **McGraw-Hill Create**™—a new, self-service website that allows you to quickly and easily create custom course materials by drawing upon McGraw-Hill Education's comprehensive, cross- disciplinary content and other third party resources.

- Select, then arrange the content in a way that makes the most sense for your course
- Combine material from different sources and even upload your own content
- Choose the best format for your students—print or eBook
- Edit and update your course materials as often as you'd like

Acknowledgments

Many people have been involved in the production of this lab manual, and I would like to thank the editorial and marketing teams at McGraw-Hill, including Marty Lange, Jessica Cannavo, Jim Connely, Amy Reed, Chloe Bouxsein, and Donna Nemmers. Thanks also goes to the McGraw-Hill production team, including Vicki Krug, Lori Hancock, and David Hash for their input and encouragement.

Reviewers and Contributors

I would like to thank the many people who have reviewed this lab manual over all past editions, contributing to its overall accuracy and scientific currency. Through the years, student and instructor comments and suggestions have been instrumental in keeping this lab manual a useful resource for the human anatomy laboratory. I also extend my appreciation to the reviewers of the Wise: *Anatomy & Physiology Laboratory Manual*; improvements to this two-semester manual also bear influence on the content changes within this *Human Anatomy Laboratory Manual*. I would also like to thank Ken Saladin for taking the time to review this edition of the lab manual.

Please feel free to write me or e-mail me with your comments, suggestions, and criticisms. I value your input and hope that your comments will lead to an even better sixth edition of this lab manual.

Eric Wise
Santa Barbara City College
721 Cliff Drive
Santa Barbara, CA 93109
wise@sbcc.edu

Student Preface

This lab manual was written to help you gain experience in the lab as you learn human anatomy. The 29 exercises explore and explain the structure of the human body. You will be asked to study the structure of the body using the materials available in your lab, which may consist of models; charts; mammal study specimens, such as cats; preserved or fresh internal organs of sheep or cows; and possibly cadaver specimens. You may also examine microscopic sections from various organs of the body. Familiarize yourself with the microscopes in your lab early, so you can take the best advantage of the information they can provide.

As a student of anatomy, you will be exposed to new and detailed information. The time it takes to learn the information will involve more than just time spent in the lab. Maximize your time in the lab by reading the assigned laboratory exercises before you come to class. At the end of each exercise are review sheets that your instructor may wish to collect. The illustrations are labeled except on the review pages. All review materials can be used as study guides for lab exams or they may be handed in to the instructor.

Although the exercises are written for use with human cadavers, there is an entire section at the end of the manual for the respective cat anatomy. Get involved in your lab experience. Don't let your lab partner do all of the dissections; likewise, don't insist on doing everything yourself. Share the responsibility and you will learn more.

Safety

Safety guidelines appear on page v of the front matter for reference. The international symbol for caution (CAUTION) is used throughout the lab manual to identify material that you should pay close and special attention to when preparing for or performing laboratory exercises.

Cleanup

Special instructions are provided for cleanup at the end of appropriate laboratory exercises and are identified by a unique icon ( ).

How to Study for This Course

Some people learn best by concentrating on the visual, some by repeating what they have learned, and others by writing what they know over and over again. In this course, you will have to adapt your learning style to different study methods. You may use one study method to learn the muscles of the body and a completely different method for understanding the structure of the nervous system. Some students need only a few hours per week to succeed in this course, while others seem to study far longer with a much less satisfactory performance. Coming to class is essential, as is coming to lab on time. The beginning of the lab is when most instructors go over the material and point out what material to omit, what to change, and how to proceed. If you do not attend lab, you do not get the necessary information. Read the material ahead of time. The subject matter is very visual, and you will find an abundance of illustrations in this manual. Record on your calendar all of the lab quizzes and exams listed on the syllabus provided by your instructor. Budget your time so you study accordingly.

Work hard! There is absolutely no substitute for hard work to achieve success in a class. Some people do math easier than others, some people remember things easier, and some people express themselves better. Most students succeed because they work hard at learning the material. It is a rare student who gets a bad grade because of a lack of intelligence. Working at your studies will get you much farther than worrying, ignoring your studies, or thinking that you can succeed by attending class some of the time and reviewing your notes the night before a lab exam. Be actively involved with the material and you will learn it better. Outline the material after you have studied it for awhile. Read your notes, go over the material in your mind, and then make the information your own. There are several ways that you can get actively involved.

Draw and doodle a lot. Anatomy is a visual science, and drawing helps. You do not have to be a great illustrator. Visualize the material in the same way you would draw a map to your house for a friend. You do not draw every bush or tree but, rather, create a schematic illustration that your friend could use to get the pertinent information. As you know, there are differences in maps. Some people need more practice than others, but anyone can do it. The head can be drawn as a circle, which can be divided into pieces representing the bones of the skull. Draw and label the illustration after you have studied the material and without the use of your text! Check yourself against the text to see if you really know the material. Correct the illustration with a colored pen, so that you highlight the areas you need work on. Go back and do it again until you get it perfect. This does take some time, but not as much as you might think.

Write an outline of the material. Take the mass of information to be learned and go from the general to the specific. Let's use the skeletal system as an example. You may wish to use these categories:

1. Bone composition and general structure
2. Bone formation
3. Parts of the skeleton
   a. Appendicular skeleton
      (1) Pectoral girdle
      (2) Upper limb
      (3) Pelvic girdle
      (4) Lower limb
   b. Axial skeleton
      (1) Skull
      (2) Hyoid
      (3) Ribs
      (4) Vertebral column
      (5) Sternum

An outline helps to organize the material in your mind and allows you to sort the information into areas of focus. Without an organizational system, this course would be a jumble of terms with no interrelationships. The outline will get more detailed as you progress, so you eventually place specific information into a framework of more general information.

Test yourself before an exam or a quiz. If you have practiced answering questions about the material you have studied, then you should do better on the real exam. As you review the material, jot down possible questions to be answered later, after you study. This compiled list of questions can be answered later to see if you have learned the material well. You can also enlist the help of friends, study partners, or family (if they are willing to do this for you). You can also study alone. Some people make flash cards for anatomy. It is a good idea to do this for the muscle section of the class, but you may be able to get most of the information down by using the preceding technique. Flash cards take time to fill out, so use them carefully.

Use memory devices for complex material. A mnemonic device is a memory phrase that has a relationship to the study material. For example, there are two bones in the wrist right next to one another, the trapezium and the trapezoid. The mnemonic device used by one student was that "trapezium" rhymes with "thumb" and it is the bone under the thumb.

Use your study group as a support group. A good study group is very effective in helping you do your best in class. Get together with people who will push you to do your best. If you get discouraged, your study partners can be invaluable support people. A good study group can help you improve your test scores, develop study hints, encourage you to do your best, and let you know that you are not the only person who is living, eating, and breathing anatomy.

Just as a good study group can really help, a bad group can drag you down farther than you might go on your own. If you are in a group that constantly complains about the instructor, that the class is too hard, that there is too much work, and so on, then you need to get out of that group and into one that is excited by the information. Don't listen to people who complain constantly and make up excuses instead of studying. There is a tendency to start believing the complaints, and that begins a cycle of failure. Get out of a bad situation early and get with a group that will move forward.

Do well in the class and you will feel good about the experience. If you set up a study time with a group of people and they spend most of the time talking about parties, sports, or personal problems, then you aren't studying. There is nothing wrong with talking about parties, sports, or helping someone with personal problems, but you also need to address the task at hand, which is learning anatomy. Don't feel bad if you must get out of your study group. It is your education, and, if your partners don't want to study, then they don't really care about your academic well-being. A good study partner is one who pays attention in class, who has prepared for the study session, and who can explain information that you may have gotten wrong in your notes. You may want to get the phone number of two or three such classmates.

Test Taking

Finally, you need to take quizzes and exams in a successful manner. Doing practice tests will help you develop confidence. Do well early in the semester. Study extra hard early—there is no such thing as overstudying! If you fail the first test or quiz, then you must work yourself out of an emotional ditch. Study early and consistently, and then spend the evening before the exam going over the material in a general way and learning those last few details. Some people do succeed under pressure and cram before exams; however, the information is temporarily stored and does not serve them well in their major field. If you study on a routine basis, then you can get up on the morning of a test, have a good breakfast, listen to some encouraging music, maybe review a bit, and be ready for the exam.

Your instructor is there to help you learn anatomy, and this lab manual was written with you in mind. Relate as much of the material as you can to your own body and keep an optimistic attitude.

Please feel free to write me or e-mail me with your comments, suggestions, and criticisms. I value your input and hope that your comments will lead to an even better sixth edition of this lab manual.

Eric Wise
Santa Barbara City College
721 Cliff Drive
Santa Barbara, CA 93109
wise@sbcc.edu

Laboratory

Exercise 1

Organs, Systems, and Organization of the Body

Introduction

Science is the study of physical phenomena and follows specific guidelines that make it unique compared to other disciplines. **Human anatomy** is the scientific study of the structure of the body. The term "anatomy" comes from Greek words meaning "to cut up" (*ana* = up, *toma* = to cut). Anatomy forms the base of study for other courses in the health sciences. It is also an important course for athletic training, dance, and visual arts. The study of the human body requires an understanding of the details of the structure of the body, the orientation of the body or organ of study, and knowledge of body regions. In this exercise, you examine the levels of organization of the body (from the subatomic level to the whole organism), organ systems, reference terms, directional terms, planes of sectioning, and major regions of the body.

Learning Objectives

At the end of this exercise you should be able to

1. list the levels of structural hierarchy from smallest to largest;
2. list all the organ systems of the body and assign organs to each system;
3. put major organs, such as the heart, lungs, and stomach, in the proper organ system;
4. describe anatomical position;
5. give directional terms that are equivalent to up, down, front, back, toward the midline, and toward the surface of the body;
6. determine from an illustration whether a section is in the frontal, transverse, or median plane;
7. assign select organs to respective body cavities; and
8. identify the quadrants and nine regions of the abdomen.

Materials

Models of human torso
Charts of human torso

Procedure

Levels of Organization

The human body can be studied at different levels from the atomic (or even smaller levels) to the organismal. The earliest study involved **gross anatomy,** examining structures visible to the naked eye. As more sophisticated equipment was developed, other levels of organization became apparent. Today, the manipulation of atomic nuclei under magnetic fields has led to magnetic resonance imaging (MRI) studies that do not depend on dissection of the body (fig. 1.1).

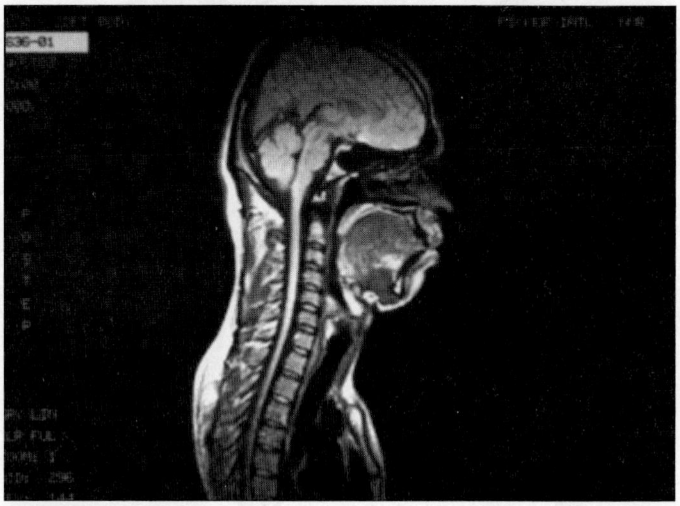

FIGURE 1.1 **MRI** Image of the neck.

Examine the following list for the levels of study along with cited examples:

Level	Examples
Atomic	Oxygen, carbon, nitrogen
Molecular/chemical	Protein, lipid, carbohydrate
Organelle	Mitochondrion, ribosome
Cellular	Fibrocyte, red blood cell
Tissue	Epithelial, muscular
Organ	Stomach, kidney
Organ system	Digestive system, urinary system
Organism	*Homo sapiens*

Organ Systems

Anatomy can be studied in many ways. **Regional anatomy** is the study of particular areas of the body such as the head or leg. Another way to study anatomy is through **systemic anatomy,** which is the study of **organ systems,** such as the skeletal system or the nervous system. Most undergraduate college courses in anatomy (and the format of this lab manual) use the systemic approach.

Although organ systems are studied separately, it is important to realize the connections between the systems. If the heart, part of the circulatory system, fails to pump blood, then the lungs (part of the respiratory system) do not receive blood for oxygenation and the intestines (in the digestive system) do not transfer nutrients to the blood as fuel. Organs throughout the body are no longer capable of functioning, and the result is death. From a clinical standpoint, the failure of one system has impacts on many other organ systems. Examine the torso models and charts in the lab and locate various organs. Using figure 1.2, find the following organ systems:

Reproductive	Respiratory	Digestive
Urinary	Skeletal	Endocrine
Nervous	Lymphatic	Circulatory
Muscular	Integumentary	

A quick way to remember all 11 systems is to remember this phrase: "Run Mrs. Lidec." Each letter of the phrase represents the first letter of one of the names of the organ systems.

Examine models or charts in lab and locate the following organs of the various organ systems.

Reproductive The gonads (testes and ovaries) contain the sex-producing cells of the body; and the accessory organs, such as

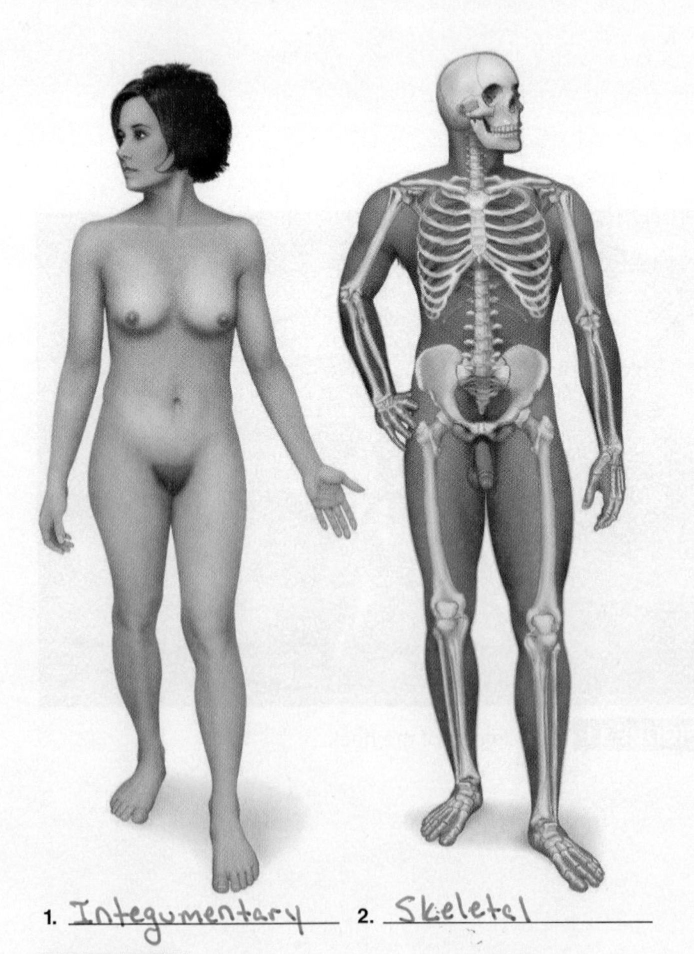

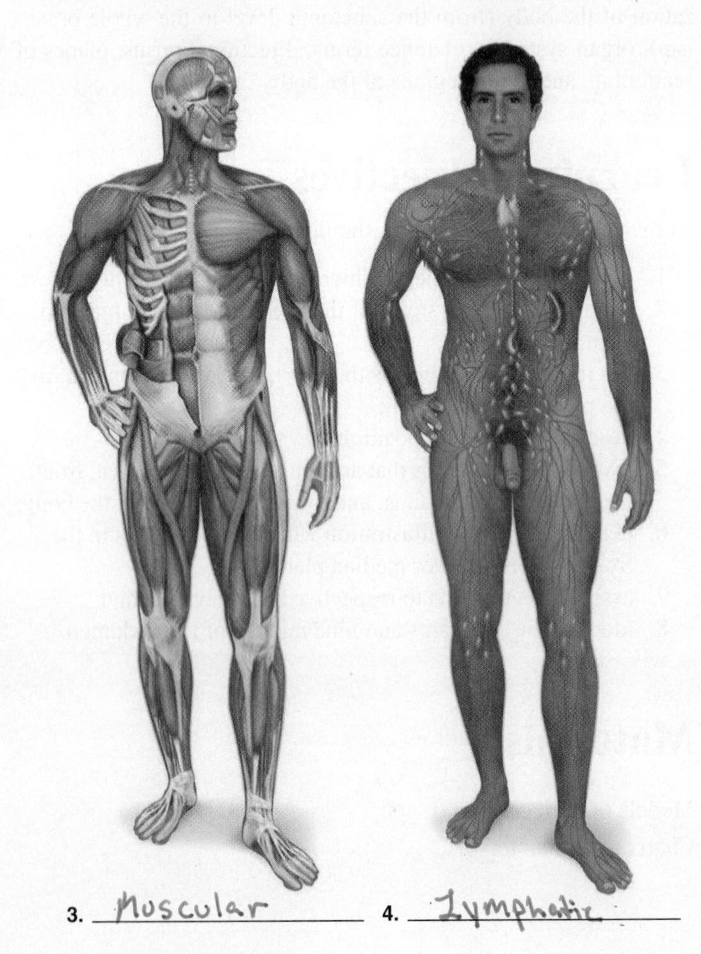

1. Integumentary 2. Skeletal 3. Muscular 4. Lymphatic

FIGURE 1.2 Organ Systems of the Human Body Fill in the names of the organ systems in the spaces below the illustrations of those systems.

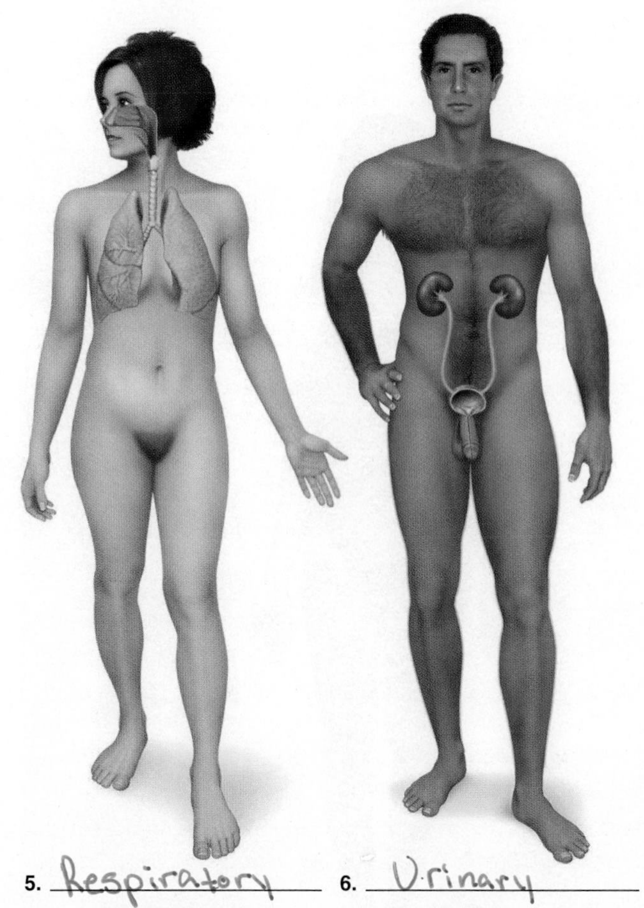

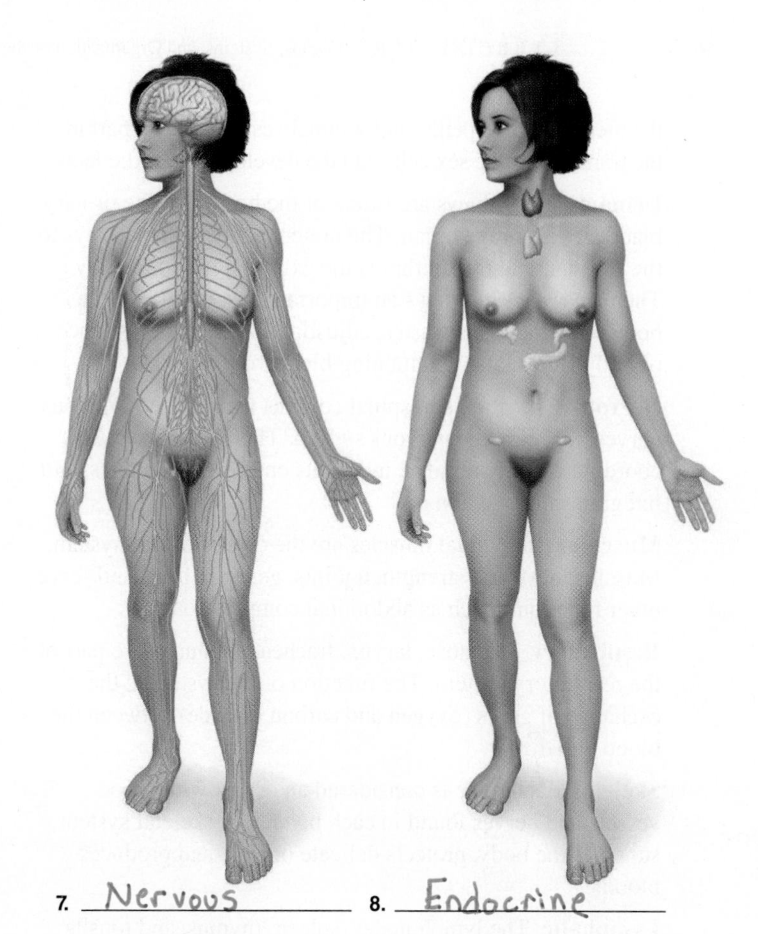

5. Respiratory 6. Urinary

7. Nervous 8. Endocrine

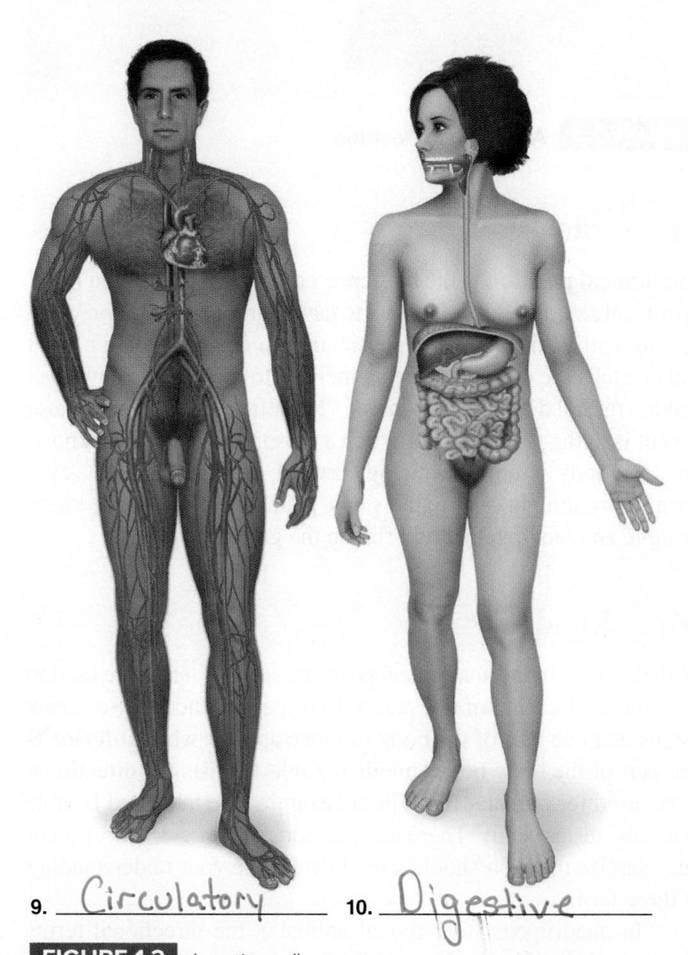

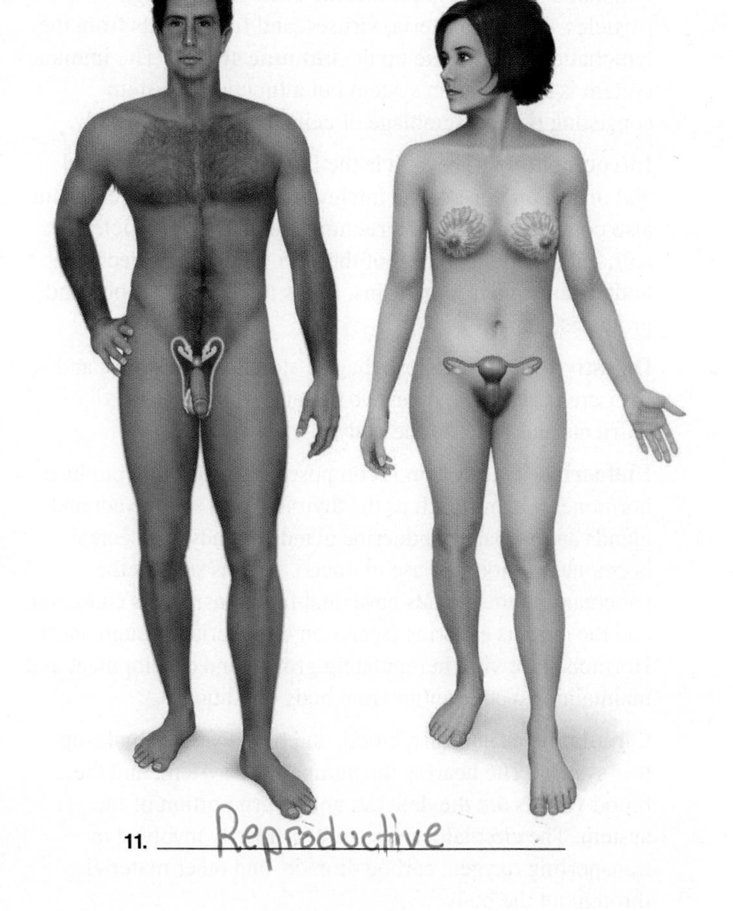

9. Circulatory 10. Digestive

11. Reproductive

FIGURE 1.2 *(continued)*

the uterus, vagina, penis, and seminal vesicles, play a part in the transport of the sex cells and the development of the fetus.

Urinary The kidneys are filters of the body, and the urinary bladder is a storage organ. The ureters connect the kidneys to the bladder, and the urethra is the exit tube from the body. The urinary system plays an important role in ridding the body of nitrogenous wastes, adjusting the chemical balance of body fluids, and maintaining blood volume.

Nervous The brain and spinal cord, as well as the numerous nerves, make up the nervous system. The nervous system coordinates body regions, interprets environmental cues, and integrates information.

Muscular Individual muscles are the organs of this system. Muscles move and strengthen joints, generate heat, and serve other functions, such as abdominal compression.

Respiratory The nose, larynx, trachea, and lungs are part of the respiratory system. The function of the system is the exchange of gases (oxygen and carbon dioxide) between the blood and the air.

Skeletal Each bone is considered an organ, with blood vessels and nerves found in each bone. The skeletal system supports the body, protects delicate organs, and produces blood.

Lymphatic The lymph nodes, spleen, thymus, and tonsils are part of the lymphatic system. A major function of the lymphatic system is to protect the body from foreign particles, such as bacteria, viruses, and fungi. Cells from the lymphatic system make up the **immune system.** The immune system is not an organ system but a functional system consisting of an assemblage of cells that defend the body.

Integumentary The skin is the largest organ of the body and makes up most of the integumentary system. The system also contains associated structures, such as hair follicles, hair, nails, and the glands of the skin. The skin protects the body against microorganisms, keeps it from drying out, and produces vitamin D.

Digestive The mouth, esophagus, stomach, intestines, and liver are parts of the digestive system, which provides nutrients and water to the body.

Endocrine This system is composed of organs that produce hormones. Organs such as the thyroid gland and the adrenal glands are primarily endocrine glands (glands that secrete hormones without the use of ducts). Organs such as the pancreas and the gonads have dual functions; one is endocrine and the other is exocrine (secretion of material through ducts). Hormones are vital in regulating growth and development and maintaining a constant internal body condition.

Circulatory The heart, blood, and blood vessels make up this system. The heart is the pump of the system, and the blood vessels are the delivery and return portion of the system. The circulatory system is primarily involved in transporting oxygen, carbon dioxide, and other materials throughout the body.

FIGURE 1.3 **Anatomical Position**

Anatomical Position

Anatomical position is the reference position for the human body. In clinical settings, it is important to have a proper orientation when dealing with patients. If two physicians are operating on a patient and one tells the other to make an incision to the left, the physician making the cut does not have to ask, "My left or your left?" because the cut is to the *patient's* left. When a person is in anatomical position, the body is upright, facing forward, the head is erect, eyes open, arms straight and by the sides, palms facing forward, knees straight, and feet together and flat on the ground (fig. 1.3).

Directional Terms

With the body in the anatomical position, specific terms are used to describe the location of one part with respect to another. **Superior** means that one part of the body is above another while **inferior** is one part of the body below another. Table 1.1 lists the directional terms used for humans. You should examine the table and be able to use the terms easily. There are questions in the review section of this exercise that you should use to determine your understanding of these terms.

In quadrupeds (four-footed animals), the directional terms are somewhat different. Note in figure 1.4 that quadrupeds do not

TABLE 1.1 Directional Terms Used in Anatomy

Term	Meaning	Example
Superior	Above	The nose is superior to the chin.
Inferior	Below	The stomach is inferior to the head.
Medial	Toward the midline	The sternum is medial to the shoulders.
Lateral	Toward the side	The ears are lateral to the nose.
Superficial	Toward the surface	The skin is superficial to the heart.
Deep	Toward the core	The lungs are deep to the ribs.
Anterior (or ventral)	To the front	The toes are anterior or ventral to the heel.
Posterior (or dorsal)	To the back	The spine is posterior or dorsal to the sternum.
Proximal	For extremities, meaning near the trunk	The elbow is proximal to the wrist.
Distal	For extremities, meaning away from the trunk	The toes are distal to the knee.
Cephalic	Toward the head	The cephalic region of the body houses the brain.
Caudal	Toward the "tail"	The tail is caudal to the chest in cats.

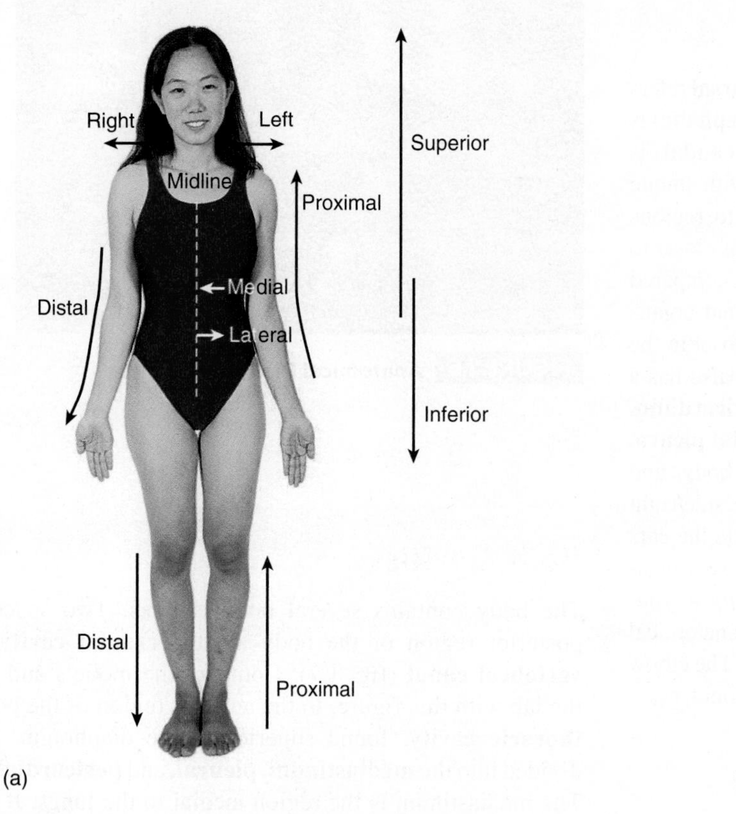

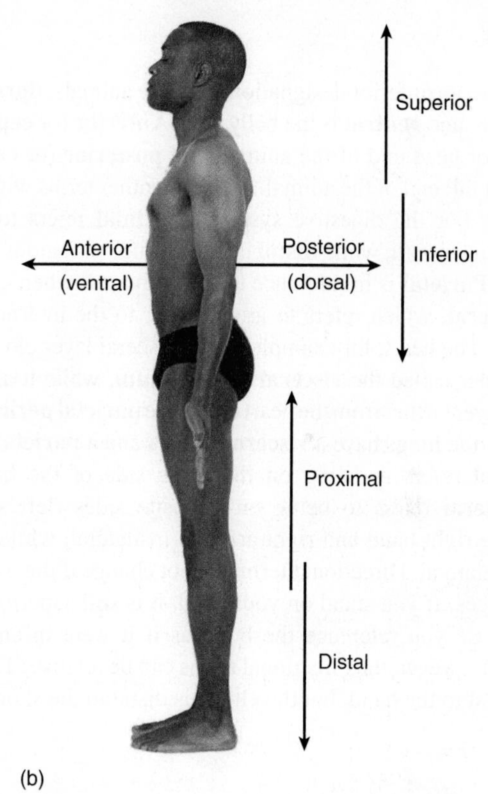

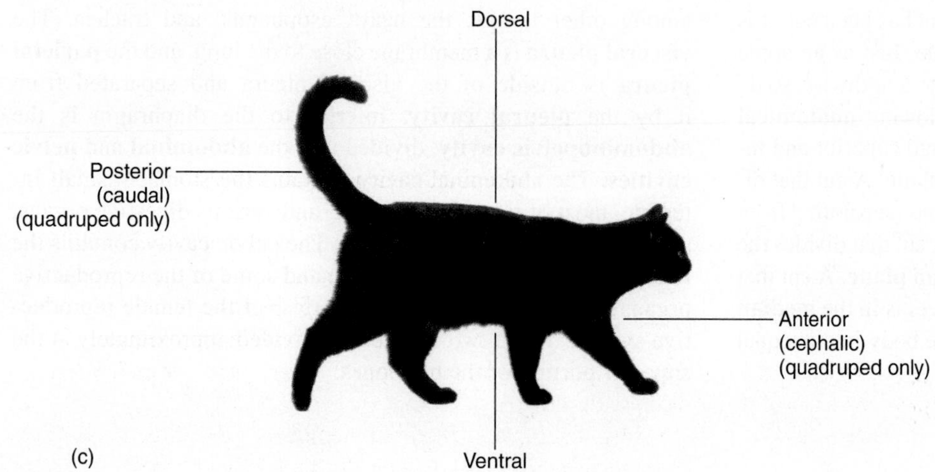

FIGURE 1.4 Directional Terms (a) For humans, anterior view; (b) for humans, lateral view; (c) for quadrupeds, lateral view.

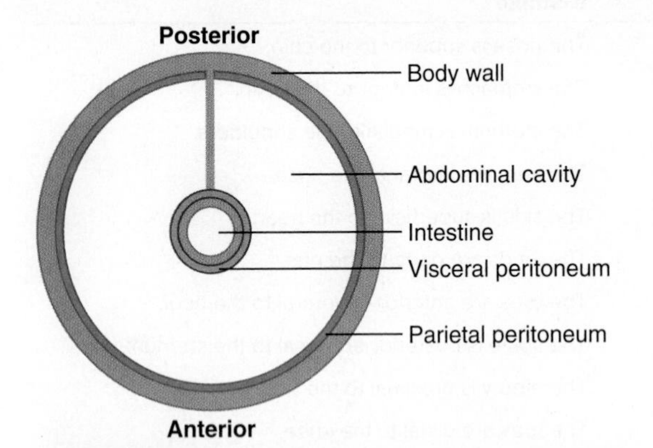

FIGURE 1.5 Idealized Cross Section of the Body with Relationships of Visceral and Parietal Terms

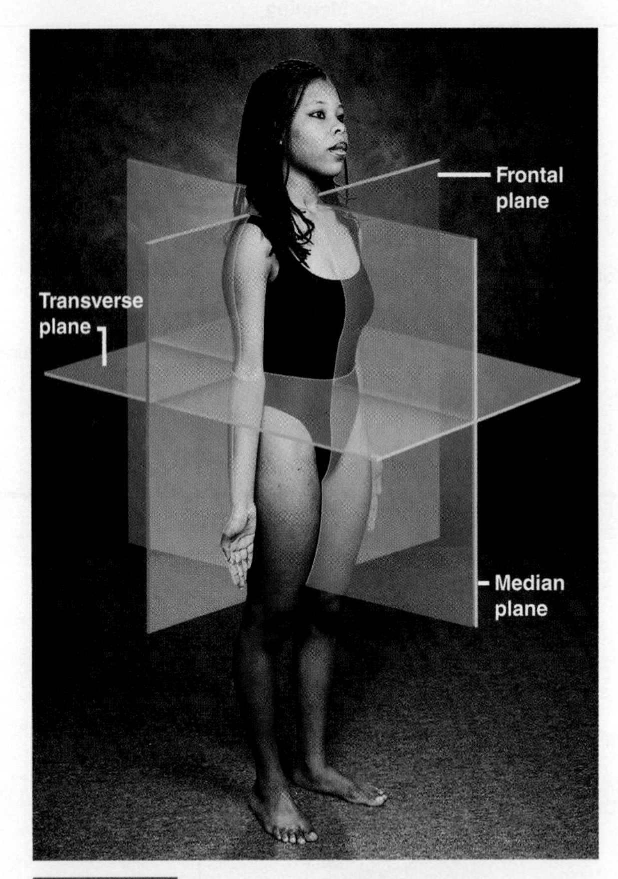

FIGURE 1.6 Anatomical Planes

have a superior/inferior designation. In these animals, **dorsal** refers to the back, and **ventral** is the belly side. **Anterior** (or **cephalic**) is the front or head end of the animal, and **posterior** (or **caudal**) is the rear or tail end of the animal. There are other terms with unique meanings. For the digestive system, **proximal** refers to regions closer to the mouth, while **distal** is in reference to regions closer to the anus. **Parietal** is in reference to the body wall when compared with **visceral,** which refers to areas closer to the internal organs (fig. 1.5). The heart, for example, has a visceral layer closer to the heart proper called the **visceral pericardium**, while it also has a parietal layer farther from the heart called the **parietal pericardium.** Likewise, the lungs have a **visceral pleura** and a **parietal pleura.** **Ipsilateral** refers to being on the same side of the body, and **contralateral** refers to being on opposite sides (left side/right side). The right hand and right arm are ipsilateral, while the ears are contralateral. Directional terms do not change if the body position changes. If you stand on your head, it is still superior to your feet because you reference the body as if it were in anatomical position. However, the directional terms can be relative. The elbow is proximal to the hand, but the elbow is distal to the shoulder.

Anatomical Planes

When you are viewing a picture of an organ that has been cut, it is important to understand how the cut was made. Just as an apple looks different when cut crosswise as opposed to lengthwise, so do some organs. Examine figure 1.6 for the following **anatomical planes.** A cut that divides the body or organ into superior and inferior parts is in the **transverse (horizontal) plane.** A cut that divides the body into anterior and posterior portions (separating front from back) is in the **frontal (coronal) plane.** A cut that divides the body into left and right portions is in the **median plane.** A cut that divides the body equally into left and right halves is in the **median (midsagittal) plane,** while one that divides the body into unequal left and right parts is in a **parasagittal plane.**

Body Cavities

The body contains several body cavities. Two spaces in the posterior region of the body are the **cranial cavity** and the **vertebral canal** (fig. 1.7). Compare the models and charts in the lab with this figure. In the anterior region of the body is the **thoracic cavity**, found superior to the diaphragm, and it is divided into the **mediastinum, pleural,** and **pericardial cavities.** The mediastinum is the region medial to the lungs. It contains, among other things, the heart, esophagus, and trachea. The **visceral pleura** is a membrane close to the lung, and the **parietal pleura** is outside of the visceral pleura and separated from it by the **pleural cavity.** Inferior to the diaphragm is the **abdominopelvic cavity**, divided into the **abdominal** and **pelvic cavities.** The abdominal cavity contains the stomach, small intestine, most of the large intestine, and various digestive organs, such as the liver and the pancreas. The pelvic cavity contains the terminal part of the large intestine and some of the reproductive organs (such as the uterus and ovaries) of the female reproductive system. These two cavities are divided approximately at the superior portion of the hip bones.

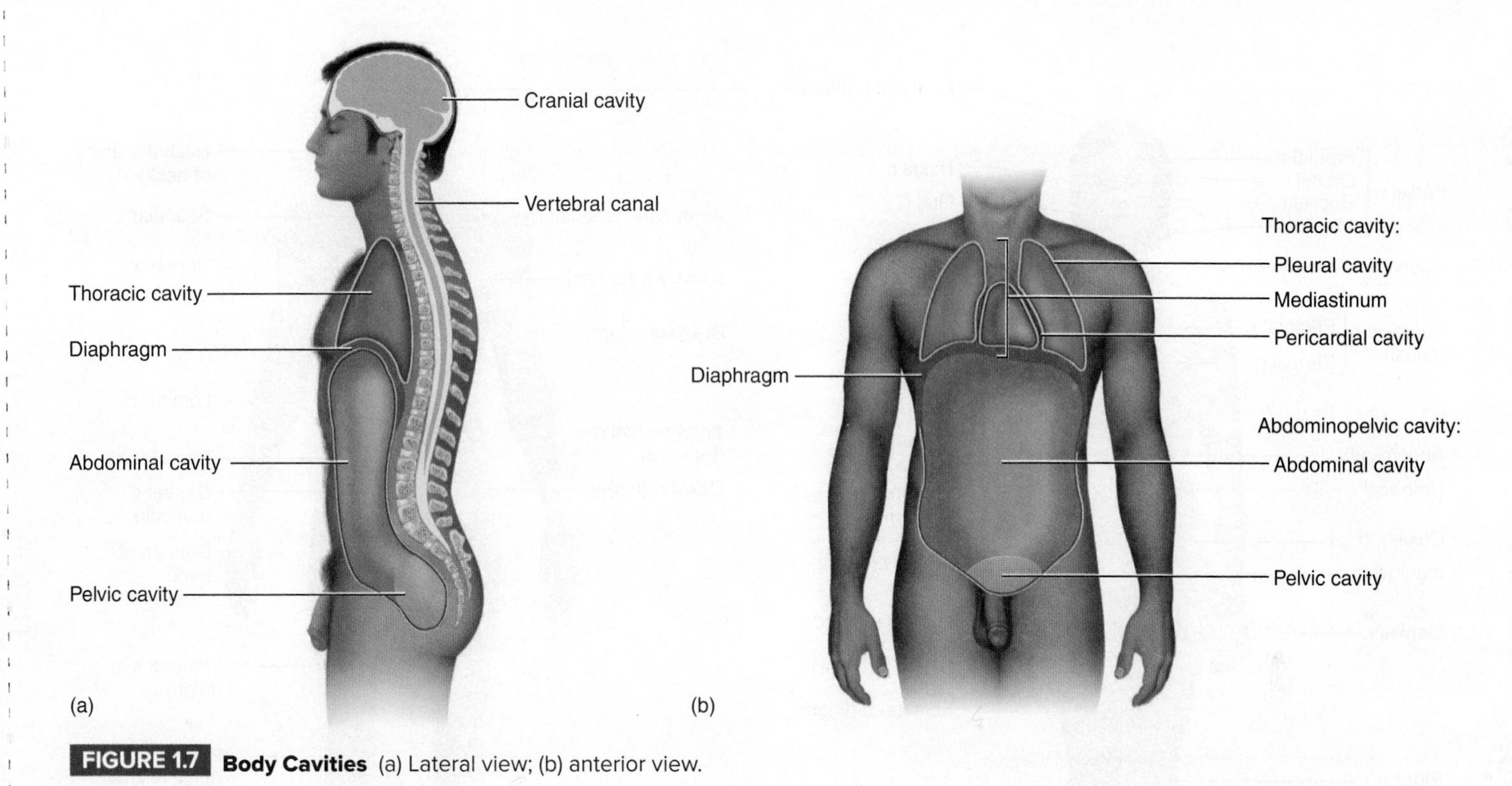

FIGURE 1.7 **Body Cavities** (a) Lateral view; (b) anterior view.

Regions of the Body

Examine figure 1.8 for the specific regions of the body. You will refer to these regions throughout this lab manual, so a complete study of them is essential. Anatomical names are listed first with the common name (if appropriate) listed in parentheses. In anatomy some regions of the body are described differently than you might expect. In anatomical usage, the **arm** is the region between the shoulder and the elbow. Distal to the elbow is the forearm. The **leg** is the region between the knee and the ankle, and from the knee to the hip is the thigh. Locate the following regions (region is abbreviated r.) on figure 1.8.

- Cephalic r. (head)
  - Frontal r. (forehead)
  - Orbital r. (eye)
  - Nasal r. (nose)
  - Buccal r. (cheek)
  - Oral r. (mouth)
  - Mental r. (chin)
- Cervical r. (neck)
- Nuchal r. (back of neck)
- Trunk
  - Thoracic r. (chest)
    - Pectoral r.
    - Sternal r.
    - Acromial r. (shoulder)
  - Abdominal r. (belly)
  - Inguinal r.
  - Genital r. (pubic)
  - Coxal r. (hip)
- Upper extremity
  - Axillary r. (armpit)
  - Brachial r. (arm)
  - Cubital r. (elbow)
  - Antebrachial r. (forearm)
  - Carpal r. (wrist)
  - Manual r. (hand)
  - Digital r. (finger)
- Lower extremity
  - Femoral r. (thigh)
  - Patellar r. (knee)
  - Popliteal r. (back of knee)
  - Crural r. (leg)
  - Tarsal r. (ankle)
  - Pedal r. (foot)
  - Digital r. (toe)

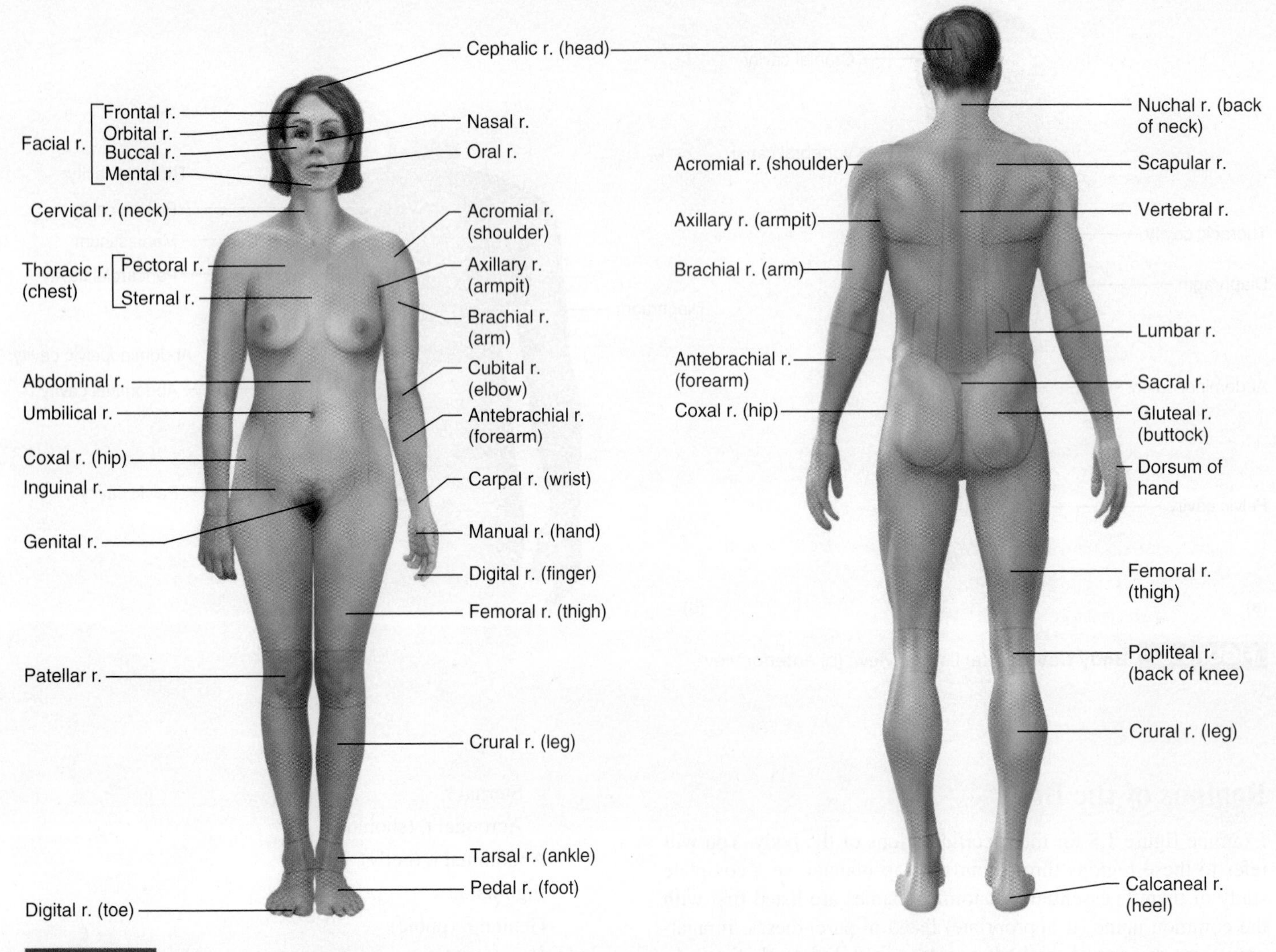

FIGURE 1.8 **Regions of the Body the indication r. stands for region.**

Find the following locations on your body and provide the appropriate anatomical description for these regions.

Shin ____________________

Elbow Cubital

Neck Cervical

Toes Digital

Shoulder Acromial

Thigh Femoral

Knee Patellar

Abdominal Regions

The abdomen can be divided into either four quadrants or nine regions. Clinicians typically use the four-quadrant terminology, while anatomists generally use nine regions. The dividing lines for the nine regions are named for anatomical structures that they pass through. A **subcostal line** is a horizontal line that occurs inferior to the ribs. An **intertubercular line** is another horizontal line that separates the umbilical region from the hypogastric region. Locate the midclavicular lines (vertical lines) that form the divisions between the medial and lateral regions. These regions can be used to pinpoint potential problems. One of the symptoms of appendicitis is pain in the lower right quadrant or right iliac region. Examine figure 1.9 and locate the lines and regions listed.

Abdominal Quadrants

Right upper quadrant
Left upper quadrant
Right lower quadrant
Left lower quadrant

Abdominal Regions

Right hypochondriac
Left hypochondriac
Epigastric
Right lumbar (lateral abdominal)
Left lumbar (lateral abdominal)
Umbilical
Hypogastric
Right iliac (inguinal)
Left iliac (inguinal)

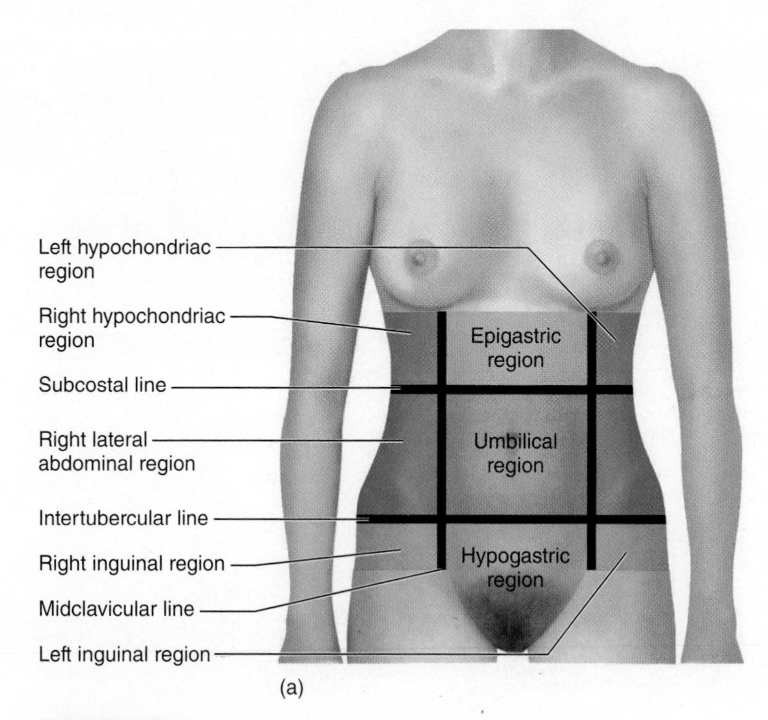

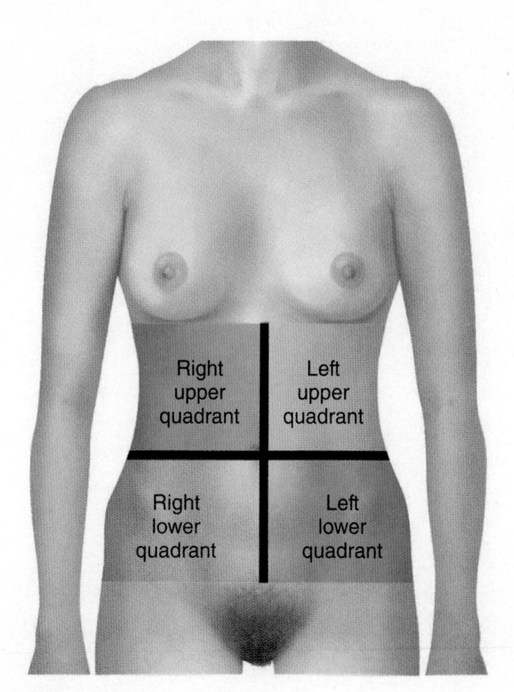

FIGURE 1.9 **Abdomen** (a) Nine regions; (b) four quadrants.

Organs, Systems, and Organization of the Body

Name Kierstyn Conyers Date ____________

1. The scientific study of the structure of the human body is known as Human Anatomy.
2. Organs are grouped into functionally related associations known as Organ systems.
3. In terms of reference, the body is placed in what position? anatomical position
4. What specific body cavity lies directly inferior to the diaphragm? pelvic Cavity
5. The kidneys belong to the Urinary system.
6. The liver belongs to the Digestive system.
7. In anatomical position, the palms of the hand are facing palms facing forward
8. In anatomical terms, referring to front and back, the pectoral region is posterior to the scapular region.
9. In terms of nearness to the trunk, the antebrachium is proximal to the carpal region.
10. The stomach is found in what specific body cavity? Abdominal Cavity
11. The region of the abdomen directly under the right side of the rib cage is the pelvic region.
12. If a hairline fracture occurred in the proximal humerus (arm bone), would the injury be closer to the shoulder or to the elbow?
    Shoulder Why? Proximal is towards the core, and the shoulder is closer to your core
13. In a clinical report, there is a note of a laceration (cut) on the posterior crural region. Where does this occur in laypersons' terms?
    Upper leg on the calf area
14. The body cavity that is enclosed by the rib cage is known as the Thoracic cavity.

15. The body cavity surrounded by the hip bones is called the pelvic Cavity.

16. The term "arm" in anatomy refers to the region between the A.

    a. shoulder and elbow b. elbow and wrist c. shoulder and wrist d. shoulder and hand

17. The term "leg" in anatomy refers to the region between the B.

    a. hip and knee b. knee and ankle c. hip and ankle d. ankle and foot

18. A mitochondrion belongs to which level of organization? Circle the correct answer.

    a. cellular b. tissue (c.) organelle d. organ system

Use correct anatomical terminology to describe the following relationships.

19. In terms of up and down, the head is Superior to the toes.

20. In terms of nearness to the trunk, the fingers are Distal to the arm.

21. In terms of nearness to the surface, the brain is Deep to the scalp.

22. In terms of front to back, the nipples are anterior to the shoulder blades.

23. The lungs belong to the respiratory system.

24. The heart belongs to the cardiovascular system.

25. If you sit on a horse's back, you are on the C/D aspect of the horse.

    a. anterior b. ventral c. posterior d. dorsal

26. What is the difference between the abdomen and the abdominal cavity? ______

27. Complete the illustration by correctly placing the following terms:

abdominal ✓	coxal ✓
acromial	crural
antebrachial ✓	femoral ✓
axillary ✓	frontal ✓
brachial ✓	genital ✓
carpal ✓	pectoral ✓
cephalic ✓	pedal ✓
cervical ✓	sternal ✓

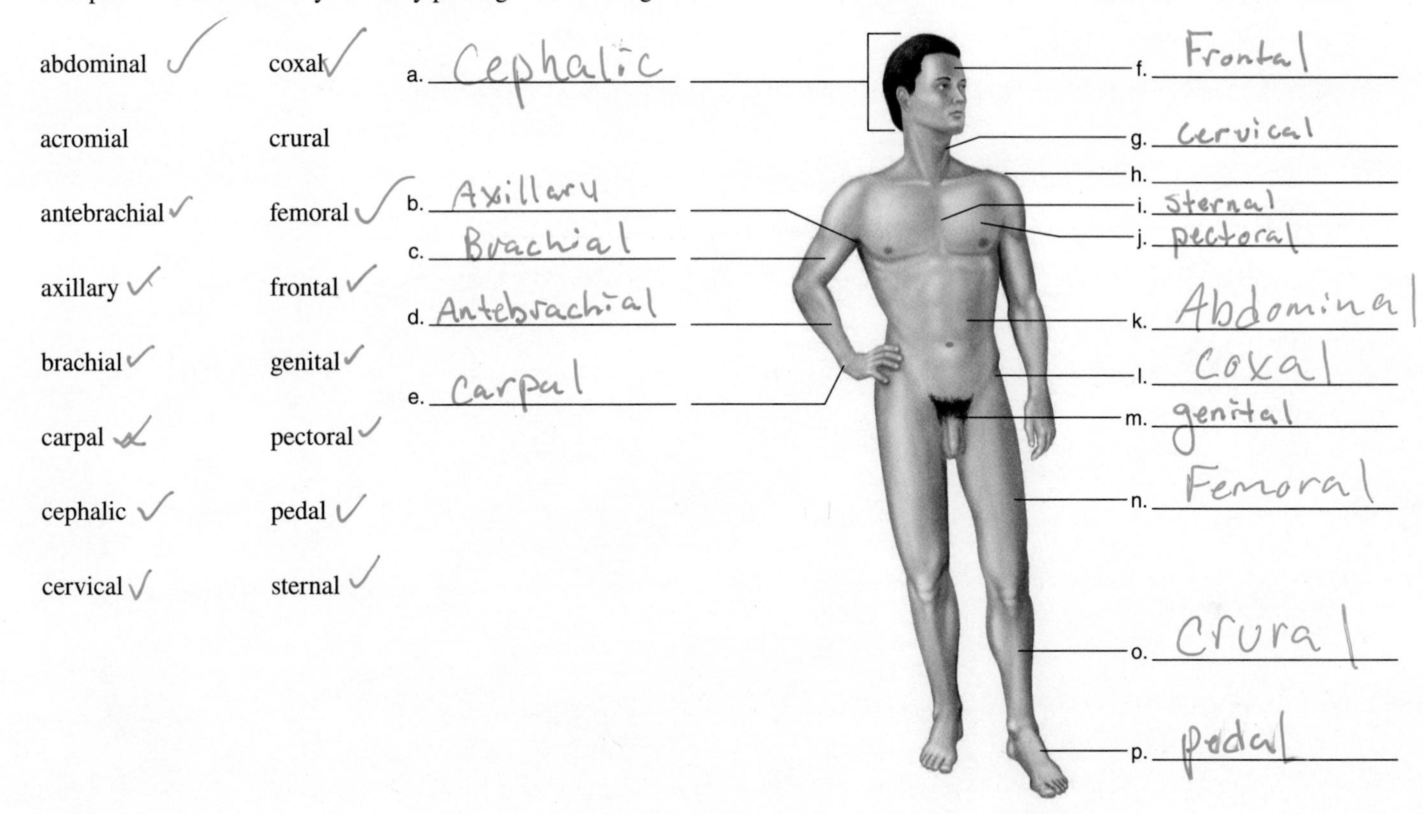

28. In the following illustrations, place the appropriate terms next to the anatomical planes that represent them.

median frontal transverse

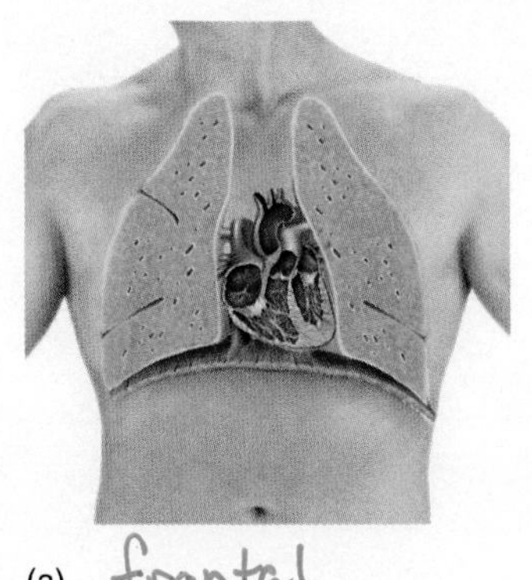

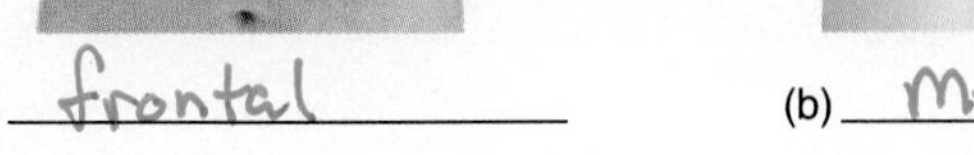
(a) frontal

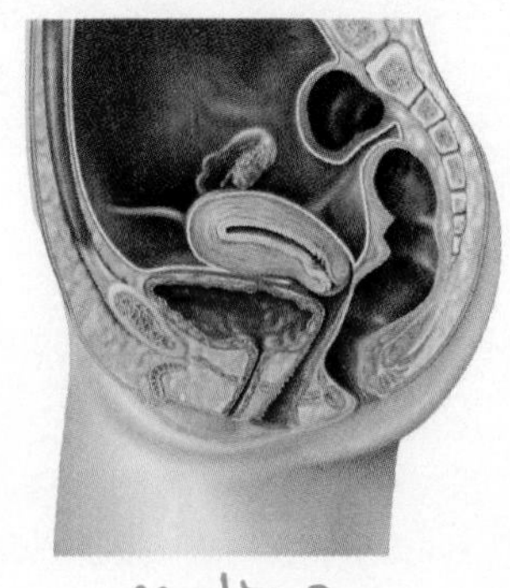

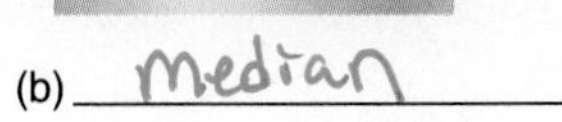
(b) median

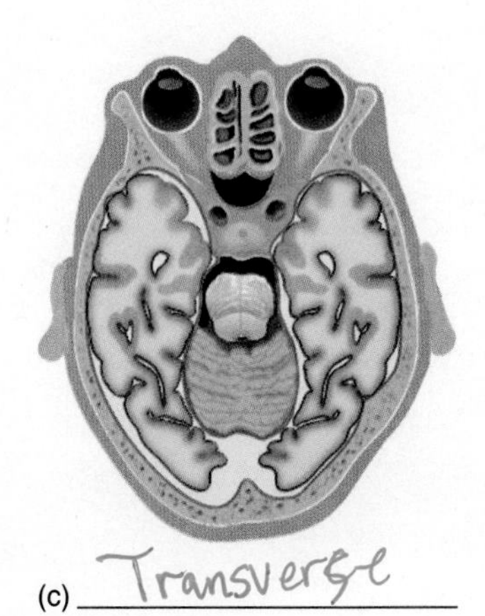

(c) Transverse

Laboratory

Exercise 2

Microscopy

Introduction

Originally, the study of anatomy and physiology was based on macroscopic, or gross, observation. This study was limited by the **resolution** of the human eye, which is the ability to distinguish two objects as separate. With the invention and use of the compound microscope (microscopes increase the resolution), much greater detail was seen, and thus began the study of cells and tissues. **Light microscopy** involves the use of visible light and glass lenses to magnify and observe a specimen. Electron microscopy, which uses electrons passing through or bouncing off of material, has revealed much greater detail than observable with light microscopy. This exercise involves the use of the compound light microscope, how to examine prepared slides under the microscope, and how to make slides of fresh material for study.

Learning Objectives

At the end of this exercise you should be able to

1. name the parts of the microscope and their functions presented in this exercise;
2. demonstrate the proper use of the compound light microscope;
3. place a microscope slide on the microscope and observe the material, in focus, under all magnifications of the microscope;
4. calculate the total magnification of a microscope based on the specific lenses used;
5. prepare a wet mount for observation; and
6. list the rules for proper microscope use.

Materials

Compound microscope
Prepared slide with the letter *e* (or newsprint and scalpel)
Transparent ruler, or slide with grid etched on it (grid slide)
Glass microscope slides
Coverslips
Lens paper
Kimwipes or other cleaning paper
Lens cleaner
Small dropper bottle of water
1% methylene blue solution
Clean, food-grade toothpicks
Histological slides of kidney, stomach, or liver
Silk threads prepared slide

Procedure

Care of the Microscope

Microscopes are expensive pieces of equipment, and you should always take great care handling them. There are a few rules concerning microscopes that you should observe:

1. When you carry the microscope, hold it securely with two hands—one hand under the base and one hand on the arm.
2. Keep the microscope upright at all times. Never tilt the microscope from an upright position, as lenses or filters may fall and break.
3. Keep microscope lenses clean with lens cleaner and softened lens paper. Do NOT use paper towels or clothing.
4. Use only the fine-focus knob when you use the high-power objective lens.
5. Remove slides from the microscope before you put it away.
6. Secure the cord with a rubber band, or wrap the cord carefully around the base of the microscope.
7. Store the microscope with the low-power objective lens in place.
8. Put the microscope away in its proper location.

Microscope Setup Procedure

1. Remove the microscope from the storage area.
2. Take the microscope to your desk. Unwrap the electrical cord and familiarize yourself with the parts of the microscope. Compare the microscope that you have in lab with the one illustrated in figure 2.1. There may be differences between the microscope on your desk and the one in the figure in this lab manual, but you should be able to locate the parts listed in the checklist. Use the checklist to make sure that you find all the listed parts. Place a check mark next to the appropriate space when you locate the part of the microscope.

Microscope Parts

____ Base	____ Arm	____ Objective lens
____ Condenser	____ Iris diaphragm lever	____ Body tube
____ Nosepiece	____ Coarse-focus knob	____ Fine-focus knob
____ Mechanical stage	____ Ocular (eyepiece) lens	____ Light source

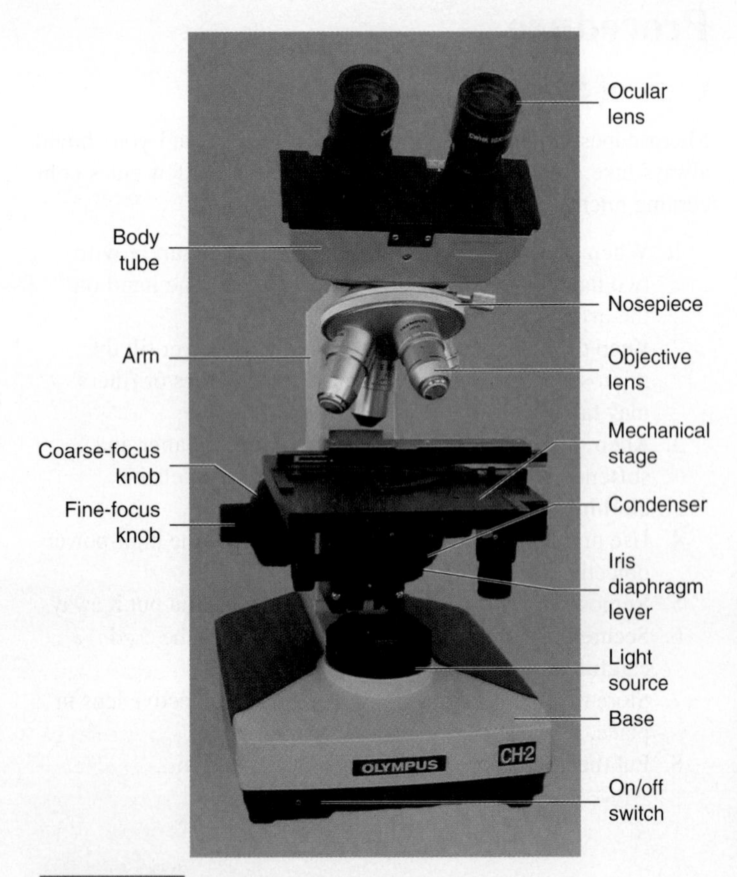

FIGURE 2.1 **Compound Light Microscope**

3. Plug in the microscope, making sure the ocular lens or lenses are facing you. The cord should not hang over the counter or in the aisle where someone might trip on the cord or pull the microscope off the counter. If the microscope has an illuminator (light source) dial, make sure that the setting is on the lowest level.
4. Rotate the nosepiece until the low-power objective lens (the one with the lowest number or the shortest barrel) clicks into place. When you first look at microscope slides, always use the low-power objective lens.
5. Examine a prepared slide with the letter *e* or take a piece of printed material and, using a scalpel, cut out a small section with some letters from the paper. Remove a glass microscope slide from the box and lay it flat on your desk. Place the piece of paper on the slide and add a drop of water to the piece of paper.
6. Place a thin coverslip on the slide by touching one edge of the coverslip to the water and lowering it slowly over the piece of printed material (fig. 2.2). If you drop the coverslip on top of the slide, you will probably trap air bubbles, which may obscure some of your specimen.
7. Locate and turn on the light switch.
8. Place the slide on the microscope stage, so that you can read the printing. Focusing the microscope requires a little bit of patience. Make sure the specimen is on the stage and centered in the open circle on the stage. There should be light coming through the specimen. The coverslip should be on top of the slide and very close to the objective lens. Look at the microscope stage from the side, and adjust the coarse-focus knob so that the coverslip is almost touching the objective lens. Look through the ocular lens and rotate the coarse-focus knob slowly, so that the objective lens and the slide begin to move away from each other. This should bring the object into focus in the field of view. The **field of view** is the circle that you see as you look into the microscope.
9. Binocular microscopes usually have an adjustable left ocular lens. Focus the coarse-focus knob so that the right eye is in focus. If this is difficult to do with both eyes open, you can place a piece of paper in front of your left eye while you focus for your right eye. Once the right eye is in

Scanning lens = 4x
Low power = 10x
High power = 40x
Oil immersion = 100x

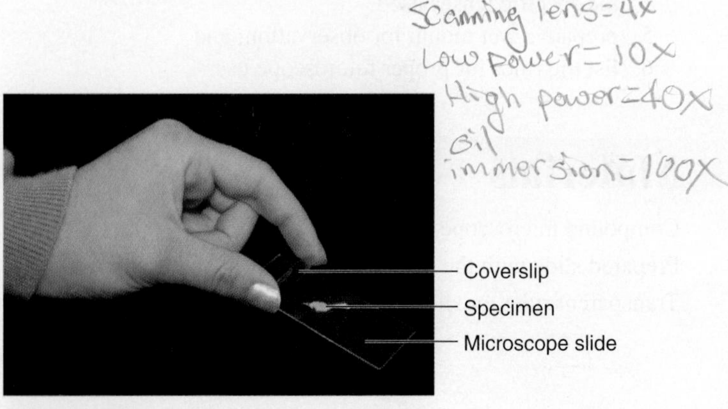

FIGURE 2.2 **Preparation of a Wet Mount**

Total magnification power = ocular lens x objective lens

focus, use the knurled ring on the ocular lens (NOT the coarse-focus knob) and adjust it so the left eye is in focus. You can block the right ocular lens with a piece of paper, if you need to, while adjusting the left ocular. It is important that both eyes be adjusted correctly so that you do not have eyestrain which can lead to headaches.

10. After you get the specimen in focus under low power, you should examine the material under higher powers. Do this by centering the image that you observe in the field of view and then switching the objective lens located on the rotating nosepiece to the next higher power. Do not touch the focus knobs at this time. The next higher objective lens should clear the slide. Once you rotate the lens, adjust the focus by using the fine-focus knob. As you move from one lens to another, the specimen should remain in the center of the field of view. This is known as being **parcentric**. You can look at the subject under high power by the same procedure by turning the objective lens to the high-dry lens. If you cannot focus on the high power or have lost the image that you were looking for, *you should return to low power and try the process again*. If you still cannot find the object under high power, you should ask your instructor to help you.
11. Adjust the light. Too much light might "wash out" the color or details of the specimen. Too little light makes it hard to see. You can change the light intensity and color by adjusting the illuminator dial. By using the iris diaphragm knob, found below the stage, you can control the aperture (or hole through which light travels), which may help you see the specimen more clearly. The condenser lens focuses the light on the specimen, and it generally is moved fairly close to the specimen.
12. Draw what the specimen looks like in the space provided.

Does the written material appear right-side-up, or is the image inverted? ______________________________

Is the material oriented correctly, or is the image flipped horizontally? ______________________________

How much of the material occupies the field of view?

____

13. Move the objective lens to the next higher power lens. Examine the specimen again. You may need to adjust the fine-focus knob so that the image becomes clearer. You may have to move the specimen a little to center the image.

As magnification increases, does the field of view increase or decrease in size?

Microscope Troubleshooting

If you are having a difficult time seeing anything or seeing your specimen in focus, there might be several reasons. Use the following troubleshooting list to help you.

Problem	Solution
Nothing is visible in the lens.	Plug in the microscope.
	Turn on the power supply.
	Rotate the objective lens so that it clicks into place.
	The bulb is burned out; replace the bulb.
You see a dark crescent.	The objective lens is not in proper position; click the lens into place.
All you see is a light circle.	The microscope is out of focus; adjust the coarse-focus knob.
	The light is up too high; turn down the light.
	The iris diaphragm is open too much; close it down slightly.

Proper Lighting

Too much or too little light makes a specimen difficult to see. One trick is to locate the edge of the coverslip and turn the coarse-focus knob up and down until the edge is in sharp focus. This lets you know that you are in the approximate focal plane for examining the material on the slide. Move the slide to where the specimen should be, adjust the light, and reexamine it.

Examination of a Prepared Slide

Examine three crossed threads of different colors under low power. As you focus the threads, which color thread is on top? __________

Which color thread is in the middle? __________

Which color thread is on the bottom? __________

Under low power, how many threads or how much of one thread appears to be in focus? As you switch to the next higher power lens, how many threads or how much of one thread appears in focus? The **depth of field** is the amount of distance that appears acceptably in focus under a particular magnification. If you take a photograph of people close to you, frequently the distant landscape is blurry. The part of the image that is out of focus is out of the depth of field. What happens to the depth of field when you increase the magnification in the microscope?

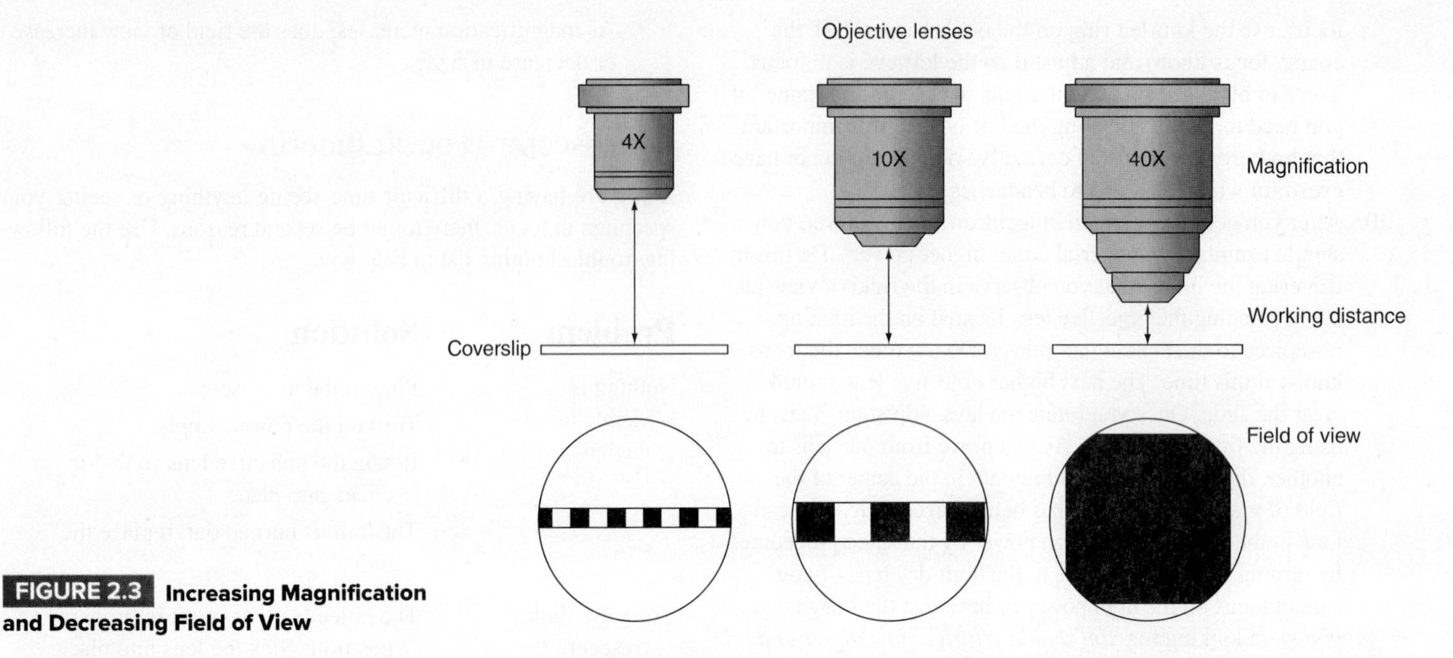

FIGURE 2.3 **Increasing Magnification and Decreasing Field of View**

Magnification and Field of View

You can determine the size of the object under observation if you know the diameter of the field of view. The field of view can be measured directly when using the low-power lens by using a clear ruler, an overhead transparency of a ruler, or a specialized slide that has a grid etched on it. If higher-power lenses are used, rulers won't work and you have to use a grid slide or calculate the field of view. Remember that a millimeter is one-thousandth of a meter and a micrometer is one-millionth of a meter. There is an inverse relationship between the diameter of the field of view and the magnification used. You can first calculate the **total magnification** using the following procedure. Look at the barrels of your microscope and determine the magnification of each lens.

Eyepiece (ocular) magnification: __________

Low-power objective lens magnification: __________

Total magnification (= ocular magnification × objective magnification): __________

Place a transparent ruler or grid slide on the stage of the microscope. The space between each dark line that runs vertically on a ruler is 1 millimeter (mm). Count the number of millimeters at the broadest part of the field of view, and enter this number as the diameter of the field of view in the following space.

Diameter of the field of view (mm): __________

You can calculate the length of an object by determining how much of the diameter of the field of view it occupies. Let's say that the diameter of the field of view is 10 mm. If an object takes up one-half of the field of view, then you can estimate its size at 5 mm. If the object takes up only one-third of the field of view, how large is it? Record your answer in the following space.

Object size (mm): __________

As the magnification increases, the field of view decreases proportionally. Thus, if the diameter of the field of view is 10 mm at one magnification and you double the magnification by changing lenses, the field of view is reduced to a diameter of 5 mm. If you switch to a new lens and increase the magnification by 10 times, then the field of view is reduced to one-tenth of the original field of view. Look at figure 2.3 for a representation of this.

Keep the clear ruler or grid slide under the microscope and increase the magnification to the next higher power by moving the next larger objective lens in place. Record the total magnification of your microscope with this objective lens.

Total magnification: __________

Examine the ruler or grid slide under the microscope and record the diameter of the field of view in millimeters.

Diameter of the field of view in millimeters: __________

Has the increase in magnification produced a decrease in the field of view? __________ If there is a decrease in the field of view, is it proportional to the magnification? __________

Now calculate the total magnification of the microscope using the high-power objective lens.

Magnification with high-power objective lens: __________

You will not be able to measure the field of view accurately under high power with a ruler; however, you should be able to use a grid slide or calculate the diameter of the field of view. For example, if the diameter of the field of view is 2.5 mm at 40 power (40×), then it is 0.25 mm at 400× (10 times more magnified yet one-tenth the field of view). Calculate or measure the diameter of the field of view under high power.

Diameter of the field of view under high power: __________

Preparation of a Wet Mount

You can make relatively quick and easy observations under the microscope as long as the material is thin enough and small enough. One technique for cell examination is to examine cells from the inside of the oral cavity. Do the following procedure.

1. With a toothpick, *gently* rub the inside of your cheek.
2. Smear the cheek material from the toothpick on a clean microscope slide.
3. Place a drop of methylene blue on the smear.
4. Place a coverslip on one edge of the drop and slowly lower it. Avoid trapping air bubbles in the process (fig. 2.2). The small, oval structures inside the cells are the nuclei.
5. Draw your observations in the space provided.

Your illustration of a cheek cell:

For another observation, remove a hair from your head (preferably one with split ends) and examine it by making a wet mount. Tie your hair in a knot, place it in the center of the slide, and add a drop of water. Draw what you see in the space below.

Observation of a Prepared Slide

Examine a prepared slide of tissue provided by your instructor. Examine the entire sample using the low-power objective lens. Scan the entire area, looking for areas you want to observe more closely. Move to the next higher power and adjust the focus using the fine-focus knob. Finally, examine the material with the high-power objective lens and draw what you see in the space provided.

Name of the sample (kidney, liver, intestine, etc.): __________

____

Your illustration of material from a prepared slide: __________

____

You should use the iris diaphragm lever and illuminator dial to adjust the light that strikes the specimen.

Oil Immersion Lens

The objective lenses you have used so far are called "dry" lenses. Your lab may be equipped with microscopes that have oil immersion lenses. Light refracts (bends) differently through air than it does through glass. Under high magnification, oil is used because it has the same level of refraction (it bends light the same) as glass. The techniques for using these lenses are somewhat different from those for dry lenses. Once you have examined the specimen using the high-power dry lens, find the spot you want to examine and center it in the field of view. Add a drop of immersion oil on top of the coverslip and carefully swing the oil immersion lens into place. Use the fine-focus knob only, or you may drive the oil immersion lens through the slide and break it. Once you have examined the slide, swing the lens away and remove the slide, carefully wiping away the immersion oil with a clean piece of lens paper (do not use your clothing or a paper towel; these can scratch the lens). Use only lens paper to clean the oil from the oil immersion lens. Use another lens paper to remove any remaining oil, if needed.

Cleaning the Microscope

Smudges on the images you view through the microscope may be due to several things. There may be makeup or dirt on the ocular lens (or lenses). There may be dirt, oil, salt, stains, or other material on the objective lenses. To clean a lens, place a small amount of lens-cleaning fluid on a clean sheet of lens paper. Make one circular pass on the lens and throw the paper away. If you continue to clean the lens with the same lens paper, you can grind dirt or dust into the lens. Use a fresh piece of lens paper and repeat the procedure if further cleaning is needed.

You may want to clean the microscope slide before you examine it. Use a cleaning paper, such as a Kimwipe, to clean oil or dust from the slide.

Finally, dust may have collected inside the microscope over the years, or the lenses may be scratched. There is nothing you can do about this, though you may want to bring this to your instructor's attention.

Putting the Microscope Away

When you are finished using the microscope, make sure that you do the following:

1. Remove the slide from the stage.
2. Rotate the nosepiece so that the low-power (4×) lens is down.
3. Make sure that the slide clip is centered so the bar is not sticking out from the side of the microscope.
4. Lower the mechanical stage (or raise the objective lenses).
5. Wrap the cord loosely around the microscope or secure the cord with a rubber band.
6. Hold the microscope with two hands, one on the arm and one on the base.
7. Put it away in the correct place.

Exercise 2

Microscopy

Name Kierstyn Convers Date ______

1. If the ocular lens is 10×, what are the total magnifications for the following objective lenses?

   a. 7× ______

   b. 15× ______

   c. 20× ______

2. What is the name of the thin piece of glass that is placed on top of a specimen? ______

______

______

______

3. The microscope that you use in lab is a(n) ______.

   a. compound light microscope b. dissecting microscope c. electron microscope

4. What is the name of the circle you see when you look through the ocular lens of the microscope?

______

______

______

5. What is the function of the iris diaphragm of the microscope? ______________________________

____

____

____

6. If the diameter of the field of view is 5.6 mm at 40×, what is the diameter at 80×? ______________________________

____

____

____

7. Label the parts of the microscope illustrated.

   arm
   base
   body tube
   coarse-focus knob
   condenser
   fine-focus knob
   light source
   mechanical stage
   objective lens
   ocular lens

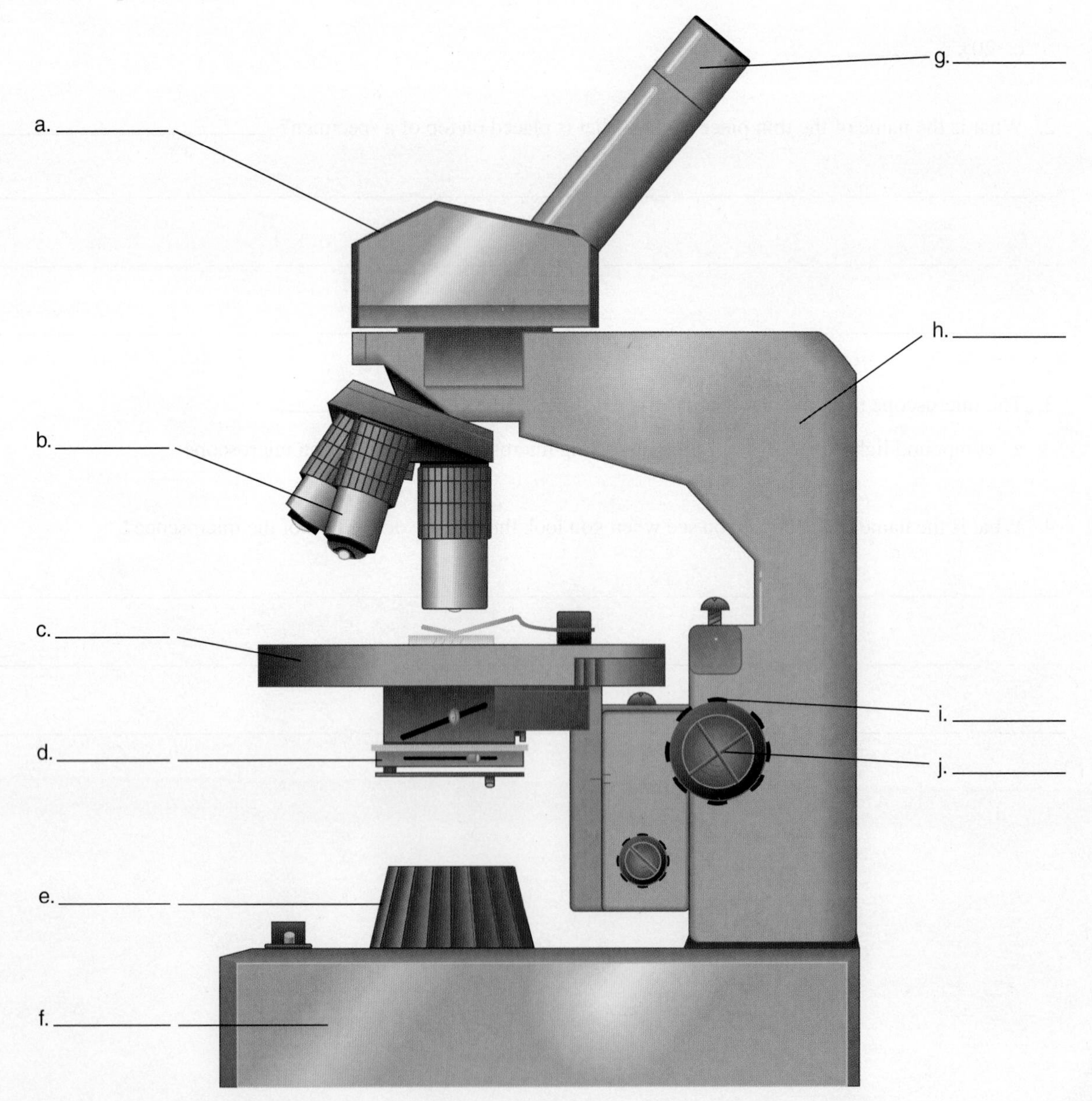

8. When you switch from a low-power objective lens (for example, 4×) to a higher-power objective lens (for example, 10×), what happens to the working distance between the lens and the coverslip? ______

9. When you change from a low-power objective lens to a high-power one, what happens to the field of view? Does it increase or decrease? ______

10. When should you use the low-power lens on the microscope? ______

11. How should you clean the lenses of a microscope? ______

12. What is the proper way to carry the microscope in lab? ______

13. Examine the following field of view. The size of the object is ______.

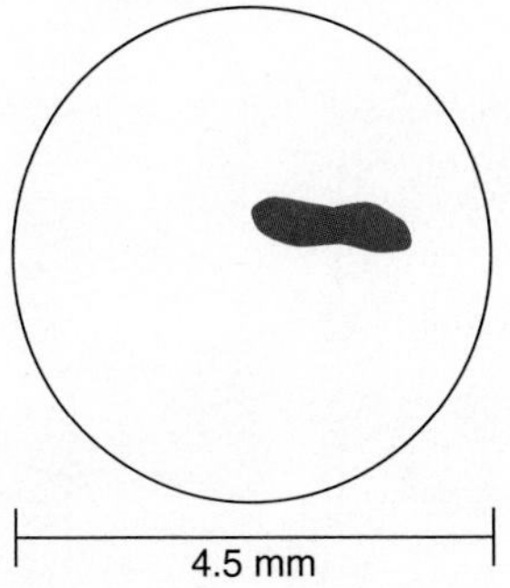

8. When you switch from a low power objective lens (for example, 4x) to a higher power objective lens (for example, 10x), what happens to the working distance between the lens and the coverslip? ______________________

____

____

____

9. When you change from a low power objective lens to a high power one, what happens to the field of view? Does it increase or decrease? ______________________

____

____

____

10. Which should you use, the low power lens or the high power lens, on the microscope? ______________________

____

____

____

11. How should you clean the lenses of a microscope? ______________________

____

____

____

12. What is the proper way to carry the microscope? ______________________

____

____

____

14. Examine the following field of view. The size of the object is ______________________

Laboratory

Exercise 3

Cell Structure

Introduction

The cell is the structural and functional unit of living organisms, including humans. **Cytology,** the scientific study of cells, is an essential part of the study of anatomy. Most diseases that produce obvious physical dysfunction can be traced to some type of cellular change. Cells grow, divide, acquire nutrients, release wastes, respond to local stimuli, and perform many functions, some of which are unique to the organ in which they are found.

The membrane of the cell **(plasma membrane)** is the dynamic interface between the internal environment of the cell and the external environment. In humans, most cells are bathed in a liquid medium called **extracellular fluid (ECF),** which provides the cells with nutrients, oxygen, hormones, water, ions, and other materials. From the cell's interior, it releases ammonia, carbon dioxide, and other metabolic products into this liquid medium. The plasma membrane is vital in the exchange of materials between the cell's interior and the environment surrounding it. The exchange of materials between the cell and the ECF maintains the homeostatic balance the cell must have in order to survive. Even small changes in the concentration of certain materials in the cell can lead to cellular death, so the constant adjustment of water, ions, and other metabolic products is extremely important. In large part, the plasma membrane actively regulates what enters and what exits the cell. This is done passively in some cases, and in others ATP is used to actively transport material.

In this exercise, you examine the structure of animal cells and learn how the cells of the body divide to make new cells.

Learning Objectives

At the end of this exercise you should be able to

1. describe the importance of cells in the makeup of the body;
2. list the functions of the plasma membrane;
3. list all the organelles and their functions;
4. describe the three main events of the cell cycle; and
5. name the four phases of mitosis and the events occurring in each phase.

Materials

Models or charts of animal cells
Electron micrographs of cells or a textbook with electron micrographs
Prepared slides of whitefish blastula
Microscopes
Modeling clay (plasticine)—two colors
Marbles

Procedure

Overview of the Cell

There are many different types of cells in the body. Some are long and thin, others are spherical, and still others are flat. We will examine a representative cell as an example, but realize that there is tremendous diversity in cell shapes and functions. Cells consist of two main parts, the plasma membrane and the **cytoplasm** (fig. 3.1). The plasma membrane is the outer boundary of the cell; although you cannot see it when examining a slide using the light microscope, its location can usually be determined by the difference in color between the **cytosol,** or **intracellular fluid (ICF),** and the extracellular fluid (ECF). The cytoplasm is the portion of the cell in which water, dissolved materials, and small cellular **organelles** (*organelle* = small organ) are found. The nucleus is one of these organelles, and it directs the cell's activities and stores its genetic information. It is visible with the light microscope and frequently appears as a spherical or oblong structure. Figure 3.1 shows the plasma membrane and cytoplasm. Locate these on the model or charts in the lab.

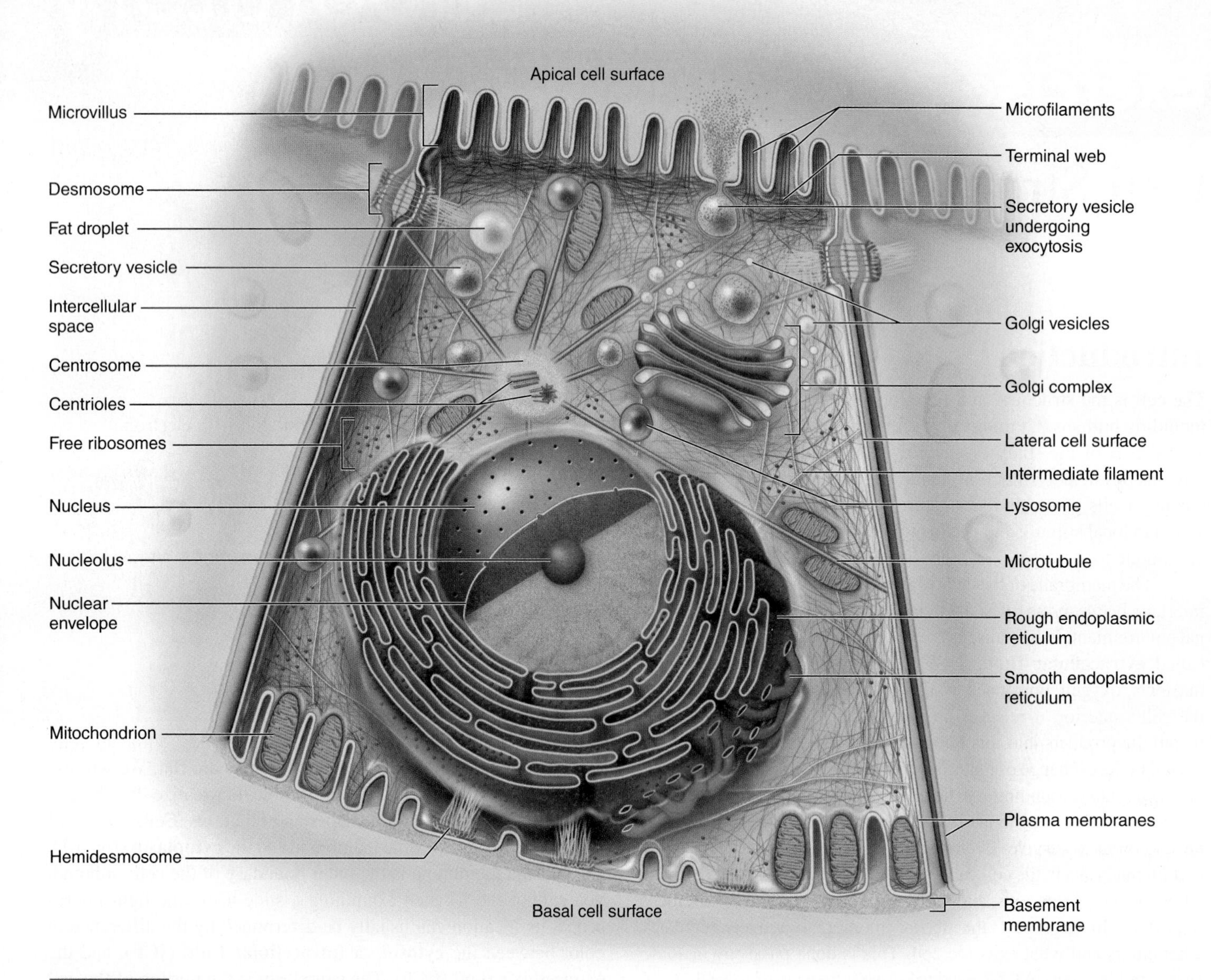

FIGURE 3.1 **Overview of the Cell**

Plasma Membrane

The plasma membrane is composed of a **phospholipid bilayer,** proteins, cholesterol, and other molecules. This bilayer consists of outer and inner phospholipid molecules, each with a peripheral phosphate head and a hydrocarbon (lipid) tail directed toward the middle of the membrane (fig. 3.2). Interspersed among the phospholipid molecules are **cholesterol molecules,** which provide stability to the membrane or, in larger concentrations, can make the membrane more fluid. Two major types of proteins are also found in the membrane. **Peripheral proteins** are found on the inner or outer surface of the membrane, while those passing through the membrane are known as **transmembrane proteins.** Transmembrane proteins may have carbohydrates or other molecules associated with them and frequently serve as cell markers. Some transmembrane proteins function as channels by which specific materials can pass through the membrane. The plasma membrane proteins are important in establishing electrochemical charge differentials, which allow for nerve impulse conduction and muscle contraction. It is also a selectively permeable membrane that provides an entrance to or exit from the cell for some materials while excluding other material from entering or exiting the cell's interior. Proteins may anchor one cell to another, provide a place for metabolic reactions to take place, act as cell markers that identify a particular cell, or act as receptors or channels. Other phospholipid bilayer membranes are found enclosing cytoplasmic structures, and these are simply known as **membranes.**

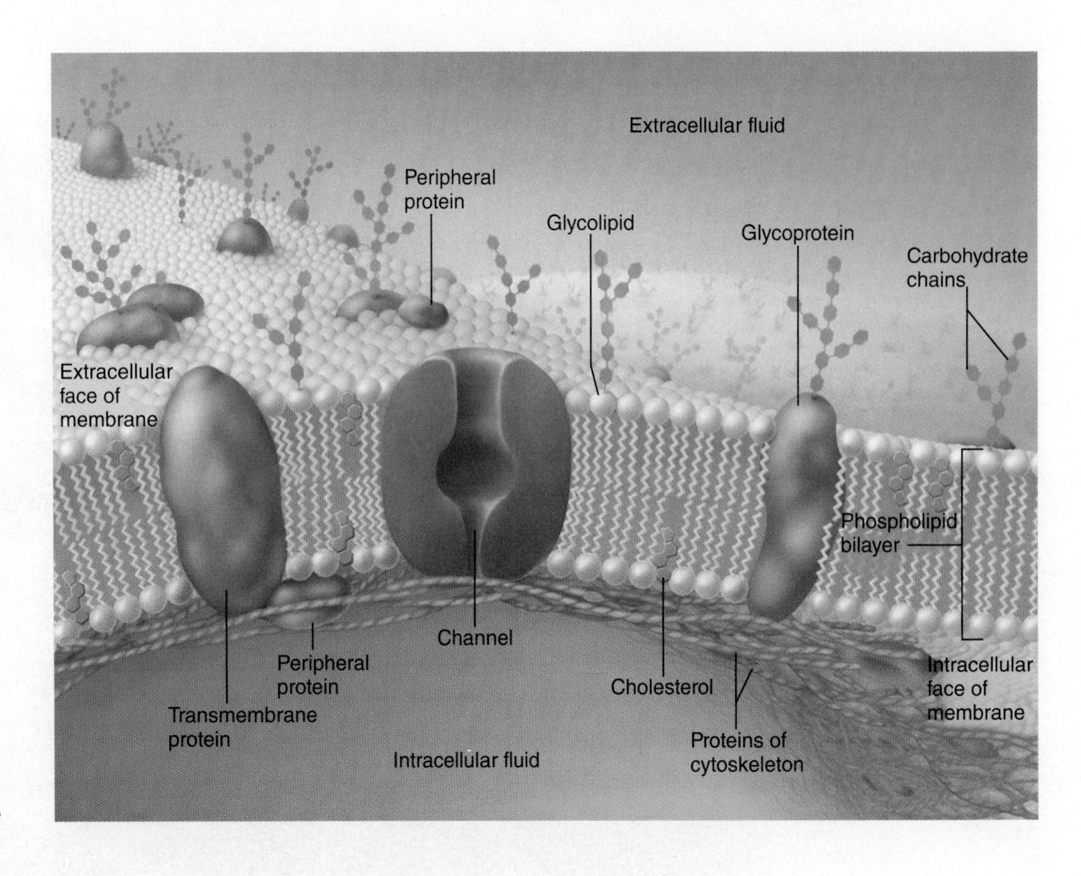

FIGURE 3.2 **Plasma Membrane**

In the following spaces, draw a sketch of the cellular structures and label them, using the listed terms. Use your lecture text as a guide for the structure of these cellular components.

Plasma membrane—phospholipid, phospholipid bilayer, transmembrane protein, peripheral protein, channel protein, cholesterol

Mitochondrion—cristae, intramembranous space, matrix, inner membrane, outer membrane

Golgi complex—cisternae, vesicle, membrane

Rough and **smooth endoplasmic reticulum**—cisternae, membrane, ribosome

Nucleus—nucleoplasm, nucleolus, nuclear membrane, nuclear pore

Centrioles—microtubules

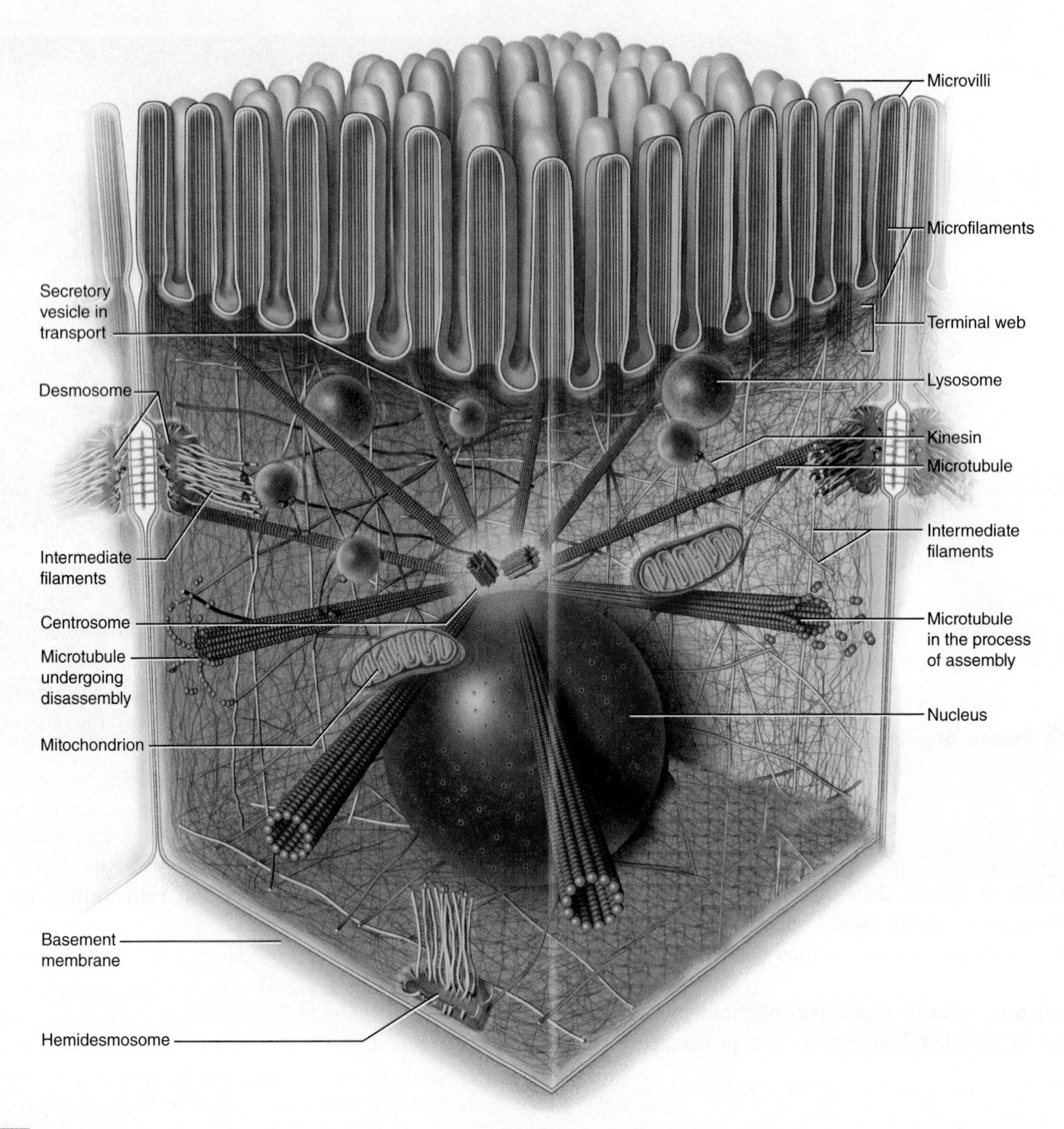

FIGURE 3.3 **Cytoskeleton** The cytoskeleton consists of microtubules, microfilaments, and intermediate filaments.

Cytoplasm

Most of the inside of the cell is cytoplasm, which consists of the inner fluid portion of the cell known as the **cytosol,** the inner framework of the cell called the **cytoskeleton,** and the small specialized units of the cell called **organelles.** The cytosol is composed of water with dissolved materials, such as sugars, ions, proteins, and amino acids. Figure 3.3 illustrates the cytoskeleton (not visible under the light microscope), which consists of microtubules, microfilaments, and intermediate filaments, all of which provide shape to the cell, a place to anchor organelles, and resistance to gravitational and other forces acting on the cell. **Microtubules** are made of the protein tubulin and are approximately 25 nanometers (nm) in diameter. **Intermediate filaments** are composed of fibrous proteins and are approximately 10 nm in diameter. **Microfilaments** are made of **actin** and are approximately 6 nm in diameter.

Other Cellular Components

Most cells have extensions on their surface such as **microvilli, cilia,** and, in human sperm cells, **flagella. Microvilli** are small extensions of the plasma membrane of some cells that increase the surface area of the cell. They are found in many cells including those of the digestive and urinary systems. In some cells, such as specialized sense organs, they serve a sensory function.

Cilia and **flagella** extend from the edges of the cell; they consist of bundles of microtubules and are covered by the plasma membrane. The general structure of cilia is the same as that of flagella, but cilia are shorter and flagella have additional cytoskeletal filaments. Cilia are seen in figure 3.4.

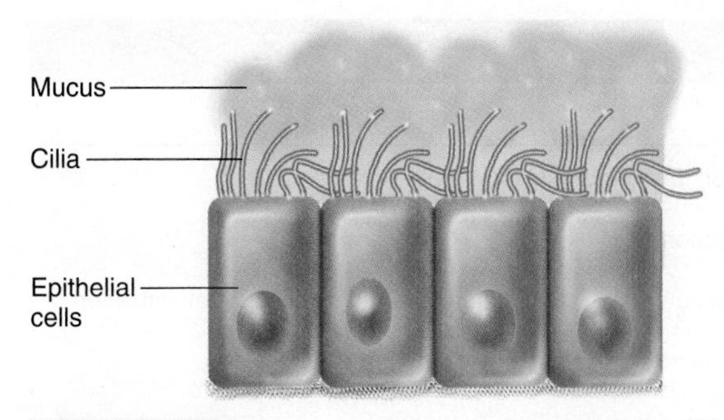

FIGURE 3.4 Cilia on Cells in the Respiratory Passage

Organelles

Organelle literally means "small organ." Each organelle serves a particular function in the cell. There are two types of organelles—membranous and nonmembranous. Organelles represent a wonderful example of specialization on a microscopic scale—individual organelles have structural characteristics reflecting their specific functions. Look at the illustrations of the various organelles as you read the following text.

Mitochondria One of the major functions of the mitochondrion (MY-toe-condree-un) (plural, *mitochondria*) is to convert the stored chemical energy in sugars, fatty acids, and amino acids to stored chemical energy in molecules of adenosine triphosphate (ATP). Mitochondria are elongated organelles approximately 0.2 to 5.0 micrometers (μm) in length. Mitochondria have two membranes similar in structure to the plasma membrane (phospholipid bilayers). Examine the illustration of a mitochondrion in figure 3.1. Note the separate outer and inner membranes. The inner membrane has folds called cristae. These cristae increase the surface area of the mitochondrion. The inner region of the mitochondrion is known as the mitochondrial matrix (stroma), and it is the area where the citric acid cycle occurs. One of the functions of the citric acid cycle is to convert food molecules into energy molecules that the cell can use.

Ribosomes Ribosomes (RYE-boh-somz) are the smallest of the organelles (about 25 nm in diameter), are nonmembranous, and produce proteins. They may be found in the cytosol, in mitochondria, on the endoplasmic reticulum, and in other locations as well. Figure 3.1 shows examples of ribosomes.

Endoplasmic Reticulum The endoplasmic reticulum (EN-do-PLAZ-mic re-TIK-ulum) is an organelle composed of a network of enclosed channels called **cisternae** (sis-TUR-nee). The name *endoplasmic reticulum* means the "little net inside the cytoplasm." There are two expressions of the endoplasmic reticulum (see fig. 3.1): rough endoplasmic reticulum (RER), which contains bound ribosomes, and smooth endoplasmic reticulum (SER), which does not have ribosomes on its surface. The rough endoplasmic reticulum is associated with the nucleus and synthesizes proteins. The endoplasmic reticulum also produces lipid and steroid compounds and detoxifies material.

Golgi Complex The Golgi (GOAL-jee) complex receives material from the endoplasmic reticulum and other parts of the cytoplasm and serves as an assembly and packaging organelle. Proteins from the endoplasmic reticulum are joined with carbohydrates, lipids, metals (such as the iron in hemoglobin), or other materials in the Golgi complex and then secreted in vesicles. Examine the Golgi complex in figure 3.1 and note the vesicles released from this organelle.

Centrioles Centrioles are unique organelles. They form microtubules and a structure known as the spindle apparatus, which is involved in cellular division.

Nucleus The nucleus is an organelle that has two major functions: one is to house the genetic information of the cell, and the other is to control the various tasks of the cell. Nucleoli (singular, *nucleolus*) make ribosomes, the protein-producing organelles in the cytoplasm of the cell.

Vesicles Vesicles are membrane-bound sacs that store and transport material within the cell, digest material brought into the cell, or digest old or damaged organelles. Vesicles can fuse with other organelles or with the plasma membrane. The endoplasmic reticulum, Golgi complex, lysosomes, peroxisomes, and the plasma membrane are involved in vesicular activity. There is a limit to the size of a molecule permitted to pass through the phospholipid bilayer or through the protein channels in the plasma membrane. Transportation of large-molecular-weight materials into and out of the cell is an important function of vesicles. They protect the integrity of the plasma membrane by maintaining the isolation of the cytoplasm from the extracellular fluid. If a substance were to exit the cell by opening a hole in the plasma membrane, the likely consequence would be that the cell would burst. Large-molecular-weight substances are enclosed in vesicles, and the vesicular membrane fuses with the plasma membrane, ejecting the material without disrupting the plasma membrane. Vesicular transport is a major form of intracellular traffic between membrane-bound organelles or between membrane-bound organelles and the plasma membrane. Transport vesicles take material from the endoplasmic reticulum to the Golgi complex (see fig. 3.1). Other vesicles have digestive or enzymatic functions. Two of these are lysosomes and peroxisomes.

Lysosomes Lysosomes (LY-so-somz) are vesicles filled with digestive enzymes. Damaged or old organelles are also removed by lysosomes.

Peroxisomes Peroxisomes use oxygen (O_2) to oxidize organic compounds. This generates hydrogen peroxide, which can subsequently be used to oxidize other molecules.

Extracellular Matrix

Humans are composed of not only cells but also nonliving material known as the **extracellular matrix (ECM).** The ECM supports cells, anchors cells, separates tissues from one another, regulates communication between cells, and assists in wound healing. The extracellular matrix consists of fibers, binding proteins, and ground substance.

ACTIVITY

Examine figure 3.1 and locate as many of the organelles listed in table 3.1 as you can. Be able to recognize the variances in structure and function of the organelles.

TABLE 3.1 Organelles and Their Functions in Cells

Organelle	Membrane	Function
Mitochondrion	Double	ATP production; fatty acid oxidation
Ribosome	None	Protein production
Endoplasmic reticulum	Single	Protein production for export; lipid and steroid synthesis; detoxification
Golgi complex	Single	Assembly of macromolecules; transport material in vesicles
Lysosome	Single	Digestion of material
Peroxisome	Single	Oxidation of organic molecules

The Cell Cycle

One of the great wonders of science is the mechanism by which a single cell, the result of the fusion of egg and sperm, develops into a complex, multicellular organism, such as a human. Various estimates put the number of cells in the human body in the trillions. All these cells came from the first cell, or zygote. In this part of the lab exercise, you examine the mechanism by which this occurs.

Most cells produce more cells by a process known as the cell cycle. The cell cycle can be divided into three general events: **interphase, mitosis,** and **cytokinesis,** illustrated in figure 3.5. Most of the time spent in the life of the average cell is in interphase.

Interphase Interphase is the time when a cell undergoes growth and duplication of DNA in preparation for the next cell division. If a cell is not going to divide any further (such as most neural cells and some muscle cells), then interphase is regarded as the time when a cell carries out normal cellular function.

Interphase has three separate phases, known as the G_1 phase, S phase, and G_2 phase. In the G_1 phase (*G* stands for "gap"), cells are in the process of growing in size and producing organelles. In the S phase (*S* stands for "synthesis"), the DNA of the cell is duplicated. The double helix of the DNA molecule unzips, and two new, identical DNA molecules are produced. In the final phase of interphase, the G_2 phase, the cell continues to grow and prepares for the process of mitosis. Some cells do not undergo further division and are said to be in the G_0 (G zero) phase. Cells in interphase have a distinct nuclear membrane, and the genetic information is dispersed in the nucleus as chromatin.

Mitosis Mitosis (my-TOE-sis) is a continuous phenomenon that has been divided into four distinct phases. Mitosis is nuclear division, and it involves the division of genetic information to produce two identical nuclei, which eventually occupy two cells. In order for mitosis to occur, the chromatin in the nucleus of the cell must condense into compact units called **chromosomes** (each species has its own number of chromosomes). Chromosomes consist of two chromatids held at the center by a centromere. Examine figure 3.6 for the structure of a chromosome. Also note the structure of chromosomes as you study the cells undergoing mitosis.

The four phases of mitosis—prophase, metaphase, anaphase, and telophase—are described next. Refer to figure 3.7 as you read the descriptions.

FIGURE 3.5 The Cell Cycle The mitotic phase has been expanded to see the distinct sections. Most of the time, a cell is in interphase. Cytokinesis is a separate event, beginning in anaphase and ending in telophase.

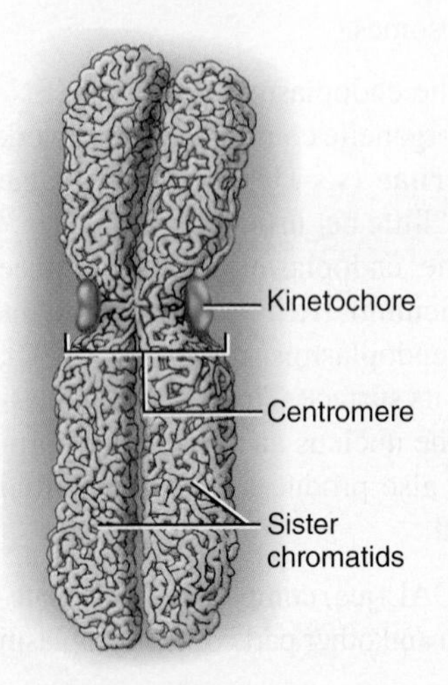

FIGURE 3.6 Structure of an Isolated Chromosome

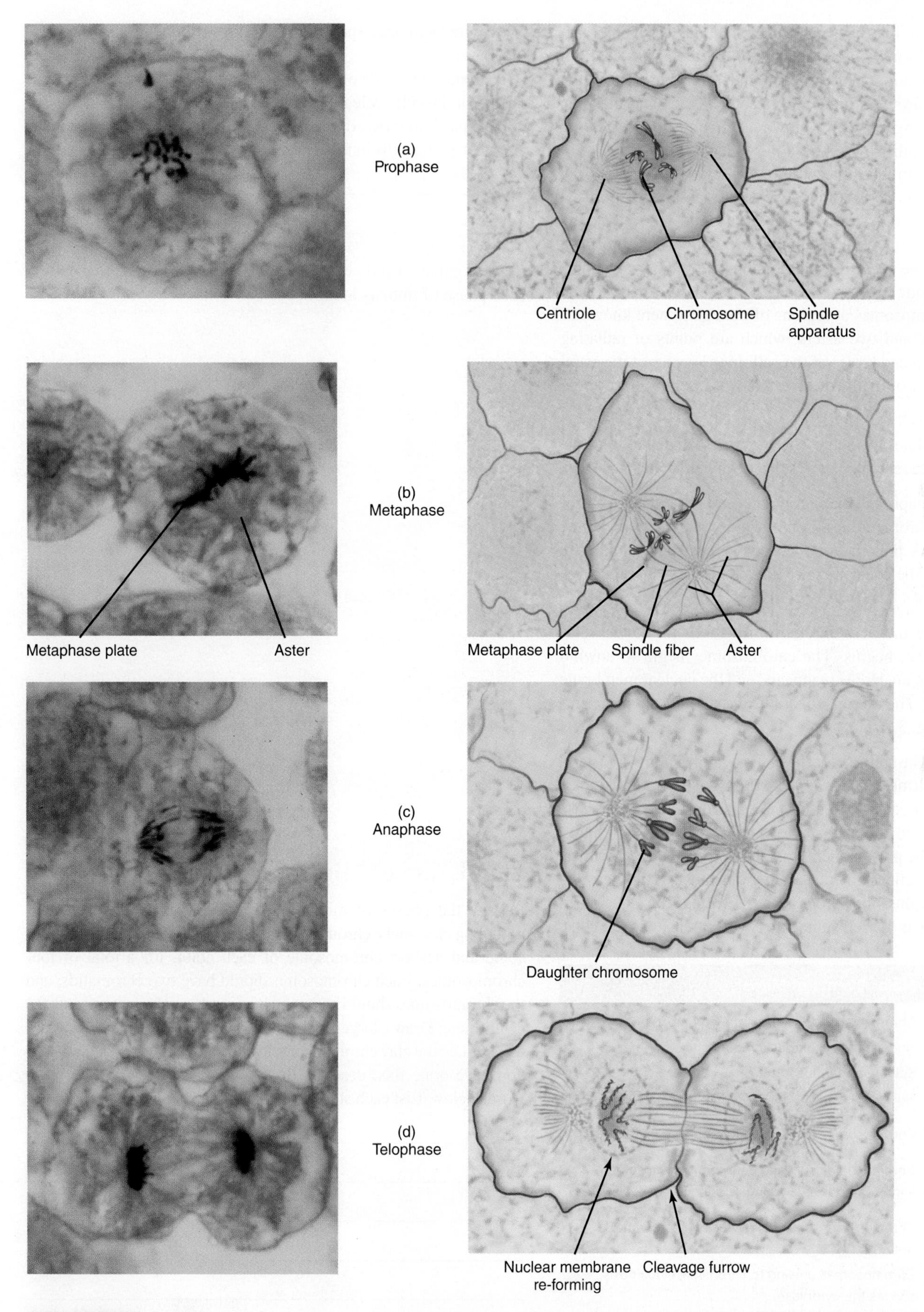

FIGURE 3.7 Phases of Mitosis: Whitefish Blastula (1,000×) (a) Prophase; (b) metaphase; (c) anaphase; (d) telophase with cytokinesis.

actually splits

Prophase Cells in interphase have a distinct nuclear membrane, and the genetic information is dispersed in the nucleus as chromatin. The first indication that a cell is undergoing mitosis is the condensation of chromatin into chromosomes. Each chromosome consists of two elongated arms known as **chromatids,** which are connected to each other by a **centromere.** See figure 3.7*a*.

In addition to the thickening of the chromosomes, the nucleolus disappears and the nuclear membrane begins to disassemble. In order for the chromosomes to separate and move away from each other, the nuclear membrane, which normally forms a barrier, must not be present. Additionally, the mitotic apparatus appears. The mitotic apparatus consists of spindle fibers, which attach to the chromosomes at regions of the centromere known as the **kinetochores,** and two asters, which are points of radiating astral fibers at each end (pole) of the cell. In the center of the astral spindle fibers are two small structures known as centrioles forming the centrosome.

Metaphase In this phase, the chromosomes align between the poles of the cell in a region known as the metaphase plate (fig. 3.7*b*).

Anaphase In anaphase, the chromatids separate at the centromere, and each chromatid is now known as a new chromosome. The spindle fibers pull the new chromosomes toward opposite poles of the cell. The centromere region moves first, and the arms of the chromosomes follow (fig. 3.7*c*).

Telophase Once the new chromosomes reach the poles, telophase (TEE-lo-faze) begins. The chromosomes begin to unwind into chromatin, the nucleolus reappears, and the nuclear membrane begins to re-form. The mitotic apparatus disassembles, thus terminating mitosis (fig. 3.7*d*).

Cytokinesis The splitting of the cell's cytoplasm into two parts is known as **cytokinesis** (SY-toe-kih-NEE-sis). Although cytokinesis is a distinct process, it frequently begins during late anaphase or early telophase. In late anaphase, as the chromosomes are moving to the poles, the plasma membrane begins to constrict, primarily by contracting actin filaments, at a region known as the **cleavage furrow.** This begins the process of dividing the cytoplasm (fig. 3.7*d*), ending as the cell splits into two separate daughter cells.

TABLE 3.2	Major Events of Mitosis
Prophase	Chromatin condenses to form chromosomes. Nuclear membrane disappears. Spindle apparatus forms. Nucleolus disappears.
Metaphase	Chromosomes align on the metaphase plate.
Anaphase	Chromosomes split and daughter chromosomes migrate to poles; cytokinesis often begins.
Telophase	Chromosomes reach poles; nuclear membrane re-forms. Chromosomes unwind to chromatin; cytokinesis divides the cytoplasm. Nucleolus reappears.

The cytoplasm and the organelles are effectively divided into two parts.

Examine a slide of whitefish blastula and look for the various phases of the cell cycle in those cells. Most of the cells that you see are in a particular part of the cell cycle. What is this phase and why are most of the cells in this phase? ______________________

____

____

Compare a slide with figure 3.7. Draw representative cells in each phase of mitosis in the space below.

Simulation of Mitosis

Review the phases of mitosis in table 3.2. Using two colors of modeling clay, make chromosomes. You should have a long chromosome and a short chromosome of each color, for a total of four chromosomes. Each chromosome should have two chromatids, and the chromosomes should be joined by a marble, which represents the centromere. Draw a large circle on a sheet of paper to represent a cell. Manipulate the clay chromosomes to show how mitosis occurs. After you have done this, describe the process of mitosis in your own words below. List each stage and what happens in that stage.

____

____

____

____

____

____

Exercise 3

Cell Structure

Name ______________________ Date __________

1. Cells in the body have a fluid surrounding them. What is the name of this fluid? Cytoplasm

2. The cytoplasm has a liquid portion. What is it called? Water?

3. What structure in a cell is mostly composed of a phospholipid bilayer? ______________________

4. Which organelle is responsible for ATP production? ______________________

5. Which organelle makes protein and lipids? ______________________

6. Which organelle in the cell has cristae? ______________________

7. Which organelle directs the activity of the cell and contains most of the genetic information of the cell?

____

8. Which cellular structure is responsible for ribosome production? ______________________

9. Name the cellular structures in the illustration, using the terms provided.

centriole	plasma membrane
Golgi complex	ribosomes
mitochondrion	rough endoplasmic reticulum
nucleolus	smooth endoplasmic reticulum

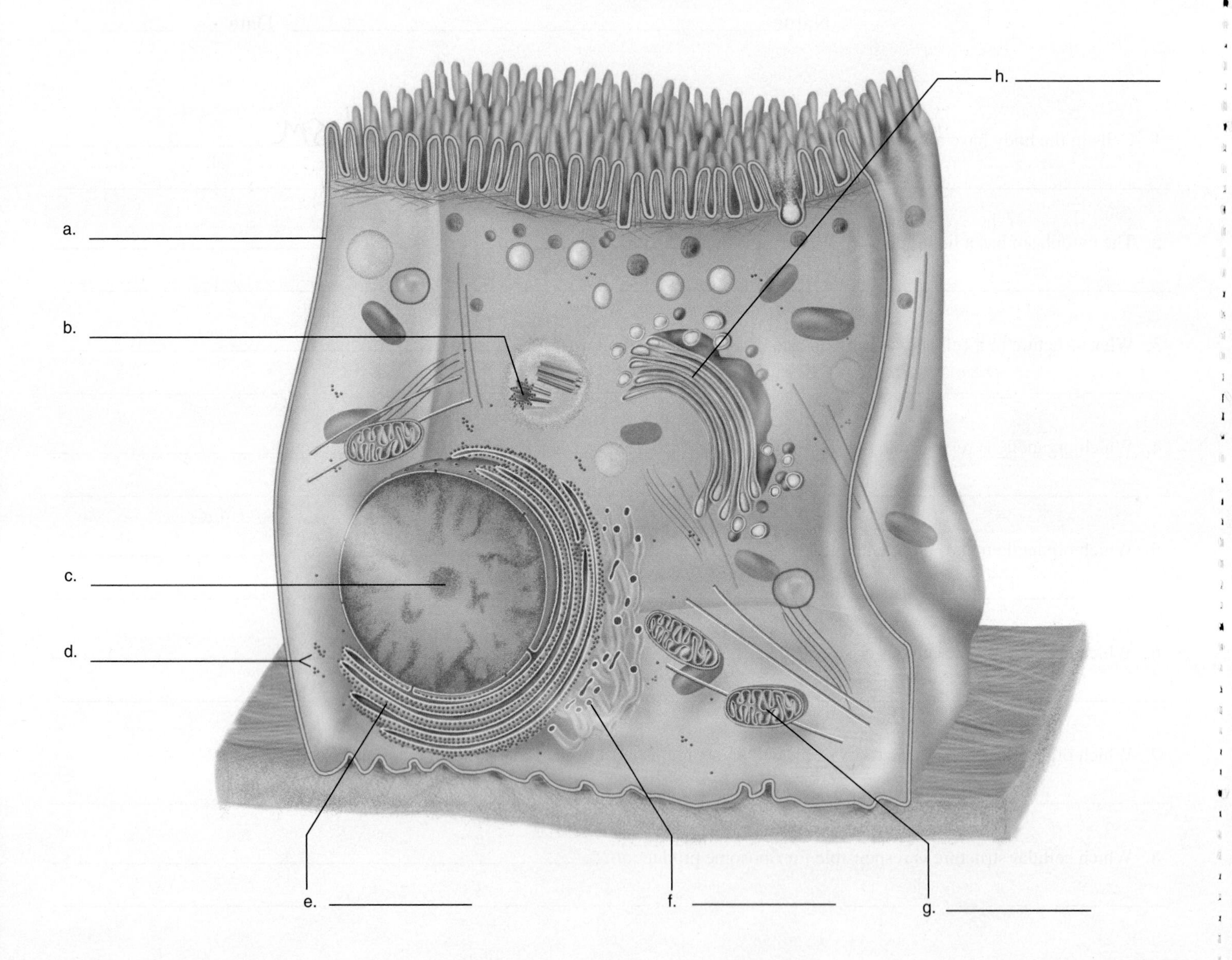

10. The G_1, S, and G_2 phases of the cell cycle are collectively known as which part of the cell cycle?

11. Considering mitosis, cytokinesis, and interphase, in which one is nuclear DNA duplicated?

12. When in the cell cycle do chromosomes first split apart?

13. The division of the cytoplasm occurs in what part of the cell cycle?

14. Describe the four phases of nuclear (mitotic) division and what occurs during those phases.

15. Name the phases of the cell cycle as illustrated.

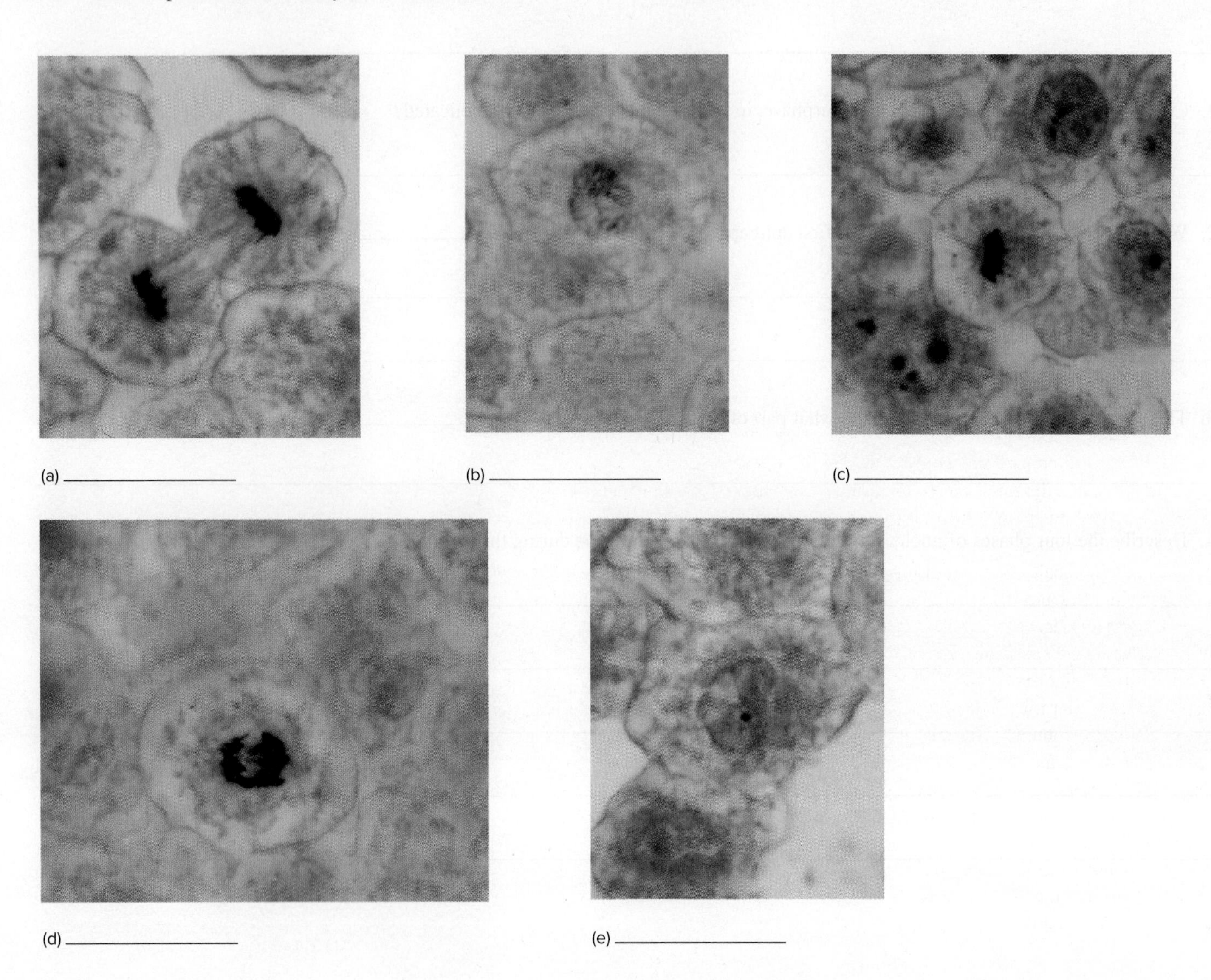

(a) ____________

(b) ____________

(c) ____________

(d) ____________

(e) ____________

16. Name the phases in interphase, and describe what happens during those phases. ____________

____

____

____

____

____

Laboratory

Exercise 4

Tissues

Introduction

The study of tissues, called **histology,** is microscopic anatomy. Individual tissues consist of cells and extracellular material that have a particular function. The study of histology is important because many organic dysfunctions of the human body are diagnosed at the tissue level. Surgical specimens are routinely sent to pathology labs, so that accurate assessment of the health of the tissue, and consequently the health of the individual, can be made.

Individual tissues have specific functions. For example, muscle contracts, and nervous tissue conducts impulses. These functions are due to the specialization of the cells that form the tissue and the extracellular material.

There are four main tissue types found in the human body—epithelial tissue, connective tissue, muscular tissue, and nervous tissue—and the organs of the body are formed by two or more of these tissues. These tissues vary by the type, function of the individual cells, and the nature of the **matrix,** or extracellular material, present.

In this exercise you examine numerous slides of tissue and begin an introduction to histology. In later exercises you revisit histology as you examine various organ systems.

Learning Outcomes

At the end of this exercise you should be able to

1. recognize the various types of epithelium;
2. associate a particular tissue type with an organ, such as kidney or bone;
3. examine a slide under the microscope or a picture of a tissue and name the tissue represented;
4. distinguish between cartilage and other connective tissues;
5. list the three parts of a neuron; and
6. describe the muscle cell types according to location and structure.

Materials

Microscope
Colored pencils
Epithelial tissue slides
- Simple squamous epithelium
- Simple cuboidal epithelium
- Simple columnar epithelium
- Pseudostratified columnar epithelium
- Stratified squamous epithelium
- Transitional epithelium

Muscular tissue slides
- Skeletal muscle
- Cardiac muscle
- Smooth muscle
- All three muscle types

Nervous tissue slide
- Spinal cord smear

Connective tissue slides
- Loose (areolar) connective tissue
- Dense regular connective tissue
- Dense irregular connective tissue
- Elastic connective tissue
- Adipose tissue
- Reticular connective tissue
- Hyaline cartilage
- Fibrocartilage
- Elastic cartilage
- Ground bone
- Cancellous bone
- Blood

Procedure

Before you begin this exercise, you should be thoroughly familiar with the microscopes in your lab. If you need a review, go back to Laboratory Exercise 2. As you examine various tissues, look for distinguishing features that will identify the tissue. Tissues consist of cells and extracellular material known as ground substance or matrix. Tissues are usually stained so that the details become visible. The most common stain is hematoxylin and eosin (H&E) stain. Hematoxylin stains the nucleus purple, and eosin colors the cytoplasm pink.

Epithelial Tissue

Epithelial tissue is a highly cellular tissue, meaning that it is composed mostly of cells with little matrix. It covers or lines parts of the body (such as the skin on the outside or the digestive tract on the inside) or is found in glandular tissue, such as the sweat glands or the pancreas. When you look at epithelial tissue under the microscope, examine the edge of the sample because epithelial tissue is frequently found as a lining. In most cases, epithelial tissue adheres to the underlying layers by way of a **basement membrane,** a noncellular adhesive layer. In the skin, the epidermis is made of epithelial tissue, and the basement membrane connects the epidermis to the underlying dermis. Epithelial tissue is classified according to the shape of the cells and the number of layers present. The cell shapes are **squamous** (SKWAY-mus = flattened), **cuboidal,** and **columnar.** The number of layers is **simple** (cells in a single layer) or **stratified** (cells stacked in more than one layer). Examine tables 4.1 and 4.2 for an overview of epithelium. Epithelial tissue is listed by cell type in the following discussion. The edge of the cell touching the basement membrane is the basal surface, and the upper edge is the apical surface.

STUDY HINT

As you examine various tissues, look for distinguishing features that will identify the tissue. It is a good idea to examine more than one slide of a particular tissue so that you see a range of samples of that tissue. Frequently, the material you see in the lab is from a slice of an organ and as such includes more than one tissue type. For example, a sample of cartilage taken from the trachea contains epithelial tissue, adipose tissue, and other connective tissues in addition to the cartilage you want to study. If you are looking for smooth muscle from the digestive tract, you may also find both epithelial tissue and connective tissue in the slide. Use the figures in this exercise to help you locate tissues on the prepared slides. Examine each slide by holding the slide up to the light and visually locating the sample. Then put the slide on the microscope and examine it on low power, scanning around the slide. Move to progressively higher powers after you have identified the tissue. If you cannot identify the tissue after some searching, then ask your lab partner or your instructor for help. When you look at epithelial tissue under the microscope, examine the edge of the sample because epithelial tissue is frequently found as a lining.

TABLE 4.1 Simple Epithelia

Simple Squamous Epithelium

Microscopic appearance: single layer of flat cells

Significant locations: lungs, inside of heart, and blood vessels

Functions: diffusion, reduction of friction and secretion of serous fluid

Simple Cuboidal Epithelium

Microscopic appearance: small cubes or wedge-shaped cells in single layer

Significant locations: kidney tubules and liver

Functions: absorption and secretion

Simple Columnar Epithelium

Microscopic appearance: tall cells in one layer, with nuclei typically in basal part of cell

Significant locations: stomach and intestines

Functions: absorption and secretion

Pseudostratified Columnar Epithelium

Microscopic appearance: looks stratified but all cells arise from basement membrane, often ciliated

Significant locations: respiratory passages

Functions: secretion of mucus and trapping dust particles, moving them away from lung

TABLE 4.2 Stratified Epithelia

Stratified Squamous Epithelium

Microscopic appearance: many layers, cells cuboidal but flattened toward surface

Significant locations: epidermis, oral cavity, esophagus, and vagina

Functions: resists abrasion, prevents microbial infection, retards water loss in skin

Transitional Epithelium

Microscopic appearance: many layers with teardrop-shaped cells that do not flatten toward surface

Significant locations: urinary bladder

Functions: allows stretching of urinary bladder

Simple Epithelium Simple epithelium is only one cell layer thick. The cells are located on the basement membrane, which may adhere to connective tissue, muscle, or other epithelial tissue. Simple epithelium is classified according to the following shapes.

Simple Squamous Epithelium This epithelial type consists of thin, flat cells that lie on the basement membrane like floor tiles. If these cells are seen from a side view, they look flat and their resemblance to floor tiles becomes apparent. Examine a prepared slide

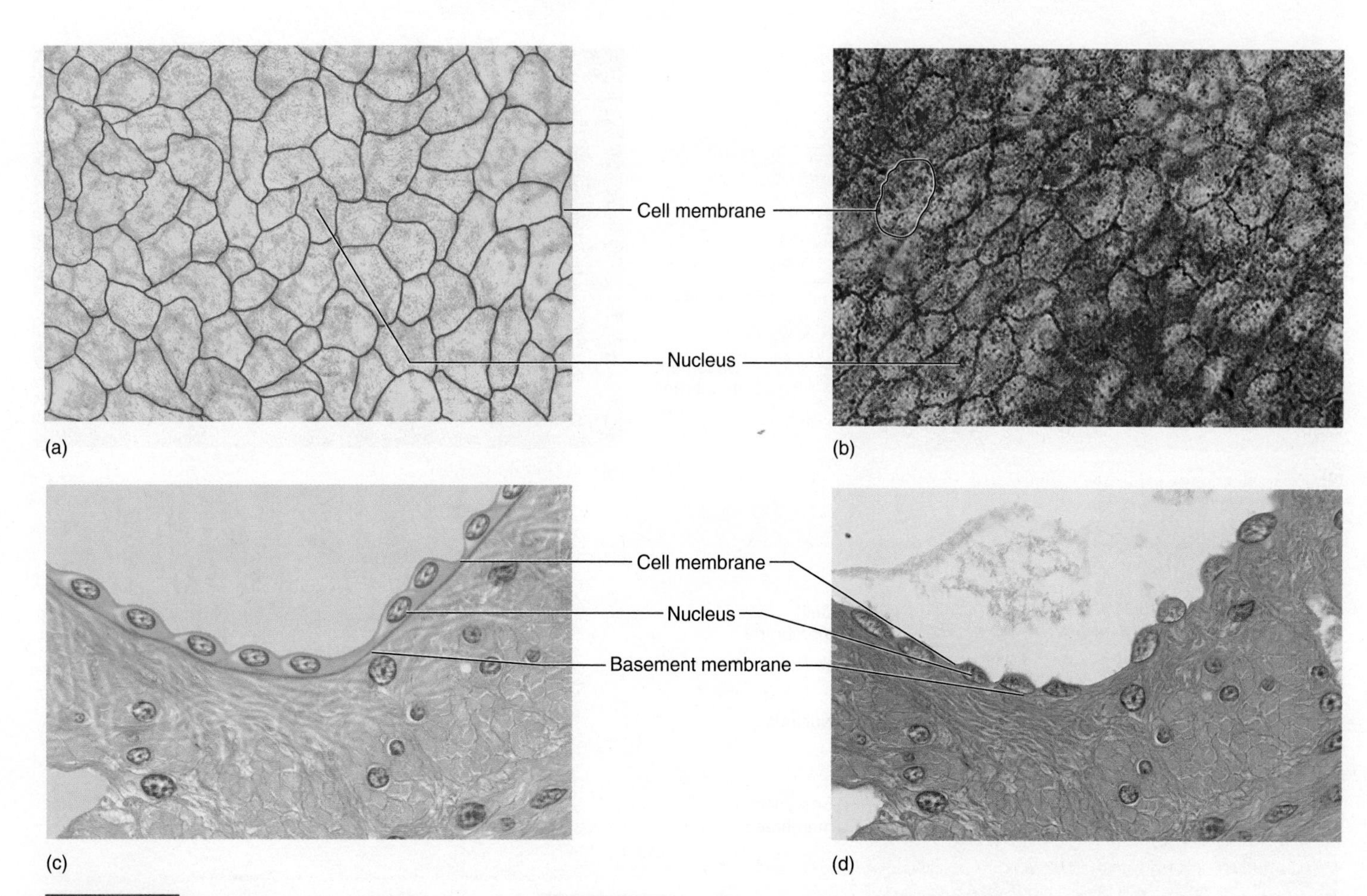

FIGURE 4.1 **Simple Squamous Epithelium** Top view of mesothelium—(a) diagram; (b) photograph (400×). Side view of serosa of small intestine—(c) diagram; (d) photograph (400×).

of simple squamous epithelium, usually seen as either a surface view or a side view. Simple squamous epithelium is found in the air sacs of lungs and in the lining of blood vessels, where it is called **endothelium.** It can be found as the surface layer of many membranes, where it is called **mesothelium.** It provides a smooth lining (as found on the inside of blood vessels), filters material in the kidney, or allows diffusion (as in the lungs). Compare your slide with figure 4.1. Draw what you see under the microscope in the space provided. Note whether your slide shows the cell as a surface or side view.

Illustration of simple squamous epithelium:

Simple Cuboidal Epithelium This tissue consists of cube-like or wedge-shaped cells mostly uniform in size. These cells form many of the glands and glandular organs of the body and form much of the kidneys. Simple cuboidal epithelium is frequently found lining tubules. It is often involved in the secretion of fluids (oil) or in reabsorption (kidneys). Examine a prepared slide of simple cuboidal epithelium and compare it with figure 4.2. Draw what you see under the microscope in the space provided, labeling the nucleus of the cell and the basement membrane.

Illustration of simple cuboidal epithelium:

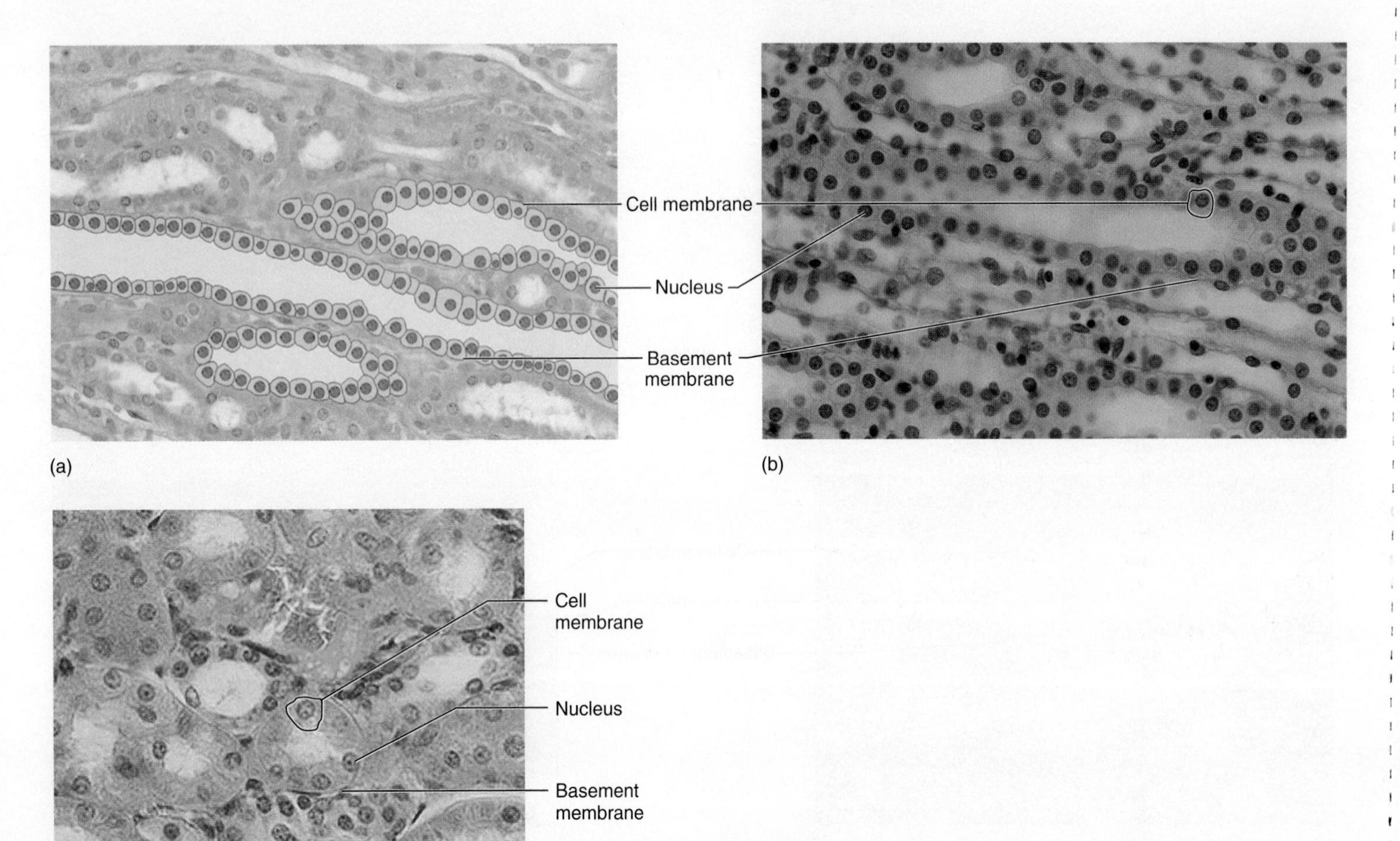

FIGURE 4.2 **Simple Cuboidal Epithelium—Kidney** Long section (a) diagram; (b) photograph (400×). Cross section (c) photograph (800×).

Simple Columnar Epithelium This epithelium resembles tall columns anchored at one end to the basement membrane. The nuclei are frequently aligned in a row. Simple columnar epithelium can be **ciliated** on the apical surface, or it may be smooth. The nonciliated type lines the inner portion of the digestive tract and provides an absorptive area for digested food. Mucus-secreting **goblet cells** are frequently found in simple columnar epithelium. Simple columnar epithelium also lines the uterine tubes and is ciliated in this case. Examine a prepared slide of simple columnar epithelium and compare it with figure 4.3. Draw a representation of what you see in the space provided.

Illustration of simple columnar epithelium:

Pseudostratified Columnar Epithelium This tissue may appear as if it occurs in a few layers, but all the cells rest on the basement membrane. The nuclei are at different levels. Pseudostratified columnar epithelium lines some portions of the respiratory passages, where it protects the lungs by trapping dust particles in a mucous sheet and by moving the particles away from them. Mucus is secreted by goblet cells. Examine a prepared slide of pseudostratified columnar epithelium and compare it with figure 4.4. Draw a representative sample of what you see in the space provided.

Illustration of pseudostratified columnar epithelium:

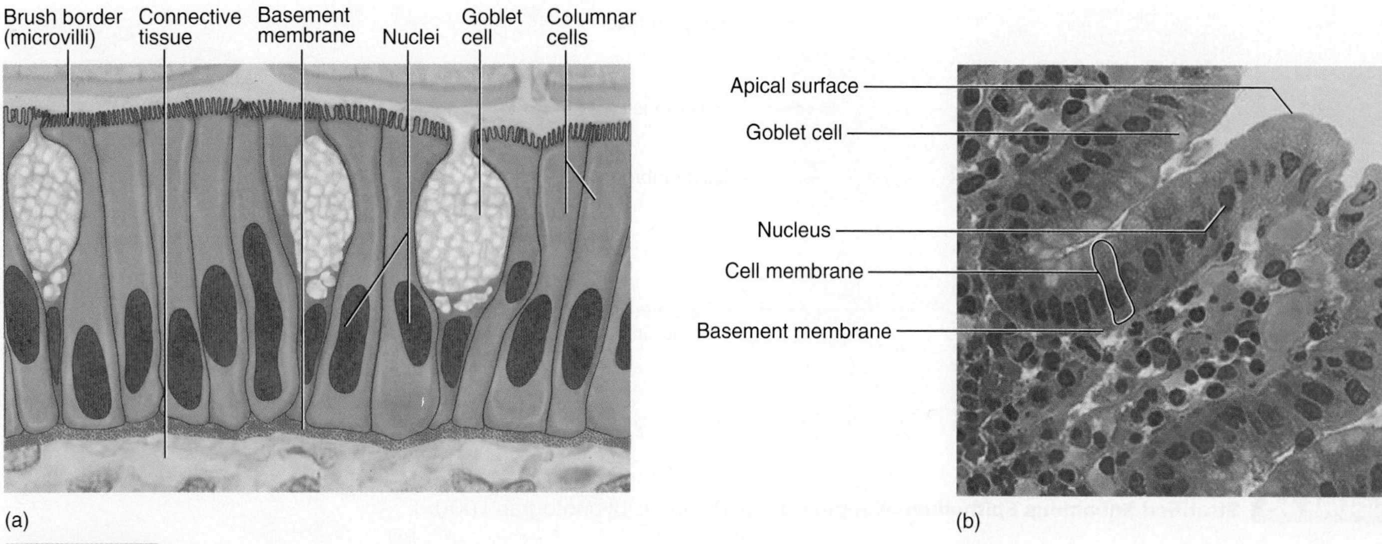

FIGURE 4.3 **Simple Columnar Epithelium—Stomach** (a) Diagram; (b) photograph (400×).

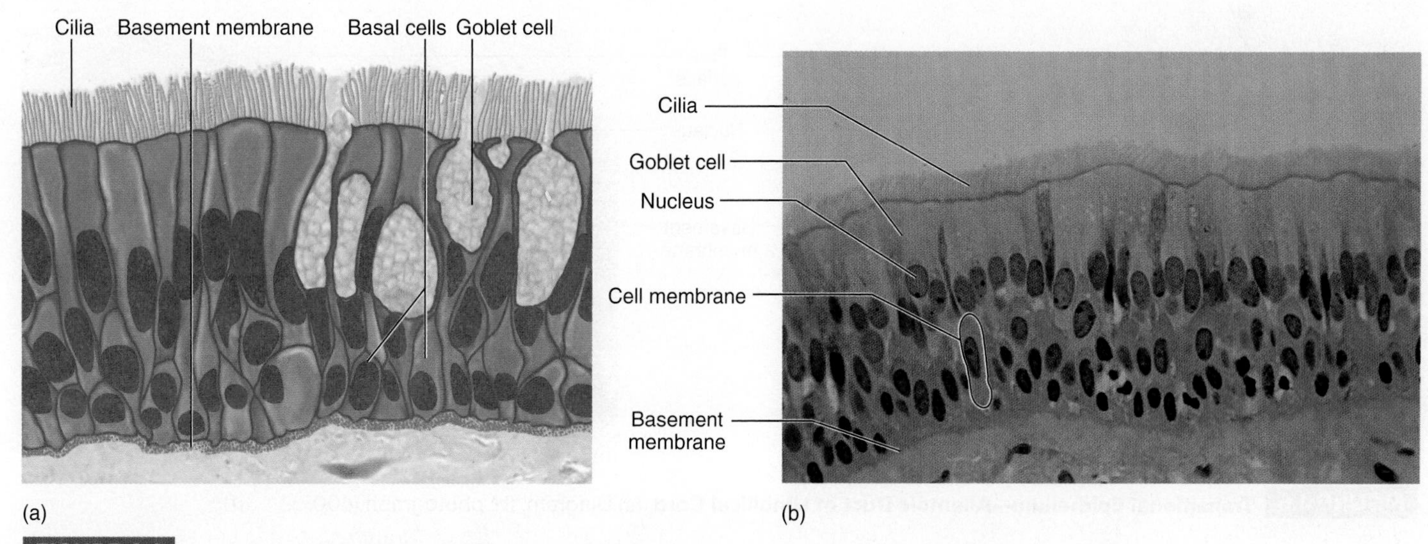

FIGURE 4.4 **Pseudostratified Ciliated Columnar Epithelium—Trachea** (a) Diagram; (b) photograph (400×).

Stratified Epithelium Stratified epithelium is so named because many epithelial cells occur in layers on a basement membrane. The term "stratified" comes from the word "strata" and refers to the layering of these cells. There are two common types belonging to this group.

Stratified Squamous Epithelium This is a tissue that covers the outside of the body and forms the outermost layer of skin. It also lines the vaginal canal, esophagus, and mouth. The multiple layers of this tissue protect the underlying tissue from mechanical abrasion. This epithelium may have cuboidal-shaped cells at the basement layer, but it derives its name from the cell shape at the free surface. Stratified squamous epithelium comes in two distinct types, **keratinized** and **nonkeratinized** (keratin is a tough protein that hardens cells in the outer layer of the skin). Examine a prepared slide of stratified squamous epithelium and locate the basement membrane. Compare your slide with figure 4.5, and draw what you see under the microscope in the space provided. Locate the basement membrane, and count the layers of cells between it and the surface of this cell type.

Illustration of stratified squamous epithelium:

Record the numbers of layers of cells here: ________

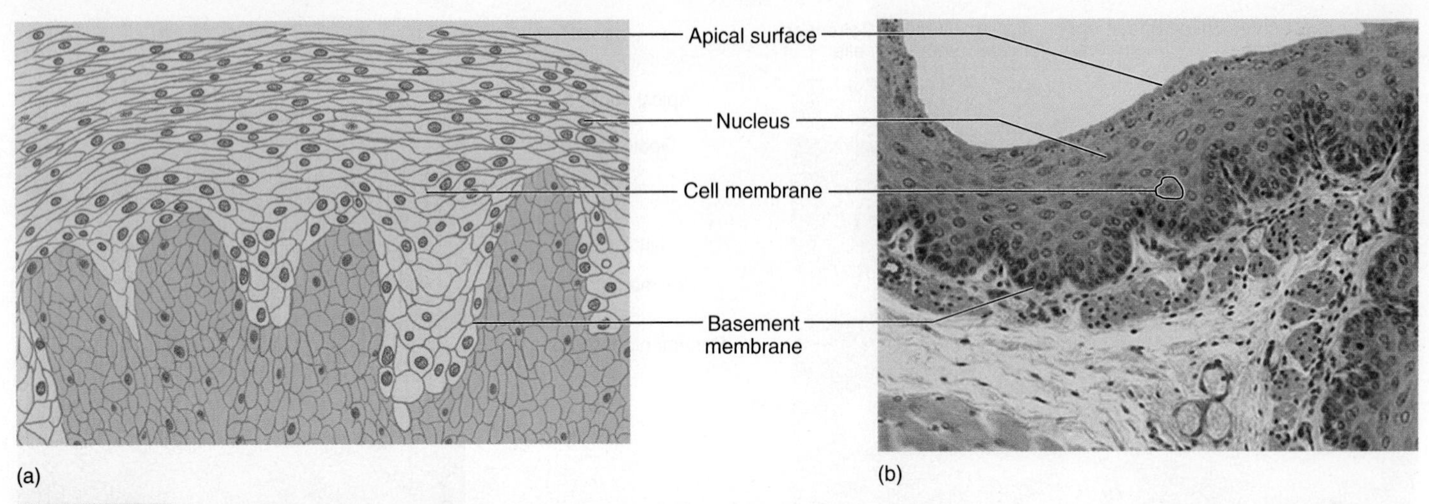

FIGURE 4.5 **Stratified Squamous Epithelium—Esophagus** (a) Diagram; (b) photograph (100×).

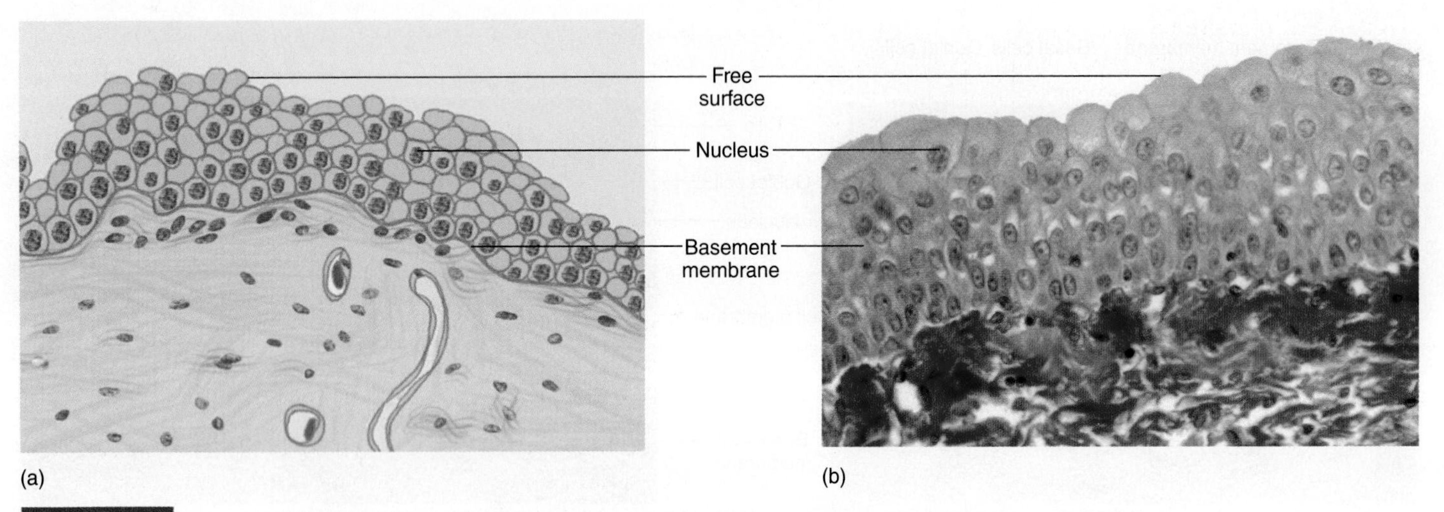

FIGURE 4.6 **Transitional Epithelium—Allantoic Duct of Umbilical Cord** (a) Diagram; (b) photograph (400×).

Transitional Epithelium This is an unusual tissue in that it has some remarkable stretching capabilities. Transitional epithelium lines the ureter, urinary bladder, and proximal urethra and allows these organs to expand as urine collects within them. It consists of teardrop-shaped cells that flatten when stretched. Examine a prepared slide of transitional epithelium. Look at figure 4.6 and draw what you see in the space provided.

Illustration of transitional epithelium:

Muscular Tissue

Like epithelial tissue, muscular tissue is a cellular tissue, with the tissue having mostly cells and little matrix. These cells are contractile and shorten due to the sliding of protein filaments across one another. There are three types of muscle: skeletal, cardiac, and smooth muscle.

Skeletal Muscle The cells of skeletal muscle are called **fibers.** When you refer to the muscles of your body, you are referring to organs made mostly of skeletal muscle. Skeletal muscle is sometimes known as striated muscle on some microscope slides because it has obvious **striations** (stry-A-shuns) (they look like stripes) in the fiber. Cardiac muscle also has striations. Skeletal muscle is **voluntary** because you have conscious control over this type of muscle. The individual muscle cells have many nuclei and

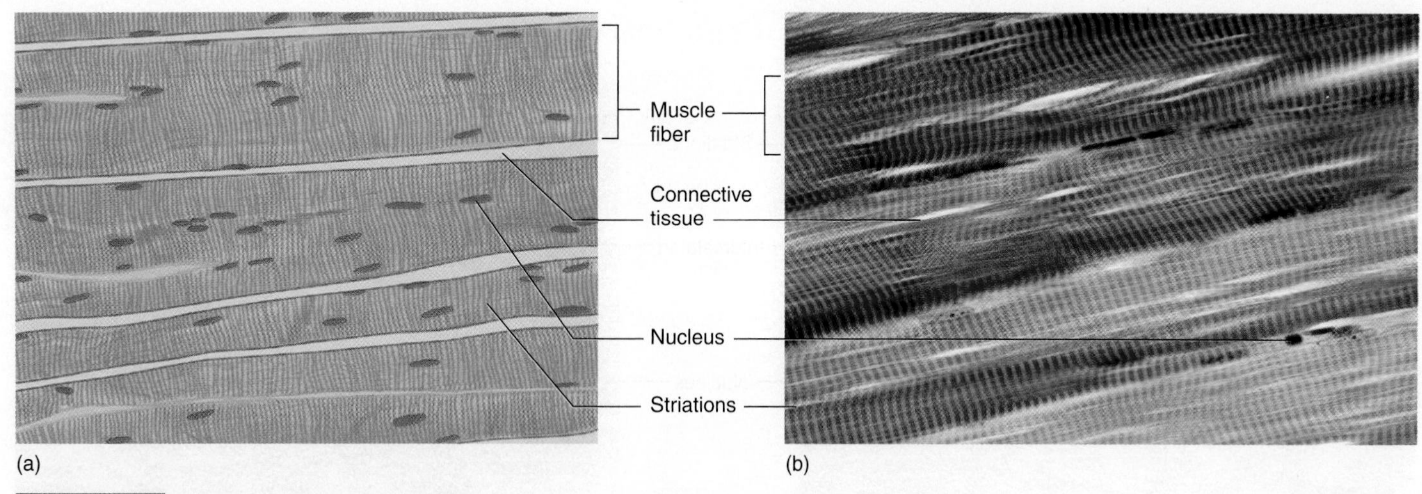

FIGURE 4.7 **Skeletal Muscle—Longitudinal Section** (a) Diagram; (b) photograph (400×).

are thus called **multinucleate** or **syncytial.** The nuclei are elongated and occur on the periphery of the cell. Examine a prepared slide of skeletal muscle under high power. Compare this slide with figure 4.7. Draw a representation of the muscle in the space provided. Examine the widths of muscle cells that fit across the diameter of your microscope when viewed at high power. How many widths of muscle cells fit across the diameter of your microscope when viewed at high power?

Illustration of skeletal muscle:

How many muscle fiber widths do you count? ________

STUDY HINT

By counting the cell widths in the slide, you should be able to see yet another way that you can distinguish among skeletal muscle, cardiac muscle, and smooth muscle.

Cardiac Muscle Cardiac muscle is found only in the heart, and it is the main tissue making up that organ. Cardiac muscle is somewhat similar to skeletal muscle in that it is striated, but the striations are much less obvious and the individual cell diameters are less than those of skeletal muscle cells. Another difference between cardiac muscle and skeletal muscle is that cardiac muscle is branched. Cardiac muscle cells are called **cardiocytes.** Place a prepared slide of cardiac muscle under the microscope, and count the number of widths of cardiocytes across the diameter of the microscope under high power.

Record the number of cell widths that occupy the diameter of your field of view at high power: ________

Cardiac muscle contracts on its own and is therefore called **involuntary muscle**. These cells are mostly uninucleate, with the nucleus centrally located and oval in shape. Cardiac muscle cells are joined by **intercalated discs** (gap junctions), which facilitate the transmission of the electrical impulses in the heart. When one muscle receives an impulse, it sends it on to the next cell. Look at a prepared slide of cardiac muscle and compare it with figure 4.8, noting the striations, intercalated discs, and nuclei of the cells. Draw the cells in the space provided.

Illustration of cardiac muscle:

Record the number of cell widths in the diameter of your field of view at high power: ________

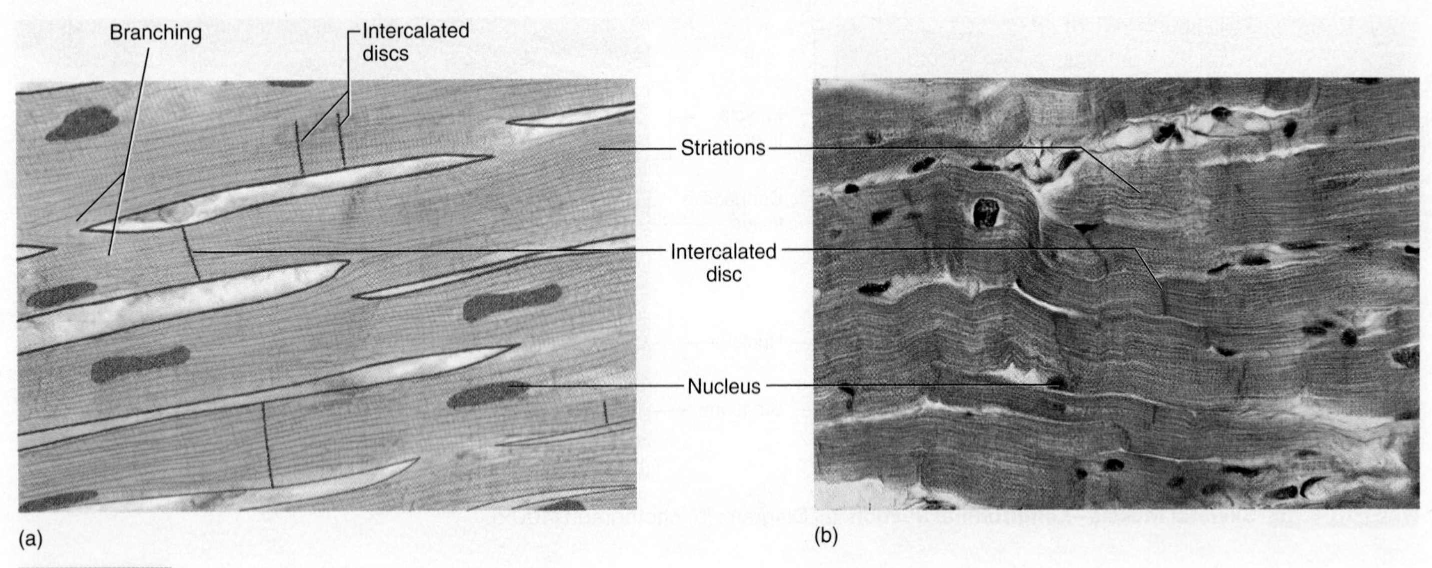

FIGURE 4.8 **Cardiac Muscle—Longitudinal Section** (a) Diagram; (b) photograph (1,000×).

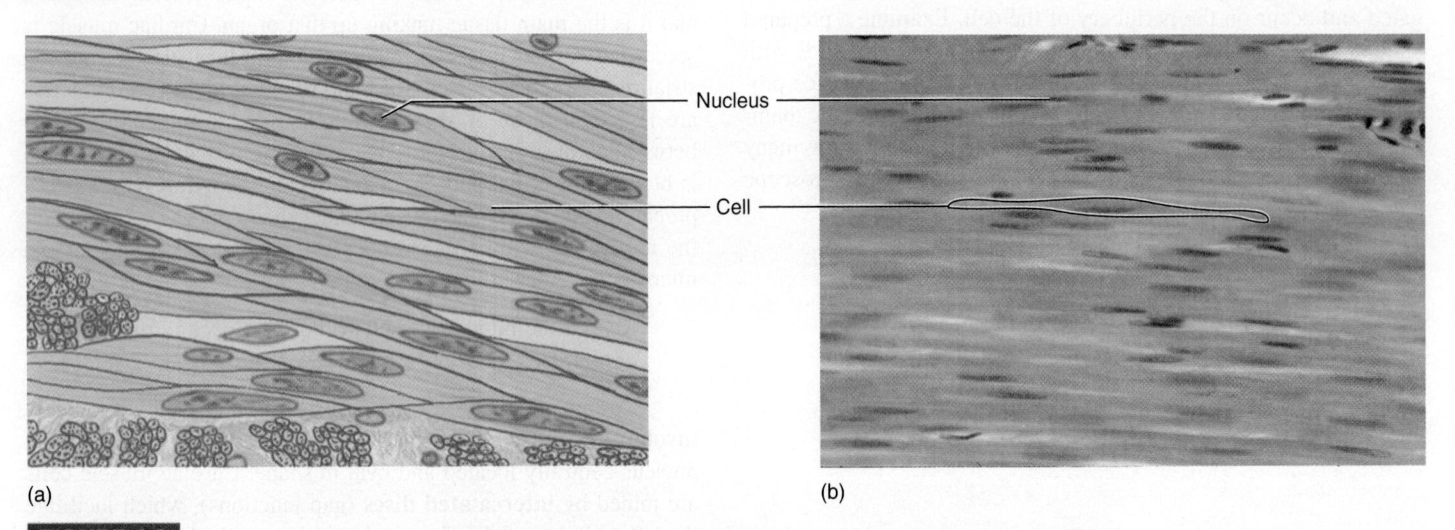

FIGURE 4.9 **Smooth Muscle—Small Intestine, Longitudinal Section** (a) Diagram; (b) photograph (400×).

Smooth Muscle This muscle is nonstriated in that the cells do not have cross-hatchings perpendicular to the length of the cell. Smooth muscle, like cardiac muscle, is involuntary. It is found in the intestine, where it propels food along by a process known as peristalsis and segmentation. The uterine contractions during labor are smooth muscle contractions. Smooth muscle is also found in the skin, where it causes hair to stand on end. In blood vessels, smooth muscle regulates the diameter of the vessels, thus maintaining blood pressure. The cells of smooth muscle are spindle-shaped and uninucleate, and the nucleus is centrally located and elongated when the muscle is relaxed. When the muscle is contracted, the nucleus appears corkscrew-shaped. Examine a prepared slide of smooth muscle and note the narrow diameter of the cells. Count the cell widths across the diameter of the field of view of your microscope under high power. Compare your slide with figure 4.9 and draw an illustration of smooth muscle in the space provided.

Illustration of smooth muscle:

Number of cell widths under high power: ________

Examine a prepared slide of all three muscle types, if available. Work with your lab partner and identify each muscle cell. Compare the cell widths in the slides and distinguish among skeletal muscle, cardiac muscle, and smooth muscle. Examine table 4.3 and compare the various muscle types and their characteristics.

TABLE 4.3	Muscular Tissue
Skeletal Muscle	
Microscopic appearance: large cell with many nuclei and obvious striations	
Significant locations: skeletal muscles of body (biceps brachii, rectus abdominis, etc.)	
Functions: voluntary contractions	
Cardiac Muscle	
Microscopic appearance: smaller, branched cell with one nucleus, intercalated discs, and less obvious striations	
Significant location: heart	
Functions: rhythmic contractions of heart	
Smooth Muscle	
Microscopic appearance: small, slender cell with one central nucleus and no striations	
Significant locations: digestive tract, uterus	
Functions: sustained contractions, propulsion of food, or delivery of infant	

Nervous Tissue

Nervous tissue, like epithelial and muscular tissues, is a cellular tissue with little matrix. Nervous tissue is found in the brain, spinal cord, ganglia, and peripheral nerves of the body. The conductive cell of this tissue is the **neuron,** which receives and transmits electrochemical impulses. A neuron is a specialized cell with three major regions—the **dendrites,** the **nerve cell body** (**perikaryon** or **soma**), and the **axon.** Examine figure 4.10 for the regions of a typical neuron and review table 4.4.

Examine the neurons of a smear of the spinal cord of an ox. Look for purple star-shaped structures, which are the nerve cell bodies. Compare them with figure 4.11. Draw what you see in the space provided.

Illustration of nervous tissue:

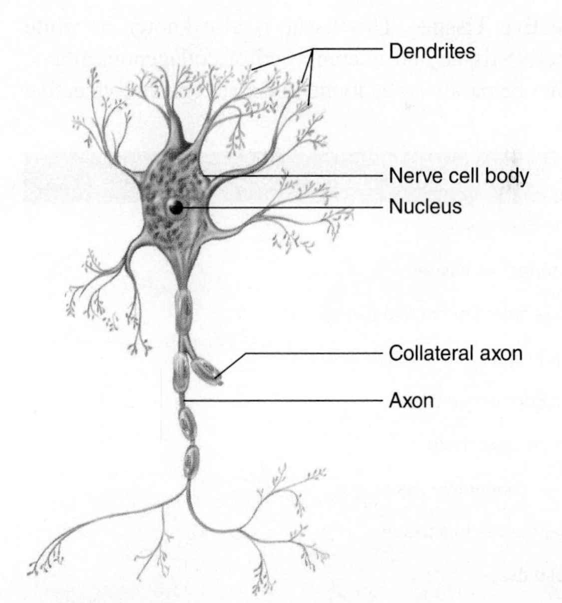

FIGURE 4.10 **Regions of a Typical Neuron**

Special cells of the nervous system are called **glial cells** or **neuroglia** (= nerve glue). These cells guide developing neurons to synapses, remove some neurotransmitters from the synapse, and perform numerous other functions. You will study glial cells in greater detail in Laboratory Exercise 15, "Introduction to the Nervous System."

TABLE 4.4	Nervous Tissue
Microscopic appearance: large, star-shaped cells (neurons in brain and spinal cord) with smaller cells (glial cells) nearby	
Significant locations: brain, spinal cord (nerves and ganglia)	
Functions: transmission of information, assimilation	

Connective Tissue

The tissues that you have seen so far, epithelial, muscular, and nervous, are composed mostly of cells with very little extracellular matrix (ECM) between them. This is not the case with connective tissue. There is usually more ECM than cells in connective tissue. ECM consists of fibers and fluid, gel, or solid ground substance. Because of the abundance of ECM, connective tissue is classified by *specific tissue.*

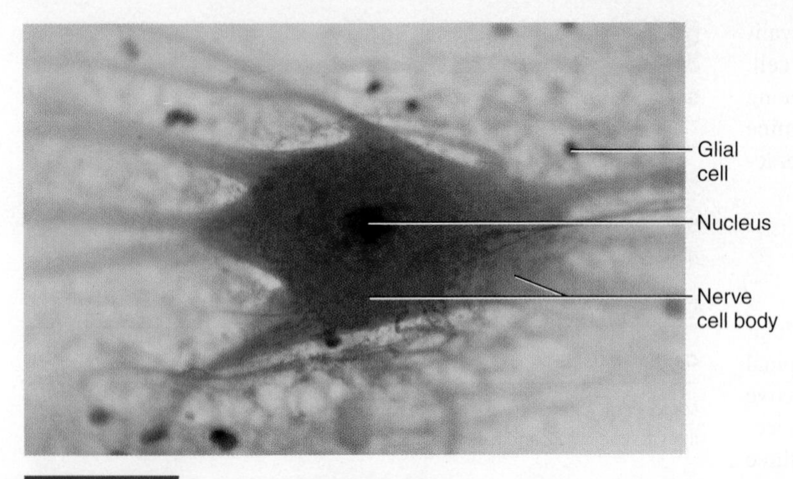

FIGURE 4.11 **Spinal Cord Smear (400×)**

Connective tissues have diverse appearances and functions, and specific tissues do not seem to have much in common with one another; however, most connective tissue arises from an embryonic tissue known as **mesenchyme.** Connective tissue can be divided into several subgroups for easier identification. These subgroups are *fibrous connective tissue, supportive connective tissue,* and *fluid connective tissue*. These are described in greater detail next and are outlined in table 4.5.

Fibrous Connective Tissue

Dense Connective Tissue This tissue is also known as white fibrous connective tissue and is composed of collagenous fibers, which can either be parallel—as found in dense regular connective tissue—or run in many directions—as found in dense irregular connective tissue. **Dense regular connective tissue** is found in tendons and ligaments. **Dense irregular connective tissue** is found in the deep layers of the skin and the white of the eye. The fibers in these tissues are made of a protein called **collagen** and are called **collagenous fibers. Fibrocytes** and **fibroblasts** are found in the tissue, in addition to the collagenous fibers. The fibroblasts actively secrete fibers and subsequently develop into fibrocytes. Examine a slide of dense connective tissue, and compare it with figure 4.12. Draw a section of dense regular or dense irregular connective tissue in the space provided.

Illustration of dense connective tissue:

TABLE 4.5 **Connective Tissue**

I. Fibrous connective tissue
  A. Dense connective tissue
    1. Dense regular connective tissue
    2. Dense irregular connective tissue
    3. Elastic connective tissue
  B. Loose connective tissue
    1. Reticular connective tissue
    2. Areolar connective tissue
    3. Adipose tissue
II. Supportive connective tissue
  A. Cartilage
    1. Hyaline cartilage
    2. Fibrocartilage
    3. Elastic cartilage
  B. Bone
III. Fluid connective tissue
  A. Blood

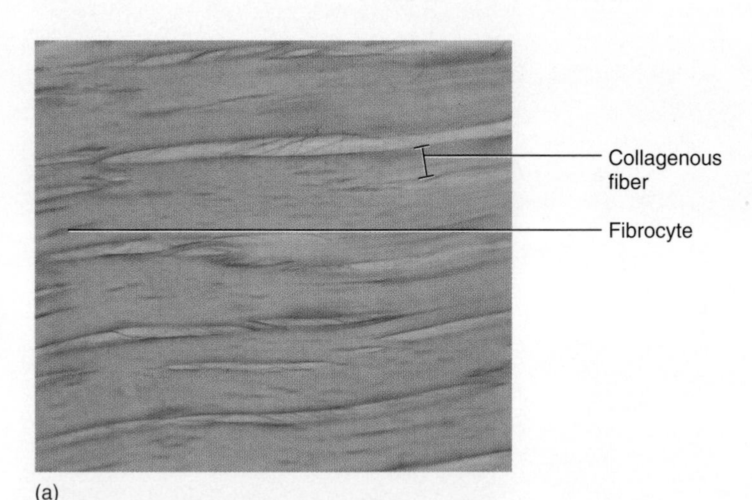

(a)

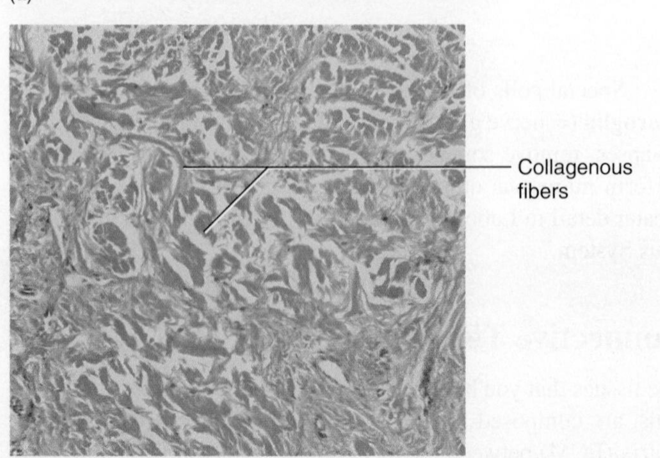

(b)

FIGURE 4.12 **Dense Connective Tissue** (a) Dense regular connective tissue—tendon (400×); (b) dense irregular connective tissue—skin (100×).

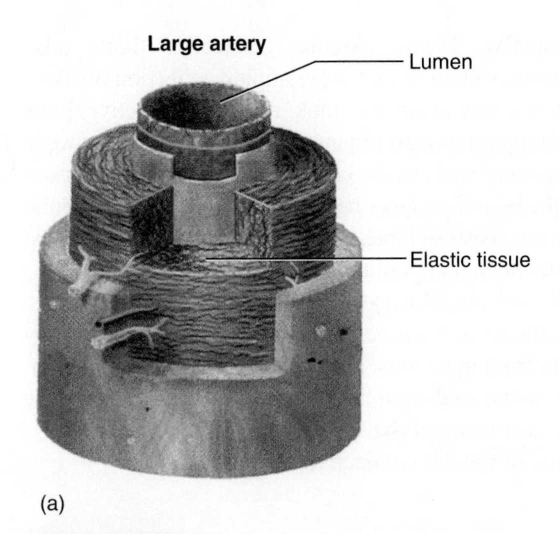

(a)

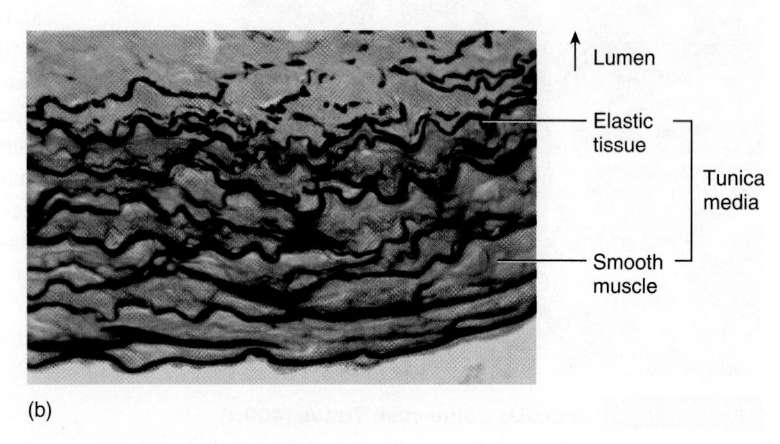

(b)

FIGURE 4.13 **Elastic Connective Tissue** (a) Overview of artery; (b) photomicrograph (400×).

Elastic Connective Tissue The cells in this tissue are *fibroblasts* and *fibrocytes*. The fibers in this tissue are made of collagenous and **elastic fibers.** Elastic fibers are made of the protein **elastin.**

Elastic tissue is found in the muscular walls of the tunica media of arteries and what appear to be dark, squiggly lines there. These are actually sheets of elastic tissue. Elastic tissue is also found in the vocal cords. Examine figure 4.13 as you look at a prepared slide of elastic tissue and review the tissues in table 4.6. Draw it in the space provided.

Illustration of elastic connective tissue:

TABLE 4.6	Dense Fibrous Connective Tissue
Dense Regular Connective Tissue	
Microscopic appearance: closely packed, wavy collagen fibers (or elastic fibers in the case of elastic connective tissue)	
Significant locations: tendons, ligaments (vocal cords in elastic connective tissue)	
Functions: binds bones together or muscle to bone (provides elastic structure)	
Dense Irregular Connective Tissue	
Microscopic appearance: randomly appearing collection of densely clustered collagen fibers	
Significant locations: dermis, sheaths around cartilage and bone	
Functions: provides strength and resists stress and strain against tearing	

Loose Connective Tissue

Reticular Connective Tissue This tissue has fibroblasts and fibrocytes with **reticular fibers** (made of collagen with a glycoprotein coat). Reticular connective tissue is found in soft internal organs, such as the liver, spleen, and lymph nodes. This tissue provides an internal framework for the organ. If your slide is stained with a silver stain, the reticular fibers appear black. If your slide is stained with Masson stain, they appear blue. In either case, look for fibers that appear branched among small, round cells, which are the cells of the organ (such as spleen, lymph node, or tonsil) that the reticular connective tissue is holding together. Examine a prepared slide of reticular tissue and compare it with

TABLE 4.7	Loose Fibrous Connective Tissue
Reticular Connective Tissue	
Microscopic appearance: reticular fibers forming meshwork around organ cells	
Significant locations: spleen, thymus, lymph nodes	
Functions: provides internal skeleton (framework) for soft organs	
Areolar Connective Tissue	
Microscopic appearance: scattered arrangement of collagenous fibers with elastic and reticular fibers along with many cells (many with protective immune functions)	
Significant locations: attachment of epithelia to lower layers, around many internal organs	
Functions: binds epithelia to lower layers, insulates organs from infections	

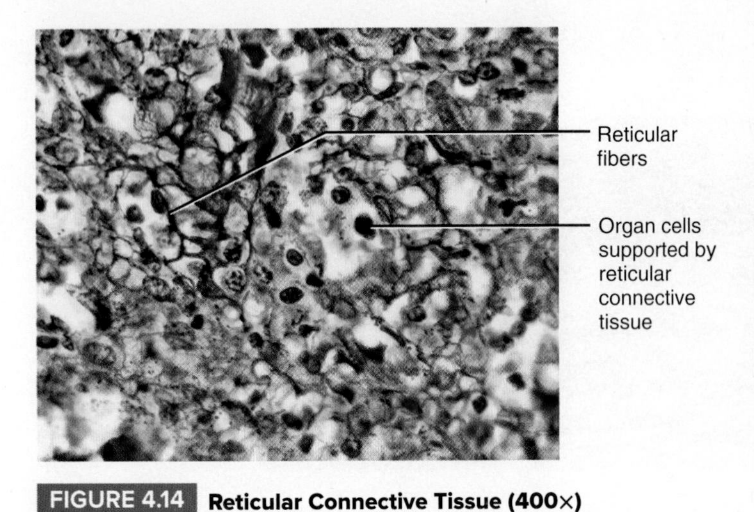

FIGURE 4.14 **Reticular Connective Tissue (400×)**

figure 4.14. Review table 4.7. Draw a sample of what you see in the space provided.

Illustration of reticular connective tissue:

Areolar Connective Tissue Areolar connective tissue (also known as loose connective tissue) is a complex collection of fibers and cells that has a very distinctive look. Areolar connective tissue is found as a wrapping around organs, as sheets of tissue between muscles, and in many areas of the body where two different tissues meet (such as the boundary layer between fat and muscle). Areolar connective tissue consists of large, pink-stained **collagenous fibers;** smaller, dark **elastic** and **reticular fibers;** and a collection of cells that includes **fibroblasts, fibrocytes, mast cells,** and **macrophages.** Mast cells and macrophages have an immune function; that is, they protect the body from infections. Examine a prepared slide of areolar connective tissue and compare it with figure 4.15. Review table 4.7. Make a drawing of the slide in the space provided.

Illustration of areolar connective tissue:

Adipose Tissue Adipose tissue (fat) is unusual as a connective tissue in that it is cellular. The cells that make up adipose tissue are called **adipocytes** *(fat cells).* Adipocytes store lipids as cellular inclusions, while the nucleus and cytoplasm remain on the outer part of the cell near the plasma membrane. In prepared slides of adipose tissue, the fat has been dissolved in the preparation process. Locate the large, empty cells, with the nuclei on the edge of some of the cells. Review table 4.8. Compare what you

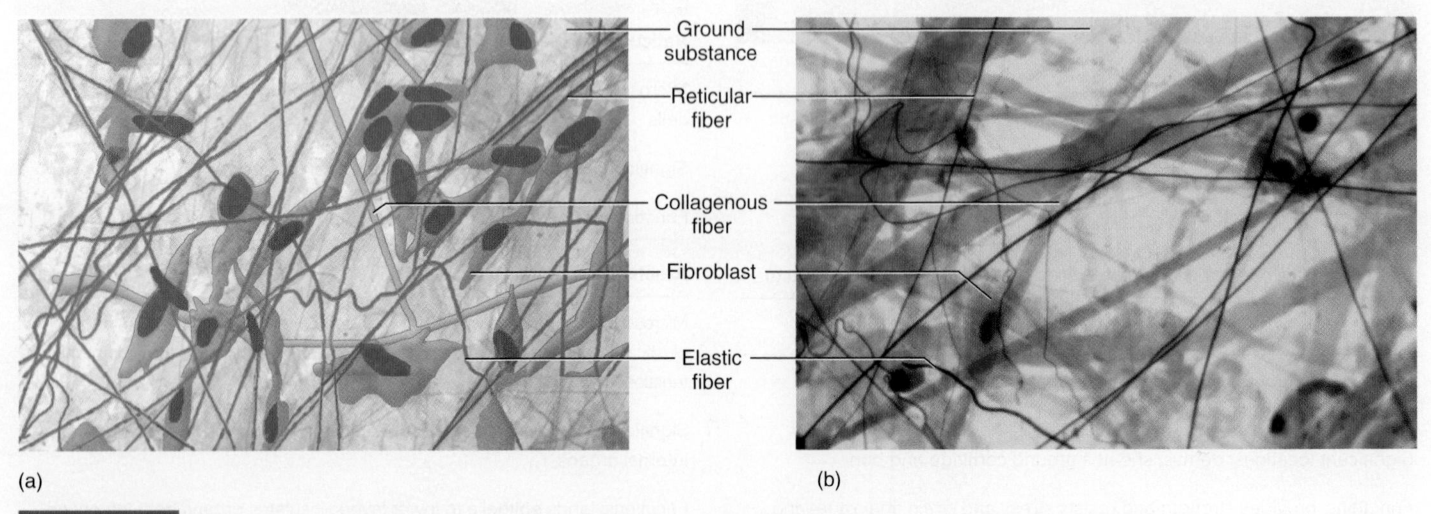

FIGURE 4.15 **Areolar Connective Tissue** (a) Diagram; (b) photograph (400×).

TABLE 4.8 Adipose Tissue

Microscopic appearance: large, pale, open cells with nuclei near periphery of cell
Significant locations: under skin, breast tissue, outside of heart and kidney
Functions: energy storage, physical protection

see in the microscope with figure 4.16 and make a drawing in the space provided. You may need to close down the iris diaphragm or reduce the amount of light shining on the specimen to see this tissue best.

Illustration of adipose tissue:

Supportive Connective Tissue

Cartilage (Cartilage and Perichondrium) Cartilage is a type of connective tissue in which the matrix is composed of a pliable material that allows for some degree of movement. The ground substance of cartilage is made of chondroitin sulfate and forms a semisolid gel enclosing both fibers and cells. In prepared slides, some of the cells (called **chondrocytes**) may have come out of the matrix and left a cavity. This cavity is known as a **lacuna** (plural, *lacunae*) and is diagnostic of cartilage tissue.

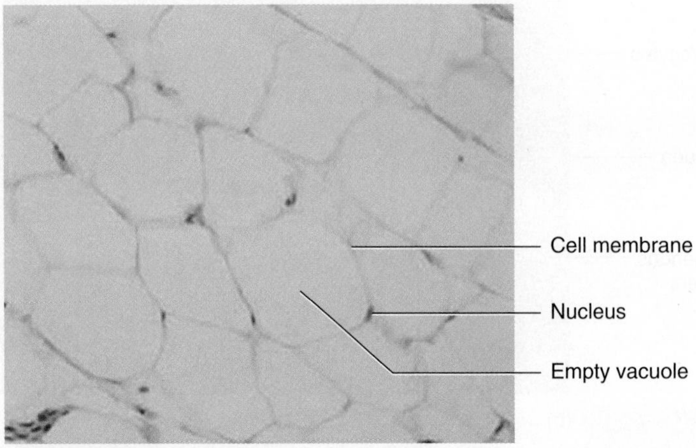

FIGURE 4.16 Adipose Tissue (400×)

TABLE 4.9 Cartilage

Hyaline Cartilage
Microscopic appearance: usually light blue or pink stained matrix, frequently with pairs of cells appearing like "eyes"
Significant locations: ends of long bones, ribs, larynx, trachea
Functions: reduces friction at joints, keeps air passages open
Fibrocartilage
Microscopic appearance: numerous collagen fibers, cells frequently in rows of four or five
Significant locations: pubic symphysis, intervertebral discs
Functions: protects from wear and tear at weight-bearing or stressed joints
Elastic Cartilage
Microscopic appearance: netlike pattern of fibers around chondrocytes
Significant locations: external ear, epiglottis
Functions: provides flexible framework

The **perichondrium** is dense irregular connective tissue on the surface of some cartilages. There are three types of cartilage in the human body, and they vary by the number of and type of protein fibers. These are hyaline cartilage, fibrocartilage, and elastic cartilage.

Hyaline Cartilage The most common cartilage in the body is hyaline (HI-ah-lin) cartilage, which is found at the apex of the nose, at the ends of many long bones at joints, between the ribs and the sternum, in the respiratory passages, and in other locations. Hyaline cartilage is clear and glassy in fresh tissue but usually appears light pink or blue in prepared, stained slides. In addition to chondrocytes there are numerous fine **collagenous fibers.** Find the

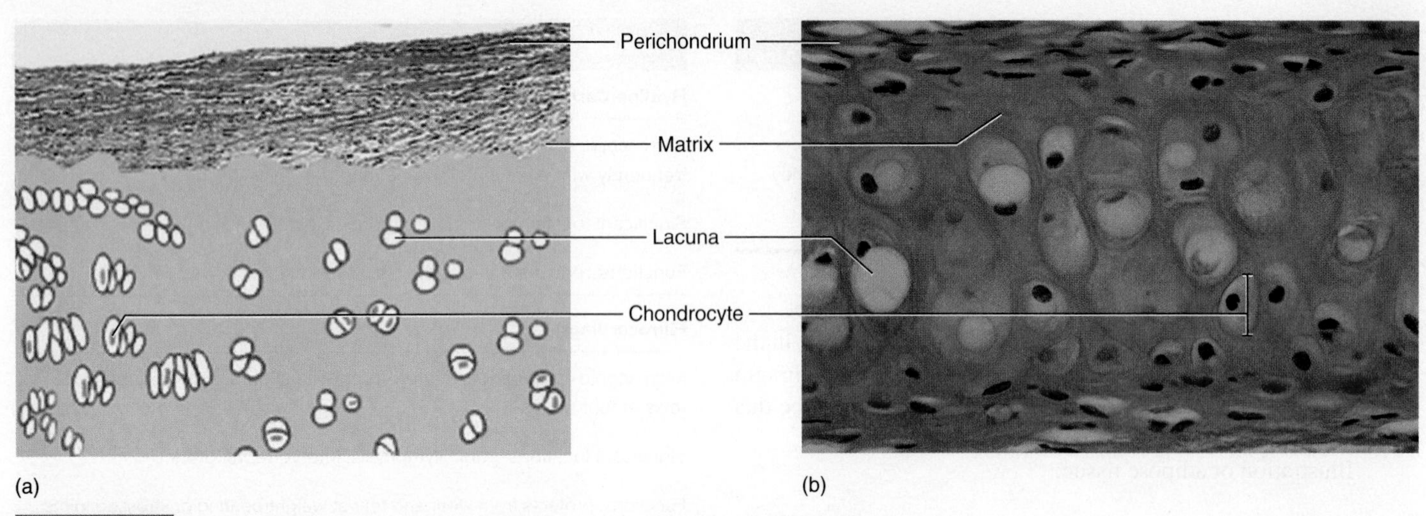

FIGURE 4.17 **Hyaline Cartilage—Trachea** (a) Overview; (b) photomicrograph (400×).

chondrocytes in a prepared slide of hyaline cartilage. The fibers do not appear distinct because they blend into the color of the ground substance. The chondrocytes of hyaline cartilage frequently occur in pairs. Some people say they look like pairs of eyes. Look around the edge of the sample of hyaline cartilage and locate the perichondrium. Review table 4.9 on the previous page. Compare the material under the microscope with figure 4.17. Draw what you see in the space provided.

Illustration of hyaline cartilage:

Fibrocartilage This tissue is similar to hyaline cartilage in that it has chondrocytes and collagenous fibers, but it differs from hyaline cartilage by having thicker collagenous fibers. These fibers are visible in the prepared slides. Fibrocartilage is found in areas where more stress is placed on the cartilage, such as in the intervertebral discs, the pubic symphysis, and the menisci of each knee. One characteristic of fibrocartilage is that the chondrocytes are frequently found in rows of four or five in the tissue. Examine a prepared slide of fibrocartilage, and locate the collagenous fibers and chondrocytes. Compare the slide with figure 4.18 and make a drawing of fibrocartilage in the space provided.

Illustration of fibrocartilage:

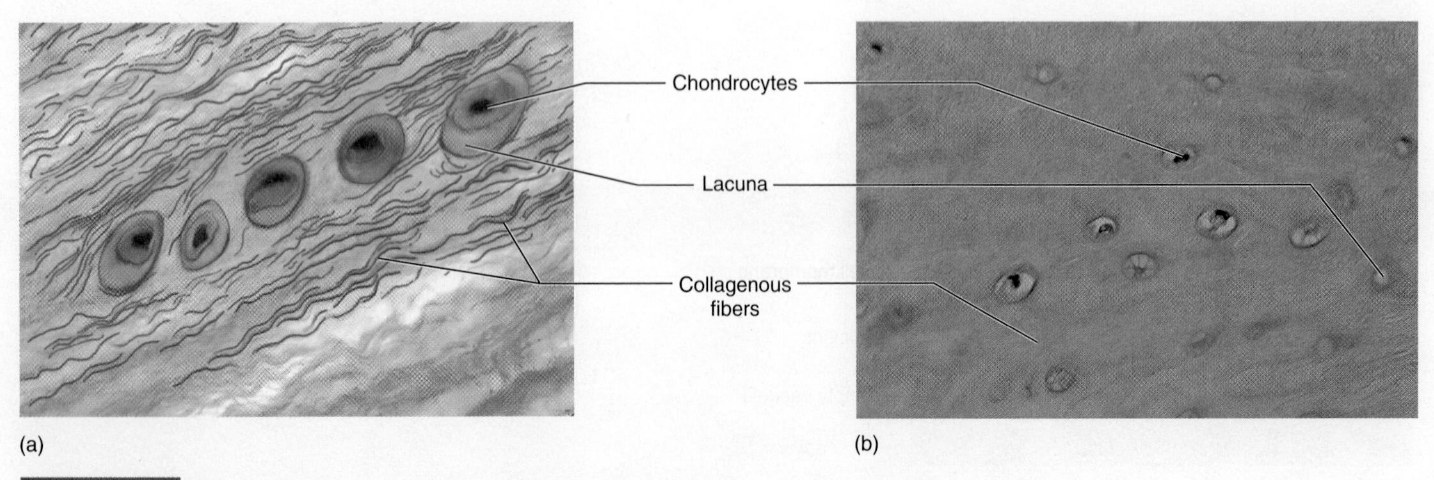

FIGURE 4.18 **Fibrocartilage** (a) Diagram; (b) photomicrograph (400×).

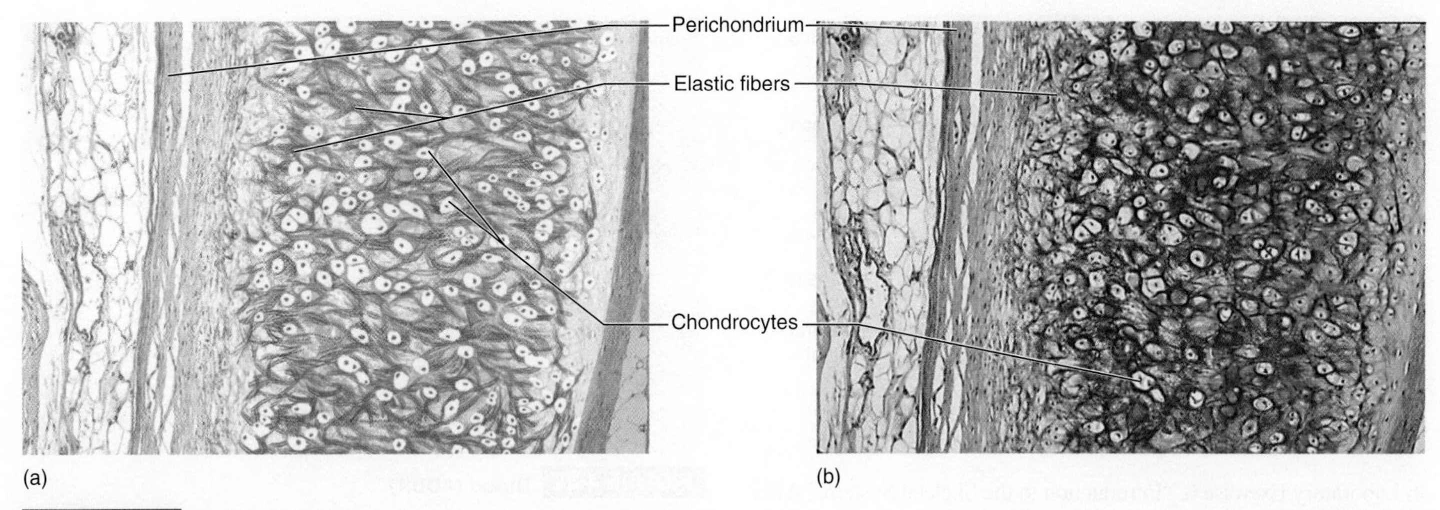

FIGURE 4.19 **Elastic Cartilage** (a) Diagram; (b) photograph (200×).

Elastic Cartilage Elastic cartilage is unique because it contains **elastic fibers,** which give the tissue its flexible nature. Elastic cartilage is found in the external ear and in the epiglottis of the larynx. If the prepared slide is stained with a silver stain, the elastic fibers appear black. If the slide is not stained in this way, the fibers may be hard to distinguish. Examine a prepared slide of elastic tissue and look for the chondrocytes and elastic fibers. Compare your slide with figure 4.19 and make a drawing of elastic cartilage in the space provided.

Illustration of elastic cartilage:

TABLE 4.10	Bone
Microscopic appearance: cells enclosed in hydroxyapatite in concentric circles	
Significant locations: skeleton	
Functions: protection of soft organs, locomotion along with muscles, scaffold for body	

Bone Bone, or osseous tissue, is a type of connective tissue in which the extracellular matrix contains a mineral component of calcium phosphate salts called hydroxyapatite. **Collagenous fibers** occur in the tissue along with osteocytes or bone cells. Two types of bone are common in the body. Compact bone is the more common one, as it makes up the dense material in a long section of bone. Cancellous (spongy or trabecular) bone is found in the end regions of long bones and has threads of bone interspersed with bone marrow. Review table 4.10. Look at a prepared slide of ground bone (which is compact bone) and compare it with figure 4.20. Locate the large, circular regions called **osteons** (which resemble a cross section of a tree trunk with the growth rings), the **central canal,** and the small spaces (lacunae) where **osteocytes** are found in the tissue. Osteons are the long, thin units of bone with the spaces in the middle of them known as the central canals. Osteocytes connect to one another through **canaliculi** and are frequently clustered in plates called **lamellae.** A more detailed examination of bone appears

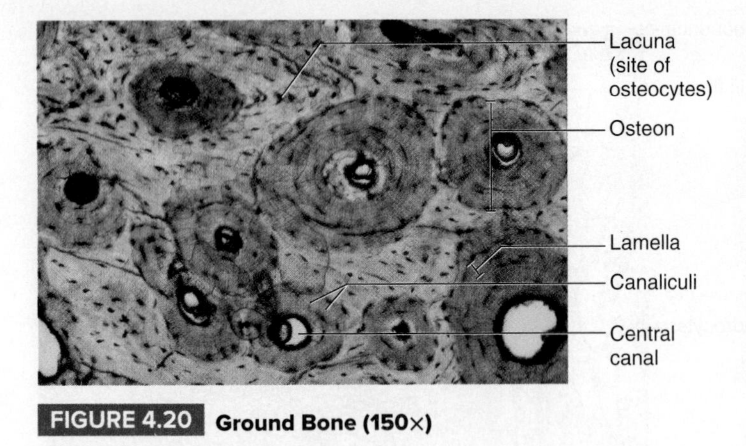

FIGURE 4.20 **Ground Bone (150×)**

in Laboratory Exercise 6, "Introduction to the Skeletal System." After you study the slide, draw an osteon in the space provided.

Illustration of ground bone:

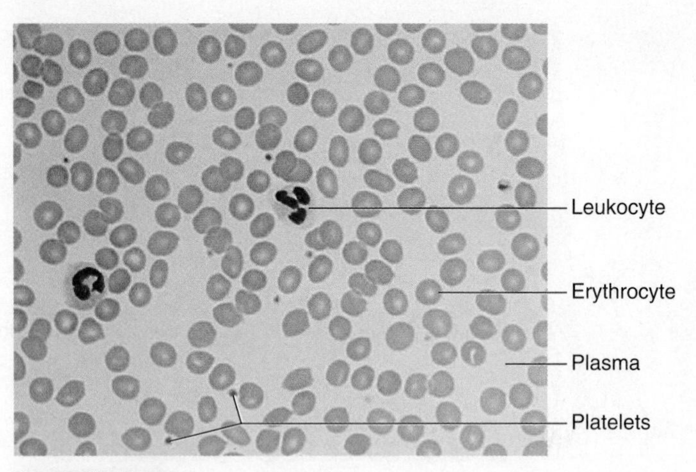

FIGURE 4.21 **Blood (400×)**

TABLE 4.11	Blood
Microscopic appearance: many cells, most with no nucleus, suspended in plasma	
Significant locations: heart, blood vessels	
Functions: transport of gases, nutrients, hormones, water, and other material throughout the body	

Fluid Connective Tissue

Blood Blood is a connective tissue in which the extracellular matrix is fluid and no fibers are visible. The matrix of blood does produce fibers when blood clots. The ECM of the blood is called **plasma,** and the cells of blood are either **erythrocytes** (eh-RITH-ro-sites) (red blood cells) or **leukocytes** (white blood cells). Examine a prepared slide of blood and look for the numerous, biconcave, pink discs in the sample. These are the erythrocytes. You may find larger cells with lobed, purple nuclei. These are the leukocytes. Numerous, small fragments of cells are in the slide. These are called platelets and they come from megakaryocytes. The details of blood are studied in Laboratory Exercise 20, "Blood Cells." Examine the slide under high power and compare it with figure 4.21. Review table 4.11. Make a drawing of blood in the space provided.

Illustration of blood:

STUDY HINT

Examine as many different images of tissues as you can. Drawing the material observed in the lab is a good memory tool. Try to make associations between the abstract image of tissue and something common. For example, you could think of bone as looking like tree rings. Adipose tissue may look like angular balloons. If you have difficulty distinguishing between two specimens, try to find a characteristic that separates the two. For example, smooth muscle and dense connective tissue may look similar in certain stained preparations. You will note that smooth muscle has nuclei inside the cells, while dense connective tissue has nuclei of the fibrocytes between adjacent fibers.

Tissues

Name ______________________________ Date ____________

1. List the four main tissues of the body. ______________________________

2. What is the name of the noncellular layer that attaches epithelial tissue to other layers? ______________________________

3. How many layers of cells are in simple epithelium? ______________________________

4. Name the three general cell shapes of epithelial tissue. ______________________________

5. A single, flattened layer of cells represents what type of epithelium? ______________________________

6. What functional problem might occur if the stomach and intestines of the digestive tract were lined with stratified squamous epithelium instead of simple columnar epithelium? ______________________________

7. Multiple layers of flattened epithelial cells represent what cell type? ______

______

8. Can you make the hairs on your arm "stand on end"? Based on your results, what kind of muscle is responsible for doing this?

______

______

______

9. Squamous cells have what general shape? ______

______

10. What tissue type lines the inside of the urinary bladder? ______

______

11. What muscle cell types have the following features?

nonstriated and involuntary ______

striated and involuntary ______

striated and voluntary ______

12. What muscle cell type has intercalated discs? ______

______

13. The intestine is mostly composed of what kind of muscle? ______

______

14. The heart is made of what type of muscle? ______

______

15. The muscles of your arm are primarily composed of what cell type? ______

______

16. What cell type is responsible for the transmission of electrochemical impulses? ______

______

17. What is the extracellular matrix in blood called? ______

______

18. Name the three types of fibers found in areolar connective tissue.

19. What type of connective tissue is found in the middle walls of arteries?

20. What is the cell type found in adipose tissue?

21. Name the outer connective tissue layer that wraps around some types of cartilage.

22. What kinds of fibers are in fibrocartilage?

23. Describe the functional difference between fibrocartilage and hyaline cartilage.

24. What is another name for calcium salts in bone?

25. In a cross section of a bone, you can usually see two types of bone tissue. What are these called?

26. How do you distinguish between epithelial cells and connective tissue?

27. Label the following photomicrographs by tissue type using the terms provided.

a. ______________________

b. ______________________

c. ______________________

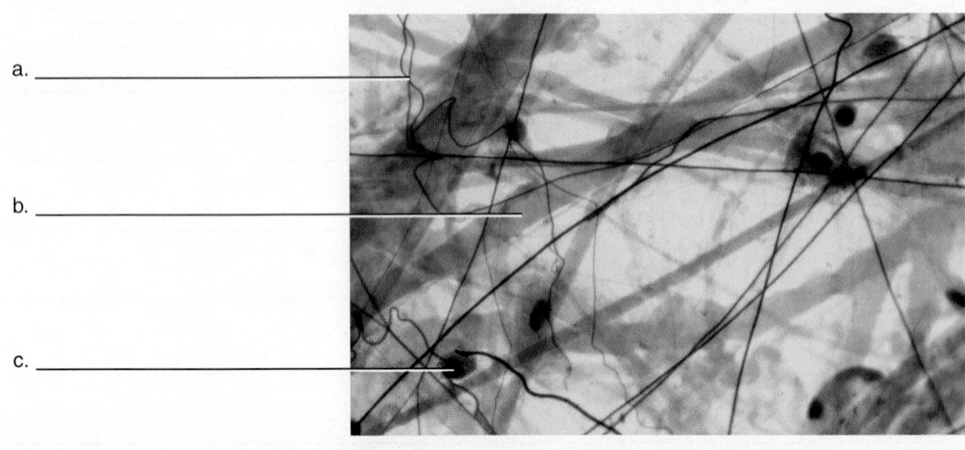

collagenous fibers, elastic fibers, fibroblast

Tissue Name ______________________

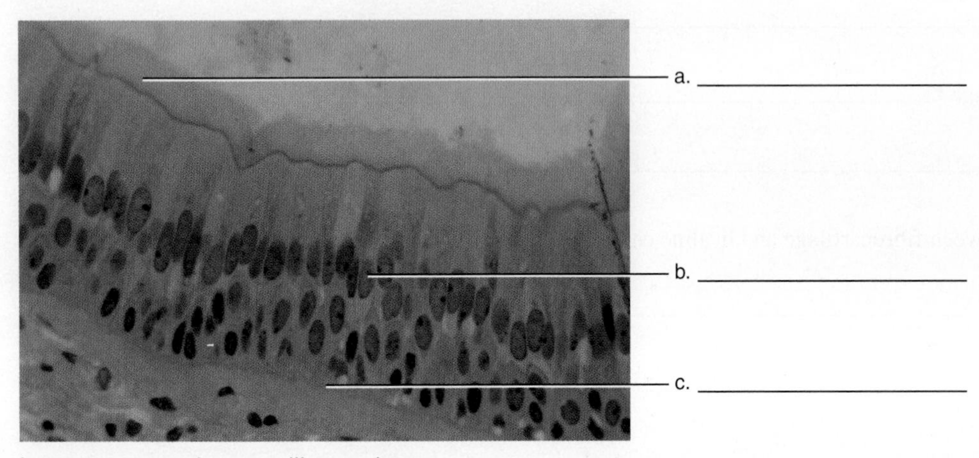

a. ______________________

b. ______________________

c. ______________________

basement membrane, cilia, nucleus

Tissue Name ______________________

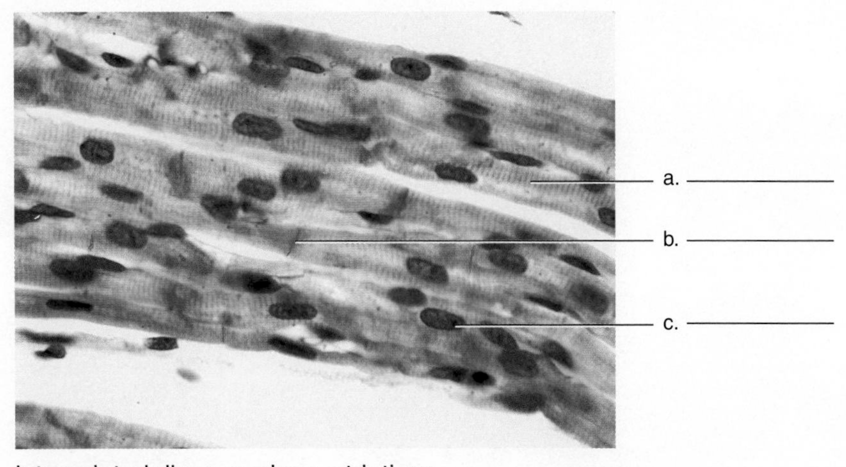

intercalated discs, nucleus, striations

Tissue Name ______________________________

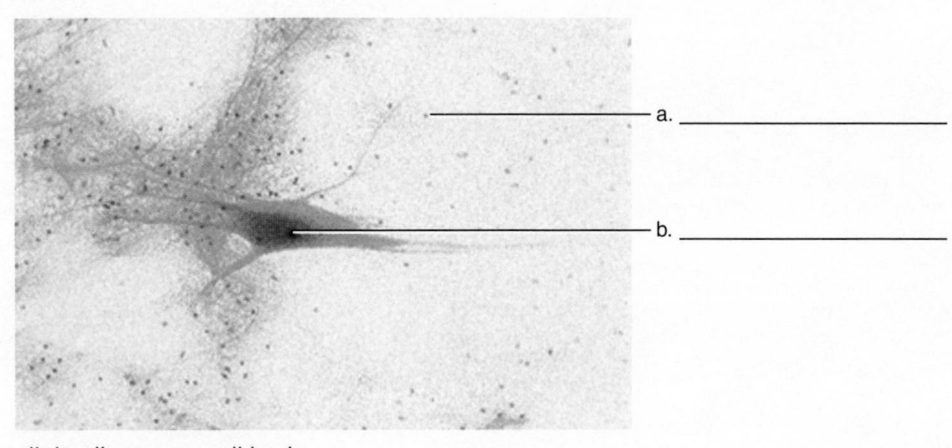

glial cells, nerve cell body

Tissue Name ______________________________

Laboratory

Exercise 5

The Integumentary System

Integumentary System

Introduction

The integumentary system consists of the skin and associated structures, such as hair, nails, and various glands. The integument consists primarily of a cutaneous membrane composed of a superficial epidermis and a deeper dermis. Below these two layers is an underlying layer called the hypodermis, which anchors the integument to deeper structures. The skin is the largest organ of the body and is vital for temperature regulation and for protection against water loss and invasion from microorganisms. In this exercise you examine the structure of skin, glands, hair, and nails and name the layers of the epidermis and dermis.

Learning Objectives

At the end of this exercise you should be able to

1. list the two main layers of the integument;
2. list all the layers of the epidermis from deep to superficial in both thick and thin skin;
3. describe the structure and function of sweat glands and sebaceous glands;
4. draw a hair in a follicle in longitudinal section; and
5. list the layers of the dermis from deep to superficial.

Materials

Models and charts of the integumentary system
Microscopes
Prepared microscope slides of
  Thick skin (with lamellar corpuscles)
  Hair follicles
  Thin skin

Procedure

Overview of the Integument

Examine models or charts of the integumentary system in the lab, and locate the two major regions of the **integument,** the **epidermis** and the **dermis.** The dermis consists primarily of collagenous fibers, and the epidermis is the most superficial layer; it can be determined by the epithelial cells of that layer. Also locate the **hypodermis,** (**subcutaneous layer,** or **superficial fascia**), not a layer of the integument but a structure anchoring the integument to underlying bone or muscle. The hypodermis can be distinguished by the presence of significant amounts of adipose tissue found there. Compare the models in the lab with figures 5.1 and 5.2.

Locate associated structures, such as the **hair follicles** in the dermis, **sebaceous (oil) glands,** and **sudoriferous (sweat) glands.** The sweat glands are connected to the surface by **sweat ducts.**

Microscopic Examination of the Integument

Examine a prepared slide of thick skin under low power. Locate the hypodermis, dermis, and epidermis. The slide will probably show three layers—one relatively clear due to a significant amount of adipose tissue, a pink layer, and a multicolored layer. The clear layer is the hypodermis, the pink layer is the dermis, and the multicolored layer is the epidermis. The epidermis may be determined by the epithelial cells of that layer. Look for cells with obvious nuclei. The colors may vary in the slides used in your lab. Compare these layers in the prepared slide to figure 5.2. Draw and label the integument in the following space, locating the epidermis, dermis, and hypodermis.

Illustration of overview of integument:

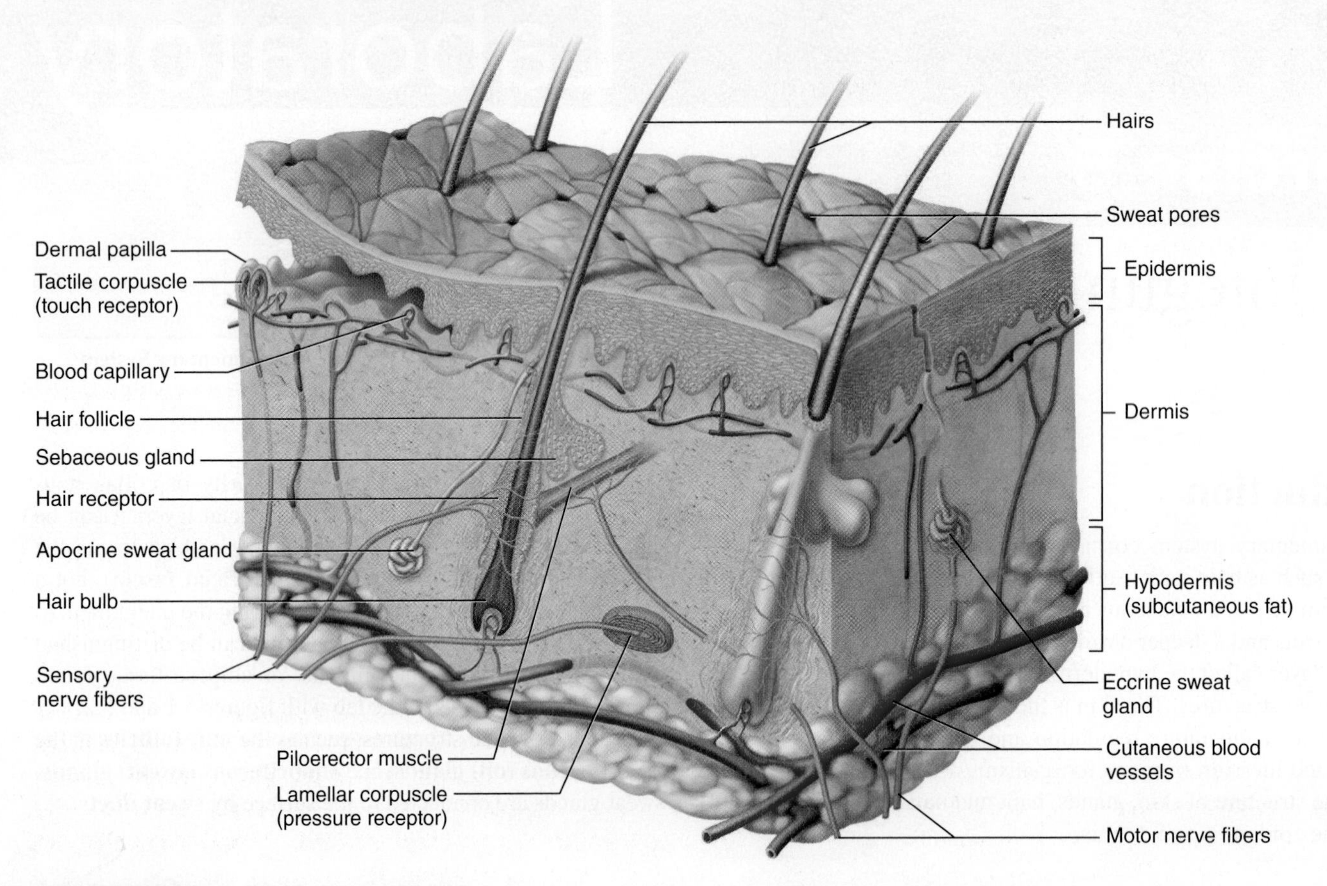

FIGURE 5.1 **Diagram of the Integument**

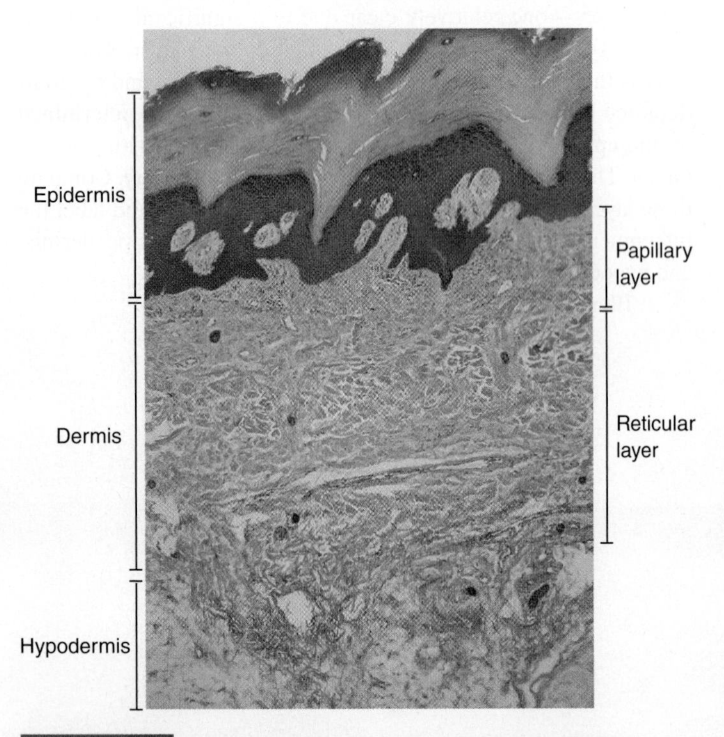

FIGURE 5.2 **Photomicrograph of the Integument (40×)**

Dermis

The dermis is responsible for the structural integrity of the integumentary system. It is composed primarily of closely apposed fibers along with blood vessels, nerves, sensory receptors, hair follicles, and glands. The dermis consists of two major regions, the superficial **papillary layer** and a deeper **reticular layer.** The papillary layer is so named for the **papillae** (bumps) that interdigitate with the epidermis, providing good adhesion between the two layers. The reticular layer is the largest layer of the dermis. It is primarily composed of irregularly arranged collagenous fibers along with some elastic and reticular fibers. Examine a slide of thick skin and locate the two layers of the dermis, as seen in figure 5.2.

The majority of the fibers of the dermis are **collagenous** fibers along with lesser numbers of **elastic** and **reticular** fibers. The collagenous fibers provide strength and flexibility. The elastic fibers, as their name implies, provide elasticity to the skin.

Blood vessels are found in the dermis, bringing nutrients to both the dermis and epidermis. The blood vessels do not enter the epidermis, but the nutrients and oxygen diffuse from the dermis into the cells of the epidermis. Blood vessels are also important because they release heat to the external environment. As the internal temperature of the body rises, vasodilation of blood vessels of the dermis occurs and allows heat to radiate from the skin. In cold weather, vessels constrict, decreasing the amount of blood flowing to the dermis, keeping heat in the body core.

There are many **nerves** and neural derivatives present in the dermis. Sensory information, such as light touch, changes in temperature, and the feeling of pain, is transmitted from receptors in the dermis to the central nervous system. These are covered in Laboratory Exercise 18, "Sensory Receptors." Draw a representation of the dermis in the following space.

Illustration of the dermis:

Epidermis

The epidermis is composed of **keratinized stratified squamous epithelium (keratinocytes)** toughened with the protein keratin. Keratin protects the lower layers of the epidermis from abrasion. The epidermis has a number of layers known as **strata** (singular, *stratum*). The epidermis does not have blood vessels between the cells but is nourished by the vessels in the dermis.

Strata of the Epidermis The deepest layer of the epidermis is known as the **stratum basale.** This single layer of cells is attached to the dermis by the **basement membrane,** a noncellular layer. Cells in the stratum basale divide repeatedly and push up toward the superficial layers as the cells age. This process is illustrated in figure 5.3.

Locate the stratum basale and draw it in the following space.
Illustration of lower epidermis:

The **stratum spinosum** is a layer of numerous cells superficial to the stratum basale. In prepared slides, the cells appear star-shaped because the cell membrane pulls away from the other cells, except in areas where they are joined at microscopic junctions called desmosomes. The stratum basale and stratum spinosum are sometimes referred to as the **stratum germinativum.** A layer of granular cells is present in the next layer. This is known as the **stratum granulosum.** The cells have purple-staining granules in their cytoplasm. The granules are precursors of keratin found in the outermost layer of the epidermis.

Superficial to the stratum granulosum is the **stratum lucidum,** so named (*lucid* = light) because this layer appears translucent in fresh material. This layer generally appears clear or pink in prepared slides stained with hematoxylin and eosin. It is found only in the skin of the palms of the hand and soles of the feet (compare figure 5.3 to figure 5.4).

The most superficial layer of the integument is the **stratum corneum,** which is a tough layer consisting of dead cells that are

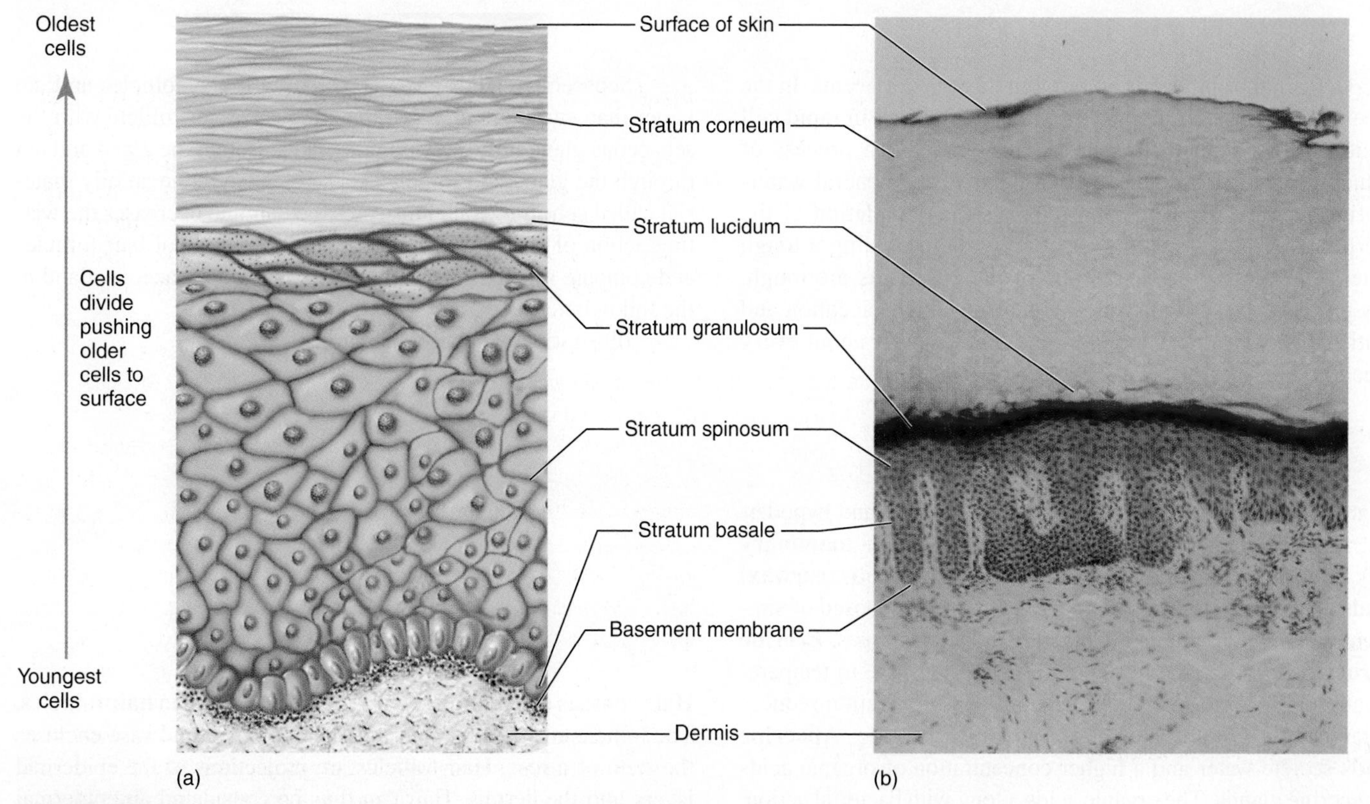

FIGURE 5.3 Proliferation of the Epidermis (a) Diagram; (b) photograph (100×).

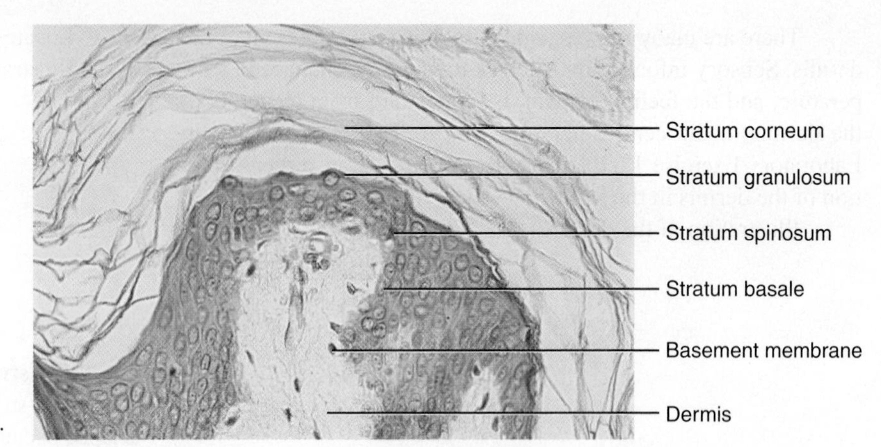

FIGURE 5.4 **Strata of the Epidermis** Thin skin.

flattened at the surface. Cells of the stratum corneum have been toughened by complete **keratinization,** and thus the epidermis is made of **keratinized stratified squamous epithelium.** Examine a prepared slide of thick skin, and locate the various strata of the epidermis as illustrated in figure 5.3. Draw an example of the superficial epidermal layers in the following space.

Illustration of superficial epidermal layers:

Cells in the epidermis go through three major events. In the deepest layer, the stratum basale, cells are involved in rapid cell division. Superficial to this layer, cells undergo a process of producing precursor molecules that leads to the general waterproofing of the cells. The final phase is the completion of the waterproofing process. Cells eventually die, providing a tough barrier. The most superficial cells of the epidermis are tough. They protect the skin from microorganisms and desiccation and eventually slough off. The epidermis renews itself about every 6 weeks.

Integumentary Glands

A significant number of glands occur in the dermis and hypodermis. These include **sudoriferous (sweat) glands, mammary (milk) glands, sebaceous (oil) glands,** and **ceruminous (earwax) glands.** The secretory parts of sweat glands are composed of simple cuboidal epithelium and come in two different types. **Eccrine (merocrine) glands,** the most common, are sensitive to temperature and produce normal body perspiration. Perspiration reduces the temperature of the body by evaporative cooling. **Apocrine glands** secrete water and a higher concentration of organic acids than eccrine glands. The organic acids, along with bacterial action, produce a pungent body odor. These glands are concentrated in the region of the axilla and groin and are associated with hair follicles. Examine a prepared slide of skin and look for clusters of cuboidal epithelial cells with purple nuclei in the dermis or hypodermis. Note that the glands are tubules and when these are cut in section on your slide, you are mostly seeing the tubules cut in cross section. These are the sudoriferous glands illustrated in figure 5.5. Draw a sudoriferous gland in the following space.

Illustration of sudoriferous gland:

Sebaceous glands are associated with hair follicles and are larger than sweat glands. You may not see a hair follicle with the sebaceous gland if the section was made through the gland and not through the gland and follicle. These glands secrete an oily material called **sebum,** which lubricates the hair and decreases the wetting action of water on the skin. Examine a slide of hair follicles and compare the slide with figure 5.5. Draw a sebaceous gland in the following space.

Illustration of sebaceous gland:

Hair

Hair consists of keratinized cells that are produced in **hair follicles.** The follicle encloses the hair in the same way a bud vase encloses the stem of a rose. Hair follicles are projections of the epidermal layers into the dermis. Hair can thus be considered an epidermal derivative. The hair consists of the **shaft,** a portion that erupts from

Sebaceous gland

Sudoriferous gland

(a)

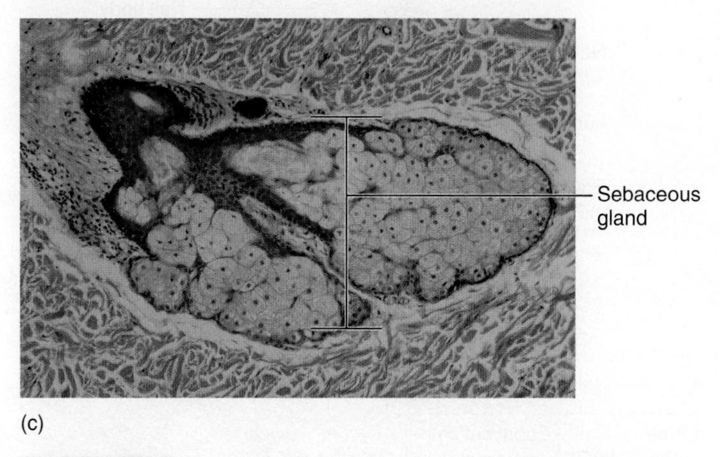

(b)

Sebaceous gland

(c)

FIGURE 5.5 **Integumentary Glands** (a) Overview of glands (20×); (b) sudoriferous gland (100×); (c) sebaceous gland (100×).

the skin surface; the **root,** which is enclosed by the follicle; and the **hair bulb,** which is the actively growing portion of the hair. There are two types of hair. **Determinate hair** grows to a specific length and then stops. Determinate hair is found in many places in the body, including the axilla, groin, arms, legs, eyelashes, and eyebrows. **Indeterminate hair** continues to grow without regard to length. It is found on the scalp and in beards in males. Examine models and charts in the lab and compare them with figure 5.6.

Look at a prepared slide of hair and locate the root, follicle, and hair bulb. At the center of the bulb is a small structure known as the **dermal papilla,** a region with blood vessels and nerves that reach the hair bulb. The follicle is composed of an outer dermal layer of connective tissue and an inner epithelial layer. These form the **root sheath** of the hair. Locate these structures and compare them with figures 5.6 and 5.7. Draw a longitudinal section of hair.

Illustration of hair:

Piloerector Muscle When the hair "stands on end," it does so by the contraction of **piloerector (arrector pili) muscles.** A piloerector muscle is a cluster of parallel smooth muscle fibers that connects the hair follicle to the upper regions of the dermis. These muscles are controlled by the autonomic nervous system. In response to cold conditions, the piloerector muscles contract, producing goose bumps. The piloerector also contracts when a person is suddenly frightened. Find a piloerector muscle in a prepared slide and compare it with figures 5.6 and 5.7. If you have trouble finding a piloerector muscle, look for slender slips of smooth muscle that angle from the follicle to the superficial regions of the dermis. You may have to look at more than one slide to find them.

Cross Section of Hair The central portion of the hair is known as the **medulla,** which is enclosed by an outer **cortex.** The cortex may contain a number of pigments, which give the hair its particular color. The brown and black pigments are due to varying types and amounts of melanin. Blond hair has pheomelanin as a pigment. Iron-containing pheomelanin pigments produce red hair.

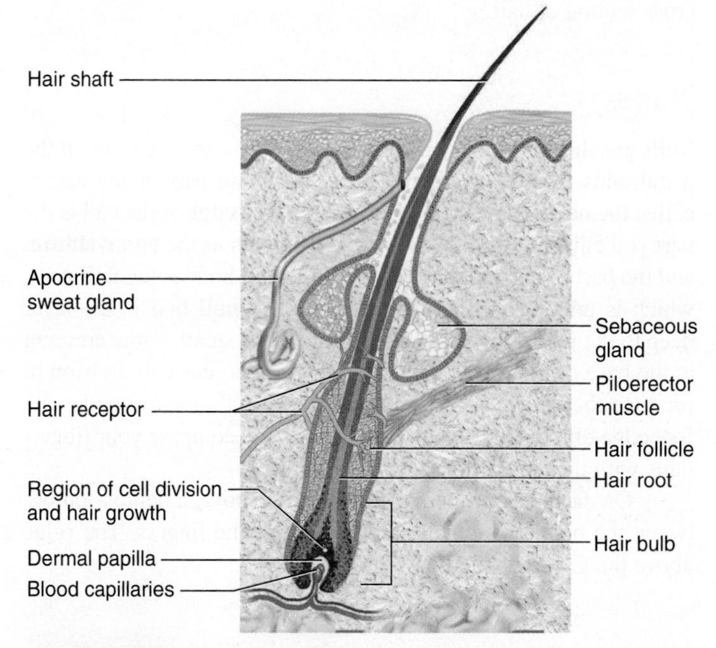

FIGURE 5.6 **Diagram of Hair in a Follicle**

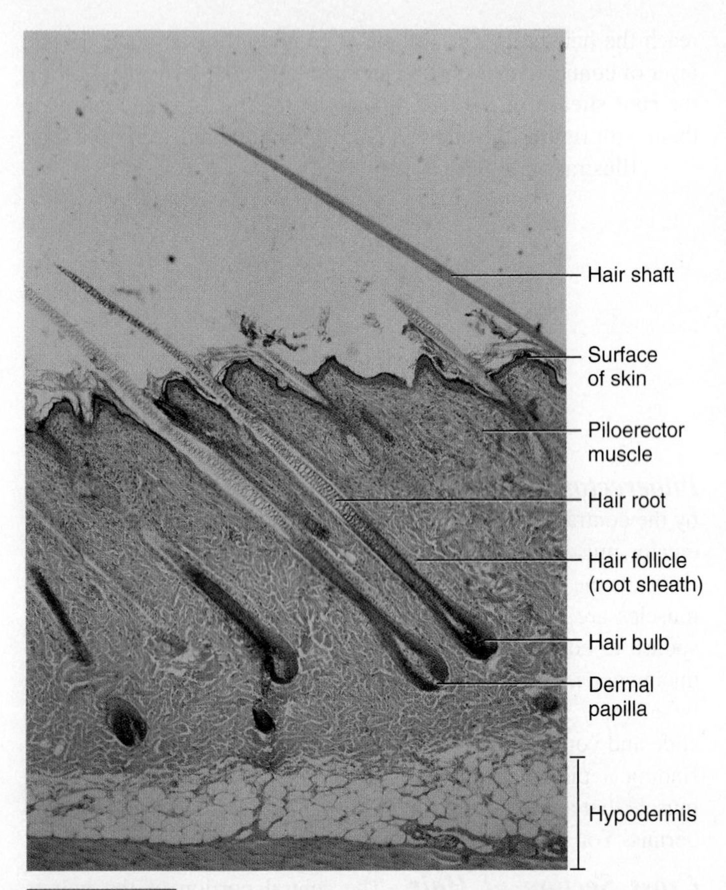

FIGURE 5.7 Photomicrograph of Hair in a Follicle (40×)

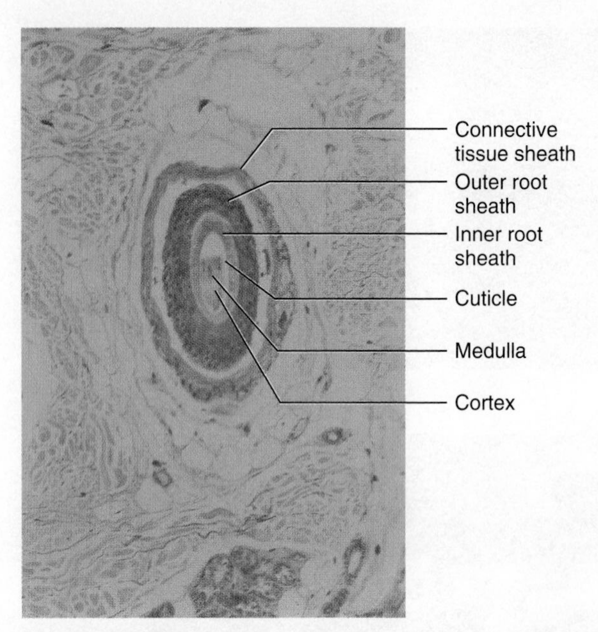

FIGURE 5.8 Hair Root and a Follicle, Cross Section (100×)

Gray hair is the result of a lack of pigment in the cortex and air in the medulla. The layer superficial to the cortex is the **cuticle** of the hair and looks rough in microscopic sections. Figure 5.8 shows a cross section of hair.

Nails

Nails are sheets of keratinized cells of the stratum corneum on the distal ends of the fingers and toes. The main part of the nail is called the **nail body,** and as it grows, the **free edge** of the nail is the part you clip. The **cuticle** of the nail is known as the **eponychium,** and the part of the nail that generates the nail body is the **nail root,** which is underneath the eponychium. The **nail bed** is the layer deep to the **nail body,** while the **lunule** is the small, white crescent at the base of the nail. The **nail matrix** undergoes cell division to produce the nail root. The **hyponychium** is the region under the free edge of the nail. Examine figure 5.9 and compare your fingernails with the diagram.

On each side of the nail is the **nail groove,** a depression between the body of the nail and the skin of the fingers. The ridge above the groove is the **nail fold.**

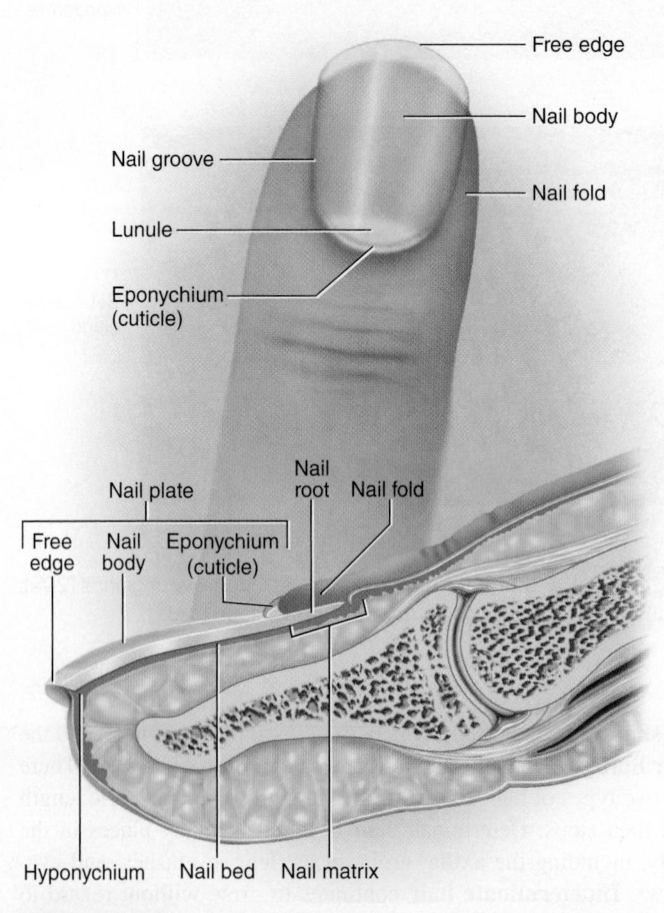

FIGURE 5.9 Fingernail, Posterior and Median Views

The Integumentary System

Name ______________________________ Date ______________

1. Label the following illustration using the terms provided.

dermis	piloerector
epidermis	sebaceous gland
hair follicle	stratum basale
hair root	stratum corneum
hair shaft	stratum spinosum
hypodermis	sweat gland

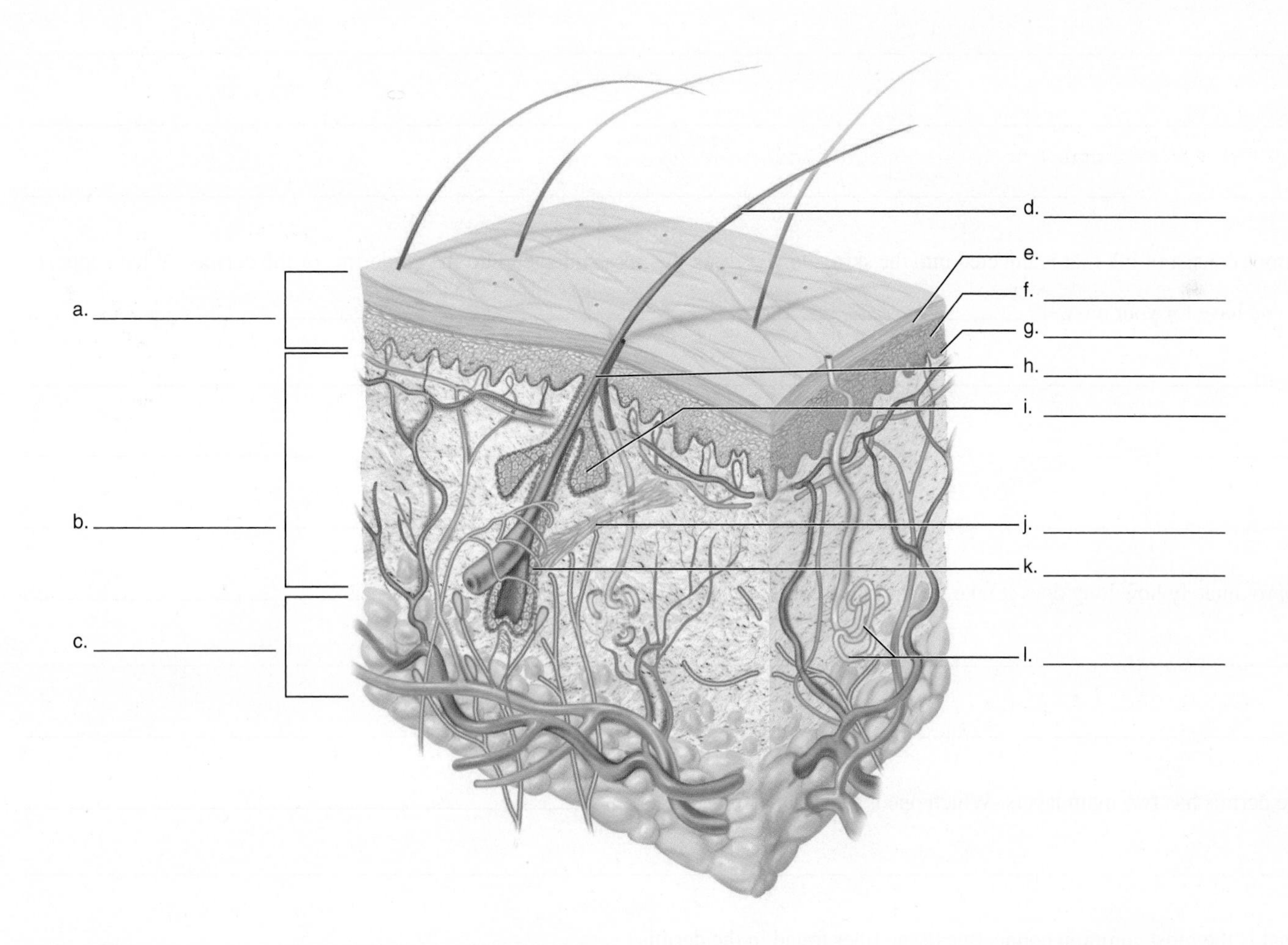

2. The main organ of the integumentary system is the skin. Name three structures that are associated structures in the integumentary system and discuss their function. __________

3. Which one of these three layers (epidermis, dermis, hypodermis) is not considered part of the integument?

4. What specific protein makes the epidermis tough? __________

5. The most superficial layer of skin (composed of keratinocytes) is dead and provides a protective barrier. If these cells were alive at the surface, how would that compromise the protective function of the integument? __________

6. Tattoos consist of ink that is injected into the skin. Do you think the ink is injected into the epidermis or the dermis? What support do you have for your answer? __________

7. Approximately how long does it take for the epidermis to renew itself? __________

8. The dermis has two main layers. Which one is more superficial? __________

9. What is the most common connective tissue fiber found in the dermis? __________

10. The release of heat from the body occurs by blood vessels in what main layer of the integument?

11. Which kind of sweat gland (eccrine or apocrine) is involved in evaporative cooling?

12. What integumentary gland secretes sebum?

13. Hair of the axilla is considered determinate/indeterminate hair (circle the correct answer). Why?

14. Electrolysis is the process of hair removal using electric current. Explain how this might destroy the process of hair growth in relation to the hair bulb.

15. Since hair color is determined by pigment in the cortex and the hair shaft is dead, explain the fallacy of claims of a person's hair turning white overnight.

16. What part of the hair is found on the outside of the skin?

17. What does the piloerector muscle do? ______________________________

____

____

____

18. The outermost portion of a cross section of hair is known as the (circle one)

a. cuticle. b. lunule. c. root. d. medulla.

19. What is another name for the cuticle of a fingernail? Circle the correct answer.

a. lunule b. eponychium c. hyponychium d. base

20. Two common blisters of the integument are watery blisters, filled with a clear fluid, and blood blisters, filled with blood. Based on your knowledge of the blood supply to the integument, describe what layers might be damaged in these two blisters.

____

____

____

Laboratory Exercise 6

Introduction to the Skeletal System

Skeletal System

Introduction

The skeletal system has numerous functions, including support of the body; protection of soft tissues, such as the brain and lungs; and provision of a structure for movement, storage of minerals, and blood cell formation.

The skeletal and muscular systems are intimately related. The interaction between the skeletal and muscular systems is one in which the skeletal system provides the lever system on which the muscles can work, resulting in movement. This is described in more detail in Laboratory Exercise 10, "Articulations." In this exercise, you examine the structure, composition, and development of the skeletal system.

Learning Objectives

At the end of this exercise you should be able to

1. describe the composition of bone tissue;
2. discuss how the skeletal system provides for efficient movement;
3. name the features on specific bones;
4. list the two major types of bone material found in long bones; and
5. draw or describe the microscopic structure of compact bone.

Materials

Chart of the skeletal system
Cut sections of bone showing compact and spongy bone
Articulated human skeleton
Disarticulated human skeleton
Prepared slide of ground bone
Prepared slide of decalcified bone
Bone heated in oven at 350°F for 3 hours or more
Bone placed in vinegar for several weeks or 1 N nitric acid for a few days
Microscopes

Procedure

Divisions of the Skeletal System

The skeletal system can be sorted into two main divisions according to location. The **axial skeleton** is so named because it is in the vertical axis of the body. It consists of the skull (including the auditory ossicles), hyoid bone, vertebral column, ribs, and sternum. The **appendicular skeleton** is that part of the skeleton that is "attached to" the axial skeleton. It can be divided further into the **pectoral girdle,** the **upper limb,** the **pelvic girdle,** and the **lower limb.** Locate the major bones in lab and compare them with figure 6.1 and table 6.1.

Major Bones of the Body

Examine figure 6.1 for the major bones or bone regions of the body. You should be familiar with the major bones of the body before you study the specifics of the individual bones in later exercises. It helps to take a bone from a disarticulated skeleton and compare it with an articulated skeleton. Locate the major bones in the figure as you look at bones in the lab.

A young adult has about 206 bones, with more in younger individuals and fewer in older individuals. Many developing bones fuse as a person ages, forming larger bones and reducing the overall number. Not everyone has the same number of bones. Bones that develop in tendons have a shape like a sesame seed and are called **sesamoid bones.**

Bone Features There are specific terms that describe bumps, hollows, or spaces of bones. **Projections** arise from the surface of bones, **depressions** occur on the surface of bones, holes traverse bones, and **cavities** are enclosed spaces in bones. Projections on bones reflect different functions. Some projections, such as condyles and heads, are surfaces where the bone articulates with another. Other projections, such as tubercles, spines, and trochanters, are attachment points for muscles or ligaments. Holes, such as foramina, canals, and fissures, are often passageways for blood vessels and/or nerves. These are presented in table 6.2. You should study these terms and be able to find them on bones in the lab.

Composition of Bone Tissue Bone consists of organic matter and inorganic matter. The **organic matter** is composed mostly of cells and collagenous fibers and makes up about one-third of the

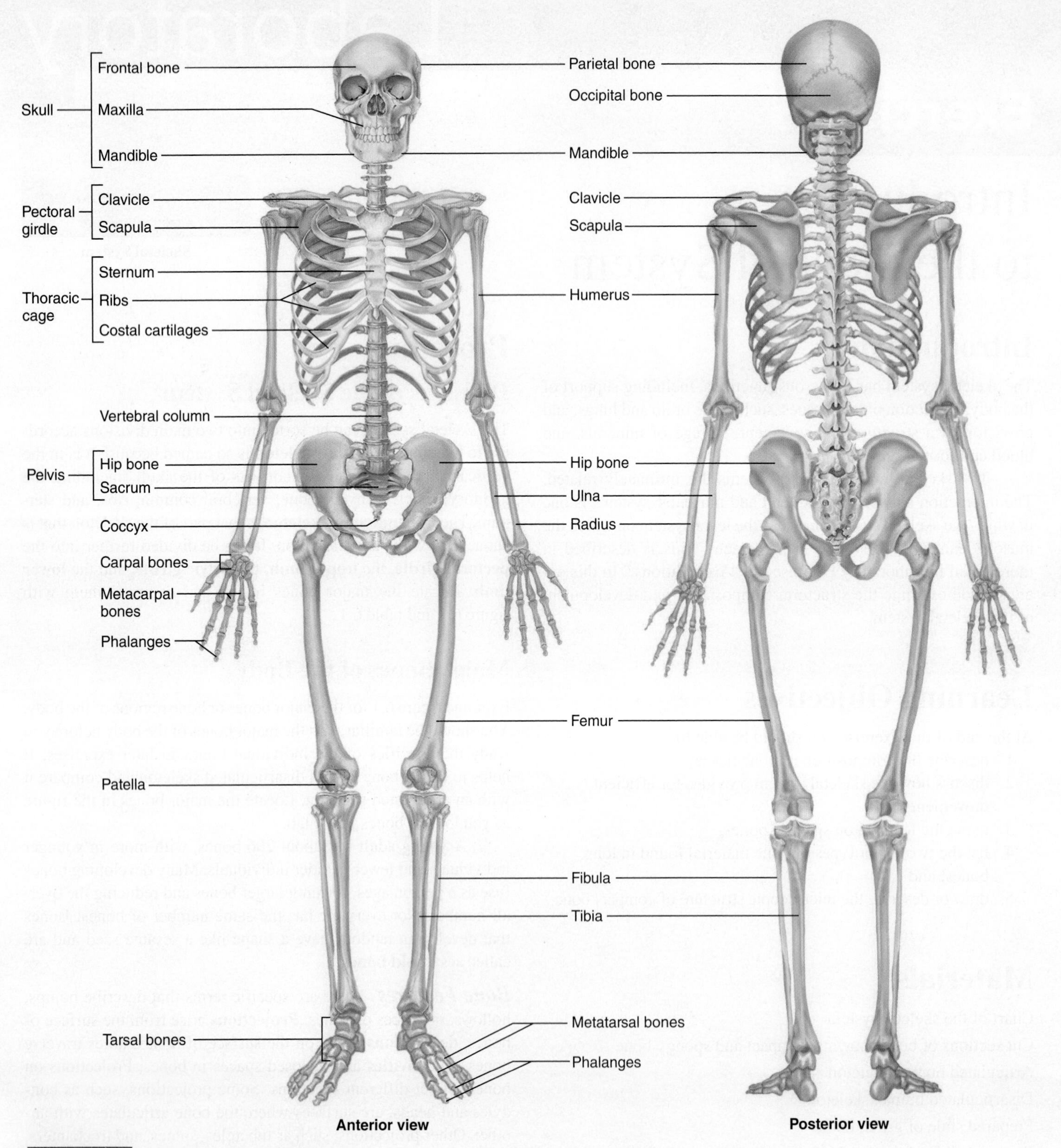

FIGURE 6.1 Major Bones of the Body Axial skeleton in beige, appendicular skeleton in green.

TABLE 6.1 Skeleton

Axial Skeleton	
Skull	Auditory ossicles
Hyoid bone	Ribs
Vertebral column	Sternum
Appendicular Skeleton	
Pectoral girdle	Pelvic girdle
Clavicle	Hip bone
Scapula	Lower limb
Upper limb	Femur
Humerus	Patella
Radius	Tibia
Ulna	Fibula
Carpals	Tarsals
Metacarpals	Metatarsals
Phalanges	Phalanges

bone by weight. The **inorganic matter** makes up the remaining two-thirds of the bone and mostly consists of **hydroxyapatite,** a complex salt consisting of calcium phosphate. There is also a lesser amount of calcium carbonate in the inorganic portion of bone. If available in the lab, examine bones that have been soaked in acid. The acid dissolves the mineral salts, leaving the collagenous fibers as a flexible framework. Your instructor may show bones that were baked in an oven for a few hours. Heating bones denatures the proteins; although the bones remain hard because of the mineral salts, they are brittle due to the destruction of the protein fibers. Your instructor may wish to demonstrate the fragility of specially prepared bones.

Bone Structure The basic anatomy of a long bone consists of the proximal and distal ends of the bone, the **epiphyses** (eh- PIF-uh-seez; singular, *epiphysis*), and the shaft of the bone, the **diaphysis** (die-AH-fuh-sis) (fig. 6.2). The epiphyses of bones have **articular cartilage** at the ends of the bone. The articular cartilage is composed of hyaline cartilage and helps reduce friction as the joint moves.

In young people, there is a hyaline cartilage plate between the epiphysis and the diaphysis. This is the **epiphyseal** (EP-ih-FIZZ-ee-ul) **plate** (fig. 6.2), commonly known as the growth plate. The cartilage increases in thickness by division of the chondrocytes. As the cartilage grows, the individual is pushed upward by the increase in the cartilage plate. Remodeling

TABLE 6.2 Bone Features

Projections from Bones	**Example**
Process—a general term for a projection from the surface of the bone	Styloid process of ulna
Tubercle—a relatively small bump on a bone	Greater tubercle of humerus
Tuberosity—a relatively large rough area on a bone	Deltoid tuberosity of humerus
Spine—a short, sharp projection	Vertebral spine
Head—a terminal projection that articulates with another bone	Head of femur
Neck—a constriction below the head	Neck of rib
Condyle—a smooth surface that articulates with another bone	Lateral condyle of femur
Epicondyle—an elevation proximal to a condyle	Medial epicondyle of humerus
Crest—an elevated ridge of bone	Crest of ilium
Line—a smaller elevation than a crest	Gluteal line of ilium
Facet—a smooth, flat face	Articular facets of vertebrae
Trochanter—a large bump (on femur)	Greater and lesser trochanter of femur
Ramus—a branch	Ramus of mandible
Depressions, Passageways, and Cavities in Bone	**Example**
Foramen—a hole	Foramen magnum of occipital bone
Meatus or canal—a tunnel	External auditory meatus of temporal bone
Sinus—a cavity in a bone	Maxillary sinus
Fossa—a surface depression in a bone	Iliac fossa
Notch—a deep cutout in a bone	Greater sciatic notch of ilium
Groove or sulcus—an elongated depression	Intertubercular groove of humerus
Fissure—a long, deep cleft in a bone	Inferior orbital fissure

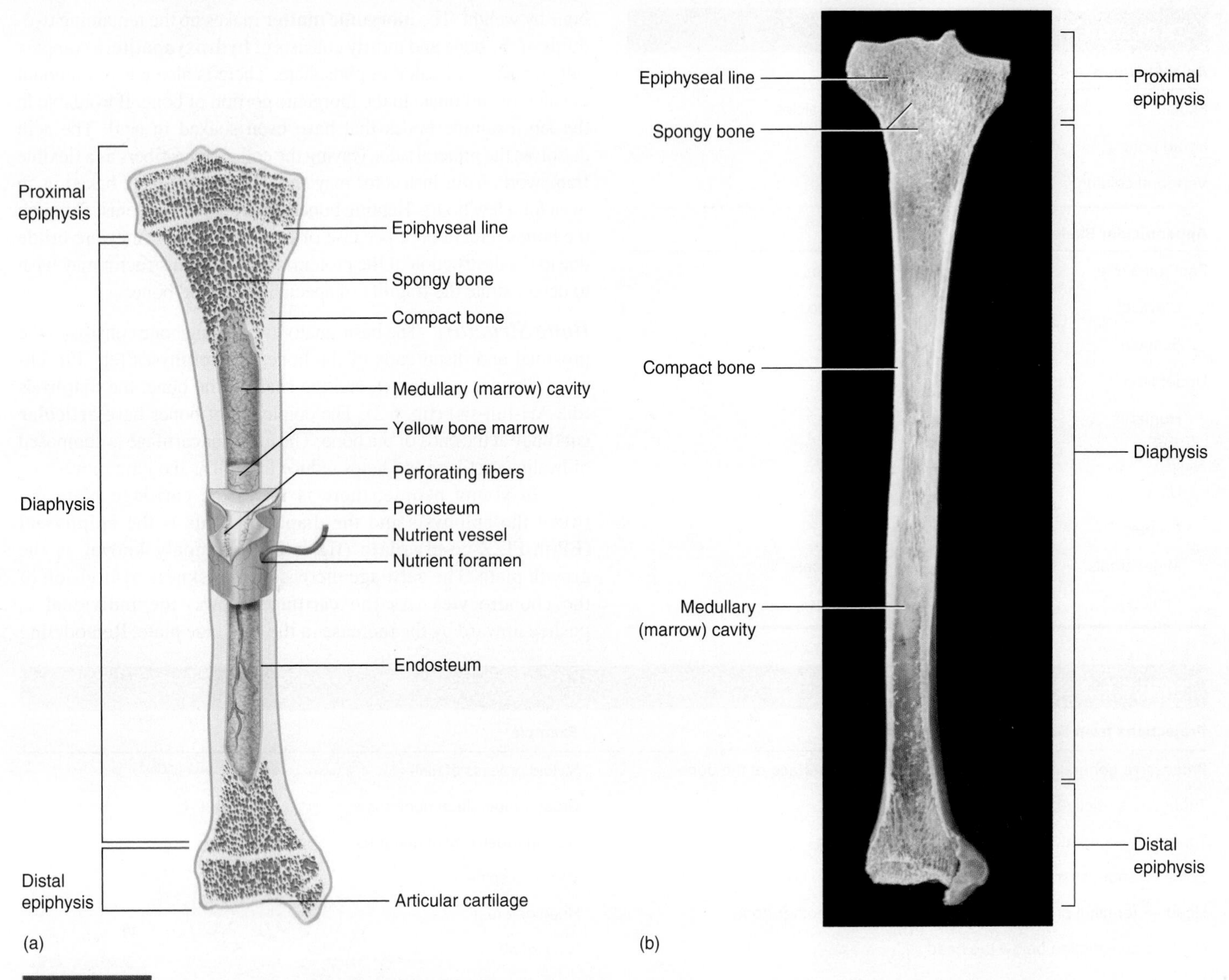

FIGURE 6.2 **Long Bone, Longitudinal Section** (a) Diagram; (b) photograph.

of the cartilage takes place, and the cartilage is eventually replaced by bone. Once a person stops growing, the epiphyseal plate is replaced by a thin line of bone called the **epiphyseal line.**

Examine a cut section of bone. Note the hard, **compact (cortical) bone** on the outside of the bone and the inner **cancellous,** or **spongy, bone.** The cancellous bone is made of **trabeculae,** which are thin rods or plates of bone that run in the same direction as the stress applied to the bone. Stress on the bone may be in the form of gravity, or it may occur due to a force applied to the limb. Trabeculae make an internal framework that strengthens the bone.

The innermost section of bone is hollow and is called the **medullary** (MED-you-lerr-ee) **(marrow) cavity. Marrow** is the material that occupies this cavity and can be of two types, a **hematopoietic** (blood cell–forming) **red marrow** and **yellow marrow** containing adipose tissue. In flat bones the inner and outer compact bone encase a material called **diploe** (DIP-low-ee), which is spongy bone found in cranial bones. Compare the cut bones in the lab with figure 6.3 and note the features listed in the discussion.

Examine a bone that has been cut lengthwise and see how the compact bone is thicker in the middle of the bone than at the ends. Many long bones have this feature and resemble an archery bow in this regard. This reinforcement decreases breakage at the region of bone that would be most prone to break, the middle.

On the surface of the diaphysis, locate the **nutrient foramina** (for-AM-ih-nuh). These are small holes that allow for the passage of blood vessels into and out of the bone. They can be found in various locations in bones. The nutrient foramina lead to **perforating (Volkmann's) canals** that pass through compact bone. The central canals typically run vertically in bones, while perforating canals carry nutrients horizontally.

The outer surface of the bone is covered with a dense connective tissue sheath called the **periosteum** (PEAR-ee-OS-tee-um). It is the site where nerves and blood vessels occur on the outer surface of the bone. The periosteum is an anchoring point for tendons and ligaments. **Tendons** attach muscles to bone at the periosteum, and this attachment is strengthened by **perforating (Sharpey's) fibers** that provide greater surface area between the tendon and the periosteum. **Ligaments** are parallel straps of connective tissue that connect one bone to another, and they are secured to the bone by the periosteum.

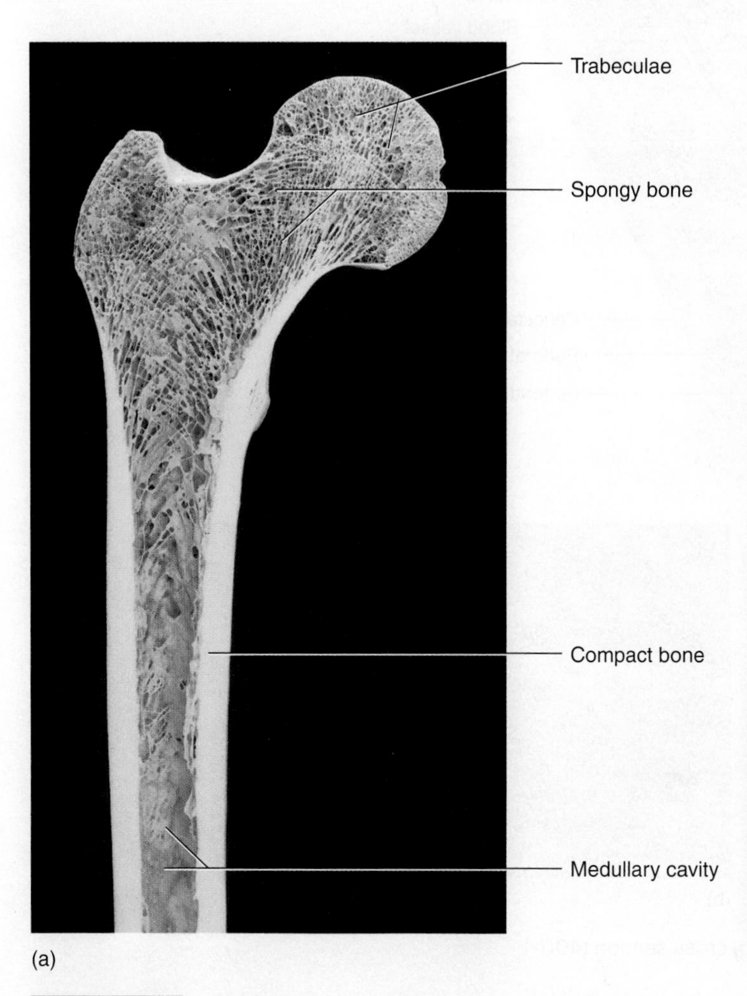

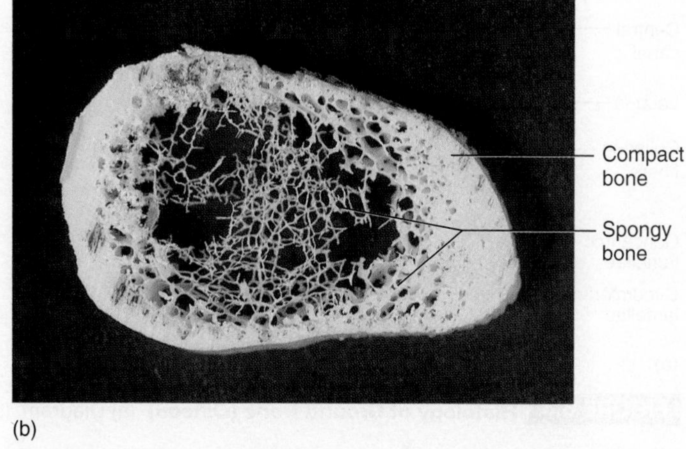

FIGURE 6.3 **Cut Bone** (a) Long section; (b) cross section.

The inner surface of long bones, near the medullary cavity, is lined with a layer known as the **endosteum** (end-OS-tee-um).

Microscopic Structure of Bone

Ground Bone Examine a prepared slide of ground bone and locate the circular structures in the field of view. These modular units of bone are called **osteons.** Each osteon has a hole in the middle called a **central** (**Haversian** or osteonic) **canal,** which houses blood vessels and nerves in the dense bone tissue. Around the central canal are rings of bone tissue known as **lamellae.** Also in the slide are dark spots. These are holes that held osteocytes in living tissue but fill with bone dust during the preparation of the slides. These spaces are the **lacunae** (singular, *lacuna*). The lacunae are connected to each other by thin tubes called **canaliculi** (CAN-uh-LIC-you-lie). The role of canaliculi is to provide a passageway through the dense bone material. Osteocytes are living and need to receive oxygen and nutrients and remove wastes. They do this by extending cellular strands through the canaliculi between osteocytes. The canaliculi connect to the **central canal.** Locate these structures in figure 6.4.

Decalcified Bone Examine a slide of decalcified bone. In this preparation the bone salts have been dissolved away and the remaining thin section of bone has been sliced, mounted, and stained. The periosteum is visible on the surface of the bone, with the compact bone in the middle and the endosteum on the inside of the bone (fig. 6.5). There are three types of bone cells in osseous tissue. These are the **osteoblasts, osteocytes,** and **osteoclasts.** *Osteoblasts* originate from stem cells called **osteogenic cells** which occur in the periosteum, endosteum, and central canals of osteons. These osteogenic cells maintain mitotic activity and produce osteoblasts, which secrete collagen fibers and calcium salts. Osteoblasts are most numerous in the endosteum and periosteum.

Osteocytes are mature bone cells; they respond to stresses placed on bone and remodel the bone in response to those stresses. For example, an increase in weight-bearing activity increases the density of bone. This is seen in the bones of athletes, which are denser than those of people who do not exercise. Osteocytes deposit or reabsorb bone matrix, regulating bone density, and control calcium and phosphate balance. Osteocytes also repair microfractures that frequently occur in bone. Examine the slide of decalcified bone and find the osteocytes.

Skeletal Anatomy of the Cat

If you are using cats, turn to section 1, "Skeletal Anatomy of the Cat," on page 408 of this lab manual.

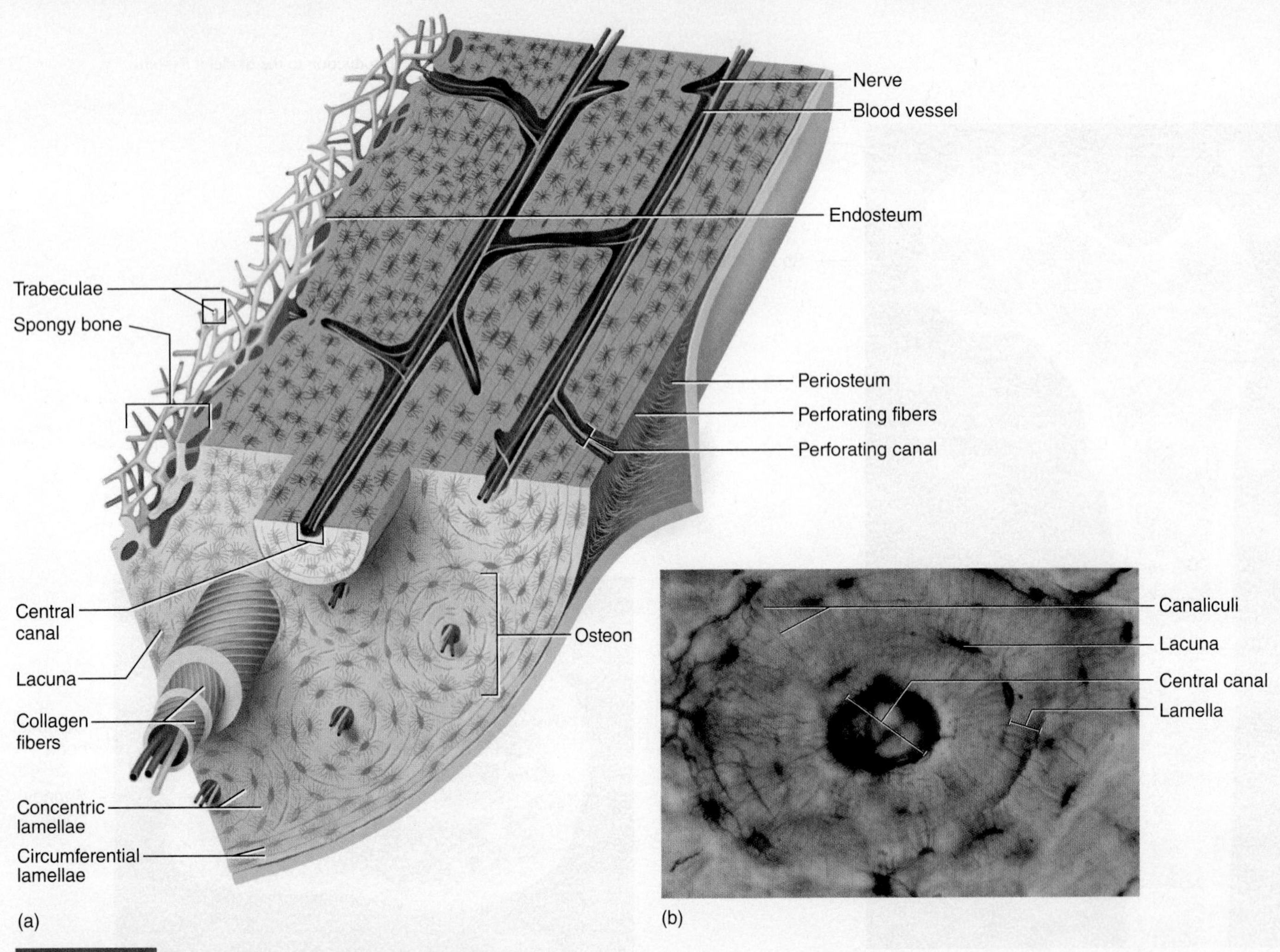

FIGURE 6.4 **Histology of Ground Bone (Osteon)** (a) Diagram; (b) cross section (400×).

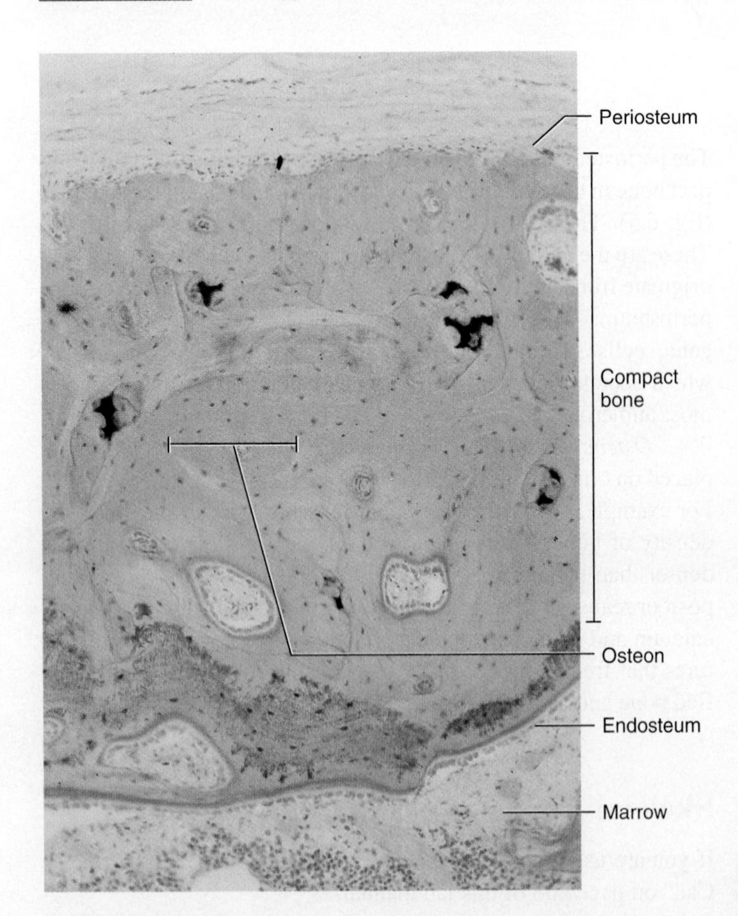

FIGURE 6.5 **Decalcified Bone (100×)**

Introduction to the Skeletal System

Name ______________________________ Date ____________

1. The hyoid bone belongs to the (circle the correct answer)

   a. appendicular skeleton. b. axial skeleton. c. upper limb. d. skull.

2. The clavicle belongs to the (circle the correct answer)

   a. axial skeleton. b. pectoral girdle. c. pelvic girdle. d. upper limb.

3. In the disease osteoporosis, there is a significant loss of cancellous bone. Explain how the loss of this specific bone material can weaken a bone. ______________________________

   ____

   ____

4. Label the following illustration using the terms provided.

   canaliculi lacuna osteon

   central canal lamella

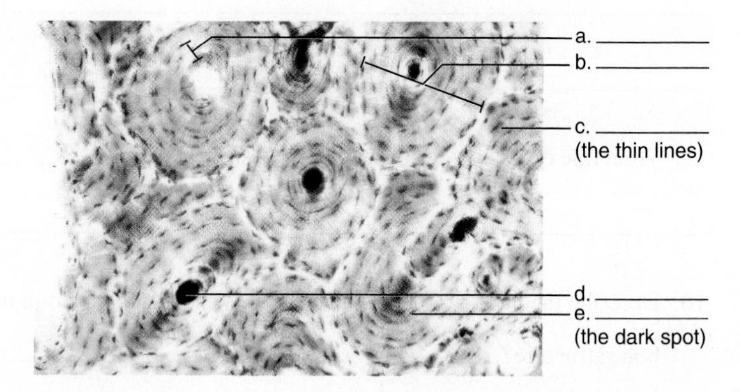

5. The ends of a long bone are known as the ______________________________.

6. The radius is part of which skeletal division? (circle the correct answer)

   a. axial b. appendicular

7. The ribs are part of which skeletal division? (circle the correct answer)

   a. axial b. appendicular

8. What role do osteoblasts have in maintaining bone tissue? ______

9. The patella is part of which skeletal division? (circle the correct answer)

   a. appendicular   b. axial

10. A young adult has how many bones, on average? ______

11. The inorganic portion of bone tissue is made of what complex mineral salt (consisting of calcium phosphate)?

______

12. How does the shape of a long bone contribute to resisting breaking when put under stress? ______

13. What is an osteon? ______

14. Synthetic bone material known as hydroxyapatite is often molded into the shape of bone. This procedure is done to replace bone that has been diseased, damaged, or surgically removed. Cells in the body remodel the synthetic material and produce new bone. What bone cells do you think are involved in this remodeling, and what role do these cells have?

______

15. What role do osteocytes have in bone tissue? ______

16. Describe the nature of decalcified bone. What was removed in the process of decalcification, and what impact did this have on the bone structure? ______

17. How does the central canal differ from a lacuna in terms of location and the material found in each respective space?

______

18. What is another name for the shoulder blade, and what two bones attach to it? ______________________________

____

19. Fill in the major bones or bony regions of the body on the illustration using the provided terms.

carpals
clavicle
femur
fibula
hip bone
humerus
metacarpals
metatarsals
patella
phalanges (foot)
phalanges (hand)
ribs
scapula
skull
sternum
tarsals
tibia
vertebral column

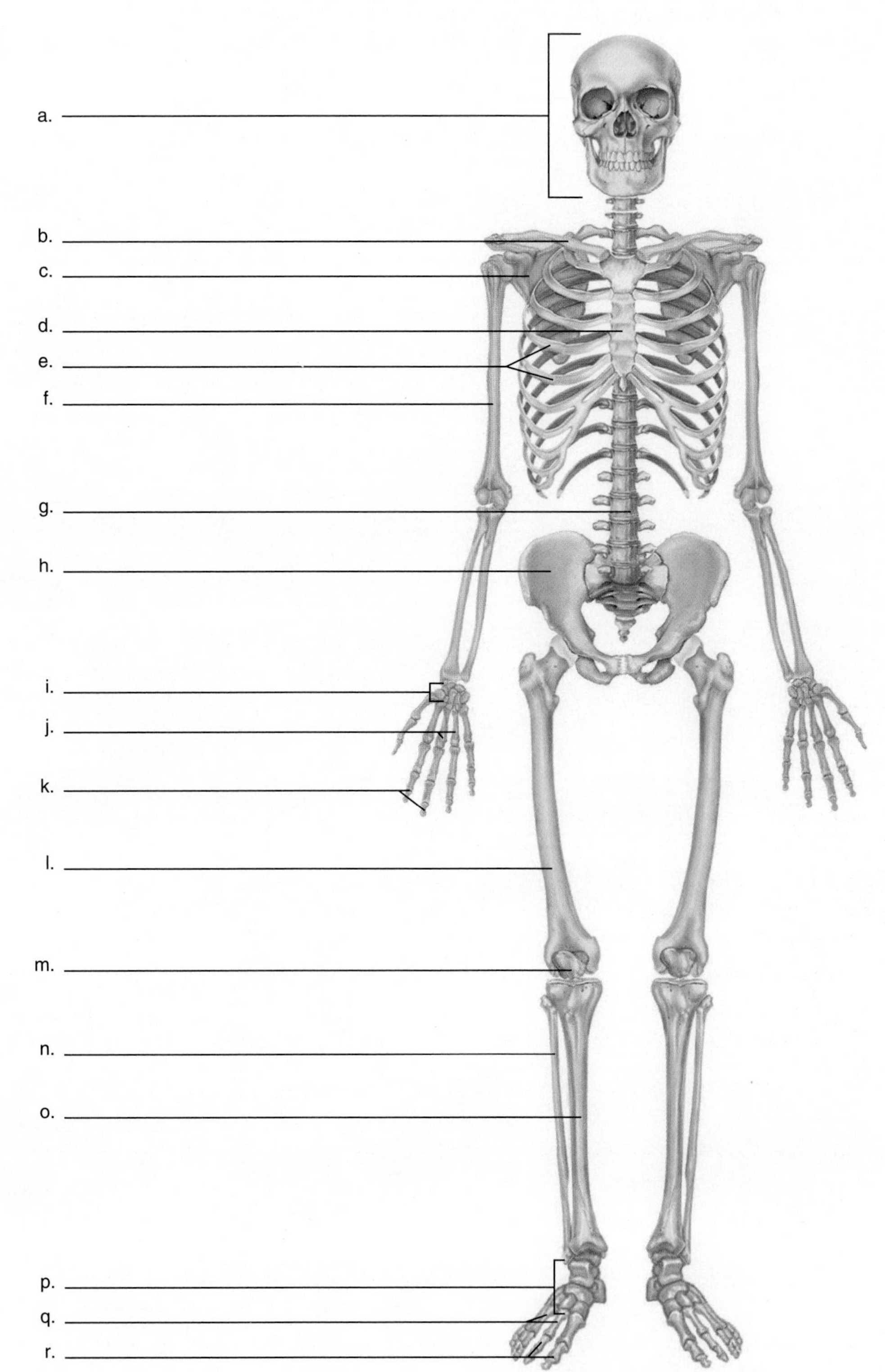

Laboratory

Exercise 7

Axial Skeleton 1: Skull

Skeletal System

Introduction

The bones of the skull can be divided into two groups: bones of the cranial vault and bones of the face. The skull is the most complex region of the skeleton and not only houses the brain but contains a significant number of sense organs as well. In this exercise you learn the details of the skull, particularly those of the cranial vault and the face. You will study the bones associated with the skull, such as those of the middle ear, in this laboratory manual in Laboratory Exercise 18, "Sensory Receptors." Be careful if you handle real bones, as they are fragile and can easily be broken.

Learning Objectives

At the end of this exercise you should be able to

1. list all the bones of the cranial vault and the major features of those bones;
2. list all the bones of the face and the major features of those bones;
3. name all the bones that occur singly or in pairs in the skull;
4. locate the major foramina of the skull;
5. find the specific bony markings of representative bones;
6. name the major sutures of the skull; and
7. list all the fontanels of the fetal skull.

Materials

Disarticulated skull, if available
Articulated skulls with the calvaria cut
Foam pads to cushion skulls from the desktop
Pipe cleaners

Procedure

Overview of the Cranial Vault

Familiarize yourself with the bones of the skull in the **cranial vault** and the **face.** The skull bones in the cranial vault are listed next with a number, in parentheses, after the name of the bone indicating whether the bone occurs singly (1) or as a pair (2). Take a skull to your lab table and locate the bones listed in figure 7.1.

Frontal (1)	Parietal (2)
Occipital (1)	Temporal (2)
Sphenoid (1)	Ethmoid (1)

You can remember the bones of the cranium with the mnemonic "of pets." Each letter represents a cranial bone (*o* for occipital, *f* for frontal, etc.).

Overview of the Face

The bones of the face are listed next. Locate these bones on a skull in the lab, as shown in figures 7.1 and 7.2.

Maxilla (2)	Nasal (2)
Mandible (1)	Palatine (2)
Vomer (1)	Zygomatic (2)
Lacrimal (2)	Inferior nasal concha (plural, *conchae*) (2)

The facial bones can be remembered by this mnemonic device: "Monkeys in New Zealand like peaches very much." In the mnemonic device, all the bones except the last two (vomer and mandible) are paired bones.

Be very careful handling skulls. As you locate the various bones on a skull in the lab, make sure you do not poke pencils, pens, or fingers into the delicate regions of the orbits (eye sockets) or nasal cavity. Do not break the delicate structures in these regions. Once you have found the major bones of the skull, use the following descriptions and illustrations to find the specific structures of the skull. The skull is described from a series of views, and you should locate the anatomical features listed for each of those views.

Anterior View

The large bone that makes up the forehead is the **frontal bone.** It is a single bone (fused from two bones in the fetus) that makes up the superior portion of the **orbits.** The frontal bone has two ridges above the eyes called **supraorbital ridges,** or **margins** (located deep to the eyebrows). There is a hole in each of these ridges called the **supraorbital foramen** that allows for nerves and arteries to reach the

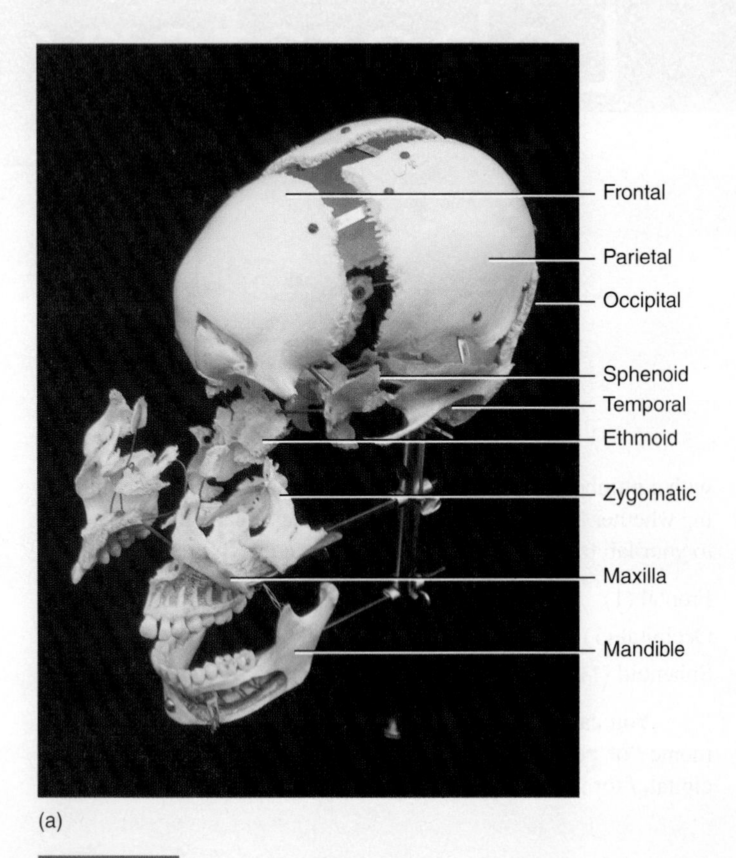

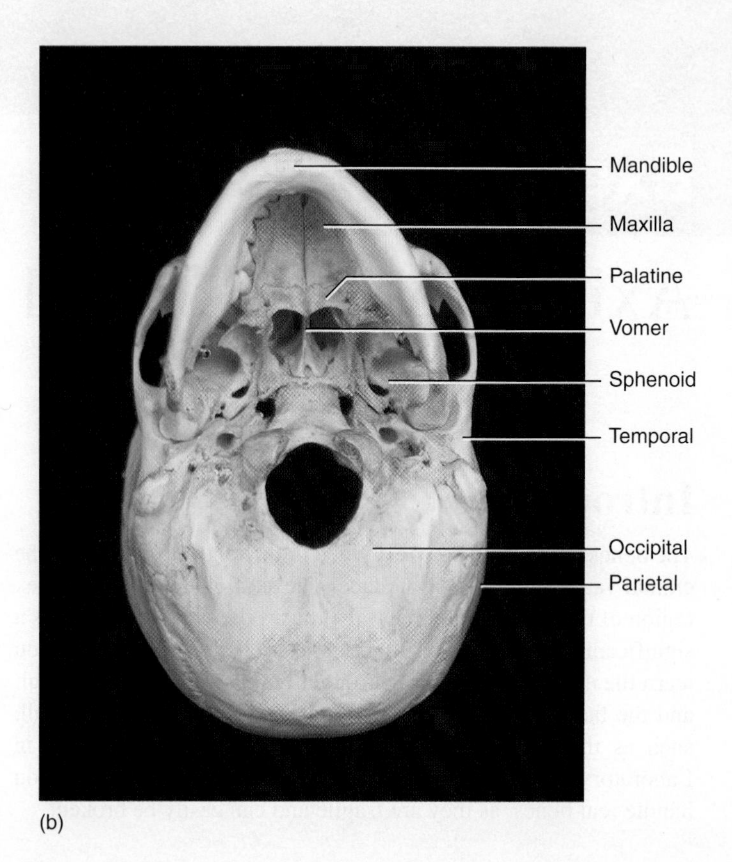

FIGURE 7.1 Overview of the Skull (a) Superolateral view; (b) inferior view.

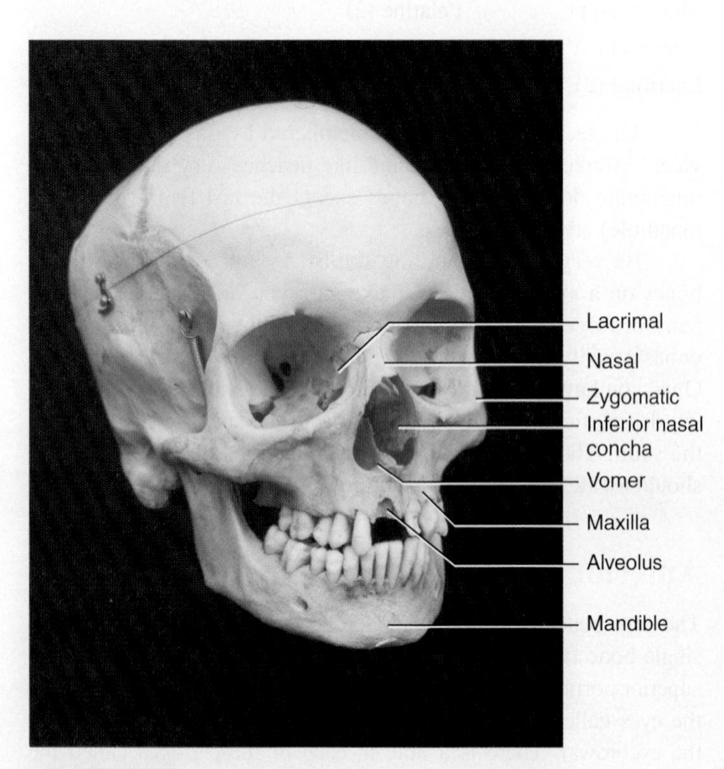

FIGURE 7.2 Bones of the Face

face. Most of what you see inferior to the frontal bone are facial bones. The bony part of the nose is composed of the paired **nasal bones,** which join with the cartilage that forms the tip of the nose. Posterior to the nasal bones are thin strips of the upper maxillae (mac-SILL-ee), bones that hold the upper teeth; posterior to the **maxilla** are the thin **lacrimal** (LACK-rih-mul) **bones,** which contain the **nasolacrimal ducts,** into which tears drain from the eyes to the nose. Locate these bones in figure 7.3. Posterior to the lacrimal bones is the **ethmoid** (ETH-moyd) **bone,** which is very delicate and frequently broken on skulls that have been mishandled. Posterior to the ethmoid is the **sphenoid bone,** which forms the posterior wall of the orbit and contains not only the **optic canal** (a passageway for the optic nerve) but also the **superior orbital fissure** and the **inferior orbital fissure.** The bones on the lateral side of the orbit are the **zygomatic** (ZY-go-MAT-ic) **bones,** commonly known as the cheekbones.

The major bones that makes up the floor of the orbit and below are the maxillae, commonly known as the upper jaw. The two maxillae have sockets called **alveoli** (singular, *alveolus*), which contain the teeth and extensions of bone between each pair of sockets called **alveolar processes.** The **infraorbital foramen** is a small hole in the maxilla below the eye; it is a passageway for nerves and blood vessels. The most inferior bone of the face is the **mandible,** commonly known as the lower jaw. The mandible also has alveoli, teeth, and alveolar processes (fig. 7.4). The mandible begins as two bones in utero and fuses at the midline of the chin. This fusion results in the **mental symphysis** (review fig. 7.3). Lateral to the mental symphysis are the **mental foramina,** which conduct nerves and blood vessels to the tissue anterior to the jaw.

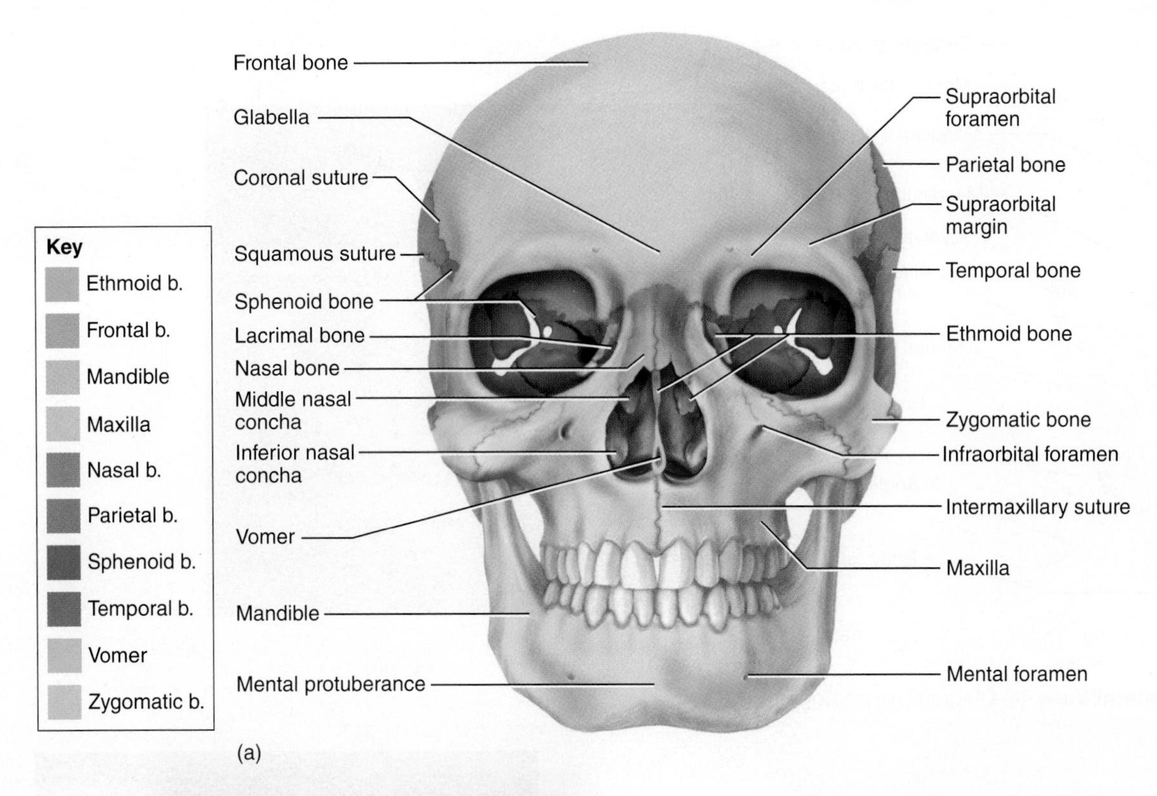

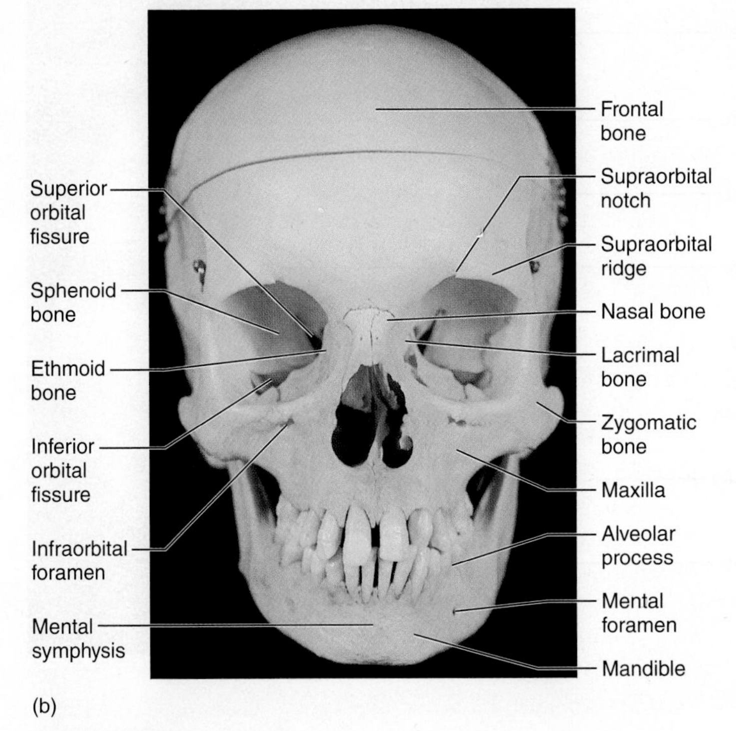

FIGURE 7.3 **Skull, Anterior View** (a) Diagram; (b) photograph.

Superior View

Locate the major suture lines of the skull from this view, as seen in figure 7.5. The frontal bone is separated from the pair of **parietal** (pa-RY-eh-tul) **bones** by the **coronal suture.** The parietal bones are separated from each other by the **sagittal suture.** The parietal bones are separated from the occipital bone by the **lambdoid** (LAM-doyd) **suture.** The lambdoid suture is named after the Greek letter lambda (λ), which looks similar to an upside-down Y. There may be small bones between the occipital bone and the parietal bones (or between other skull bones), and these are known as **sutural,** or **Wormian, bones.**

Lateral View

The coronal and lambdoid sutures can also be seen from the lateral view along with the **squamous** (SKWAY-mus) **suture,** which separates the **temporal bone** from the parietal bone. Locate the cranial bones from a lateral view. These are the **frontal, parietal, occipital, temporal, sphenoid,** and **ethmoid bones.** You may be able to see a bump at the back of the occipital bone from this angle. This is known as the **external occipital** (oc-SIP-ih-tul) **protuberance.** The hole on the side of the head where the ear attaches is the **external acoustic meatus.** The large process posterior and inferior to this opening is the **mastoid** (MAS-toyd) **process.** A long, thin spine medial to the mastoid process is the **styloid process,** which has muscles that connect it to the hyoid bone, tongue, and larynx. The sphenoid bone can be seen from this view just adjacent to the temporal bone. Anterior to the sphenoid bone is the zygomatic bone forming the lateral wall of the orbit. The zygomatic bone has a **temporal process** that connects with the temporal bone. The temporal bone has a **zygomatic process,** and both of these processes, along with a part of the maxilla, make up the **zygomatic arch.** The inner wall of the orbit is made up of the ethmoid bone, lacrimal bone, maxilla, and nasal bone. Examine these structures in figure 7.6.

The mandible has a **condylar process** with a terminal **mandibular condyle** articulating with the temporal bone; a

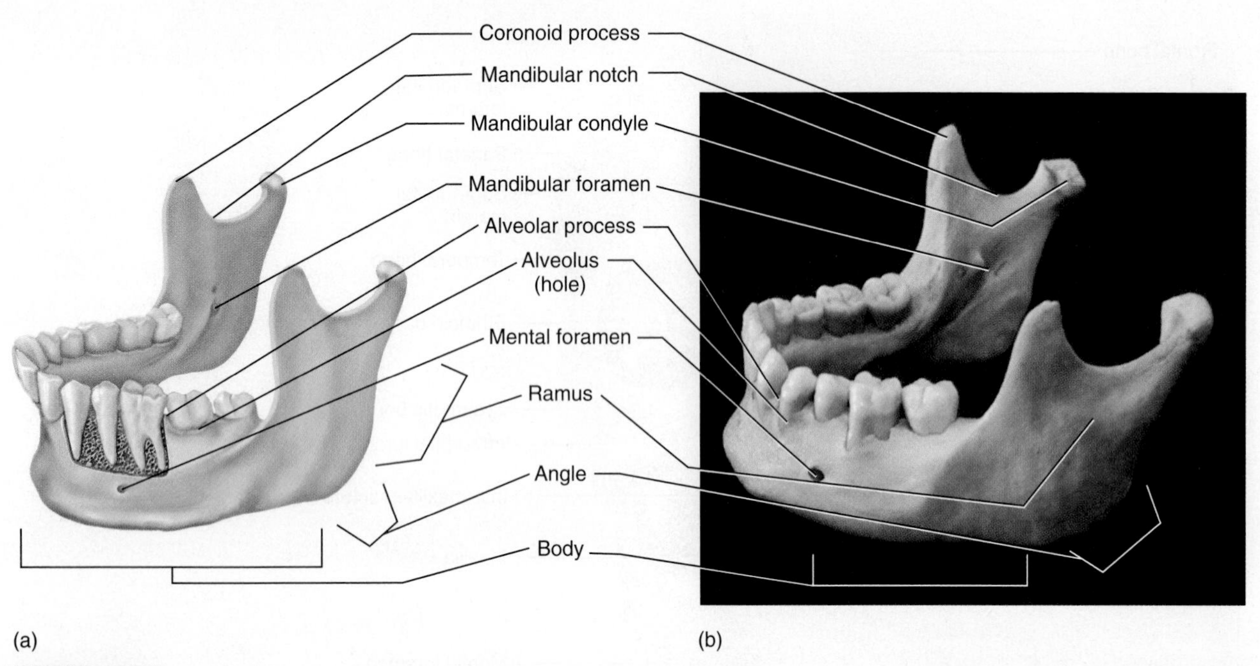

FIGURE 7.4 **Mandible, Lateral View** (a) Diagram; (b) photograph.

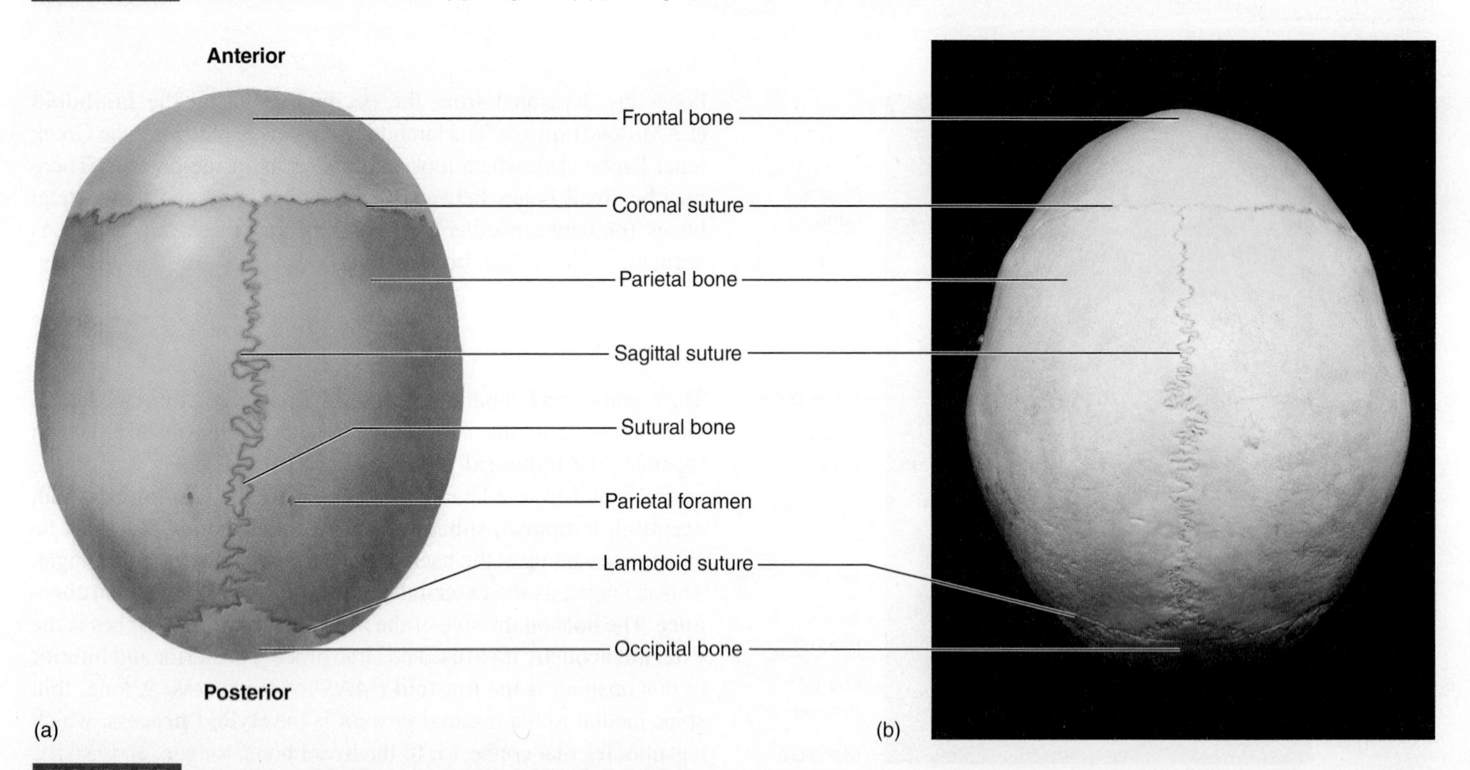

FIGURE 7.5 **Skull, Superior View** (a) Diagram; (b) photograph.

coronoid process, medial to the zygomatic arch; and a **mandibular notch,** a depression between the condyloid process and coronoid process. The vertical section of the mandible is the **mandibular ramus** (RAY-mus; *ramus* = branch), and the horizontal portion of the mandible is the **body.** The **angle** of the mandible is at the posterior part of the bone at the junction of the body and the ramus. On the inside of each ramus of the mandible is the **mandibular foramen,** a conduit for a nerve. The parts of the mandible can be identified in figures 7.4 and 7.6.

Inferior View

Place the skull in front of you with the mandible and the top of the skull removed (fig. 7.7). The largest hole in the skull, the **foramen magnum,** should be close to you. The foramen magnum is in the occipital bone and is the dividing line between the brain and the spinal cord. Lateral to the foramen magnum are the **occipital condyles,** processes that articulate with the superior articular facets of the first cervical vertebra. A small bump at the posterior part of

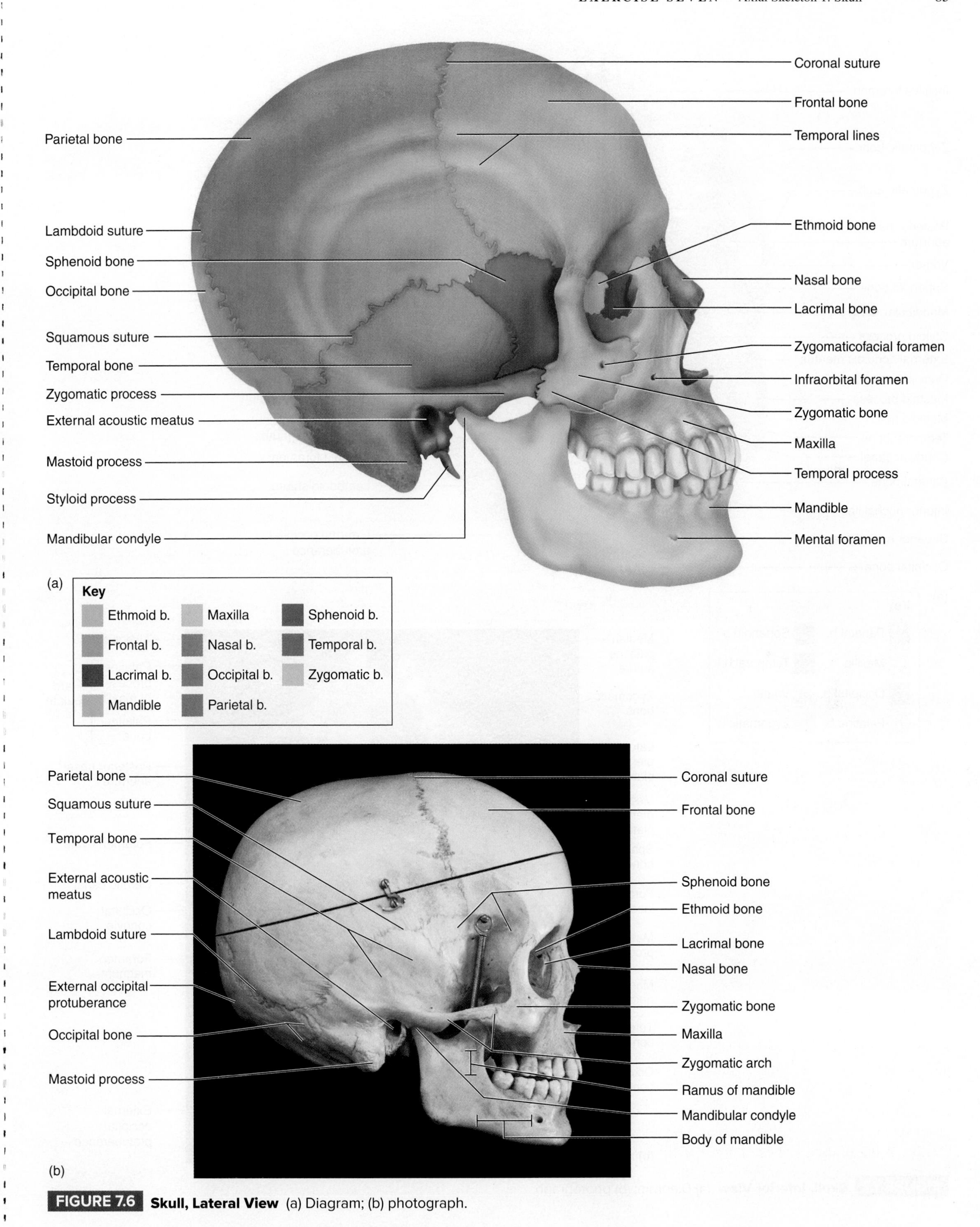

FIGURE 7.6 **Skull, Lateral View** (a) Diagram; (b) photograph.

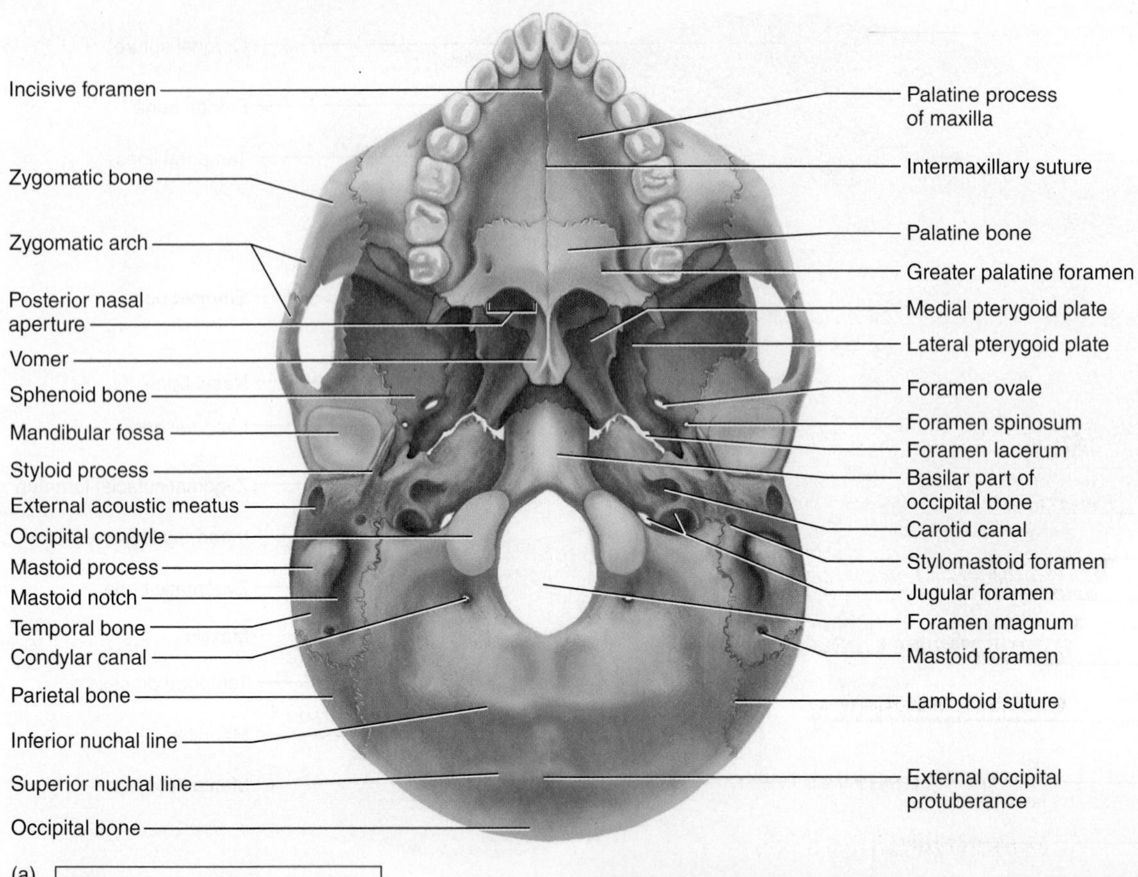

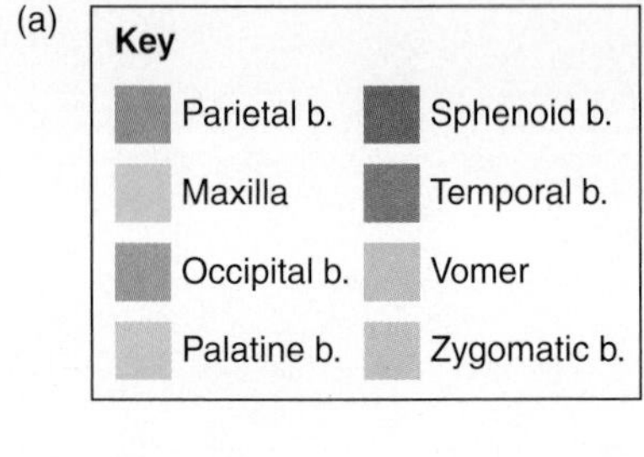

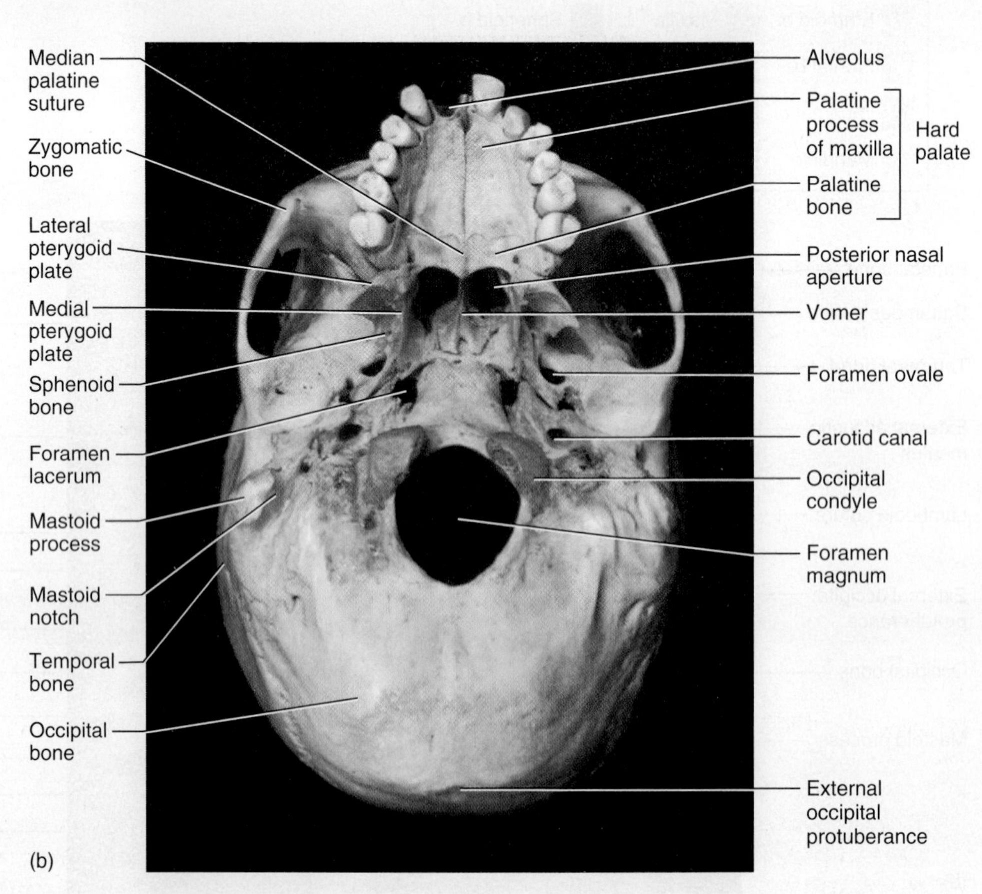

FIGURE 7.7 **Skull, Inferior View** (a) Diagram; (b) photograph.

the occipital bone is the external occipital protuberance. Two small openings are found near the foramen magnum: the **hypoglossal canals,** which allow the passage of the hypoglossal nerve. At the junction of the occipital bone and the temporal bone is the **jugular foramen,** a hole where the jugular vein begins. If you carefully insert a pipe cleaner into the jugular foramen and turn the skull over, you will see that the foramen leads to the posterior part of the skull.

You can see the mastoid process of the temporal bone and a depression just medial to the process known as the **mastoid notch.** Find the styloid processes in your specimen, though they may be hard to locate, since they are frequently broken in lab specimens. Medial to the styloid process is the **carotid canal,** through which passes the internal carotid artery to the brain. If you *carefully* insert a pipe cleaner into the carotid canal of a skull, you will notice that the canal bends at about a 90° angle; if you turn the skull over, you will note that the opening occurs in the middle of the skull. You cannot do this with most plastic casts of skulls. At the junction of the temporal bone and the sphenoid bone is the **foramen lacerum,** which is adjacent to the carotid canal. The temporal bone also has a **mandibular fossa,** which is the articulation site of the mandible. The zygomatic process of the temporal bone can also be seen from this view.

From the inferior view, the sphenoid bone can be seen as a bone that runs from one side of the skull to the other. Two pairs of flattened shelves can also be seen, the **lateral pterygoid plates** and the **medial pterygoid plates** (*pterygoid* = winglike). These are attachments for muscles that extend from the sphenoid to the mandible. Just posterior to the pterygoid plate is the **foramen ovale,** a hole that conducts one of the branches of the trigeminal nerve to the mandible.

In the midline of the skull, and sometimes looking like a part of the sphenoid, is the **vomer.** This is a single bone of the face that forms part of the **nasal septum.** The two large holes on each side of the vomer are the **posterior nasal apertures.** Connected to the vomer and forming part of the **hard palate** is the **palatine bone.** The palatine bones are L-shaped bones with a horizontal plate and a vertical plate. The horizontal plates normally join at the **median palatine suture.** If this suture does not fuse completely at birth (along with the **intermaxillary suture** which occurs between the two maxillae), an individual has a cleft palate. The anterior portion of the hard palate is made of the **palatine processes of the maxillae.**

The major openings of the skull are presented in table 7.1. Locate the openings and note the number of structures that pass through these holes.

Interior of the Cranium

With the superior portion of the skull removed, you can see the **cranial cavity** divided into three major regions. These are the **anterior cranial fossa,** a depression anterior to the lesser wings of the sphenoid; the **middle cranial fossae,** which lie between the **lesser wings of the sphenoid** and the petrous portion of the temporal bone; and the **posterior cranial fossa,** posterior to the petrous portion of the temporal bone. Examine figure 7.8 for a view of the interior of the cranium.

TABLE 7.1 Openings of the Skull

Opening	Function or Structure in Opening
Carotid canal	Internal carotid artery, nerves
External acoustic meatus	Opening for sound transmission
Foramen lacerum	Internal carotid artery
Foramen magnum	Spinal cord, vertebral arteries, accessory nerves
Foramen ovale	Mandibular branch of trigeminal nerve
Foramen rotundum	Maxillary branch of trigeminal nerve
Foramen spinosum	Meningeal blood vessels
Hypoglossal canal	Hypoglossal nerve
Inferior orbital fissure	Maxillary branch of trigeminal nerve
Infraorbital foramen	Infraorbital nerve and artery for the face
Internal acoustic meatus	Vestibulocochlear nerve and facial nerve
Jugular foramen	Internal jugular vein, vagus, and other nerves
Mandibular foramen	Mandibular branch of trigeminal nerve and blood vessels
Mental foramen	Mental nerve and blood vessels
Optic canal	Optic nerve
Stylomastoid foramen	Facial nerve exits skull
Superior orbital fissure	Nerves to the eye and face
Supraorbital foramen	Supraorbital nerve and artery for the face

Beginning with the anterior cranial fossa, you should find the centrally located **ethmoid bone.** A sharp ridge known as the **crista galli** is part of the ethmoid and projects from the main portion of this bone. The small, horizontal plate of bone with numerous holes lateral to the crista galli is the **cribriform** (CRIB-rih-form) **plate.** It is also part of the ethmoid bone. The holes in this plate, called **cribriform (olfactory) foramina,** transmit the sense of smell from nerves in the nose to the brain. If the skull is cut close to the orbit, you can see the **frontal sinus** as a hollow space in the anterior portion of the frontal bone.

The dividing line between the anterior and middle cranial fossae is the sphenoid bone. Locate the lesser wings of the sphenoid and the **sella turcica** (SEL-la TUR-sih-ca; turk's saddle) just posterior to it. The **greater wings of the sphenoid** are more inferior than the lesser wings, and each one contains the **foramen rotundum,** which takes a branch of the trigeminal nerve to the maxilla. The **foramen ovale** can be seen from this view as well.

Posterior to the sphenoid bone is the temporal bone, having a flattened lateral section known as the **squamous portion**

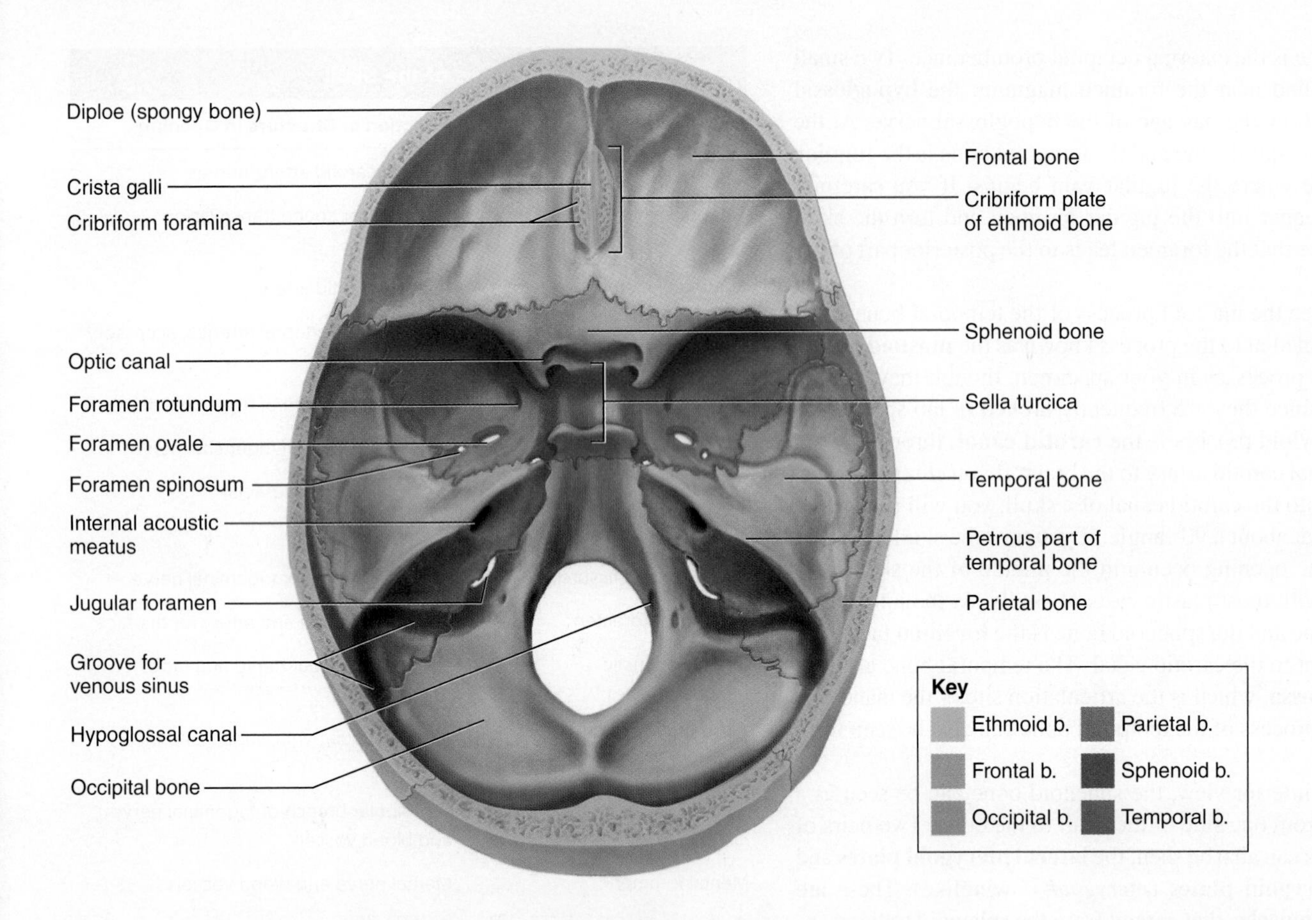

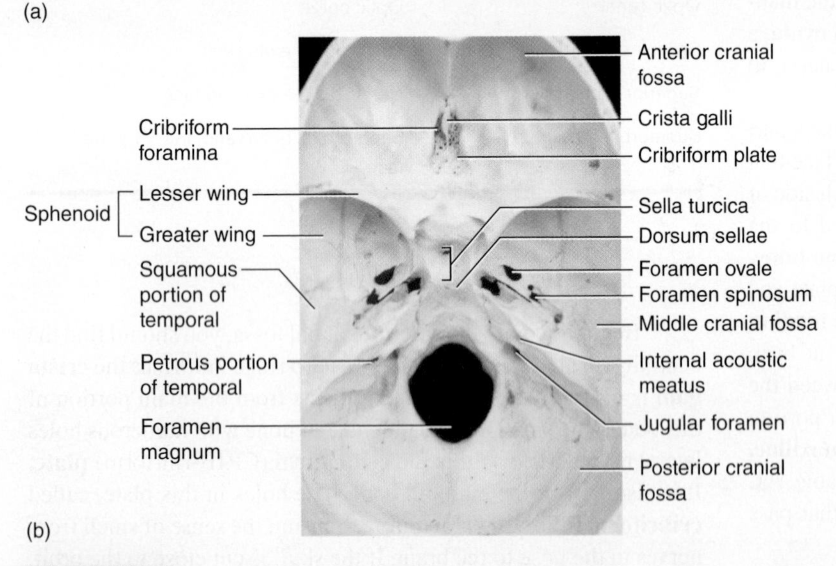

FIGURE 7.8 **Interior of the Cranium** (a) Diagram; (b) photograph.

and a heavier mass of bone known as the petrous portion. The **petrous** (rocklike) **portion** divides the middle and posterior cranial fossae. The petrous portion also has a hole in the posterior surface, which is the **internal acoustic meatus.** This is a passageway for the nerves that come from the inner ear. The posterior cranial fossa is located dorsal to the petrous portion of the temporal bone. It contains the foramen magnum and the jugular foramina. Most of this fossa is formed by the occipital bone.

Median Section of the Skull

Examine the structures seen in a median section. Be extremely careful with the specimen, since many of the internal structures are fragile. Use figure 7.9 as a guide to the median section of the skull, whether or not one is present in the lab. Locate the **nasal septum,** which is composed of the **vomer,** the **perpendicular plate of the ethmoid bone,** and the **nasal cartilage** (absent in skull preparations). If the nasal septum is removed, you can see

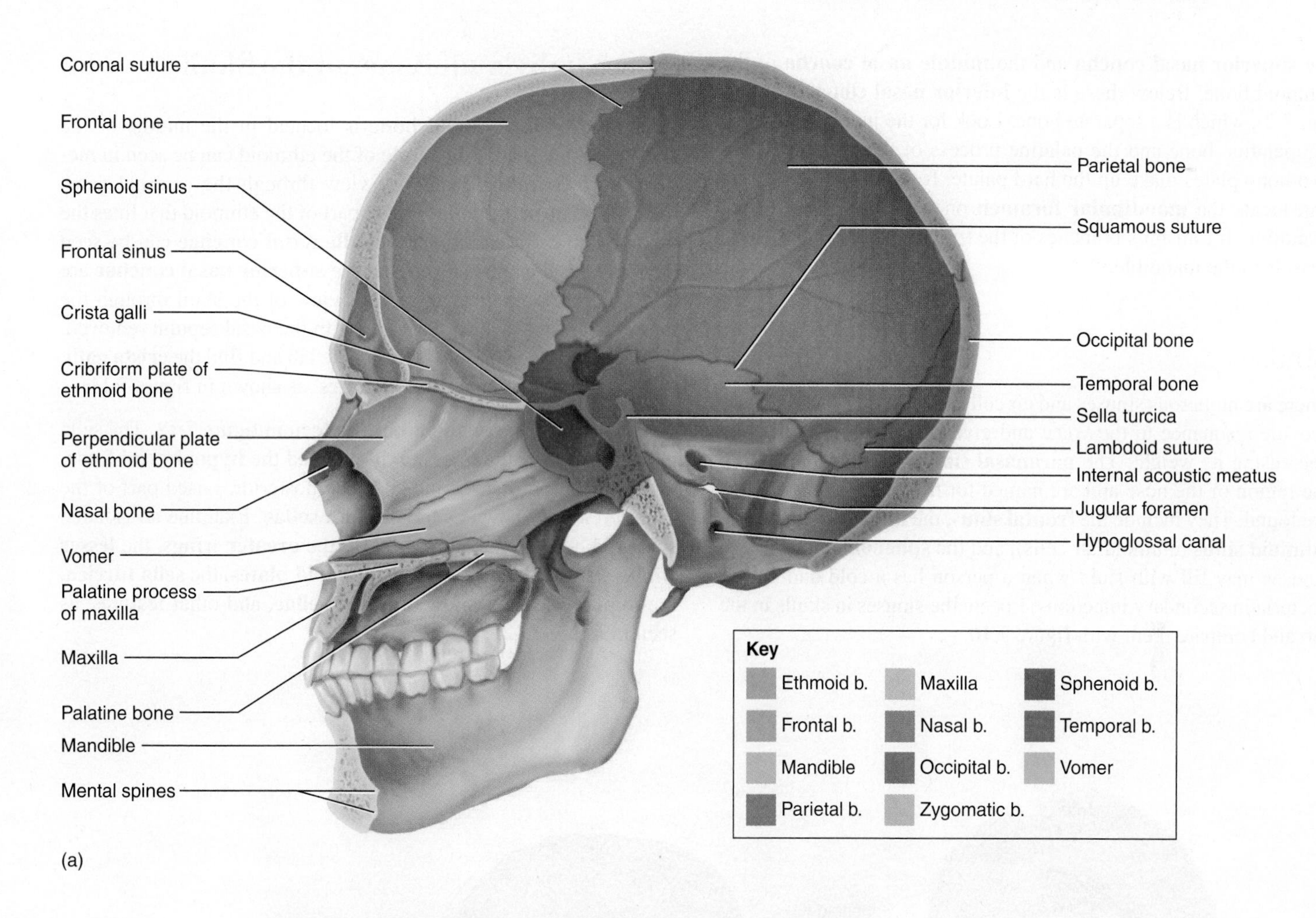

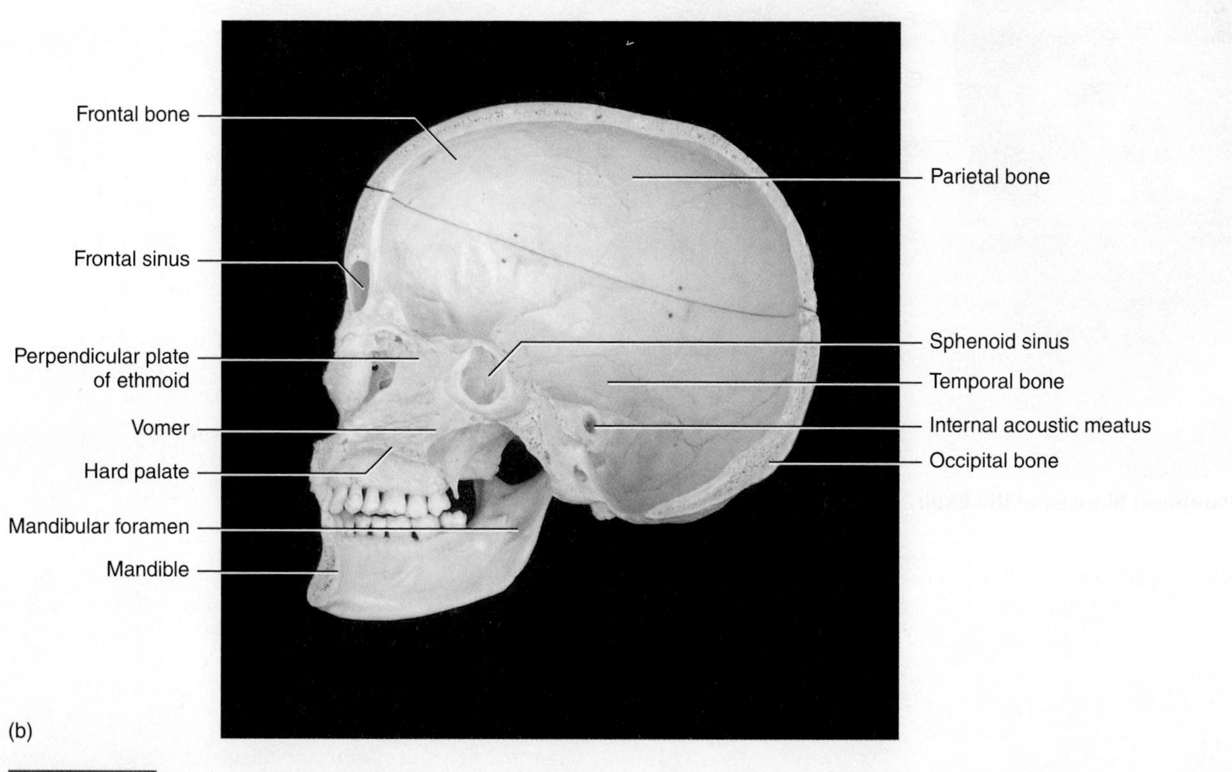

FIGURE 7.9 **Skull, Median Section** (a) Diagram; (b) photograph.

the **superior nasal concha** and the **middle nasal concha** of the ethmoid bone. Below these is the **inferior nasal concha** (review fig. 7.2), which is a separate bone. Look for the junction between the palatine bone and the palatine process of the maxilla. These two bony plates make up the hard palate. If the mandible is present, locate the **mandibular foramen** on the inner aspect of the mandible. It transmits branches of the trigeminal nerve and blood vessels to the mandible.

Sinuses

There are numerous sinuses and air cells in the skull. These sinuses provide resonance to the voice and give shape to the skull while decreasing its weight. The **paranasal sinuses** are located around the region of the nose and are named for the bones in which they are found. They include the **frontal sinus,** the **maxillary sinus,** the **ethmoid sinus** (ethmoid air cells), and the **sphenoid sinus.** These sinuses may fill with fluid when a person has a cold and harbor bacteria in secondary infections. Locate the sinuses in skulls in the lab and compare them with figure 7.10.

Select Individual Bones of the Skull

Ethmoid The **ethmoid bone** is located in the middle of the skull. The **perpendicular plate** of the ethmoid can be seen in median view or from the anterior view through the external nares (nostrils). The **orbital plate** is the part of the ethmoid that lines the medial wall of the orbit. The **middle nasal conchae** can be seen from the nasal cavity as well, but the **superior nasal conchae** are best seen by looking at an inferior view of the skull through the internal nares or in a median view with the nasal septum removed. Examine isolated ethmoid bones in the lab and find the **crista galli, cribriform plate,** and other structures, as shown in figure 7.11.

Sphenoid The **sphenoid bone** is seen in figure 7.12. The sella turcica has a small depression in it called the **hypophyseal fossa,** in which the pituitary gland sits. The posterior, raised part of the sella turcica is known as the **dorsum sellae.** Examine an isolated sphenoid bone in the lab and locate the **greater wings,** the **lesser wings,** the **medial** and **lateral pterygoid plates,** the **sella turcica,** the **hypophyseal fossa,** the **dorsum sellae,** and other features, as seen in figure 7.12.

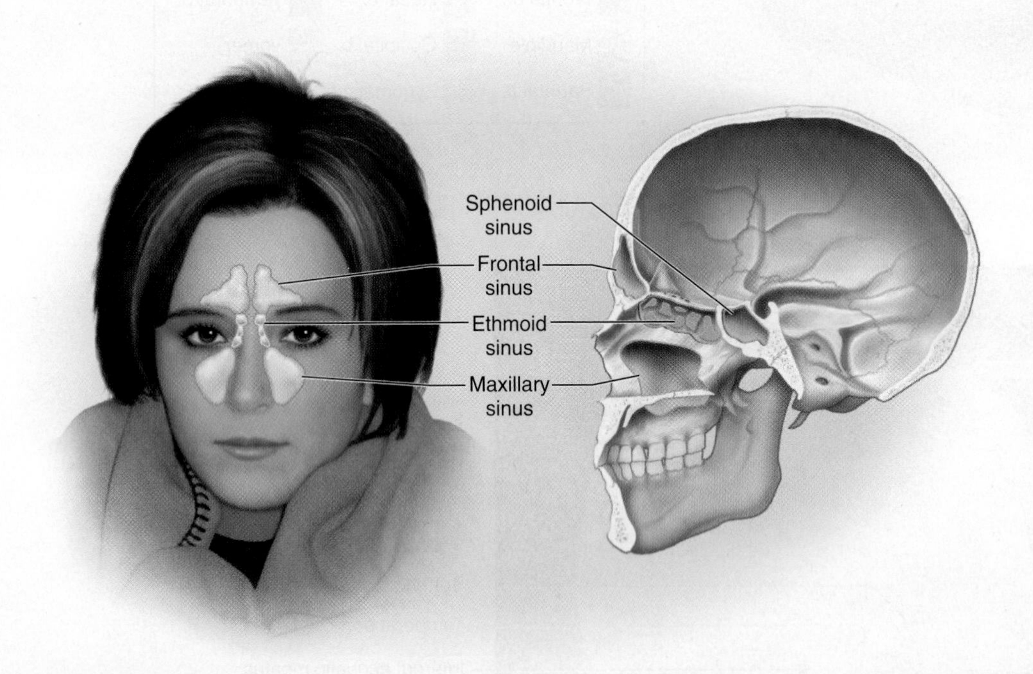

FIGURE 7.10 Paranasal Sinuses of the Skull

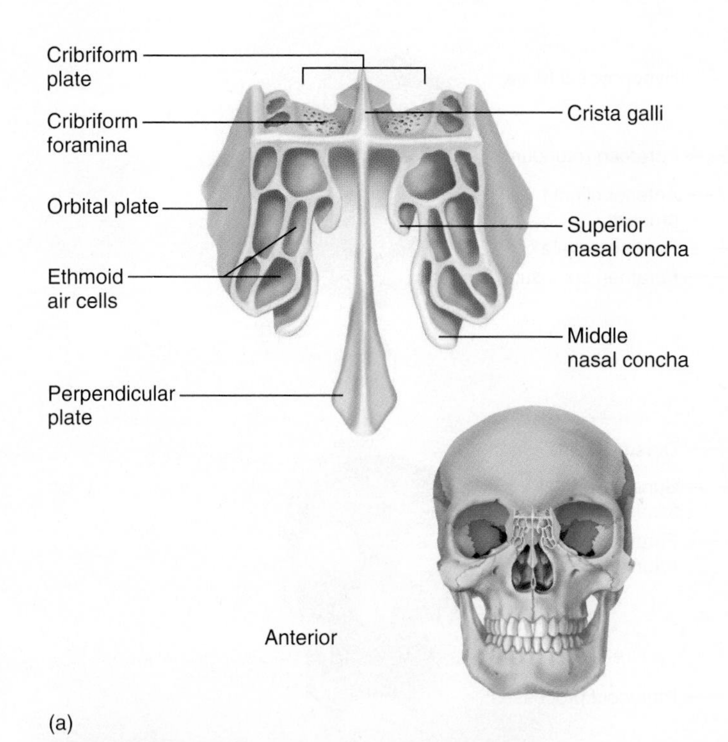

(a)

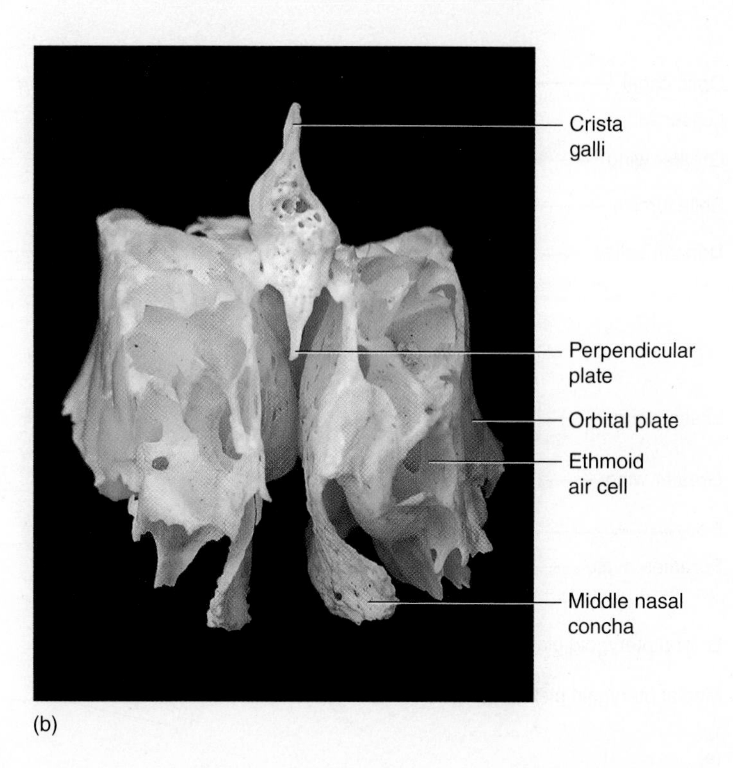

(b)

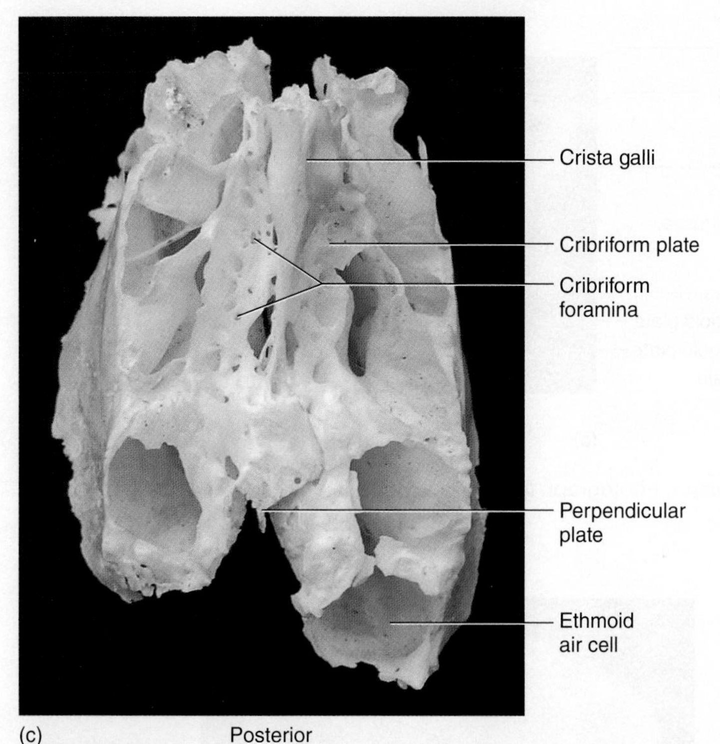

(c)

FIGURE 7.11 **Ethmoid Bone** Diagram (a) anterior view. Photograph (b) anterior view; (c) superior view.

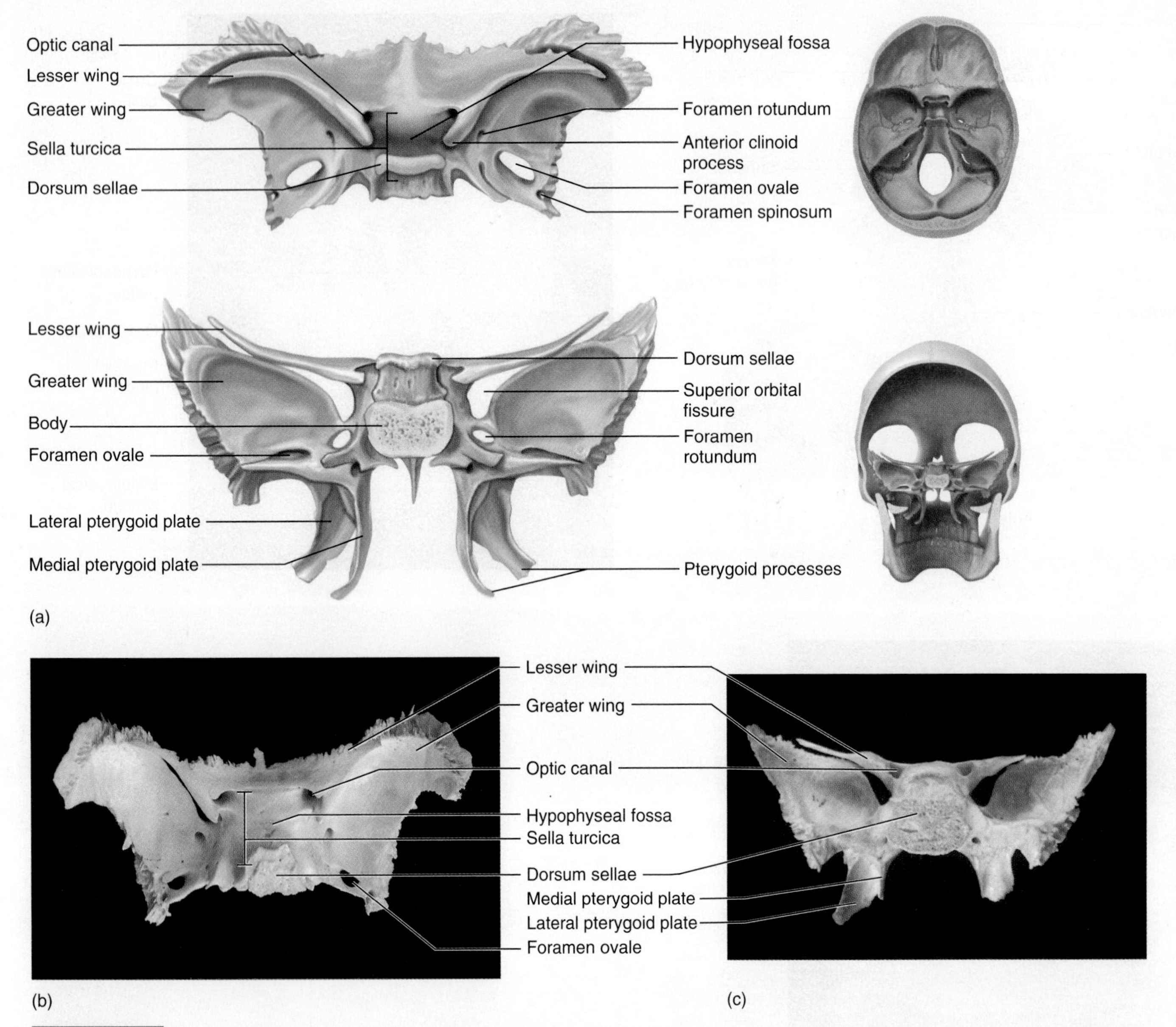

FIGURE 7.12 **Sphenoid Bone** Diagram (a) superior and posterior views. Photograph (b) superior view; (c) posterior view.

Temporal There are two **temporal bones** in the cranium. Each has a flat, superior **squamous portion,** which forms part of the cranial vault, and a medial part called the **petrous portion.** The petrous portion contains the ear ossicles and the opening of the **internal acoustic meatus,** as seen in the medial view of the temporal bone in figure 7.13. Examine an isolated temporal bone in lab and locate these features, as well as the **zygomatic process,** which articulates with the zygomatic bone, and the **mastoid process,** which can be palpated (felt) as a bump posterior to the ear.

Fontanels The fontanels are the "soft spots" of an infant's skull. There are four types of fontanels, and they allow for the passage of the skull through the birth canal by enabling the bones of

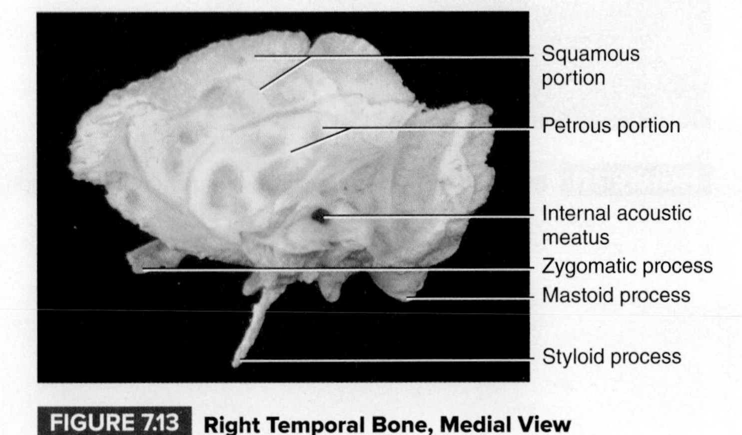

FIGURE 7.13 **Right Temporal Bone, Medial View**

the cranium to slide over one another in a process called molding. After birth the fontanels allow for further expansion of the skull. The **anterior (frontal) fontanel** is an area between the frontal bone and the parietal bones. The **posterior (occipital) fontanel** is between the occipital bone and the parietal bones. The **anterolateral (sphenoid) fontanels** are paired structures on each side of the skull and are located superior to the sphenoid bone, and the **posterolateral (mastoid) fontanels** are also paired structures posterior to the temporal bone. Most fontanels fuse typically before 1 year of age, though the anterior fontanel may fuse as late as age 2. Locate these structures in figure 7.14 and on the material available in the lab.

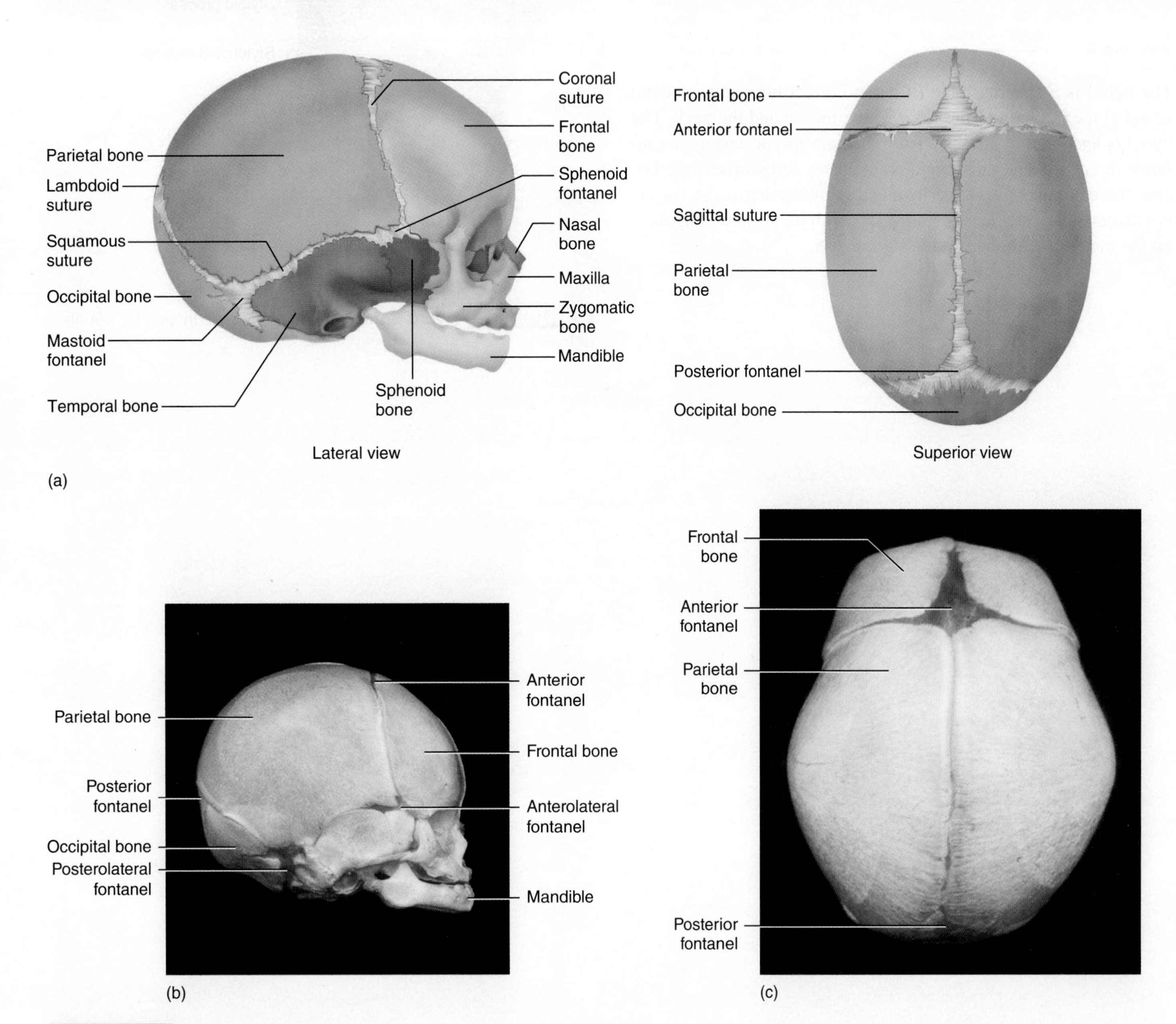

FIGURE 7.14 **Fetal Skull and Fontanels** (a) Diagram of lateral and superior views. Photograph (b) lateral view; (c) superior view.

Bones Associated with the Skull

There are seven bones that are not part of the skull but are closely associated with it. These are the auditory ossicles which are discussed in Exercise 18 later on in this lab manual and the hyoid bone.

Hyoid

The hyoid is a floating bone (it has no direct bony attachments) found at the junction of the floor of the mouth and the neck. The hyoid is anchored by muscles from the anterior, posterior, and inferior directions and aids tongue movement and swallowing. Locate the central body of the hyoid, the greater horns, or cornua (COR-new-uh), and the lesser horn, or cornu (COR-new; singular), on the material in the lab and in figure 7.15.

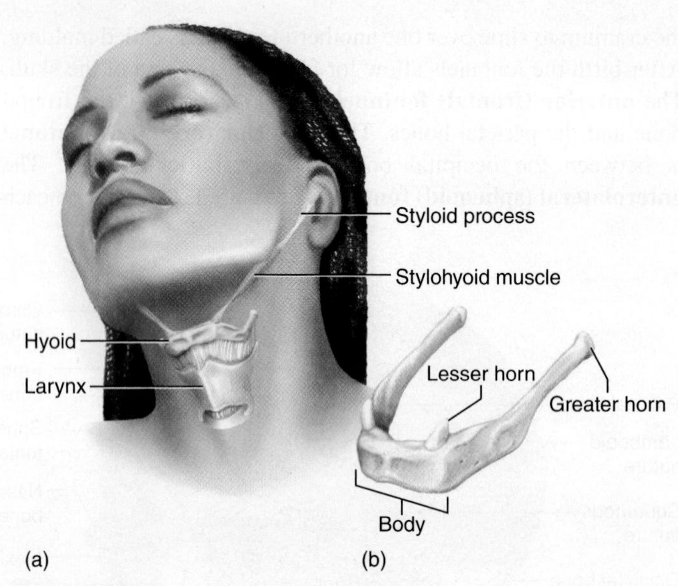

FIGURE 7.15 Hyoid Bone, Anterior View (a) In situ; (b) details of hyoid.

Exercise 7

Axial Skeleton: Skull

Name ______________________________ Date ______________

1. The eyebrows are superfical to what bone? ______________________________

2. What is the common name for the zygomatic bone? ______________________________

3. What is the name of the bony process directly posterior to the earlobe? ______________________________

4. The hard palate is made up of what bones? ______________________________

5. What are the names of the major paranasal sinuses? ______________________________

6. The mandible fits into what part of the temporal bone to form the jaw joint? ______________________________

7. What bone is found just posterior to the ethmoid bone? ______________________________

8. The sella turcica is part of what bone? ______________________________

9. What occupies the depression in the sella turcica? ______________________________

10. What are the names of the bones that surround the anterior opening of the nose? ______________________________

11. The upper teeth are held in place by what bones? ______________________________

12. In what bone would you find the foramen magnum? ______________________________

13. What are the two bony structures that make up the nasal septum? ______________________________

____

____

14. The sagittal suture separates the _____ from the _____. (Circle the correct answer.)

a. sphenoid, ethmoid b. left parietal, right parietal c. frontal, parietal d. parietals, occipital

15. Which bone is *not* located in the orbit? (Circle the correct answer.)

a. maxilla b. zygomatic c. ethmoid d. sphenoid e. temporal

16. Which bone is *not* a paired bone of the skull? (Circle the correct answer.)

a. zygomatic b. temporal c. lacrimal d. vomer

17. Label the following illustration using the terms provided.

carotid canal	maxilla	posterior nasal aperture
foramen magnum	occipital bone	temporal bone
jugular foramen	occipital condyle	vomer
mastoid process	palatine bone	zygomatic bone

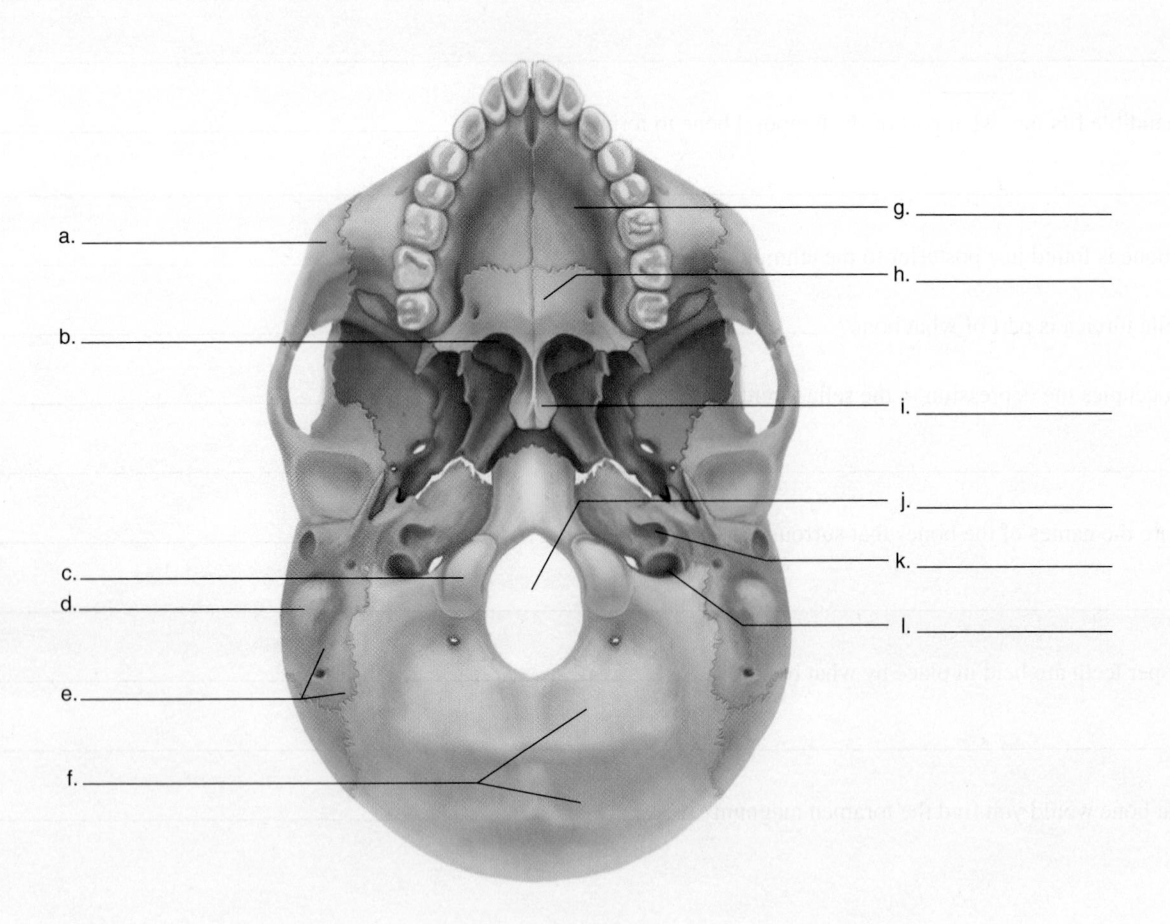

18. Label the following illustration using the terms provided.

angle
body
condylar process
coronoid process
mandibular notch
mental foramen
ramus

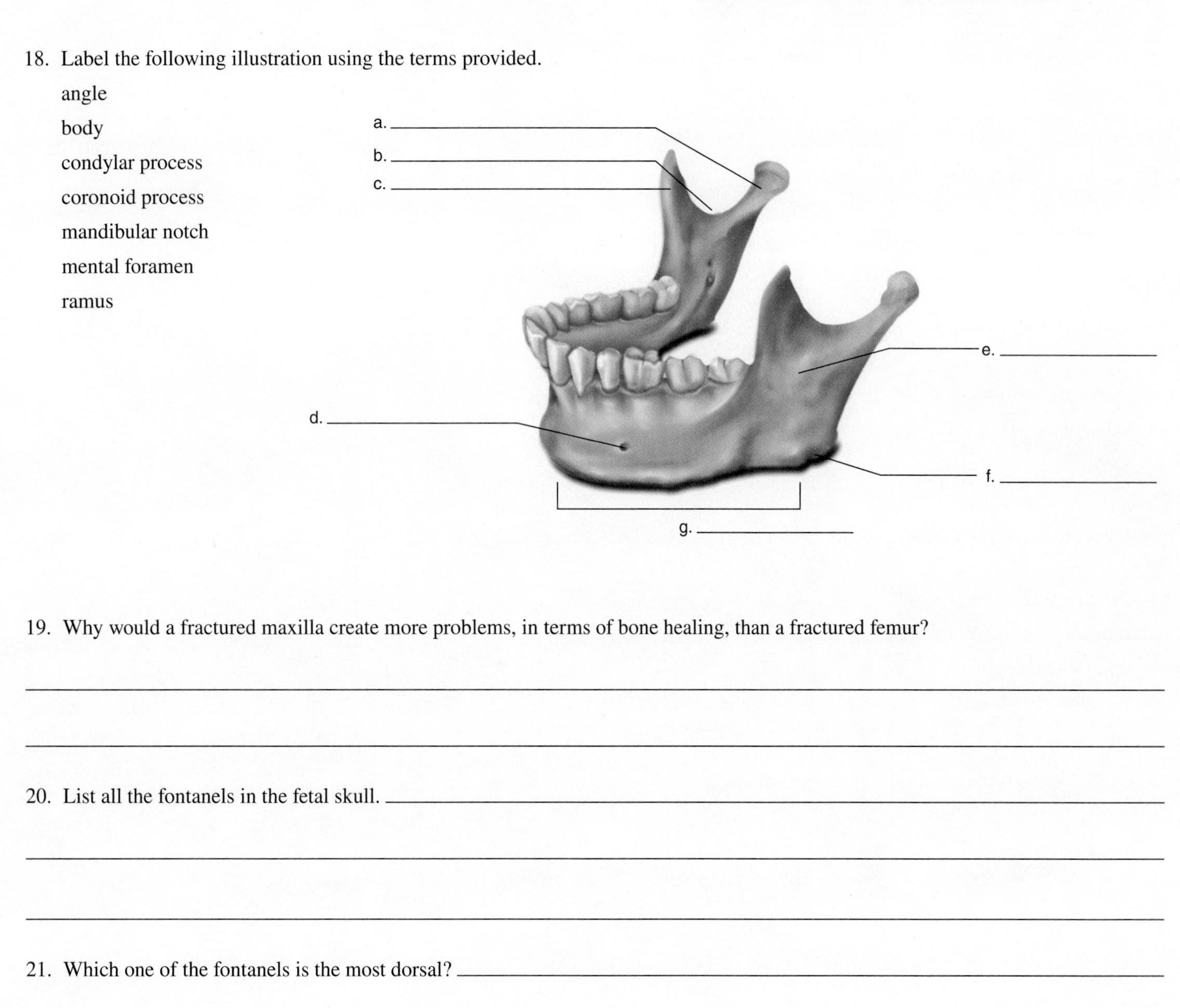

19. Why would a fractured maxilla create more problems, in terms of bone healing, than a fractured femur?

____

____

20. List all the fontanels in the fetal skull. ____

____

____

21. Which one of the fontanels is the most dorsal? ____

____

Laboratory

Exercise 8

Axial Skeleton 2: Vertebrae, Ribs, Sternum

Skeletal System

Introduction

The axial skeleton consists of 80 bones, including the skull, hyoid, vertebrae (singular, *vertebra*), sternum, auditory ossicles, and ribs. In this exercise you examine all parts of the axial skeleton except the skull and hyoid, which were covered in laboratory exercise 7. There are five regions of the vertebral column. These are the cervical (SERVE-ik-ul), thoracic (thor-AH-sic), and lumbar vertebrae and vertebrae that make up the sacrum and the coccyx. There are 33 individual vertebrae, which form the vertebral column. Some fuse into larger structures, such as the sacrum.

The thoracic cage consists of the 12 pairs of ribs, thoracic vertebrae, and sternum, protecting the lungs and heart yet providing for flexibility during breathing.

Learning Objectives

At the end of this exercise you should be able to

1. find a specific vertebra (such as T6) on an articulated vertebral column;
2. name selected vertebrae on an articulated vertebral column;
3. name the bony features of an individual vertebra and determine the region of the vertebral column to which it belongs;
4. identify the atlas and the axis as specific vertebrae;
5. describe the normal and abnormal spinal curvatures;
6. distinguish between the different types of ribs, the markings of the individual ribs, and whether a rib comes from the left or right side of the body; and
7. demonstrate the location of the three major regions of the sternum.

Materials

Articulated skeleton or plastic casts of a skeleton

Disarticulated skeleton or plastic casts of bones

Charts of the skeletal system

Plastic drinking straws cut at an angle or pipe cleaners for pointer tips

Foam pads of various sizes (to protect bone from hard countertops)

A cardboard box or piece of wood about 1 foot square and 3 inches deep

Procedure

Review the bones of the axial skeleton in the lab and in laboratory exercise 6 (fig. 6.1). In this exercise you study the bones of the axial skeleton by examining both the disarticulated bones and the articulated skeleton. Hold each bone up to the skeleton to see how it is positioned in relation to the other bones of the body. When you examine bones in the lab, do not use your pen or pencil to locate a structure. Use a cut drinking straw or a pipe cleaner to point out structures. Your instructor may want you to place real bone material on foam pads to cushion the bone from the tabletop.

Vertebrae

Overview of the Vertebrae The **vertebral column** of humans is significantly different from that of all other mammals in that humans are bipedal. In humans the vertebrae increase in size from the cervical to the lumbar vertebrae. This is due to the increase in weight on the lower vertebrae. The diameter of the vertebral canal increases from the sacral region to the cervical region. This reflects the greater number of neural fibers in the spinal cord as sensory information is transmitted to the brain and motor information is transmitted from the brain. Between the vertebrae are fibrocartilaginous pads known as **intervertebral discs.**

The vertebrae are listed according to their location. The cervical vertebrae are located in the neck with C1 (the first cervical vertebra) being closest to the skull. The thoracic vertebrae are listed as T1 through T12, and they are defined because they are attached to the ribs. The lower the number in a vertebral sequence, the more superior it is. There are five lumbar vertebrae (L1–L5), and these are found between the thoracic vertebrae and the sacrum. The sacral vertebrae (S1–S5) are fused in adults and form the singular sacrum. The inferiormost vertebrae are the coccygeal vertebrae, and there are generally four of them (Co1–Co4).

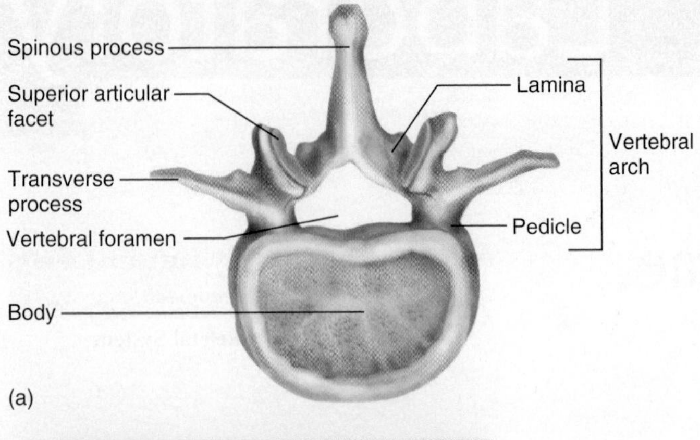

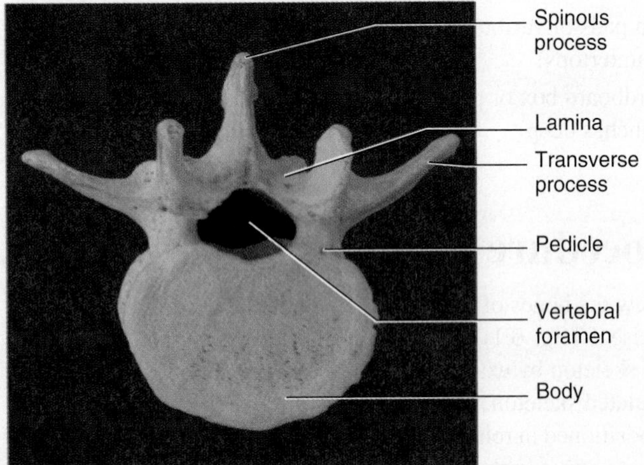

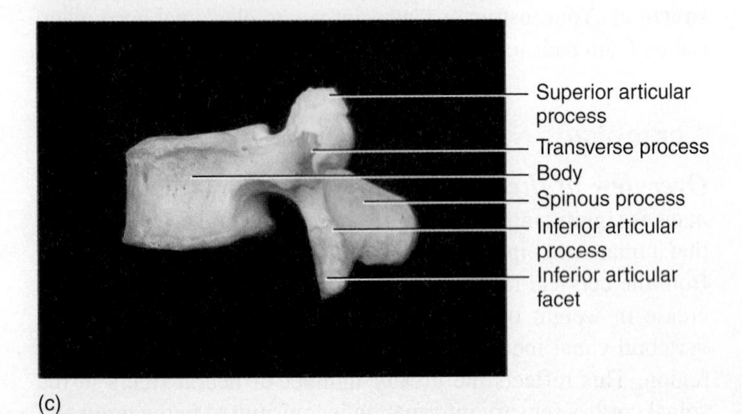

FIGURE 8.1 Features of a Typical Vertebra (a) Diagram, superior view. Photograph (b) superior view; (c) lateral view.

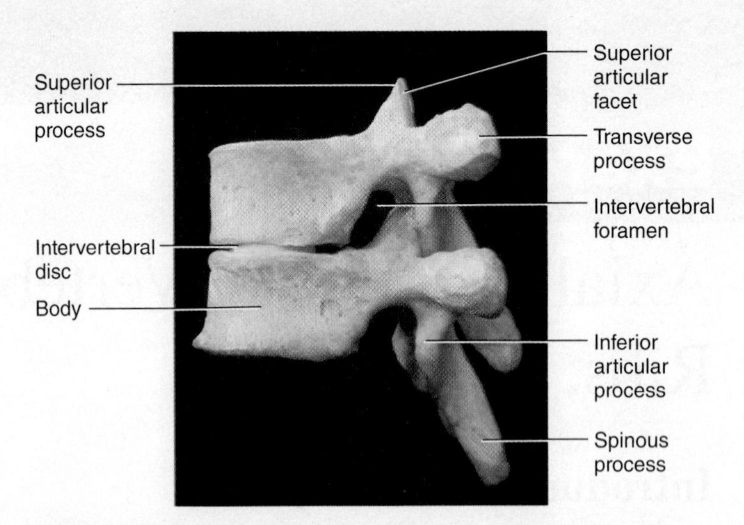

FIGURE 8.2 Articulated Vertebrae, Lateral View

Typical Vertebrae Obtain a vertebra and bring it to your table. Compare it to figure 8.1. Place the vertebra in front of you, so that you can see through the large hole known as the **vertebral foramen,** where the spinal cord is located. Note the large **body** of the vertebra, which supports the weight of the vertebral column and is in contact with the intervertebral discs. Posterior to the body is the **vertebral,** or **neural, arch,** which consists of two pedicles (PED-ik-uls) and two laminae (LAM-i-nee). The **pedicles** are the parts of the arch that extend from the body of the vertebra to the two lateral projections, called the **transverse processes.** Each **lamina** (LAM-in-uh) is a broad, flat structure between the transverse process and the dorsal **spinous process.**

If you rotate the vertebra as illustrated in figure 8.1*c*, you should be able to see the vertebral body in lateral view with the **superior articular process** and the **superior articular facet.** The superior articular facet is a flat surface that articulates with the vertebra above. There is also an **inferior articular process** with an **inferior articular facet** that articulates with the vertebra below. If you put two adjacent vertebrae together and view them from the side, you can see the **intervertebral foramen,** a hole where the spinal nerve is located. This is illustrated in figure 8.2.

Spinal Curvatures The spine has four curvatures, which alternate from superior to inferior. The **cervical** (SERVE-ik-ul) **curvature** is convex (bowed forward) when seen from the anatomical position. The **thoracic** (thor-AH-sic) **curvature** is concave, while the **lumbar curvature** is convex and the **sacral (pelvic) curvature** is concave. These curvatures allow for balance in an upright posture and provide room for the lungs and abdominal organs. Locate them in figure 8.3. In addition to the spinal curvatures described, there are abnormal spinal curvatures. **Scoliosis** is a lateral curvature. **Kyphosis** (KY-foh-sis) is an exaggerated thoracic curvature, and **lordosis** is an exaggerated lumbar curvature commonly seen during pregnancy.

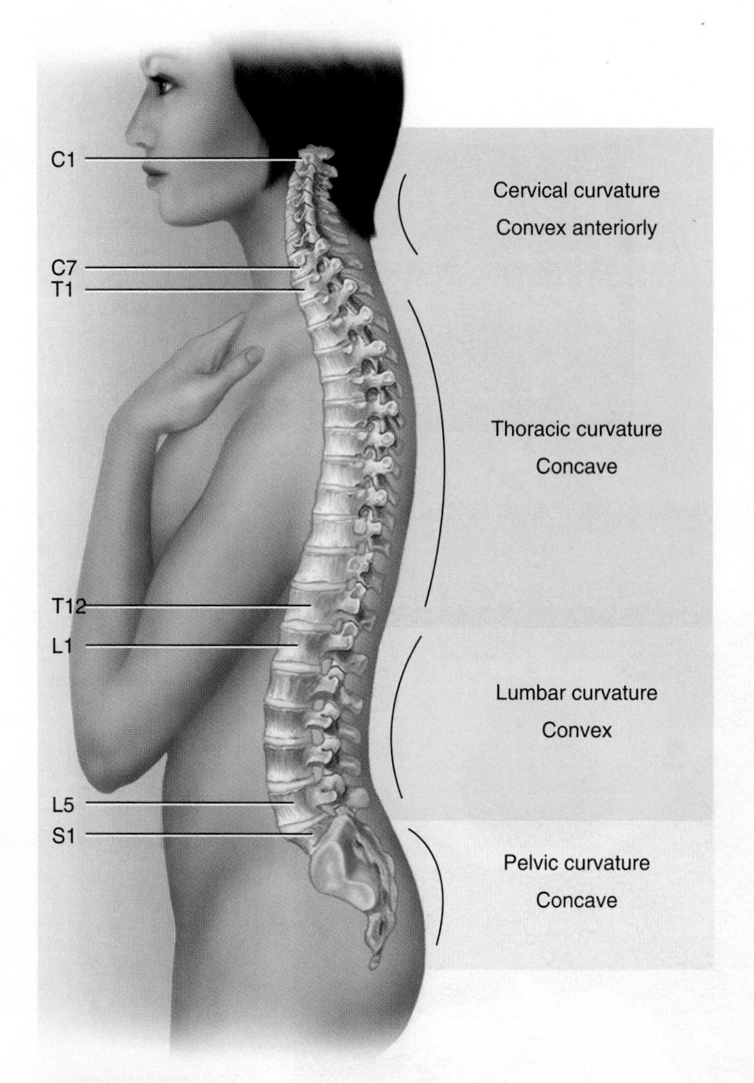

FIGURE 8.3 **Spinal Curvatures, Lateral View**

Cervical Vertebrae There are seven **cervical vertebrae** (C1–C7), and these can be distinguished from all other vertebrae in that each vertebra has three foramina. The **vertebral foramen** is the largest opening, and the two **transverse foramina** are found only in cervical vertebrae. The transverse foramina house the vertebral arteries and vertebral veins. Some of the cervical vertebrae (C2–C6) have **bifid** (BY-fid) **spinous processes** (*bifid* = split in two), and the bodies of the cervical vertebrae are less massive than those of the inferior vertebrae. Examine the isolated cervical vertebrae in the lab and compare them with figure 8.4. Three cervical vertebrae are unique enough to warrant special attention. The first cervical vertebra (C1) is known as the **atlas** and is the only cervical vertebra without a body. Atlas was a titan in Greek mythology who held up the heavens. The atlas joins with the head and allows you to nod your head to indicate "yes." The second cervical vertebra (C2) is the **axis,** and it has a process called the **dens,** or **odontoid process,** which runs superiorly through the atlas. The odontoid process allows the atlas to move on the axis and allows you to rotate your head to indicate "no." The seventh cervical vertebra (C7) is known as the **vertebra prominens.** It has a spinous process that projects sharply in a posterior direction and can be palpated (felt) as a significant bump at the junction of the neck and the back. Look at an atlas, an axis, and a vertebra prominens in the lab and compare them with figure 8.5.

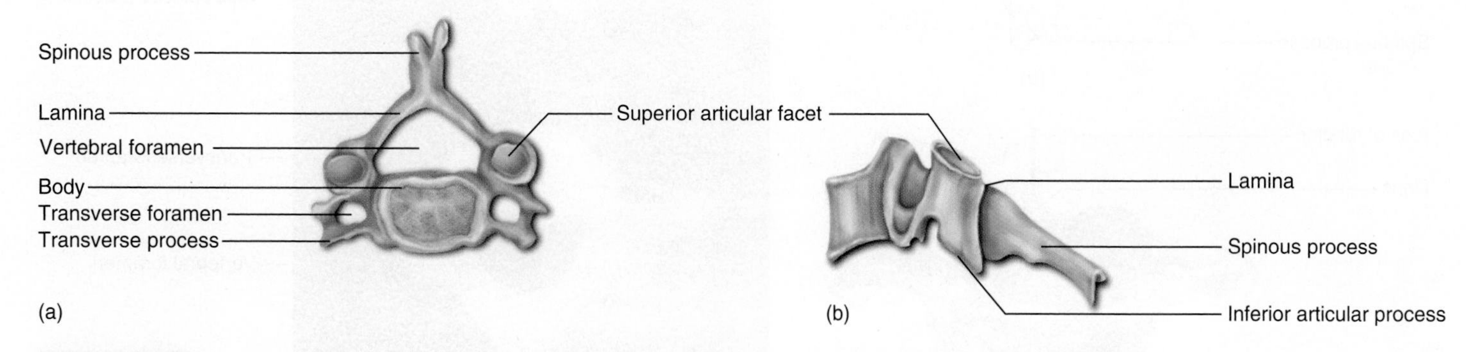

FIGURE 8.4 **Cervical Vertebra** Diagram (a) superior view; (b) lateral view. *(continued)*

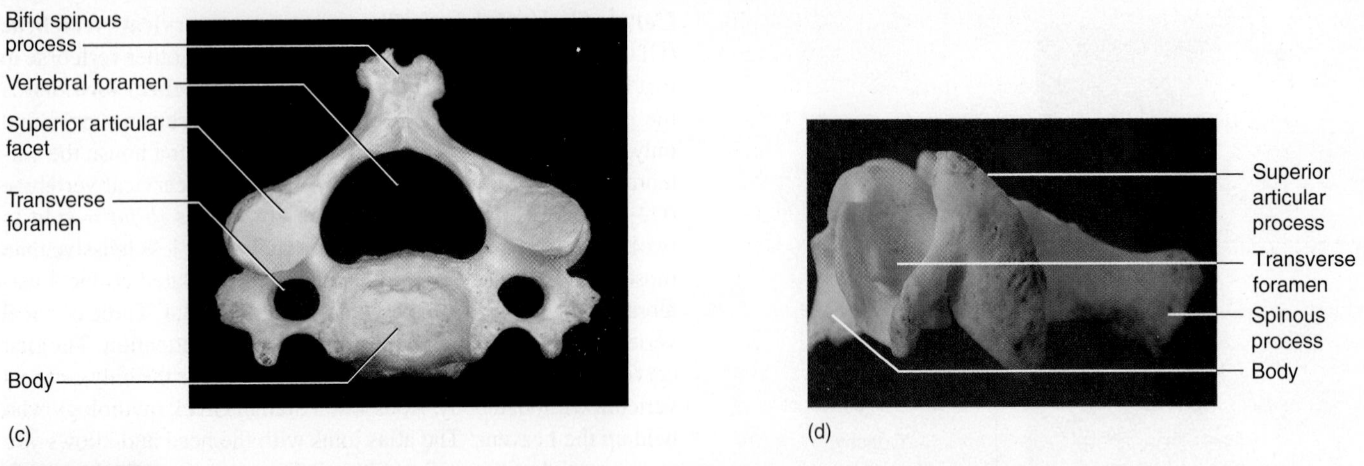

FIGURE 8.4 *(continued)* **Cervical Vertebra** Photograph (c) superior view; (d) lateral view.

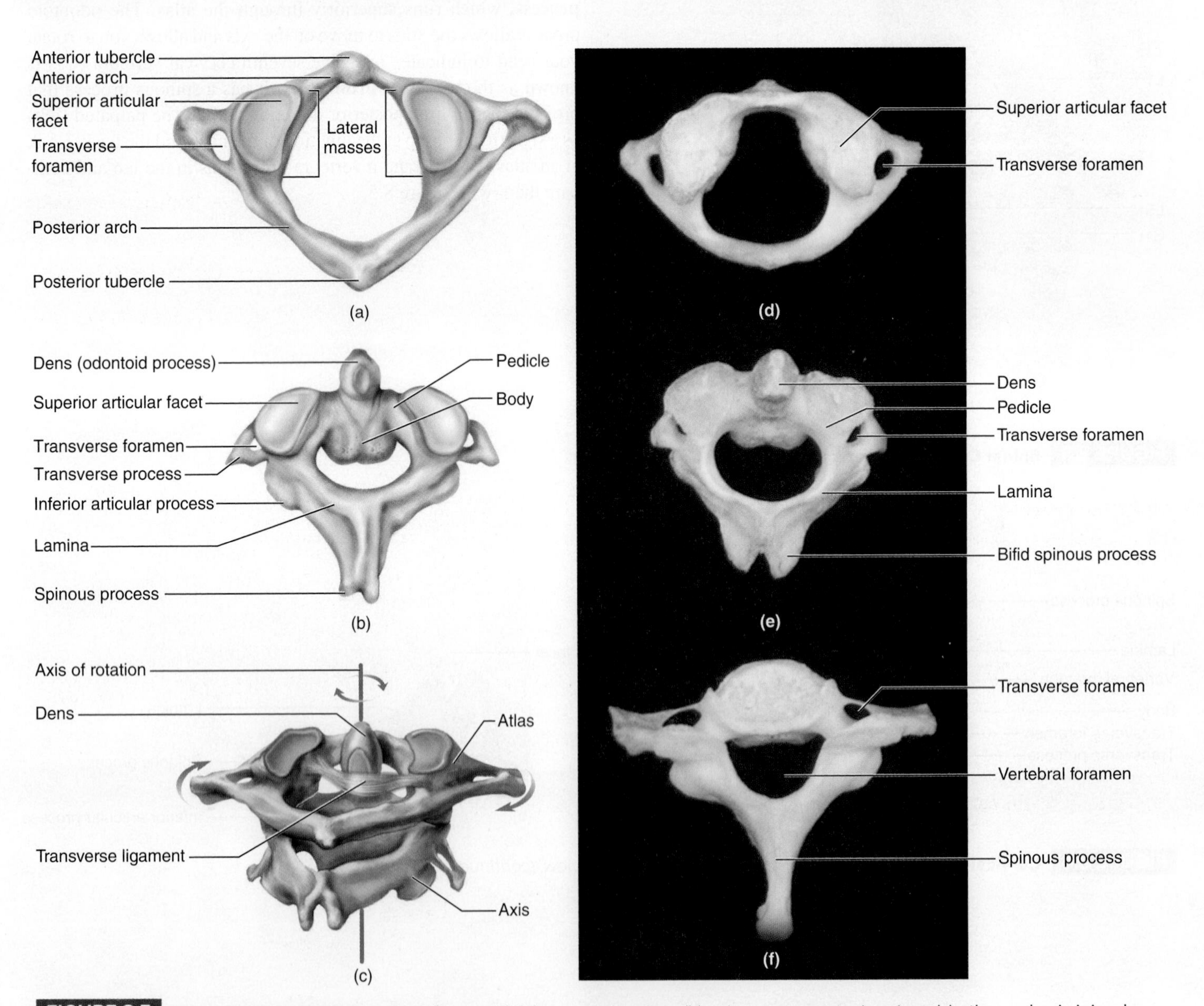

FIGURE 8.5 **Atlas, Axis, and Vertebra Prominens** (a) Atlas, superior view; (b) axis, superoposterior view; (c) atlas and axis joined. Photographs of three vertebrae, superior view (d) atlas; (e) axis; (f) vertebra prominens.

Thoracic Vertebrae The 12 **thoracic vertebrae** are distinguished from all other vertebrae by markings on their lateral posterior bodies, which are attachment points for ribs. In some areas, a rib attaches to just 1 vertebra, leaving a mark on the vertebral body known as a **complete costal facet** (FASS-et). In other areas, the head of a rib attaches to 2 vertebrae. The point of attachment at the superior part of the vertebra is called the **superior costal facet,** and the attachment of the rib to the inferior part of the vertebra is called the **inferior costal facet.** The head of the rib spans both of the vertebrae and articulates with the facet of the superior and inferior vertebrae. Vertebrae T1 through T10 have **transverse costal facets** on the terminal portions of the transverse processes.

The thoracic vertebrae also have longer spinous processes than the cervical vertebrae, and the spinous processes of the thoracic vertebrae tend to angle in a more inferior direction. Thoracic vertebrae do not have transverse foramina. Examine figure 8.6 and note the characteristics of the vertebrae as seen in the lab.

Lumbar Vertebrae There are five lumbar vertebrae. The **lumbar vertebrae** are distinguished from the other vertebrae by having neither transverse foramina nor rib facets. The spinous processes of the lumbar vertebrae tend to be more horizontal than those of the thoracic vertebrae, and the bodies of the lumbar vertebrae are larger, as they carry more weight than the thoracic vertebrae; yet the twelfth thoracic and the first lumbar vertebrae are remarkably similar. The last thoracic vertebra has rib facets on it, and the first lumbar does not. Look at the lumbar vertebrae in the lab and compare them with figure 8.7.

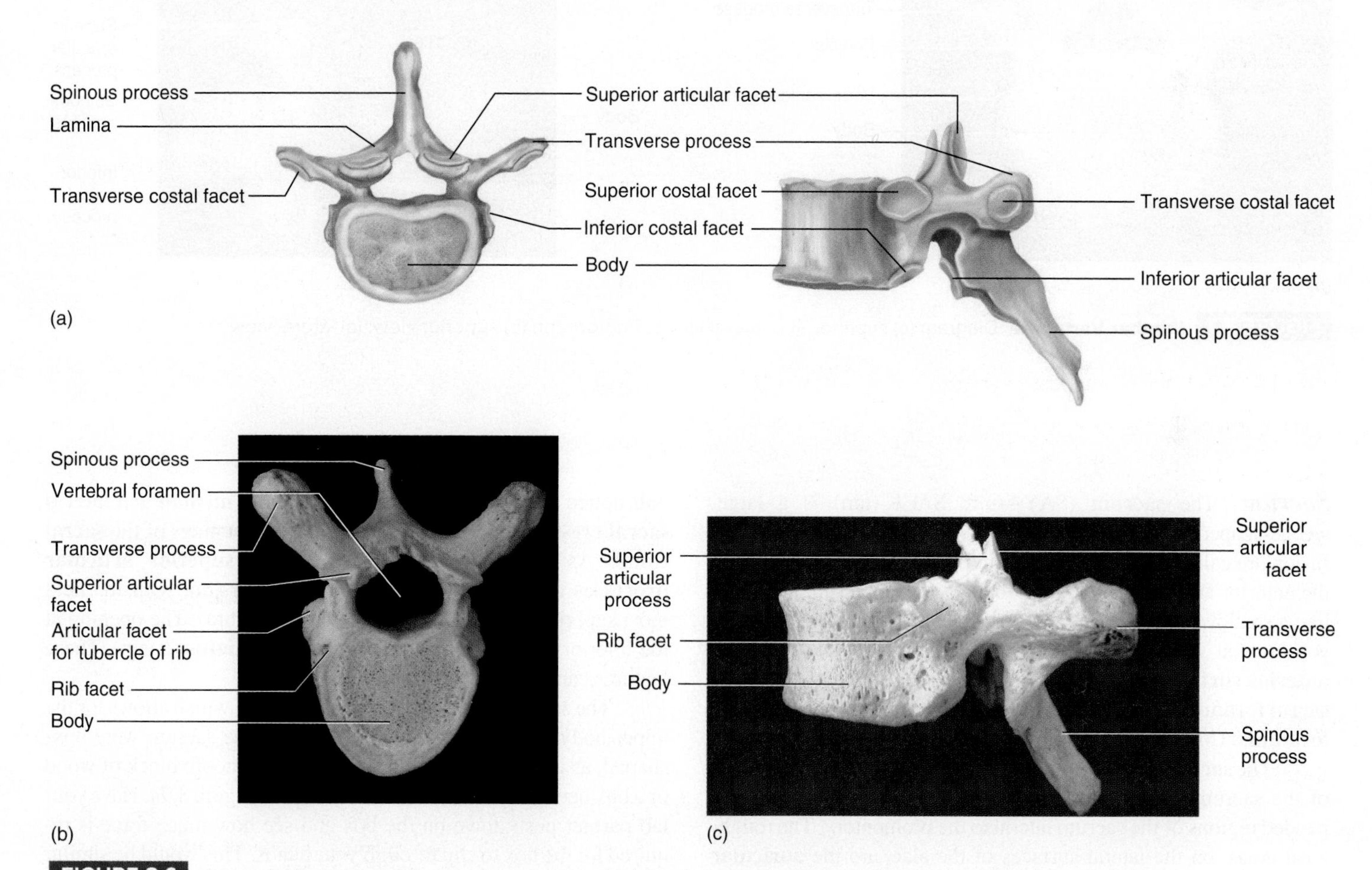

FIGURE 8.6 **Thoracic Vertebrae** Diagram (a) superior and lateral views. Photograph (b) superior view; (c) lateral view.

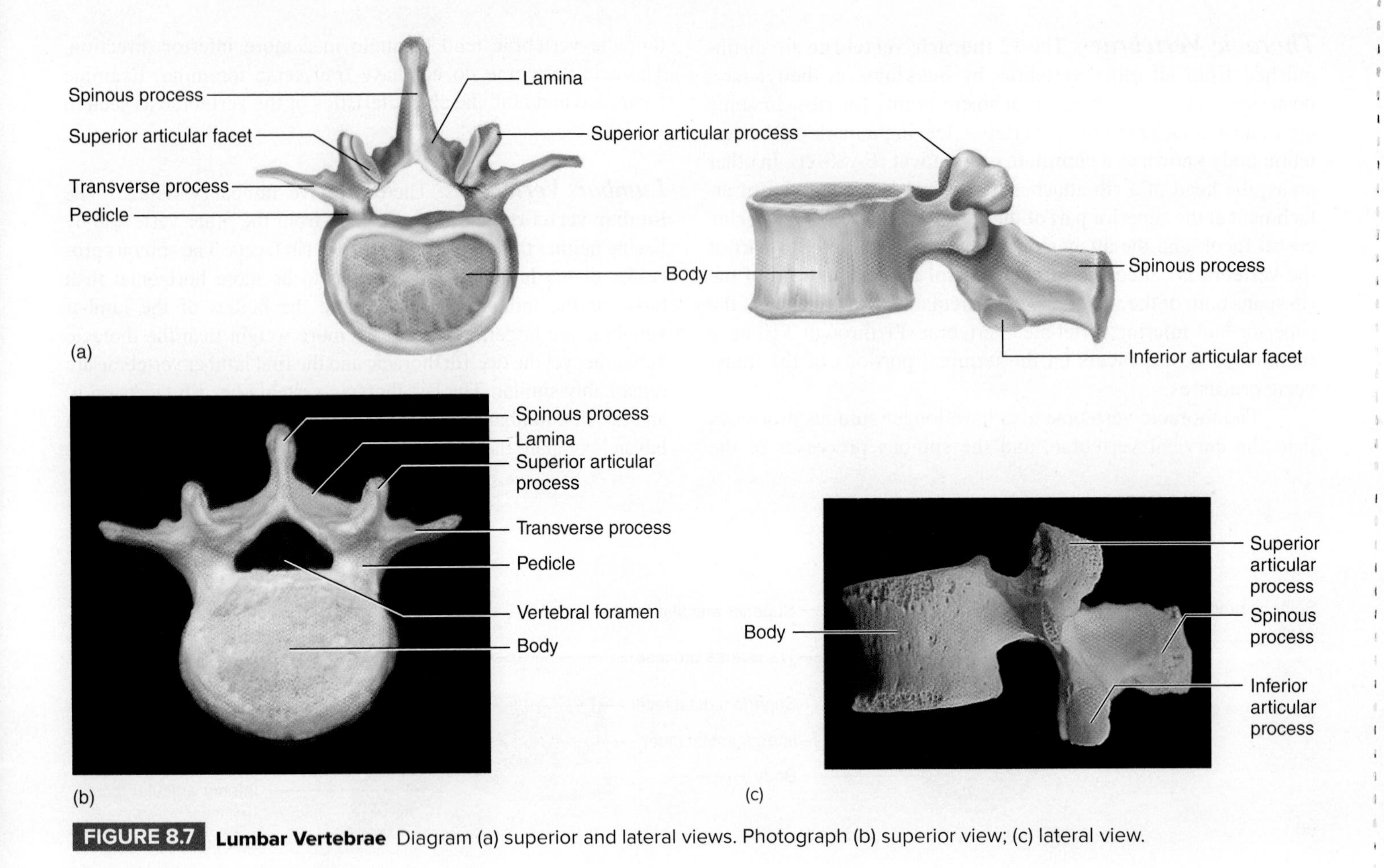

FIGURE 8.7 **Lumbar Vertebrae** Diagram (a) superior and lateral views. Photograph (b) superior view; (c) lateral view.

Sacrum The **sacrum** (SAY-krum, SACK-rum) is a large, wedge-shaped bone composed of five fused vertebrae. The lines of fusion are called **transverse lines,** and these may be seen on both the anterior and posterior sides. Notice how the sacrum is shaped like a shallow, triangular bowl. If you place the sacrum in front of you, as you would a cereal bowl, the shallow depression is the **anterior surface.** The two rows of holes you see are the **anterior sacral foramina.** On the posterior surface are the **posterior sacral foramina.** Compare a sacrum from your lab with figure 8.8.

The **sacral promontory** is a rim on the anterior superior part of the sacrum, and the **alae** (AIL-ee; singular, *ala*) are two expanded regions of the sacrum lateral to the promontory. The roughened areas, on the lateral surfaces of the alae, are the **auricular** (aw-RIC-you-lur; ear-shaped) **surfaces** of the sacrum; each joins with the ilium to form the **sacroiliac** (SACK-ro-ILL-ee-ak) **joint.** As additional force is applied to the sacrum, it wedges itself into the ilium. If you examine the posterior surface of the sacrum, you will notice the posterior sacral foramina, the **median** and **lateral sacral crests,** and the superior and inferior openings of the **sacral canal.** As with the other vertebrae, the **superior articular processes** and the **superior articular facets** join with the next most superior vertebra (the fifth lumbar vertebra). The opening at the inferior part of the sacrum is the **sacral hiatus** (a gap). These features can be seen in figure 8.8.

The sacrum in humans is wedge-shaped, which allows for the upper body to carry greater weight than if the sacrum were box-shaped, as it is in many quadrupeds. Hold a smooth block of wood or a box between your hands, as illustrated in figure 8.9*a*. Have your lab partner push down on the box and see how much force is required for the box to slip through your hands. This would be similar to the forces that would act on a rectangular sacrum. Now turn the box or block of wood so that it forms a wedge in your hands, as illustrated in figure 8.9*b*. Have your lab partner push down on the box. Is it easier or harder to dislodge the "sacrum" in this way?

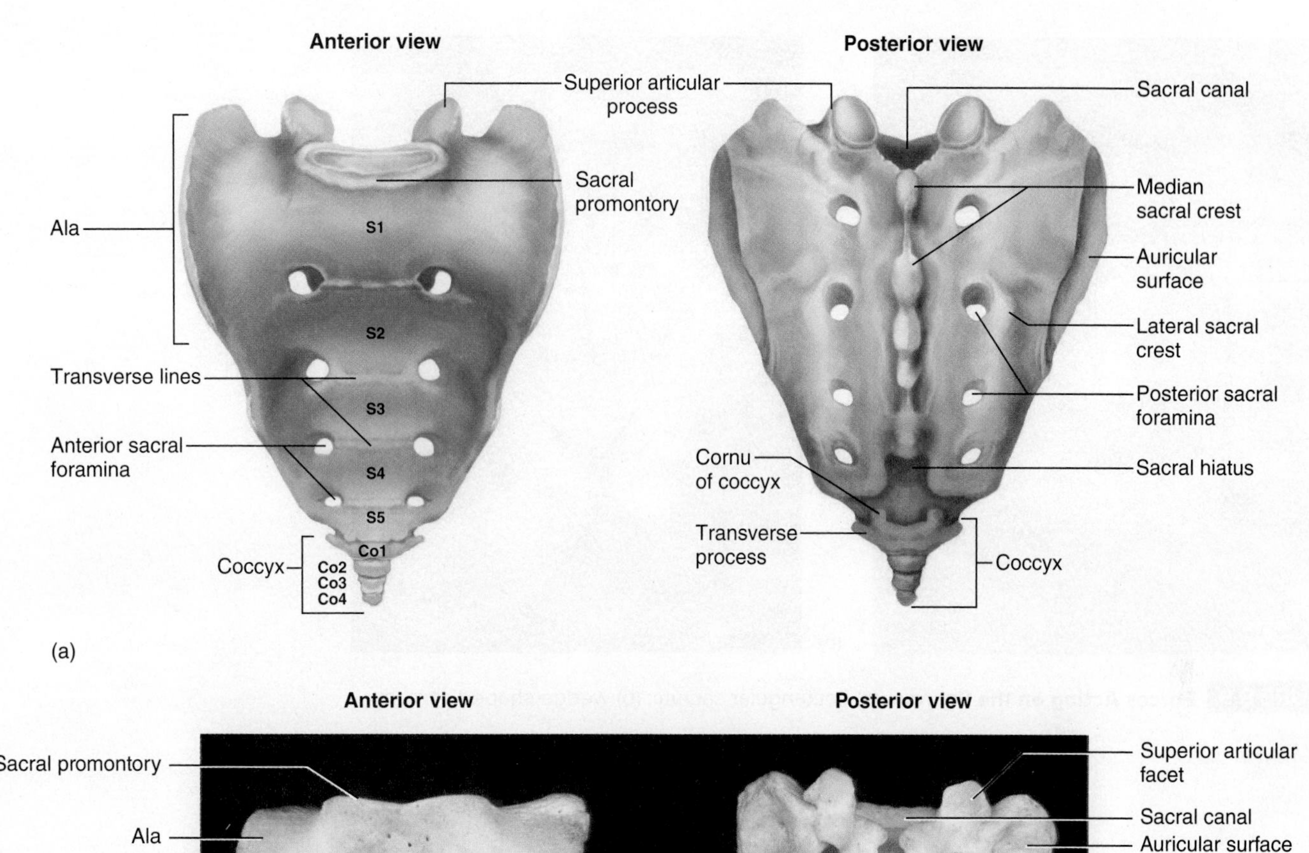

FIGURE 8.8 **Sacrum and Coccyx** (a) Diagram; (b) photograph.

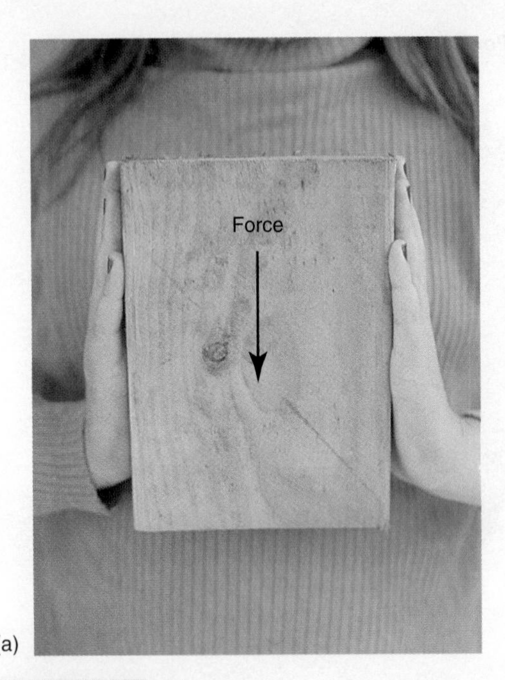

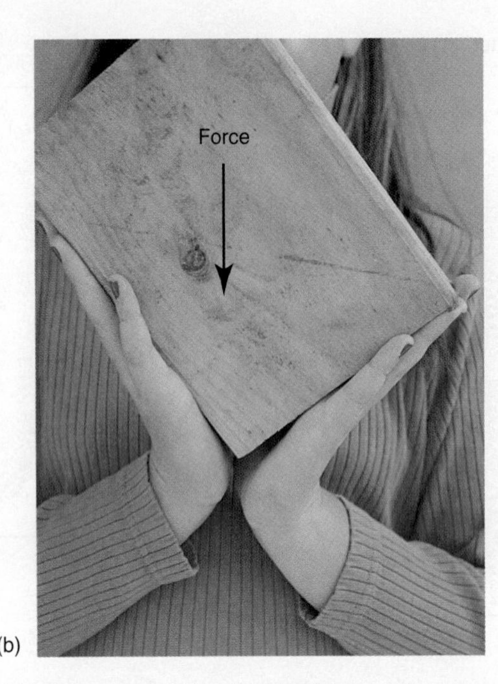

FIGURE 8.9 **Forces Acting on the Sacrum** (a) Rectangular sacrum; (b) wedge-shaped sacrum.

Coccyx The **coccyx** is the terminal portion of the vertebral column, and it usually consists of four fused vertebrae. The coccyx may be fused with the sacrum in some females who disarticulated the joint during childbirth and in individuals who have fallen backward and landed on a hard surface. Examine the coccyx in the lab and compare it with figure 8.8.

Ribs

There are 12 pairs of **ribs** in the human; along with the sternum and the thoracic vertebrae, these structures make up the **thoracic cage.** Each rib has a **head** that articulates with the body of one or two vertebrae. On the head is a **facet,** or two. Each facet is the site of articulation with the vertebral body. A constricted region near the head is the **neck** of the rib. Locate these features on figure 8.10. The process near the neck on ribs 1–10 is the **tubercle** of the rib, which articulates with the transverse process of the vertebra. Examine the articulated skeleton from the back and notice that the **shaft** of the ribs bends at about the same distance from the midline of the body as the inferior angle of the scapula. This bend can be seen on ribs 2–10 and is known as the **angle** of the rib, or the **costal angle.** Some ribs have a truncated **sternal end** that attaches to a costal cartilage prior to joining the sternum. The superior edge of the rib is more rounded than the inferior edge. The depression that runs along the inferior edge of each rib is known as the **costal groove.** The intercostal blood vessels and nerves are nestled in the space provided by the costal groove. With the costal groove in an inferior position and the blunt sternal end toward the midline, determine whether you are looking at a left rib or a right rib.

The first 7 pairs of ribs are **true ribs.** A true rib is one that attaches to the sternum by its own cartilage. These are known also as **vertebrosternal ribs,** as they attach to the vertebra posteriorly and to the sternum anteriorly. Ribs 8 through 12 are **false ribs,** because they do not attach to the sternum by their own cartilage. Ribs 8 through 10 attach to the sternum by way of the cartilage of rib 7. These ribs are also called **vertebrochondral ribs** for their posterior attachment to the vertebrae and their anterior attachment to the cartilage of rib 7. Ribs 11 and 12 do not attach to the sternum at all and are known as **floating ribs** (a specific type of false rib) or **vertebral ribs.** Examine the articulated skeleton in the lab and compare it with figure 8.11.

Sternum

The **sternum** is composed of three fused bones. The superior segment is the **manubrium** (ma-NOO-bree-um) and the depression at the top of the manubrium is a **jugular,** or median **suprasternal, notch.** The lateral indentations on the manubrium are sites of articulation with the clavicles known as **clavicular notches.** The main portion of the sternum is the **body,** or **gladiolus,** and between the body and the manubrium is the **sternal angle.** The sternal

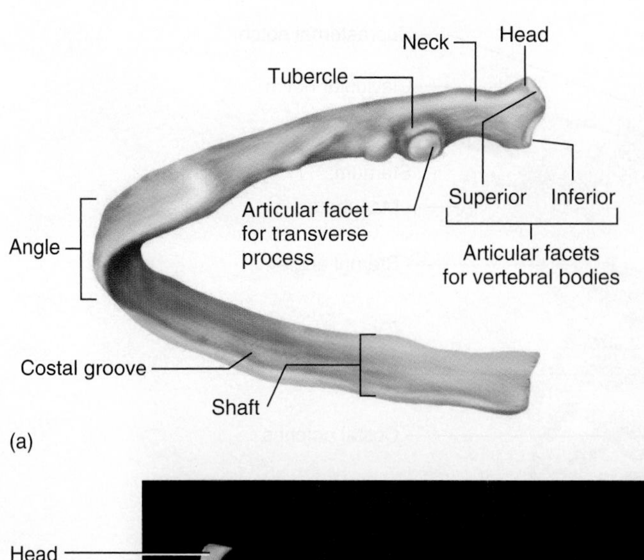

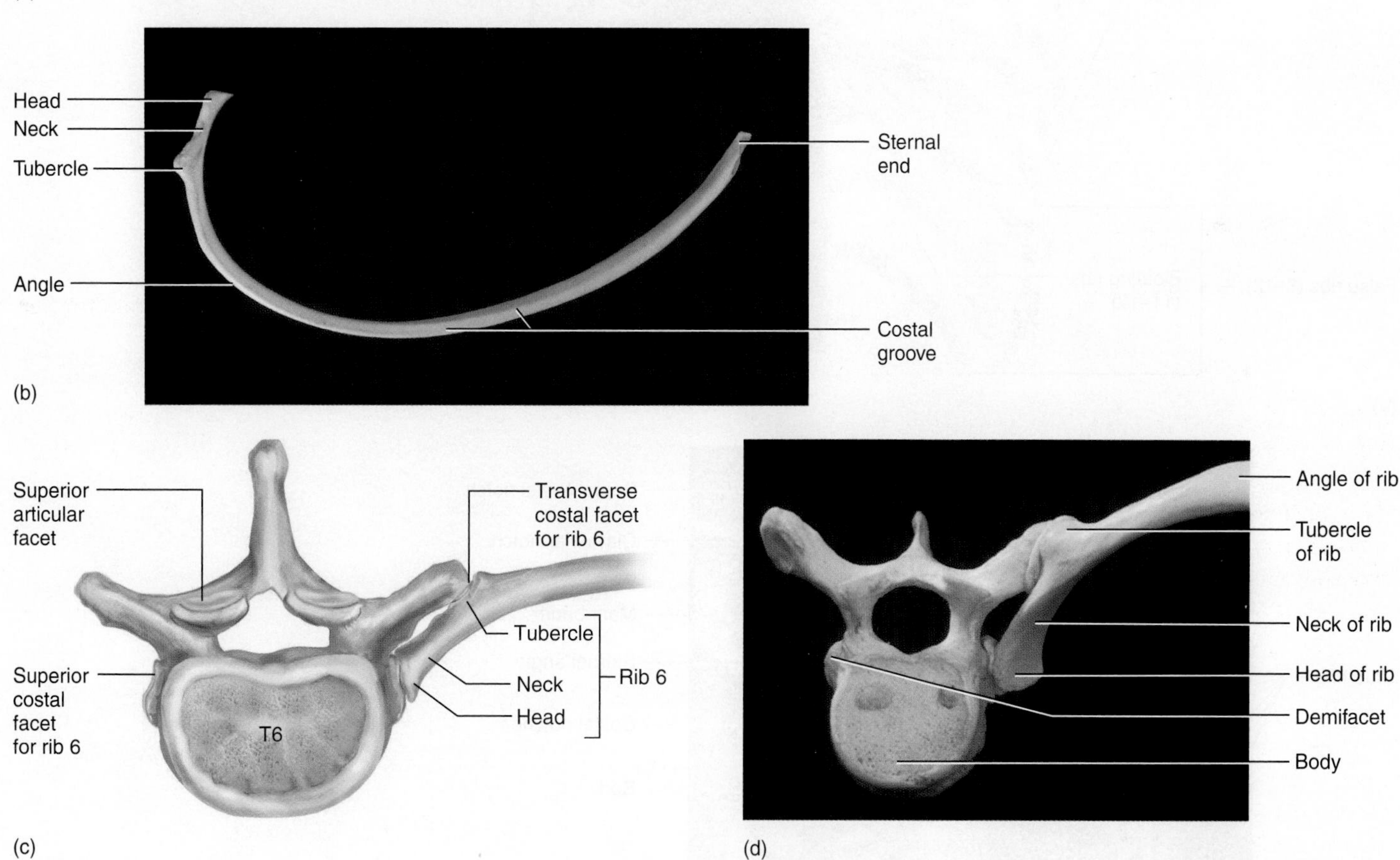

FIGURE 8.10 Individual and Articulated Rib (a) Diagram of an individual rib, posterior view; (b) photograph of an individual rib, inferior view; (c) diagram of an articulated rib; (d) photograph of an articulated rib.

angle is a landmark for finding the second rib when using a stethoscope to listen to heart sounds. Locate the **costal notches** on the body of the sternum. These are where the cartilages of the ribs attach.

The narrow, bladelike part that is the most inferior segment of the sternum is the **xiphoid** (ZYE-foyd) **process.** Care must be taken when performing CPR (cardiopulmonary resuscitation) so that pressure is applied to the body of the sternum and not to the xiphoid process. If the force is applied to the xiphoid process, it could fracture and penetrate the liver. Examine the structures of the sternum on lab specimens and in figure 8.11.

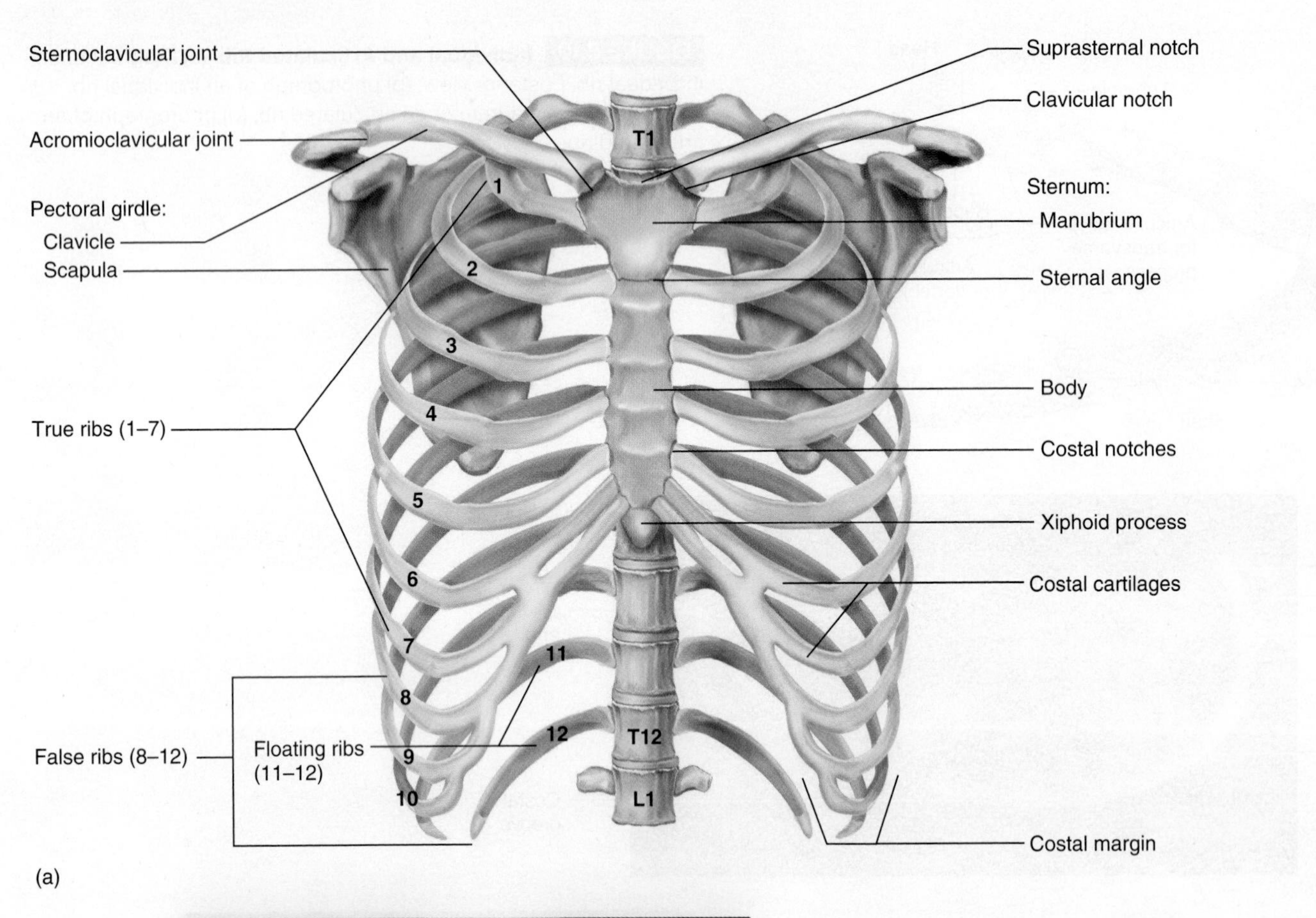

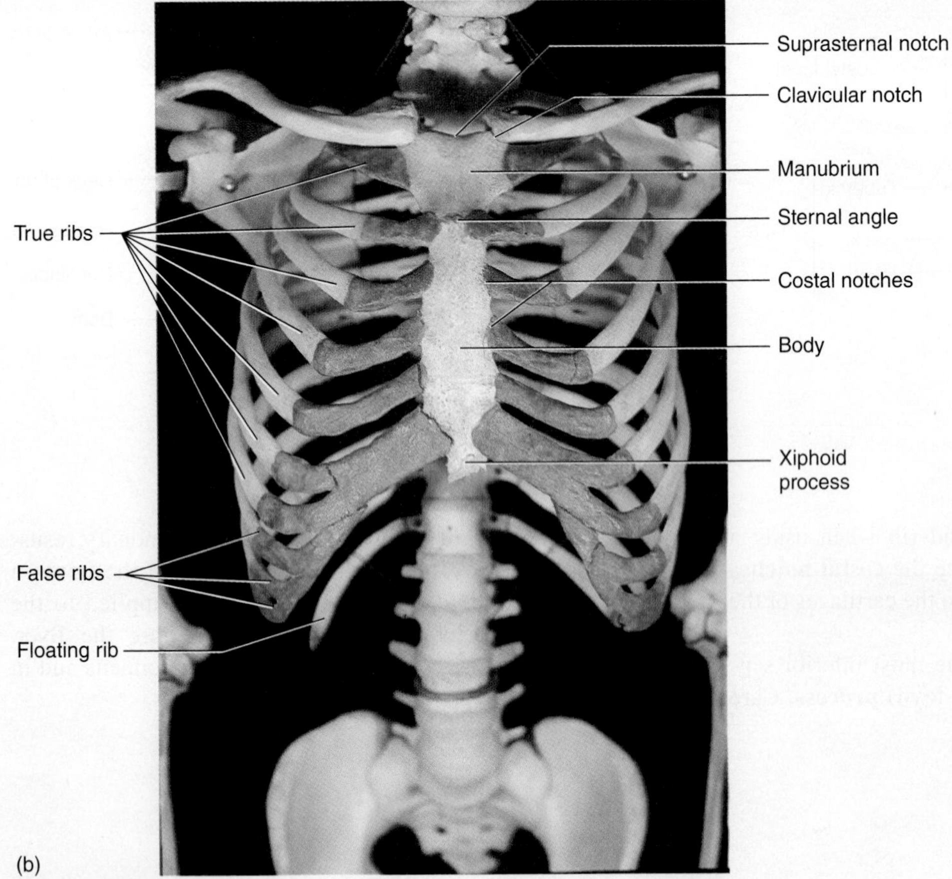

FIGURE 8.11 **Thoracic Cage** (a) Diagram; (b) photograph.

Exercise 8

Axial Skeleton 2: Vertebrae, Ribs, Sternum

Name ______________________________ Date ____________

1. Label the following illustration, using the terms provided.

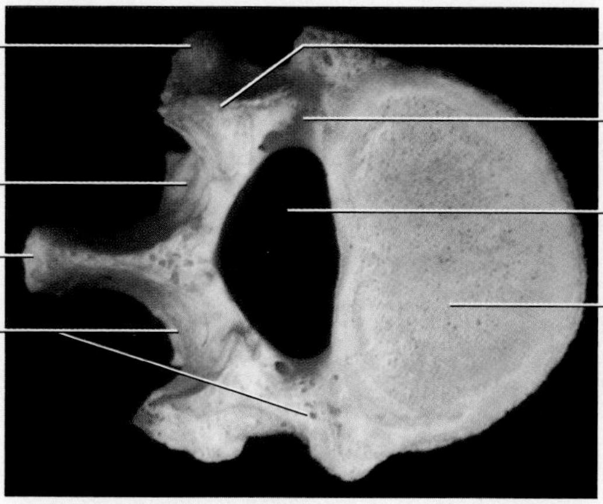

a. ______________________

b. ______________________

c. ______________________

d. ______________________

e. ______________________

f. ______________________

g. ______________________

h. ______________________

lamina	transverse process
pedicle	vertebral arch
spinous process	vertebral body
superior articular process	vertebral foramen

2. What are the names of the fibrocartilage pads found between adjacent bodies of the vertebrae? ______________________

3. What structures make up the vertebral arch? ______________________

4. The superior articular facet of a vertebra articulates with what specific structure? ______________________

5. Which spinal curvature is the most superior one? ____________________

____

6. On what vertebra would you find the odontoid process? ____________________

____

7. Which vertebra has no body? ____________________

____

8. What two features do cervical vertebrae have that no other vertebrae have? ____________________

____

____

9. Match the cervical vertebrae with their numeric names (draw a line connecting each pair).

axis	C1
vertebra prominens	C2
atlas	C7

10. Determine from what part of the spinal column the vertebra in the following illustration comes. ____________________

____

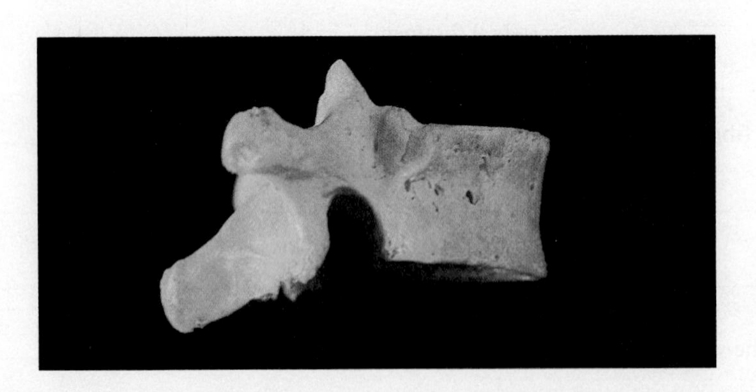

11. How many lumbar vertebrae are in the human body? ____________________

____

12. Distinguish between the posterior sacral foramina and the sacral canal. ____________________

____

____

13. Name the horizontal lines that result from the fusion of the sacral vertebrae. ____________________

____

14. What is the joint between the sacrum and the hip bones called? ____________________

____

15. Match the vertebrae from a region with the features or structures found there.

Vertebrae	**Features or Structures**
____ cervical	a. ala
____ coccyx	b. vertebrae with the largest vertebral bodies
____ lumbar	c. vertebrae with rib facets
____ sacral	d. typically four fused vertebrae
____ thoracic	e. transverse foramina

16. A rib that attaches to the sternum by the cartilage of rib 7 has what name? ____________________

____

17. Is the angle of the rib on the anterior or posterior side of the body? ____________________

____

18. Which ribs (by number) are the floating ribs? ____________________

____

19. What part of a rib articulates with the transverse process of a vertebra? ____________________

____

20. The most inferior portion of the sternum is the (circle the correct answer)

a. body. b. manubrium. c. angle. d. xiphoid.

21. The superior portion of the sternum has what name? ____________________

____

Laboratory

Exercise 9

Appendicular Skeleton

Skeletal System

Introduction

The appendicular skeleton consists of 126 bones in four major groups. These are the pectoral girdle, upper limb, hip bones, and lower limb. The **pectoral girdle** is composed of a scapula and clavicle on each side of the body. The **upper limb** consists of the humerus, radius, ulna, and bones of the hand including the carpal bones, metacarpal bones, and phalanges. The **hip bones** consist of two sets of fused bones. Each hip bone is composed of an ilium, ischium, and pubis. The **lower limb** consists of the femur, patella, tibia, fibula, and bones of the foot including the tarsal bones, metatarsal bones, and phalanges.

Learning Objectives

At the end of this exercise you should be able to

1. locate and name all the bones of the appendicular skeleton;
2. name the significant surface features of the major bones;
3. place a bone into one of the four major groups (pectoral girdle bones, upper limb bones, hip bones, lower limb bones);
4. determine whether a selected bone is from the left side or right side of the body;
5. locate the major surface features of the scapula and hip bone; and
6. name the individual carpal bones and tarsal bones in articulated hands and feet, respectively.

Materials

Articulated skeleton or plastic cast of articulated skeleton

Disarticulated skeleton or plastic casts of bones

Charts of the skeletal system

Plastic drinking straws (cut on a bias) or pipe cleaners for pointer tips

Foam pads of various sizes (to protect real bones from hard countertops)

Procedure

Locate the bones of the appendicular skeleton on the material in the lab and in laboratory exercise 6 (fig. 6.1). When you study a bone, compare an isolated (or disarticulated) bone to the articulated skeleton (one that is joined together). Hold the individual bone up to the skeleton to see how it is positioned in relation to the other bones of the body. When you examine bones in the lab, do not use your pen or pencil to locate a structure. They leave marks on the bone. Use a cut drinking straw, a wooden applicator stick, or a pipe cleaner to point out structures. Your instructor may want you to place real bone material on foam pads to cushion the bone from the tabletop. Use your time in lab to find the material at hand. Once you are at home, draw or trace bone images from your manual and be able to name all the parts.

Pectoral Girdle

The pectoral, or shoulder, girdle consists of the right and left **scapulae** (singular, *scapula*) and the right and left **clavicles.** The pectoral girdle provides a movable yet stable support for the upper limb.

Scapula The scapula, commonly known as the shoulder blade, is found on the posterior, superior portion of the thorax. Because it has few bony attachments, it provides a great range of motion for the arm. The scapula is roughly triangular and has three borders—a **superior border,** a **vertebral (medial) border,** and an **axillary (lateral) border.** On the superior border of the scapula is an indentation known as the **scapular notch,** which contains a nerve that innervates shoulder muscles.

On the anterior surface of the scapula is a smooth, hollowed depression known as the **subscapular fossa.** The posterior surface of the scapula is divided by the **scapular spine** into a superior depression known as the **supraspinous fossa** and an inferior depression known as the **infraspinous fossa.** The spine of the scapula is a bony crest that runs from the vertebral border to the lateral edge of the scapula, where it expands to form the **acromion** (ah-CRO-me-on). Another projection from the scapula is the

coracoid (COR-uh-coyd) **process,** which projects anteriorly. The scapula also has three angles known as the **lateral angle,** the **inferior angle,** and the **superior angle.** The features of the scapula are important because numerous muscles attach to them. The humerus moves in the **glenoid cavity,** a depression on the lateral edge of the scapula. Above the cavity is the supraglenoid tubercle, a process on which the biceps brachii muscle attaches. Just inferior to the glenoid cavity is a bump known as the **infraglenoid tubercle,** an attachment point for the triceps brachii muscle.

You should also be able to distinguish a right scapula from a left one. This can be done if you recognize that the spine of the scapula is posterior, the inferior angle is less than 90°, and the glenoid cavity is lateral. Locate the features of the scapula on figure 9.1.

Clavicle Commonly known as the collarbone, the **clavicle** is a small bone that serves as a strut between the scapula and the sternum. The blunt end of the scapula is the **sternal end,** and the horizontally flattened end is the **acromial end.** Note the **conoid** (CON-oyd) **tubercle** on the inferior surface of the clavicle. The arm

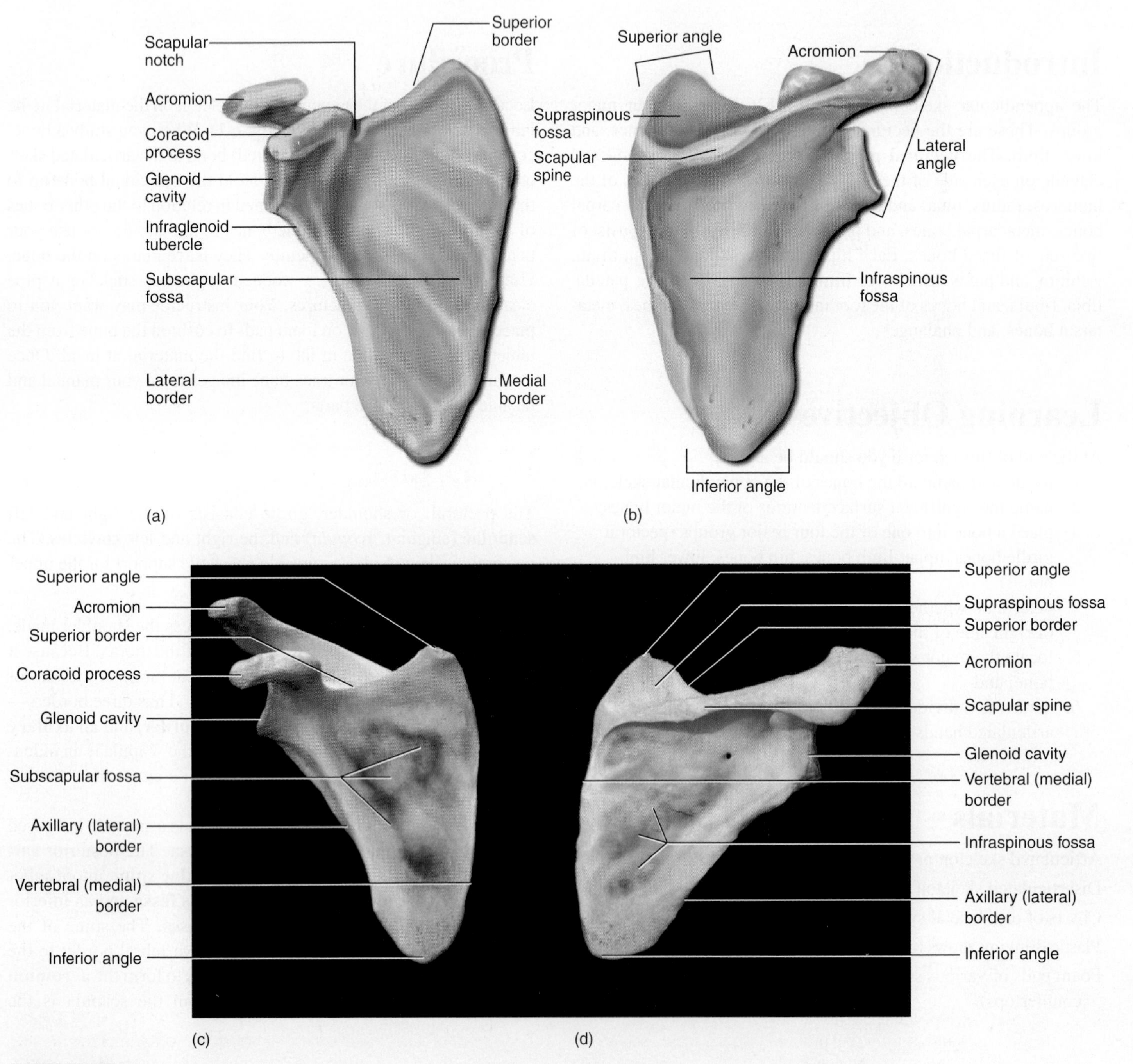

FIGURE 9.1 **Right Scapula** Diagram (a) anterior view; (b) posterior view. Photograph (c) anterior view; (d) posterior view.

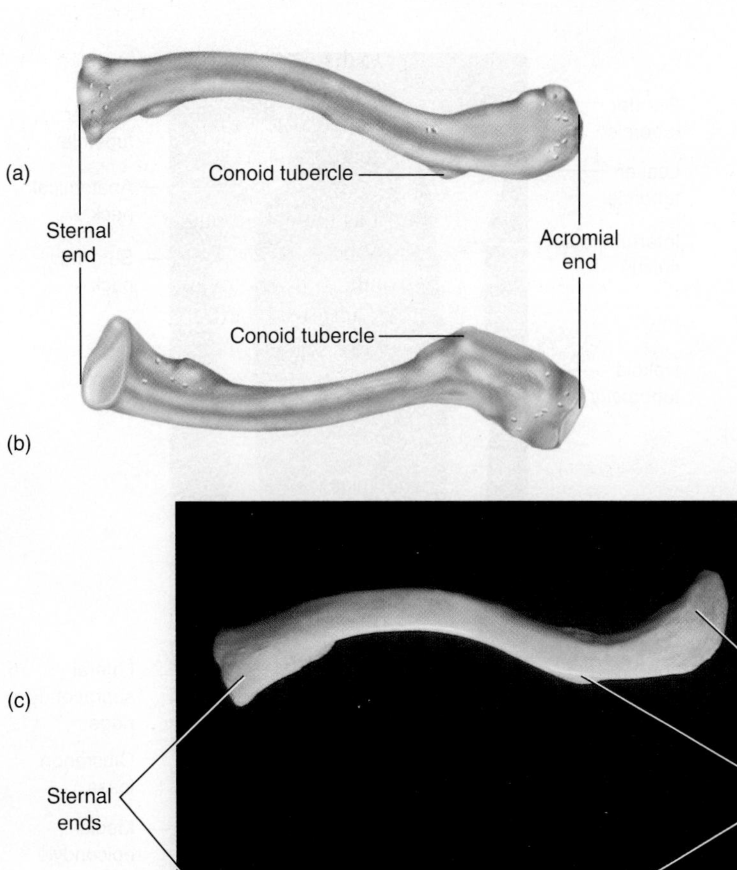

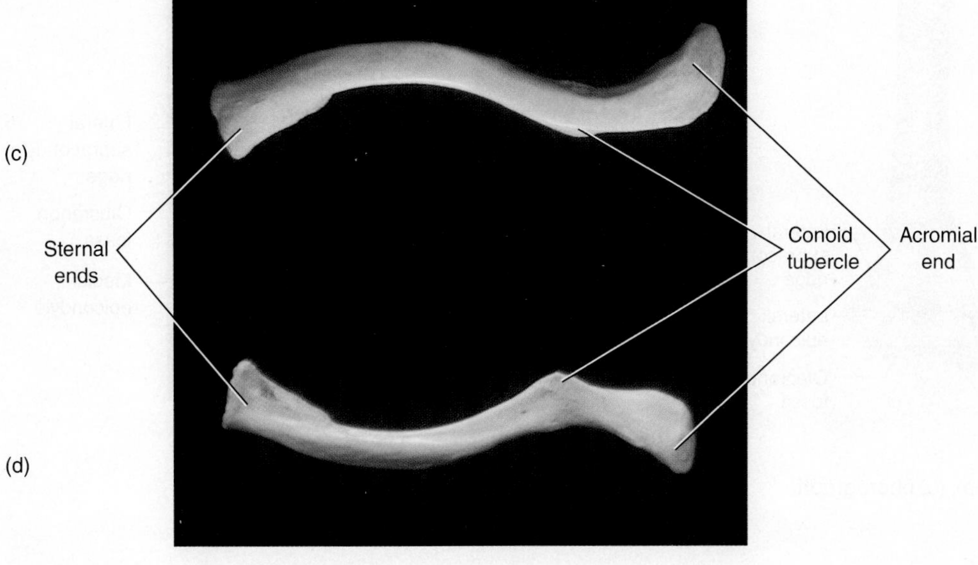

FIGURE 9.2 Clavicle Diagram (a) superior view; (b) inferior view. Photograph (c) superior view; (d) inferior view.

transmits force to the clavicle, which is sandwiched between the scapula and the sternum. The clavicle is the most frequently fractured bone in the body. Examine the clavicles in the lab and compare them with figure 9.2.

Upper Limb

Bones of the Arm

Humerus The **humerus** is a long bone that articulates with the scapula at its proximal end and with the radius and ulna at its distal end. The majority of the length of the humerus is the **diaphysis,** or **shaft,** and small holes penetrating the shaft are the **nutrient foramina.** Nutrient foramina are holes that occur in many bones and conduct blood vessels into these bones. The **head** of the humerus is a hemispheric structure that ends in a rim known as the **anatomical neck.** The anatomical neck is the site of the epiphyseal line. The humerus has two large processes—the **greater tubercle,** which is the largest and most lateral process, and the **lesser tubercle,** which is medial and smaller. These processes are attachment points for muscles originating from the scapula. Between the tubercles is the **intertubercular,** or **bicipital, sulcus,** which is an elongated depression through which one of the biceps brachii tendons passes. The intertubercular sulcus is anterior, and the head of the humerus is medial. These two features can orient you to the right or left humerus. Examine figure 9.3 and locate these features. Below the tubercles of the humerus is a constricted region known as the **surgical neck.** The surgical neck is so named because it is a frequent site of fracture. Locate the surgical neck of the humerus on your arm and name a muscle located nearby in the space provided.

Muscle near surgical neck ____________________

A roughened area on the lateral surface of the humerus is the **deltoid tuberosity,** named for the attachment of the deltoid muscle.

At the distal end of the humerus are the regions of articulation with the bones of the forearm. On the lateral side is a round hemisphere known as the **capitulum** (ca-PIT-you-lum). This is where the head of the radius fits into the humerus. On the medial side is an hour glass–shaped structure called the **trochlea** (TROCK-lee-uh). The trochlea is where the ulna attaches to the humerus. The capitulum and trochlea represent the **condyles** (KON-dials) of the humerus. To the sides of these condyles are the

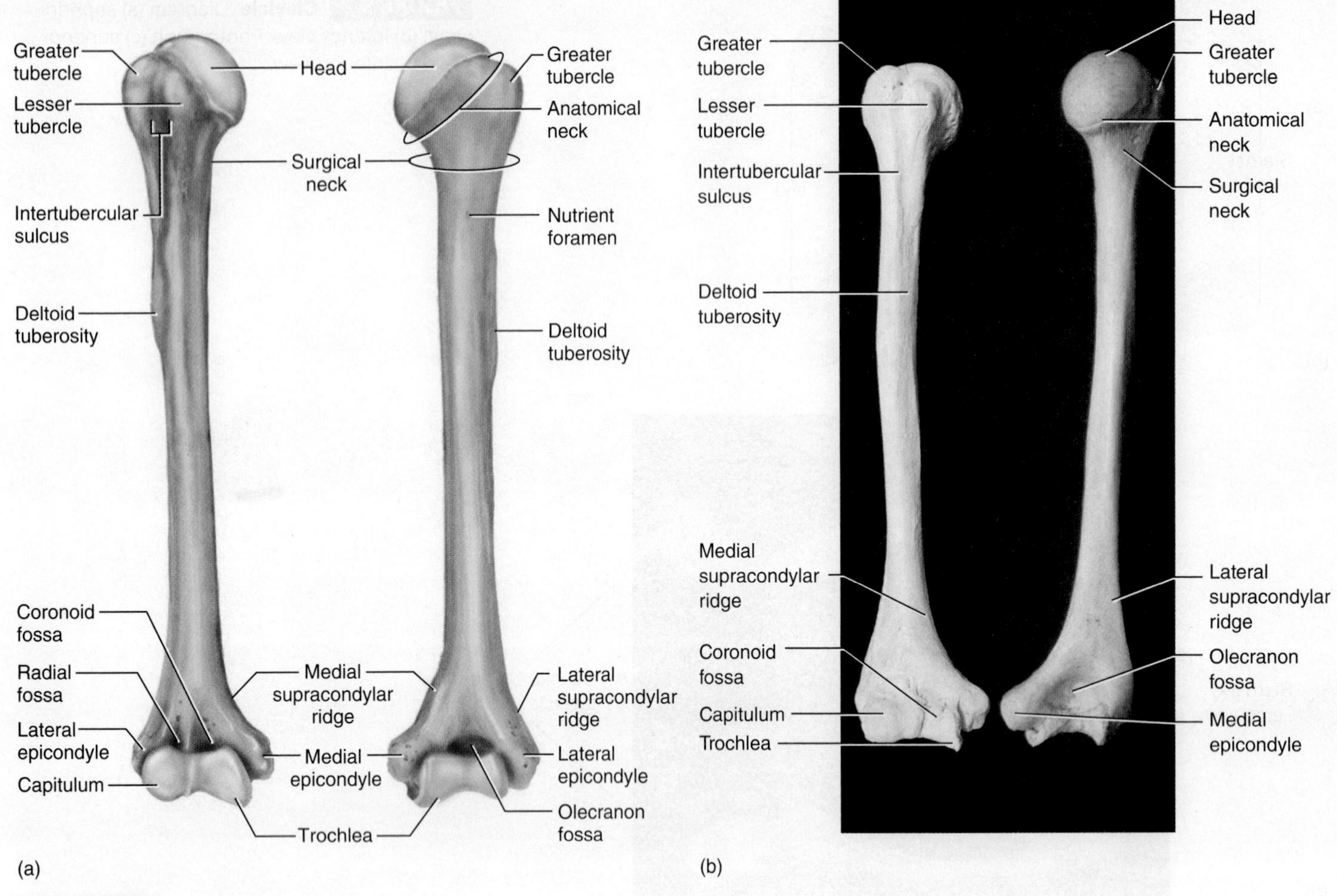

FIGURE 9.3 **Right Humerus** (a) Diagram; (b) photograph.

epicondyles (EP-ee-Kon-dials). The **medial epicondyle** of the humerus is a process commonly known as the "funny bone," and the **medial supracondylar ridge** forms a small wing of bone extending from the epicondyle to the distal shaft of the humerus. The medial epicondyle is the bump that you can palpate (feel) medial to the elbow. The **lateral epicondyle** also has a **lateral supracondylar ridge.** The epicondyles and supracondylar ridges are points of attachment for muscles that manipulate the forearm and hand.

At the distal end are three depressions in the humerus. The ones on the anterior surface are the **coronoid fossa** and the **radial fossa,** and the one on the posterior surface is the **olecranon** (uh-LEC-ruh-non) **fossa.** These depressions accommodate the ulna as the forearm is flexed or extended. Locate these structures on specimens in the lab and on figure 9.3.

Bones of the Forearm Two bones make up the skeletal portion of the forearm. These are the **radius,** lateral (on the thumb side), and the **ulna,** a medial bone. The **radius** has a proximal **head,** resembling a wheel that articulates with the capitulum of the humerus. Distal to the head is the **tuberosity** of the radius, where the biceps brachii muscle inserts, and at the most distal and posterior end of the radius is the **styloid** (STY-loyd) **process.** On the distal end of the radius is a medial depression called the **ulnar notch.** This notch joins with the medial bone of the forearm, the ulna. Examine figure 9.4 for these structures and compare them with the bones in the lab.

The **ulna** has a U-shaped depression (some students remember this as "U for ulna") on the proximal portion. This depression is the **trochlear** (TROK-lee-ur), or **semilunar, notch** that articulates with the trochlea of the humerus. The most proximal portion of the ulna is the **olecranon,** commonly known as the bony tip of the elbow. The process distal to the trochlear notch is the **coronoid process,** which provides a relatively tight fit with the humerus and, along with the **tuberosity of the ulna,** provides an area for muscle attachment. On the proximal part of the ulna, where the head of the radius fits into the ulna, is a depression known as the **radial notch.** If you know that the trochlear notch is anterior and the radial notch is lateral, then you can determine whether you are looking at a left or right ulna. In the following space, state what side of the body the bones of the forearm you are studying come from.

Side of the body: ____________________

The ulna has a *distal* **head,** unlike the radius (the head of the radius is proximal), and a **styloid process** on the distal, posterior side. Locate these structures in figure 9.4.

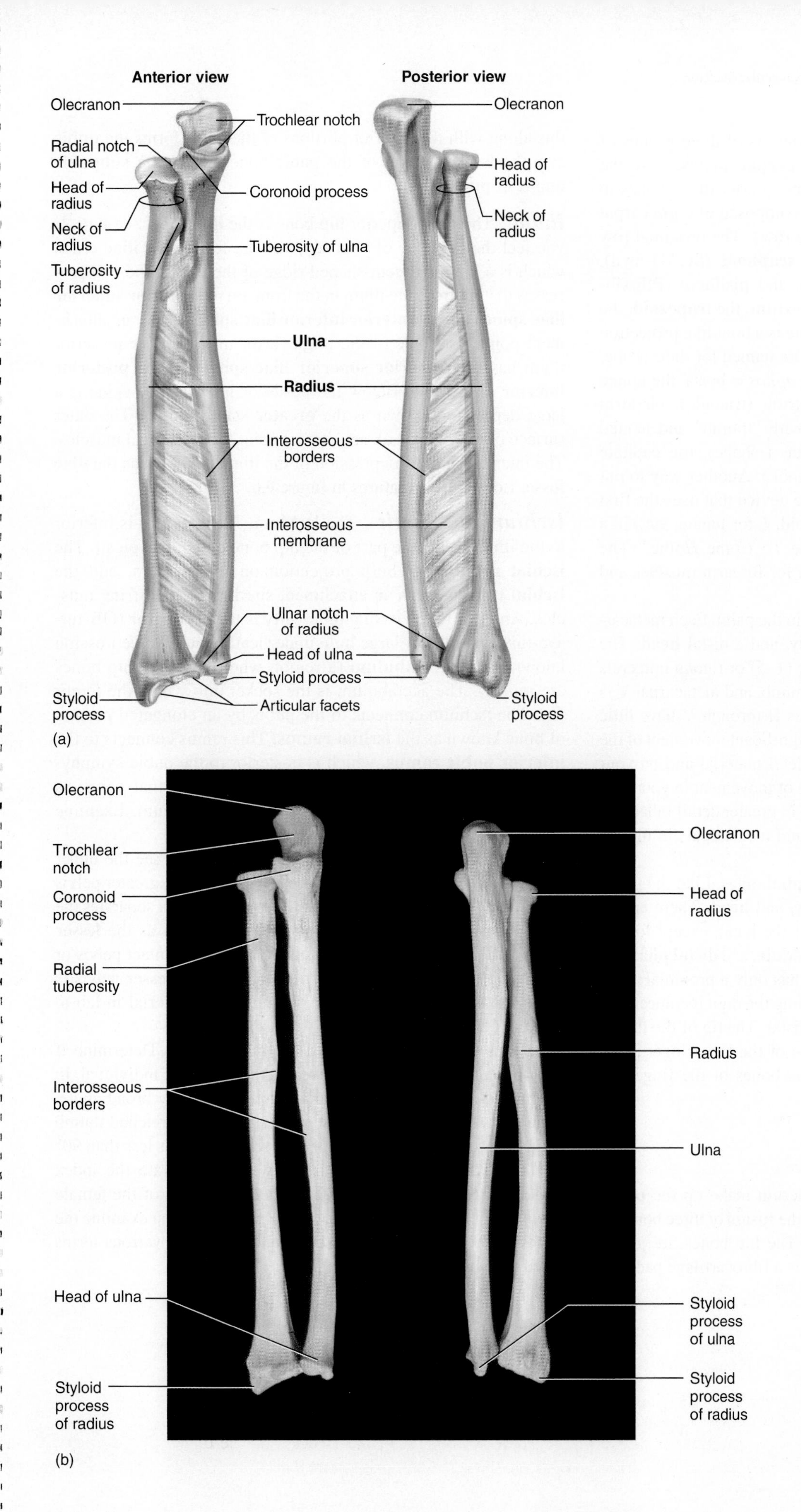

FIGURE 9.4 Right Radius and Ulna (a) Diagram; (b) photograph.

Bones of the Hand Each hand consists of three groups of bones—the **carpal bones,** the **metacarpal bones,** and the **phalanges** (singular, *phalanx*). Over one-quarter of all bones in the body are in the hands. The hand is composed of eight carpal bones (fig. 9.5) that are aligned in two rows. The proximal row from lateral to medial includes the **scaphoid** (SCAH-foyd), **lunate, triquetrum** (tri-KWEE-trum), and **pisiform** (PIE-sih-form). The distal row includes the **trapezium,** the **trapezoid,** the **capitate,** and the **hamate.** On the hamate is a hooklike projection called the **hamulus.** The carpal bones are named for their shape, with the scaphoid resembling a boat (*scaphos* = boat), the lunate named for the crescent moon, triquetrum (triangle), pisiform (pea-shaped), the trapezium (rhymes with "thumb" and named for a shape), the trapezoid (named for a shape), the capitate (headlike), and the hamate (*hamus* = hook). Another way to put the bones in order is to use a mnemonic device that uses the first letter of each carpal bone (*S* for scaphoid, *L* for lunate, etc.) in a sentence—"*S*ay *L*oudly *T*o *P*am, *T*ime *T*o *C*ome *H*ome." The carpal bones are regions of attachment for forearm muscles and for intrinsic muscles of the hand.

The **metacarpal bones** are bones in the palm. Each metacarpal consists of a proximal **base,** a **body,** and a distal **head.** The metacarpals are labeled by either arabic (1–5) or roman numerals (I–V). Metacarpal I is proximal to the thumb, and metacarpal V is proximal to the little finger. Metacarpals II through V have little movement, but metacarpal I allows for significant movement of the thumb. Locate metacarpal I on the skeletal material and on your own hand, and note the substantial range of movement in your own hand. This type of movement is covered in greater detail in laboratory exercise 10. Examine the carpals and metacarpals in the lab and in figure 9.5.

Distal to the metacarpals are the **phalanges.** Like the metacarpals, each phalanx has a **base, body,** and **head.** There are 14 phalanges in each hand. Each finger of the hand, except for the thumb, has 3 phalanges—a **proximal, middle,** and **distal phalanx.** The thumb is known as the **pollex** and has only a proximal and a distal phalanx. Identify each bone by noting the digit it comes from and whether it is proximal, middle, or distal. The tip of the thumb is the distal phalanx I, whereas the base of the little finger is the proximal phalanx V. Locate the various bones of the fingers in material in the lab and in figure 9.5.

Hip Bones

The hip bones (ossa coxae) and the sacrum make up the **pelvic girdle.** Each adult hip bone results from the fusion of three bones—the ilium, the ischium, and the pubis. The hip bones are joined anteriorly by the interpubic disc, which is a fibrocartilage pad, and this along with the anterior portions of the pubis forms the **pubic symphysis.** The union of the pubic bones forms the **subpubic angle,** or **pubic arch.**

Ilium The most superior hip bone is the **ilium** (ILL-ee-um). If you feel the top edge of your hip, you are feeling the **iliac crest,** which is a long, crescent-shaped ridge of the ilium. The two processes that jut from the ilium in the front are the **anterior superior iliac spine** and the **anterior inferior iliac spine.** These are attachment points for some of the thigh flexor muscles. The posterior ilium has the **posterior superior iliac spine** and the **posterior inferior iliac spine.** Below the posterior inferior iliac spine is a large depression known as the **greater sciatic notch.** The outer surface of the ilium is an attachment point for the gluteal muscles. The interior, shallow depression of the ilium is known as the **iliac fossa.** Locate these features in figure 9.6.

Ischium and Pubis The **ischium** (ISS-kee-um) is inferior to the ilium and is the part of the hip bone on which you sit. The **ischial spine** is a sharp projection on the ischium, and the **ischial tuberosity** is an attachment site for the hamstring muscles. Anterior to the ischial tuberosity is the **obturator** (OB-tur-aye-tur) **foramen,** a large hole underneath a cuplike depression known as the **acetabulum** (a region where all three hip bones are joined). The acetabulum is the socket into which the femur fits. The ischium connects to the pubis by an elongated portion of bone known as the **ischial ramus.** This ramus connects to the **inferior pubic ramus,** which is posterior to the pubic symphysis. The pubis (PYOU-bis) is a U-shaped bone that connects inferiorly to the ischium and superiorly to the ilium. Examine these bones in lab.

If the two hip bones are articulated, you can see the upper basin of the pelvis, which is the **greater pelvis.** The greater pelvis is medial to the iliac fossa. There is a rim of bone that separates the greater pelvis from a deeper, smaller basin known as the **lesser pelvis.** This rim is known as the **pelvic brim** in an intact pelvis or the **arcuate line** on an isolated pelvic bone. The lesser pelvis is medial to the obturator foramen. Compare the material in lab to figure 9.6.

Examine the disarticulated hip bones in lab. Determine if you have a left or right bone as well as the sex of the individual. In a female pelvis the greater sciatic notch has a fairly broad angle, which approximates the arc inscribed by your outstretched thumb and index finger. In a male the greater sciatic notch is less than 90° and approximates the angle if you were to separate the index finger and the middle finger. Changes in the shape of the female pelvis occur due to the influence of hormones. As you examine the bones in the lab, look at figure 9.7 and locate the various terms listed on the illustration.

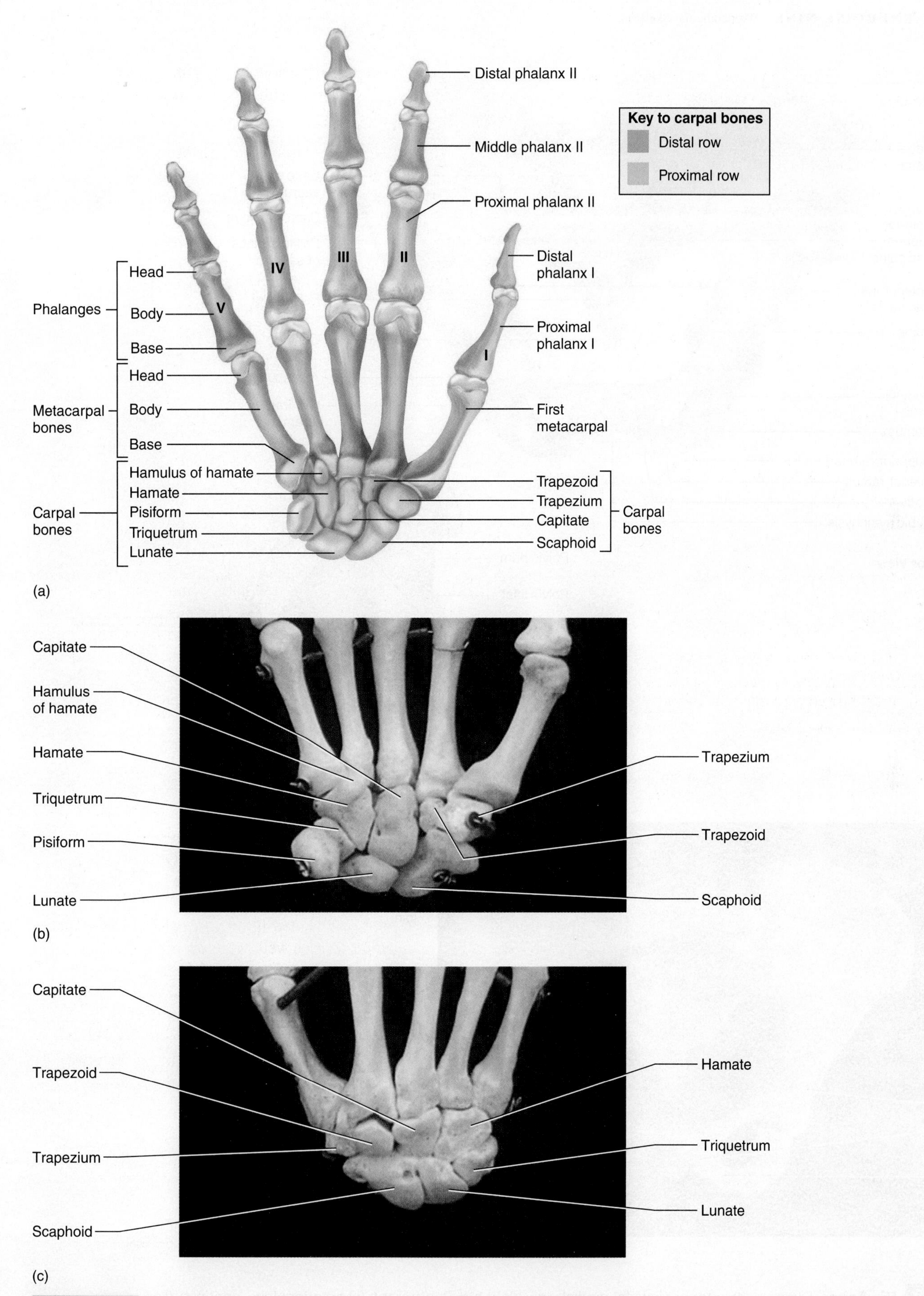

FIGURE 9.5 **Bones of the Right Hand** Diagram (a) anterior view. Photograph (b) anterior view; (c) posterior view.

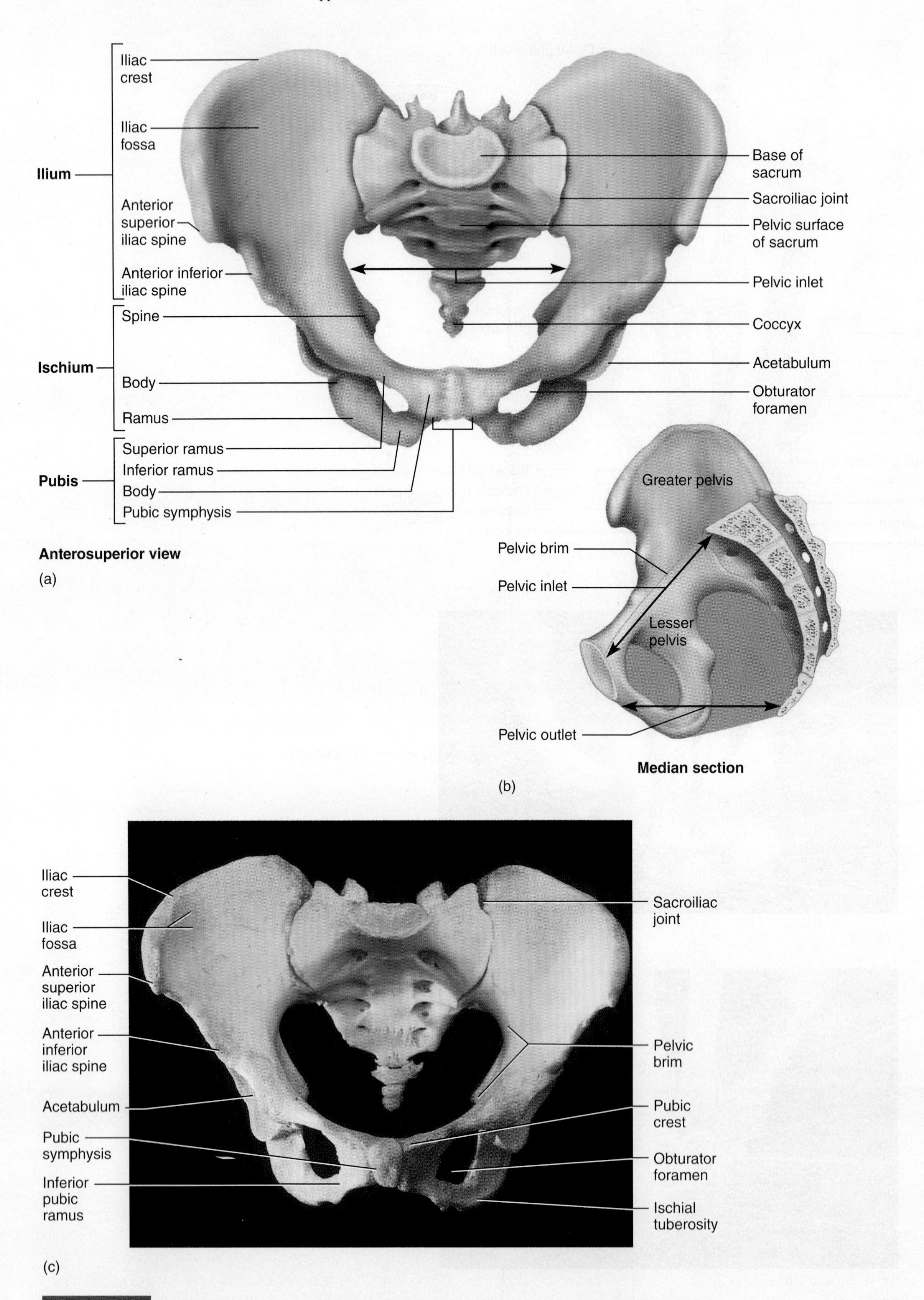

FIGURE 9.6 **Hip Bones** Diagram (a) anterior view; (b) medial view, right hip. Photograph (c) anterior view. *(continued)*

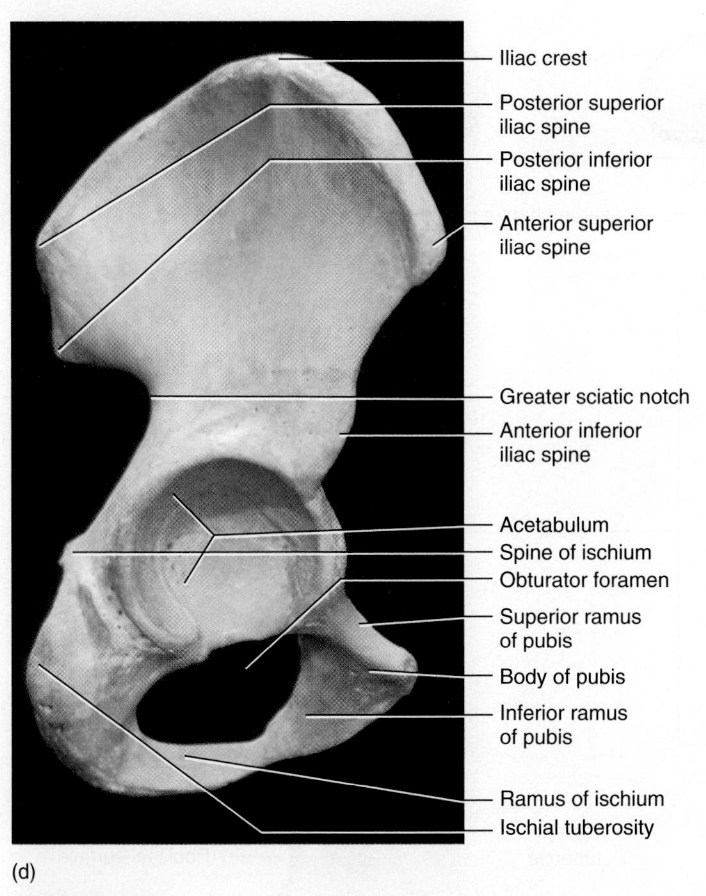

FIGURE 9.6 *(continued)* **Hip Bones** Diagram (d) lateral view, right hip.

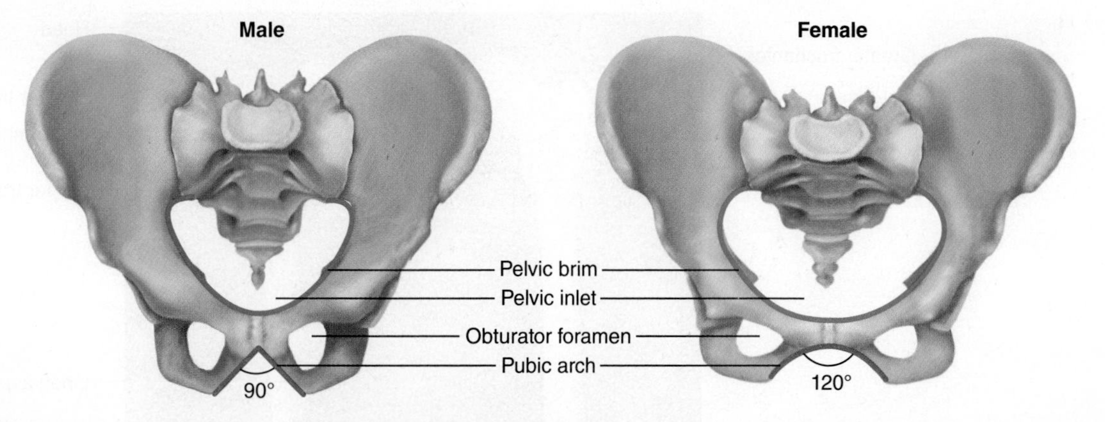

FIGURE 9.7 **Comparison of the Male and Female Hip Bones** The pelvic brim is the margin highlighted in blue; the pubic arch is highlighted in red.

Lower Limb

Bones of the Thigh

Femur The **femur** is the longest and heaviest bone in the body, and it is the only bone of the thigh. It articulates proximally with the hip and inferiorly with the tibia. The femur has a proximal, medial, spherical **head,** which inserts into the acetabulum of the hip bone. The **fovea capitis** is a depression in the head where a ligament attaches. Inferior to the head is an elongated, constricted **neck.** Inferior to the neck are two large processes called the **trochanters,** which are areas of muscle attachment. The larger, proximal process is the **greater trochanter,** and the smaller, distal process is the **lesser trochanter.** Locate these structures in figure 9.8. You should also find the **intertrochanteric crest,** a posterior ridge between the two trochanters and the anterior **intertrochanteric line.** In the following space,

FIGURE 9.8 Right Femur
(a) Diagram; (b) photograph.

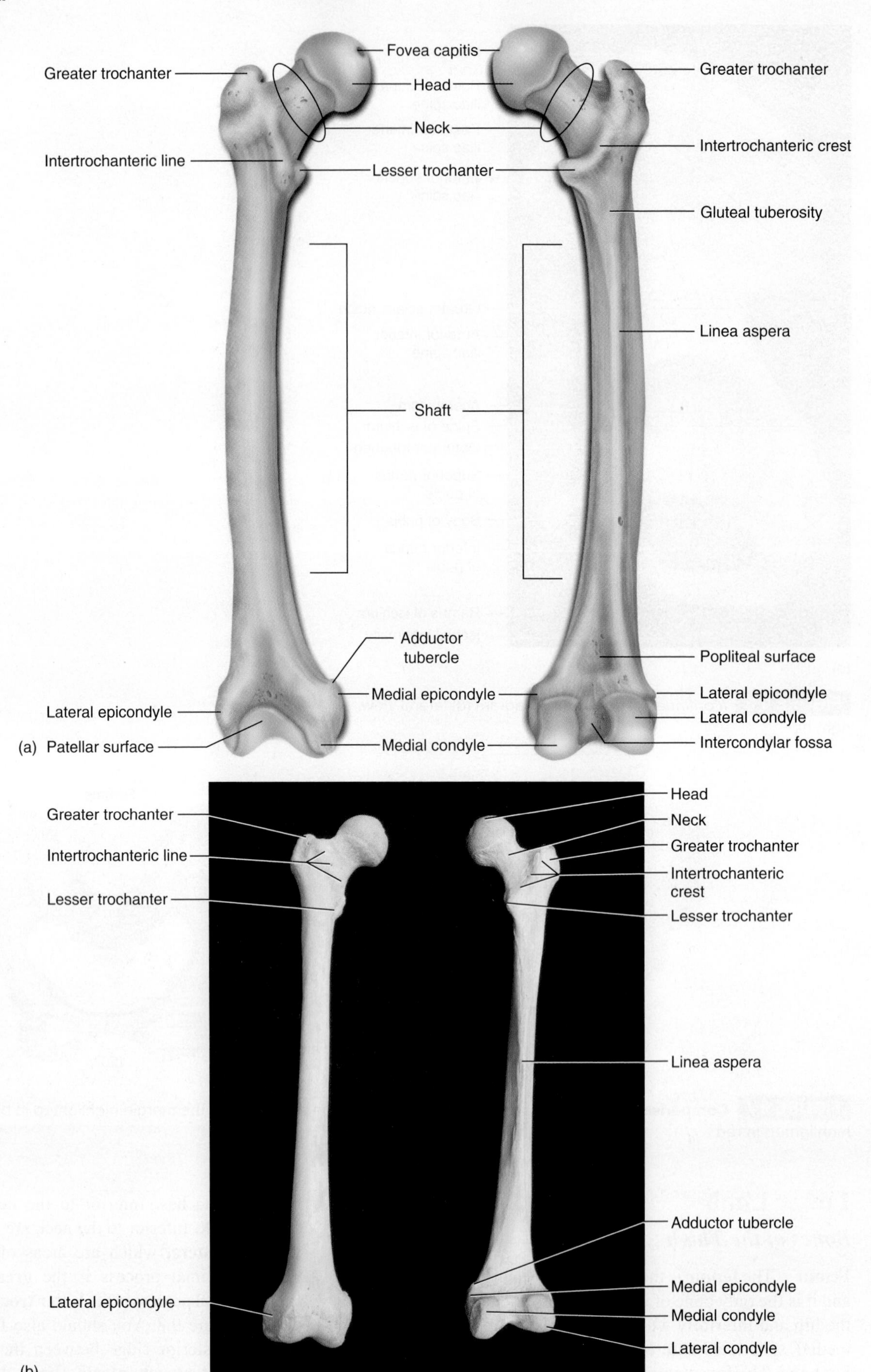

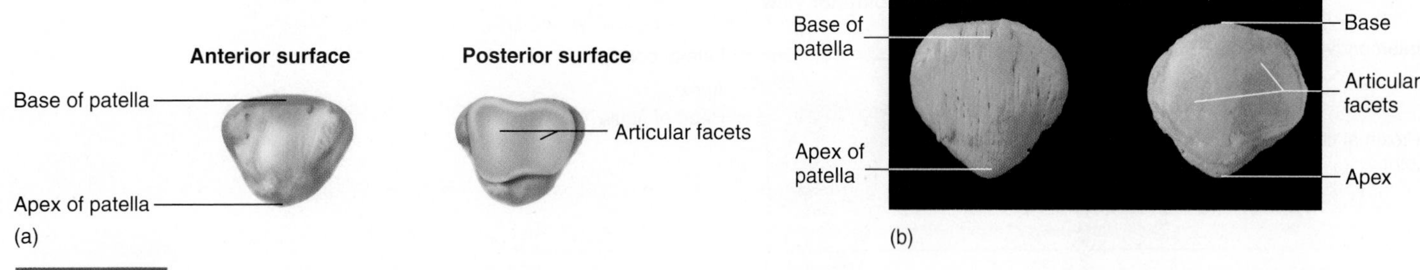

FIGURE 9.9 **Patella, Anterior and Posterior Views** (a) Diagram; (b) photograph.

determine which is broader—the intertrochanteric crest or intertrochanteric line.

Broader surface feature ______________________________

The **shaft** of the fêmur is curved and bows anteriorly with the posterior shaft marked by the **linea aspera** (meaning rough line). At the proximal and lateral portion of the linea aspera is a roughened area called the **gluteal tuberosity,** an attachment site for the gluteus maximus muscle. The distal portion of the femur consists of **medial** and **lateral condyles.** These two smooth processes articulate with the condyles of the tibia. To the sides of the condyles are bulges known as the **epicondyles.** Locate the **lateral epicondyle** and the **medial epicondyle** on the femur. Also note the location of the **adductor tubercle,** a small, triangular process proximal to the medial epicondyle. This is one of the attachment points for one of the adductor muscles. The head of the femur is medial, and the linea aspera is posterior. These features help distinguish whether you have a left or right femur.

Patella The patella, commonly known as the kneecap, is a bone formed in the tendon of the quadriceps femoris muscles on the anterior thigh. The patella is a specific type of bone known as a **sesamoid bone.** Sesamoid bones develop in the tendons. The patella ossifies between ages 3 and 6. When the leg is flexed, as when kneeling, the patella protects the ligaments of the knee joint. Examine the patella in the lab and compare it with figure 9.9.

Bones of the Leg

Tibia The **tibia** is the largest bone of the leg. It is the weight-bearing bone, inferior to the femur. The tibia is roughly triangular in cross section and is medial to the fibula. The **tibial condyles** are separated by the **intercondylar eminence,** and they articulate with the condyles of the femur. There is a rough area on the proximal, anterior surface of the tibia known as the **tibial tuberosity.** This is the attachment point for the patellar ligament. At the distal end of the tibia is an extension of bone known as the **medial malleolus** (MAH-lee-OH-lus). This process is one-half of the ankle joint that articulates with the talus of the foot. Along the length of the tibia is the **anterior border,** a ridge that runs from superior to inferior. This is the border that is bruised when you hit your shin. Locate the structures of the tibia in the lab and in figure 9.10.

Fibula The **fibula** is smaller than the tibia (a way to remember that it is smaller than the tibia is this mnemonic: "tell a little *fib*"). The fibula is lateral to the tibia and is not a weight-bearing bone but is an attachment point for muscles. The fibula has a proximal **head** and a distal process called the **lateral malleolus.** The lateral malleolus is the other half of the ankle that forms a joint with the talus. Examine the bones in the lab and compare the fibula with figure 9.10.

Bones of the Foot The feet have 26 bones each, and if you add up all the bones of the feet along with the hands, you have 106 bones, or more than 50% of all the bones in the body!

There are seven **tarsal bones** of the foot, including the **talus,** which articulates with the tibia and fibula. Directly inferior to the talus is the **calcaneus** (cal-CAY-nee-us), commonly known as the heel bone. The bone anterior to both the talus and the calcaneus is the **navicular.** The foot has three **cuneiform** (cue-NEE-ih-form) **bones,** which are wedge-shaped bones (*cuneiform* = wedge-shaped). These are the **medial,** or **first, cuneiform;** the **intermediate,** or **second, cuneiform;** and the **lateral,** or **third, cuneiform.** Locate these on specimens in the lab and in figure 9.11. The lateral cuneiform bone is not the most lateral bone of the foot. The cuboid is lateral to the lateral cuneiform. One way to remember the tarsal bones is with the mnemonic "Children that never march in line cry." The beginning letter of each word is the same as a tarsal bone (calcaneus; talus; navicular; medial, intermediate, and lateral cuneiform; and cuboid).

The pattern of the **metatarsal bones** is similar to that of the metacarpal bones of the hand. The first metatarsal bone is under the largest digit, the **hallux** (big toe), and the fifth metatarsal bone is under the smallest digit. The pattern of the phalanges of the foot is the same as in the hand, with toes two to five each having a **proximal, middle,** and **distal phalanx** and the big toe having only a proximal and a distal phalanx. Examine the material in the lab and compare the metatarsal bones and phalanges with figure 9.11.

When you have finished locating the major features of the bones of the appendicular skeleton, test yourself and your lab partner by pointing to various structures and see how well you identify the structures.

STUDY HINT

Use your time in lab to *find* the material at hand. Once you are at home, draw or trace material from your laboratory manual and name all the parts.

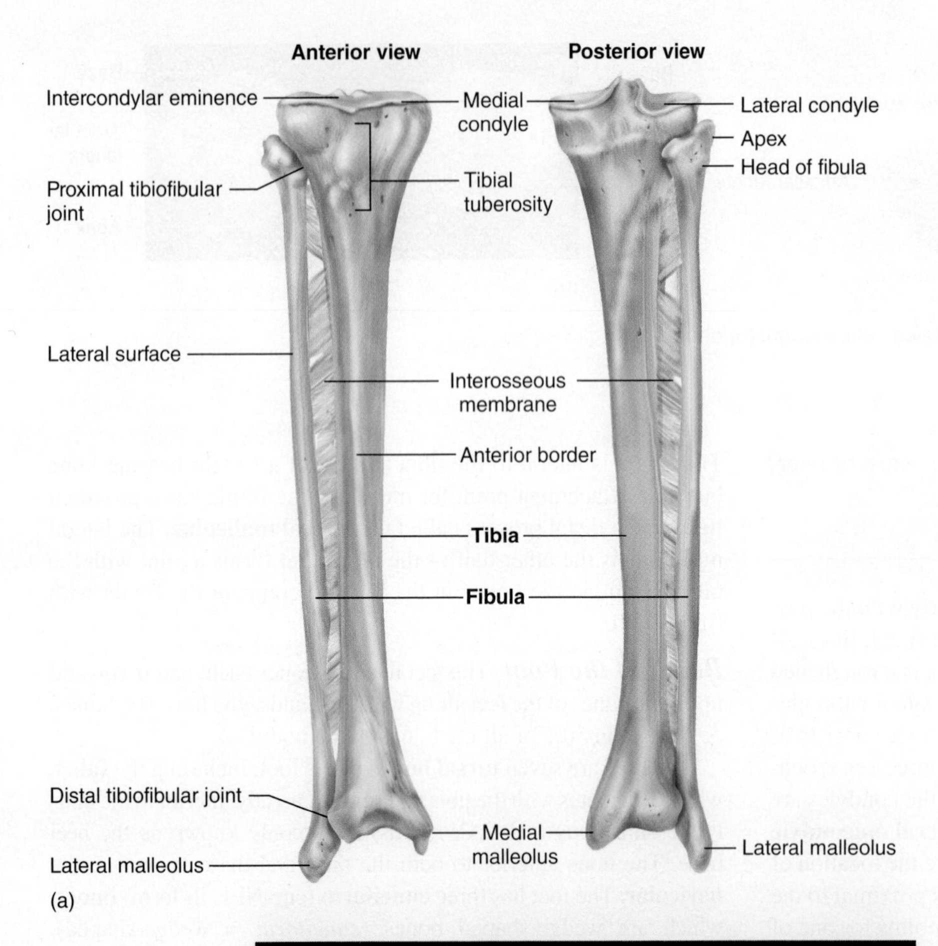

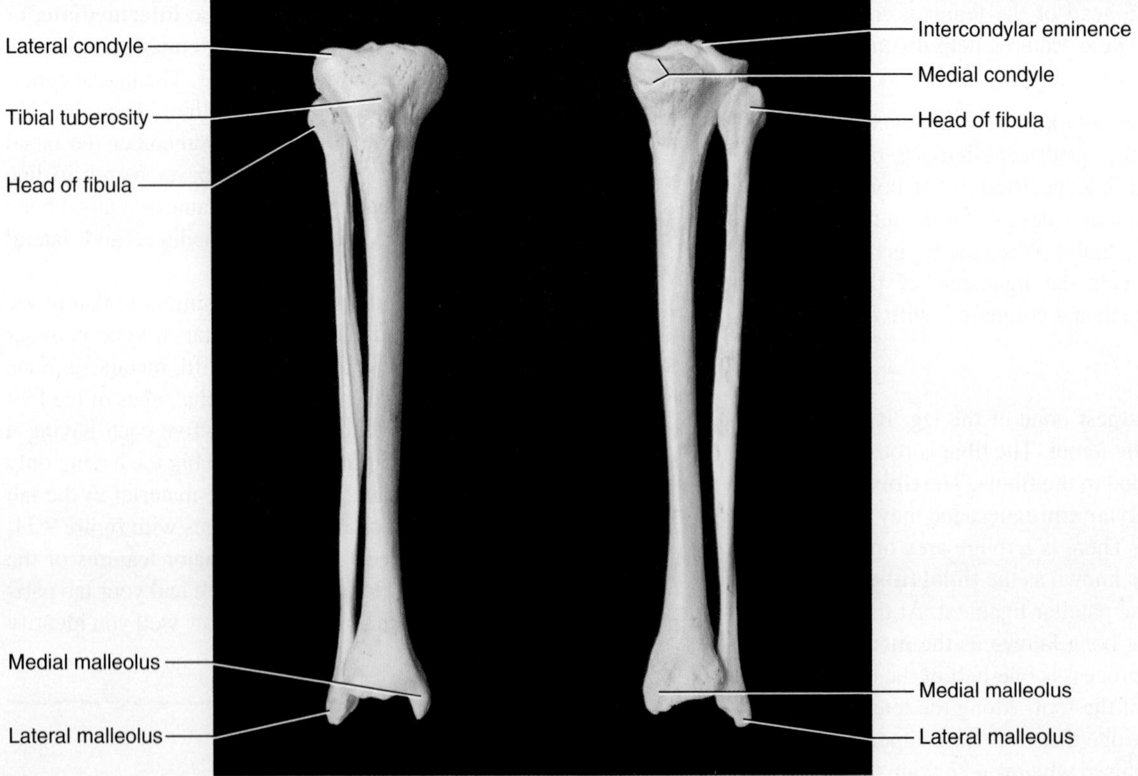

FIGURE 9.10 Right Tibia and Fibula (a) Diagram; (b) photograph.

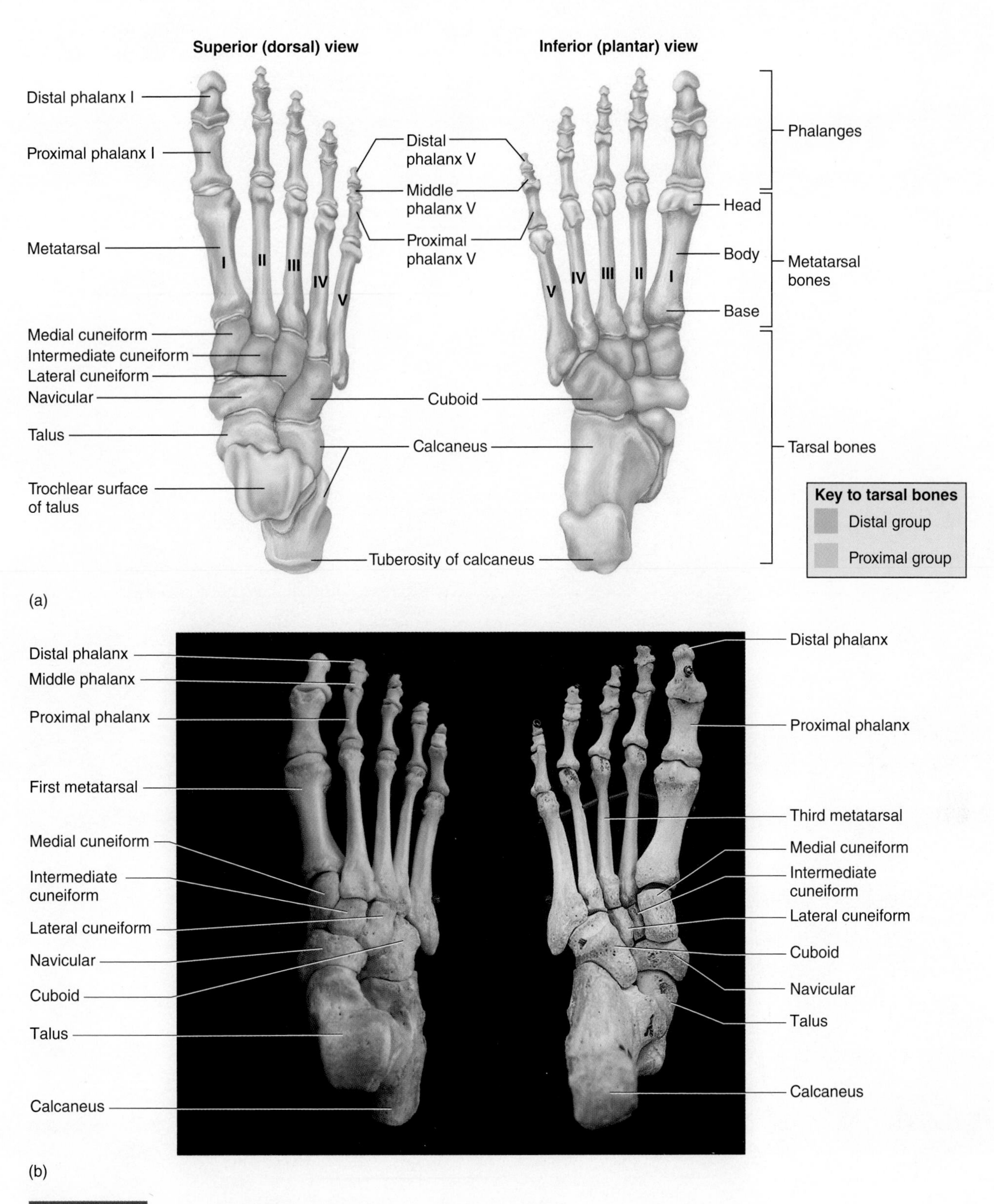

FIGURE 9.11 **Bones of the Right Foot** (a) Diagram; (b) photograph.

Exercise 9

Appendicular Skeleton

Name ______________________________ Date ______________

1. Name the anterior depression on the scapula. ______________________________

2. The humerus fits into what specific part of the scapula? ______________________________

3. Draw a line connecting the bone with the region it comes from.

clavicle	foot
hallux	forearm
ilium	hand
radius	hip bone
scaphoid	pectoral girdle

4. What specific part of the clavicle attaches to the scapula? ______________________________

5. Frequently, the clavicle is broken when the arms are extended to brace a fall. Explain how hitting the ground with your hands can fracture the clavicle. ______________________________

6. The proximal epiphyseal line on the humerus has what other name? ______________________________

7. The medial and lateral condyles of the humerus have specific names. What are they? ______________________________

8. Name the depression in the ulna into which the humerus inserts. ______________________________

9. Name the bony process that extends distally from the head of the ulna. ______________________________

10. How does the head of the ulna differ in position from the head of the radius? ______________________________

11. Each metacarpal bone consists of three major regions. What are they? ______________________________

12. Name the carpal bone at the base of the thumb. ______________________________

13. What is the name of the joint at the anterior junction of the pubic bones? ______________________________

14. Name the notch found just inferior to the posterior inferior iliac spine. ______________________________

15. Explain the difference between the greater pelvis and the lesser pelvis. ______________________________

16. Name the roughened line that runs along the length of the posterior femur. ______________________________

____

17. Name the sesamoid bone that forms in the quadriceps femoris tendon. ______________________________

____

18. Name the weight-bearing bone of the leg. ______________________________

____

19. What is the name of the process on the distal portion of the tibia? ______________________________

____

20. What is the function of the fibula? ______________________________

____

21. What is another name for the heel bone? ______________________________

____

22. What is the name of the bone of the foot that joins with the tibia and fibula? ______________________________

____

23. What is another name for the big toe? ______________________________

____

24. How many ankle bones are there versus the number of wrist bones? ______________________________

____

25. Label the parts of the scapula in the following illustration, using the terms provided.

acromion	lateral border
coracoid process	medial border
inferior angle	scapular spine
infraspinous fossa	supraspinous fossa

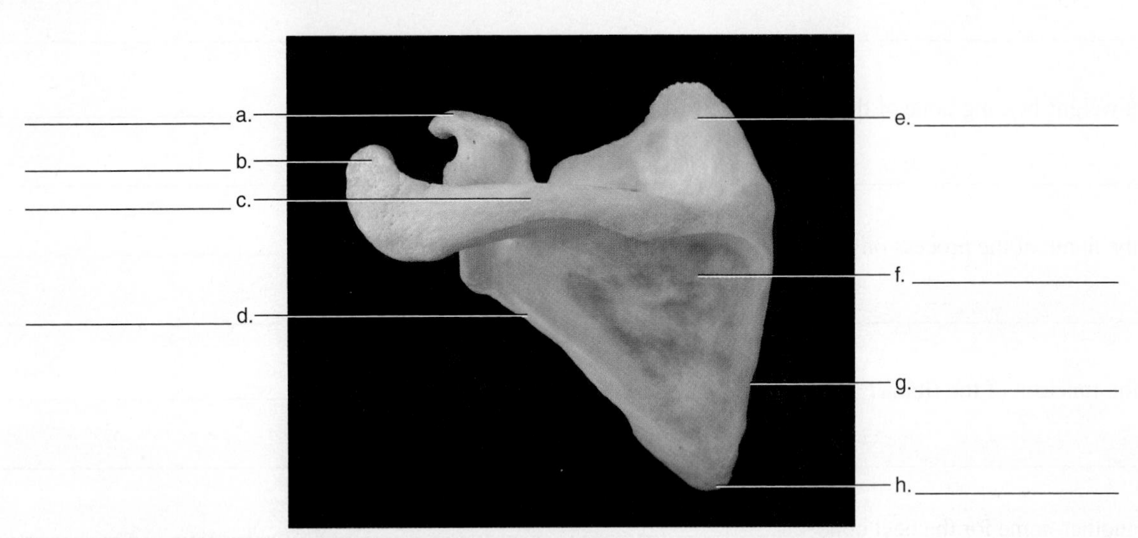

26. Label the following illustration of the foot using the terms provided.

calcaneus	medial cuneiform
cuboid	navicular
intermediate cuneiform	talus
lateral cuneiform	

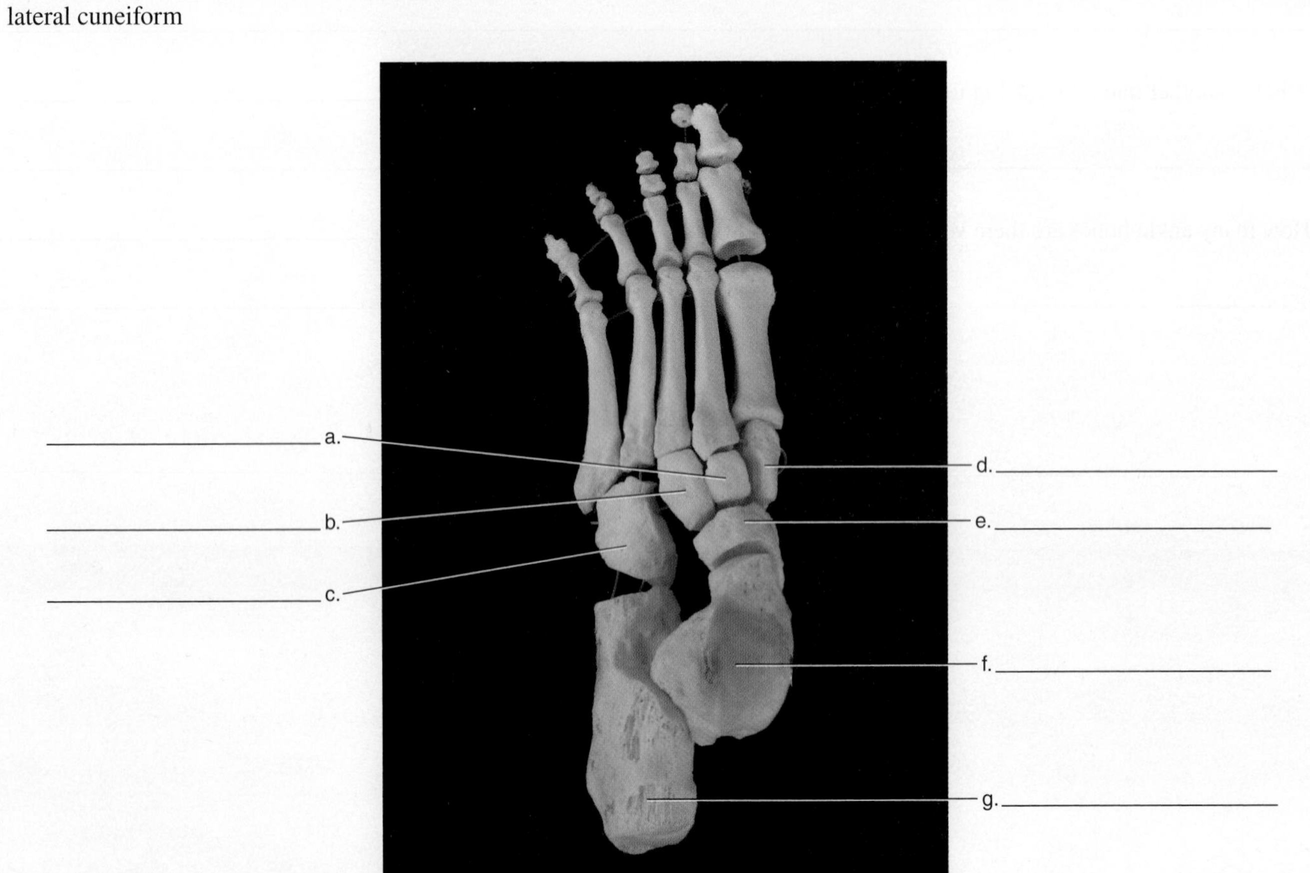

27. Label the following illustration of the hand, using the terms provided.

capitate
hamate
lunate
pisiform
trapezoid

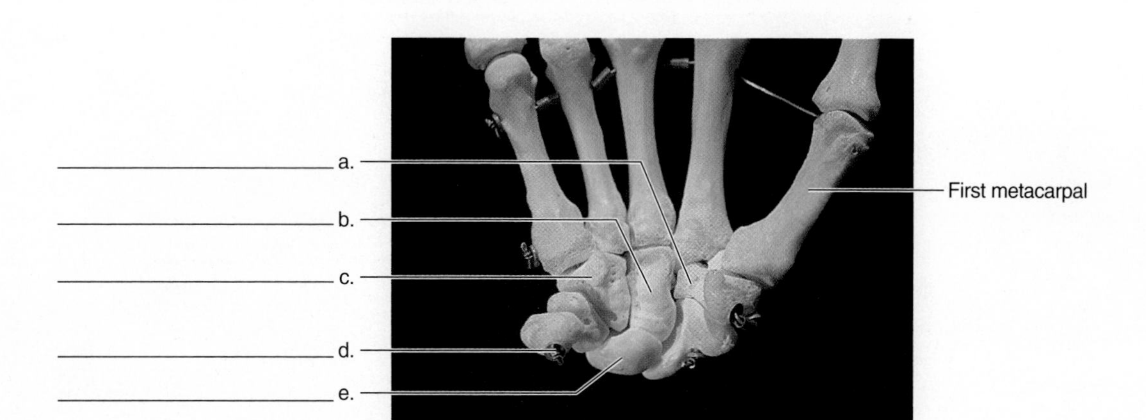

Laboratory Exercise 10

Joints

Skeletal System

Introduction

The study of **joints** or **articulations** is known as **arthrology.** Articulations are the areas of interplay between the skeletal system and the muscular system. They are important because of the significant trauma and disease that occur in them. Diseases such as arthritis affect millions of people worldwide. The development of artificial joints is continually advancing to replace specific joints of the body that can no longer provide either a comfortable range of movement or the stability required for body support.

In this exercise you study the structure and function of various types of joints in the body. Joints are classified according to their physical composition. As you study the joints in the lab, use the joints in your body as references and mimic the actions of specific articulations.

Learning Objectives

At the end of this exercise you should be able to

1. distinguish among bony, fibrous, cartilaginous, and synovial joints;
2. explain the structure of the knee, hip, jaw, elbow, and shoulder;
3. discuss the nature of a synovial joint;
4. locate the fibrous capsule, synovial membrane, synovial fluid, and articular cartilage in a dissected synovial joint;
5. list five types of synovial joints; and
6. describe actions at joints, such as flexion, extension, medial rotation, and lateral rotation.

Materials

Vertebrate joint with intact synovial capsule
Dissection tray with scalpel, blunt probe, and protective gloves
Waste container
Model or chart of joints, including those of the shoulder, elbow, knee, hip, and jaw
Articulated skeleton

Procedure

Types of Joints

The four major groups of joints classified according to composition are bony, fibrous, cartilaginous, and synovial joints. Bony joints occur when two bones fuse to form one. Fibrous joints are typically composed of connective tissue fibers between two bones. They permit little movement. Cartilaginous joints consist of cartilage between two bones and are generally more movable than fibrous joints, although some cartilaginous joints may have no movement at all. Synovial joints have the most complex structure, including a joint capsule, an inner membrane, and synovial fluid, and they are the most movable of the joints (table 10.1).

Bony Joints In **bony joints,** or **synostoses** (SIN-oss-TOE-sees; singular, *synostosis*), two separate bones fuse and become immovable joints. Developing bones in young individuals fuse and become synostoses. As a person approaches the age of 35 or so, some of the sutures of the skull begin to fuse from the region closest to the brain toward the superficial surface of the skull. This fusion leads to the complete union of two bones. The two frontal bones, present as separate bones in the fetal skeleton, fuse as a synostosis and form the single frontal bone. When a growth plate fuses in a long bone, a synostosis forms.

Fibrous Joints **Fibrous joints** are those with two bones held together by connective tissue fibers; they may provide some movement or no movement. If the bones are bound closely together, the joint does not allow for movement between the bones. One example of this type of joint is called a **suture;** it occurs between adjacent bones in the cranium. In these joints the bones are tightly held together by dense fibrous connective tissue. Examine a skull in the lab and locate the sutures represented in figure 10.1. These sutures may be **serrated, lap,** or **plane sutures.**

TABLE 10.1 Classification of Joints

Bony Joints—Joints That Result from the Fusion of Two Bones	
Synostosis	Fusion of parietal bones in older individuals
Fibrous Joints—Joints Held Together by Collagenous Fibers	
Suture	Frontal and parietal bones
Gomphosis	Teeth and mandible
Syndesmosis	Distal tibia and distal fibula
Cartilaginous Joints (Synchondroses)—Joints Held Together by Cartilage	
Epiphyseal plate	Humeral head and shaft
Costal cartilages	Ribs and sternum
Symphysis	Joint between two vertebral bodies
Synovial Joints—Joints Enclosed by Synovial Capsule	
Plane	Between carpal bones
Hinge	Humerus and ulna
Pivot	Atlas and axis, radius and ulna
Condylar	Radius and scaphoid, atlas and occipital
Saddle	Trapezium and first metacarpal
Ball-and-socket	Acetabulum and femur

Another type of fibrous joint is a **gomphosis** (gom-FOE-sis; plural, *gomphoses*), which is the joint of the teeth in the alveoli of the maxilla and the mandible. This type of joint is like a peg in a socket. A gomphosis connects the bone of the jaw to the tooth by fibrous connective tissue called a **periodontal ligament.** Examine a jaw or complete skull in the lab and locate a gomphosis. Compare your specimen with figure 10.2.

A **syndesmosis** (SIN-dez-MO-sis; plural, *syndesmoses*) is a fibrous joint where the fibrous connective tissue is longer than in sutures or gomphoses, allowing for some movement. An example of a syndesmosis is the connection between interosseus membranes of the radius and ulna or the tibia and fibula. Examine an articulated skeleton in the lab and compare it with figure 10.3.

Cartilaginous Joints If bones are held together by cartilage, the articulation is known as a **cartilaginous joint.** Cartilaginous joints have cartilage between the bones and may be either immovable or slightly movable. The articulation is known as a **synchondrosis** (plural, *synchondroses*) if it consists of hyaline cartilage and a **symphysis** if it is composed of fibrocartilage. If the cartilage is thin between the bones, the joint is immovable (a synarthrosis). An example of this is the **epiphyseal plate** (fig. 10.4). This joint eventually fuses to form a single bone. Another kind of synchondrosis is the **costal cartilage** between the first rib and the sternum. The cartilage in this joint is longer than one in an epiphyseal plate, and the joint is more movable (amphiarthrotic), allowing the ribs limited movement. Examine these joints in lab and compare them with figure 10.5.

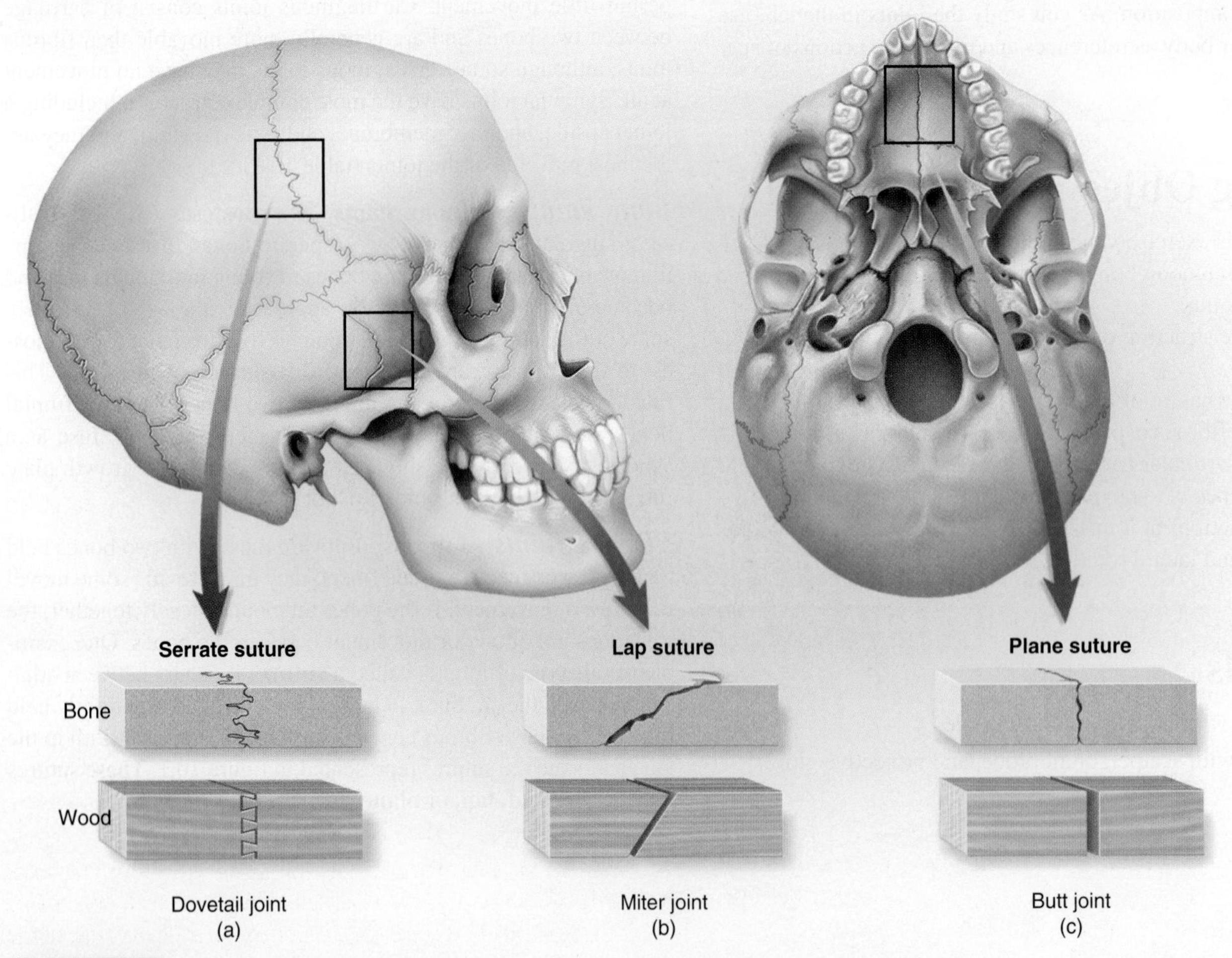

FIGURE 10.1 **Sutures** (Top) seen in position in the skull; (middle) a magnified view of the joint in bone; and (bottom) analogous wood joints.

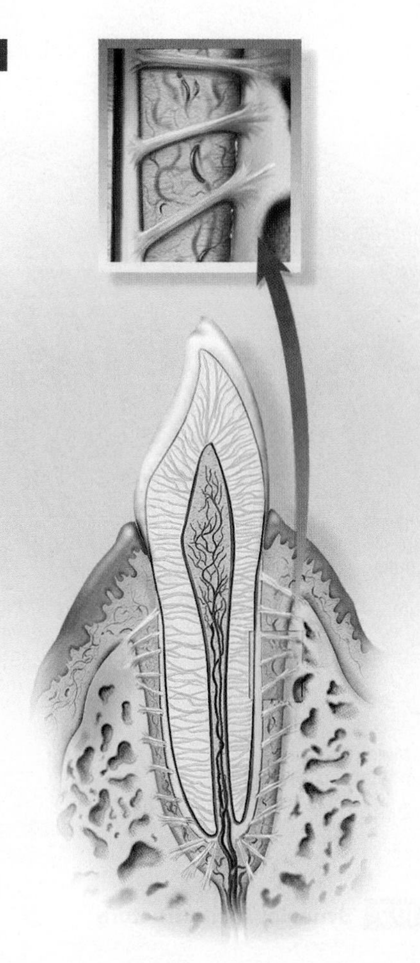

FIGURE 10.2 **Gomphosis**

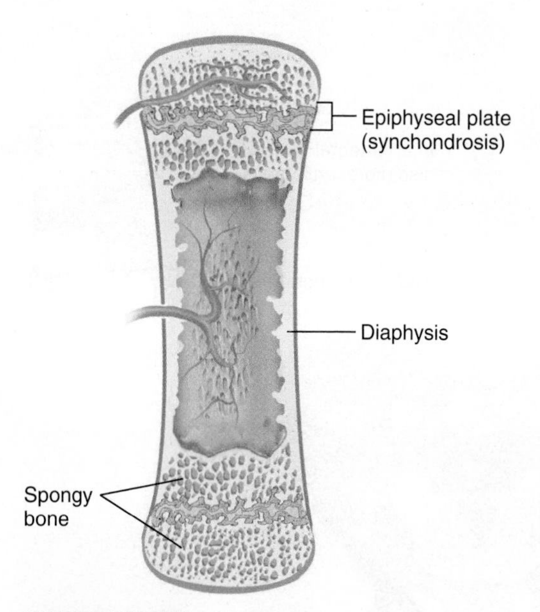

FIGURE 10.4 **Cartilaginous Joint** Epiphyseal plate.

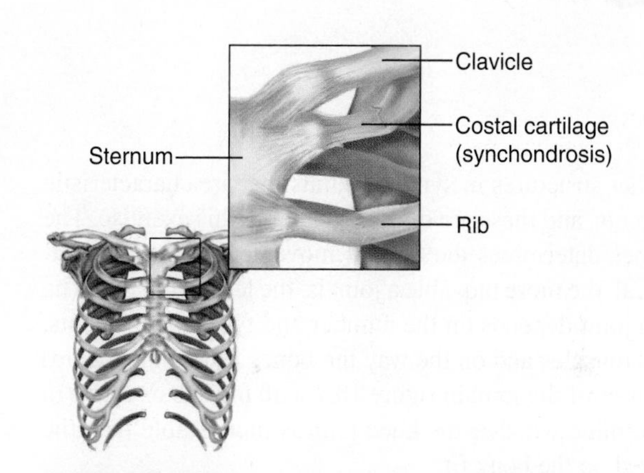

FIGURE 10.5 **A semimovable synchondrosis**

FIGURE 10.3 **Syndesmosis**

A **symphysis** (SIM-fih-sis) is a fibrocartilaginous pad and the two bones connecting it (for example, the pubic symphysis) or in the intervertebral discs. These joints allow for limited movement. The physical stress in a symphysis is greater than in other cartilaginous joints, such as the costal cartilages, and the joint reflects this with the presence of fibrocartilage. Fibrocartilage endures much greater stress than hyaline cartilage. Locate a symphysis in lab and compare it with figure 10.6.

Synovial Joints These joints consist of two bones in a capsule that contains synovial fluid. Synovial joints allow for extensive movement. The outer part of the synovial joint is the **joint capsule,** which is made of an outer **fibrous capsule** and an inner **synovial membrane.** The synovial membrane secretes **synovial fluid,** a lubricating liquid that reduces friction inside the joint. The space inside the joint is called the **synovial cavity,** and each bone of the joint ends in a hyaline cartilage cap called the **articular cartilage.**

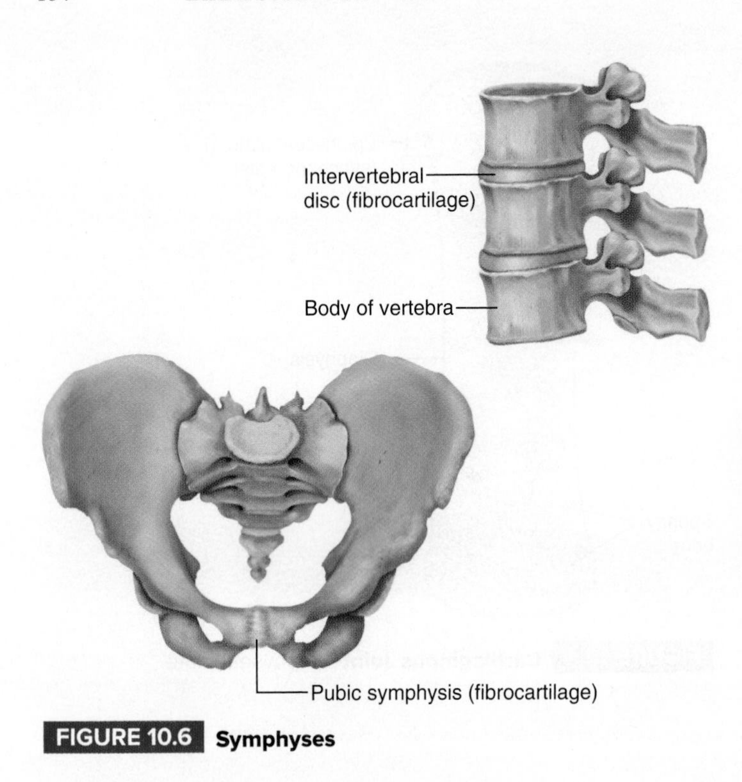

FIGURE 10.6 **Symphyses**

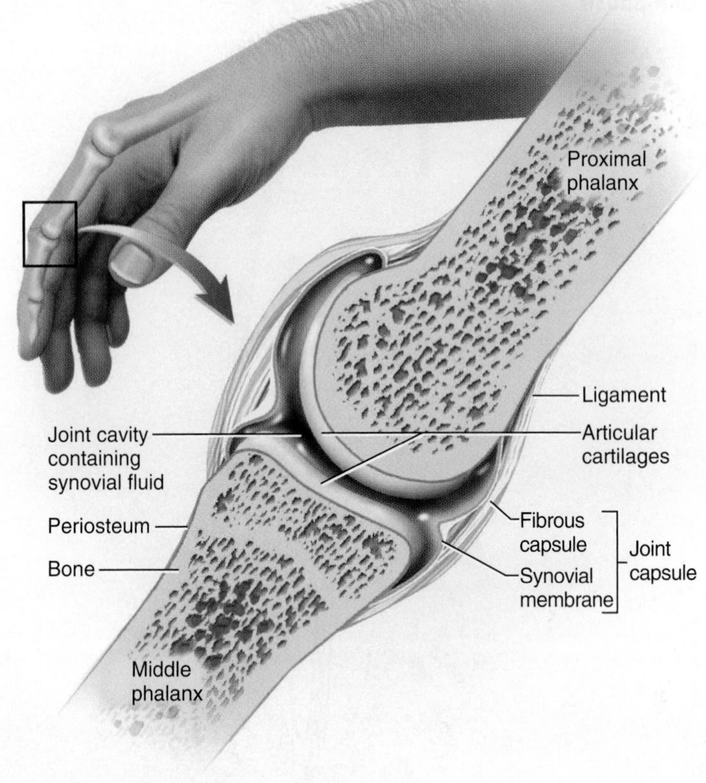

FIGURE 10.7 **Synovial Joint Structure**

There are other structures in synovial joints that are characteristic of specific joints, and these are discussed later in this exercise. The shape of bones determines the type of movement at the articulation. In general, the more movable a joint is, the less stable it is. The stability of a joint depends on the number and types of ligaments, tendons, and muscles and on the way the bones fit together. Compare the features of the joint in figure 10.7 with models or charts in the lab. Determine whether the knee joint is more stable than the hip joint based on the bony fit.

Dissection of a Synovial Joint To understand the structure of a synovial joint, examine a fresh or recently thawed vertebrate joint such as a beef or chicken joint. Wear protective gloves as you dissect the joint, and wash your hands thoroughly with soap and water after the dissection. Place the joint in front of you on a dissecting tray, and cut into the joint capsule with a scalpel. Note the tough, white material that surrounds the joint. This is the joint capsule, and it may be fused with ligaments that bind the bones of the joint together. Notice the synovial fluid, which is a slippery substance that provides a slick feel to the inside of the capsule, reducing friction between the bones. Once you have cut into the joint, examine the articular cartilage on the ends of the bones. *Carefully* cut into this cartilage with a scalpel and notice how the material chips away from the bone. When you have finished with the dissection, make sure to rinse off the dissection equipment, dispose of the joint in the appropriate animal waste container, and wash your hands.

Other Synovial Structures Bursae and tendon sheaths are synovial structures. **Bursae** (singular, *bursa*) are small synovial sacs between tendons and bones or other structures. The bursae cushion the tendons as they pass over the other structures. **Tendon sheaths** are synovial structures encircling tendons. There are numerous tendon sheaths in the palm of the hand. Tendon sheaths protect the tendons as they slide past one another. Examine figure 10.8 for these structures.

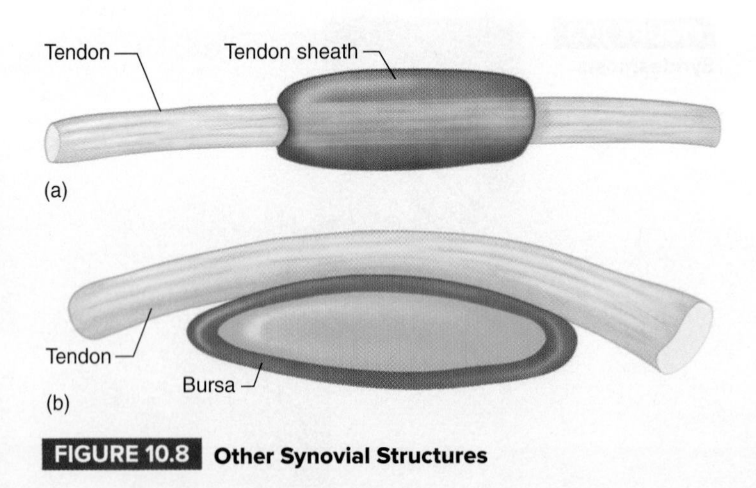

FIGURE 10.8 **Other Synovial Structures**

Joints Classified by Movement

Joints are classified as immovable, semimovable, or freely movable. Immovable joints are known as **synarthrotic joints.** The bones in these joints are tightly bound by connective tissue. Synarthrotic joints may be fibrous or cartilaginous. In fibrous joints, collagenous fibers bind the bones. In cartilaginous joints, hyaline cartilage joins the two bones. Semimovable joints are known as **amphiarthrotic joints,** and they may be fibrous or cartilaginous. The cartilage of amphiarthrotic joints may be hyaline cartilage or fibrocartilage. Freely movable joints are **diarthrotic joints,** and they are always synovial joints.

Synovial Joints Classified by Movement Synovial joints are classified according to the type of movement they allow between the articulating bones. The joints are listed here in the general order of least movable to most movable. Examine the joints on an articulated skeleton in lab and compare them with figure 10.9.

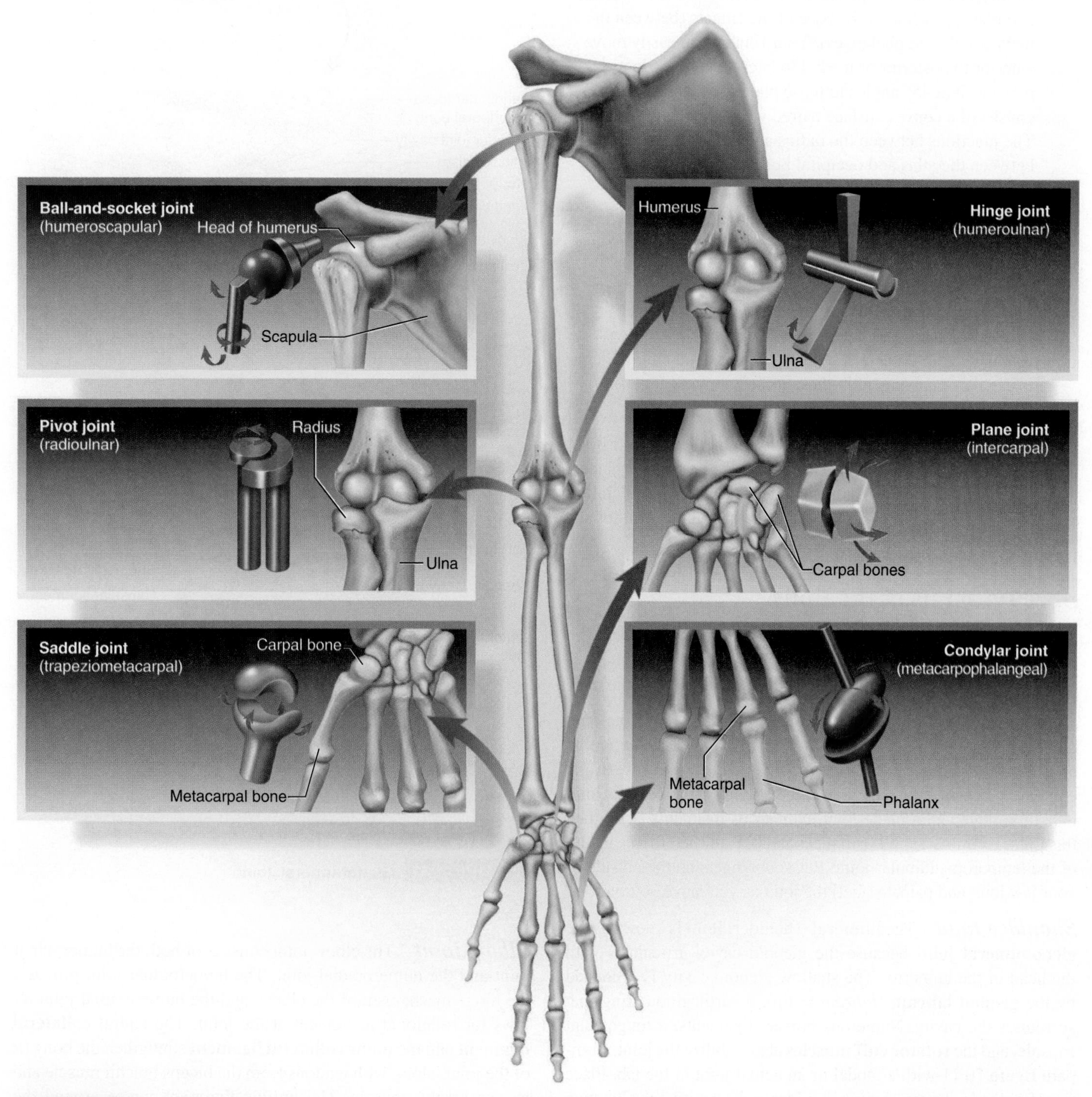

FIGURE 10.9 **Synovial Joints** Six different synovial joints are unique due to their bony fit.

1. **Plane joints** allow for movement between two flat surfaces, such as between the superior and inferior facets of adjacent vertebrae or intertarsal or intercarpal joints.
2. **Hinge joints** allow for angular movement, such as in the elbow, in the knee, or between the phalanges of the fingers. You can increase or decrease the angle of the two bones with this joint.
3. **Pivot joints** allow for rotational movement between the atlas and the axis when a person is moving the head to indicate "no." They are also located at the proximal radius and ulna.
4. **Condylar** (ellipsoid) **joints** allow significant movement in two planes, such as at the base of the fingers (between the metacarpals and phalanges). Your fingers can easily move anterior to posterior or medial to lateral, yet they do not move well at 45° angles to these planes. Condylar joints consist of a convex surface paired with a concave surface. The junctions between the radius and scaphoid bone and between the atlas and occipital bone are good examples of condylar joints.
5. **Saddle joints** have two concave surfaces that articulate with each other. An example of a saddle joint is between the trapezium and the first metacarpal of the thumb. This provides for greater movement in the thumb than the condylar joint of the wrist.
6. **Ball-and-socket joints** consist of a spherical head in a round concavity, such as in the shoulder and the hip. There is extensive movement in these joints, yet they are less stable due to the freedom of movement they afford.

Monaxial joints move in only one plane. Hinge and pivot joints are monaxial joints. **Biaxial joints** move in two planes. Condylar and saddle joints are biaxial joints. **Multiaxial joints** move in many planes. Ball-and-socket joints are multiaxial joints.

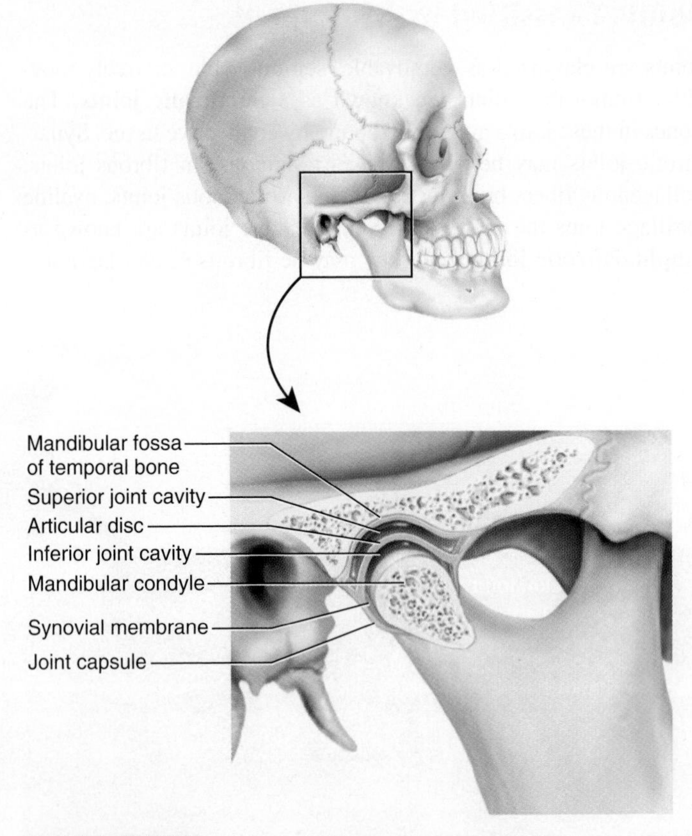

FIGURE 10.10 **Temporomandibular Joint (TMJ)**

Specific Joints of the Body

There are several types of specific joints in the body; some are primarily hinge joints, such as the jaw, elbow, and knee, while others are ball-and-socket joints, such as the shoulder and hip joints.

Jaw Joint The **temporomandibular joint (TMJ)** is the only diarthrotic joint of the skull (fig. 10.10). This joint has an **articular disc,** a pad of fibrocartilage that provides a cushion between the condylar process of the mandible and the temporal bone. This joint is both a hinge and a plane joint. Numerous ligaments strengthen this joint. Examine a skull with the mandible attached for the nature of the temporomandibular joint. Put your fingers on the outside of your jaw joint and palpate (feel) the joint as you move your jaw.

Shoulder Joint The **humeral** (shoulder) **joint** is known as the **glenohumeral joint** because the glenoid cavity articulates with the head of the humerus. The shallow glenoid cavity is deepened by the **glenoid labrum** (*labrum* = lip), a cartilaginous ring that surrounds the cavity. Numerous bursae, ligaments, a tough joint capsule, and the **rotator cuff muscles** also stabilize the joint. Compare figure 10.11 with a model or an actual joint in the lab. Place your fingers on the top of your shoulder and determine the location of the shoulder joint as you elevate your arm.

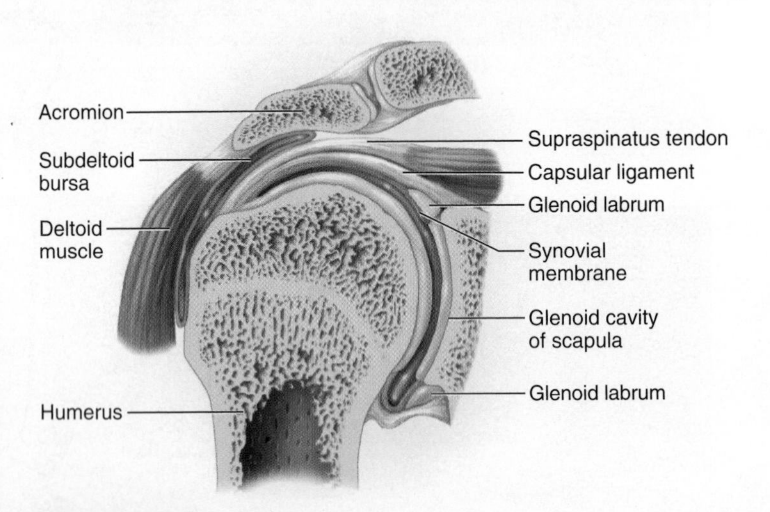

FIGURE 10.11 **Glenohumeral Joint**

Elbow Joint The elbow joint consists of both the humeroulnar joint and the humeroradial joint. The humeroulnar joint provides the hinge mechanism of the elbow, and the humeroradial joint allows for rotational movement at the joint. The **radial collateral ligament** and the **ulnar collateral ligament** strengthen the bony fit of the joint, along with tendons from the biceps brachii muscle and triceps brachii muscle. The **anular ligament** wraps around the head of the radius and is subject to dislocation when the forearm is

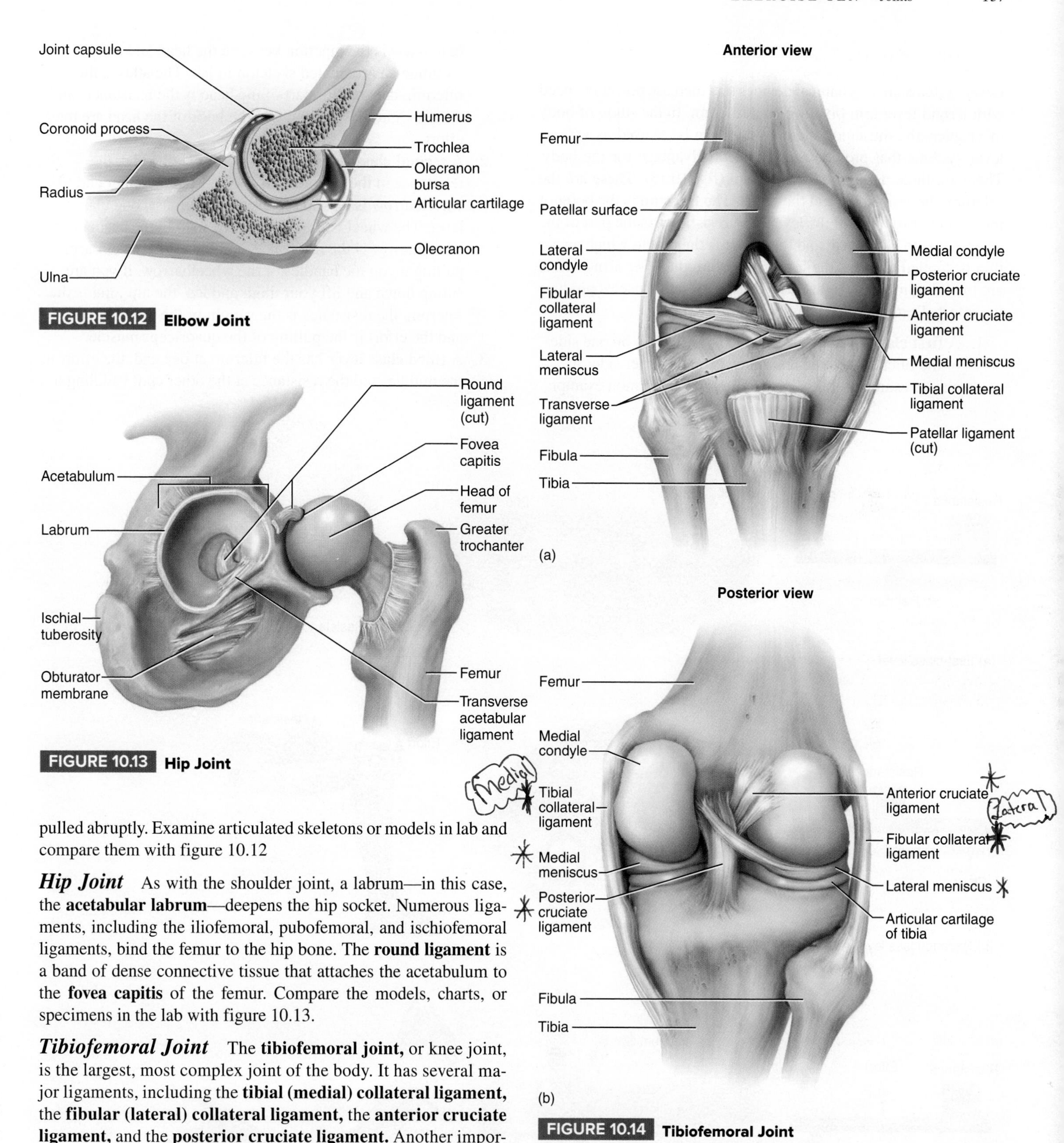

FIGURE 10.12 Elbow Joint

FIGURE 10.13 Hip Joint

FIGURE 10.14 Tibiofemoral Joint

pulled abruptly. Examine articulated skeletons or models in lab and compare them with figure 10.12

Hip Joint As with the shoulder joint, a labrum—in this case, the **acetabular labrum**—deepens the hip socket. Numerous ligaments, including the iliofemoral, pubofemoral, and ischiofemoral ligaments, bind the femur to the hip bone. The **round ligament** is a band of dense connective tissue that attaches the acetabulum to the **fovea capitis** of the femur. Compare the models, charts, or specimens in the lab with figure 10.13.

Tibiofemoral Joint The **tibiofemoral joint,** or knee joint, is the largest, most complex joint of the body. It has several major ligaments, including the **tibial (medial) collateral ligament,** the **fibular (lateral) collateral ligament,** the **anterior cruciate ligament,** and the **posterior cruciate ligament.** Another important structure is the **patellar tendon,** which runs from the quadriceps femoris muscle to the tibial tuberosity. The **patellar ligament** is a specific connection between the patella and the tibial tuberosity. The other major structures of the tibiofemoral joint are the **medial** and **lateral menisci** (singular, *meniscus*), wedge-shaped pads of fibrocartilage that provide a cushion between the femur and tibia. The knee is primarily a hinge joint with a little lateral movement allowed. Compare specimens in the lab with figure 10.14. Palpate the tibial collateral ligament, the fibular collateral ligament, and the patellar tendon on yourself.

Lever Systems

Lever systems are mechanical devices that increase power or speed with a rigid lever arm pivoting on a fulcrum. In the study of body mechanics, the musculoskeletal system can be viewed as various lever systems that provide a mechanical advantage for the body. There are three parts to a lever system (fig. 10.15). These are the fulcrum, the resistance, and the effort. The **fulcrum** is the point of movement or rotation of the lever. The **resistance** is the part of the lever that is to be moved, and the **effort** is the part to which an action is applied to move the lever. Bones are the lever arms, joints are the fulcrum, the weight is the resistance, and muscles produce the effort. There are three classes of levers:

1. A **first class lever** is one in which the effort is on one side of the fulcrum and the resistance is on the other side. A classic example of this is a seesaw, and a common example in humans is the junction between the head and atlas. Examine an articulated skeleton in lab. The atlas is the fulcrum, the anterior part of the head is the resistance, and the neck muscles attaching to the back of the head are the effort.
2. A **second class lever** has the fulcrum at one end, the resistance in the middle, and the effort at the other end. A wheelbarrow is a good representative of a second class lever. The wheel is the fulcrum, the load in the basin of the wheelbarrow is the resistance, and the effort is your arms pulling up on the handles of the wheelbarrow. If you are sitting down and lift your thigh and leg, the hip joint is the fulcrum, the resistance is the weight of the thigh and leg, and the effort is the pulling of the quadriceps muscle.
3. A **third class lever** has the fulcrum at one end, the effort in the middle, and the resistance at the other end. Paddling a

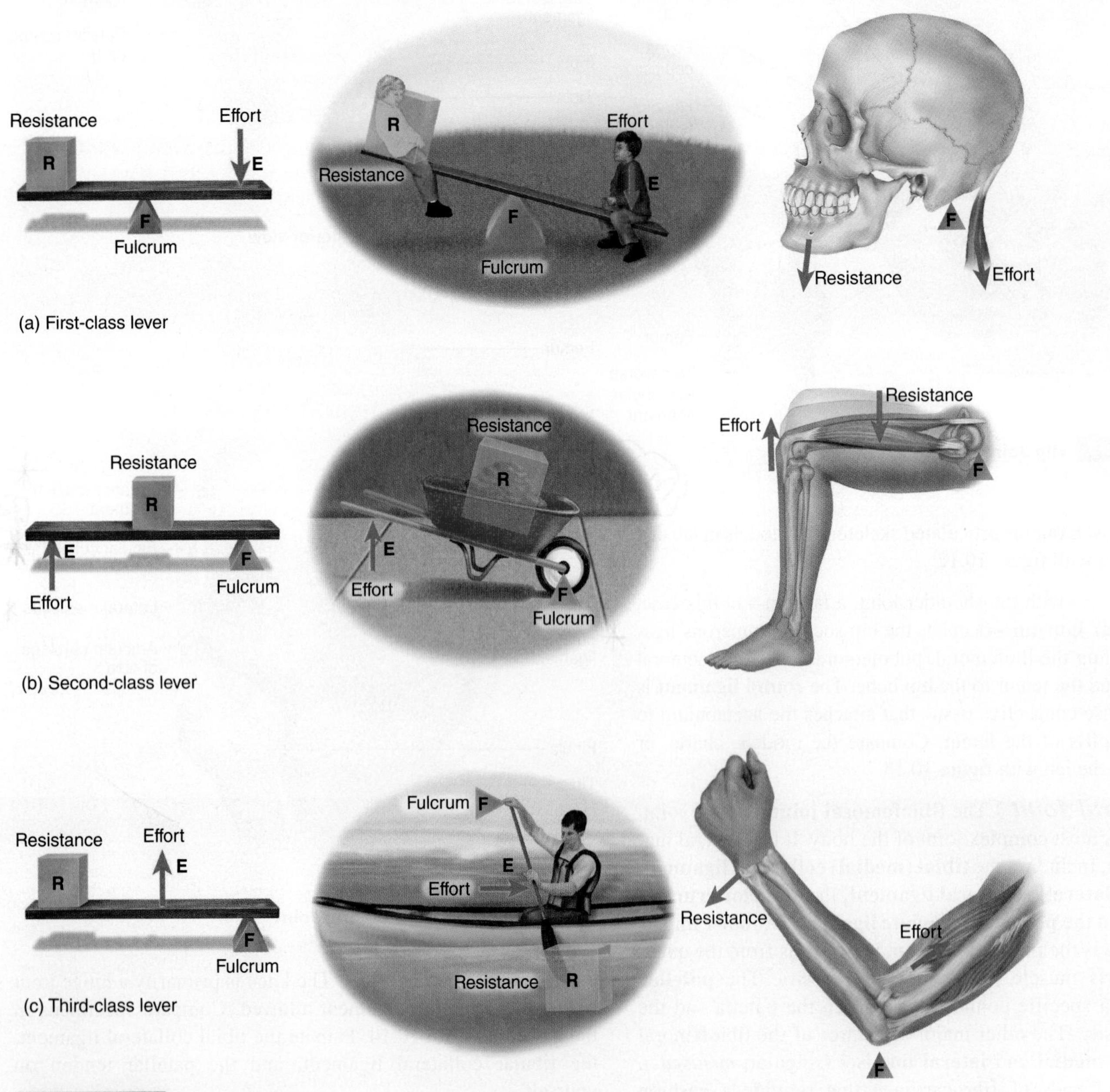

FIGURE 10.15 **Lever Systems**

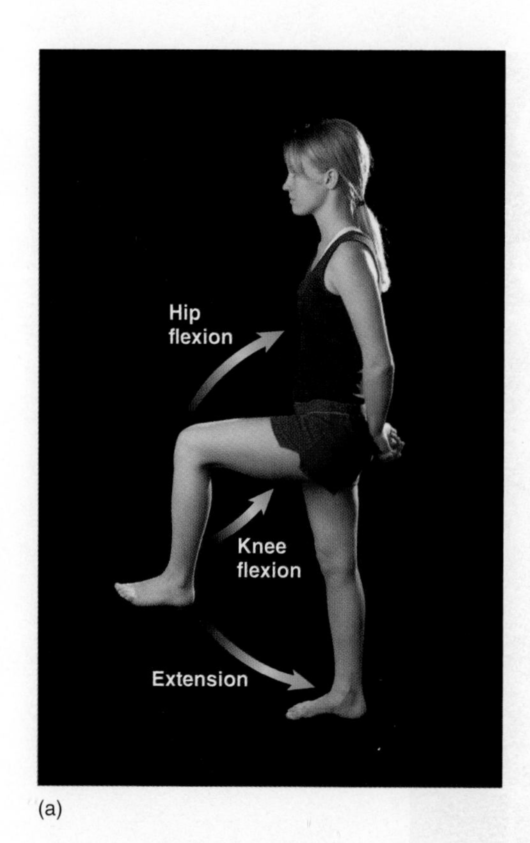

(a)

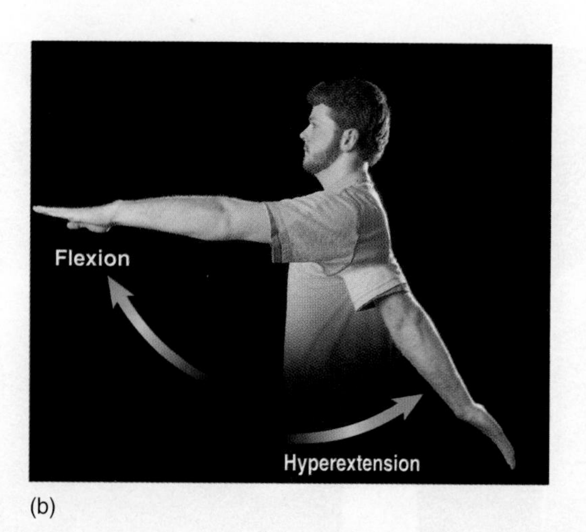

(b)

FIGURE 10.16 Flexion, Extension, and Hyperextension of Selected Joints (a) Flexion and extension of the hip and knee; (b) flexion and hyperextension of the arm.

canoe is an example of a third class lever. Most of the joints of the body involve third class levers and a common example is the elbow joint. The fulcrum is the elbow, the effort is the biceps brachii muscle, and the resistance is the weight of the forearm. Examine this joint in an articulated skeleton.

*Movement at Joints

Many kinds of movements occur at joints. These movements, controlled by muscles, are called actions. Actions can decrease a joint angle, increase a joint angle, and cause rotation at a joint, among other movements. The specific types of action at a joint are as follows.

Flexion is a decrease in the joint angle from anatomical position. If you bend your elbow, you are flexing your forearm. Flexion of the thigh is in the anterior direction, yet flexion of the leg is in the *posterior* direction. Bending forward at the waist is flexion of the vertebral column. Looking at your toes is flexion of the head. Examine figure 10.16 for examples of flexion.

Extension is a return to anatomical position of a part of the body that was flexed. If you are looking at your toes and lift your head back to anatomical position, you are extending your head. If you straighten your knee after it is bent (flexed), you are extending the leg. Examine figure 10.16 for examples of extension. Extension of the part of the body beyond anatomical position is known as **hyperextension.** When you are about to roll a bowling ball and your arm reaches the very back of the arc, you are hyperextending your arm.

Abduction (*abduct* = to take away) is movement of the limbs in the coronal plane away from the body. Abducting the fingers is spreading the fingers apart in the frontal plane. Abduction is taking away a part of the body in a lateral direction, as seen in figure 10.17.

(a) Abduction

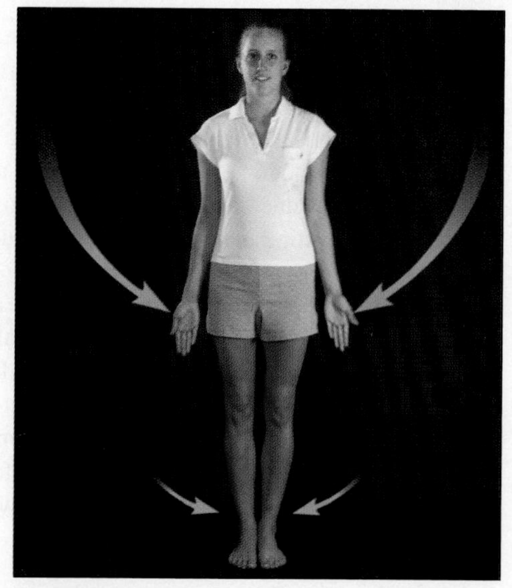

(b) Adduction

FIGURE 10.17 Abduction and Adduction of the Arms and Thighs

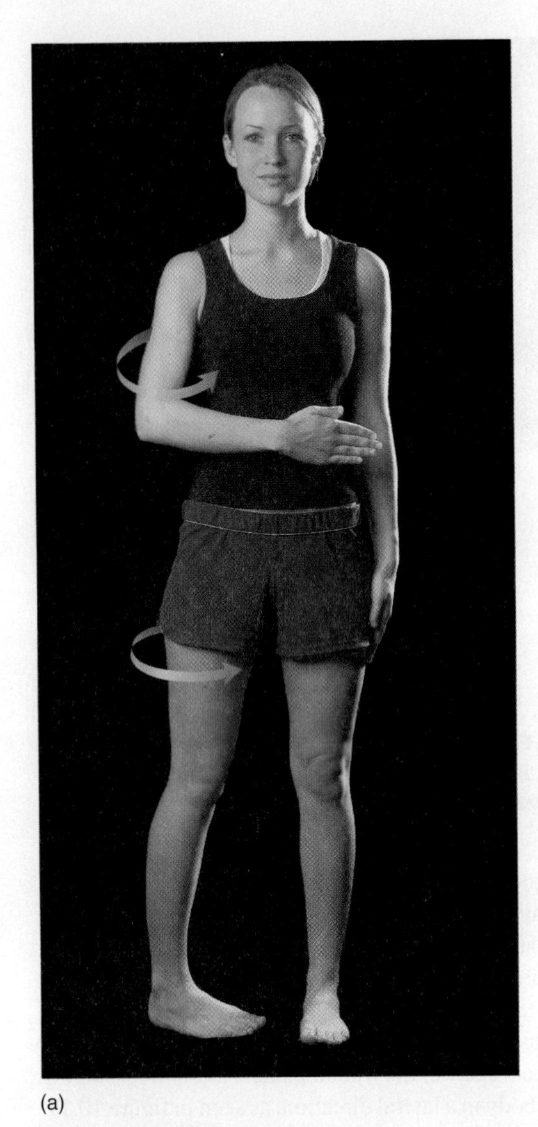

(a)

(b)

FIGURE 10.18 Rotation of Joints (a) Medial rotation; (b) lateral rotation.

Adduction is the return of the part of the body to anatomical position after abduction. When your fingers are straight and together, they are adducted. In doing "jumping jacks," you are abducting and adducting in series. Think of adduction as "adding" a limb back to the body, as seen in figure 10.17.

Rotation is the circular movement of a part of the body around an axis, as seen in figure 10.18. **Lateral rotation** of a limb moves the anterior surface of the limb toward the lateral side of the body. **Medial rotation** of a limb turns the anterior surface of the limb toward the midline. The trunk and neck may also be rotated.

Supination is lateral rotation of the forearm and consequently the hand. The hands are supinated when the body is in anatomical position.

Pronation is medial rotation of the forearm and consequently the hand as well. When you turn your palms posteriorly from anatomical position, you are pronating your hands.

Circumduction is the movement of a muscle in a conical shape, with the point of the cone being proximal.

Elevation is movement in a superior direction. Elevation occurs when you shrug your shoulders or close your mouth.

Depression is the opposite of elevation. It is movement in the inferior direction. Opening your mouth is depression of the mandible. Elevation and depression are seen in figure 10.19.

Protraction is a horizontal movement in the anterior direction, as in jutting the chin forward.

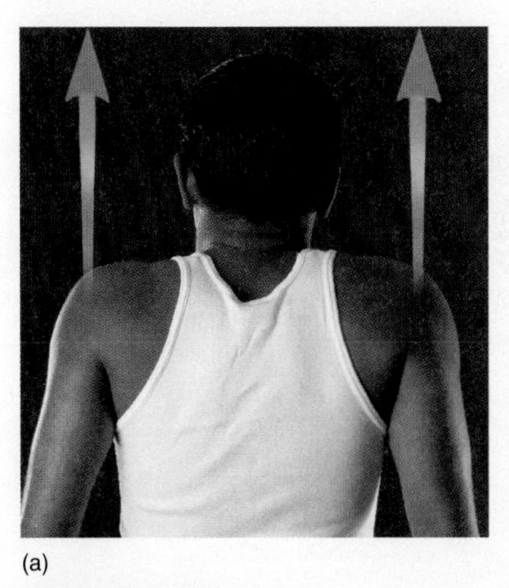

(a)

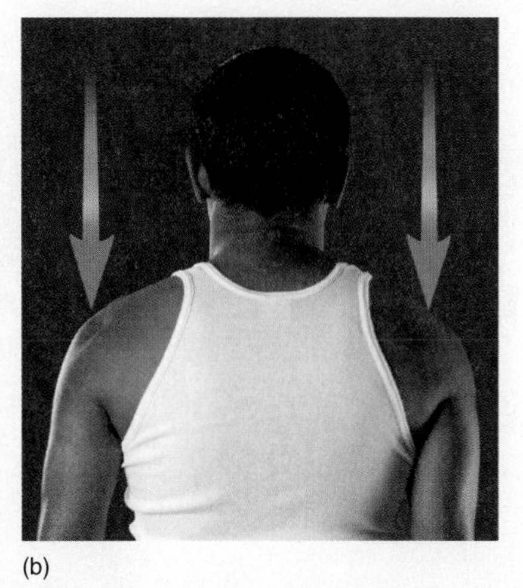

(b)

FIGURE 10.19 **Elevation and Depression of the Shoulders** (a) Elevation; (b) depression.

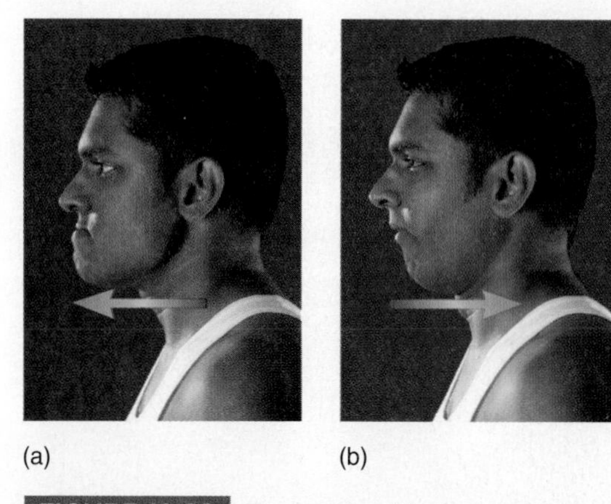

(a) (b)

FIGURE 10.20 **Protraction and Retraction of the Mandible** (a) Protraction; (b) retraction.

Retraction is the reverse of protraction. A jaw that moves from anterior to posterior is retracted. These two actions are illustrated in figure 10.20.

Inversion is movement of the feet—turning the soles of the feet medially so they face each other. Sometimes inversion of the foot is known as *supinating* the foot.

Eversion means turning the soles of the feet laterally. This is also known as *pronating* the foot. Inversion and eversion can be seen in figure 10.21.

Fixing a muscle prevents motion in either direction. This is done with opposing muscles contracting simultaneously. The muscle that has the main force on a joint is called the **prime mover,** or **agonist.** Muscles that assist with the prime mover are **synergists,** while those that oppose the muscle are **antagonists.**

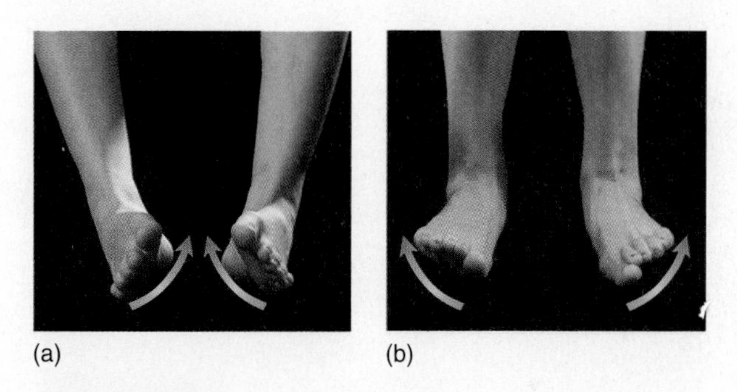

(a) (b)

FIGURE 10.21 **Inversion and Eversion of the Feet** (a) Inversion; (b) eversion.

Don't worry about levers

Look @ conversions
Canvas practice page

Review

Exercise 10

Joints

Name ______________________ Date __________

1. The study of articulations, or joints, is known as ______________________.
2. In what kind of joint (based on joint structure) are the bones held together by collagenous fibers? ______________________.

____

3. When two bones fuse into a single bone, this union is called a(n) ______________________.
4. The teeth are held into the jaw by what kind of joint? ______________________

____

5. What kind of joint is found between the distal tibia and fibula? ______________________

____

6. Bones held together by cartilage (cartilaginous joints) are also known as ______________________ joints.
7. The epiphyseal plate is a cartilaginous joint. In terms of movement, it is also called a(n) ______________________.
8. Label the following illustration, using the terms provided.

   articular cartilage
   bone
   joint capsule
   synovial cavity
   synovial membrane

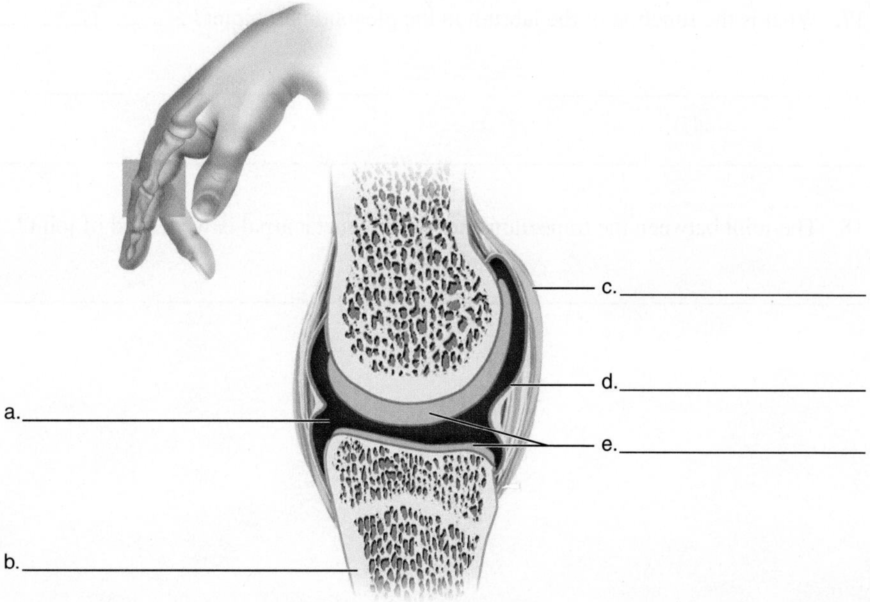

a. ______________________

b. ______________________

c. ______________________

d. ______________________

e. ______________________

9. What is the name of a joint that is held together by a joint capsule and contains a slippery fluid? ______________________________

____

10. Synovial fluid is secreted by what structure? ______________________________

____

11. How much movement occurs in a suture between two bones of the skull? ______________________________

____

12. Rank the following joints in terms of least movable to most movable, with 1 being the least movable and 5 being the most movable.

plane ________ saddle ________ suture ________ syndesmosis ________ ball-and-socket ________

13. Which one of the following joints has the greatest range of movement? (Circle the correct answer.)

a. gomphosis b. suture c. synchondrosis d. hinge

14. In which of these joints would you find a meniscus? (Circle the correct answer.)

a. cartilaginous b. fibrous c. synovial

15. What is the function of the meniscus in the knee? ______________________________

____

____

16. Match the joint in the left column with the type of joint in the right column.

_____ acetabulofemoral	a. ball-and-socket
_____ intercarpal	b. condylar
_____ radiocarpal	c. hinge
_____ tibiofemoral	d. plane

17. What is the function of the labrum in the glenohumeral joint? ______________________________

____

____

18. The joint between the trapezium and the first metacarpal is what kind of joint? ______________________________

____

19. The joint between the femur and the tibia is what type of joint? ______________________

____

20. What kind of joint is found at the wrist (between the radius and the carpal bones)? ______________________

____

Next test

Axial Muscles 1: Muscles of the Head and Neck

Muscular System

Introduction

Laboratory exercises 11 to 14 focus on skeletal muscles. There are more than 600 skeletal muscles in the human body. The muscles are grouped according to their location, with most muscles occurring as pairs. In the following exercises, you will learn about some of the major muscles of the body. The attachment points, action, and innervation are listed for about 100 muscles in this lab manual. Your instructor may wish to customize the list, so that you learn specific muscles or specific things about each muscle. The study of muscles can be both fun and challenging. Begin your study of muscles early and do not wait to learn them until the night before the lab exam!

Skeletal muscle is voluntary muscle. It contracts when consciously stimulated by specific nerves. Muscles pull on bones or other muscles. They do not push. The muscular system functions in movement, maintenance of posture, generation of heat (shivering), and compression of the abdomen, among other functions. In this exercise you are introduced to some muscles of the axial skeleton, starting with the muscles of the head and neck. These muscles can be grouped into a few functional classifications. Some of the head muscles provide the strength for chewing food or help manipulate the food in the mouth for chewing and swallowing. Other muscles control facial expression or the closing of the eyes or mouth. The neck muscles are those that move the neck or head or those that move the hyoid, larynx, or tongue during speech or swallowing.

Learning Objectives

At the end of this exercise you should be able to

1. locate the muscles of the head and neck on a torso model, chart, or cadaver (if available);
2. list the attachment points and action of each muscle presented;
3. describe what nerve controls (innervates) each muscle;
4. name all the muscles that have an action on a joint, such as all the muscles that flex the head; and
5. reproduce the actions of select muscles on your own head or neck.

Materials

Human torso model and head and neck model
Human muscle charts
Articulated skeleton
Cadaver (if available)

Procedure

The study of human musculature is seen as a daunting task by some students, while other students recall the muscle section of the course as their favorite part of the study of human anatomy. Studying muscles is more satisfying if attainable goals are set. This is best accomplished by choosing small numbers of muscles to study at a sitting and using flash cards as study aids. It is vital that you know the bones and bony markings before studying muscles. Look over the exercises on the skeletal system if you need to review the skeleton. You may also want to make a quick sketch of bones and how muscles attach to them to help you visualize points of attachment. A thorough study of muscles involves knowing not only the name of the muscle but also the attachment points, action, and innervation. Traditionally muscles have been listed by their **origin,** which was perceived as the most stable part of the muscle, and the **insertion,** the more movable attachment of the muscle. Since some muscles have movable attachment points (depending on what part of the body is stable), some authors are moving away from the concept of the origin and the insertion. For the purpose of this lab manual, the origins of the muscle are listed as attachment point 1 and the insertions are listed as attachment point 2 so that either can be used per the wishes of your instructor.

Examine a model or chart of the human musculature and cadaver (if available) as you read the following descriptions. These muscles are listed in table 11.1. Examine a skull or an articulated skeleton and review the bony markings as you study the origins and insertions of the muscles in this exercise. Once you know the attachment points, the actions should be more comprehensible. Each muscle has an **action,** which is the effect the muscle has on a part of the body (such as *flexion* of the head). There are two schools of thought on action. One is that the action has an impact on a joint, such as flexion of the wrist. The other is that an action moves the region beyond a joint, such as flexion of the hand. This lab manual will mostly follow the former. In addition to moving a joint, muscles can fix a joint. Fixing means to prevent motion in either direction. This is done with opposing muscles contracting simultaneously. The muscle that has the main force on a joint is called the prime mover. Muscles that assist the prime mover are synergists, while those that oppose the muscle are antagonists. Finally, the

TABLE 11.1 Axial Muscles 1

Muscles of Facial Expression

Name	Attachment 1 (Origin)	Attachment 2 (Insertion)	Action	Innervation
Frontalis	Galea aponeurotica	Subcutaneous tissue of eyebrows	Raises eyebrows, draws scalp anteriorly	Facial nerve (VII)
Occipitalis	Occipital and temporal bone	Galea aponeurotica	Retracts scalp	Facial nerve (VII)
Orbicularis oculi	Frontal and maxilla on medial margin of orbit	Skin of eyelid	Closes eyelid	Facial nerve (VII)
Corrugator supercilii	Medial portion of frontal bone	Skin of eyebrows	Adducts eyebrows (pulls them medially and inferiorly)	Facial nerve (VII)
Orbicularis oris	Modiolus of mouth	Skin of lips	Closes and protrudes lips	Facial nerve (VII)
Levator labii superioris	Maxilla and zygomatic bones	Muscles of upper lips	Elevates upper lip, flares nostril	Facial nerve (VII)
Zygomaticus (major, minor)	Zygomatic bone	Superolateral angle of mouth and muscles of upper lip	Elevates corners of mouth (in smiling and laughing)	Facial nerve (VII)
Risorius	Fascia of masseter zygomatic arch	Modiolus at angle of mouth	Abducts corner of mouth (draws edge of mouth laterally)	Facial nerve (VII)
Depressor labii inferioris	Mandible	Muscles and skin of lower lip	Depresses lower lip	Facial nerve (VII)
Mentalis	Anterior mandible	Skin of chin below lower lip	Protrudes lower lip	Facial nerve (VII)
Buccinator	Maxilla and mandible near molar teeth	Orbicularis oris, lips	Compresses cheek	Facial nerve (VII)
Platysma	Fascia covering pectoralis major and deltoid	Mandible and skin of lower region of face	Depresses lower lip and angle of mouth, opens jaw	Facial nerve (VII)

Muscles of Chewing and Swallowing

Name	Attachment 1 (Origin)	Attachment 2 (Insertion)	Action	Innervation
Temporalis	Temporal fossa	Coronoid process and ramus of mandible	Elevates (closes) mandible	Trigeminal nerve (V)
Masseter	Zygomatic arch	Angle and ramus of mandible	Elevates (closes) mandible	Trigeminal nerve (V)
Pterygoids (medial, lateral)	Pterygoid processes of sphenoid bone	Medial ramus of mandible	Medial excursion of mandible (for chewing)	Trigeminal nerve (V)
Digastric	Interior, distal margin of mandible, mastoid notch of temporal bone	Hyoid bone	Elevates hyoid and mandible	Trigeminal (V) and facial (VII) nerves
Mylohyoid	Inferior margin of mandible	Hyoid bone	Elevates hyoid and tongue	Trigeminal nerve (V)
Omohyoid	Superior border of scapula	Hyoid bone	Depresses hyoid	Spinal nerves C1–3
Sternohyoid	Manubrium of sternum	Hyoid bone	Depresses hyoid	Spinal nerves C1–3
Sternothyroid	Manubrium of sternum	Thyroid cartilage of larynx	Depresses thyroid cartilage	Spinal nerves C1–3

Muscles Acting on the Head

Name	Attachment 1 (Origin)	Attachment 2 (Insertion)	Action	Innervation
Sternocleidomastoid	Sternum, clavicle	Mastoid process of temporal	Abducts, rotates, and flexes head	Accessory nerve (XI) Spinal nerves C2–3
Scalenes (anterior, middle, posterior)	Transverse process of all cervical vertebrae	Ribs 1 and 2	Flexes and rotates neck, elevates ribs 1 and 2	Spinal nerves C3–8
Trapezius	Posterior occipital bone Nuchal ligament, C7-T4	Clavicle, acromion, spine of scapula	Extends and abducts head, rotates and adducts scapula, fixes scapula	Accessory nerve (XI) Spinal nerves C2–4

innervation of a muscle is the specific nerve that controls it. Nerves are important in that if the motor portion of a nerve is damaged, then the muscle innervated by that nerve cannot function.

In this exercise you are introduced to some muscles of the axial skeleton, starting with muscles of facial expression, muscles of chewing and swallowing, and muscles acting on the head (moving the head in particular directions). Examine the charts, models, or cadavers in the lab; read the material presented in this lab manual; and compare the illustrations in this book with the material in the lab.

Muscle Nomenclature

Another aid to learning muscles is to understand how they are named. Muscles are named by a number of criteria: Location: Some muscles are named by their location, such as the tibialis anterior, the frontalis, or the temporalis.

Size: The adductor magnus is the largest of the adductor muscles. The gluteus minimus is the smallest of the gluteal muscles.

Shape: The deltoid muscle is shaped like a delta, or triangle.

Orientation: The transversus abdominis muscle has fibers that are oriented horizontally, or in a transverse direction.

Attachments: The infraspinatus muscle attaches to the infraspinous fossa of the scapula, while the flexor hallucis longus muscle attaches to the hallux, or big toe.

Number of heads: The triceps brachii muscle has three heads.

Action: The extensor digitorum is a muscle that extends the fingers.

Length: Muscles with the term *longus* or *brevis* are named for the overall length of the muscle.

Muscles of Facial Expression

Humans are social animals and communicate in many different ways including moving parts of the face such as sneering, frowning, or smiling. These can be obvious movements such as opening the mouth widely when laughing or more subtle such as the slight elevation of an eyebrow. The muscles in this exercise are described below, and the specifics of each muscle are listed in table 11.1.

The **frontalis** (fron-TAL-is) muscle attaches to a broad, flat tendinous sheet on the superior aspect of the skull known as the **galea aponeurotica** (GALE-ee-uh AP-oh-nu-ROT-ih-KAH). Another attachment of the frontalis is on the subcutaneous tissue deep to the eyebrow. The **occipitalis** (AUK-sip-ih-TAL-us) attaches to the posterior part of the galea aponeurotica. If both muscles contract, the eyebrows are raised. If just the frontalis contracts, the scalp is brought forward, as in frowning. These muscles are sometimes listed as one muscle, the **occipitofrontalis.**

The **orbicularis oculi** is a sphincter muscle that closes the eye. Sphincter muscles act as the strings of a drawstring purse. The orbicularis oculi attaches to the medial portion of the orbit and on the eyelid. The muscles ring the eye and close the eyelids. Examine these muscles in figures 11.1 to 11.3.

The **corrugator supercilii** (SOO-pur-SIL-ee-ee) muscle attaches to the frontal bone between the eyebrows and inserts laterally. It furrows the eyebrows as in frowning. Examine this muscle in lab and compare it to figures 11.2 and 11.3.

The **orbicularis oculi** attaches to the corners of the mouth in a complex arrangement of muscles and connective tissue known as

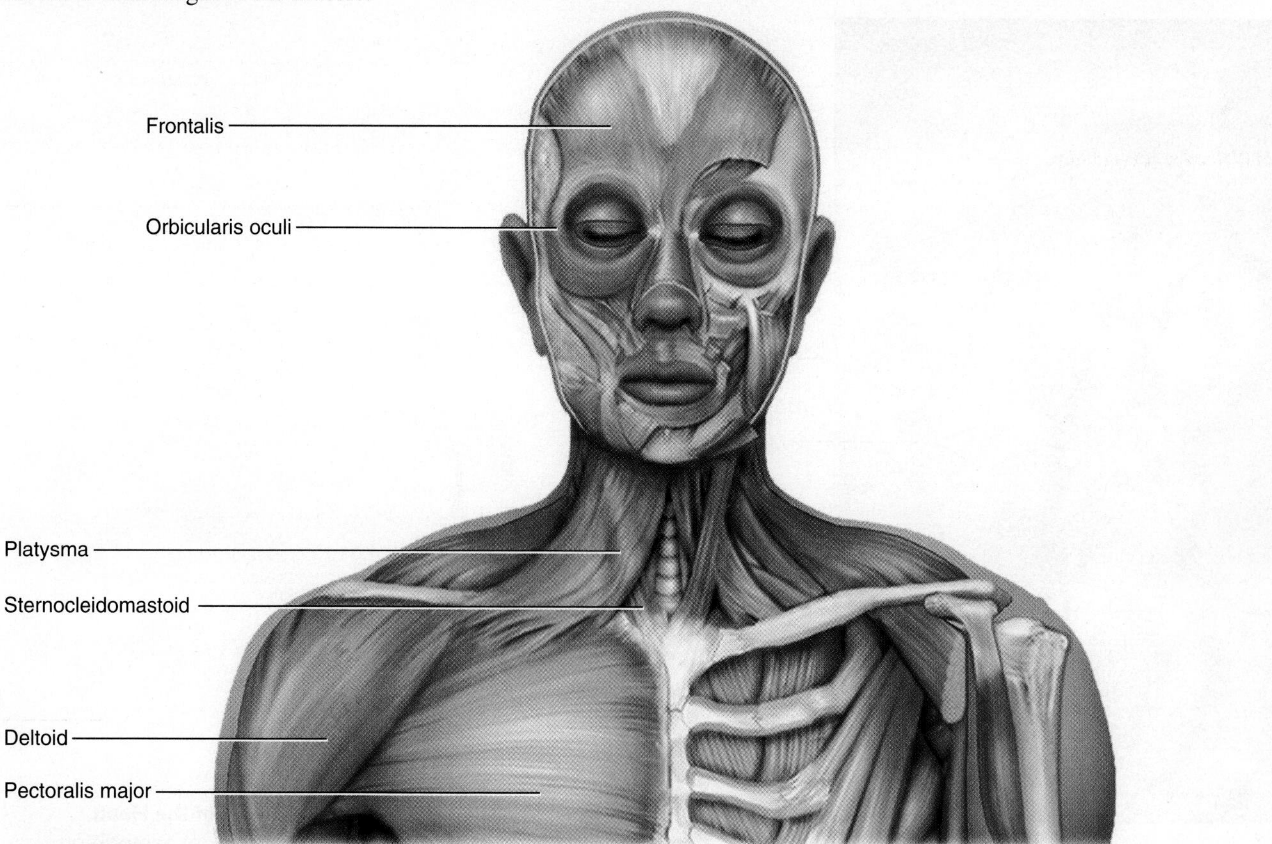

FIGURE 11.1 Platysma, Anterior View

Remember

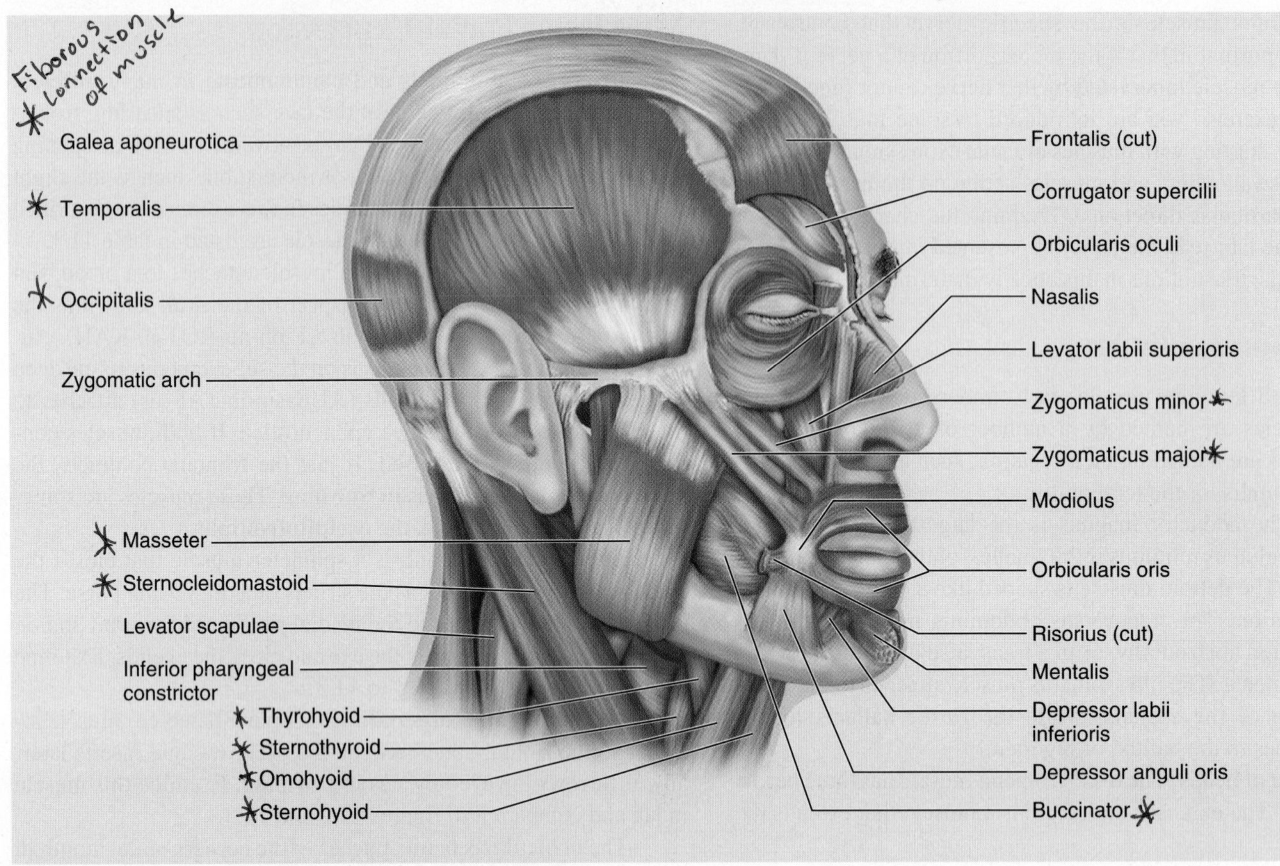

(a)

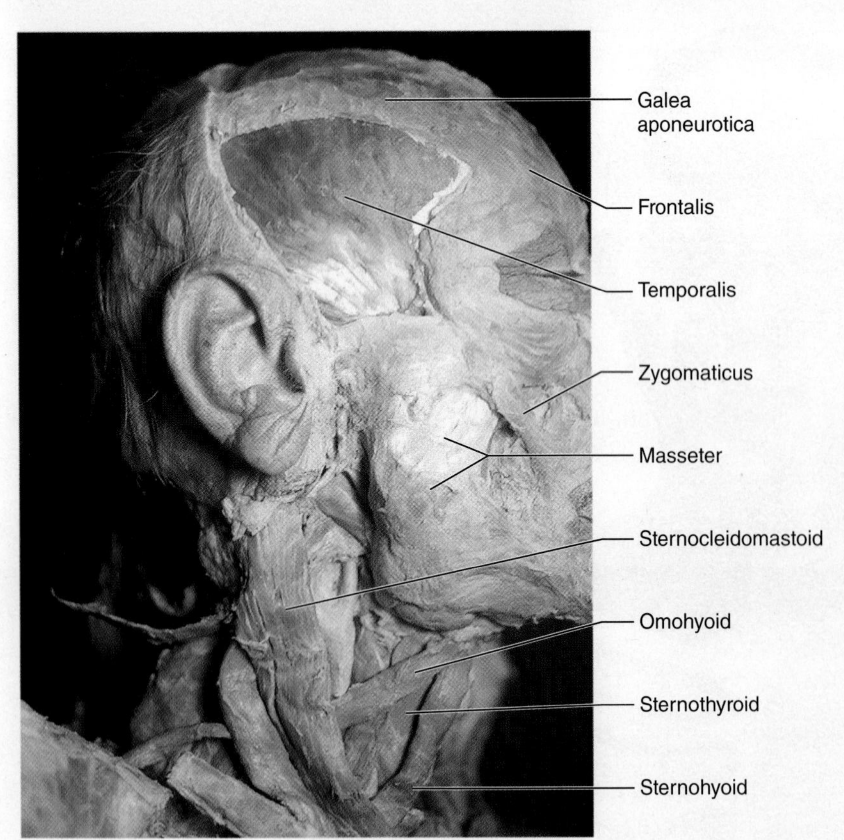

(b)

FIGURE 11.2 Muscles of the Head, Lateral View (a) Diagram; (b) photograph.

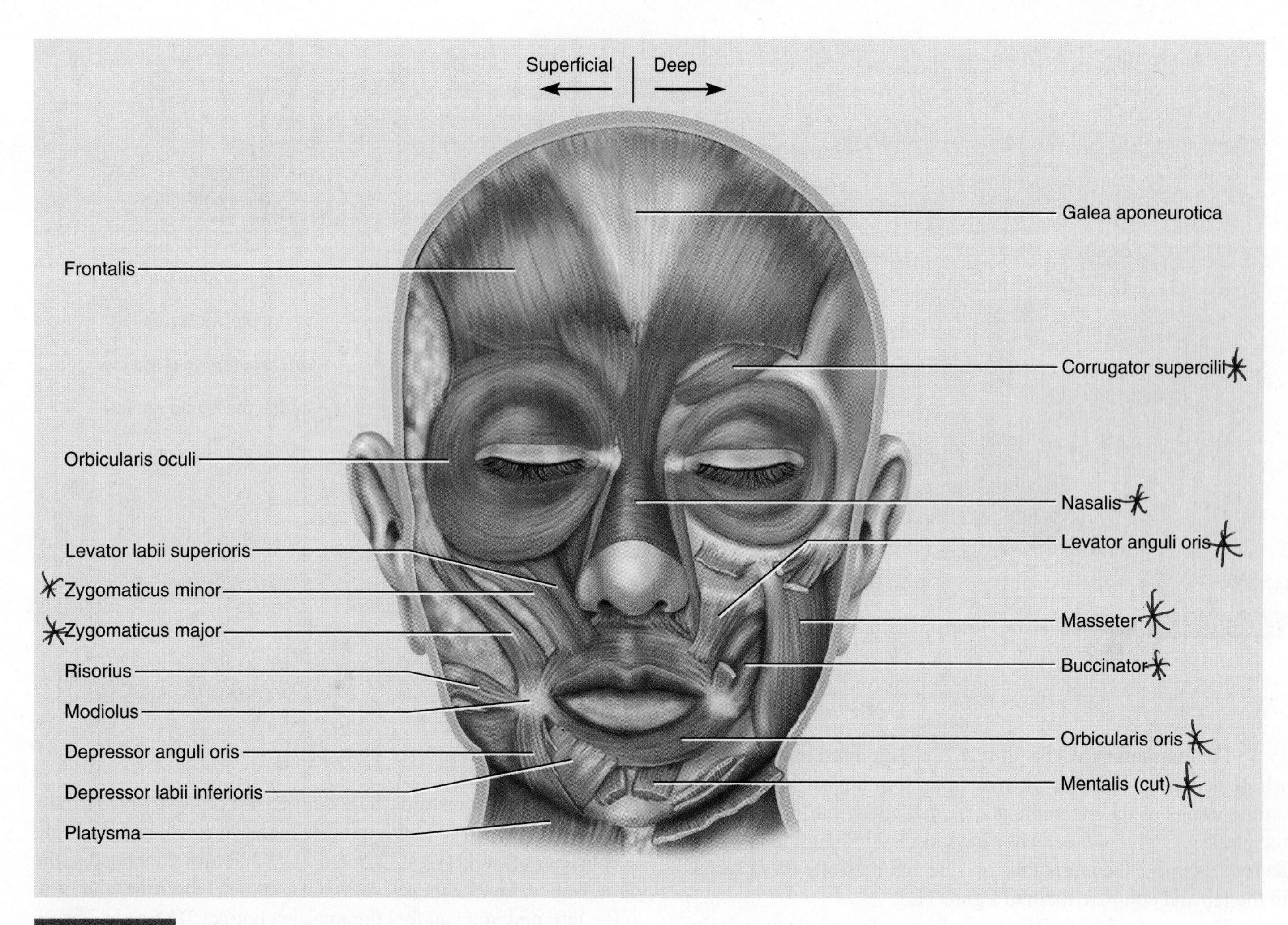

FIGURE 11.3 **Muscles of the Face, Anterior View**

the **modiolus.** The orbicularis oris attaches to the skin of the lips, thus closing the lips. Examine the orbicularis oris in lab and in figures 11.2 and 11.3.

The **levator labii superioris** muscle has a bony attachment on the maxilla and zygomatic bones, has a soft attachment on the muscles of the upper lip, and raises the skin of the lips and expands the nostrils, as in the expression of showing extreme disgust. Find the levator labii superioris in lab and in figures 11.2 and 11.3 and notice how it attaches to the upper lip.

The **zygomaticus** (ZYE-go-MAH-tih-cus) **major** and **zygomaticus minor** (figures 11.2 and 11.3) elevate the corners of the mouth by pulling them superiorly and laterally, as in smiling or laughing. They are named for their attachment on the zygomatic bone. The **risorius** (Rih-ZOR-ee-us) (figure 11.3) is known as the laughing muscle because it pulls the lips laterally. It does not have bony attachments, but is fixed to fascia and the modiolus.

The **depressor labii** (LAYB-ee-eye) inferioris (figures 11.2 and 11.3) pulls the lower corners of the mouth inferiorly when pouting. It is named for its action (depressing the lower lip).

The **mentalis** (men-TAL-is) has a stable attachment (origin) on the chin (anterior mandible) and is another of the pouting muscles. The **buccinator** (BUX-in-aye-tur) muscle of the cheek runs in a horizontal direction. It puckers the cheeks, as in trumpet playing, and pushes food toward the molars in chewing. These are seen in figures 11.2 and 11.3.

The **platysma** (plah-TISS-mah) is a broad, thin muscle that has a soft attachment (on the fascia of the pectoral and deltoid muscles). It also attaches to the mandible and skin of the lips and can be seen if you elevate your chin and subsequently pout. The thin wings that stick out on the side of your neck are the edges of the platysma muscle. Examine the platysma in figure 11.1 and in the models or on the cadaver in the lab.

Muscles of Chewing and Swallowing

Chewing and swallowing are complex actions that involve several muscles. As for all skeletal muscles, their contractions are coordinated by the brain and directed by the nerves that innervate them. The **temporalis** (TEMP-oh-RAL-us) is a powerful muscle that elevates the mandible. The temporal fossa is so named because it is a depression superior to the zygomatic arch. Recall that a fossa is a depression although when you examine the skull, the temporal fossa appears slightly domed. The temporalis muscle is deep to the zygomatic arch and inserts on the coronoid process and on the superior and medial ramus of the mandible.

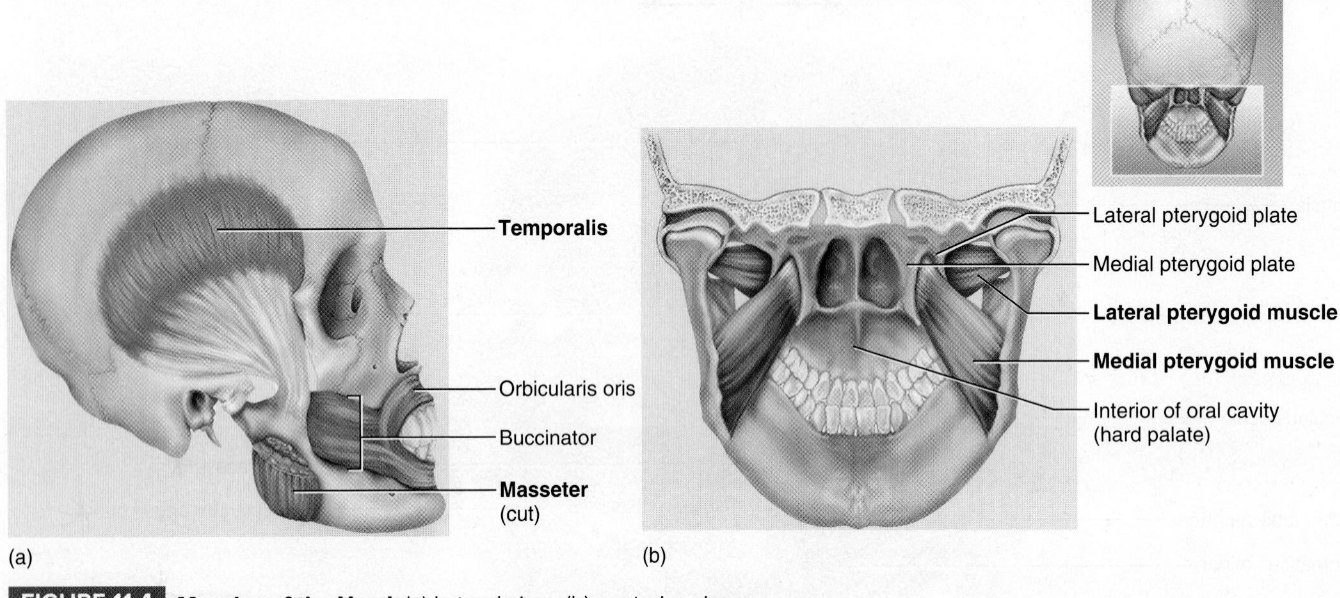

FIGURE 11.4 **Muscles of the Head** (a) Lateral view; (b) posterior view.

The **masseter** (MASS-ih-tur) is a large muscle of the head whose action is to elevate the mandible. If you place your fingers on the ramus of the mandible and clench your teeth, you can feel the masseter tighten. It is a chewing muscle, or a muscle of *mastication*. Examine the temporalis muscle and masseter on materials in the lab and compare them to figure 11.2.

The **pterygoid** (TARE-ih-goyd) muscles are deep muscles that originate on the sphenoid bone and insert laterally on the mandible. They pull the jaw horizontally, which helps in rotatory chewing. The temporalis and masseter close the jaw, while the pterygoids provide the sideways movement characteristic of a person chewing gum. These muscles can be seen in figure 11.4.

The **digastric** (di-GAS-trik) is a muscle with two bellies. The digastric is one of a few muscles that depresses the mandible. The digastric also moves the hyoid, which is important in tongue movement for speech and swallowing. Deep to the digastric is the **mylohyoid,** a broad muscle of the floor of the mouth that aids in pushing the tongue superiorly when swallowing. These muscles can be seen in figure 11.5.

The **omohyoid, sternohyoid,** and **sternothyroid** are all named for their attachment points. The sternum anchors the stable part of the muscle (the origin) for the sternohyoid and the sternothyroid. The thyroid cartilage of the larynx is depressed when the sternothyroid contracts, and the hyoid bone is depressed when the sternohyoid contracts. The term "omo" means shoulder, and the scapula anchors the stable part of the omohyoid. When the omohyoid contracts, the hyoid is depressed. Examine material in the lab and compare it with figures 11.5 and 11.6 for muscles of the anterior neck.

Muscles Acting on the Head

The **sternocleidomastoid** (STER-no-KLY-doh-MAS-toyd) muscle rotates the head in a unique way. Place your hand on the right sternocleidomastoid (figs. 11.5 and 11.6) and turn your head to the right. Notice how the muscle does not contract. Now turn your head to the left, and you can feel the muscle contract. The right sternocleidomastoid turns the head to the left, and the left sternocleidomastoid turns the head to the right.

The **scalenes** (skah-LEENS) are a group of muscles on the lateral side of the neck. They are bordered by the sternocleidomastoid in the front and the levator scapulae muscle in the back. They rotate the neck or elevate the ribs. Locate these muscles in the lab and examine figure 11.5.

The trapezius is a muscle that attaches medially to the skull, a tough, posterior ligament of the neck called the **nuchal ligament,** and the seventh cervical vertebra and thoracic vertebrae 1 through 4. It can extend the head when it is flexed or move the scapula. Examine the material in the lab and compare it to figure 11.5.

Activities

1. When you turn your head to the left, which sternocleidomastoid muscle contracts?

   ____

2. Tilt your head so your chin is elevated and pout your lower lip. What muscle forms a thin membrane along the anterolateral neck? ______________________

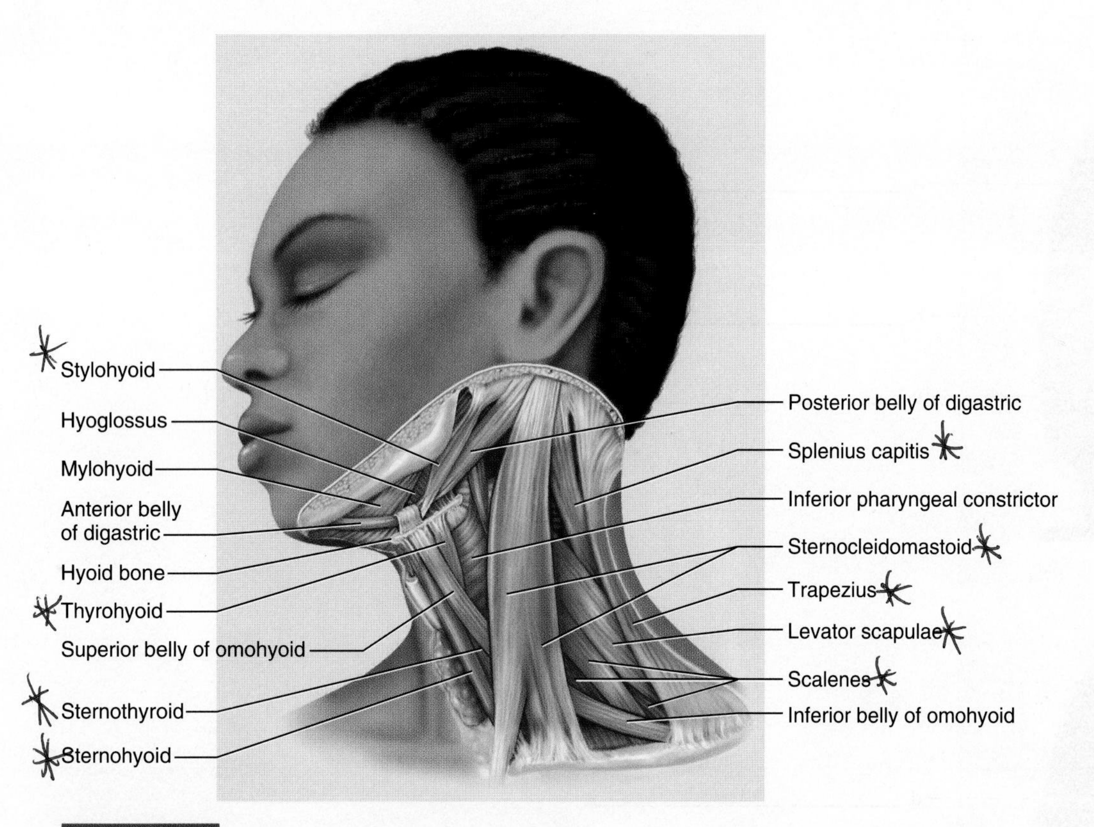

FIGURE 11.5 **Muscles of the Neck, Lateral View**

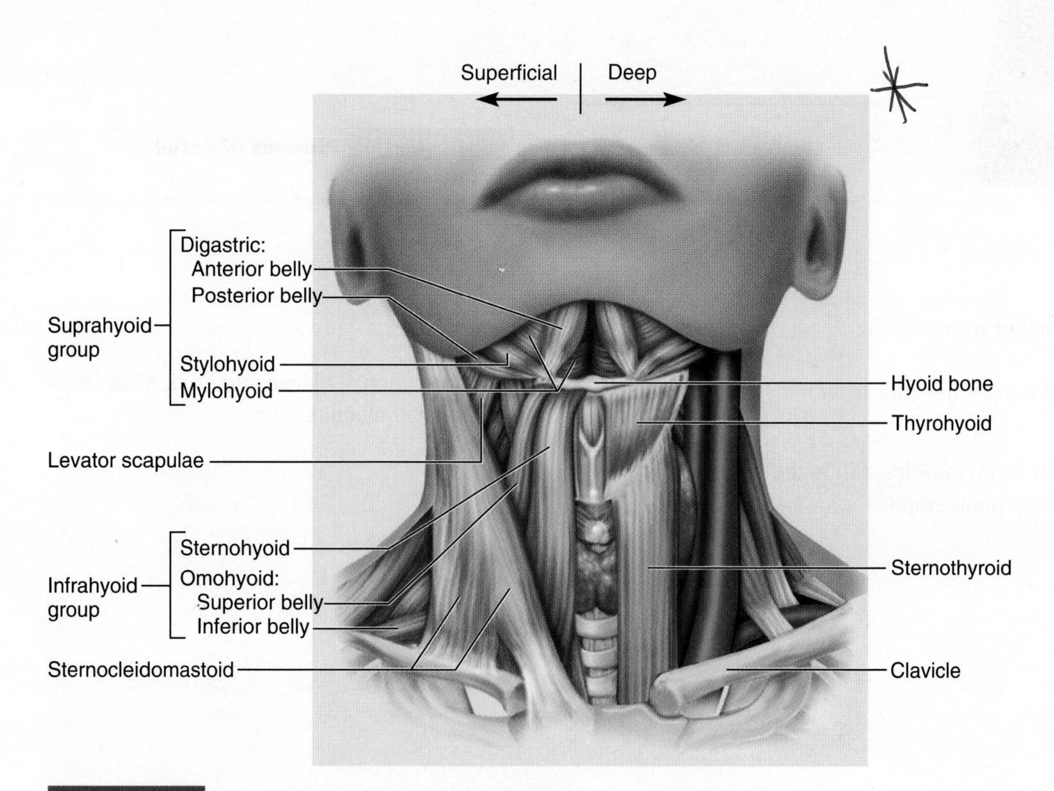

FIGURE 11.6 **Muscles of the Neck, Anterior View**

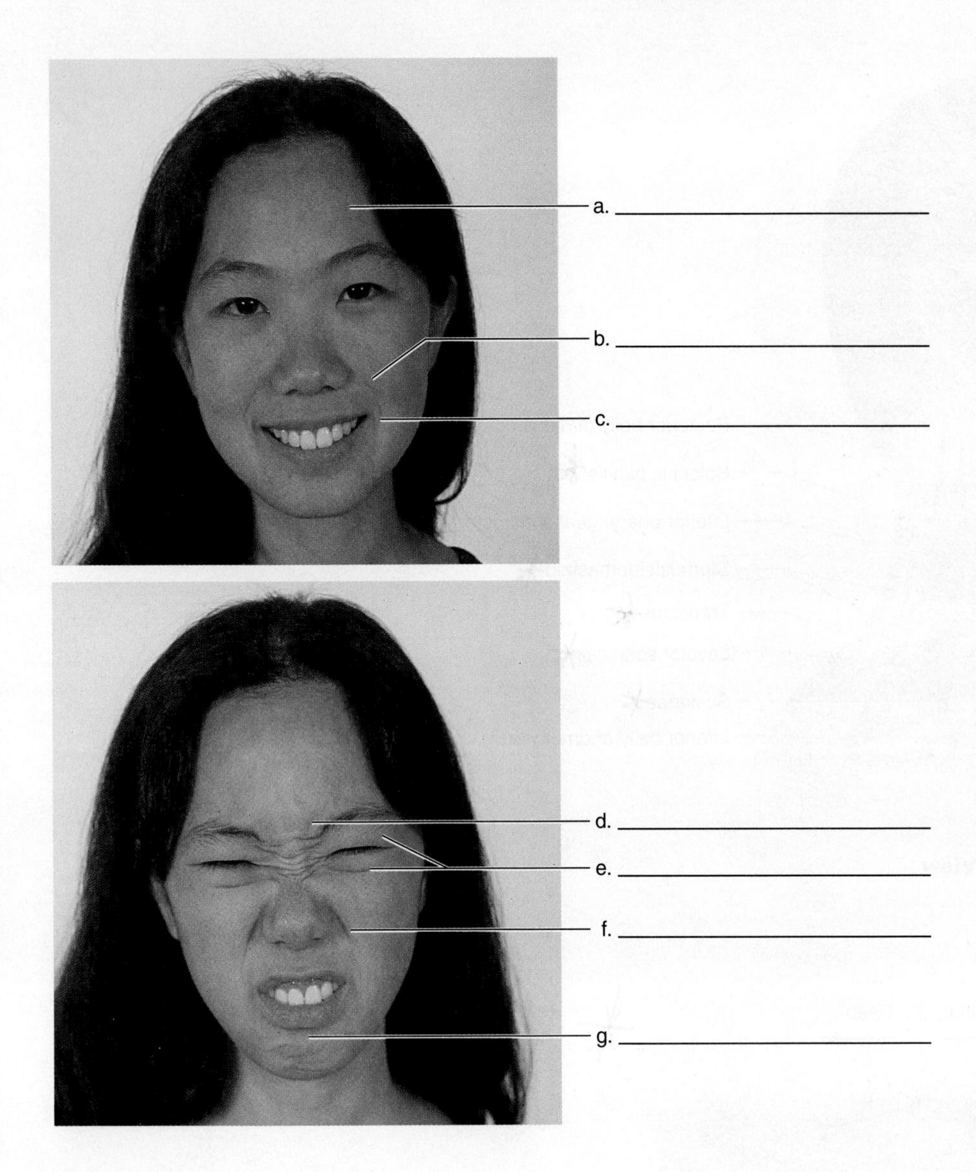

FIGURE 11.7 **Muscles of Facial Expression**

3. Purse your lips and feel the buccinator muscle as it contracts in the cheek.
4. Clench your teeth and palpate the temporalis muscle and the masseter.

Examine figure 11.7 for surface views of facial muscles. Fill in the responses that represent the muscles in the photographs.

Cat Anatomy

If you are using cats, turn to section 4, "Head and Neck Muscles of the Cat," on page 421 of this lab manual.

Review

Exercise 11

Axial Muscles 1: Muscles of the Head and Neck

Name ______________________________ Date ______________

1. What is the origin of the masseter muscle? ______________________________

____

2. What kind of muscle is the orbicularis oculi or orbicularis oris muscle in terms of action? ______________________________

____

3. Fill in the following illustration for the muscles of the head, using the terms provided.

buccinator	occipitofrontalis	orbicularis oris
masseter	orbicularis oculi	temporalis

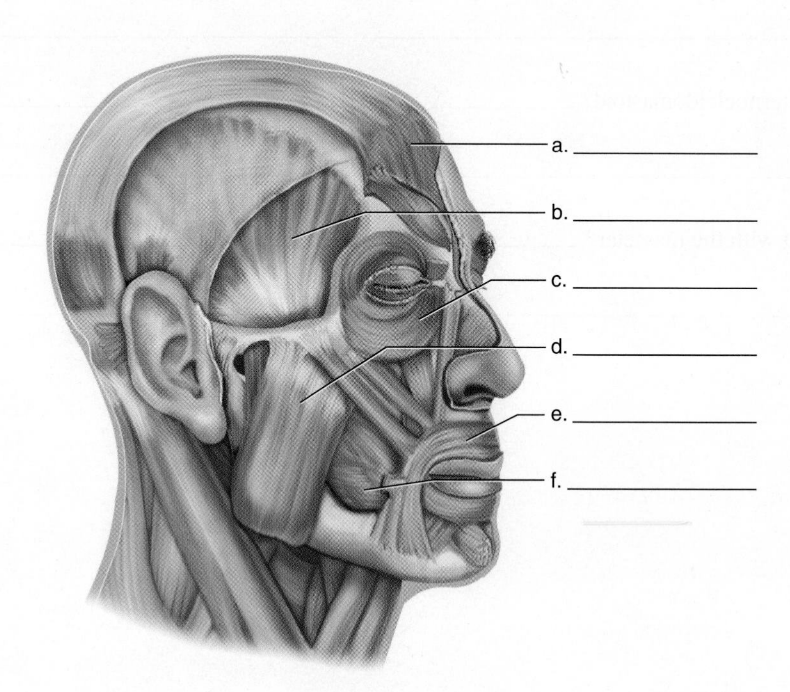

4. What muscle attaches to/originates on the temporal fossa? ______________________

5. Name two muscles that close the jaw. ______________________

6. Where does the sternocleidomastoid muscle attach superiorly/insert? ______________________

7. What muscle closes and protrudes the lips? ______________________

8. Where does the orbicularis oculi insert? ______________________

9. What is the inferior attachment/insertion of the temporalis? ______________________

10. Name a muscle that closes the eye lid. ______________________

11. What is the action of the sternocleidomastoid? ______________________

12. What muscle is a synergist with the masseter? ______________________

Laboratory

Exercise 12

Axial Muscles 2: Muscles of the Trunk

Muscular System

Introduction

The muscles of the trunk can be grouped into a few functional areas. These are the abdominal muscles, which tighten the abdomen; the respiratory muscles, which assist in breathing; the postural muscles of the back; and the muscles that act on the scapula or on the head and neck.

Learning Objectives

At the end of this exercise you should be able to

1. locate the muscles of the trunk on a torso model, chart, or cadaver (if available);
2. list the origin, insertion (attachments), and action of each muscle;
3. describe what nerve controls (innervates) each muscle; and name all the muscles that have a particular action on the trunk, such as all the muscles that compress the abdomen.

Materials

Human torso model
Human muscle charts
Articulated skeleton
Cadaver (if available)

Procedure

Review the actions as outlined in laboratory exercise 10. Examine a torso model or chart in the lab and locate the muscles described in this exercise and in table 12.1. Correlate the shape or fiber direction of the muscle with the name and visualize the muscles as you study the models or charts. You may want to look at an articulated skeleton as you review the attachment points of the muscles, so that you can better see them. The descriptions in the text portion of this exercise can help you understand the nature of the muscle, while table 12.1 gives you specific information about the muscle. Once you have learned the attachment points, you should be able to understand the action of the muscle. This is done, in part, by visualizing how the muscle is pulled between the attachment points when the muscle contracts.

Muscles of Respiration

The diaphragm and **intercostal** muscles are respiratory muscles. Normally the diaphragm is responsible for about 60% of the resting breath volume, while the intercostal muscles contribute to the remaining volume. The **diaphragm** is a domed muscle that has a peripheral attachment. The central attachment of the diaphragm occurs deep on a sheet of fascia called the central tendon at the base of the mediastinum. If you think of the diaphragm as a trampoline, the outer springs represent the first attachment (origin), while the center (where you jump) represents the second attachment (insertion). Examine the models in the lab and compare them with the diaphragm in figure 12.1.

The intercostal muscles contribute to the breathing volume at rest, but they also contribute to a greater increase in the movement of the thorax during times of exercise. The **external intercostal** muscles are involved in inhalation, and the **internal intercostal** muscles are involved in both inhalation and exhalation. Locate the intercostal muscles and compare them with figure 12.2.

Muscles of the Anterior Abdominal Wall

The muscles of the abdomen compress the viscera, which aids in urination, childbirth, and bowel movements. This is accomplished by the **valsalva maneuver,** which consists of taking a breath and contracting these abdominal muscles. The **external abdominal oblique** is a broad, superficial muscle of the abdomen with fibers that run from a superior direction to an inferior, medial direction. The **internal abdominal oblique** is deep to the external abdominal oblique and has fiber directions that run perpendicular to the external abdominal oblique.

The deepest of the abdominal muscles is the **transverse abdominal,** which has fibers running in a horizontal direction. The **rectus abdominis** (*rectus* = straight) runs vertically up the abdomen. The rectus abdominis has small connective tissue bands called **tendinous intersections,** located horizontally across the muscle and dividing it into small segments, and the muscle is enclosed by the **rectus sheath.** If the abdominal fat is minimal and

TABLE 12.1 Axial Muscles 2: Muscles of the Trunk

Name	Attachment 1 (Origin)	Attachment 2 (Insertion)	Action	Innervation
Muscles of Respiration				
Diaphragm	Xiphoid process, ribs 7–12, upper lumbar vertebrae	Central tendon	Inspiration	Phrenic nerve
External intercostals	Inferior border of ribs 1–11	Superior border of rib below	Elevates ribs (increases volume in thorax)	Intercostal nerves
Internal intercostals	Superior border of rib	Inferior border of next higher rib (2–12)	Elevates and depresses ribs	Intercostal nerves
Muscles of the Anterior Abdominal Wall				
External abdominal oblique	Ribs 5–12	Iliac crest, pubis	Compresses abdominal wall, laterally rotates waist	Intercostal nerves from T7–T12
Internal abdominal oblique	Inguinal ligament, iliac crest	Pubis, ribs 10–12	Compresses abdominal wall, laterally rotates waist	Intercostal nerves from T7–T12 and spinal nerve L1
Transverse abdominal	Inguinal ligament, iliac crest, costal cartilages 7–12	Linea alba, pubis	Compresses abdominal wall, laterally rotates waist	Intercostal nerves from T7–T12 and spinal nerve L1
Rectus abdominis	Crest of pubis, pubic symphysis	Costal cartilages 5–7 xiphoid process	Flexes vertebral column, compresses abdominal wall	Intercostal nerves from T6–12
Muscles of the Back				
Erector spinae: Iliocostalis, longissimus, spinalis	Nuchal ligament, thoracic and lumbar vertebrae, sacrum, ilium ribs 3–12	All ribs, all thoracic vertebrae, temporal bones	Extends and rotates vertebral column and head	Dorsal rami of spinal nerves
Splenius	Nuchal ligament, C7–T6	C2–4, occipital and temporal bone	Extends and rotates head	Middle and lower cervical nerves
Semispinalis	C4–T10	Occipital bone and spinal processes of C2–C5	Extends and rotates head and vertebral column	Cervical and thoracic spinal nerves
Quadratus lumborum	Iliac crest	L1–4, rib 12	Extends and abducts vertebral column	T12, L1–4
Multifidus	Vertebrae C4–L5, sacrum, ilium	Vertebral column	Extends and rotates vertebral column	Dorsal rami of spinal nerves

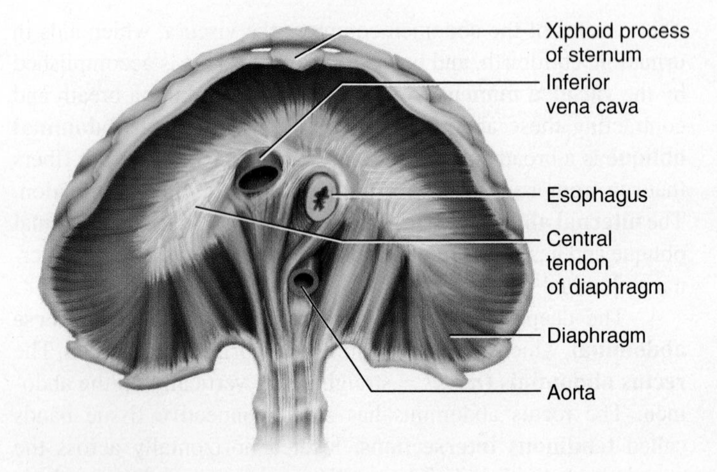

FIGURE 12.1 Diaphragm, Inferior View

the muscles are well developed, the "washboard stomach" (or "six-pack") is clearly seen because the muscle fibers increase in girth while the tendinous intersections remain undeveloped. The rectus abdominis not only compresses the abdominal viscera, as do the other abdominal muscles, but also flexes the vertebral column. Examine the abdominal muscles in the lab and compare them with figure 12.2.

Muscles of the Back

The postural muscles of the back of the trunk mostly consist of the **erector spinae** muscles. These are located deep to the trapezius (fig. 12.3). The erector spinae muscles are actually many muscles found between individual vertebrae or between the vertebrae and the ribs. These separate muscles are grouped conveniently into long strap muscles known collectively as the erector spinae. There

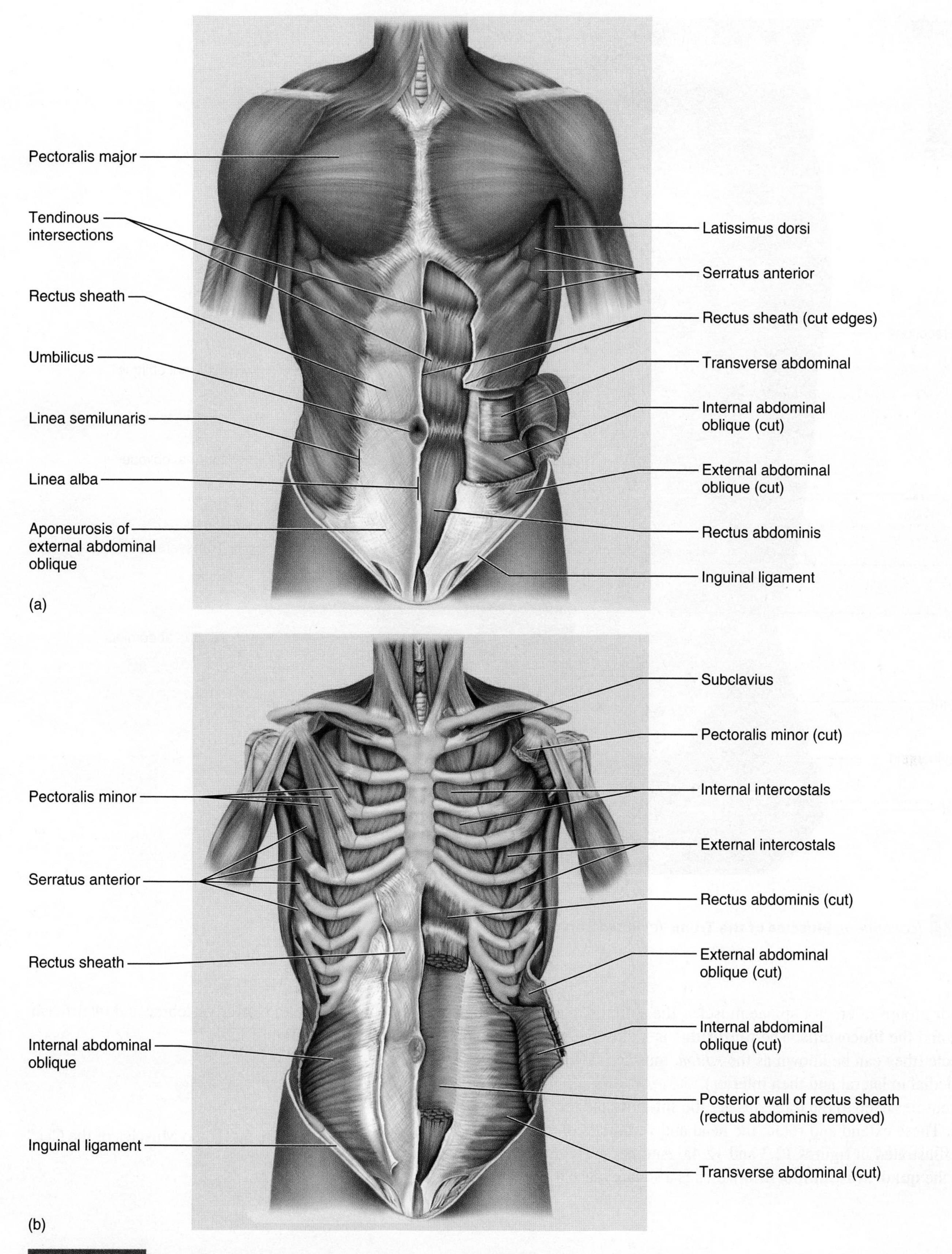

FIGURE 12.2 Muscles of the Trunk, Diagram of Anterior View (a) Superficial; (b) deep.

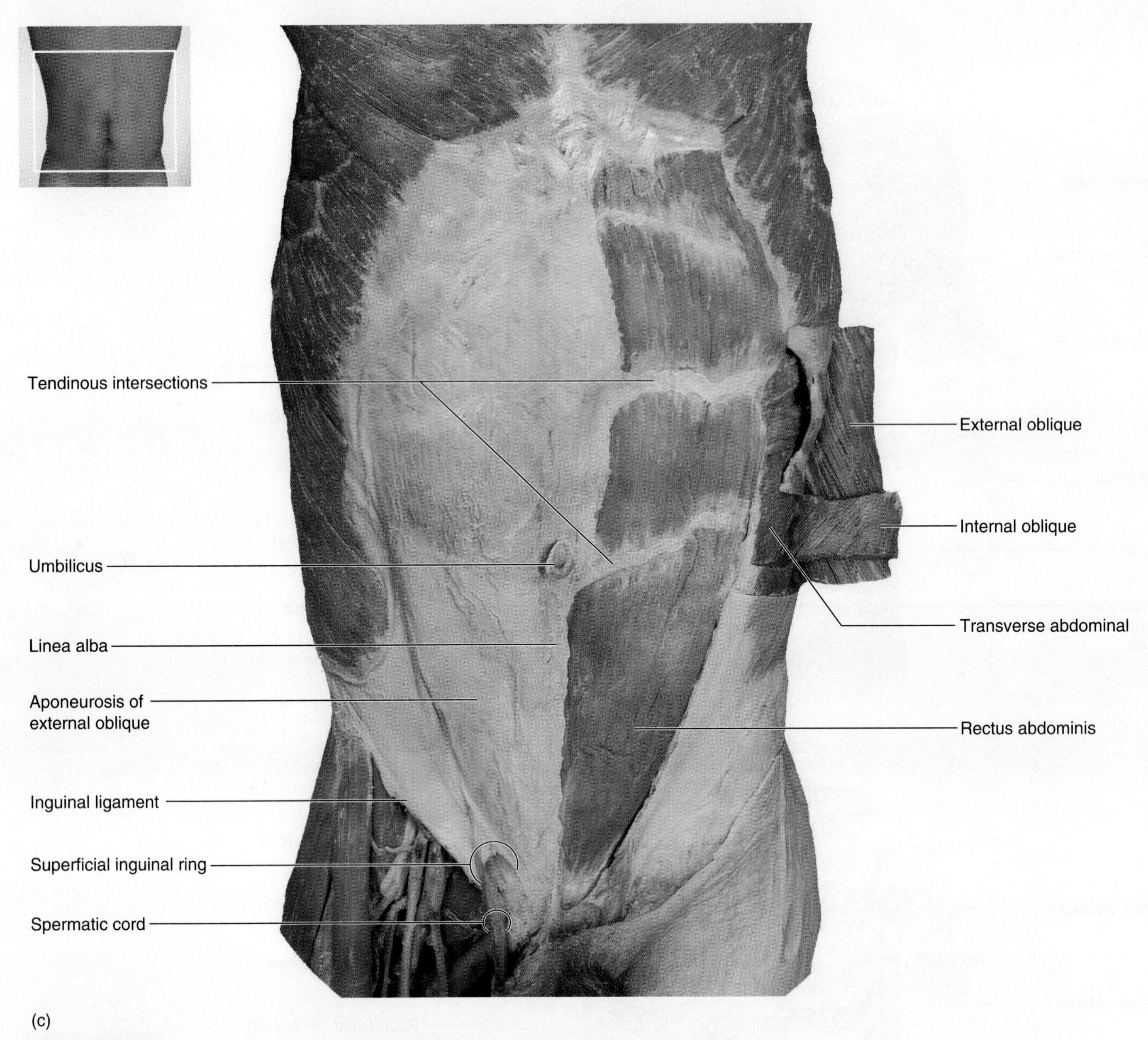

FIGURE 12.2 *(continued)* **Muscles of the Trunk** (c) Photograph.

are three major groups of erector spinae muscles, the **spinalis,** the **longissimus,** and the **iliocostalis.** The **multifidus** is a closely associated muscle (they can be known as the *s.l.i.m.* muscles as you move from medial to lateral and then inferior).

Other muscles deep to the trapezius are the **splenius** and the **semispinalis.** These extend and rotate the head and vertebral column and are illustrated in figures 12.3 and 12.4a. Another muscle in the area is the **quadratus lumborum,** which is a square muscle that runs from the iliac crest to the lower vertebrae and twelfth rib. These muscles can be seen in figure 12.4.

Cat Anatomy

If you are using cats, turn to section 5, "Torso Muscles of the Cat," on page 423 of this lab manual.

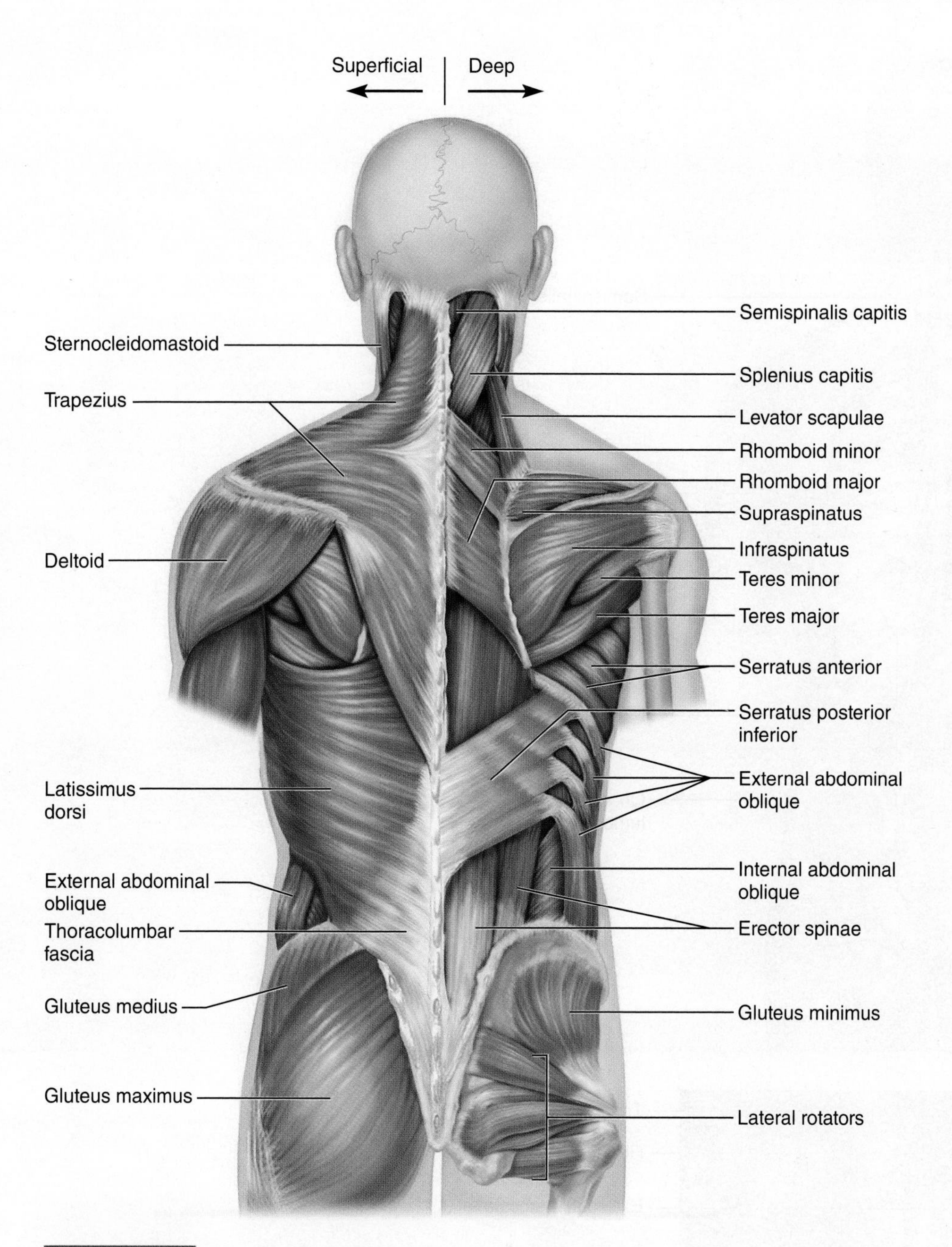

FIGURE 12.3 **Muscles of the Back, Posterior View**

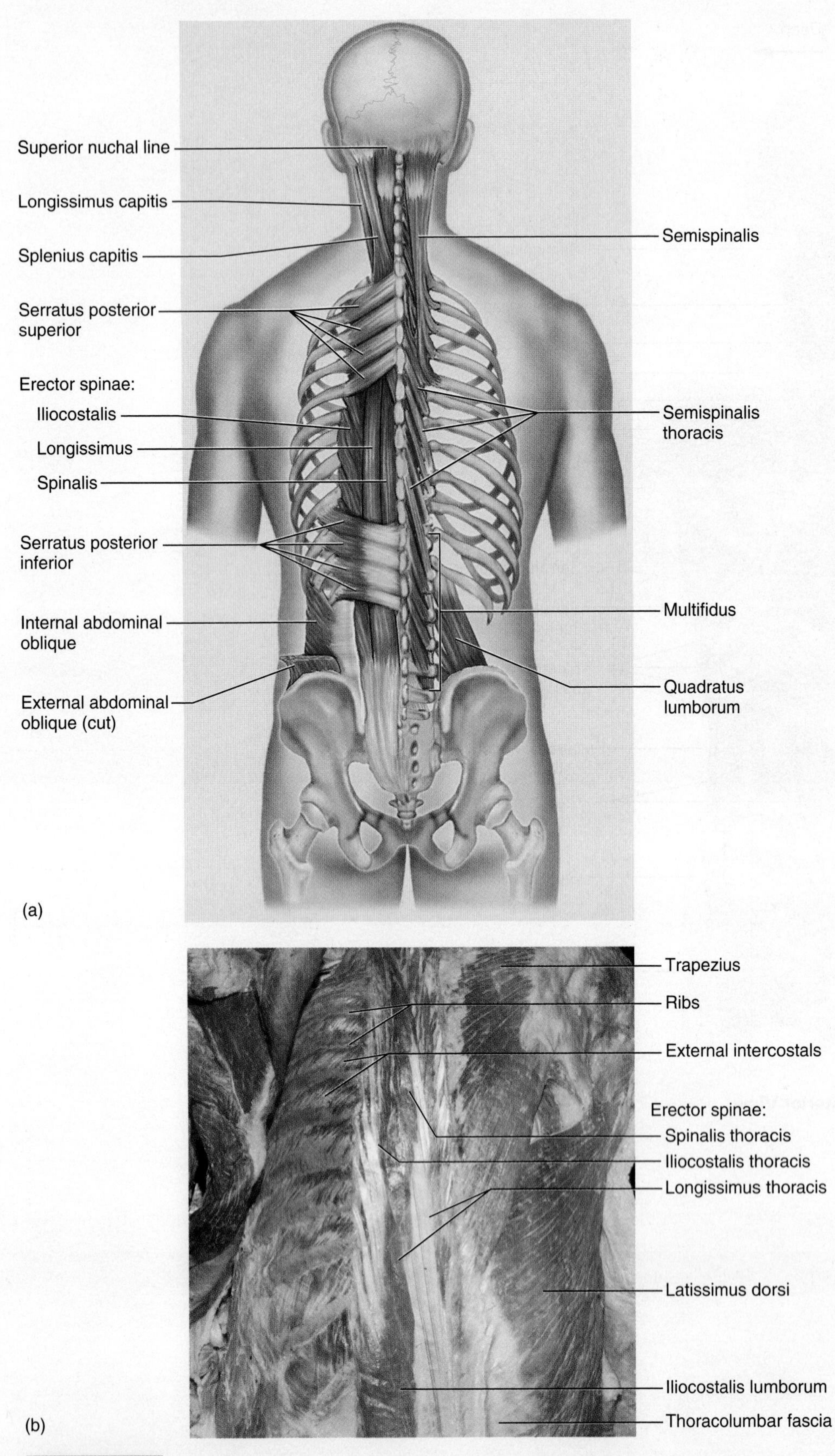

FIGURE 12.4 **Deep Muscles of the Back, Posterior View** (a) Diagram; (b) photograph.

Review

Exercise 12

Axial Muscles 2: Muscles of the Trunk

Name ______________________________ Date ____________

1. What is the central innervation of the diaphragm? ______________________________

2. What is the action of the quadratus lumborum? ______________________________

3. What action does the multifidus have in common with the erector spinae? ______________________________

4. How does the action of the rectus abdominis differ from that of the other abdominal muscles? ______________________________

5. What is the physical relationship of the intercostal muscles to one another? ______________________________

6. Compression of the abdominal wall occurs by what four muscles? ______________________________

7. Extension and rotation of the vertebral column occur by what group of muscles? ______________________________

8. What muscle is responsible for most of the air inhaled during relaxed breathing? ______________________

____

9. What is the action of the external intercostal muscles? ______________________

____

10. What muscle attaches to/inserts on the central tendon? ______________________

____

11. The tendinous intersections are found in what muscle? ______________________

____

12. Flexion of the vertebral column occurs by what abdominal muscle? ______________________

____

13. Which is the deepest anterior abdominal muscle? ______________________

____

14. Label the muscles in the following illustration.

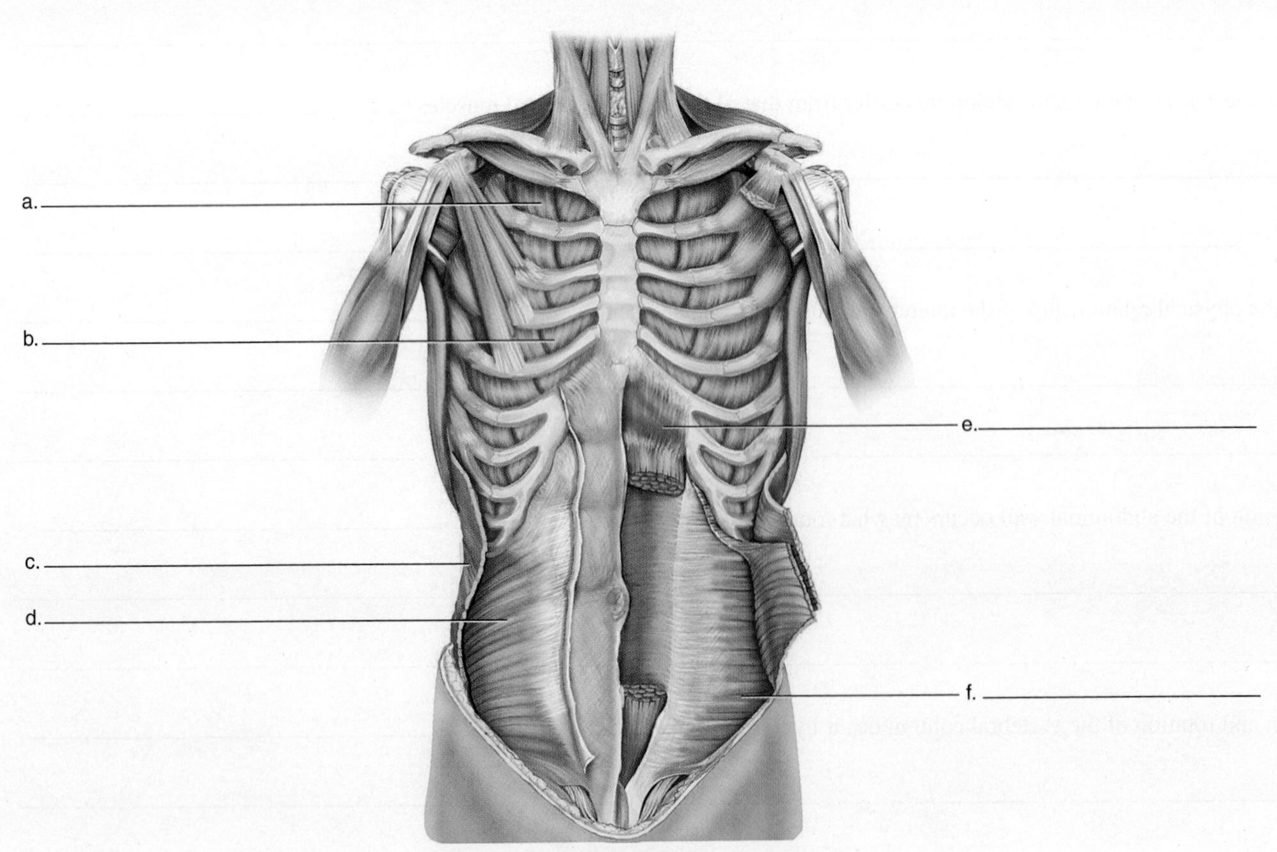

15. Label the muscles in the following illustration.

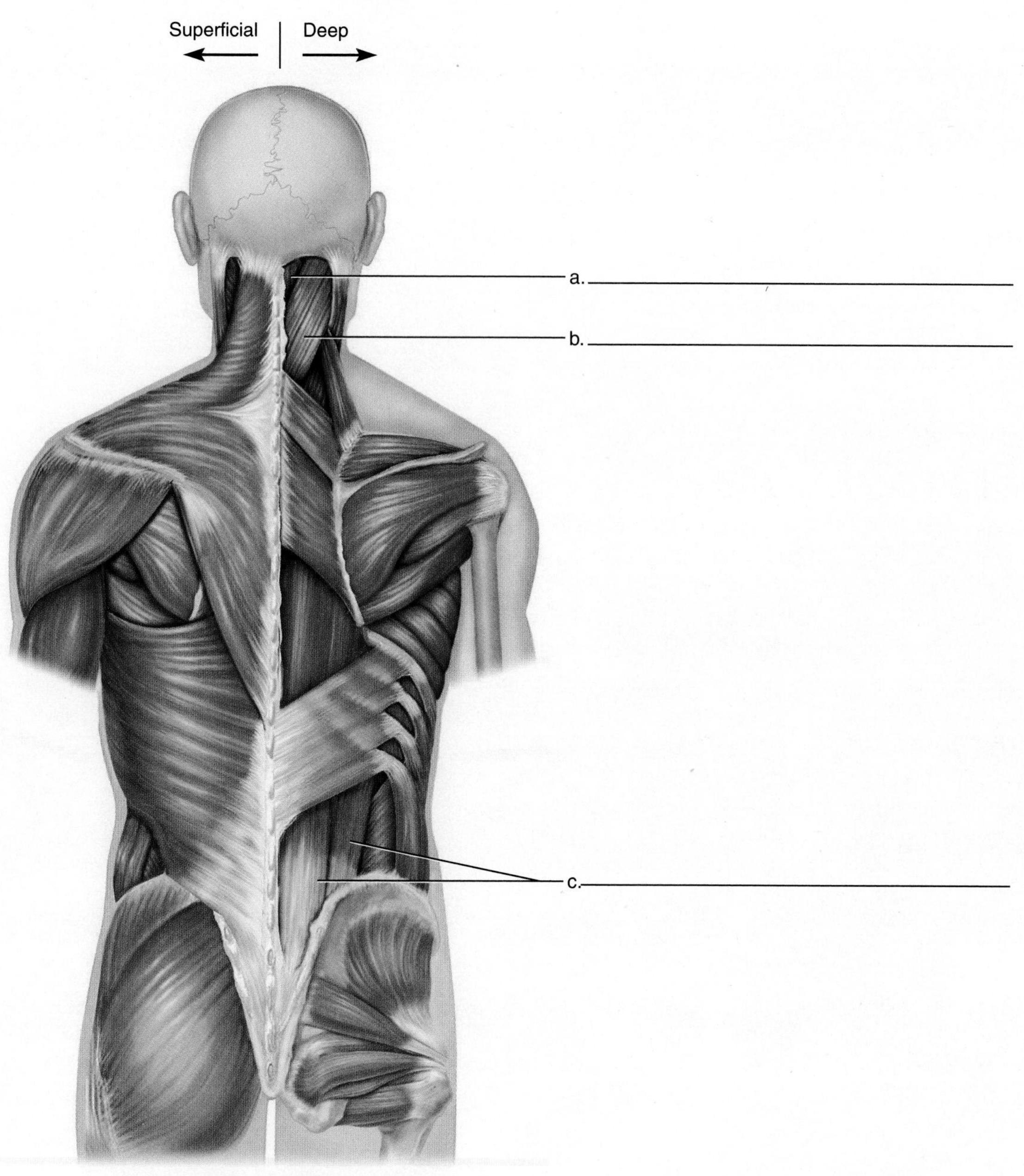

Laboratory

Exercise 13

Appendicular Muscles 1: Muscles of the Shoulder and Upper Limb

Muscular System

Introduction

In this exercise, you study the major muscles of the shoulder and upper limb. Some of these muscles attach to the vertebrae or the scapula and move or stabilize the humerus. Others attach to the humerus and insert on the forearm.

Other muscles in this exercise are those of the forearm and hand, and these allow us to grasp materials and finely manipulate things, which in turn allows us to do things such as create fine art and construct precision tools that help define us as being human. You should learn the attachment points as you learn the muscles of the forearm, since many of these muscles look quite similar. By knowing the attachments of a muscle, you will not easily mistake one muscle for another.

Learning Objectives

At the end of this exercise you should be able to

1. locate the muscles of the shoulder and upper limb on a torso model, chart, or cadaver (if available);
2. list the attachment points, and action of each muscle presented;
3. describe what nerve controls (innervates) each muscle;
4. list what muscles function as synergists or antagonists to the prime mover;
5. name all the muscles that have an action on a joint, such as all the muscles that laterally rotate the arm or muscles that flex the hand; and
6. reproduce the actions of each listed muscle by moving the appropriate limbs to demonstrate the actions.

Materials

Human torso model and upper limb model
Human muscle charts
Articulated skeleton
Cadaver (if available)

Procedure

Examination of Muscles

Look at the charts, models, and cadaver (if available) in the lab and locate the muscles presented in this section. Review the bones of the hand, as well as the bones of the arm and forearm, in Laboratory Exercise 9, "Appendicular Skeleton," to locate precisely the bony origins and insertions. Refer to the following descriptions as you examine the muscles. The specific details of the muscles are provided in table 13.1.

Muscles Acting on the Shoulder Locate the diamond-shaped **trapezius** (tra-PEE-zee-us) muscle, which has an attachment in the midline of the vertebral column and head and has other attachments laterally. If the scapula is fixed, the head moves; yet, if the vertebral column and head are fixed, then the trapezius moves the scapula. The **levator scapula** (le-VAY-tur SCAP-u-lay) is named for what it does, elevate the scapula. The levator scapula has one attachment on the lateral side of the neck and another on the upper scapula. If the neck is fixed, the levator scapula elevates the scapula, as in shrugging. If the scapula is fixed, the muscle rotates or abducts the neck. Figure 13.1(a) illustrates the trapezius and levator scapulae.

The rhomboid muscles are deep to the trapezius and reflection of the trapezius is necessary to see them. The rhomboid muscles attach to the vertebral column and also on the medial border of the scapulae. As they contract they pull the scapulae together, adducting the scapulae. The rhomboid major is a larger and more inferior muscle than the rhomboid minor. Compare the material in lab with figure 13.1.

The **pectoralis** (PEC-tur-AL-is) **major** muscle has fibers that run horizontally across the chest region, and it is a superficial muscle of the chest. It flexes, adducts, and medially rotates the arm. It is seen in figure 13.1b. Another superficial muscle, but on the posterior surface, is the **latissimus dorsi** (la-TISS-ih-muss DOR-sye) muscle, which arises from the vertebrae by way of a broad, flat **thoracolumbar fascia.** The stable part of the latissimus dorsi is on the back, and the movable part is on the anterior aspect of the humerus. It is a powerful extensor of the arm and is called the swimmer's muscle (fig. 13.1*a*).

TABLE 13.1 Muscles of the Shoulder and Upper Limb

Name	Attachment 1 (Origin)	Attachment 2 (Insertion)	Action	Innervation
Muscles Acting on the Shoulder and Arm				
Trapezius	Posterior occipital bone Nuchal ligament, C7–T4	Clavicle, acromion, spine of scapula	Extends and abducts head, rotates and adducts scapula, fixes scapula	Accessory nerve (XI) Spinal nerves C2–4
Levator scapulae	C1–4	Superior angle and medial border of scapula	Elevates scapula, abducts and rotates neck	Spinal nerves C3–5 and dorsal scapular nerve
Rhomboid minor	Nuchal ligament and spines C7–T1	Medial border of scapula at scapular spine	Adducts scapula (draws scapulae together)	Dorsal scapular nerve
Rhomboid major	Spines of T2–5	Inferior one-third of medial border of scapula	Adducts scapula (draws scapulae together)	Dorsal scapular nerve
Pectoralis major	Clavicle, sternum, cartilages of ribs 1–7	Crest of greater tubercle of humerus	Flexes, adducts, and medially rotates humerus	Medial and lateral pectoral nerves
Latissimus dorsi	T7–T12, L1–5, crest of ilium, ribs 10–12	Intertubercular sulcus of humerus	Extends, adducts, and medially rotates arm; draws shoulder inferiorly	Thoracodorsal nerve
Deltoid	Clavicle, acromion, spine of scapula	Deltoid tuberosity of the humerus	Abducts arm; flexes, extends, medially and laterally rotates arm	Axillary nerve
Pectoralis minor	Ribs 3–5	Coracoid process of scapula	Depresses glenoid cavity	Medial pectoral nerve
Teres major	Inferior angle of scapula	Crest of lesser tubercle of humerus	Extends and medially rotates arm	Subscapular nerve
Coracobrachialis	Coracoid process of scapula	Midmedial shaft of humerus	Flexes and medially rotates arm	Musculocutaneous nerve
Supraspinatus	Supraspinous fossa	Greater tubercle of humerus	Abducts arm, helps stabilize shoulder joint	Suprascapular nerve
Infraspinatus	Infraspinous fossa	Greater tubercle of humerus	Laterally rotates arm, stabilizes shoulder joint	Suprascapular nerve
Teres minor	Lateral border of scapula	Greater tubercle of humerus	Laterally rotates and adducts arm, stabilizes shoulder joint	Axillary nerve
Subscapularis	Subscapular fossa	Lesser tubercle of humerus	Medially rotates arm, stabilizes shoulder joint	Subscapular nerve
Muscles Acting on the Forearm				
Brachialis	Anterior, distal surface of humerus	Coronoid process of ulna	Flexes forearm	Musculocutaneous nerve
Biceps brachii	Long head: superior margin of glenoid cavity. Short head: coracoid process of scapula	Radial tuberosity	Flexes arm, flexes forearm, supinates hand	Musculocutaneous nerve
Triceps brachii	Infraglenoid tuberosity of scapula, lateral and posterior surface of humerus	Olecranon of ulna	Extends and adducts arm, extends forearm	Radial nerve
Brachioradialis	Lateral supracondylar ridge of humerus	Near styloid process of radius	Flexes forearm	Radial nerve
Pronator quadratus	Distal part of anterior ulna	Distal radius	Pronates forearm (and hand)	Median nerve
Pronator teres	Medial epicondyle of humerus, coronoid process of ulna	Lateral, middle shaft of radius	Pronates and flexes forearm	Median nerve
Supinator	Lateral epicondyle of humerus, anterior ulna	Proximal radius	Supinates forearm	Radial nerve

(continued)

TABLE 13.1 Muscles of the Shoulder and Upper Limb *(continued)*

Muscles Acting on the Wrist and Hand

Name	Attachment 1 (Origin)	Attachment 2 (Insertion)	Action	Innervation
Flexor carpi radialis	Medial epicondyle of humerus	Second and third metacarpals	Flexes and abducts wrist	Median nerve
Flexor carpi ulnaris	Medial epicondyle of humerus, olecranon process, and dorsal surface of ulna	Pisiform, hamate, fifth metacarpal	Flexes and adducts wrist	Ulnar nerve
Flexor digitorum superficialis	Medial epicondyle of humerus, proximal ulna, proximal radius	Middle phalanges of digits 2–5	Flexes proximal and middle phalanges, flexes wrist	Median nerve
Palmaris longus	Medial epicondyle of humerus	Palmar aponeurosis	Anchors skin and fascia of palmar region	Median nerve
Flexor digitorum profundus	Anterior, proximal surface of ulna; interosseous membrane	Distal phalanges of digits 2–5	Flexes phalanges, flexes wrist	Median and ulnar nerves
Flexor pollicis longus	Anterior portion of radius and interosseous membrane	Distal phalanx of pollex (thumb)	Flexes thumb	Median nerve
Extensor carpi radialis longus	Lateral supracondylar ridge of humerus	Second metacarpal	Extends and abducts wrist	Radial nerve
Extensor carpi radialis brevis	Lateral epicondyle of humerus	Third metacarpal	Extends and abducts wrist	Radial nerve
Extensor digitorum	Lateral epicondyle of humerus	Middle and distal phalanges of digits 2–5	Extends phalanges, extends wrist	Radial nerve
Extensor carpi ulnaris	Lateral epicondyle of humerus, proximal ulna	Fifth metacarpal	Extends and adducts wrist	Radial nerve
Abductor pollicis longus	Posterior radius and ulna, interosseous membrane	First metacarpal, trapezium	Abducts thumb	Radial nerve
Extensor pollicis longus and brevis	Posterior radius and ulna, interosseous membrane	Proximal and distal phalanges of pollex (thumb)	Extends thumb	Radial nerve
Abductor pollicis brevis	Scaphoid and trapezium flexor retinaculum	Proximal phalanx of first digit	Abducts thumb	Median nerve
Flexor pollicis brevis	Trapezium, trapezoid, capitate	Proximal phalanx of first digit	Flexes thumb	Median and ulnar nerves
Opponens pollicis	Trapezium, flexor retinaculum	First metacarpal	Opposes thumb	Median nerve
Abductor digiti minimi	Pisiform	Proximal phalanx of fifth digit	Abducts fifth digit	Ulnar nerve
Flexor digiti minimi brevis	Hamulus of hamate	Proximal phalanx of fifth digit	Flexes fifth digit	Ulnar nerve
Opponens digiti minimi	Hamulus of hamate	Fifth metacarpal	Opposes fifth digit	Ulnar nerve

The **deltoid** is located on top of the shoulder and abducts the arm. Place your hand on your shoulder, abduct your arm, and feel the deltoid move. It also flexes, extends, medially rotates, and laterally rotates the arm. This diverse action occurs because different parts of the deltoid can contract independently. Locate the deltoid on material in the lab and compare it with figure 13.1. It is a fan-shaped muscle whose insertion partially covers the insertion of the pectoralis major. Deep to the pectoralis major muscle is the **pectoralis minor** muscle, whose fibers run in a more vertical direction, as seen in figure 13.1*b*. The **teres major** is a round muscle (*teres* means round). It attaches to the scapula and also to the anterior humerus. When it contracts, it extends and medially rotates the arm (figure 13.2).

The **coracobrachialis** (COR-uh-co-BRAYkee-AL-iss) is a short muscle that attaches to the scapula and also to the humerus. It is found deep to the pectoralis major on the anterior side of the body. Examine material in the lab and find the coracobrachialis as illustrated in figure 13.3.

Deep to the deltoid and the trapezius muscle are muscles that are located on the scapula proper. The **supraspinatus** (SOO-pra-spy-NAY-tus) is named for its attachment on the supraspinous fossa, and it is a synergist to the deltoid. The **infraspinatus** (IN-fra-spy-NAY-tus) is a muscle located on the infraspinous fossa. It laterally rotates the arm. The **subscapularis** (sub-SCAP-you-LAR-is) is named for it being found on the subscapular fossa. The other attachment of the subscapularis is on the anterior surface of the humerus; therefore, when it contracts, it medially rotates the arm. The subscapularis is located on the anterior surface of the scapula, between the scapula and the ribs, and cannot be seen in a posterior view. Locate these muscles in the lab and compare them with figure 13.2.

The **teres** (TARE-eez) **minor** is another muscle that attaches to the scapula, and it appears like a slip of the infraspinatus.

The **rotator cuff** muscles stabilize the shoulder joint by holding the head of the humerus into the glenoid cavity. Muscles composing the cuff are the supraspinatus, infraspinatus,

FIGURE 13.1 Overview of the Muscles of the Body
(a) Posterior view

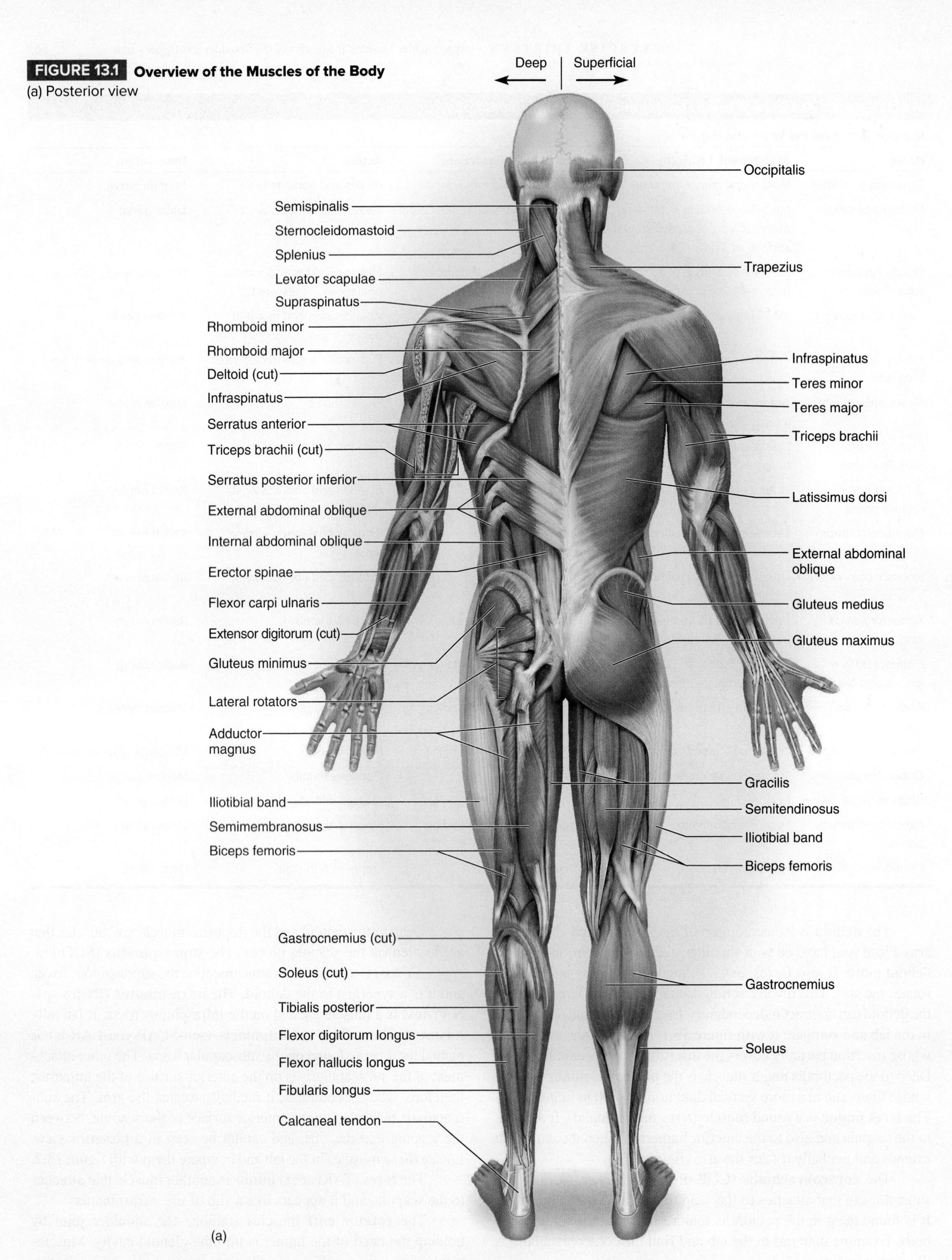

FIGURE 13.1 *(continued)* **Overview of the Muscles of the Body** (b) Anterior view.

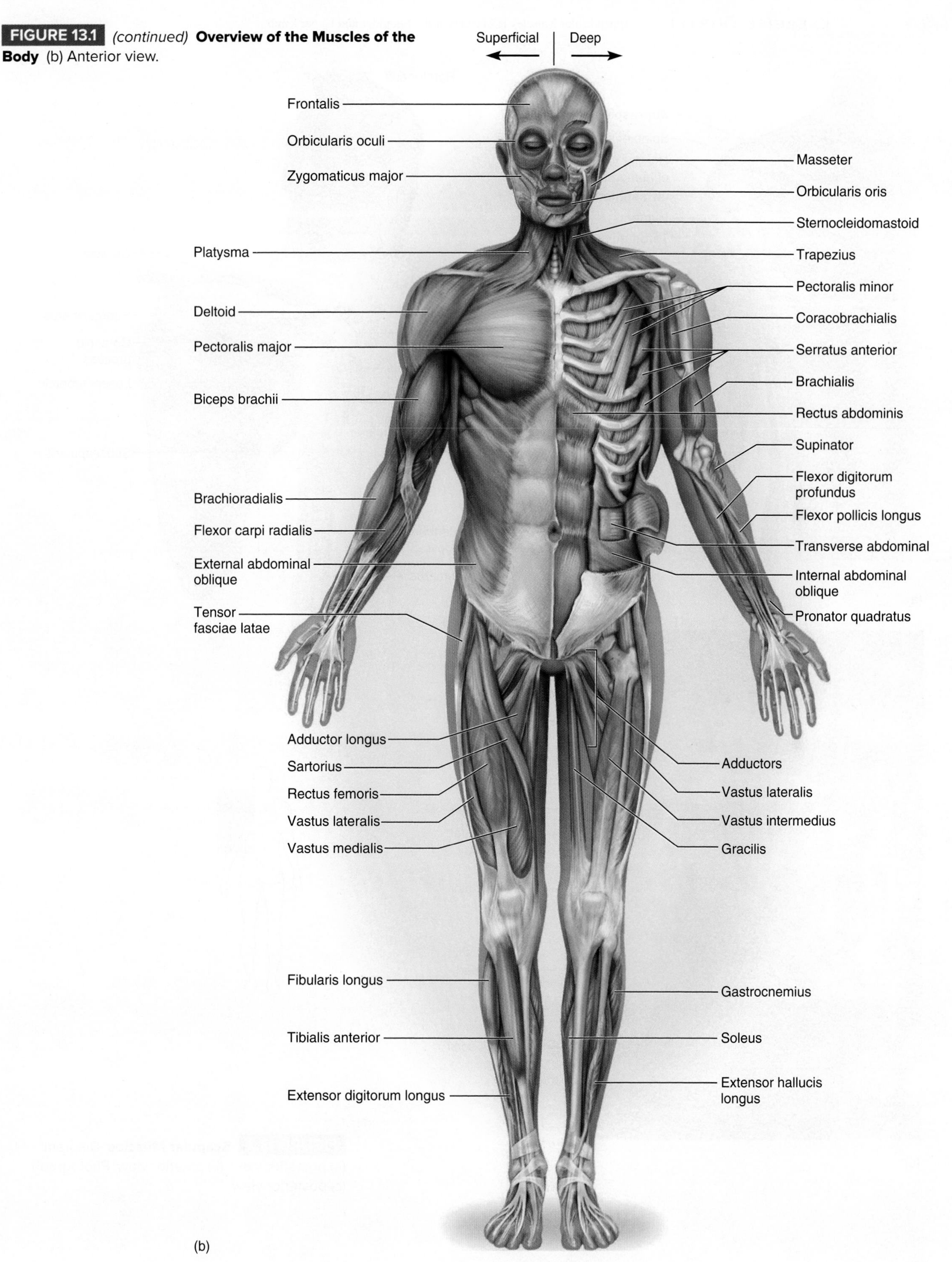

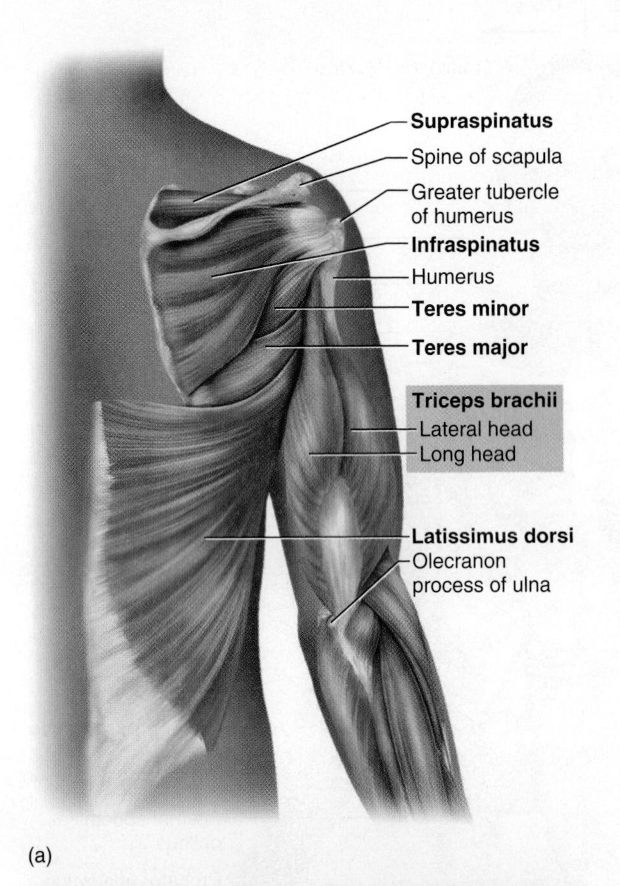

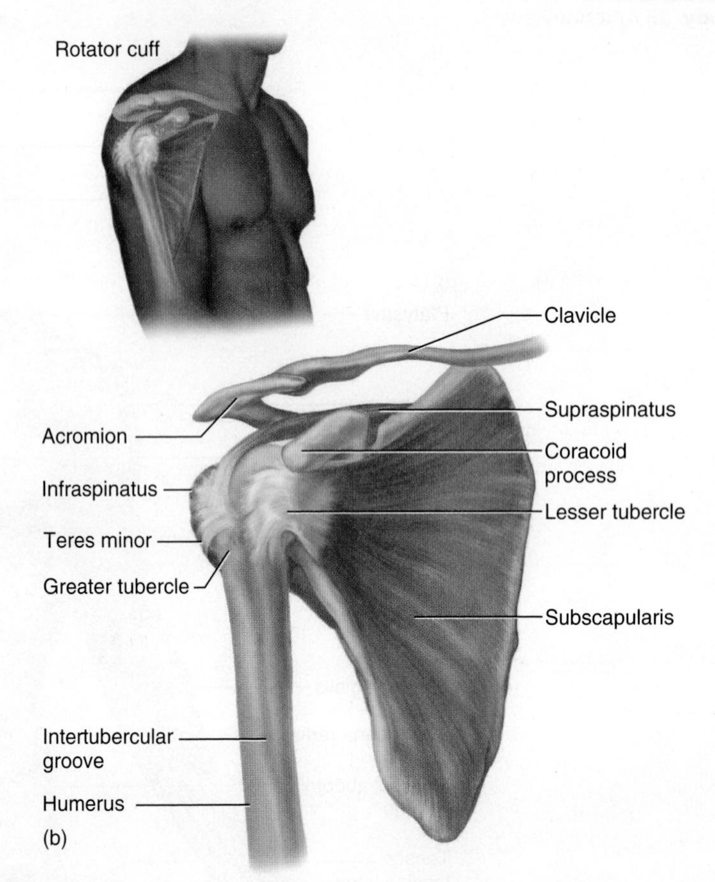

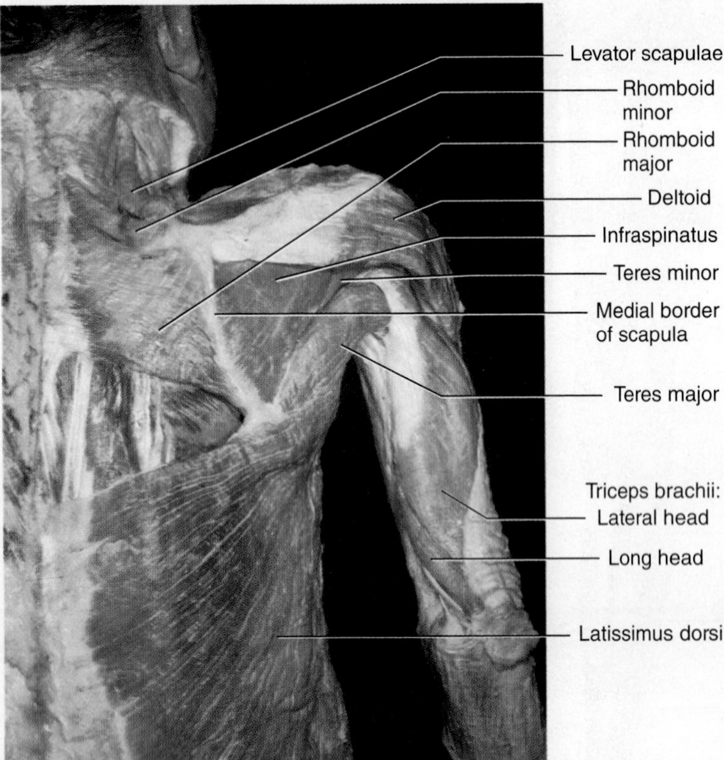

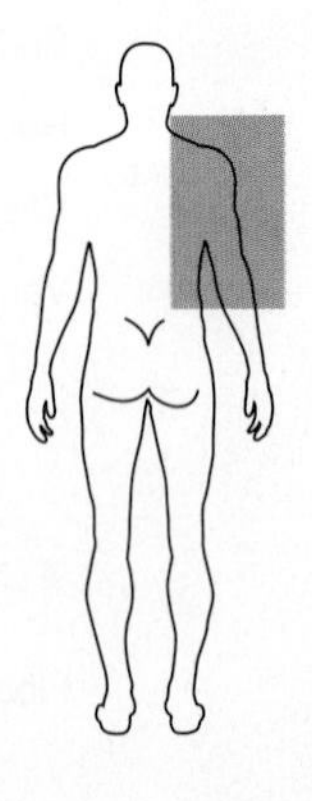

FIGURE 13.2 **Scapular Muscles** Diagram (a) posterior view; (b) anterior view. Photograph (c) posterior view.

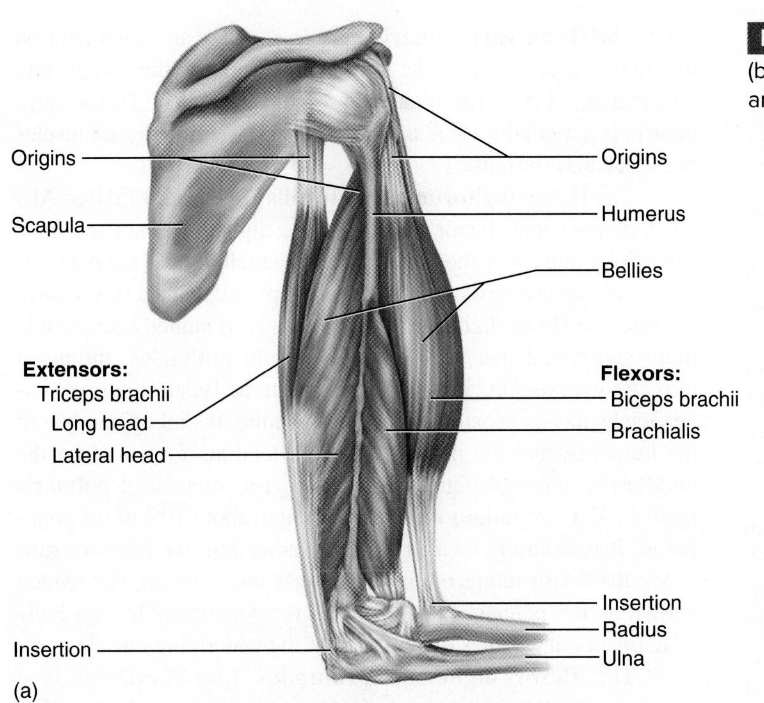

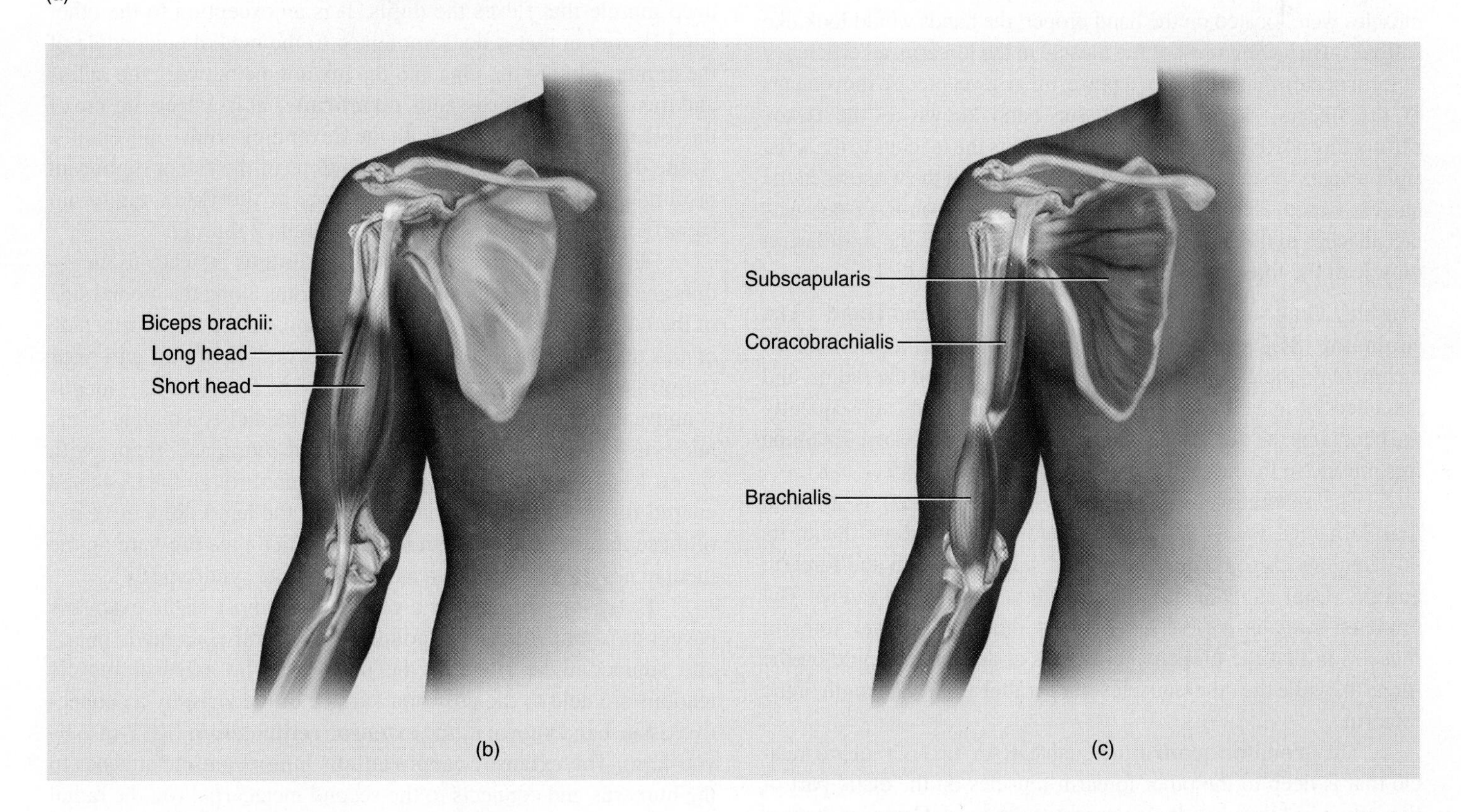

FIGURE 13.3 Muscles of the Arm Diagram (a) lateral view; (b) superficial muscles, anterior view; (c) deep muscles, anterior view.

subscapularis, and teres minor. You can remember these muscles because the humerus "SITS" in the rotator cuff. A rotator cuff injury occurs if there is damage to any of these muscles.

Muscles Acting on the Forearm The **biceps brachii** (BY-seps BRAY-kee-eye) muscle is a two-headed muscle of the arm. It has no attachment to the humerus, yet it has an action on the arm. Flexion of the forearm occurs primarily from contraction of this muscle. The biceps brachii has a long tendon that originates on the scapula and runs between the intertubercular sulcus of the humerus and a short tendon that originates on the coracoid process. The insertion of the biceps brachii is on the radius.

Underneath the biceps brachii on the anterior surface of the arm is the **brachialis** (BRAY-kee-AL-iss) muscle. The brachialis crosses the anterior surface of the elbow joint and flexes the forearm. The **triceps** (TRY-ceps) **brachii** is the only major muscle on the posterior surface of the humerus. The triceps brachii extends the arm and forearm and is an antagonist to the biceps brachii. Examine the material in the lab and compare these muscles to figure 13.3 and to descriptions found in table 13.1.

Muscles Acting on the Wrist and Hand The muscles found on the forearm, due to their similar appearance, provide greater challenges than do the muscles of the shoulder and arm. As you study these muscles, it is important that you locate them and determine their attachment points. These will help you determine if you are looking at the correct muscle. The muscles of the forearm and hand are generally named for their action (*pronate* the hand or *flex* the digits) or for their distal attachment (*carpi* for attaching to the carpals or metacarpals, *digitorum* for fingers, and *pollicis* for the thumb). In a few instances, they are named for the shape of the muscle (*teres* for round or *quadratus* for square).

Many of the anterior muscles are innervated by the median nerve. As you examine these muscles, look for the median nerve.

The fingers are controlled by muscles that attach to the humerus or the proximal radius or ulna and extend by long tendons to the digits. The fingers move by a force applied from a distance, analogous to a puppet controlled by puppet strings. If the hand muscles were located on the hand proper, the hands would look like softballs. By having most of the muscle in the forearm, an efficiency of form occurs that allows for a powerful grip yet precise movements of the fingers. A connective tissue band known as the **flexor retinaculum** (RET-in-AK-you-lum) anchors the tendons to the wrist and prevents the tendons from pulling away from the wrist when the hand is flexed. The **brachioradialis** (BRAY-kee-oh-RAY-dee-AL-iss) attaches to the distal part of the humerus and is the most lateral muscle of the forearm, as seen in figures 13.1*a* and 13.4.

Muscles Tnat Supinate and Pronate the Wrist and Hand The **supinator** (SOO-pin-AY-tor) is a muscle that has a proximal attachment on the arm and forearm, wrapping around the radius, and is named for its action of supinating the forearm and subsequently the hand. It is the deepest proximal muscle of the forearm. Examine this muscle in the lab and compare it with figure 13.5.

The **pronator teres** (PRO-nay-tur TARE-eez) is a round muscle named for its action of pronating the wrist and hand. Its proximal attachment is on the medial side of the arm and forearm and the distal attachment is on the lateral side of the radius. The pronator teres is different from the other superficial forearm muscles in that the pronator teres runs at an oblique angle on the forearm, while the other muscles run parallel along the length of the forearm.

The **pronator quadratus** (quah-DRAY-tus) is a square muscle that is deep to the other forearm muscles on the distal part of the radius and the ulna. It also pronates the hand. Compare the two pronator muscles in the lab with figures 13.4 and 13.5.

Flexor Muscles The flexor muscles are grouped by their insertion on the hand. The **flexor carpi** (CAR-pie) **radialis** muscle attaches to the metacarpals on the radial side of the hand. The pulse of the radial artery is taken at the wrist just lateral to the tendon of the flexor carpi radialis muscle. Most of the flexor muscles of the hand, including the flexor carpi radialis, have a proximal attachment on the medial epicondyle of the humerus. The flexor carpi radialis runs underneath the **flexor retinaculum.**

The **flexor carpi ulnaris** muscle also has an attachment on the medial epicondyle of the humerus and one on the carpals and on metacarpal 5 of the ulnar side of the hand. The flexor carpi ulnaris is a medial muscle of the forearm. Examine these muscles in the lab and in figure 13.4.

The **flexor digitorum superficialis** (SOO-per-FISH-ee-AL-is) is a superficial flexor muscle of the digits. It is not the most superficial muscle of the forearm but is actually under the palmaris longus (described next) and the flexor carpi radialis and flexor carpi ulnaris. The flexor digitorum superficialis is so named because it is more superficial than the flexor digitorum profundus, discussed next. As with the previous flexor muscles, the flexor digitorum superficialis has an proximal attachment on the medial epicondyle of the humerus, and the distal attachment tendons form a V on the middle phalanges of digits 2 through 5. The superficial **palmaris** (pahl-MARE-us) **longus** muscle is absent in about 10% of the population. It is centrally located in the middle, anterior forearm; runs above the flexor retinaculum; and inserts into a broad, flat tendon known as the palmar aponeurosis. This aponeurosis has no bony attachment but attaches to the fascia of the underlying muscles.

The **flexor digitorum profundus** (pro-FUND-us) is a deep muscle that flexes the digits. It is an exception to the other hand flexors in that it does *not* attach to the medial epicondyle of the humerus but to the ulna and the membrane between the radius and the ulna (the **interosseous membrane**). It is a deep muscle of the forearm that runs underneath the flexor digitorum superficialis. At the distal attachment point, the tendons of the flexor digitorum profundus run through the split tendons of the flexor digitorum superficialis to the distal phalanges of digits 2 through 5.

The **flexor pollicis** (PAUL-ih-sis) **longus** attaches to the radius and the interosseous membrane and runs along the medial side of the radius, attaching to the distal phalanx on the fingerprint side of the thumb. Flexion of the thumb occurs when you curl your thumb as if you were ready to flip a coin. The flexor pollicis longus is unusual in that it does not attach to the medial epicondyle of the humerus. Examine these muscles and compare them with figure 13.6. Many of the flexor muscles run through the U-shaped **carpal tunnel** at the proximal portion of the hand. Repetitive use of these muscles causes them to swell, which puts pressure on the median nerve. This is known as carpal tunnel syndrome.

Extensor Muscles As a general rule, most of the extensors have a proximal attachment point on the lateral epicondyle or lateral supracondylar ridge of the humerus. The extensor muscle tendons are held to the posterior surface of the wrist by a connective tissue band known as the **extensor retinaculum** (RET-in-AK-you-lum). The **extensor carpi radialis longus** muscle attaches to the humerus and connects to the second metacarpal (on the radial side) of the hand. The **extensor carpi radialis brevis** (BREV-is) is deep to the longus, and it attaches to the dorsum of the third metacarpal. Both of these muscles extend and abduct the hand. Examine these muscles in figure 13.7.

The **extensor digitorum** is a singular muscle on the back of the hand (remember that there are two flexor digitorum muscles). It has a proximal attachment to the lateral epicondyle of the humerus, as do the other extensors, and connects to the middle and distal phalanges of the second through fifth digits. The tendons of this muscle can be seen on the dorsal surface of the hand and in figure 13.7.

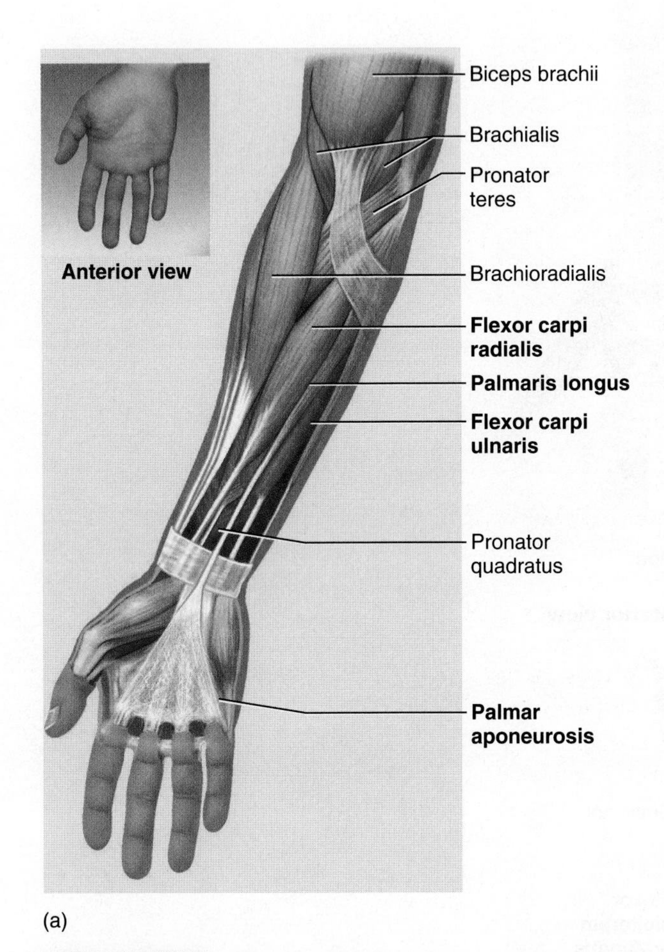

(a)

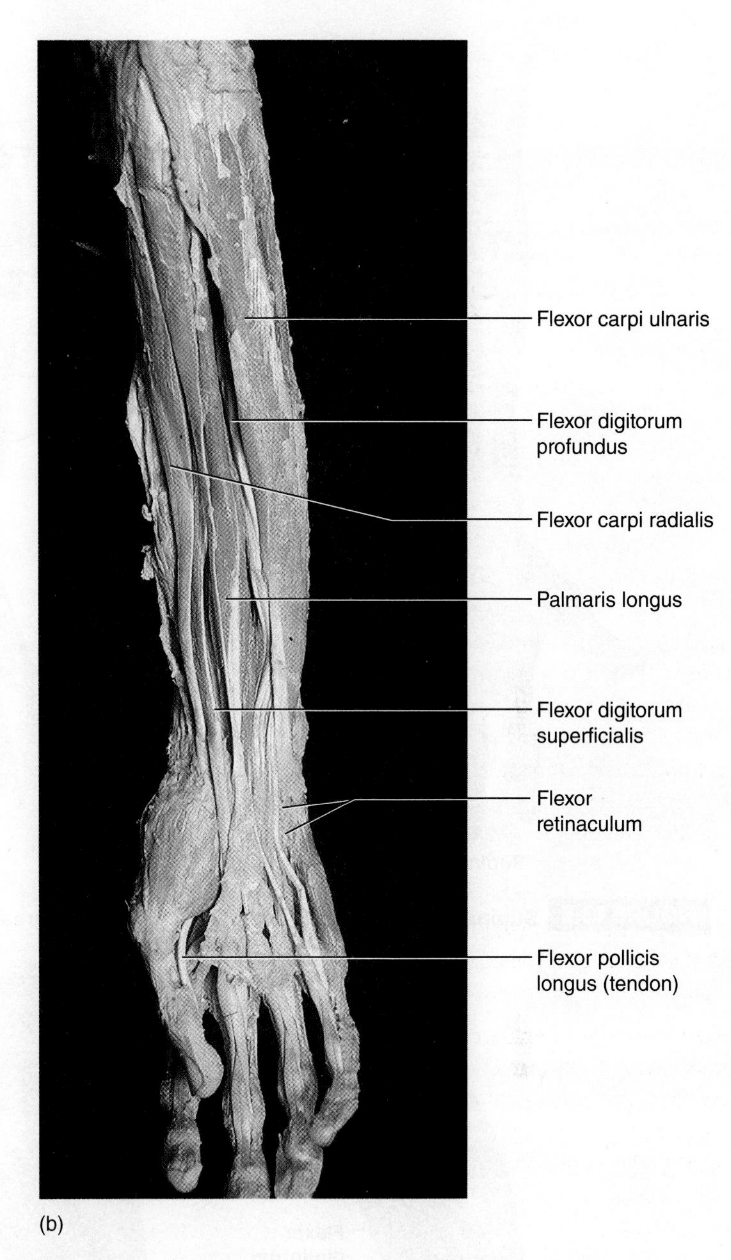

(b)

FIGURE 13.4 **Superficial Flexor Muscles of the Right Forearm, Anterior View** (a) Diagram; (b) photograph.

The **extensor carpi ulnaris** has a proximal attachment to the lateral epicondyle and a distal attachment to the fifth metacarpal (on the ulnar side) of the hand. It extends and adducts the hand.

The **abductor pollicis longus** muscle runs to the metacarpal of the thumb. By pulling your thumb in an anterior direction away from your index finger, you are abducting the thumb. There are two extensor pollicis muscles, the **extensor pollicis longus** and the **extensor pollicis brevis.** Extension of the thumb is done when you flip a coin. At the end of the flip, the thumb is extended. The tendons of the extensor pollicis muscles and the abductor pollicis muscles make a depression in the form of a triangle at the base of the thumb. This depression is known as the anatomical snuff box. These muscles can be seen in figures 13.7 and 13.8.

Intrinsic Muscles of the Hand Many muscles have both attachment points on the hand and have an action on the hand. Some of these arise on the pad of muscles at the base of the thumb known as the **thenar eminence.** Thenar muscles are named for their action on the thumb, including the **abductor pollicis brevis,** the **flexor pollicis brevis,** and the **opponens** (up-POH-nenz) **pollicis.** The pad of tissue at the base of the fifth digit is known as the hypothenar eminence, and the muscles that arise there are the **abductor digiti minimi,** the **flexor digiti minimi brevis,** and the **opponens digiti minimi.** Find these muscles on models or cadavers in the lab and compare them with figure 13.9.

Cat Anatomy

If you are using cats, turn to section 2, “Dissection, Overview, and Forelimb Muscles of the Cat,” on page 409 of this lab manual.

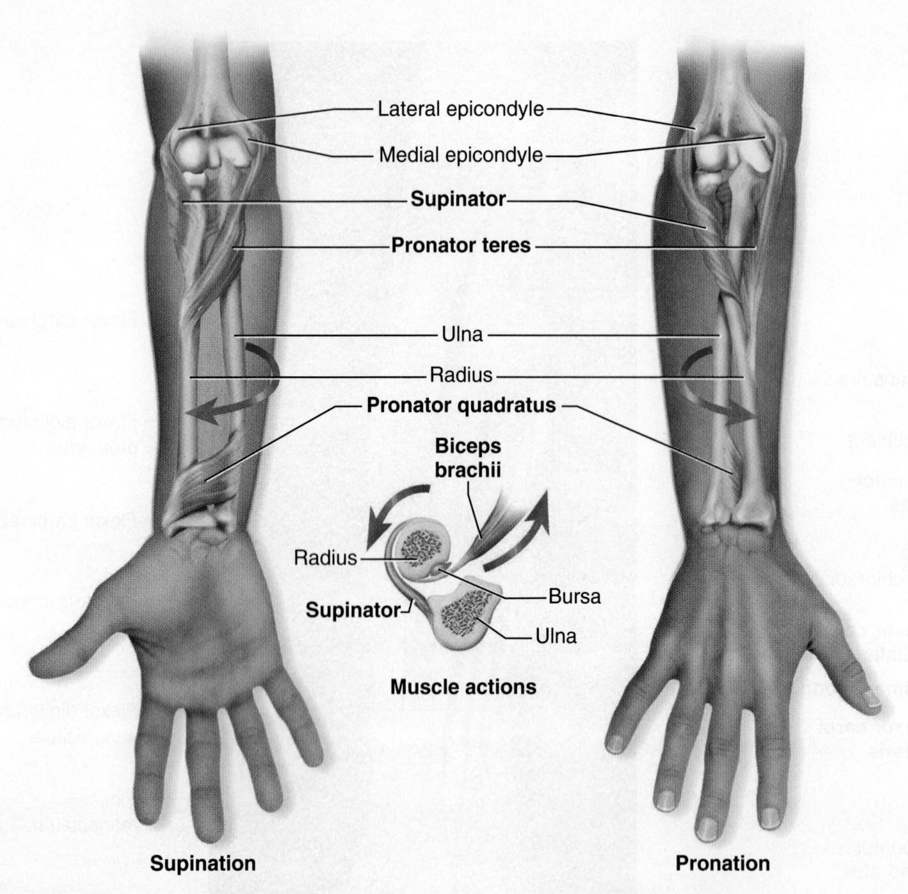

FIGURE 13.5 Supinator and Pronator Muscles of the Right Forearm, Anterior view

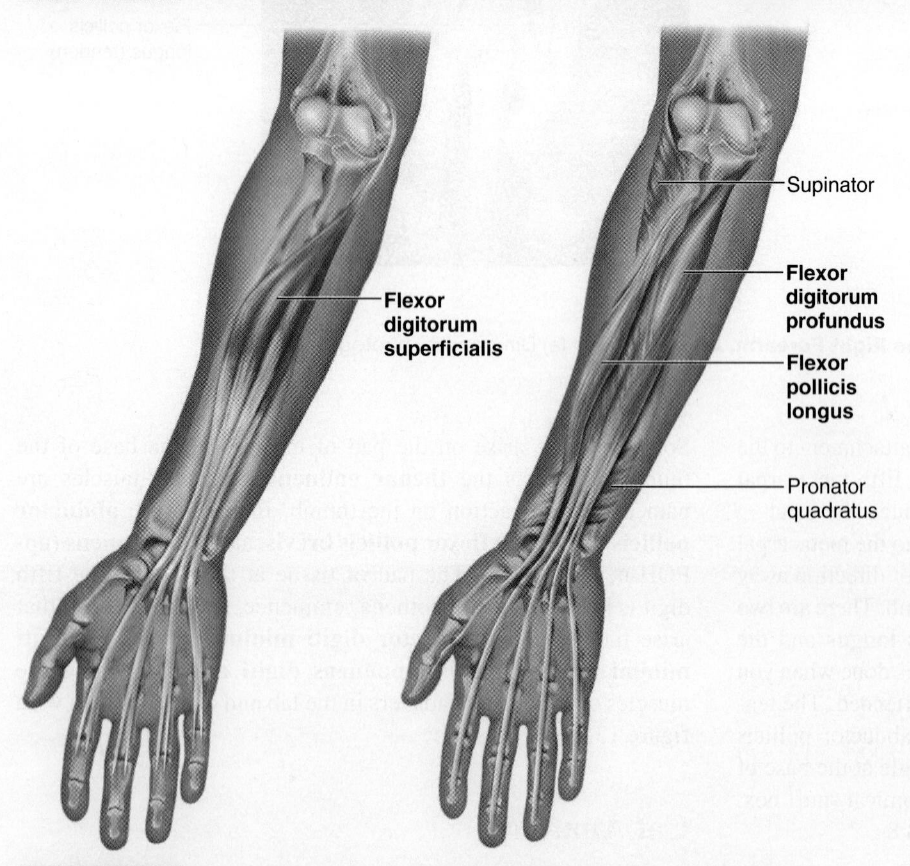

FIGURE 13.6 Deep Flexor Muscles of the Right Forearm, Anterior View

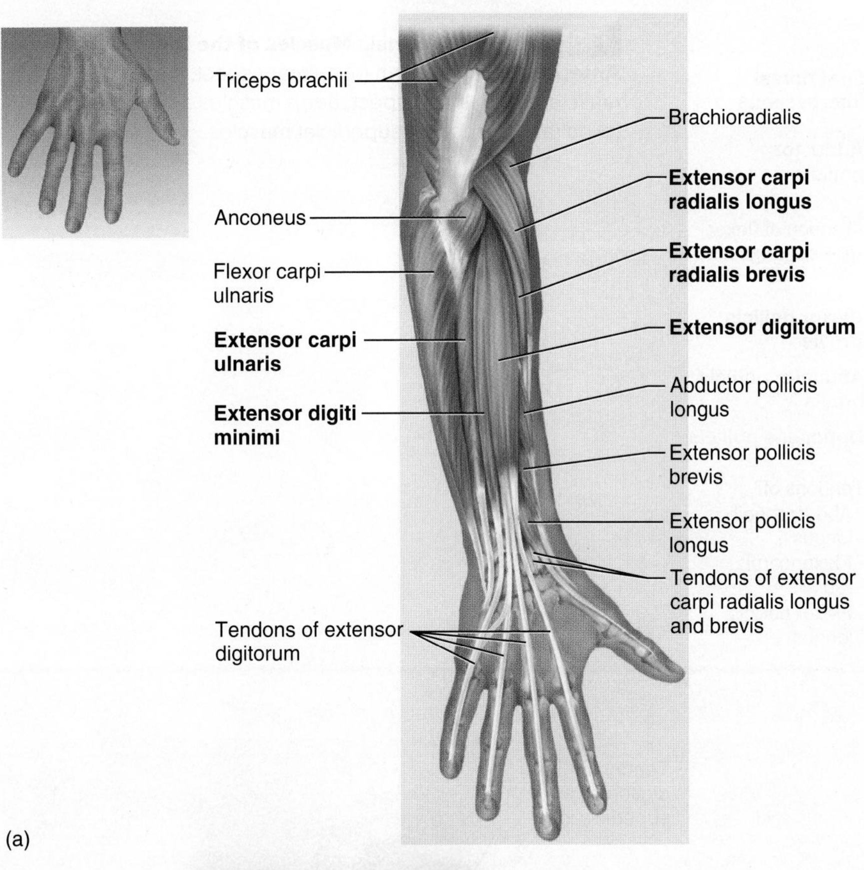

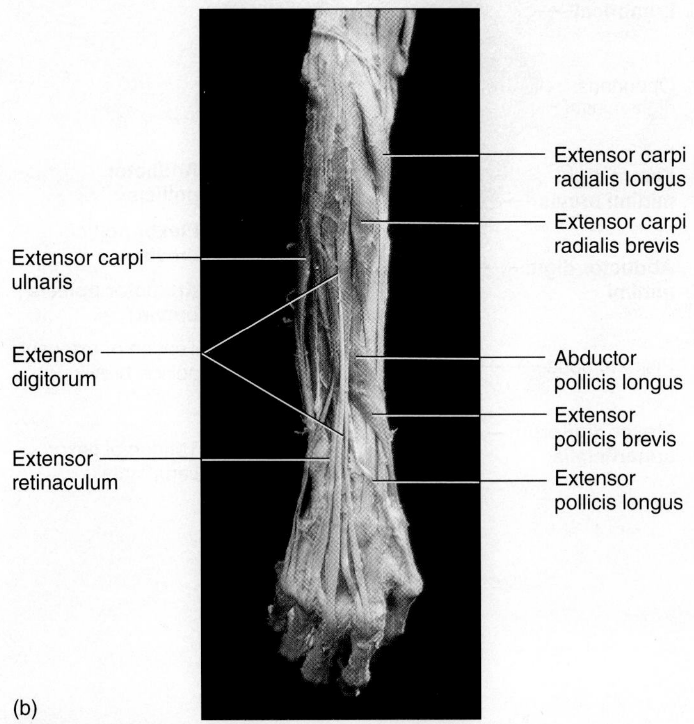

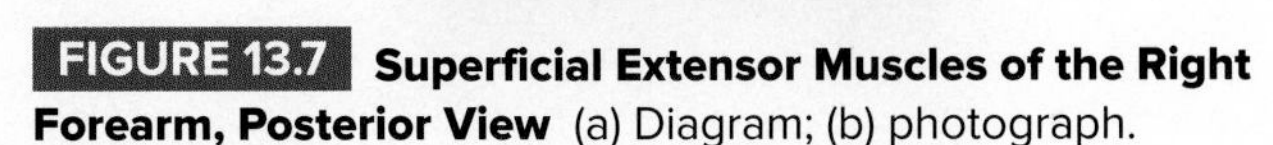
FIGURE 13.7 Superficial Extensor Muscles of the Right Forearm, Posterior View (a) Diagram; (b) photograph.

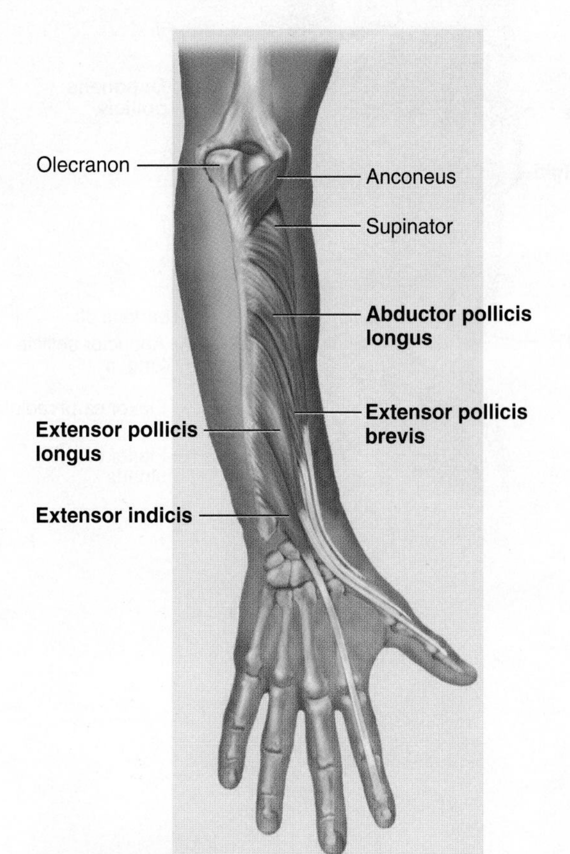

FIGURE 13.8 Deep Extensor Muscles of the Right Thumb, Posterior View

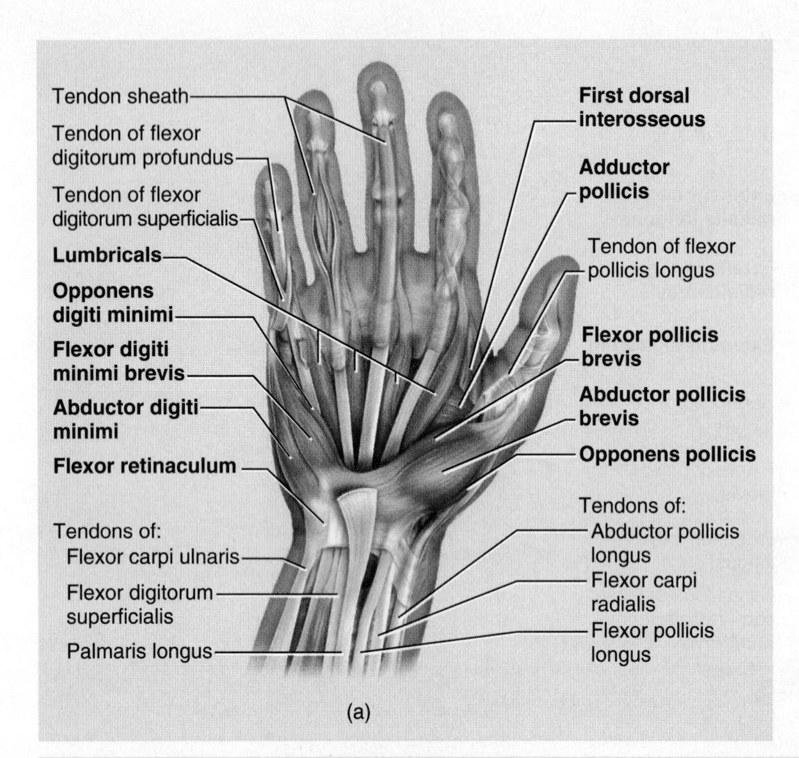

FIGURE 13.9 Intrinsic Muscles of the Right Hand, Anterior View Diagram (a) palmar aspect, superficial muscles; (b) palmar aspect, deep muscles. Photograph (c) palmar dissection, superficial muscles.

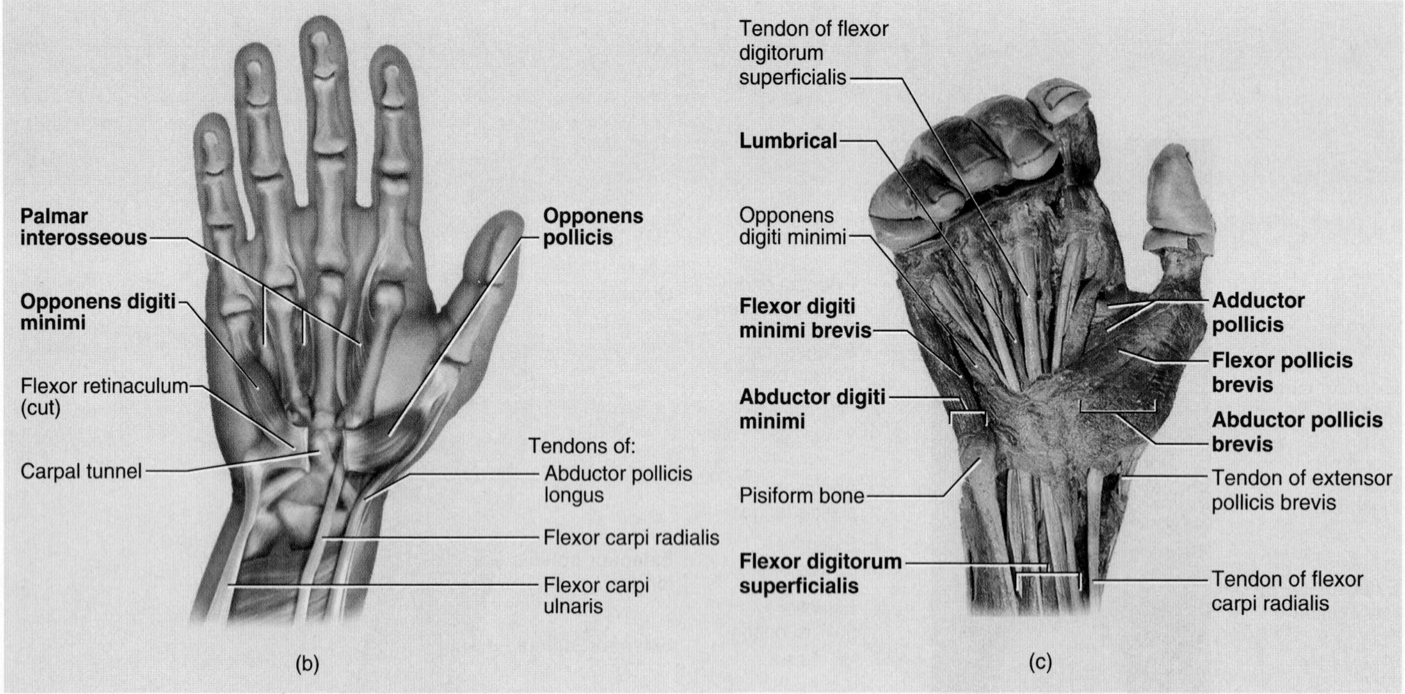

Exercise 13

Appendicular Muscles 1: Muscles of the Shoulder and Upper Limb

Name ______________________________ Date ______________

1. What is the lateral attachment of the trapezius muscle? ______________________________

2. What is the action of the deltoid muscle? ______________________________

3. What is the attachment point on the limb of the latissimus dorsi muscle? ______________________________

4. Where on the scapula is the attachment point of the supraspinatus muscle? ______________________________

5. What is the superior attachment point of the pectoralis minor? ______________________________

6. What is the inferior attachment point of the pectoralis minor? ______________________________

7. Does the biceps brachii muscle attach to the humerus? ______________________________

8. Name the action of the pectoralis major. ______________________________

9. What is the lateral attachment point of the subscapularis? ______________________________

10. What is the action of the triceps brachii? ______________________________

____

11. What is the proximal attachment point of the brachialis? ______________________________

____

12. Name all the muscles that flex the arm. ______________________________

____

____

13. Which muscles are antagonists to the triceps brachii? ______________________________

____

14. What is the proximal attachment point of the flexor carpi ulnaris? ______________________________

____

15. What is an antagonist to the supinator muscle? ______________________________

____

16. Where does the flexor digitorum superficialis distally attach? ______________________________

____

17. What is the distal attachment point of the extensor carpi ulnaris muscle? ______________________________

____

18. What is the action of the flexor carpi radialis muscles? ______________________________

____

19. What muscle extends both the hand and the phalanges? ______________________________

____

20. What muscles flex the hand? ______________________________

____

21. What muscles extend the thumb? ______________________________

____

22. Where are the extensor carpi muscles found, on the anterior or posterior side of the forearm? ______________________________

____

Laboratory Exercise 14

Appendicular Muscles 2: Muscles of the Hip, Thigh, Leg, and Foot

Muscular System

Introduction

The muscles of the hip, thigh, leg, and foot are primarily muscles of locomotion. They can be divided into muscles that have an action on the thigh, the leg, or the foot.

The size and shape of the muscles of the hip and thigh in humans are different from those of other mammals in that humans are bipedal. This two-legged walking habit is different from a quadruped's locomotion in terms of balance and forward movement. When a quadruped walks, typically three legs remain on the ground. When a biped walks, the body is placed in an unstable position and the support limb must be stable to maintain the center of gravity. Think about what happens to the mechanics of movement and the center of balance when you move a leg to walk or run.

The muscles of the leg originate typically from the femur, tibia, or fibula and insert on the bones of the foot. These muscles, involving movement of the foot, use some terms of action specific to the foot. The term **dorsiflexion** describes a decrease in the angle between the shin and dorsum (top of the foot). **Plantar flexion** is an increase in the angle between the shin and the dorsum of the foot (seen as you stand on your toes to reach something high). Review other actions of the leg and foot in laboratory exercise 10. Some of the muscles of the feet are similar to those of the hand, such as the extensor digitorum muscles.

Learning Objectives

At the end of this exercise you should be able to

1. locate the muscles of the hip, thigh, leg, and foot on a model, chart, or cadaver (if available);
2. list the origin, insertion, and action of each muscle presented;
3. describe what nerve controls each muscle;
4. list what muscles function as synergists or antagonists to the prime mover; and
5. name all the muscles that move a joint, such as all the muscles that flex the thigh or those that dorsiflex the foot.

Materials

Human torso model
Human leg models
Human muscle charts
Articulated skeleton
Cadaver (if available)

Procedure

Review the muscle nomenclature and the action of muscles as outlined in laboratory exercises 10 and 11. Examine muscle models or charts in the lab, and locate the muscles described in the following section and in table 14.1. Associate the shape of the muscle with the name and begin to visualize the muscles as you study the models or charts. You may want to look at an articulated skeleton as you review the origins and insertions of the muscles, so that you can better see the muscle attachment points. Once you have learned the origin and the insertion, you should be able to understand the action of the muscle. The action of the muscles will be easier to understand when you have identified the origins and the insertions. This is done, in part, by imagining how the bony attachments of the origin and insertion would come together if the muscle pulled them closer to one another. If you have a cadaver, pay particular attention to the origins and the tendons of insertion of the muscles of the leg and foot.

Anterior Muscles of the Hip and Thigh

The **iliopsoas** (IL-ee-oh-SO-az) muscle is a major flexor of the thigh. It attaches inside the abdominopelvic cavity and crosses over the pelvic brim to the posterior aspect of the femur. The iliopsoas is actually two muscles, the **iliacus** (ILL-ee-AK-us) and **psoas** (SO-az) **major.** A companion muscle to the iliopsoas is the **psoas minor.** This exercise treats these muscles as one. The iliopsoas is seen in figure 14.1.

The **tensor fasciae latae** (TEN-sur FASH-ee-ee LAY-tee) is a lateral muscle of the thigh that attaches to a thick band of connective tissue (the **fasciae latae**) that runs down the lateral aspect of the thigh similar to a stripe on a pair of tuxedo pants. This is known as the **iliotibial tract,** a thickened portion of the broader fasciae latae. This muscle abducts the thigh and is illustrated in figures 14.1 and 14.2.

TABLE 14.1 Appendicular Muscles 2: Muscles of Hip, Thigh, Leg, and Foot

Name	Attachment 1 (Origin)	Attachment 2 (Insertion)	Action	Innervation
Muscles Acting on the Hip and Thigh				
Iliopsoas	T12, L1–5, sacrum, iliac crest, iliac fossa	Lesser trochanter of femur	Flexes thigh and lumbar vertebrae	Femoral nerve, spinal nerves L1–3
Tensor fasciae latae	Iliac crest, anterior iliac surface	Iliotibial tract of fasciae latae to lateral condyle of tibia	Flexes, abducts, and medially rotates thigh	Superior gluteal nerve
Gluteus maximus	Outer iliac surface, iliac crest, sacrum, coccyx	Gluteal tuberosity of femur, iliotibial tract of fasciae latae	Extends and adducts thigh	Inferior gluteal nerve
Gluteus medius	Lateral surface of ilium	Greater trochanter of femur	Abducts and medially rotates thigh	Superior gluteal nerve
Gluteus minimus	Lateral surface of ilium	Greater trochanter of femur	Abducts and medially rotates thigh	Superior gluteal nerve
Medial (Adductor) Compartment of the Thigh				
Adductor brevis	Pubis	Linea aspera of femur	Adducts thigh	Obturator nerve
Adductor longus	Pubis	Linea aspera of femur	Adducts, flexes, and medially rotates thigh	Obturator nerve
Adductor magnus	Pubis, ischium	Gluteal tuberosity, linea aspera, adductor tubercle of distal femur	Adducts, flexes, extends, and medially rotates thigh	Obturator nerve
Gracilis	Pubis, ischium	Proximal, medial tibia	Flexes and medially rotates leg	Obturator nerve
Pectineus	Pubis	Femur inferior to lesser trochanter	Adducts and flexes thigh	Femoral nerve
Muscles Acting on the Knee and Leg				
Quadriceps femoris				
Rectus femoris	Anterior inferior iliac spine, margin of acetabulum	Tibial tuberosity by the patellar tendon, lateral and medial condyles of tibia	Flexes thigh, extends leg	Femoral nerve
Vastus lateralis	Greater trochanter of femur linea aspera of femur	Tibial tuberosity by the patellar tendon, lateral and medial condyles of tibia	Extends leg	Femoral nerve
Vastus medialis	Linea aspera, medial side	Tibial tuberosity by the patellar tendon, lateral and medial condyles of tibia	Extends leg	Femoral nerve
Vastus intermedius	Proximal, anterior femur	Tibial tuberosity by the patellar tendon, lateral and medial condyles of tibia	Extends leg	Femoral nerve
Sartorius	Anterior superior iliac spine	Proximal, medial tibia	Flexes and laterally rotates thigh, flexes knee	Femoral nerve
Popliteus	Lateral condyle of femur	Proximal tibia	Flexes leg, thus "unlocking" knee	Tibial nerve
Hamstrings				
Biceps femoris	Ischial tuberosity, linea aspera	Head of fibula	Extends thigh, flexes knee	Sciatic nerve
Semitendinosus	Ischial tuberosity	Proximal, medial tibia	Extends thigh, flexes knee	Sciatic nerve
Semimembranosus	Ischial tuberosity	Medial condyle of tibia	Extends thigh, flexes knee	Sciatic nerve
Muscles Acting on the Foot				
Fibularis tertius	Medial, distal fibula	Dorsum of fifth metatarsal	Dorsiflexes and everts foot	Deep fibular nerve
Extensor digitorum longus	Lateral condyle of tibia, shaft of fibula	Middle and distal phalanges of second through fifth digits	Extends toes, dorsiflexes foot	Deep fibular nerve
Extensor hallucis longus	Anterior shaft of fibula	Distal phalanx of hallux (big toe)	Extends hallux, dorsiflexes foot	Deep fibular nerve
Tibialis anterior	Lateral condyle and proximal tibia	First metatarsal, medial cuneiform	Dorsiflexes and inverts foot	Deep fibular nerve
Gastrocnemius	Condyles of femur	Calcaneus	Flexes leg, plantar flexes foot	Tibial nerve
Soleus	Posterior, proximal tibia and fibula	Calcaneus	Plantar flexes foot	Tibial nerve
Flexor digitorum Longus	Posterior tibia	Distal phalanges of second through fifth digits	Flexes toes, plantar flexes and inverts foot	Tibial nerve
Flexor hallucis longus	Inferior shaft of fibula	Distal phalanx of hallux (big toe)	Flexes hallux, inverts foot	Tibial nerve
Tibialis posterior	Proximal tibia and fibula	Second through fourth metatarsals and some tarsals	Plantar flexes and inverts foot	Tibial nerve
Fibularis brevis	Shaft of fibula	Fifth metatarsal	Plantar flexes and everts foot	Superficial fibular nerve
Fibularis longus	Head and shaft of fibula, lateral condyle of tibia	First metatarsal, medial cuneiform	Plantar flexes and everts foot	Superficial fibular nerve

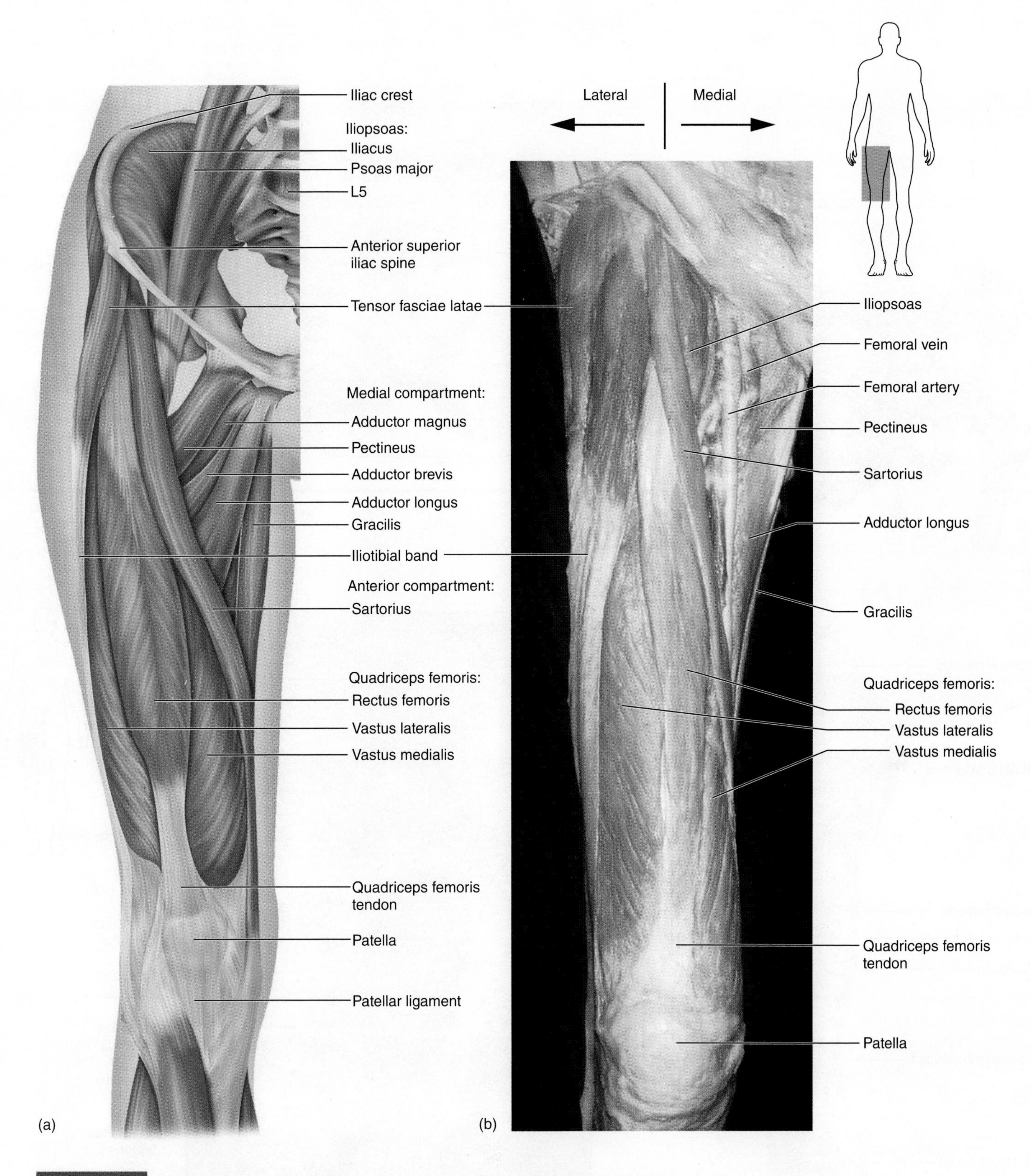

FIGURE 14.1 **Muscles of the Right Thigh, Anterior View** (a) Diagram; (b) photograph.

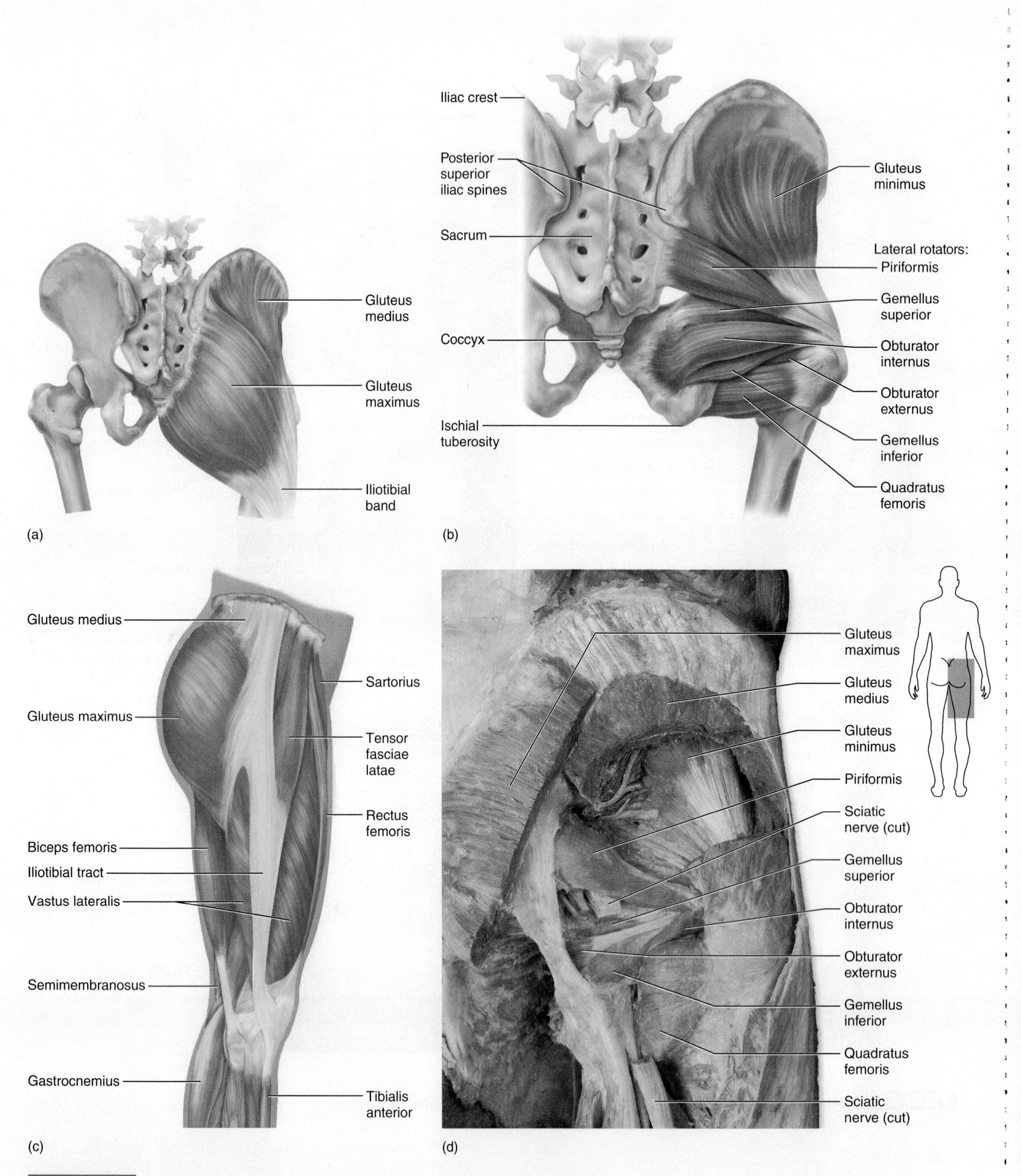

FIGURE 14.2 **Muscles of the Right Hip and Thigh** Diagram (a) superficial muscles, posterior view; (b) deep muscles, posterior view; (c) lateral muscles. Photograph (d) superficial muscles of the hip.

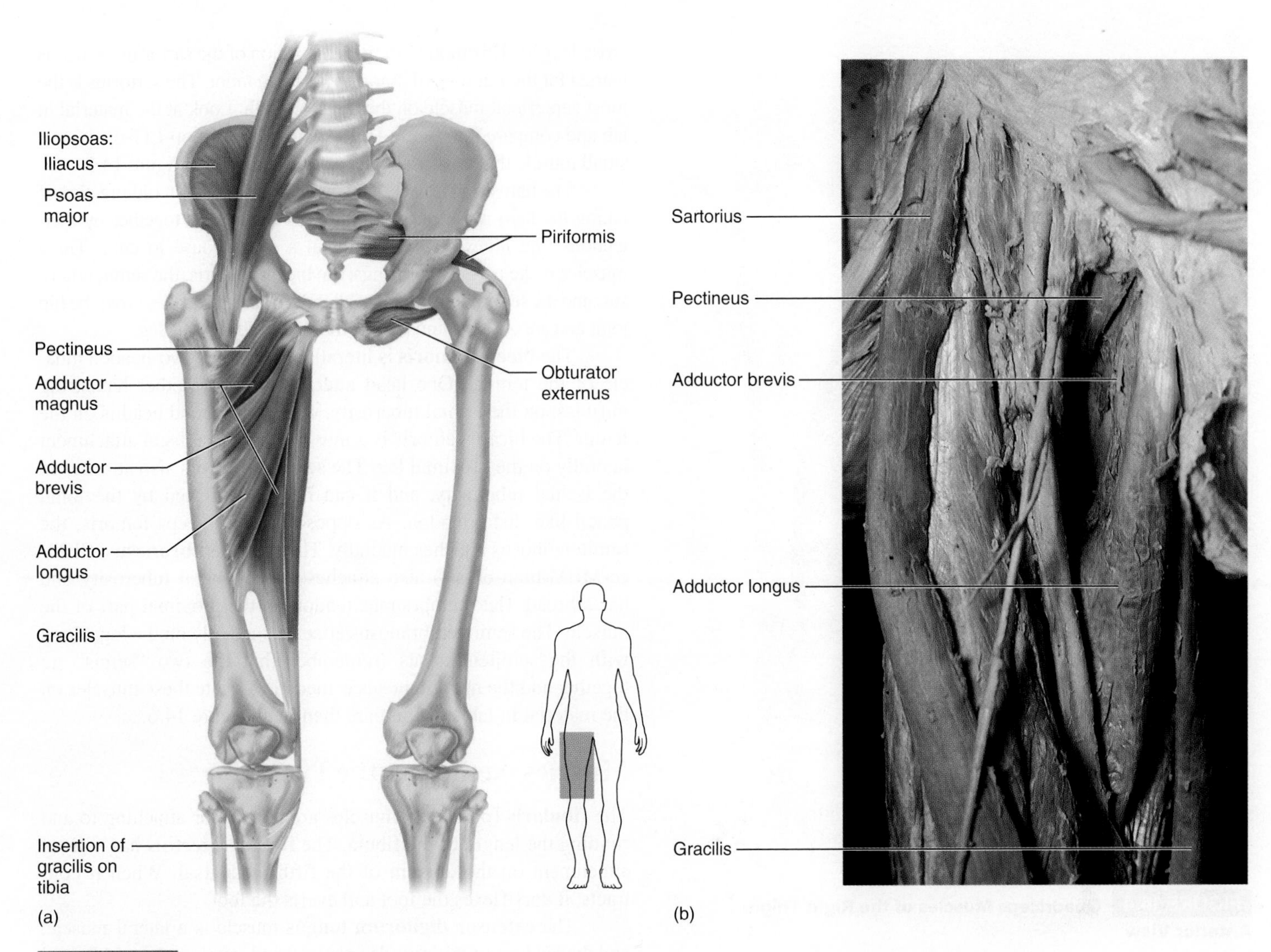

FIGURE 14.3 **Iliopsoas and Adductor Muscles of the Right Thigh, Anterior View** (a) Diagram; (b) photograph.

The gluteus muscles in humans are different than in other mammals because we are bipedal. The largest of the gluteal muscles is the **gluteus maximus**. It extends the thigh when standing from a sitting position or when climbing up stairs. The **gluteus medius** and the **gluteus minimus** both aid in keeping balance during walking because they maintain the center of gravity. The gluteus medius is superficial to the gluteus minimus but both of these muscles originate on the outer surface of the ilium. When you raise one leg to walk, gravity pulls your body in the direction of the lifted leg. The gluteus medius and minimus on the opposite side of the body contract to maintain upright posture. Compare these muscles with figure 14.2. Deep to the gluteal muscles are lateral rotators and abductors of the thigh. These are the **piriformis, obturator internus, superior gemellus, inferior gemellus,** and **quadratus femoris.** Find these muscles in lab and compare them with figure 14.2*b*.

The muscles in the medial aspect of the thigh are adductor muscles. The gracilis is an example of an adductor muscle, as are the pectineus and the three adductor muscles proper. The **adductor brevis** is the shortest adductor muscle and is found deep in the medial aspect of the thigh. It can be seen if the pectineus (discussed shortly) is reflected. The adductor brevis has a stable attachment on the pubis and connects to the proximal third of the thigh. The **adductor longus** also has a pubic attachment, but it goes to the middle third of the thigh. The largest of the adductor muscles is the **adductor magnus,** and it attaches along the length of the inferior hip bone (pubis and ischium) and also along the length of the femur from the gluteal tuberosity, along the length of the linea aspera to the adductor tubercle.

The most superficial muscle of the medial aspect of the thigh is the gracilis. The **gracilis** (GRASS-ih-lis) is a thin muscle that adducts the thigh. The **pectineus** (pec-TINee-us) is a small, quadrangular muscle that is the most proximal of the group. It is superficial to the adductor brevis muscle and proximal to the adductor longus muscle. The gracilis is medial, the adductors are more lateral, and the pectineus is more lateral still. This sequence of the first letters of the muscles spells the word "gap." Examine the material in the lab and compare these muscles with figures 14.1 and 14.3.

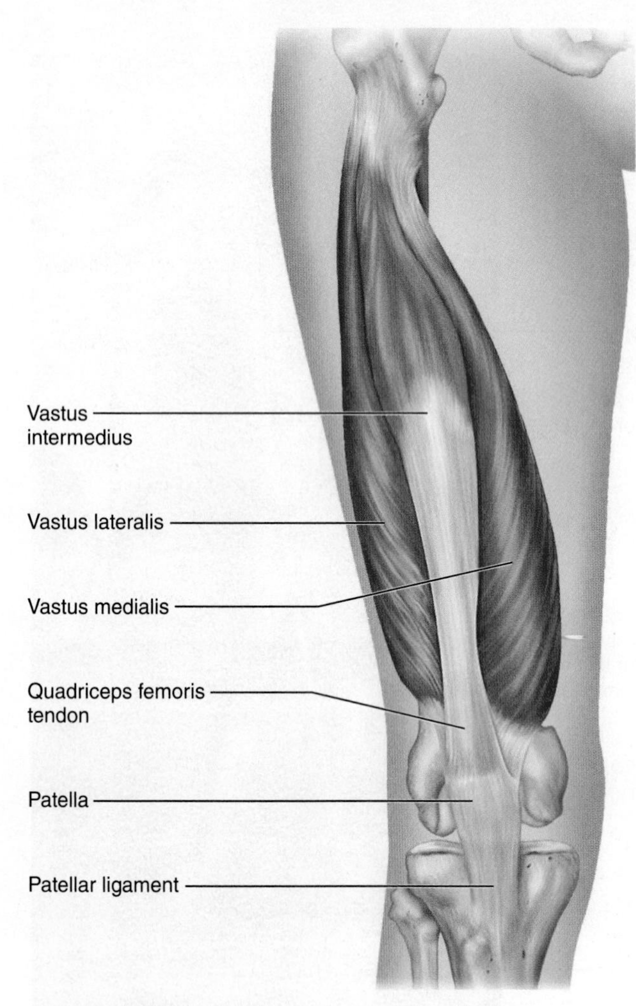

FIGURE 14.4 Quadriceps Muscles of the Right Thigh, Anterior View

The muscles of the anterior aspect of the thigh belong to the **quadriceps femoris** group. All these muscles extend the leg, while only one member of the group flexes the thigh. The quadriceps muscles all have a common ligament called the patellar ligament. This ligament attaches to the tibial tuberosity which attaches to the medial and lateral condyles of the tibia. The most superficial muscle of the group is the **rectus femoris** (FEM-or-us) (fig. 14.1). It is a muscle that has a proximal attachment directly inferior to the sartorius. The rectus femoris runs straight down the femur (*rectus* = straight). Because it crosses the hip joint, the rectus femoris is the only quadriceps muscle to flex the thigh.

The three other quadriceps muscles are the vastus muscles and are illustrated in figures 14.1 and 14.4. The word "vastus" means large. The **vastus lateralis** is named due to its lateral position. The **vastus intermedius** is deep to the rectus femoris, and the **vastus medialis** is the most medial of these muscles. These muscles attach to the femur, so they do not have an action on the hip. They also attach to the tibia and extend the leg.

The **sartorius** (sar-TOR-ee-us) is a slender muscle that has a superior attachment on the superior, lateral hip and attaches in an inferior, medial location. When hemming a pair of pants, a tailor sits cross-legged. This motion mimics the action of the sartorius, which is named for the Latin word "sartor," meaning tailor. The sartorius is the most superficial muscle on the anterior thigh. Look at the material in lab and compare it to figure 14.1. The **popliteus** (pop-LIT-ee-us) is a small muscle that crosses the knee joint. It is seen in figure 14.5.

The hamstring muscles are named because of an old practice of taking the ham muscles from a pig and tying them together by their tendons (the *hamstrings*) to hang in a smokehouse to cure. Three muscles make up the hamstrings: the biceps femoris, the semitendinosus, and the semimembranosus. All three of these muscles cross the hip joint and are extensor muscles of the thigh and flex the leg.

The **biceps femoris** is literally named the "two-headed muscle of the femur." One head attaches, with the other hamstring muscles, on the ischial tuberosity, while the second head is on the femur. The biceps femoris is a muscle that has a distal attachment laterally on the proximal leg. The **semitendinosus** also attaches to the ischial tuberosity, and it can be distinguished by the long, pencil-like distal tendon. As opposed to the biceps femoris, the semitendinosus attaches medially. The **semimembranosus** (SEM-ee-MEM-bran-oh-sis) also attaches to the ischial tuberosity and has a broad, flat membranous tendon on the proximal part of the muscle. The semimembranosus attaches medially on the leg, along with the semitendinosus (remember that the two "semis" go together and the membranosus is medial). Locate these muscles on the material in lab and compare them with figure 14.6.

Muscles Acting on the Foot

The **fibularis** (peroneus) muscles are named for attaching to and running the length of the fibula. The **fibularis tertius** has a distal attachment on the dorsum of the fifth metatarsal. When it contracts, it dorsiflexes the foot and everts the foot.

The **extensor digitorum longus** muscle is a lateral muscle, and the tendons of this muscle splay out and attach to the middle and distal phalanges of all the digits of the foot except the hallux. Extend your toes and feel these tendons. The **extensor hallucis** (HAL-uh-sis) **longus** is the muscle that extends the hallux and acts as a synergist to the extensor digitorum longus in dorsiflexion of the foot. Examine these muscles in the lab and compare them with figures 14.7 and 14.8. The muscles on the dorsum of the foot are held down by the **extensor retinaculum.** This is a connective tissue band that prevents the extensor muscles from bowstringing. These are major muscles of the foot. There are also smaller muscles such as the extensor digitorum brevis and the extensor hallucis brevis.

The **tibialis anterior** is located just lateral to the crest of the tibia on the anterior side of the leg, as seen in figure 14.8. Palpate the tibia. Just lateral to the tibia is a fleshy mass of muscle which is the tibialis anterior. The tendon of the tibialis anterior crosses to the medial side of the foot and attaches to the first metatarsal and medial cuneiform. As the tibialis anterior contracts, it decreases the angle between the anterior tibial crest and the dorsum of the foot in an action known as dorsiflexion of the foot. Because the distal attachment of this muscle is medial, it also inverts the foot.

The **gastrocnemius** (GAS-trock-NEE-meus) is a calf muscle with a medial and lateral head; it is the most superficial muscle of the posterior leg group. The gastrocnemius crosses the knee joint and

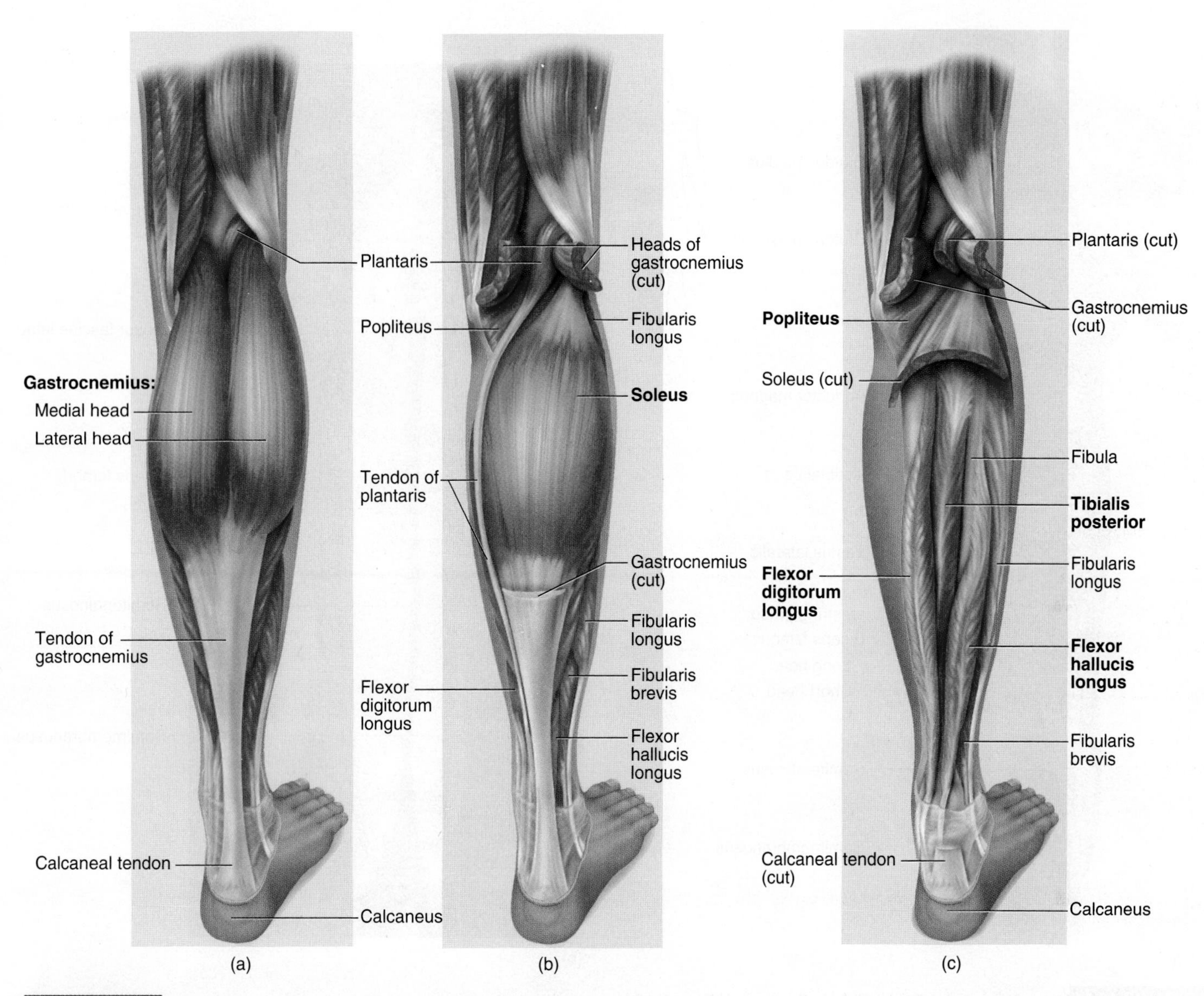

FIGURE 14.5 **Muscles of the Right Leg and Foot, Posterior View** (a) Superficial muscles of the leg; (b) middle-level muscles of the leg; (c) deep muscles of the leg.

flexes the leg. It attaches to the calcaneus by way of the **calcaneal tendon,** plantar flexing the foot as well. The calcaneal tendon is also known as the Achilles tendon. Deep to the gastrocnemius is the **soleus** (SO-lee-us) muscle. Unlike the gastrocnemius, the soleus does not attach to the femur; therefore, it does not cross the knee and has no action on the leg. It inserts on the calcaneus, sharing the calcaneal tendon with the gastrocnemius, and it plantar flexes the foot. Palpate the lateral side of the gastrocnemius and feel the junction between it and the soleus. These are seen in figure 14.5.

The **flexor digitorum longus,** is a posterior muscle that attaches to the distal phalanges of all the digits of the foot except for the hallux (big toe). The flexor digitorum longus flexes (curls) the toes in addition to plantar flexing and inverting the foot. The **flexor hallucis longus** flexes the hallux and aids the flexor digitorum longus in inverting the foot. Locate these muscles in the lab and compare them with figure 14.5. The **tibialis posterior** is a large muscle deep to the soleus that plantar flexes and inverts the foot. The tendon of the muscle runs along the medial aspect of the ankle and is seen in figure 14.5.

The **fibularis longus** has a tendon that travels from the lateral side of the foot underneath to the medial side, crossing under the arch of the foot. The **fibularis brevis** parallels the fibularis longus, except that the tendon stops short and attaches to the lateral side of the foot at the fifth metatarsal. Both of these muscles have tendons that hook posterior to the lateral malleolus, and they both plantar flex and evert the foot. Examine these muscles in figure 14.5.

Cat Anatomy

If you are using cats, turn to section 3, "Hindlimb Muscles of the Cat," on page 416 of this lab manual.

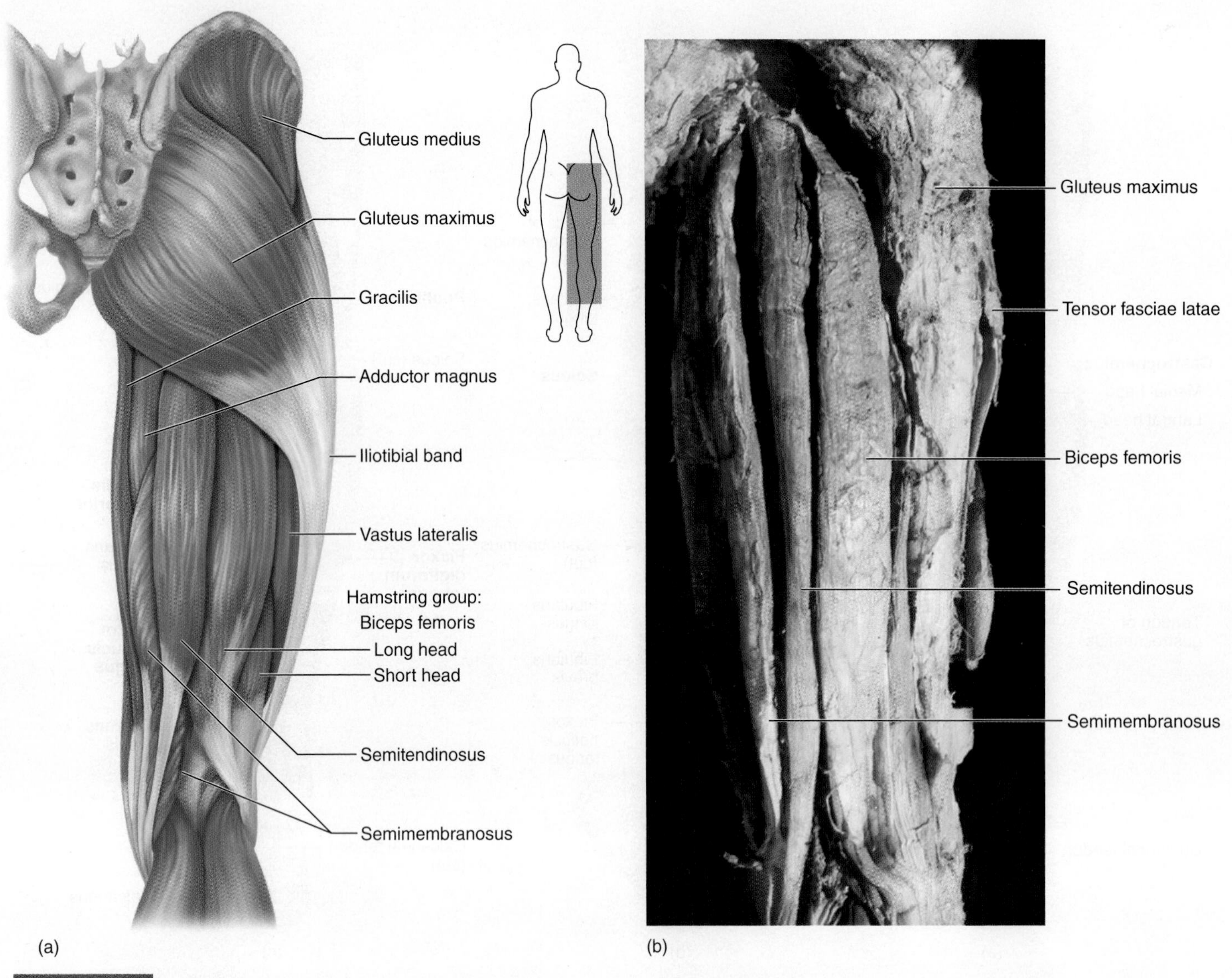

FIGURE 14.6 **Muscles of the Right Thigh, Posterior View** (a) Diagram; (b) photograph.

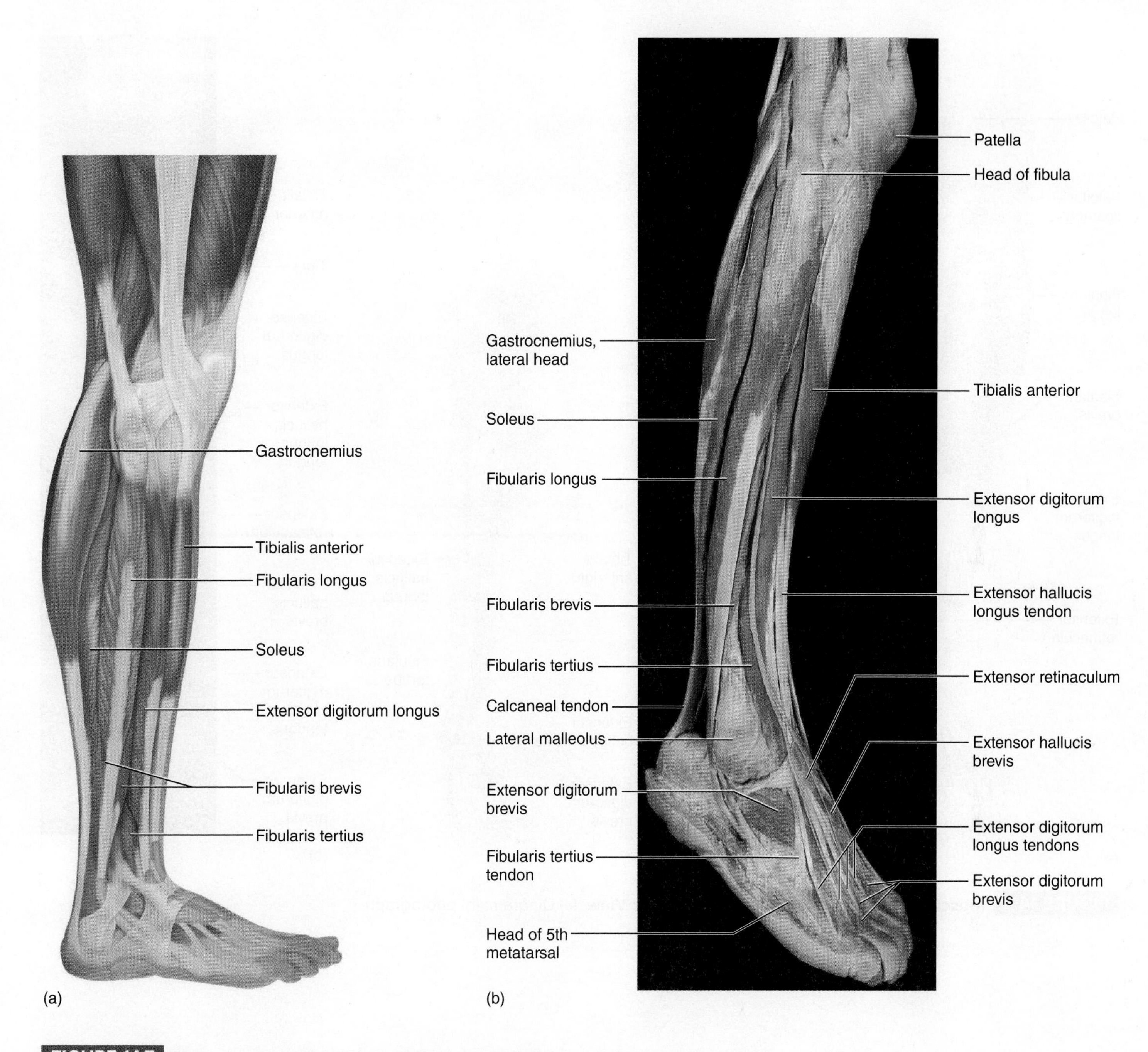

FIGURE 14.7 **Muscles of the Right Leg and Foot, Lateral View** (a) Diagram; (b) photograph.

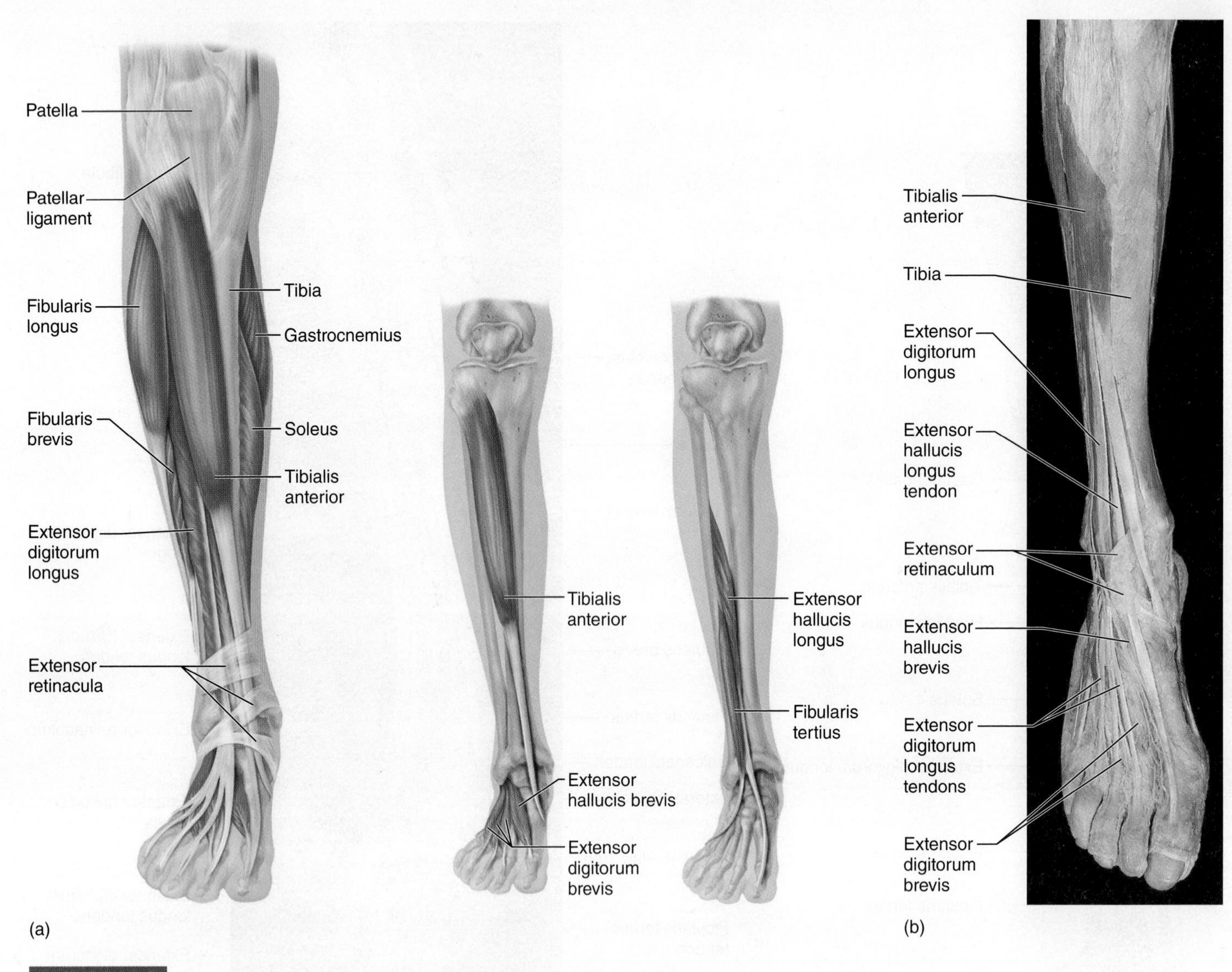

FIGURE 14.8 **Muscles of the Right Leg and Foot, Anterior View** (a) Diagram; (b) photograph.

Review

Exercise 14

Appendicular Muscles 2: Muscles of the Hip, Thigh, Leg, and Foot

Name ______________________________ Date ______________

1. If you were to ride a horse, what muscles would you use to keep your seat out of the saddle as you ride?

2. How do the gluteus medius and gluteus minimus prevent you from toppling over as you walk? ______________

3. What muscle is an antagonist to the biceps femoris muscle? ______________

4. What are two muscles that are synergists with the biceps femoris muscle? ______________

5. Are all the hamstring muscles identical in action? ______________

   What is the action of the hamstring muscles? ______________

6. What is the distal attachment of all the muscles of the quadriceps group? ______________

7. How does the action of the rectus femoris differ from that of the other quadriceps muscles? ______________

8. How many adductor muscles are there? ______________

9. List two muscles in this exercise that are responsible for thigh flexion. ______

10. Where do the hamstring muscles have their superior attachment as a group? ______

11. What is the action of the vastus lateralis? ______

12. Which muscle group is found on the anterior part of the thigh? ______

13. Is abduction of the thigh movement toward or away from the midline? ______

14. What muscle flexes the lumbar vertebrae as part of its action? ______

15. What is the proximal attachment of the gastrocnemius? ______

16. What is the distal attachment of the tibialis anterior? ______

17. How does the action of the fibularis longus differ from that of the fibularis tertius? ______

18. What is the action of the extensor hallucis longus? ______

19. The calf is made of what two major muscles? ______

20. What muscle extends the toes? ______

21. Name a muscle in this exercise that dorsiflexes the foot. ______

22. Plantar flexion and eversion of the foot occur by what muscles? ______________________________

23. What is the distal attachment of the soleus? ______________________________

24. What is the action of the tibialis anterior? ______________________________

25. What is the distal attachment of the fibularis tertius muscle? ______________________________

26. What is the distal attachment of the flexor digitorum longus? ______________________________

27. Label the muscles in the following illustration using the terms provided.

extensor digitorum longus	soleus
extensor hallucis longus	tibialis anterior
fibularis longus	

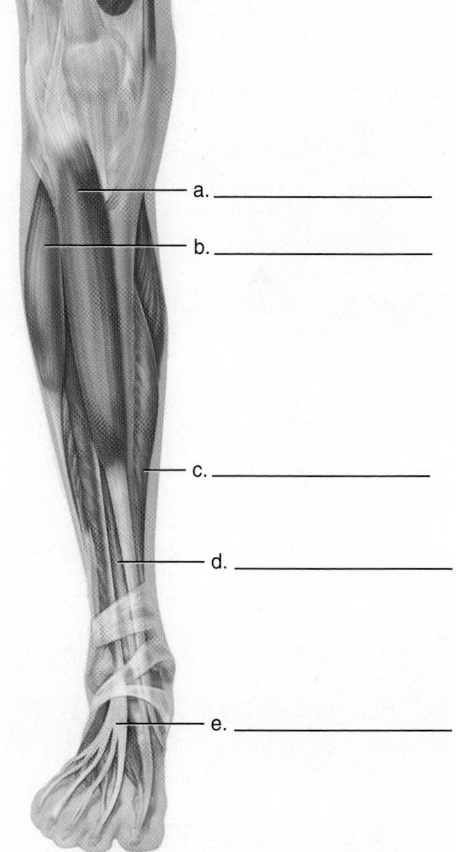

Laboratory

Exercise 15

Introduction to the Nervous System

Nervous System

Introduction

The function of the nervous system, among other things, is **communication** between the various regions of the body, **coordination** of body functions (as in digestion or walking), **orientation** to the environment, and **assimilation** of (learning) information. The functional unit of the nervous system is the neuron. It is the cell that carries out the activity of nervous tissue. Neurons comprise the nerves of the body, the spinal cord, and the brain. The processes of neurons form nerves. Neuroglia, or glial cells, make up about 50% of the volume of nervous tissue. They aid the neurons in terms of increasing the speed of neuron transmission, providing nutrients to the neurons, and protecting the neurons. In this exercise you learn about the basic structure of the nervous system and the cells that are part of the nervous system.

Learning Objectives

At the end of this exercise you should be able to

1. describe the three parts of the neuron;
1. list the main divisions of the nervous system;
2. group the organs of the nervous system into the main divisions; and
3. describe the functions of the various neuroglia.

Materials

Charts or models of the nervous system
Charts or models of neurons
Microscopes
Prepared slides:
- Spinal cord smear
- Longitudinal section of a nerve
- Neuroglia

Procedure

Divisions

The nervous system can be divided into two anatomical subdivisions, the **central nervous system (CNS)** and the **peripheral nervous system (PNS).** The CNS consists of the brain and the spinal cord. The PNS consists of cranial nerves (for example, the facial nerve), spinal nerves (nerves that are adjacent to the spinal cord), ganglia (nerve cell bodies outside of the central nervous system), and somatic nerves (such as the sciatic nerve or radial nerve). The peripheral nervous system is functionally subdivided into a sensory division (one that transmits sensations to the CNS) and a motor division (one that carries information away from the CNS). Each one of these divisions has a somatic division (involving the muscles and/or skin, joints, and bones) and a visceral division (usually involving internal organs). The **afferent (sensory) division** conducts impulses *from* the regions of the body to the central nervous system. Sensations, such as touch or sight, travel to the CNS by the afferent division. The **efferent (motor) division** conducts impulses from the central nervous system *to* the various regions of the body. The movement of muscles occurs when **somatic motor nerves** take impulses from the CNS to the skeletal muscles of the body. Other bodily functions are controlled by **visceral motor nerves.** These are part of the **autonomic nervous system,** and they regulate functions that occur automatically, such as pupil constriction when you move from a shady area to a bright one.

Visceral motor nerves carry information to the thoracic and abdominal regions. The **visceral motor division** sends impulses to glands, smooth muscle, and cardiac muscle. It can be divided into a **sympathetic division,** which prepares the body for active responses, and a **parasympathetic division,** which slows the body down or allows certain functions, such as kidney function and digestion, to take place. The visceral motor division functions independently and provides automatic controls for activities normally under subconscious direction.

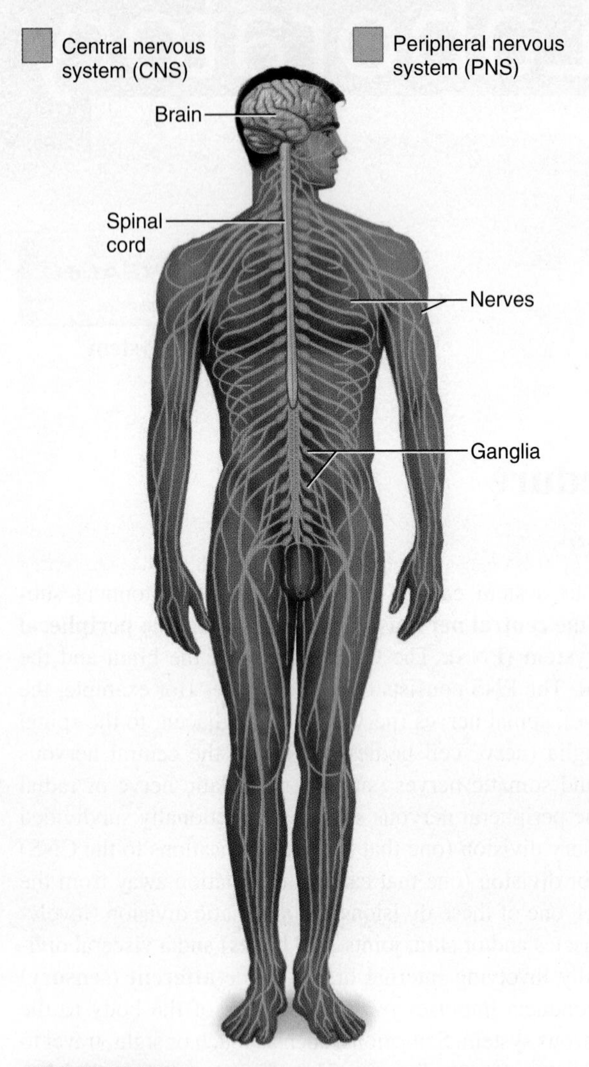

FIGURE 15.1 Divisions of the Nervous System

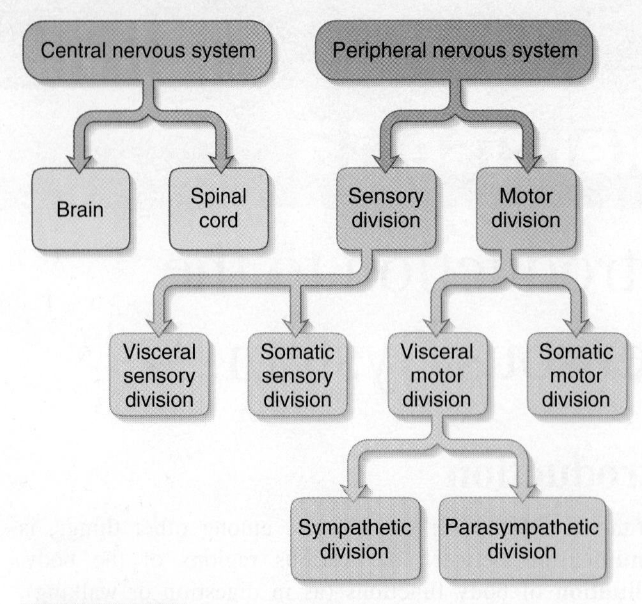

FIGURE 15.2 Schematic Representation of the Nervous System

Somatic nerves receive sensory information from the skin, muscles, joints, and bones and take motor information to skeletal muscles. The term "somatic nerves" generally refers to the nerves in the trunk and extremities. Nerves that are associated with the brain are called **cranial nerves.** They may be sensory, motor, or both.

Examine the models or charts in the lab for the central and peripheral divisions of the nervous system. Compare the material in the lab with figures 15.1 and 15.2.

Histology of the Neuron

The **neuron,** or **nerve cell,** is a remarkable cell not only for its functional nature but also for the anatomical extremes it exhibits. The nerves in the thigh and leg are composed of many neurons that may be more than a meter in length. When you look at prepared slides of neurons in the microscope in this exercise, remember that some of these neurons are of great length.

Neurons consist of three main parts: the **dendrites;** the **nerve cell body,** or **neurosoma;** and the **axon.** Neurons communicate with one another at junctions called **synapses.** Examine figure 15.3*a* and models or charts in the lab for the structure of neurons. Dendrites are so named because they have branching structures that resemble a tree (*dendros* = tree). The nerve cell body is the metabolic center of the neuron. Nerve cell bodies consist of the cytoplasm, which contains **Nissl bodies,** or **chromatophilic substances** (rough endoplasmic reticulum of the neuron), and many organelles, including the **nucleus.** The triangular region of the nerve cell body that is devoid of Nissl bodies is the **axon hillock,** and it leads to the axon that exits the nerve cell body. Axons consist of long strands of neurofibrils. In both the PNS and CNS, they may be wrapped in myelin sheaths. These sheaths are discussed later in the exercise. The axon in figure 15.3 is much shorter than what would normally be found in the body.

Examine a prepared slide of a spinal cord smear under the microscope and locate the purple, star-shaped structures under low power. These are the nerve cell bodies of multipolar neurons. Switch to high power and locate the darker-stained Nissl bodies, the nucleus, and the clear, triangular axon hillock. If you find the axon hillock, you should be able to see the axon leading away from this structure. All the other processes that are attached to the nerve cell body are dendrites. The small nuclei that are scattered throughout the smear belong to glial cells in the spinal cord. Compare your slide with figure 15.3*b*.

Functions of Neurons

There are three types of neurons based on function. **Afferent,** or **sensory, neurons** conduct impulses *to* the central nervous system. They convey information from receptors in the external or internal

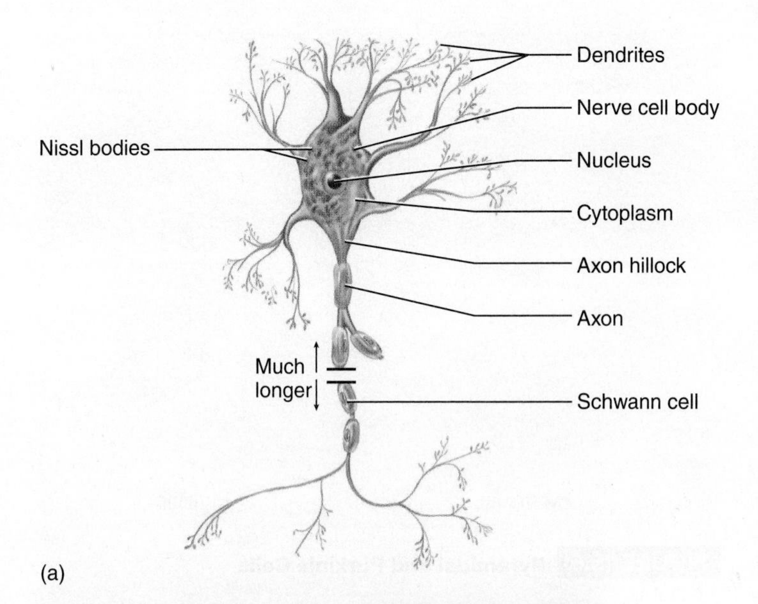

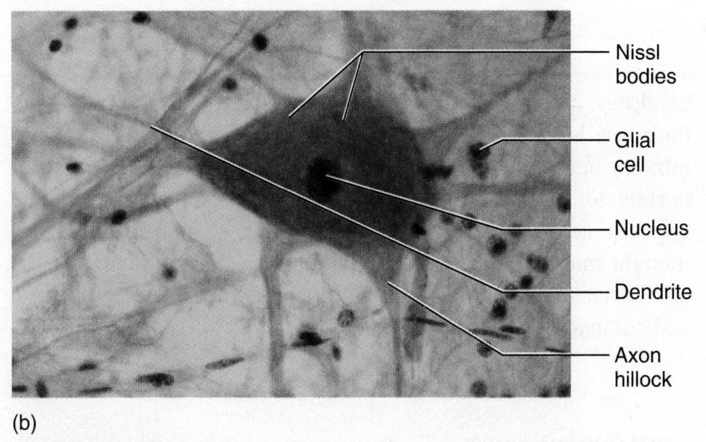

FIGURE 15.3 **Parts of a Multipolar Neuron** (a) Diagram; (b) photomicrograph.

body environment *to* the spinal cord and/or the brain. **Efferent,** or **motor, neurons** conduct impulses *away from* the central nervous system to organs or glands that carry out an activity. **Interneurons,** are located *between* sensory and motor neurons, transmitting information to the brain for processing. The interplay between neurons constitutes **reflex arcs.** When there are only two neurons in an arc (an afferent and an efferent), this is known as a monsynaptic reflex. If there is one or more interneurons in the arc, then it is known as a polysynaptic reflex arc. Examine the different types of neurons in a polysynaptic reflex arc in figure 15.4.

Neuron Shapes

Neurons can be classified according to shape. A **multipolar neuron** consists of several dendritic processes, a single nerve cell body, and a single axon. The majority of the neurons of the body (such as those in the brain and spinal cord) are multipolar neurons. Compare the material in the lab with figure 15.5. **Bipolar neurons** are so named because the nerve cell body has two poles. Dendrites receive information and conduct it to one pole of the nerve cell body. At the other pole, an axon leaves the nerve cell body and transmits the impulse away from the cell body. Bipolar neurons are found in the nose, eye, and inner ear.

Unipolar, or **pseudounipolar, neurons** are cells with just one process attaching to the nerve cell body. Dendrites receive the stimulus and conduct the impulse to the axon, which bypasses the nerve cell body, taking the impulse to the spinal cord. Most of the sensory nerves of the body are composed of unipolar neurons. How does this arrangement of axons differ from both multipolar neurons and bipolar neurons in terms of conduction of impulses with respect to the nerve cell body?

Specialized Neurons

The cerebral cortex of the brain, the amygdala, and the hippocampus contain **pyramidal cells** that have extensive dendritic branches.

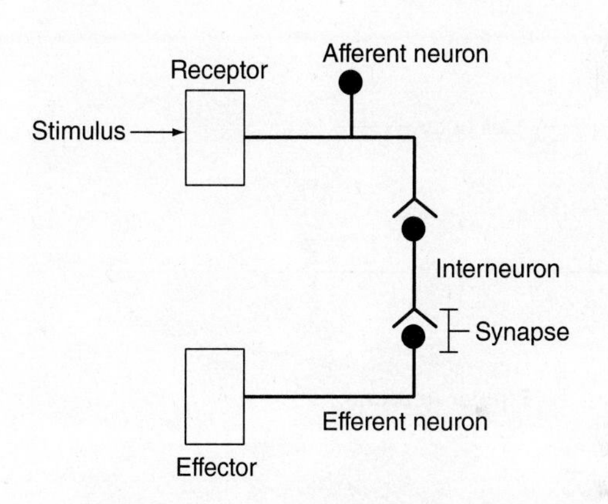

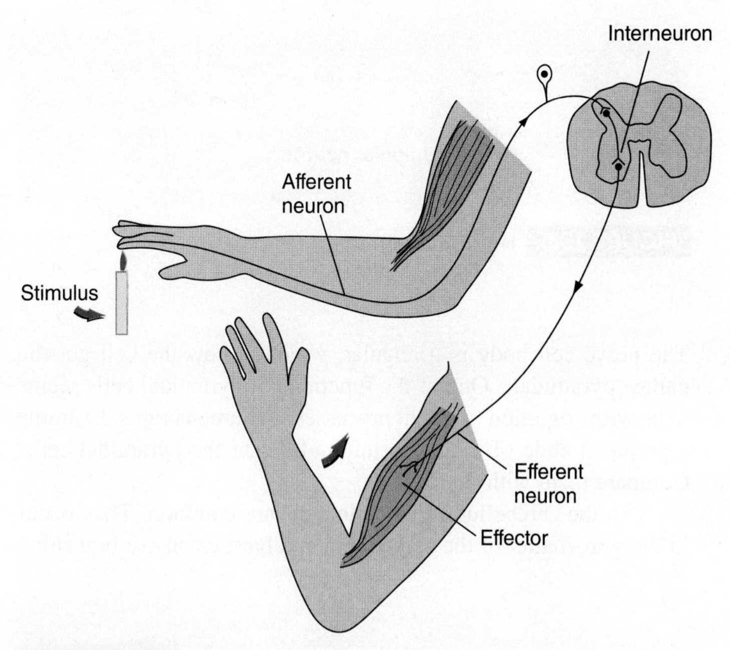

FIGURE 15.4 **Afferent, Efferent, and Interneurons in a Polysynaptic Reflex Arc**

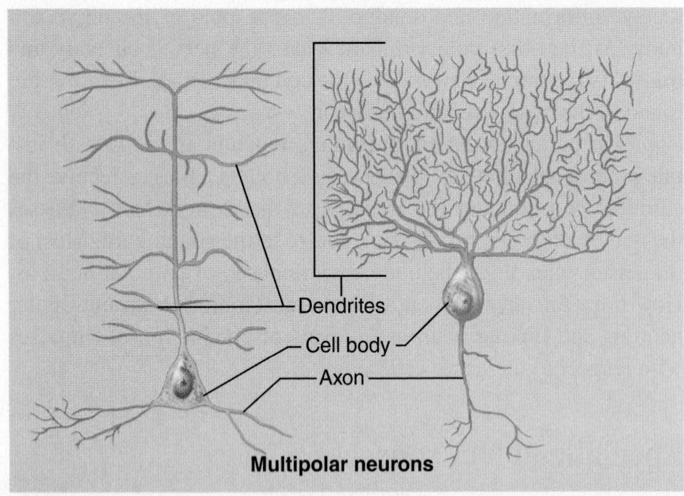

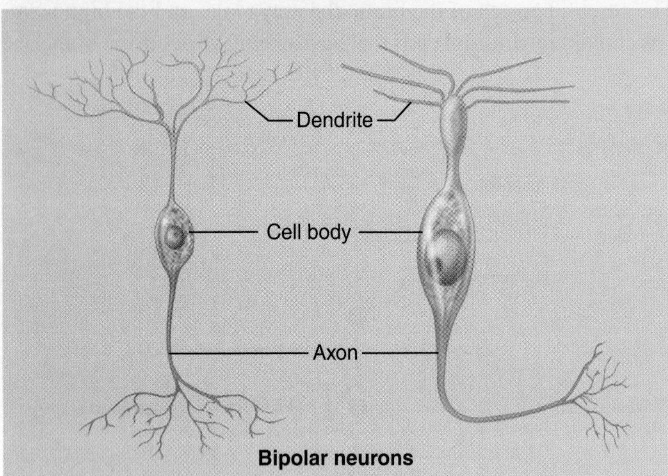

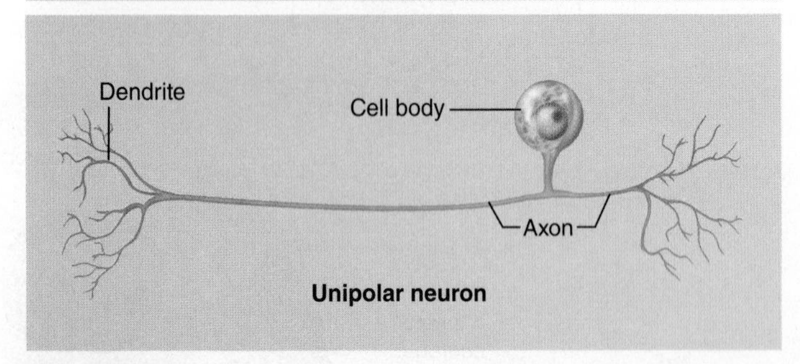

FIGURE 15.5 **Neuron Shapes**

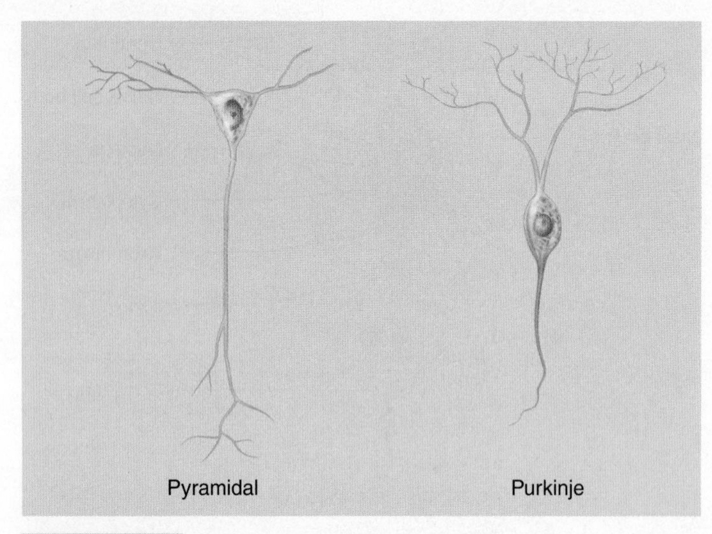

FIGURE 15.6 **Pyramidal and Purkinje Cells**

The nerve cell body is triangular, which is how the cell got the name "pyramidal." One of the functions of pyramidal cells seems to be with cognition (thought processes) in human brains. Examine a prepared slide of the cerebrum and locate the pyramidal cells. Compare them with figure 15.6.

In the cerebellum, **Purkinje cells** are common. They occur in the gray matter of the cerebellum and have extensive branching dendrites. They are some of the largest neurons in the brain, and their role is largely inhibitory by the secretion of GABA, an inhibitory neurotransmitter. When a muscle is contracting, it is important to relax the antagonists (the muscles that perform the opposite action of the contracting muscle) to that muscle. It is thought that the Purkinje cells inhibit muscle contraction of the antagonist muscles. Examine a prepared slide of the cerebellum and compare it with figure 15.6.

Neuroglia

Numerous cells aid the functioning of the neuron. These cells are called **neuroglia** (*neuro* = nerve, *glia* = glue), or **glial cells.**

PNS Neuroglia The common glial cell of the peripheral nervous system is the **Schwann cell,** or **neurolemmocyte.** These glial cells wrap around the axon, much as a thin strip of paper can be wrapped around a pencil, leaving small gaps between successive cells called the **nodes of Ranvier** (neurolemmocyte nodes). The gaps allow for the nerve transmission to be transmitted from node to node, increasing the transmission speed of the neuron. This type of jumping transmission is called **saltatory conduction.** The Schwann cell consists of a significant amount of a lipoprotein material called **myelin,** and the series of Schwann cells produces a **myelin sheath. Myelinated nerve fibers** appear white. Small axons that are not enclosed by myelin sheaths are called unmyelinated fibers. Nerve cell bodies, unmyelinated fibers, dendrites, and other parts of the neurons form the portion of the nervous tissue known as **gray matter.**

Histology of the Neurolemmocyte Examine a prepared slide of a longitudinal section of nerve under high power. You should see the axon fibers as long, dark threads in the microscope. What appear to be clear areas on each side of the axon fibers is the myelin sheath. If you scan the slide closely, you can see the junction of two Schwann cells and the node of Ranvier between them. Compare your slide with figure 15.7.

CNS Neuroglia Schwann cells are located in the PNS. Other types of neuroglia produce myelin in the CNS (fig. 15.8). These are known as **oligodendrocytes.** Unlike Schwann cells, oligodendrocytes frequently wrap around several neurons. The oligodendrocytes have the same general function as the neurolemmocytes, which is to increase the speed of neural transmission. Nervous tissue with significant myelinated fibers is called **white matter.** Nervous tissue with a large percent of unmyelinated fibers has a darker color and is called gray matter.

Astrocytes are branched glial cells that provide a barrier between the nervous tissue and the blood. They are the most common glial cells in the CNS, and they have numerous functions. They stimulate the capillaries in the brain to form the blood-brain barrier, which limits the entrance of microbes into the brain tissue. They feed the neurons, release nerve growth factors, play a role in synaptic transmission, and form scar tissue when neurons are damaged. They also inhibit some medications from reaching the brain.

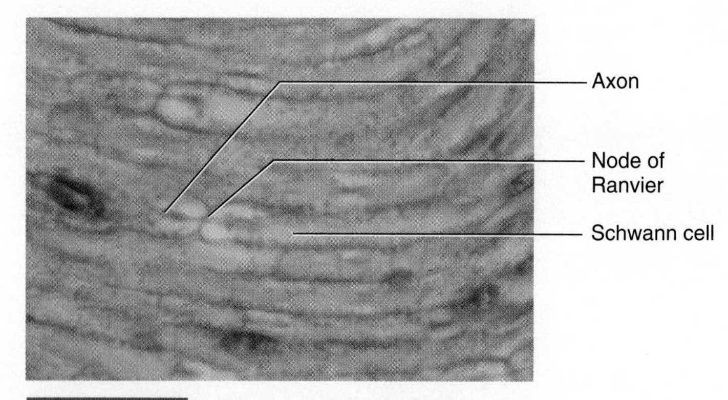

FIGURE 15.7 **Nerve, Longitudinal Section (400x)**

Microglia are small phagocytic glial cells—they digest foreign particles that invade the nervous tissue and remove dead or damaged neurons. They are the brain's resident immunocompetent cells. When they are in their resting phase, they are actively sampling the brain for tissue damage or infection. In their reactive phase, they resemble macrophages as they engulf microbes or damaged tissue. Examine figure 15.8 for microglial cells.

The last of the glial cells covered in this exercise are the **ependymal cells.** These cells line the spaces of the brain and spinal cord and secrete and circulate cerebrospinal fluid.

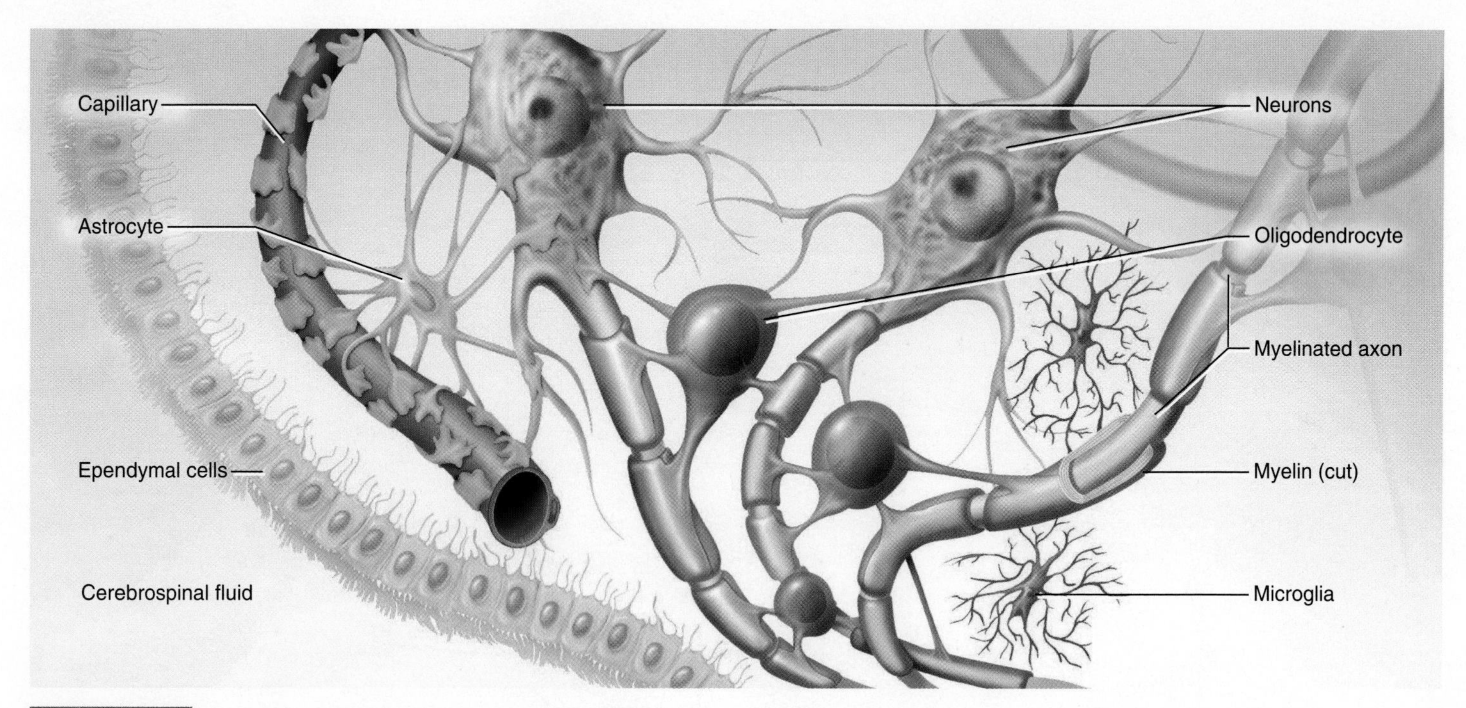

FIGURE 15.8 **Neuroglia of the Central Nervous System**

Introduction to the Nervous System

Name ______________________ Date __________

1. What type of neuron (multipolar, bipolar, or unipolar) is represented by the drawing? Label the parts with the terms provided.

axon	nerve cell body
dendrite	Nissl body

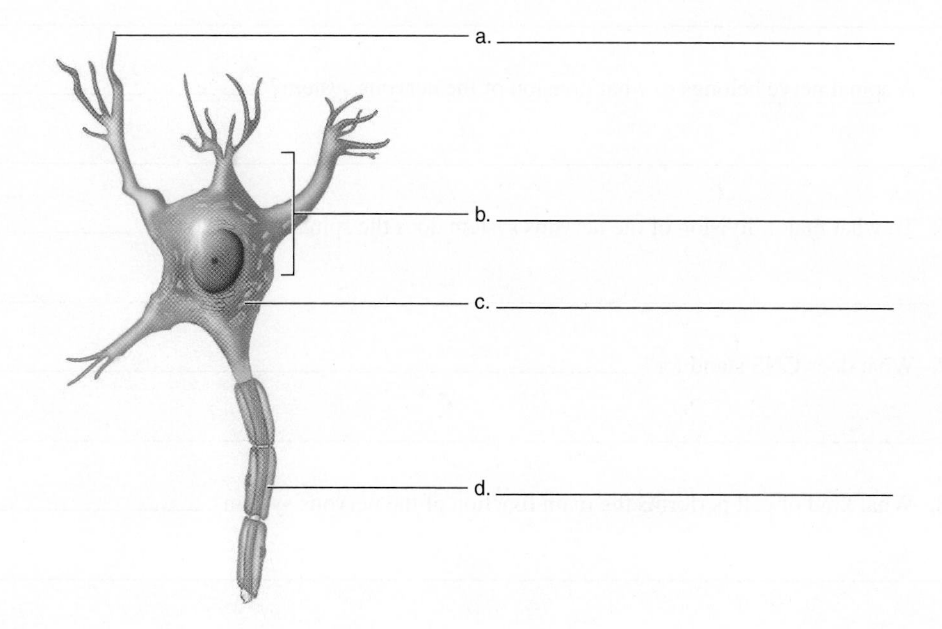

Neuron type ______________

2. What type of neuron (multipolar, bipolar, or unipolar) is represented by the drawing? Label the parts with the terms provided.

axon	nerve cell body
dendrite	nucleus

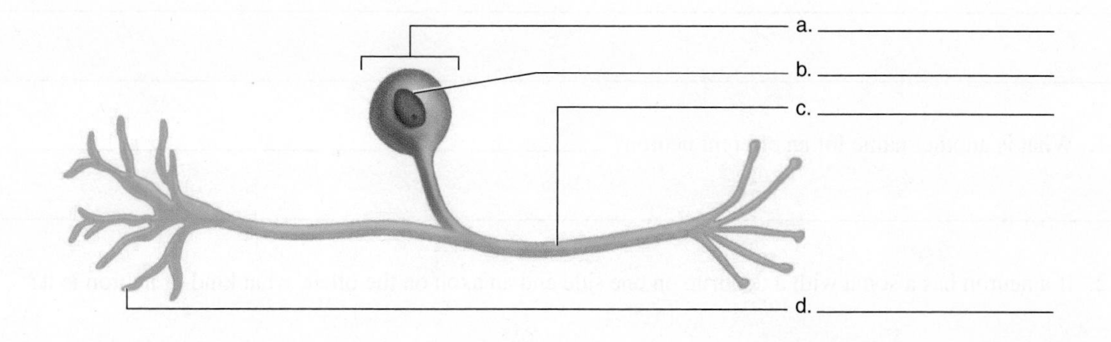

Neuron type ______________

3. Describe the function of

   a. an astrocyte: ______

   b. an ependymal cell: ______

   c. an oligodendrocyte: ______

   d. a microglial cell: ______

   e. a neurolemmocyte: ______

4. The brain belongs to what division of the nervous system? ______

5. A spinal nerve belongs to what division of the nervous system? ______

6. To what major division of the nervous system does the spinal cord belong? ______

7. What does CNS stand for? ______

8. What kind of cell performs the main function of the nervous system? ______

9. The nucleus is found in what specific part of the neuron? ______

10. A neuron has three main parts. What are they? ______

11. What is another name for an efferent neuron? ______

12. If a neuron has a soma with a dendrite on one side and an axon on the other, what kind of neuron is it?

______

13. Two adjacent neurons that communicate with each other are separated by a space. What is this space called?

____

14. How are neuroglia different from neurons in terms of function? _______________________

____

____

15. In which one of the nervous system divisions are neurolemmocytes found? _______________

____

16. Myelin is made of what kind of material? _______________________________

____

Laboratory

Exercise 16

Spinal Cord and Spinal Nerves

Nervous System

Introduction

The spinal cord is part of the central nervous system (CNS); it begins at the foramen magnum of the skull and ends at about vertebra L1 or L2. The spinal cord stops growing early in life, yet the vertebral bodies continue to grow, causing the spinal cord to be shorter than the vertebral canal. The spinal cord receives sensory information from, and transmits motor information to, the spinal nerves, which radiate into the body as peripheral nerves.

The peripheral nervous system consists of the cranial nerves and the spinal nerves. Cranial nerves are associated with the brain and are covered in this lab manual in laboratory exercise 17. Spinal nerves take both sensory and motor information to and from the spinal cord. Peripheral nerves **innervate** (IN-ur-vate; functionally connect to) muscles and other parts of the body. Nerves such as the radial nerve and the femoral nerve are spinal nerves. These nerves travel throughout the body, receiving sensory information and sending it to the CNS or taking motor information from the CNS to skeletal muscles. In this exercise, you learn the major features of the spinal cord in longitudinal aspect and in cross section and the major nerves and plexuses of the body.

Learning Objectives

At the end of this exercise you should be able to

1. demonstrate the major regions in a cross section of spinal cord;
2. describe the structure of the longitudinal aspect of the spinal cord;
3. list the major nerves that arise from each plexus;
4. name all the major nerves of the upper and lower limbs; and
5. list the structures that carry the impulses to and away from the spinal cord.

Materials

Models or charts of the central and peripheral nervous systems

Cadaver (if available)

Prepared slide of a spinal cord in cross section

Model or chart of a spinal cord in cross section and longitudinal section

Procedure

Spinal Cord

Longitudinal Aspect of the Spinal Cord Examine a model or chart in the lab of a longitudinal view of the spinal cord, and locate the major features illustrated in figure 16.1. The spinal cord terminates inferiorly as the **medullary cone** (MED-you-LAR-ee CONE) at approximately vertebra L1 and is attached to the coccyx by a continuation of the pia mater known as the **terminal filum** (FYE-lum). The neural continuation of the spinal cord exists as an extension of parallel nerve fibers in the lumbar and sacral regions. These parallel fibers resemble a horse's tail and are called the **cauda equina** (CAW-duh ee-KWY-nah). The spinal cord is expanded in two locations. The **cervical enlargement** occurs at vertebrae C3 through T2 and represents a bulge in the spinal cord that has increased the number of neurons going to or coming from the upper limbs. The **lumbar enlargement** occurs at about vertebrae T7 through T11, and this expanse is also due to the number of neurons going to or coming from the lower limbs. Compare the material in the lab with figure 16.1.

Cross Section of Spinal Cord Examine a model or chart in the lab of a cross section of the spinal cord. Locate the **gray matter,** which appears as an H pattern or a butterfly pattern in the middle of the spinal cord, with the **white matter** located on the periphery of the cord. Note how the distribution of gray and white

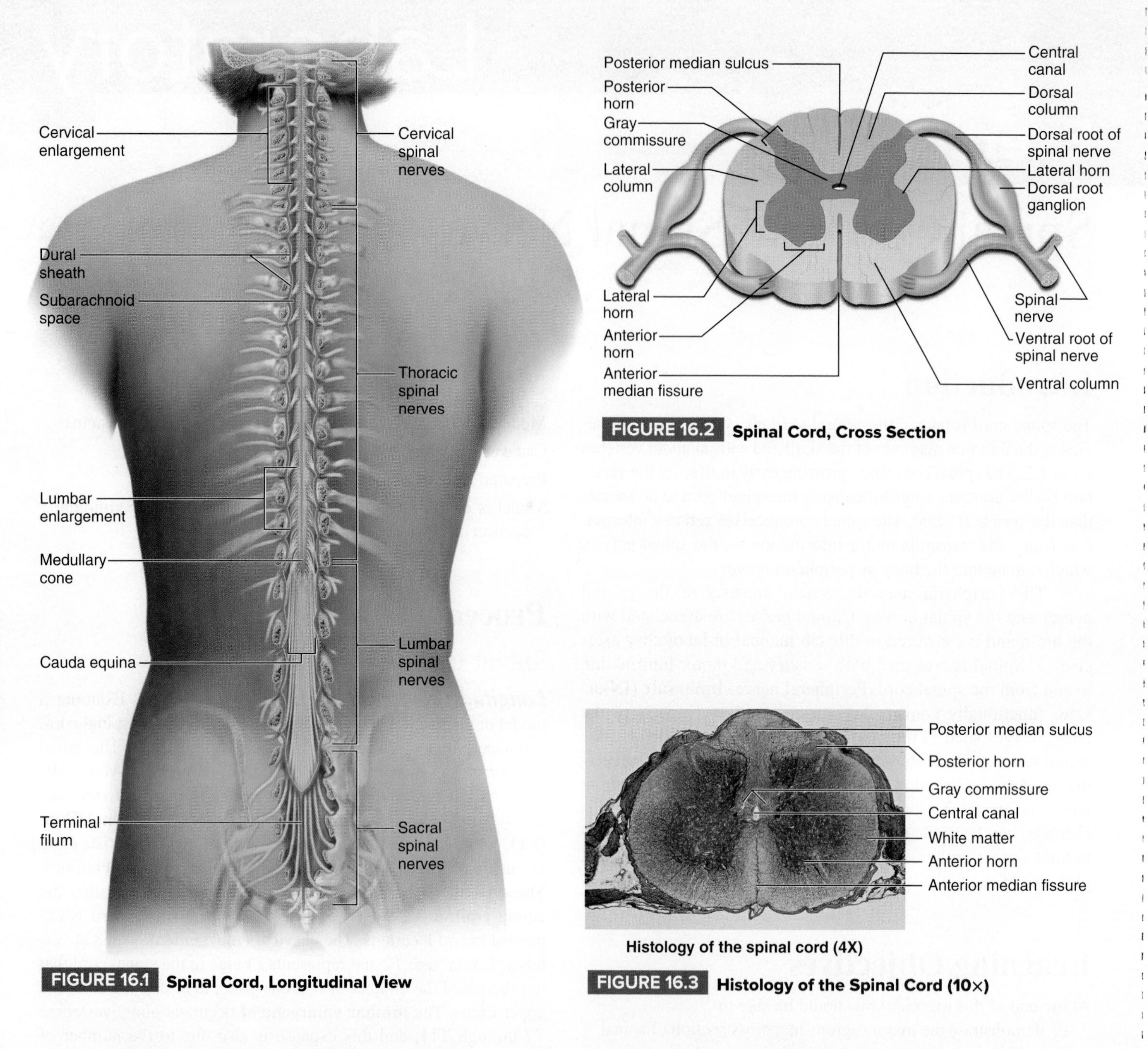

FIGURE 16.1 Spinal Cord, Longitudinal View

FIGURE 16.2 Spinal Cord, Cross Section

FIGURE 16.3 Histology of the Spinal Cord (10×)

matter in the spinal cord is opposite to that in the brain. In a cross section of the spinal cord, you should see the gray matter divided in two narrow horns and two rounded horns. The narrow horns are known as the **posterior horns,** and these areas receive sensory information from the spinal nerves. The rounded horns are the **anterior horns,** and these send motor signals to the spinal nerves. In some parts of the spinal cord, there are additional sections of gray matter known as the **lateral horns,** which contain neurons of the sympathetic division.

Each side of the gray matter is connected to the other by a crossbar known as the **gray commissure.** In the middle of the gray commissure is the **central canal,** which runs the length of the spinal cord and contains CSF. The white matter of the cord is divided into **tracts,** or **funiculi**—ascending tracts that carry sensory information to the brain and descending tracts that carry motor information down from the brain. **Tracts** are parallel nerve fibers in the CNS, while **nerves** are parallel fibers in the PNS. You should also see a depression in the posterior surface of the spinal cord. This is the **posterior median sulcus,** while the deeper depression on the anterior side is known as the **anterior median fissure.** Examine a model or chart in the lab and compare it with figure 16.2.

Examine a prepared slide of the spinal cord under low power, and locate the **anterior** and **posterior horns,** the **gray commissure,** the **central canal,** the **posterior median sulcus,** and the **anterior median fissure.** Examine the anterior horn of the spinal cord, and look for the nerve cell bodies of the **multipolar neurons** there. Compare your slide with figure 16.3.

Meninges The spinal cord is covered by **meninges,** which are non-neural sheaths. The outer covering of the cord consists of the **dura mater** (dural sheath). Between the dura mater and the vertebra is the **epidural space,** a site for the injection of anesthetics. The next layer deeper to the dura mater is the **arachnoid mater.** Deep to the arachnoid is the **pia mater,** the innermost of the meninges and a thin cover on the spinal cord proper. Between the pia mater and the arachnoid is the **subarachnoid space,** which contains the **cerebrospinal fluid.** Examine figure 16.4 for an illustration of the meninges of the spinal cord.

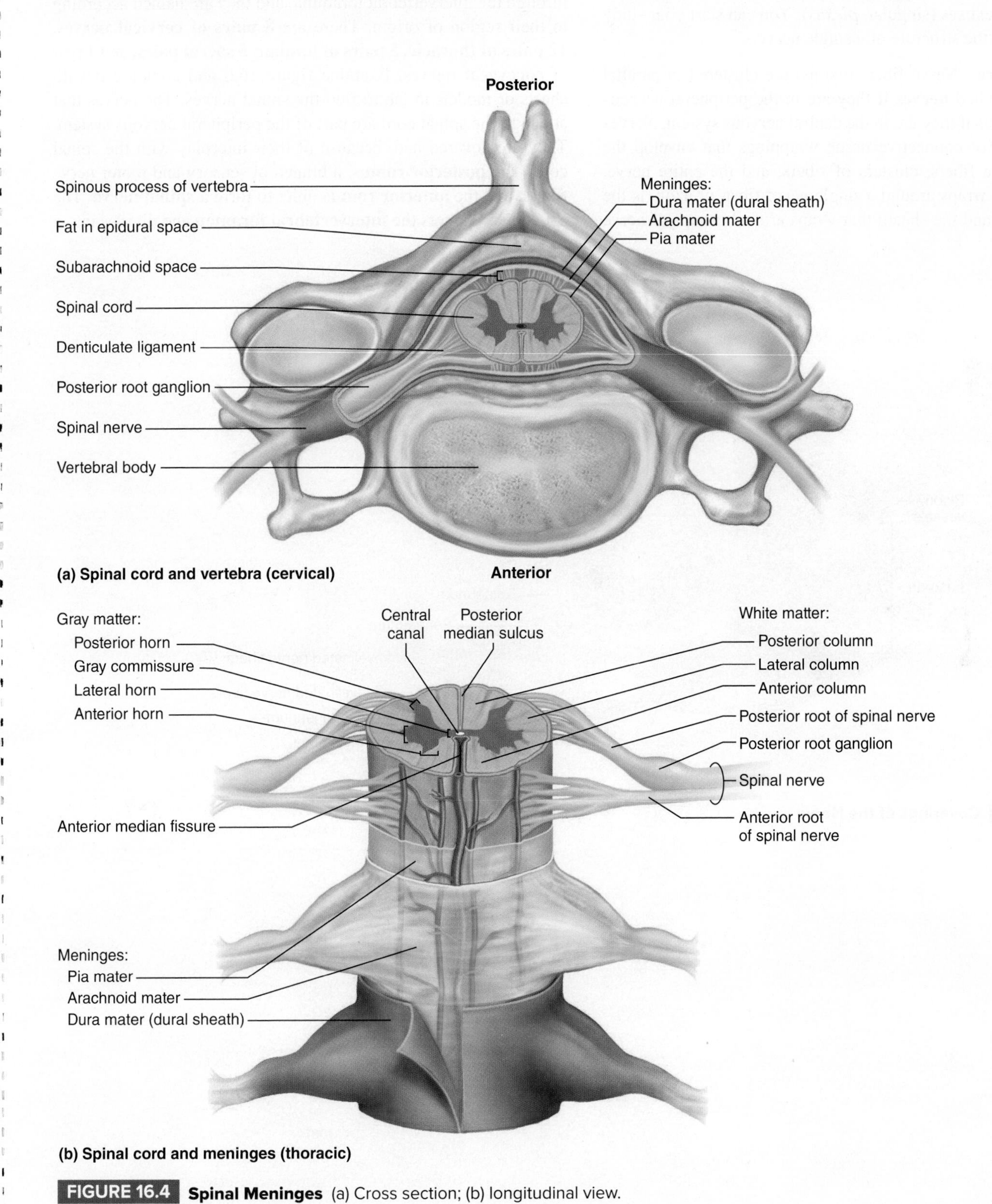

FIGURE 16.4 **Spinal Meninges** (a) Cross section; (b) longitudinal view.

Nerves Associated with the Spinal Cord The anatomy of the peripheral nervous system near the spinal cord is complicated. Some nerves, such as those in the middle of the thorax, leave the spinal cord and travel to the ribs. In the neck, arm, thigh, and sacral region, the nerve fibers split, intertwine, and form complex networks called **plexuses** (singular, *plexus*). You can start your study with examining the structure of a single nerve.

Nerve Structure Nerve fibers (axons) are clustered in parallel arrangements called **nerves** if they are in the peripheral nervous system and **tracts** if they are in the central nervous system. Nerves have a number of connective tissue wrappings that envelop the individual nerve fibers, clusters of fibers, and the entire nerve. The sheath that wraps around a single nerve fiber, or axon, is the **endoneurium,** and the sheath that wraps around groups of nerve fibers (nerve fascicles) is the **perineurium.** The wrapping that covers the entire nerve is called the **epineurium.** Examine figure 16.5 for these layers.

Spinal Nerves There are 31 pairs of **spinal nerves,** which pass through the intervertebral foramina, and they are named according to their region of origin. There are **8 pairs of cervical nerves, 12 pairs of thoracic, 5 pairs of lumbar, 5 sacral pairs,** and **1** pair of **coccygeal nerves.** Examine figure 16.6 and compare it with charts or models in lab to find the spinal nerves. The nerves that attach to the spinal cord are part of the peripheral nervous system. They are covered here because of their interplay with the spinal cord. The **posterior ramus,** a branch of sensory and motor nerve fibers, and the **anterior ramus** unite to form a **spinal nerve.** The spinal nerve enters the **intervertebral foramen** and divides into a

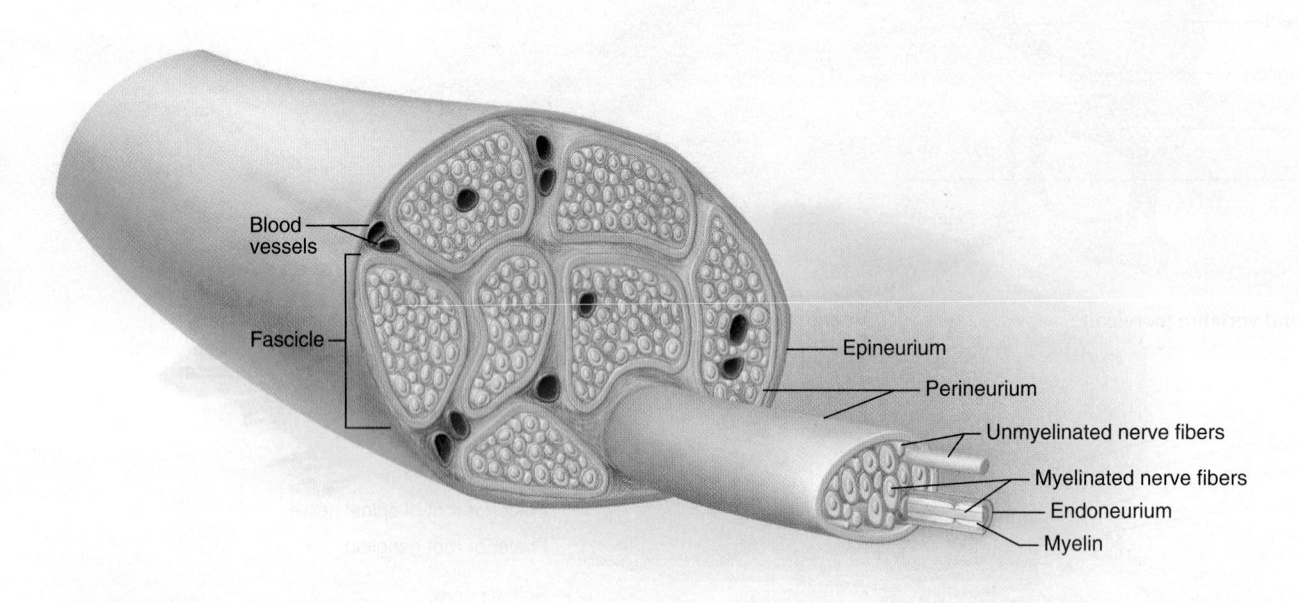

FIGURE 16.5 **Coverings of the Nerve**

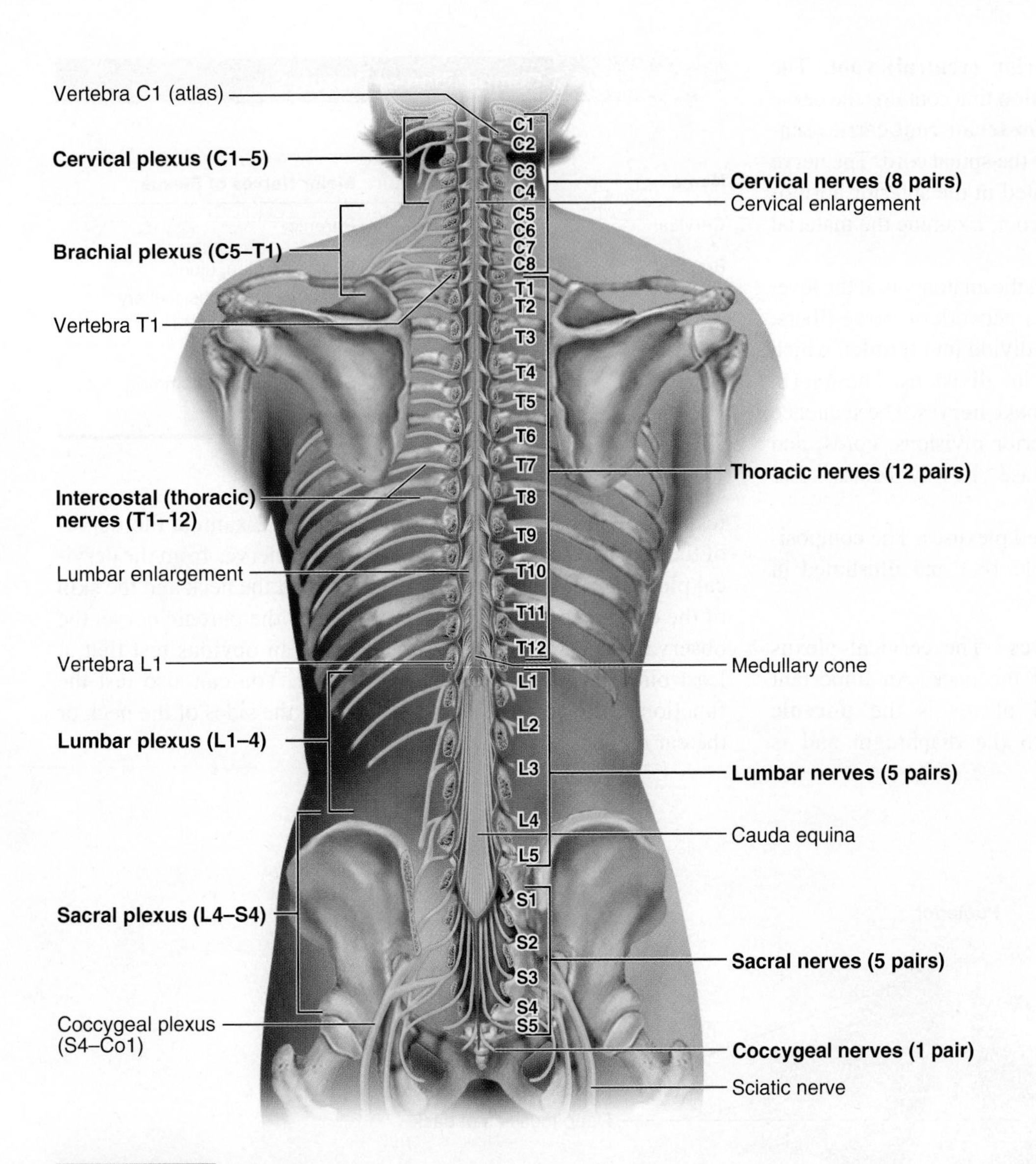

FIGURE 16.6 **Spinal Nerves and Plexuses**

posterior (dorsal) root and an **anterior (ventral) root.** The posterior root has a **posterior root ganglion** that contains the nerve cell bodies of the sensory nerves. The **posterior root** carries sensory information to the posterior horn of the spinal cord. The nerve cell bodies of the motor nerves are located in the anterior horn of the spinal cord and exit via the anterior root. Examine the material in the lab and compare it to figure 16.7.

Where things get complicated with the anatomy is at the level of the plexus. A plexus is made up of a network of nerve fibers. Spinal **roots (spinal nerves)** sometime divide into **trunks,** which split into **anterior divisions** and **posterior divisions.** These form **cords** in some plexuses, which unite to make **nerves.** The sequence (roots, trunks, anterior divisions, posterior divisions, cords, and nerves) can be remembered by the phrase "Rowdy Tourists Are Playing Cards Now."

There are four generally recognized plexuses. The composition of the plexuses is outlined in table 16.1 and illustrated in figure 16.6.

TABLE 16.1 Composition of Plexuses

Name	Spinal Nerves Contributing to Plexus	Major Nerves of Plexus
Cervical	C1–5	Phrenic
Brachial	C5–T1	Radial, median, ulnar, musculocutaneous, axillary
Lumbar	L1–4	Femoral, obturator
Sacral	L4–S4	Sciatic (tibial and common fibular)

Cervical and Brachial Plexus Nerves The cervical plexus exits from the upper spinal nerves of the neck. An important nerve that comes from the cervical plexus is the **phrenic** (FREN-ic) **nerve.** This nerve runs to the diaphragm and is responsible for its contraction in breathing. Examine the nerves of the **cervical plexus** in figure 16.8. Many nerves from the cervical plexus innervate the muscles and skin of the neck and the skin of the ear. Since breathing is controlled by the phrenic nerve, the observation that a person is breathing is an obvious test that at least one nerve of the plexus is working. You can also test the function of the nerves by lightly pinching the sides of the neck or the ear for sensory perception.

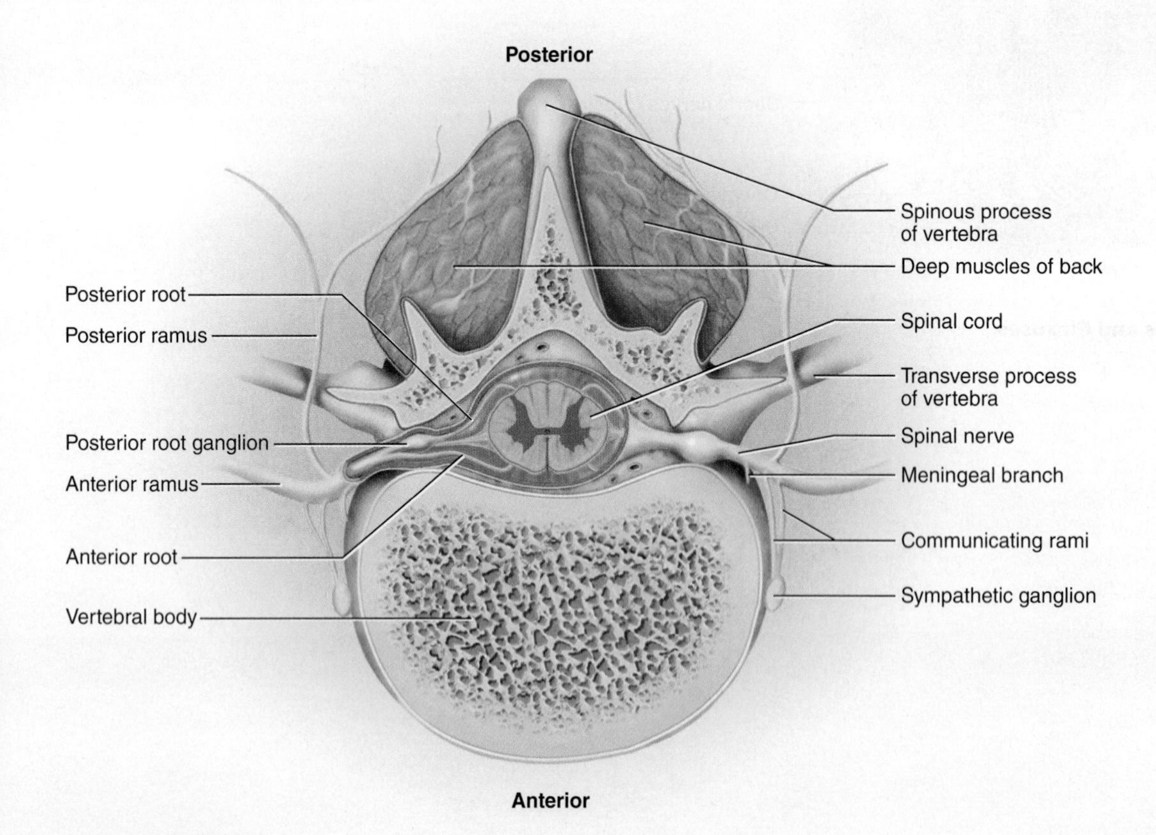

FIGURE 16.7 Nerves Associated with the Spinal Cord

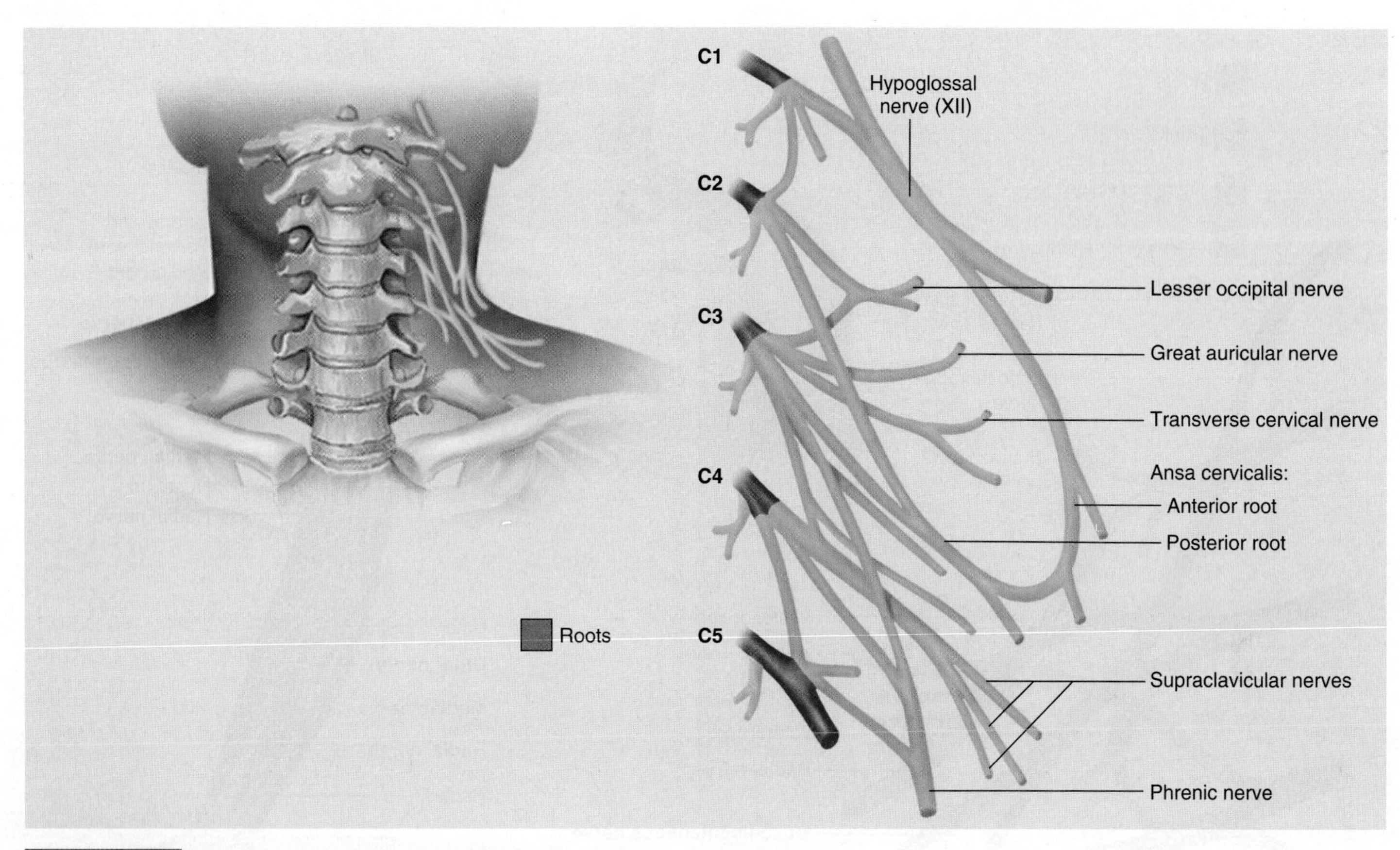

FIGURE 16.8 **Nerves of the Cervical Plexus**

The **brachial plexus** has branches from C5 to T1 and forms nerves that primarily innervate the upper limbs. The plexus is illustrated in figure 16.9. The muscles innervated from this plexus were covered in laboratory exercise 13. The major nerves are divided into anterior and posterior divisions. In the anterior division is the **musculocutaneous** (MUS-cue-lo-cyu-TANE-ee-us) **nerve,** innervating many of the muscles of the arm; the **median nerve,** running the length of each upper limb and serving important hand and forearm flexors; and the **ulnar nerve,** serving other forearm and hand flexors. The ulnar nerve crosses behind the medial epicondyle of the humerus and is commonly known as the "funny bone." In the posterior division is the **axillary nerve,** innervating the upper shoulder, and the **radial nerve,** predominantly innervating the triceps brachii and the extensors of the hand. The nerves of this plexus can be tested by pinching the fingers, the medial and lateral aspects of the forearm, and the anterior and posterior aspects of the arm. Sensations from these areas are conducted to the brain via the brachial plexus.

Thoracic Nerves There are numerous nerves not associated with a plexus. The thoracic nerves are a good example. Many of the **thoracic nerves** exit through the intervertebral foramina of the vertebral column and innervate the ribs, muscles, and other structures of the thoracic wall. Look at models or charts in lab, and compare these with the nerves illustrated in figure 16.6.

Lumbar and Sacral Plexus Nerves The **lumbar** and **sacral plexuses** take sensory information from and motor information to the lower limb. Some authors combine the two plexuses into the **lumbosacral plexus.** The lumbar plexus is composed of spinal nerves L1–4. One of the nerves originating from the lumbar plexus is the **obturator nerve.** It innervates the adductor

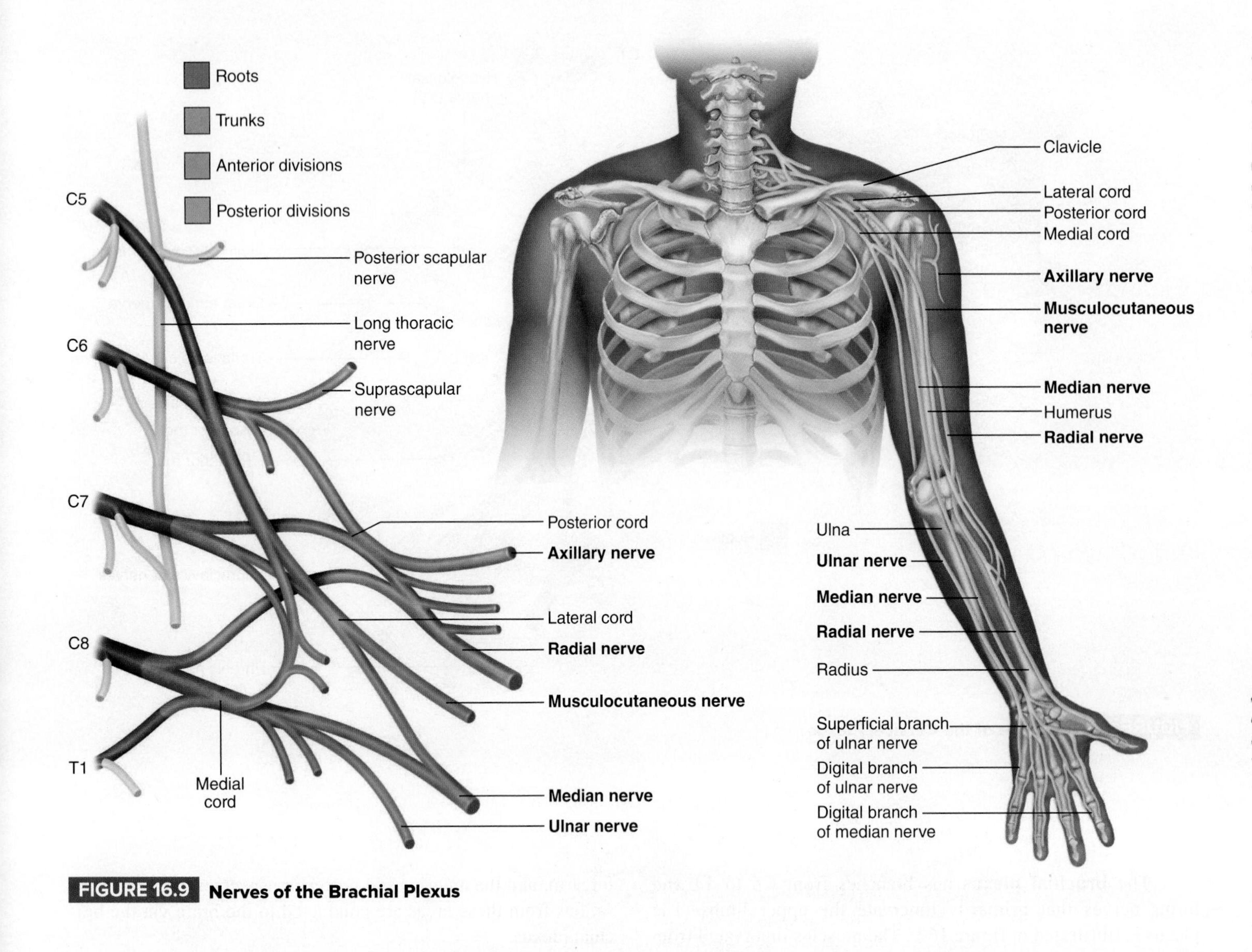

FIGURE 16.9 **Nerves of the Brachial Plexus**

muscles of the thigh, as seen in figure 16.10. The **femoral nerve** is another nerve arising from the **lumbar plexus.** This large nerve passes posterior to the inguinal ligament and mostly innervates the muscles of the anterior thigh. The femoral nerve is seen in figure 16.10. To test for the nerves of this plexus, you can lightly pinch the anterior thigh for the femoral nerve and the medial thigh for the obturator nerve.

The **sacral plexus** consists of fibers from L4–S4. Many innervate the pelvis and muscles that move the hip, thigh, and leg. Two of the nerves from this plexus, the **tibial** and **common fibular (peroneal) nerves,** unite proximally in one sheath to form the **sciatic nerve** (fig. 16.11). At the distal thigh, the tibial nerve and the common fibular nerve separate. These nerves innervate the posterior thigh, leg, and foot. Test for the sciatic nerve by lightly pinching the posterior aspect of the thigh. Examine the nerves in the lab and compare them with the illustrations.

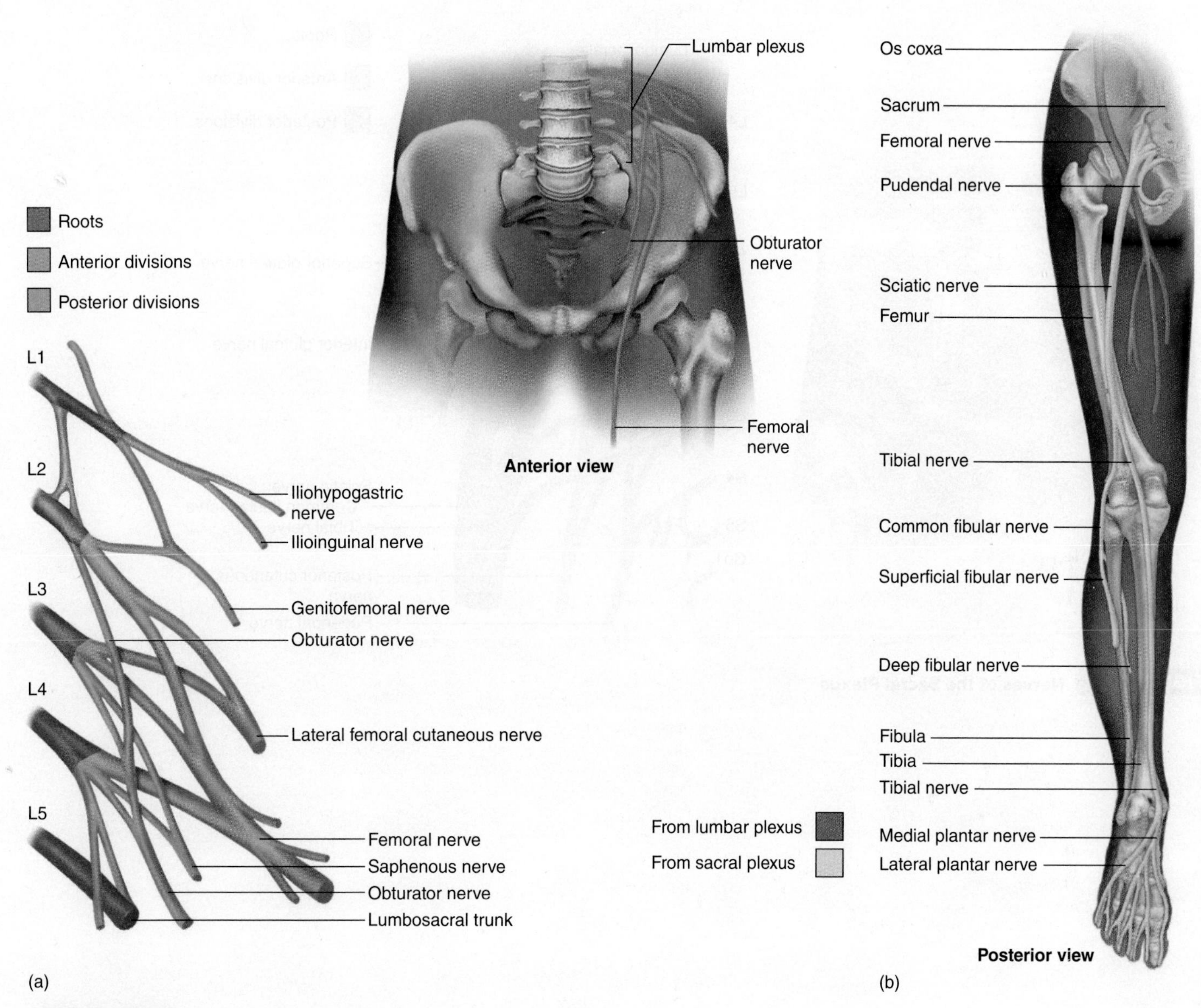

FIGURE 16.10 **Nerves of the Lumbar Plexus** (a) Lumbar plexus; (b) nerves of the lower extremity.

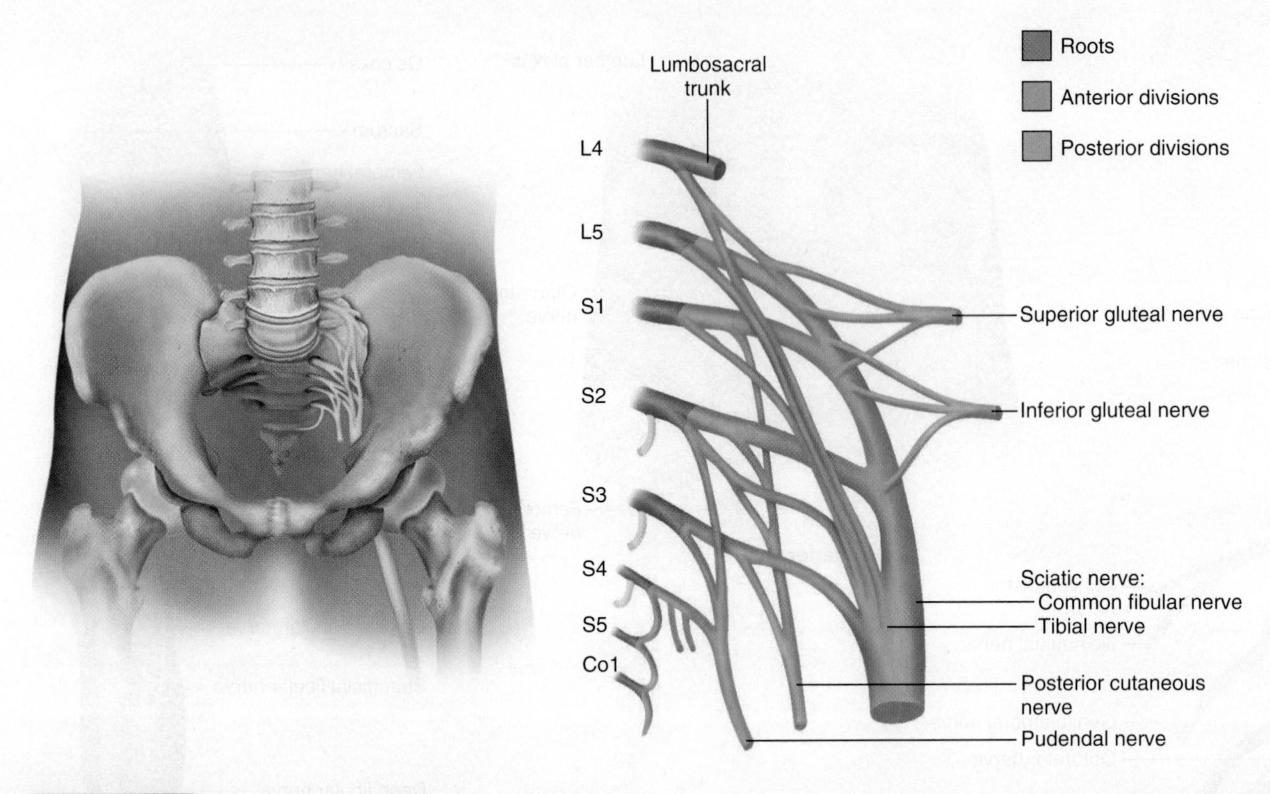

FIGURE 16.11 **Nerves of the Sacral Plexus**

Exercise 16

Spinal Cord and Spinal Nerves

Name ______________________________ Date ____________

1. What anatomical feature is responsible for the cervical enlargement of the spinal cord? ____________

2. Where is the terminal filum found? ____________

3. What is the medullary cone? ____________

4. What is the cauda equina? ____________

5. In the spinal cord, which is deep, the white matter or the gray matter? ____________

6. What is the area of gray matter between the lateral halves of the spinal cord? ____________

7. *In the spinal cord,* what type of impulse (sensory/motor) travels through the

   a. anterior horn? ____________

   b. posterior horn? ____________

   c. ascending spinal tracts? ____________

   d. descending spinal tracts? ____________

8. In function, how does the posterior spinal root vary from the anterior spinal root? ______

9. What is the endoneurium? ______

10. How do tracts differ from nerves? ______

11. What is a mixed nerve? ______

12. What major nerves arise from the following plexuses? ______

a. cervical ______

b. brachial ______

c. lumbar ______

d. sacral ______

13. The diaphragm contractions are regulated by what nerve? ______

14. The muscles of the arm, such as the biceps brachii, have what innervation? ______

15. The extensor muscles of the hand are controlled by what nerve? ______

16. The sciatic nerve is composed of two nerves. What are they? ______

17. Label the major plexuses in the following figure.

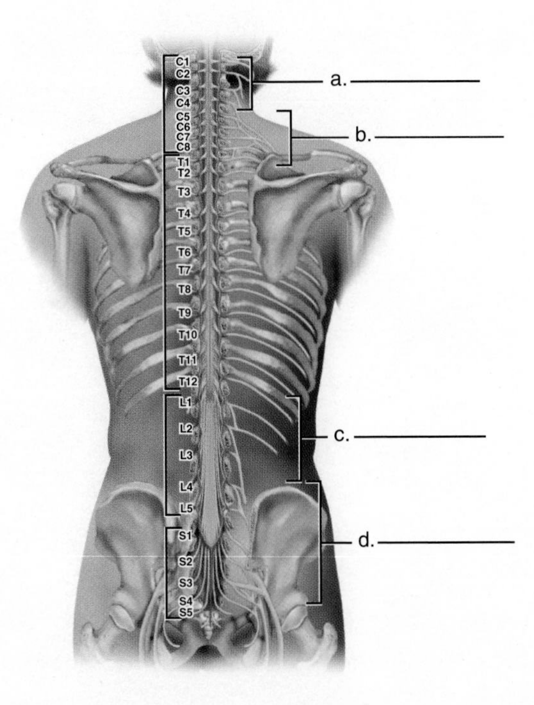

18. A person has feeling from the deltoid and biceps brachii region but no feeling from the wrist extensors. Where on the spinal cord has injury occurred?

___

Laboratory

Exercise 17

Brain and Cranial Nerves

Nervous System

Introduction

Two specific traits distinguish humans from other animals. One is our upright posture, and the other is the extensive development of the brain. In this exercise you examine the anatomy of the human brain and the cranial nerves. You also compare the human brain to a sheep brain and note similarities and differences. There is a difference not only in the overall size of the brains between humans and sheep but also in the relative size of various structures of the brain and the position of some of the anatomy of the brain. You may want to review the planes of sectioning in Exercise 1. Sheep brains are commonly used as dissection specimens because of their cost, availability, and similarity in structure to human brains. The anatomical differences in the sheep brain will be described in this exercise.

The brain begins development, as does the rest of the nervous system, in the third week of pregnancy as a **neural groove** in the ectoderm. By the fourth week, the brain has folded into a **neural tube** that contains the **central canal** (fig. 17.1). The posterior portion of the central canal becomes the central canal of the spinal cord. The anterior portion of the canal becomes the ventricles of the brain in adults. By the sixth week of development, the cerebral hemispheres have begun to form and continue their development throughout pregnancy.

Learning Objectives

By the end of this exercise you should be able to

1. name the three meninges of the brain and their location relative to one another;
2. locate the three major regions of the brain;
3. name the main structures in each of the three regions of the brain;
4. describe the function of specific areas of the brain, such as the thalamus, Broca area, the occipital lobe, the temporal lobe, the cerebellum, and the medulla oblongata;
5. trace the path of cerebrospinal fluid through the brain;
6. identify each of the 12 pairs of cranial nerves on an illustration, a model, or a real brain;
7. describe whether a cranial nerve is sensory, motor, or both; and
8. identify the structure that a particular cranial nerve innervates.

Materials

Models and charts of the human brain
Preserved human brains (if available)
Cast of the ventricles of the brain
Sheep brains
Dissection trays
Scalpels
Protective gloves
Blunt (Mall) probe
First aid kit in lab or prep area
Sharps container
Animal waste disposal container

Procedure

Overview of the Brain

The brain is located in the cranial cavity of the skull and weighs approximately 1.4 kilograms (3 pounds). The brain is derived from three embryonic regions, each of which further develops into more specific areas. The embryonic brain consists of the **forebrain,** or **prosencephalon** (PROZ-en-SEF-uh-lon); the **midbrain,** or **mesencephalon** (MES-en-SEF-uh-lon); and the **hindbrain,** or **rhombencephalon** (ROMB-en-SEF-uh-lon). These regions are illustrated in figure 17.1. Examine table 17.1 for an overview of specific areas of the brain and the regions to which they belong. Figure 17.2 is an illustration of the major

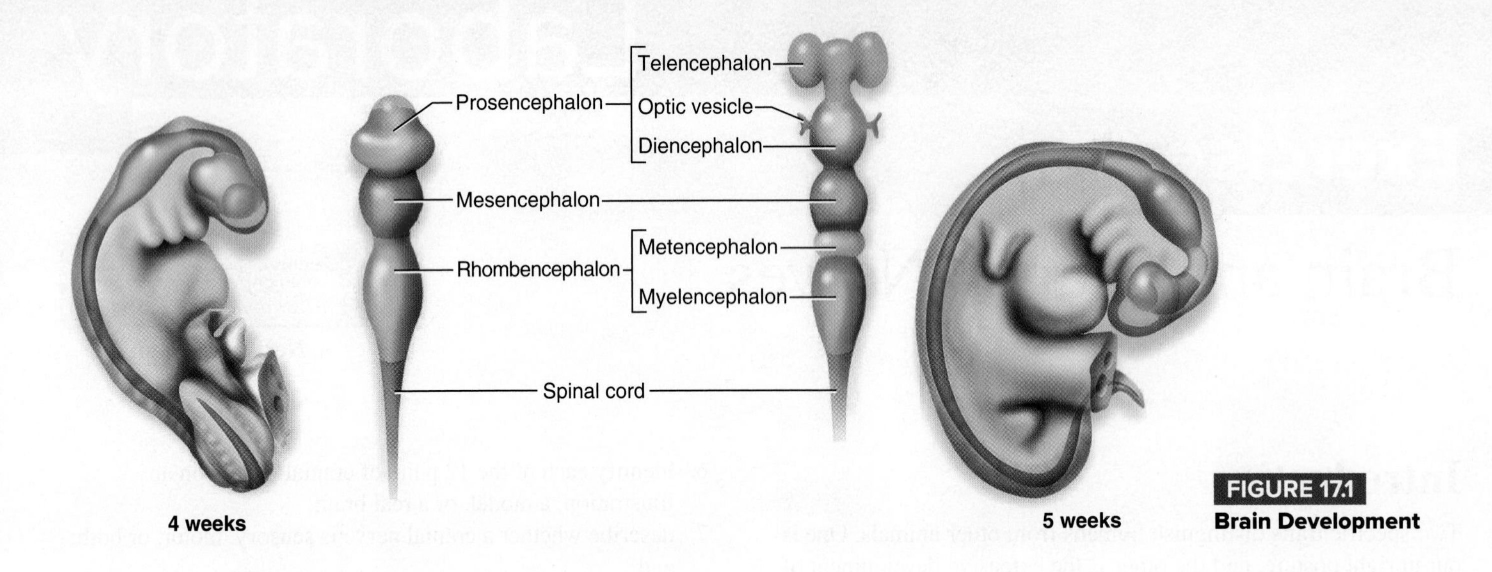

FIGURE 17.1 Brain Development

TABLE 17.1 Regions of the Brain

Prosencephalon
Telencephalon
Cerebrum (cerebral hemispheres)
Cerebral cortex (gray matter)
Basal nuclei (gray matter)
Corpus callosum
Diencephalon
Pineal body
Thalamus
Hypothalamus
Pituitary gland
Mammillary bodies
Mesencephalon
Peduncles
Tectum
Corpora quadrigemina
Superior colliculus
Inferior colliculus
Rhombencephalon
Metencephalon
Pons
Cerebellum
Myelencephalon
Medulla oblongata

areas in an adult brain. Compare this illustration with a model of the brain in lab and locate the **forebrain.** It is the largest region of the brain, consisting primarily of the **cerebrum** (seh-REE-brum) and the **diencephalon** (DY-en-SEF-uh-lon). The **midbrain** is the smallest region of the brain, located between the forebrain and the hindbrain. The **hindbrain** is composed of the **pons,** the **medulla oblongata** (meh-DULL-ah OB-long-GAH-ta), and the **cerebellum** (SER-eh-BEL-um). The hindbrain is the most inferior portion of the brain; it connects to the spinal cord at the foramen magnum.

Meninges

Three membranes surround the brain. These are called the meninges. The outermost membrane is the **dura mater** (DOO-rah MAH-tur; Latin for "tough mother"), a tough dense connective tissue sheath that encircles the brain and has sections that extend into the brain. The dura resembles a swim cap but is tough and tendinous. The dura mater is divided into an outer, **periosteal layer** and an inner, **meningeal layer.** The dural sinus is a space between these two layers. Deep to the dura mater is the **arachnoid** (uh-RAK-noyd; Greek for "spider web–like") **mater,** which is a soft, thin membrane that resembles a spider's web. Between the dura mater and the arachnoid membrane is the **subdural space.** Blood vessels are in the subdural space. Deep to the arachnoid is the **subarachnoid space,** which contains **cerebrospinal fluid (CSF).** The function of the cerebrospinal fluid is to keep the brain buoyant in the skull, cushion the brain, and remove waste from the central nervous system (CNS). CSF is produced in the subarachnoid space, the ependymal cells, and the choroid plexuses. There is approximately 150 milliliters of CSF in the central nervous system, and it takes about 6 hours to circulate through the system. The deepest layer is the **pia** (PEE-uh; Latin for "tender") **mater,** a membrane that adheres directly to the outer surface of the brain and appears like the surface of the brain. Locate the meninges in preserved brains in the lab.

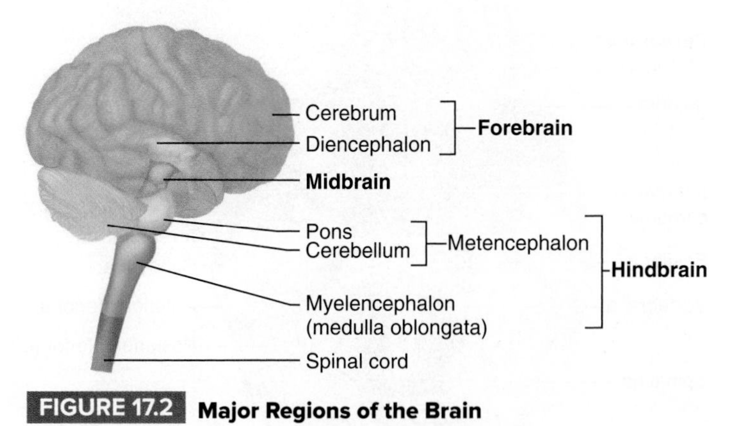

FIGURE 17.2 Major Regions of the Brain

Ventricles of the Brain

Part of the hollow neural tube that develops in the first trimester of pregnancy becomes the ventricles of the brain in the adult. The two ventricles that occupy the center of each cerebral hemisphere are known as the **lateral ventricles.** These ventricles receive cerebrospinal fluid from tufts of capillaries called **choroid** (CO-royd) **plexuses.** You can see the choroid plexuses in preserved brains as small, brown areas in the superior portions of the ventricles. Fluid from the lateral ventricles flows through the **interventricular foramina** and into the **third ventricle.** The third ventricle lies between the walls of the thalamus and receives CSF from choroid plexuses in that area. The third ventricle drains into the **fourth ventricle** by way of the **cerebral** (**mesencephalic,** MEZ-en-suh-FAHL-ik) **aqueduct.** If this duct becomes occluded, then CSF accumulates in the lateral and third ventricles, causing a condition known as **hydrocephaly.** Locate the ventricles of the brain in figure 17.3.

Normally the cerebral aqueduct is open and passes through the midbrain. Posterior to the cerebral aqueduct is the fourth ventricle, which occupies a space anterior to the cerebellum. The fourth ventricle also has a choroid plexus that secretes CSF. Cerebrospinal fluid flows from the fourth ventricle into the central canal of the spinal cord and the subarachnoid space of the spinal cord and brain. CSF is absorbed by the arachnoid villi under the dura mater of the skull. The villi penetrate the dural sinuses, and the CSF flows to the venous sinuses and returns to the cardiovascular system primarily by the **internal jugular veins** (fig. 17.4).

Blood Supply to the Brain

The blood vessels that supply nutrients and oxygen to the brain, and generate CSF, form a meshwork around the brain and into the ventricles. The brain receives blood from four major arteries. These are the **left** and **right vertebral arteries** and the **left** and **right internal carotid arteries.**

Locate the vertebral arteries as illustrated in figure 17.5. These arteries pass through the transverse foramina of the cervical vertebrae and join to form the **basilar artery** at the base of

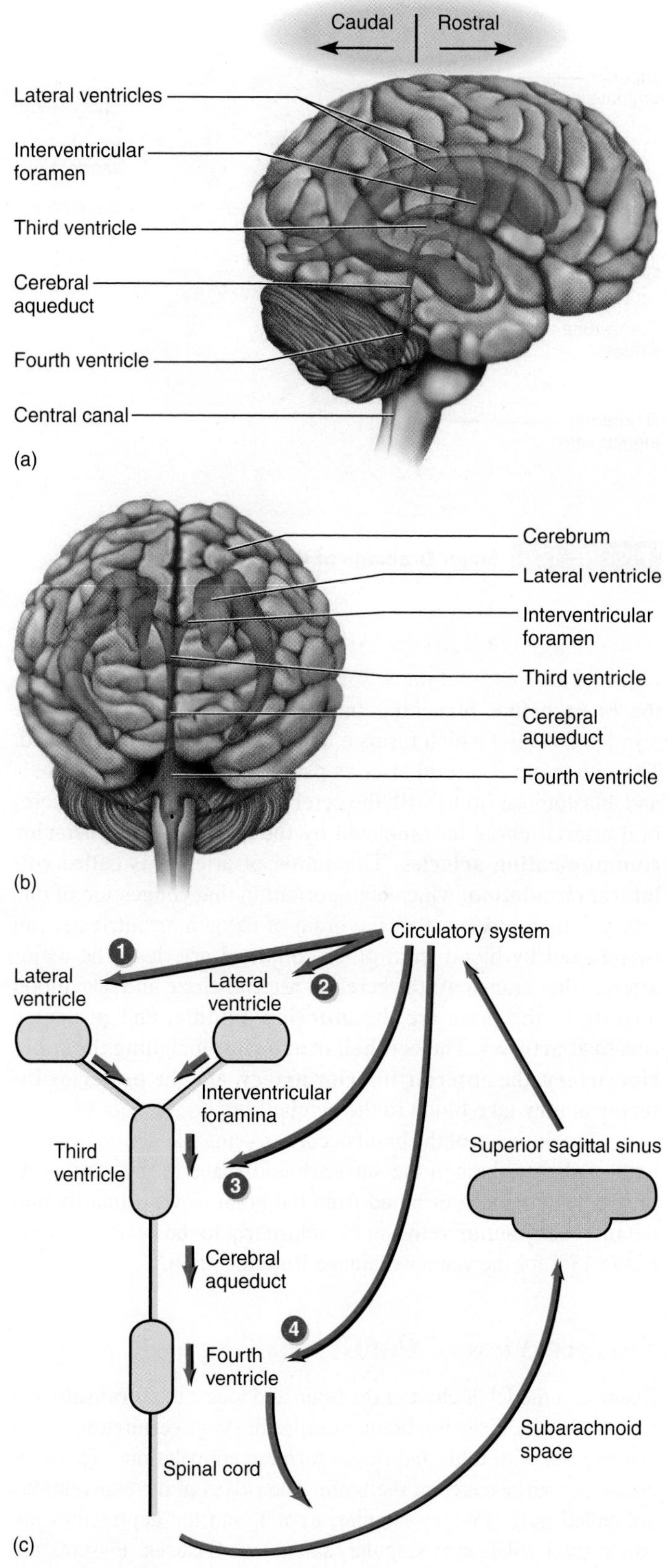

FIGURE 17.3 Ventricles of the Brain (a) Lateral view; (b) anterior view; (c) schematic presentation of the flow of cerebrospinal fluid. Numbers 1–4 represent flow through choroid plexuses.

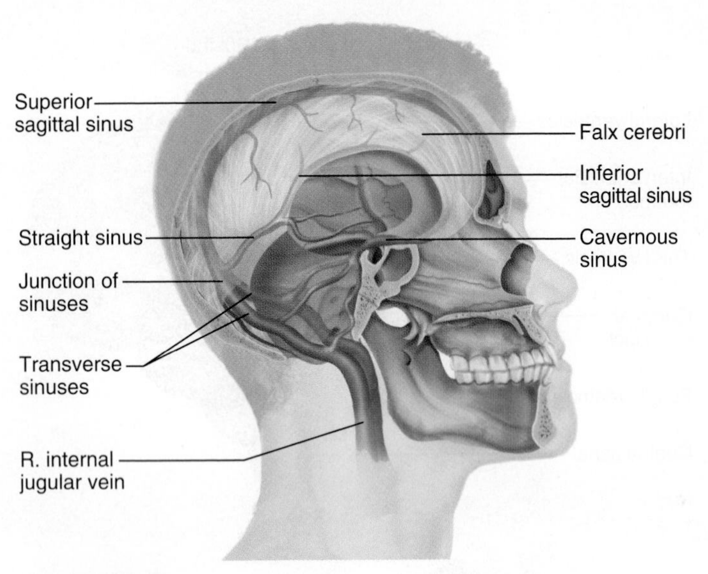

FIGURE 17.4 **Major Drainage of the Brain**

the brain before branching into the **cerebral arterial circle** (circle of Willis) which forms a loop around the pituitary gland. The two internal carotid arteries pass through the carotid canals and anastomose (join) with the cerebral arterial circle. The cerebral arterial circle is completed by the **anterior** and **posterior communicating arteries.** This union of arteries is called **collateral circulation,** which is important in that congestion of one artery, which might starve the brain of oxygen or nutrients, can be relieved by blood from other collateral arteries. The major arteries that branch off the cerebral arterial circle and take blood directly to the brain are the **anterior, middle,** and **posterior cerebral arteries.** The **cerebellar arteries,** including the **superior artery,** the **anterior inferior artery,** and the **posterior inferior artery** take blood to the cerebellum.

The drainage of the brain occurs as veins take blood from the brain and pass through the subarachnoid space to the **venous sinuses.** The drainage of blood from the brain flows primarily into the **internal jugular veins** on the return trip to the heart. Examine figure 17.4 for the venous drainage from the brain.

Surface View of the Brain

Examine a model or chart of the brain and locate the **forebrain** and the **hindbrain.** In the forebrain, examine the large **cerebrum,** which can be seen with folds and ridges forming convolutions. These increase the surface area of the brain. The ridges of the convolutions are called **gyri** (JY-rye; singular, *gyrus*), and the depressions are either **sulci** (SUL-sye; singular, *sulcus*) or **fissures.** Fissures are deeper than sulci. The lateral view of the brain allows you to see the major lobes of each cerebral hemisphere. These lobes are named for the bones of the skull under which they lie. They are the **frontal, parietal, occipital,** and **temporal lobes.** Deep to the lateral sulcus of the brain is a small lobe called the **insula.** The **lateral sulcus**

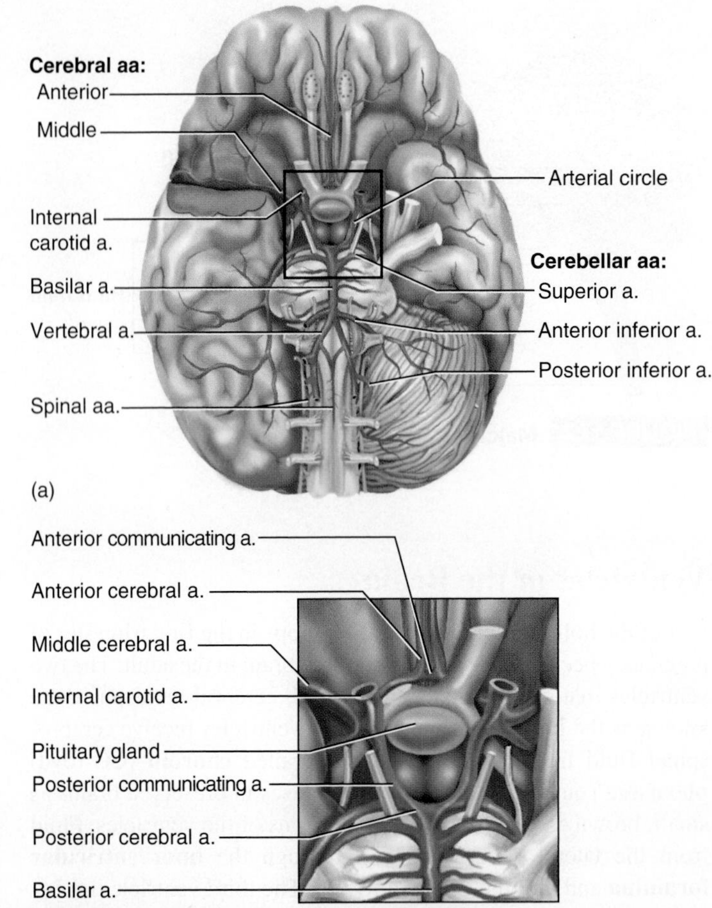

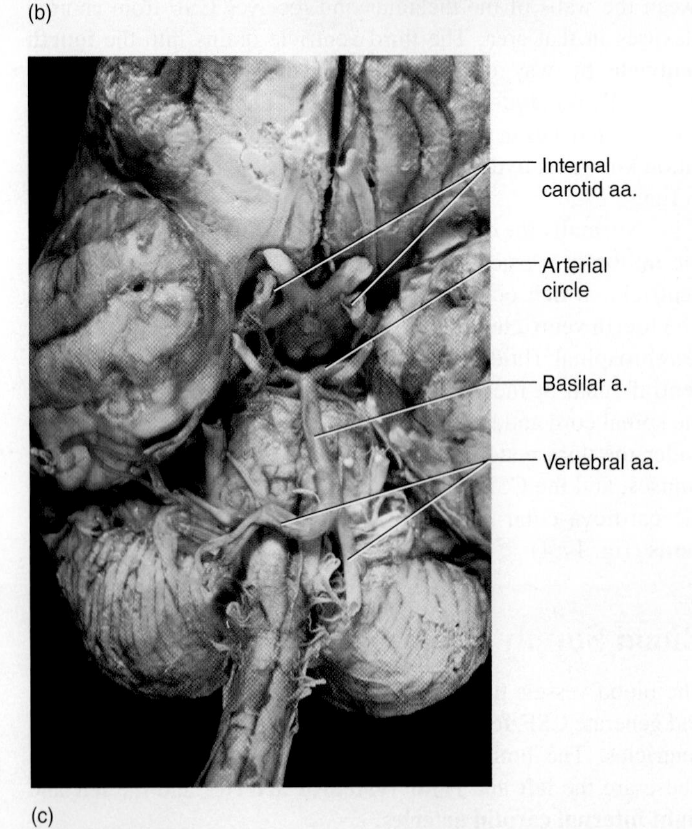

FIGURE 17.5 **Brain with Arteries, Inferior View** (a) Overview of arterial supply to brain; (b) close-up of arterial circle; (c) photograph. In this illustration a = artery and aa = arteries.

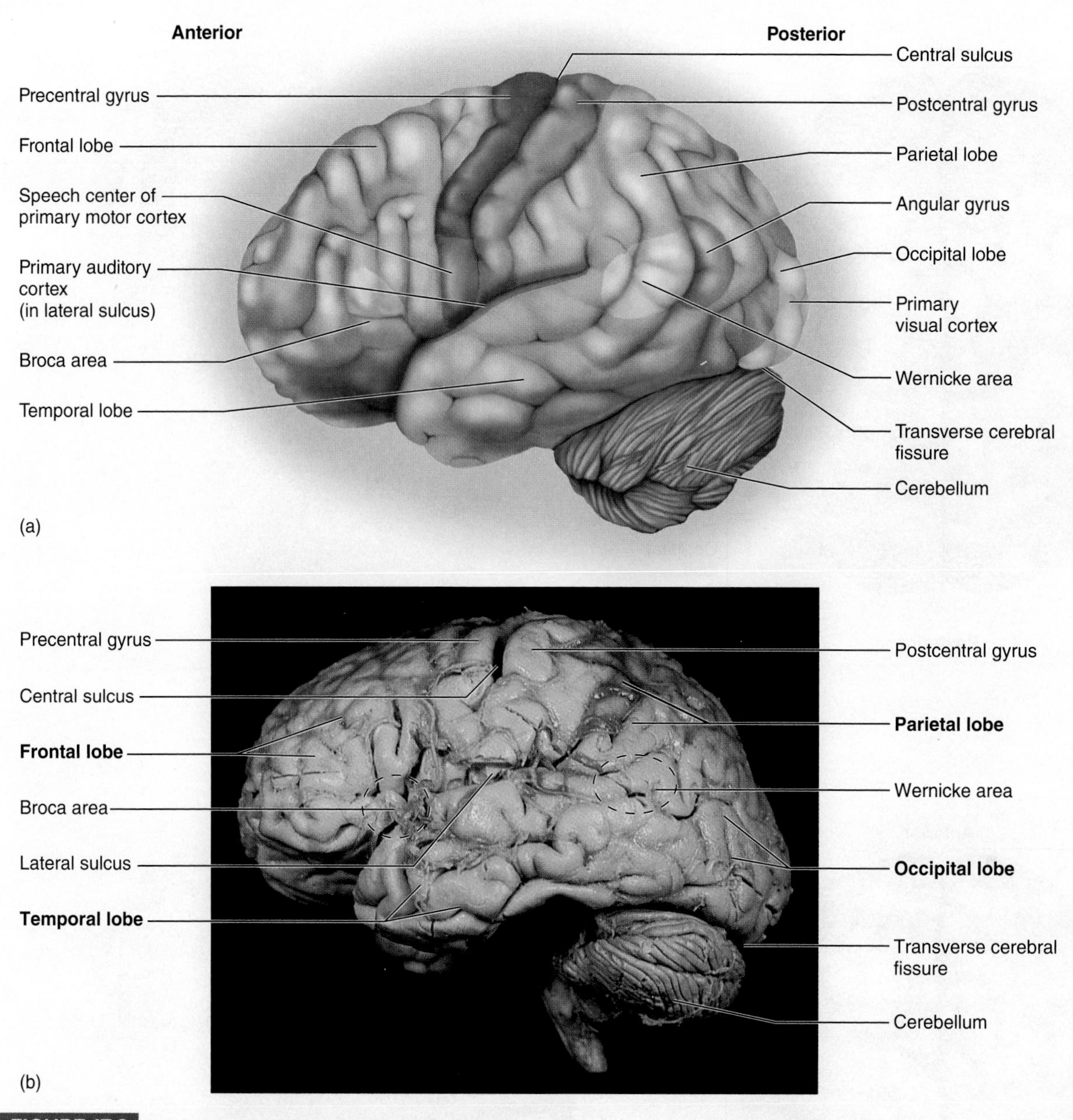

FIGURE 17.6 **Brain, Lateral View** (a) Diagram; (b) photograph.

separates the temporal lobe from the frontal and parietal lobes of the brain. These regions can be seen in figure 17.6.

Frontal Lobe The **frontal lobe** is responsible for many of the higher functions associated with being human. Humans have a six-layered neocortex, which is characteristic of primates. The frontal lobe is involved in intellect, abstract reasoning, creativity, social awareness, and language. An important area responsible for contributing to the formation of speech is called the **Broca area,** or the **motor speech area.** This region is located on the left side of the brain in most people, whether they are right-handed or left-handed. It is located in the lateral aspect of the left frontal lobe. Locate the frontal lobe in figure 17.6. The posterior border of the frontal lobe is defined by the **central sulcus.**

To find the central sulcus, look for two convolutions that run from the superior portion of the cerebrum to the lateral fissure, more or less continuously. The gyrus anterior to the central sulcus is part of the frontal lobe and is known as the **precentral gyrus,** or the **primary motor cortex.** This cortex is important for directing a part of the body to move and has been mapped, as in figure 17.7*a*. When you construct a figure of a human that reflects this map, it is called a homunculus. A **motor homunculus** (hoh-MUNK-you-lus) is seen in figure 17.7*a*.

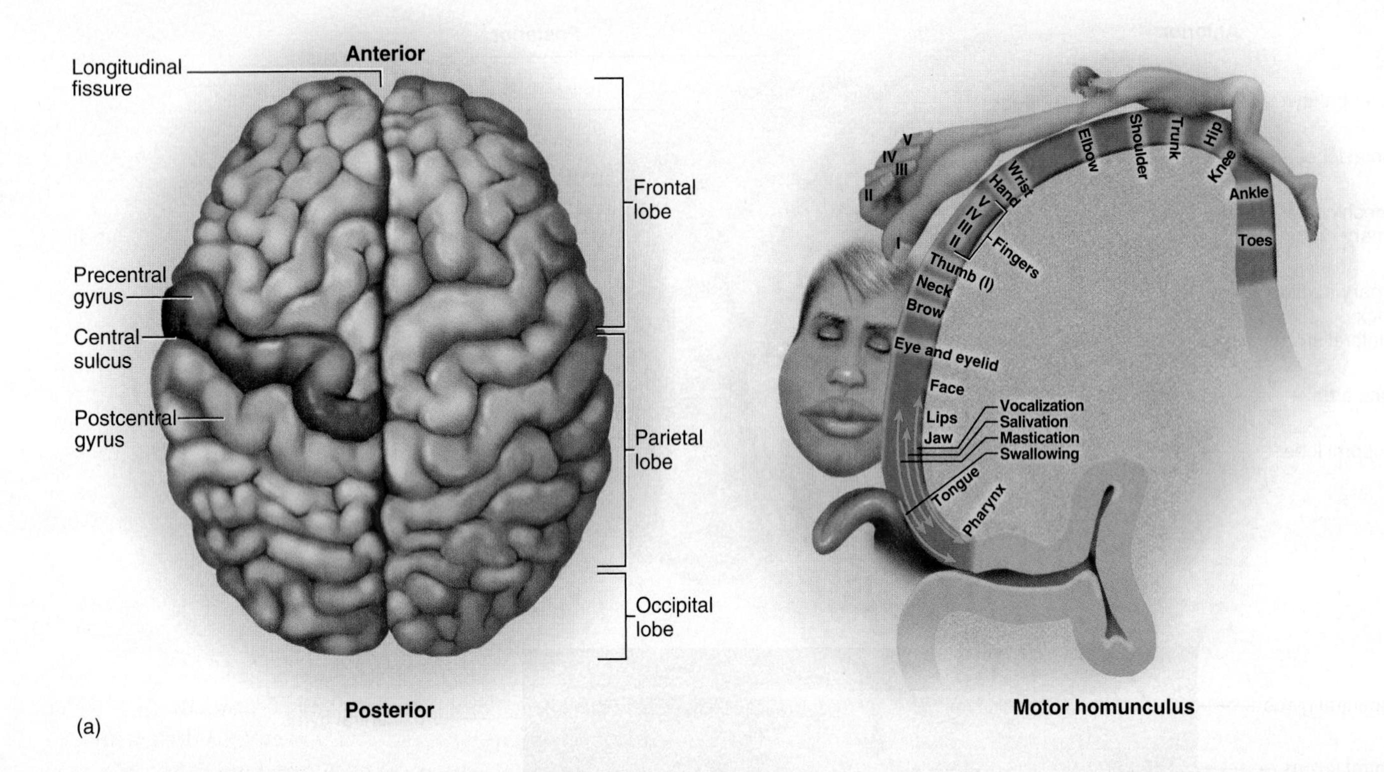

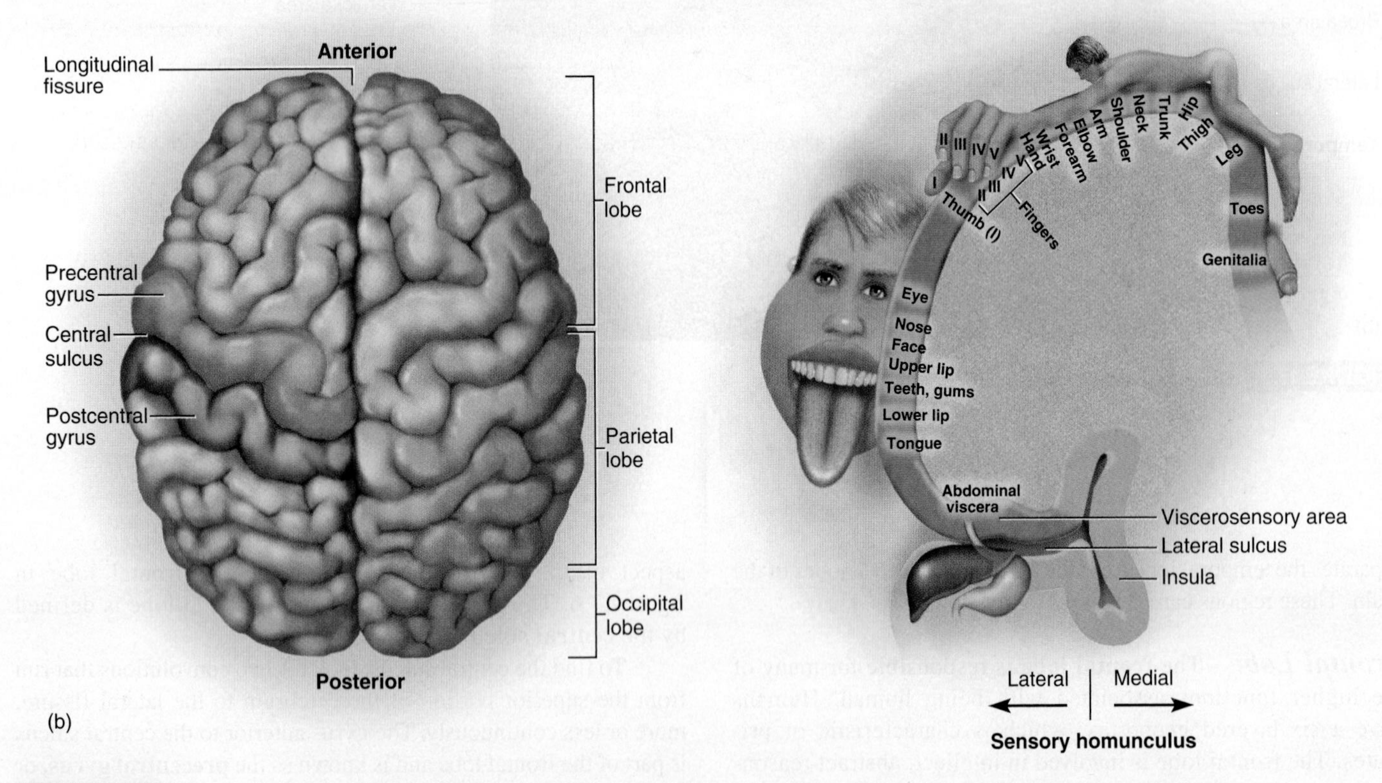

FIGURE 17.7 Primary Motor and Somatosensory Cortex (a) Motor cortex (precentral gyrus); (b) somatosensory cortex (postcentral gyrus).

What reason could be ascribed to having so much of the precentral gyrus dedicated to the face and hands?

What reason might there be for having so little of the precentral gyrus dedicated to the trunk?

Parietal Lobe The gyrus posterior to the central sulcus is known as the **postcentral gyrus,** or the **primary somatosensory cortex.** This is part of the parietal lobe and is involved in receiving sensory information from the body. This area has also been mapped, and you can see the **sensory homunculus** represented in figure 17.7*b*. The primary somatosensory cortex receives information, yet the material is integrated just posterior to the sensory cortex in the **association areas.** The primary somatosensory cortex pinpoints the part of the body affected, and the association area interprets the sensation (pain, heat, cold, and so on).

Occipital Lobe Posterior to the parietal lobe is the **occipital lobe** of the cerebrum. The occipital lobe is considered the **primary visual area** of the brain, and damage to this lobe can cause blindness. The shape, color, and distance of objects are perceived here, as is the recollection of past visual images. As you are reading these words, your occipital lobe is receiving the information and transferring it to other regions, which are converting the words to thought. Between the occipital lobe and the cerebellum is a **transverse cerebral fissure,** which separates these two regions of the brain. Locate the occipital lobe and the cerebellum in figure 17.6.

Temporal Lobe The **temporal lobe** is separated from the frontal and parietal lobes by the **lateral sulcus,** as seen in figure 17.6. The temporal lobe contains an area, known as the **primary auditory cortex,** which receives neural impulses sent from the inner ear. The nearby auditory association area distinguishes the nature of the sound (music, noise, speech) as well as the location, distance, pitch, and rhythm. The **Wernicke** (WUR-ni-keh) **area** is involved in the formation of language, such as the recognition of written and spoken language and in the forming of coherent sentences. The temporal lobe also has centers for the sense of smell **(olfactory centers)** and taste **(gustatory centers).**

Cerebral Hemispheres Rotate the brain so that you are looking at it from a superior view, and find the **longitudinal fissure** that separates the cerebrum into the left and right cerebral hemispheres. The **left cerebral hemisphere** in most people is involved in language and reasoning. For example, the Broca area is on the left side of the brain (figs. 17.6 and 17.8).

The **right cerebral hemisphere** of the brain in most people is involved in space and pattern perceptions, artistic awareness, imagination, and music comprehension. This specialization, in which one hemisphere of the cerebrum is involved in a particular task, is known as cerebral lateralization. Examine the surface features of the brain, seen in figure 17.8.

Inferior Aspect of the Brain

Forebrain Examine the inferior aspect of the brain (fig. 17.9), and find the frontal lobes of the cerebrum as well as the temporal lobes. You may be able to see the **pituitary gland** if it has not been removed. The **optic chiasm** (KYE-az-um; *chiasm* = cross) is anterior to the pituitary and transmits visual impulses from the optic nerves to the brain. Two small processes posterior to the pituitary are the **mammillary bodies** (so named because they resemble little breasts), which function in olfactory reflexes and memory formation.

Hindbrain From an inferior view of the brain, locate the hindbrain, which includes the **pons, medulla oblongata,** and

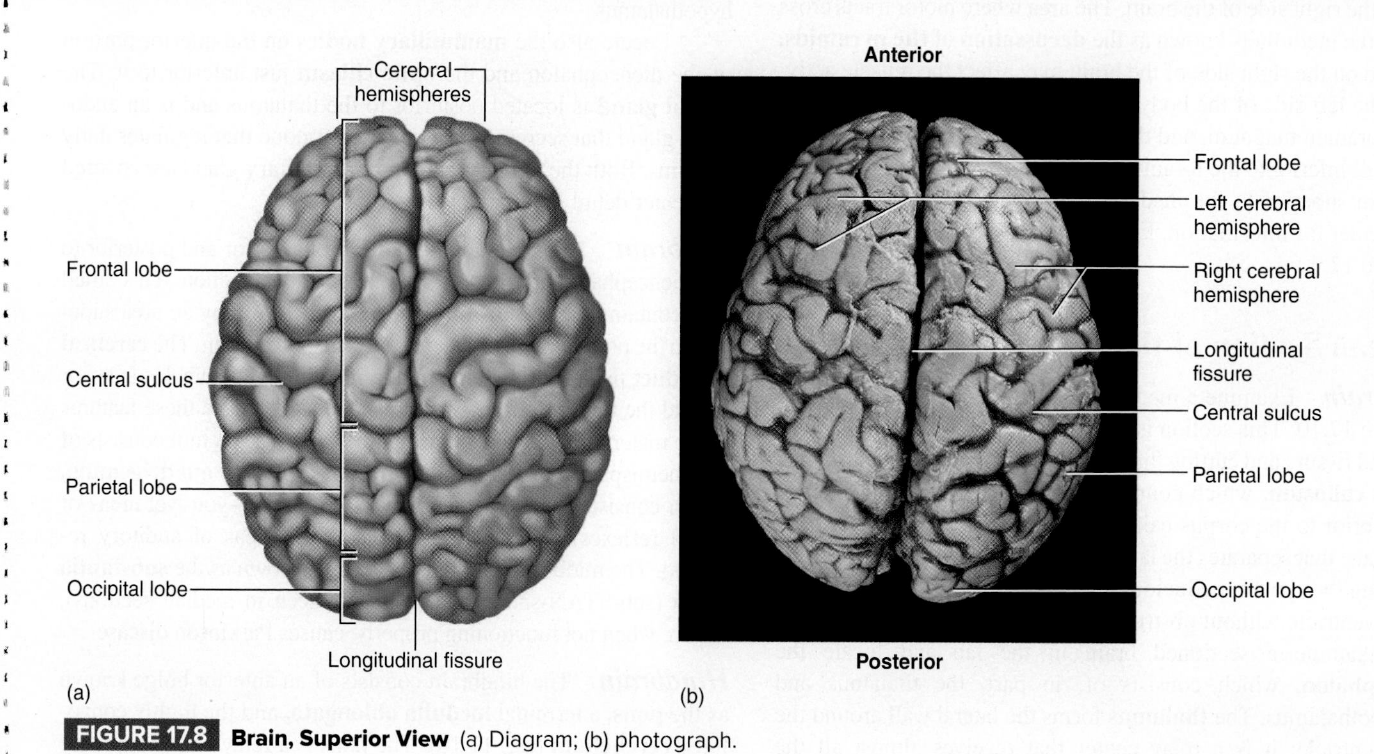

FIGURE 17.8 Brain, Superior View (a) Diagram; (b) photograph.

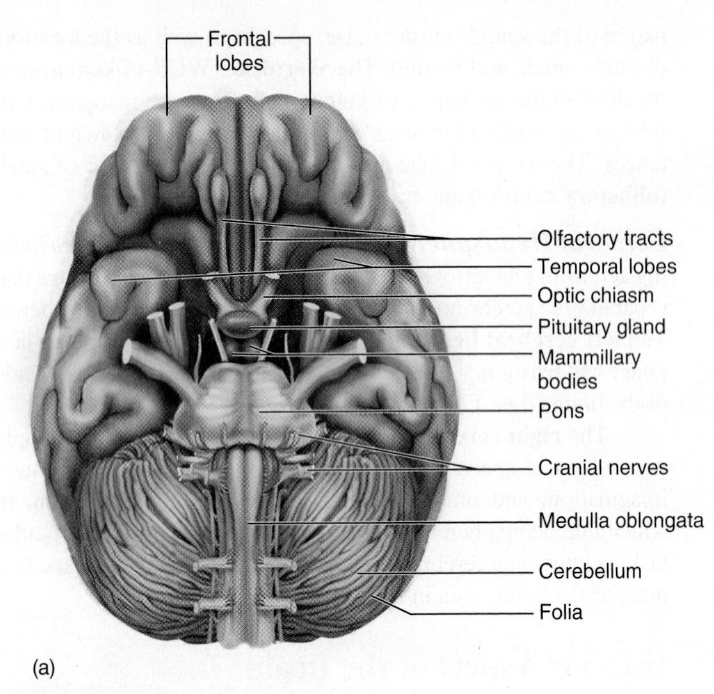

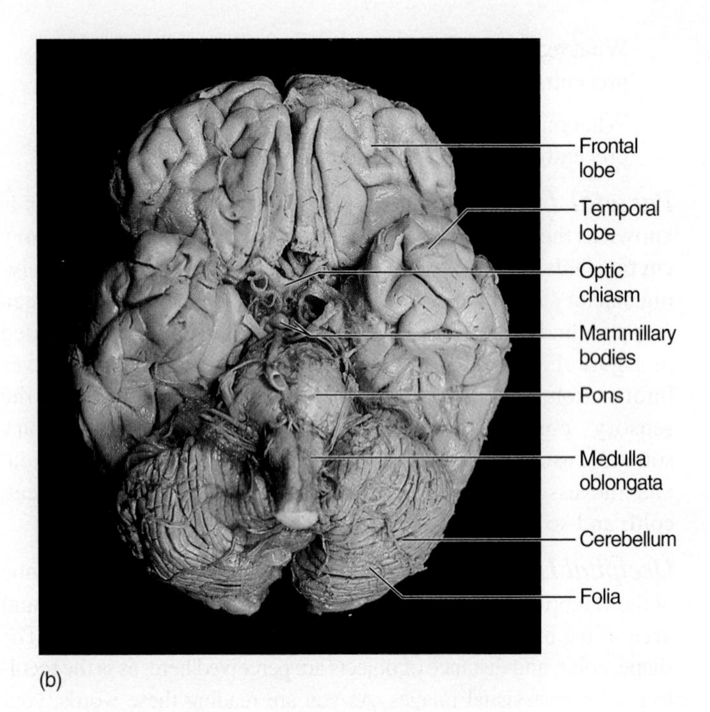

FIGURE 17.9 **Brain, Inferior View** (a) Diagram; (b) photograph.

cerebellum (SER-eh-BEL-um). The cerebellum has fine folds of neural tissue called **folia,** which are seen in an inferior view of the brain. The medulla oblongata is located anterior and inferior to the cerebellum and connects the brain to the spinal cord. The medulla oblongata has centers for control of respiratory rate and blood pressure as well as other vital centers. Some information from the right side of the body crosses over to the left brain in the medulla oblongata. Information coming from the left side of the body crosses over to the right side of the brain. The area where motor tracts cross over in the medulla is known as the **decussation of the pyramids.** A stroke on the right side of the brain may affect the muscle activity on the left side of the body. The medulla oblongata terminates at the foramen magnum, and the cervical region of the spinal cord continues inferior to the foramen magnum. The enlarged portion of the brain superior to the medulla is the pons, which serves as a relay center for information. Examine these structures of the brain in figure 17.9.

Median Section of the Brain

Forebrain Examine a median section of a brain, as illustrated in figure 17.10. This section is made by placing a knife in the longitudinal fissure and cutting through the brain. Locate the C-shaped **corpus callosum,** which connects the two cerebral hemispheres. Just inferior to the corpus callosum is the **septum pellucidum,** a membrane that separates the lateral ventricles from one another. By removing the septum pellucidum, you will be able to look into the lateral ventricle without obstruction.

Examine a sectioned brain in the lab and locate the **diencephalon,** which consists of, in part, the thalamus and the hypothalamus. The **thalamus** forms the lateral wall around the third ventricle; it is a relay center that receives almost all the sensory information from various tracts in the CNS and sends it to the cerebral cortex.

Below the thalamus is the **hypothalamus,** which has numerous autonomic centers. The hypothalamus, in part, directs the autonomic nervous system (ANS) and is involved with the **pituitary gland** (in the hypothalamopituitary axis) in many endocrine functions. Centers for thirst, water balance, pleasure, rage, sexual desire, hunger, sleep patterns, temperature, homeostasis, and aggression are located in the hypothalamus.

Locate also the **mammillary bodies** on the inferior portion of the diencephalon and the **optic chiasm** just anterior to it. The **pineal gland** is located posterior to the thalamus and is an endocrine gland that secretes melatonin, a hormone that regulates daily rhythms. Both the pineal gland and the pituitary gland are covered in greater detail in laboratory exercise 19.

Midbrain The midbrain is a small area inferior and posterior to the diencephalon and is best seen in a median section. This small area contains the **cerebral peduncles,** which occupy an area superior to the pons and on the anterior surface of the brain. The **cerebral aqueduct** passes through the midbrain, with the peduncles anterior to, and the **tectum** posterior to, the aqueduct. Locate these features in the material in the lab and in figure 17.10. The tectum consists of four hemispheric processes known as the **corpora quadrigemina,** which consist of the **superior colliculi** (col-LIC-you-lye; areas of visual reflexes) and the **inferior colliculi** (areas of auditory reflexes). The midbrain also houses a center known as the **substantia nigra** (sub-STAN-she-uh NY-gruh; not seen in median sections), which, when not functioning properly, causes Parkinson disease.

Hindbrain The hindbrain consists of an anterior bulge known as the **pons,** a terminal **medulla oblongata,** and the highly convoluted **cerebellum** (fig. 17.10). The pons is a relay center shunting

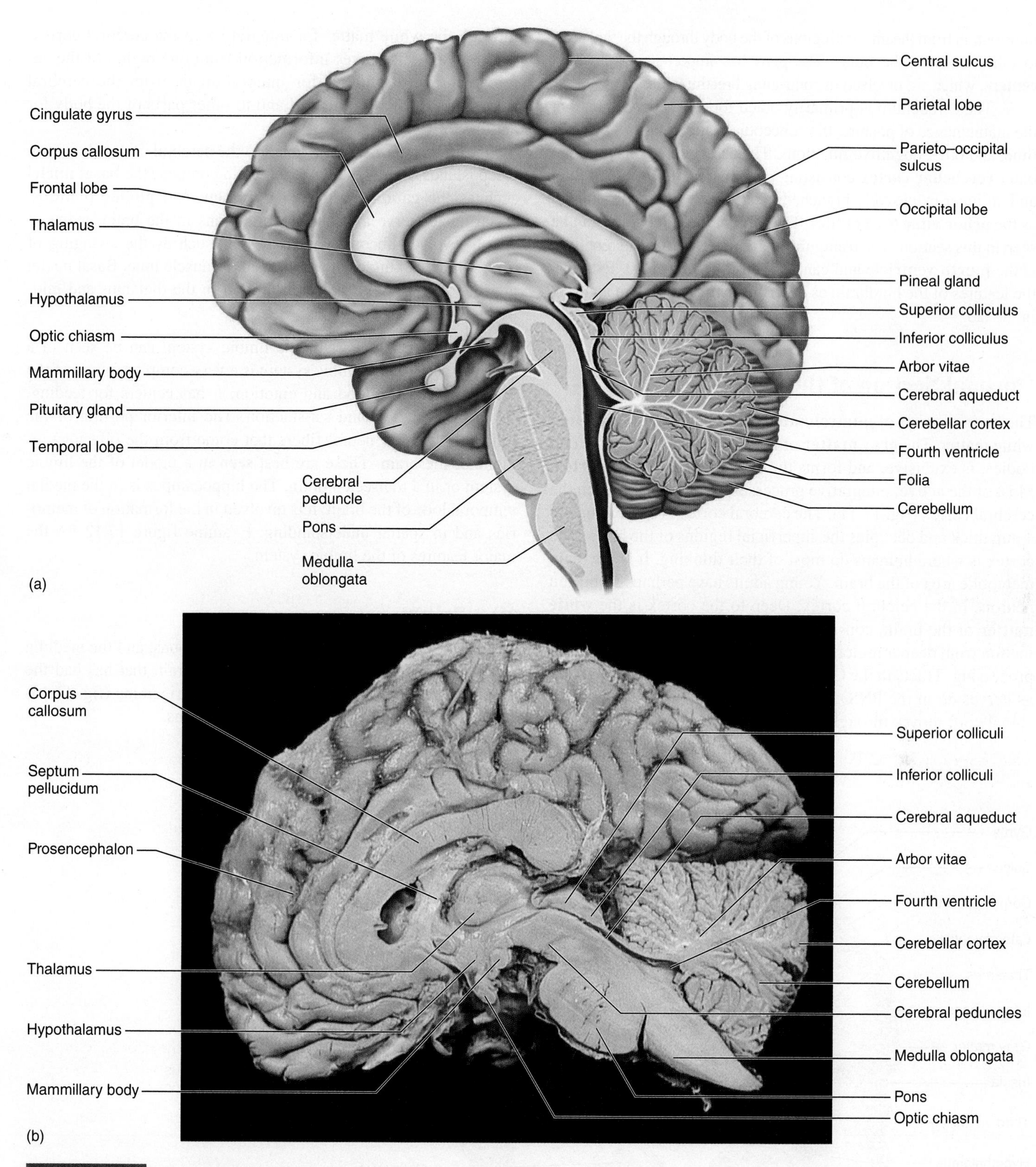

FIGURE 17.10 **Brain, Median Section** (a) Diagram; (b) photograph.

information from the inferior regions of the body through the thalamus to other areas of the brain. The pons has important **respiratory centers,** which are involved in controlling breathing rate.

The cerebellum is primarily noted for muscle coordination, the maintenance of posture, the conceptualization of the passage of time, and other cognitive functions. The cerebellum consists of an outer **cerebellar cortex** consisting of numerous folds called folia and an inner extensively branched pattern of white matter known as the **arbor vitae** (tree of life). The **folia** of the cerebellum can be seen in this section. The triangular space anterior to the cerebellum is the **fourth ventricle** and can be seen in figure 17.10. Examine the features of the hindbrain as described here and seen in material in the lab.

Coronal Section of the Brain

The brain consists of **unmyelinated** gray matter and **myelinated** white matter. The **gray matter** of the brain consists of nerve cell bodies, is extensive, and forms the superficial **cerebral cortex.** Most of the active, integrative processes of the brain occur in the cerebral cortex (fig. 17.11). The cerebral cortex is approximately 4 mm thick and occupies the superficial regions of the brain. The cortex is where humans do most of their thinking. It is the main metabolic area of the brain. Young adults have perhaps 50 billion neurons in the cerebral cortex. Deep to the cortex is the **white matter** of the brain, consisting mostly of axons that take information from deeper regions of the brain to the cerebral cortex for processing. Tracts in the CNS are myelinated fibers and function as nerves do in the PNS. Sensory information coming from the spinal cord moves through the inferior regions of the brain and through the white matter for integration in the cerebral cortex. White matter also takes information from one region of the cerebral cortex to another for integration or from the cerebral cortex back to the spinal cord and to other parts of the body for action.

Gray matter is not restricted to the cerebral cortex, however. Deep islands of gray matter in the brain compose the **basal nuclei** such as the **caudate nucleus, putamen,** and **globus pallidus.** Basal nuclei serve a number of functions in the brain, many of which involve subconscious processes, such as the swinging of arms while walking or the regulation of muscle tone. Basal nuclei are found not only in the cerebrum but in the thalamus and midbrain as well.

Limbic System Part of the limbic system can be seen in a coronal section. The limbic system is a very complex region of the brain involved in mood and emotion; it has centers for feeding, sexual desire, fear, and satisfaction. The inferior portion of the limbic system has neural fibers that come from the olfactory regions of the brain. These are best seen in a model of the limbic system or in a transected brain. The hippocampus is in the medial temporal lobe of the brain; it is involved in the formation of memories and in spatial understanding. Examine figure 17.12 for the major features of the limbic system.

Brainstem

The brainstem consists of the midbrain, the pons, and the medulla oblongata. Look at a model or section of brain that has had the cerebrum removed. Locate the corpora quadrigemina (fig. 17.13) along with the medulla oblongata and the pons.

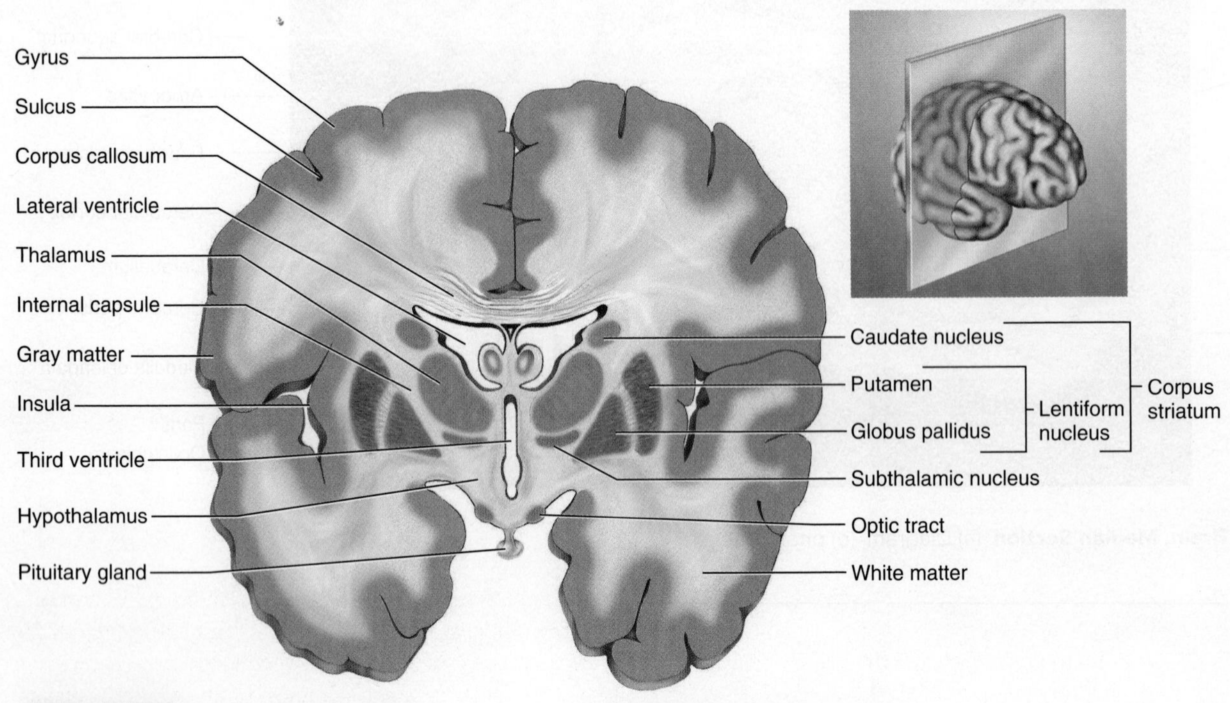

FIGURE 17.11 Brain, Coronal Section

Medial prefrontal cortex
Corpus callosum
Cingulate gyrus
Orbitofrontal cortex
Basal nuclei
Amygdala
Temporal lobe
Fornix
Thalamic nuclei
Mammillary body
Hippocampus

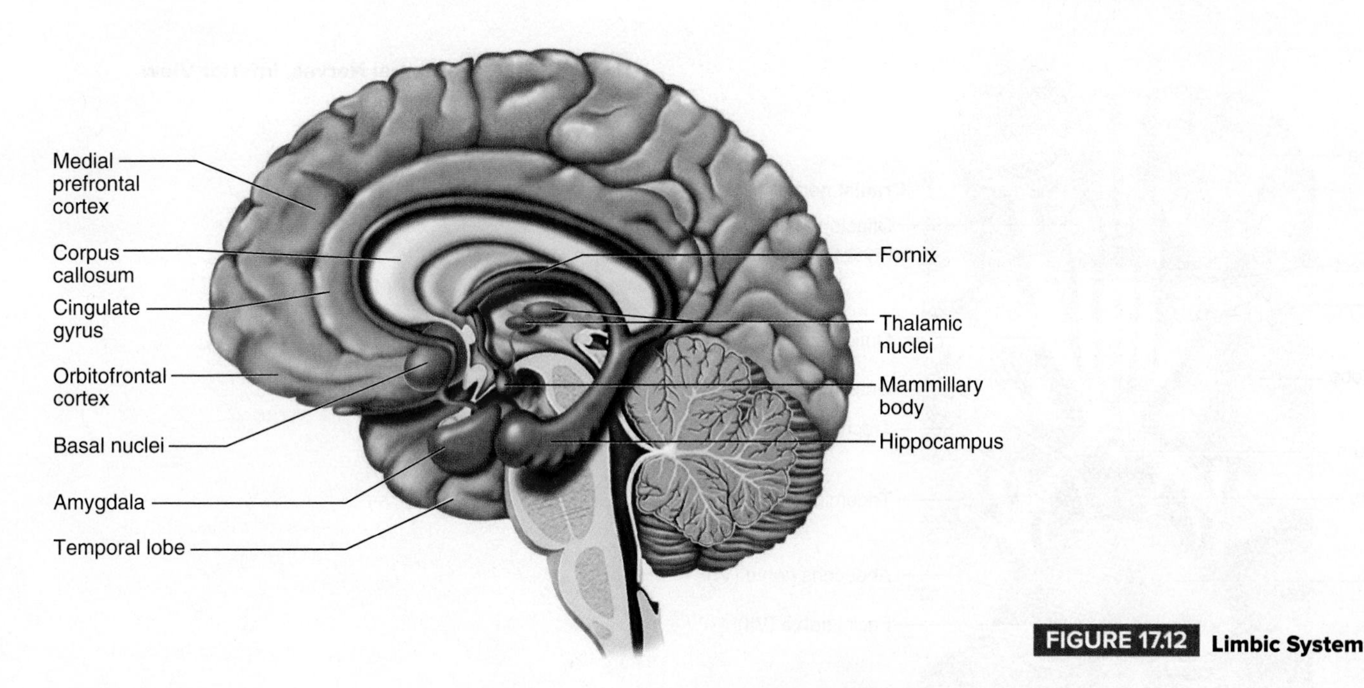

FIGURE 17.12 Limbic System

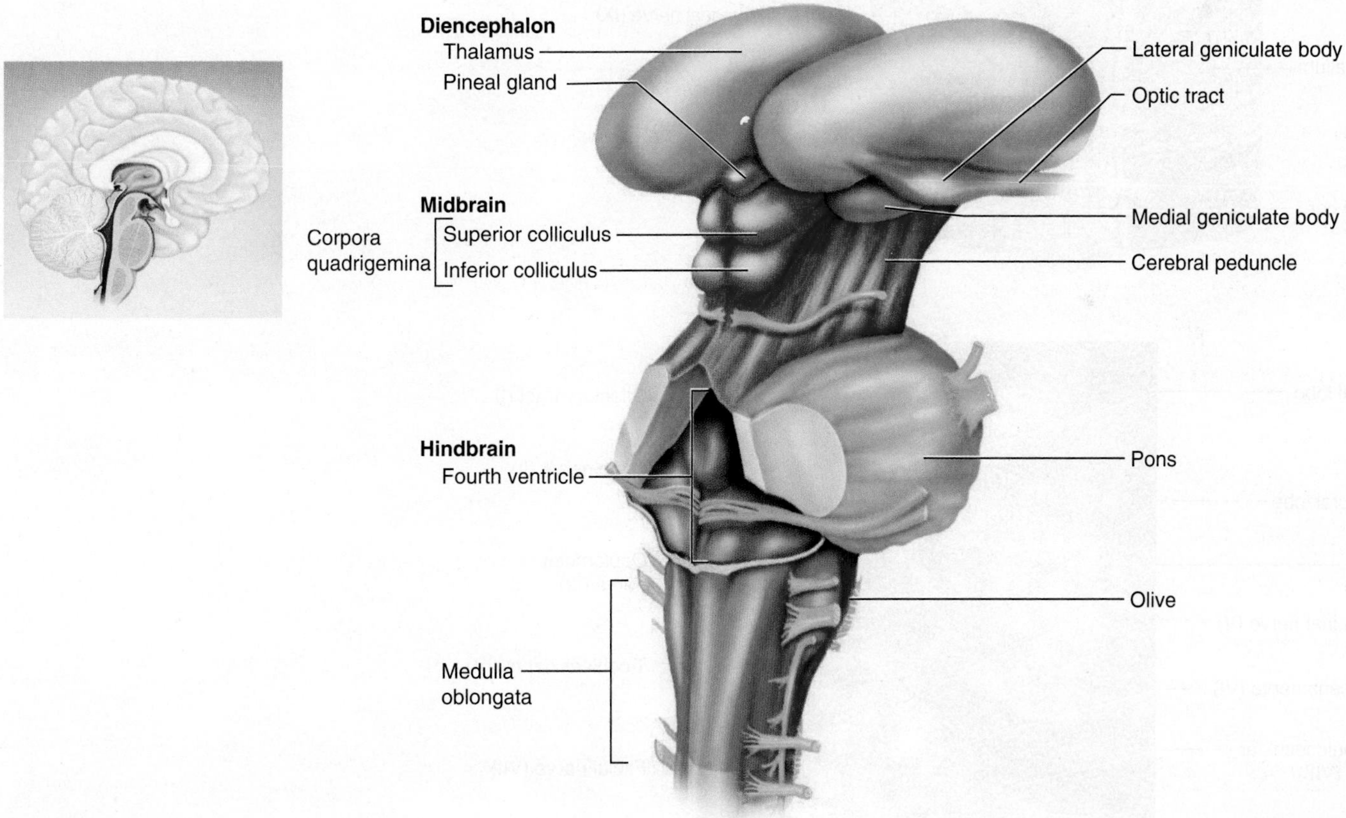

Diencephalon
Thalamus
Pineal gland
Midbrain
Corpora quadrigemina
Superior colliculus
Inferior colliculus
Hindbrain
Fourth ventricle
Medulla oblongata
Lateral geniculate body
Optic tract
Medial geniculate body
Cerebral peduncle
Pons
Olive

FIGURE 17.13 Brainstem, Posterolateral

Cranial Nerves

There are 12 pairs of cranial nerves. The cranial nerves are part of the PNS, but they are frequently studied along with the brain. The cranial nerves are listed by roman numeral (I–XII), and you should know the nerve by name *and* by number. Examine a model of the brain along with figure 17.14, and note that all the cranial nerves except nerve XII are in sequence from anterior to posterior. Nerves may be sensory, motor, or mixed (both sensory and motor). If the predominant function of the nerve is to receive

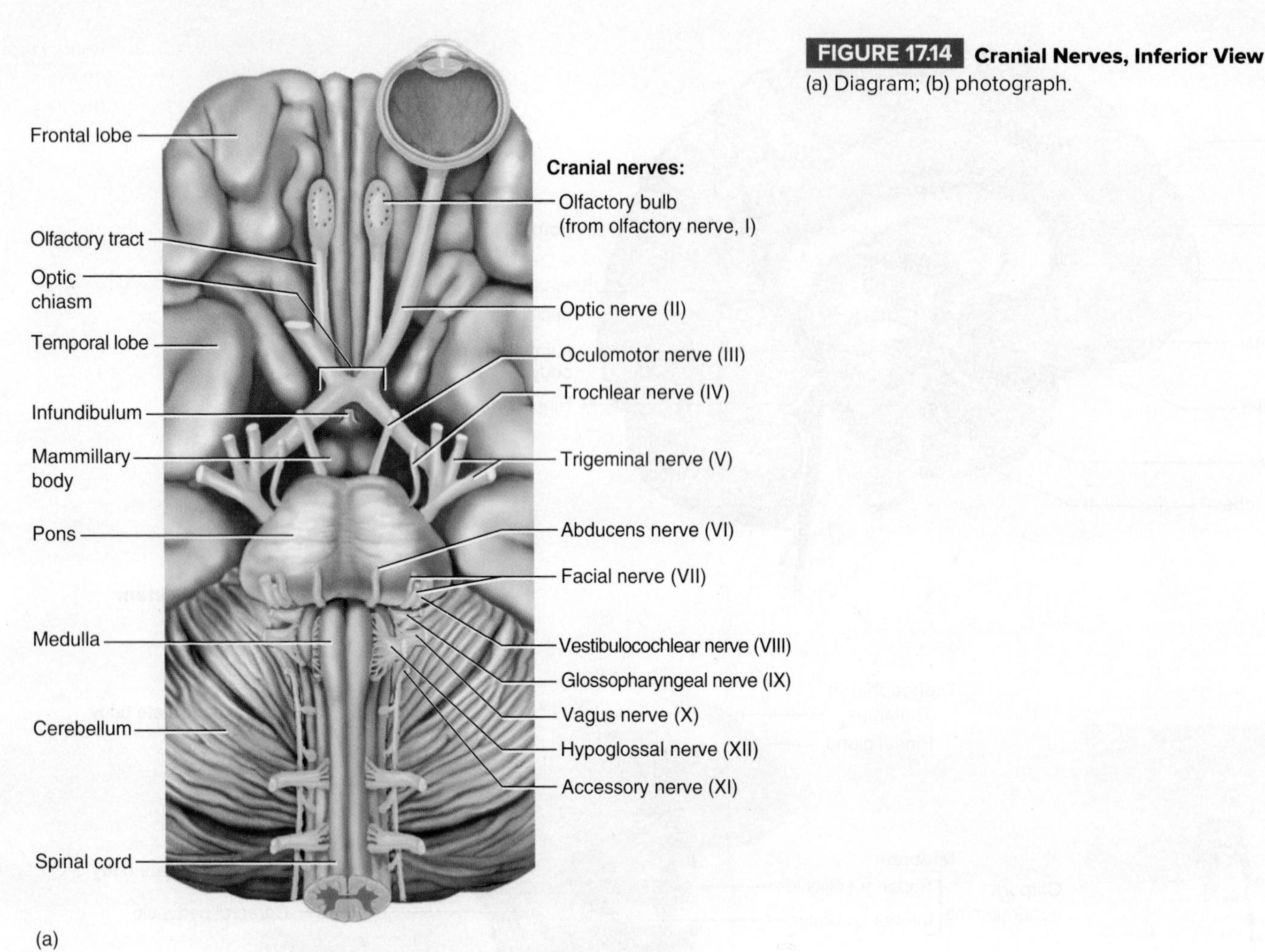

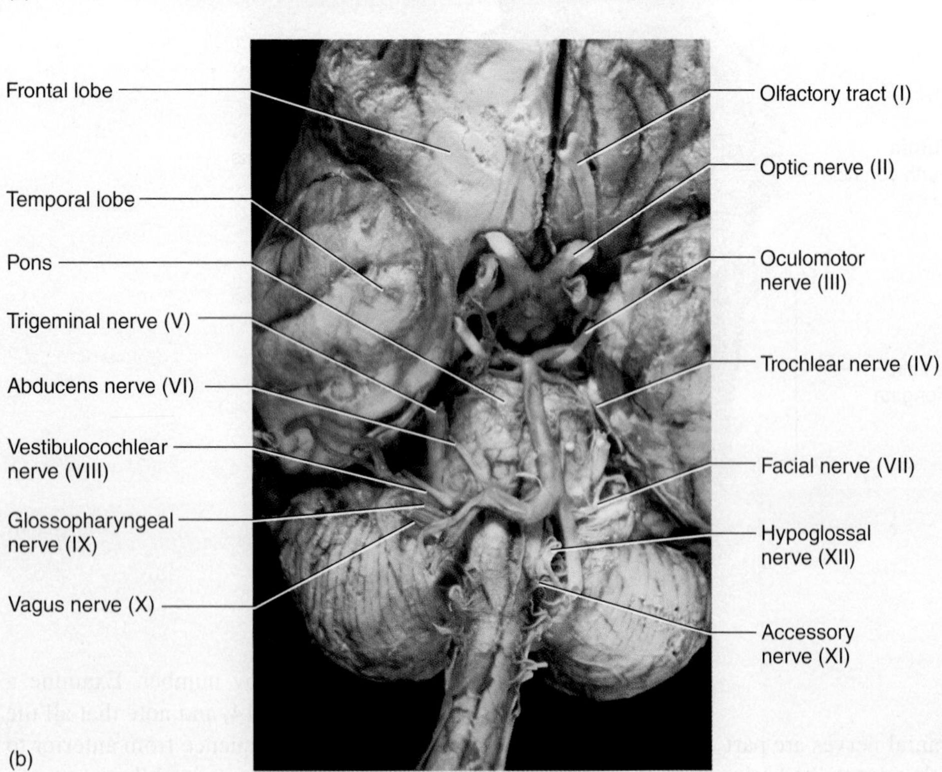

FIGURE 17.14 Cranial Nerves, Inferior View
(a) Diagram; (b) photograph.

neural input, then the nerve is considered a **sensory nerve.** If the predominant function of the nerve has some kind of action (such as skeletal muscle contraction or glandular secretion), then the nerve is considered a **motor nerve.** Some nerves have significant sensory and motor functions, and these are **mixed nerves.** When you study nerves, you should examine the anatomy of the brain to see where the nerve emerges from the brain. If you know, for example, that the abducens is found between the pons and the medulla oblongata, you can use that nerve as a way to locate other nerves.

The **olfactory nerves** (I) (*olfaction* = smell) pass through the cribriform plate of the ethmoid, synapsing in the olfactory bulbs. Impulses from the bulbs pass into the olfactory tracts that run along the anterior base of the brain at the inferior aspect of the frontal lobe and transmit the information to the brain. The **optic nerve** (II) (*optic* = sight) is a sensory nerve from the eye that leads to the base of the brain and forms the **optic chiasm.** Some fibers of the optic nerve lead to one side of the brain, while others cross to the other side. The **oculomotor nerves** (III) (*oculo* = eye; *motor* = muscle) control eye muscles and can be found anterior to the pons, close to the midline of the brain, while the **trochlear nerve** (IV) (*trochlea* = hoop that the superior oblique muscle passes through) is found at about a 45° angle from midline on the lateral aspect of the pons. The large **trigeminal nerve** (V) (*trigeminal* = triplets as the nerve divides into three parts) is found at a 90° angle to the pons and is located on the lateral aspect of the pons. It is a mixed nerve (both sensory and motor) that innervates much of the face. The **abducens nerve** (VI) (*abduce* = to pull away; the muscle controlled by this nerve pulls the eye laterally) is found close to the midline of the brain at the junction of the pons and medulla oblongata, while the **facial nerve** (VII) (named for its innervation of the face) is more lateral. Lateral to the facial nerve is the **vestibulocochlear nerve** (VIII) (the vestibule and cochlea are parts of the ear), and the nerve inferior to that is the **glossopharyngeal nerve** (IX) (*glosso* = tongue; *pharyngeal* = pharynx). The **vagus nerve** (X) (*vagus* = wandering) is a large nerve or large cluster of fibers on the lateral aspect of the medulla oblongata. The vagus nerve travels to the thorax and abdomen. Inferior to the vagus nerve is the **accessory nerve** (XI), which controls the muscles of the neck and back and arises from the superior spinal cord. Toward the midline of the medulla oblongata is the **hypoglossal nerve** (XII) (*hypo* = below; *glosso* = tongue), which innervates tongue muscles. Locate these nerves in figure 17.14 and note their details in table 17.2. There have been many mnemonic devices constructed to remember the sequence of cranial nerves. One such mnemonic is "Old Oliver Ogg Traveled To Africa For Very Good Vacations And Holidays." The first letter of each word of the mnemonic represents the first letter of the name of each cranial nerve.

TABLE 17.2 Cranial Nerves—Function

Number	Name	Function	Location
I	Olfactory	Receives sensory information from the nose, conducting the sense of smell to the brain	Begins in the upper nasal cavity and passes through the cribriform plate of the ethmoid bone. It synapses in the olfactory bulb on either side of the longitudinal fissure of the brain. The fibers take information on the sense of smell and pass via the olfactory tracts to be interpreted in the temporal lobe of the brain.
II	Optic	Receives sensory information from the eye, conducting the visual impulses to the brain	Takes sensory information from the retina at the back of the eye and transmits the impulses through the optic canal in the sphenoid bone. Some axons cross at the optic chiasm and pass via the optic tracts to the occipital lobe, where vision is interpreted. Other axons do not cross over the optic chiasm.
III	Oculomotor	Conducts motor information to move the medial, superior, and inferior rectus muscles and the inferior oblique muscle of the eye	Emerges from the surface of the brain near the midline and just superior to the pons. It passes through the superior orbital fissure and innervates the inferior oblique muscle and the medial, superior, and inferior rectus muscles and carries parasympathetic fibers to the lens and iris.
IV	Trochlear	Conducts motor information to move the superior oblique muscle of the eye	Seen at the sides of the pons at about a 45° angle from the midline of the brain. It passes through the superior orbital fissure to the superior oblique muscle.

(continued)

TABLE 17.2 Cranial Nerves—Function *(continued)*

Number	Name	Function	Location
V	Trigeminal	A three-branched nerve; conducts both sensory information from, and motor information to, the face, also important for mastication (chewing)	Seen at a 90° angle from the midline at the lateral sides of the pons. The trigeminal has three branches: (1) The ophthalmic branch passes through the superior orbital fissure; (2) the maxillary branch passes through the foramen rotundum of the sphenoid bone; (3) the mandibular branch passes through the foramen ovale of the sphenoid bone and enters the mandible by the mandibular foramen and exits by the mental foramen.
VI	Abducens	A motor nerve to move the lateral rectus muscle of the eye	Begins at the midline junction between the pons and the medulla oblongata and passes through the superior orbital fissure to carry motor information to the lateral rectus muscle of the eye.
VII	Facial	A large nerve that receives sensory information of taste from the anterior tongue and takes motor information to the facial muscles. Innervates salivary and other glands.	Begins as the first of a cluster of nerves on the anterolateral part of the medulla oblongata. It passes through the internal auditory meatus and near the inner ear to the stylomastoid foramen of the temporal bone to innervate facial muscles and glands. It carries sensory information from the anterior tongue. Sensory information of the tongue is interpreted in the parietal and temporal lobes of the brain.
VIII	Vestibulocochlear	Receives sensory information, from the ear; the vestibular part transmits equilibrium information, and the cochlear part conducts acoustic information	Comes from the inner ear and passes through the internal auditory meatus. The conduction passes to the pons, and hearing and balance are interpreted in the temporal lobe.
IX	Glossopharyngeal	A mixed nerve of the tongue and throat, including motor functions, also receives information on taste	Passes through the jugular foramen to innervate muscles of the throat (pharyngeal branches) and sensory receptors of the tongue. Motor portions of the nerve control a muscle of swallowing and a salivary gland, while sensory nerves carry information from the posterior tongue and from baroreceptors and chemoreceptors of the carotid artery.
X	Vagus	Receives sensory information from the abdomen, thorax, neck, and root of the tongue; conducts motor information to the pharynx and larynx and controls autonomic functions of heart, digestive organs, spleen, and kidneys	Passes through the jugular foramen and along the neck to the larynx, heart, lungs, and abdominal region. The sensory impulses travel in this nerve from the viscera in the abdomen, thorax, neck, and root of the tongue to the brain.
XI	Accessory	A motor nerve to the muscles of the neck and back also important for swallowing	Multiple fibers arise from the lateral sides of the superior spinal cord and pass through the jugular foramen to numerous muscles of the neck and back.
XII	Hypoglossal	A motor nerve to the tongue	Begins at the anterior surface of the medulla and passes through the hypoglossal canal to innervate the muscles of the tongue.

TABLE 17.3 Types of Cranial Nerves

Number	Name	Name Mnemonic	Type	Type Mnemonic
I	Olfactory	Old	Sensory	Sally
II	Optic	Oliver	Sensory	Sells
III	Oculomotor	Ogg	Motor*	Many
IV	Trochlear	Traveled	Motor	Mangoes
V	Trigeminal	To	Both	But
VI	Abducens	Africa	Motor	My
VII	Facial	For	Both	Brother
VIII	Vestibulocochlear	Very	Sensory	Sells
IX	Glossopharyngeal	Good	Both	Bigger
X	Vagus	Vacations	Both	Better
XI	Accessory	And	Motor	Mega
XII	Hypoglossal	Holidays	Motor	Mangoes

*Many of the motor nerves have sensory fibers that come from proprioceptors in the muscles they innervate. Information about the tension of the muscle is sent back to the brain to make adjustments in amount of muscle contraction. Since the main function of these nerves is motor, they are listed as motor nerves, even though they have some sensory capabilities.

The cranial nerves are listed in table 17.3 as sensory nerves, motor nerves, or both sensory and motor nerves. A mnemonic device is also listed in the table to help you remember the name and function of each nerve.

Dissection of a Sheep Brain

Work in pairs during the dissection of a sheep brain. Take a sheep brain back to your table, along with a dissecting tray and appropriate dissection tools. Work with another group, so that you can dissect two different sheep brains. One of the brains should be sectioned in the median plane and the other in the coronal plane. If the brains still have the **dura mater,** examine this tough connective tissue coat on the outside of the brain. Cut through this layer to examine the other meninges that are underneath it. Deep to the dura mater is a filmy layer of tissue that contains blood vessels. This is known as the **arachnoid mater.** If you tease some of the membrane away from the brain, you will see that it has a cobweblike appearance in the **subarachnoid space.** In life, the subarachnoid space contains the **CSF.** Underneath this layer and adhering directly to the brain convolutions is the **pia mater,** the surface lining of the convolutions of the brain.

Find the major lobes of the cerebrum, along with the cerebellum, pons, and medulla oblongata (fig. 17.15). Because sheep are

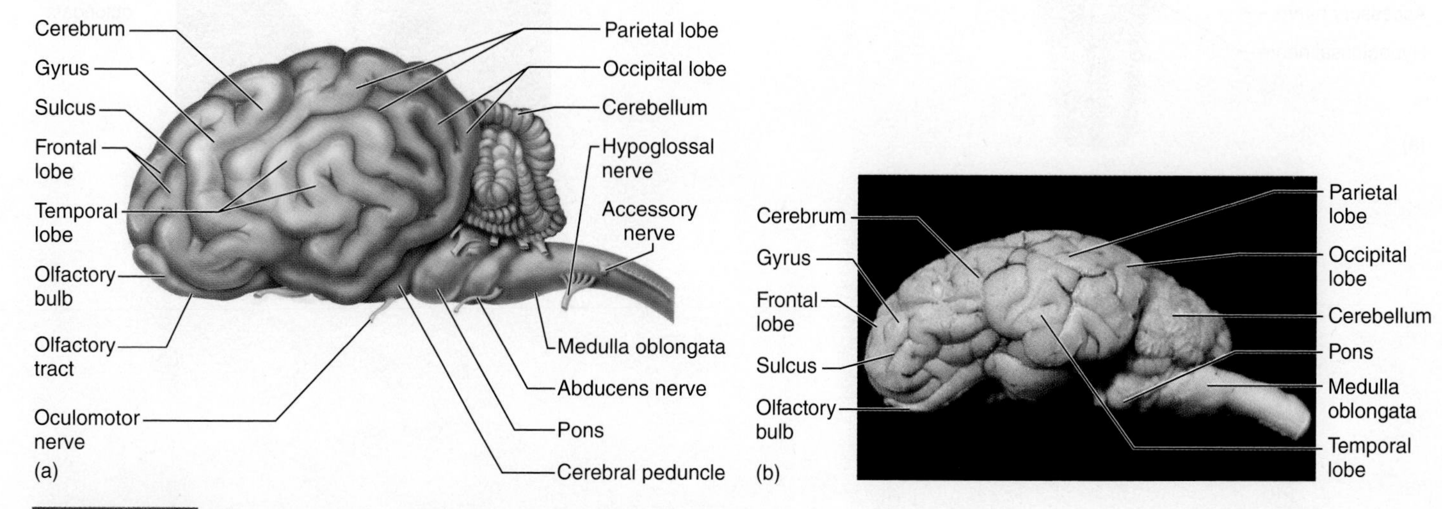

FIGURE 17.15 Sheep Brain, Lateral View (a) Diagram; (b) photograph.

quadrupeds, the flexure of the brain does not occur in them as it does in humans. Sheep have a horizontal spinal cord, while humans have a vertical one. Sheep also have a reduced cerebrum. The average weight of the sheep brain is about one-tenth that of humans. The greatest difference between the two is the larger cerebrum in humans, which reflects our greater cognitive abilities. The pons in humans is larger due to increased neural transmission to and from the enlarged cerebrum. Sheep have an excellent sense of smell, and this is reflected in their larger olfactory bulbs. Examine the inferior surface of the sheep brain. Locate the olfactory bulbs, tracts, optic nerve, and optic chiasm. The pituitary gland will probably not be attached, but you should locate the infundibulum, posterior (caudad) to the optic chiasm. Locate these structures in figure 17.16. Work with another group so that you can dissect two sheep brains. If the sheep brains are intact, you will need to decide which brain will be sectioned in the median plane and which will be sectioned in the coronal plane. For the median section, divide the brain in the plane of the longitudinal fissure. Your cut should resemble the brain in figure 17.17. Note that the sheep brain has an enlarged corpora quadrigemina, compared with humans'. Locate the corpus callosum, lateral ventricles, third ventricle, hypothalamus, pineal gland, superior and inferior colliculi, cerebellum, arbor vitae, pons, medulla oblongata, cerebral aqueduct, and fourth ventricle. The coronal section of a sheep brain is illustrated in figure 17.18. After making a coronal section about midway through the cerebrum, you should locate the cerebral cortex, cerebral medulla, lateral ventricles, corpus callosum, third ventricle, thalamus, and hypothalamus.

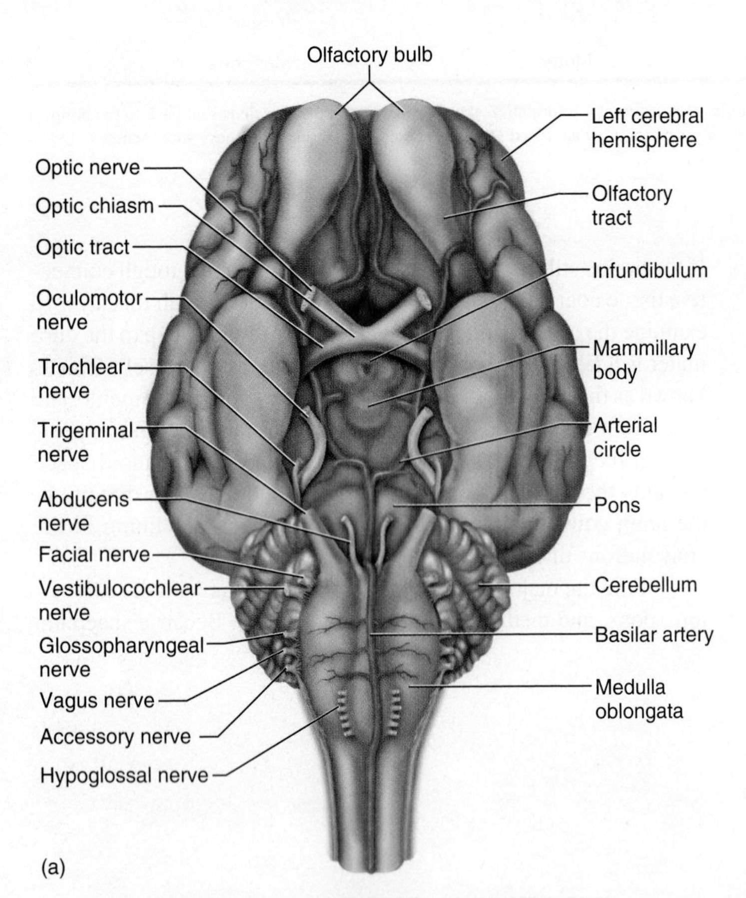

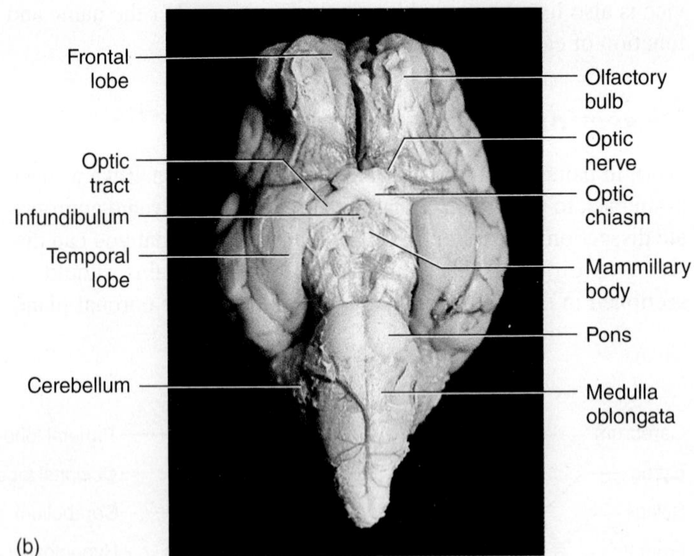

FIGURE 17.16 Sheep Brain, Inferior View (a) Diagram; (b) photograph.

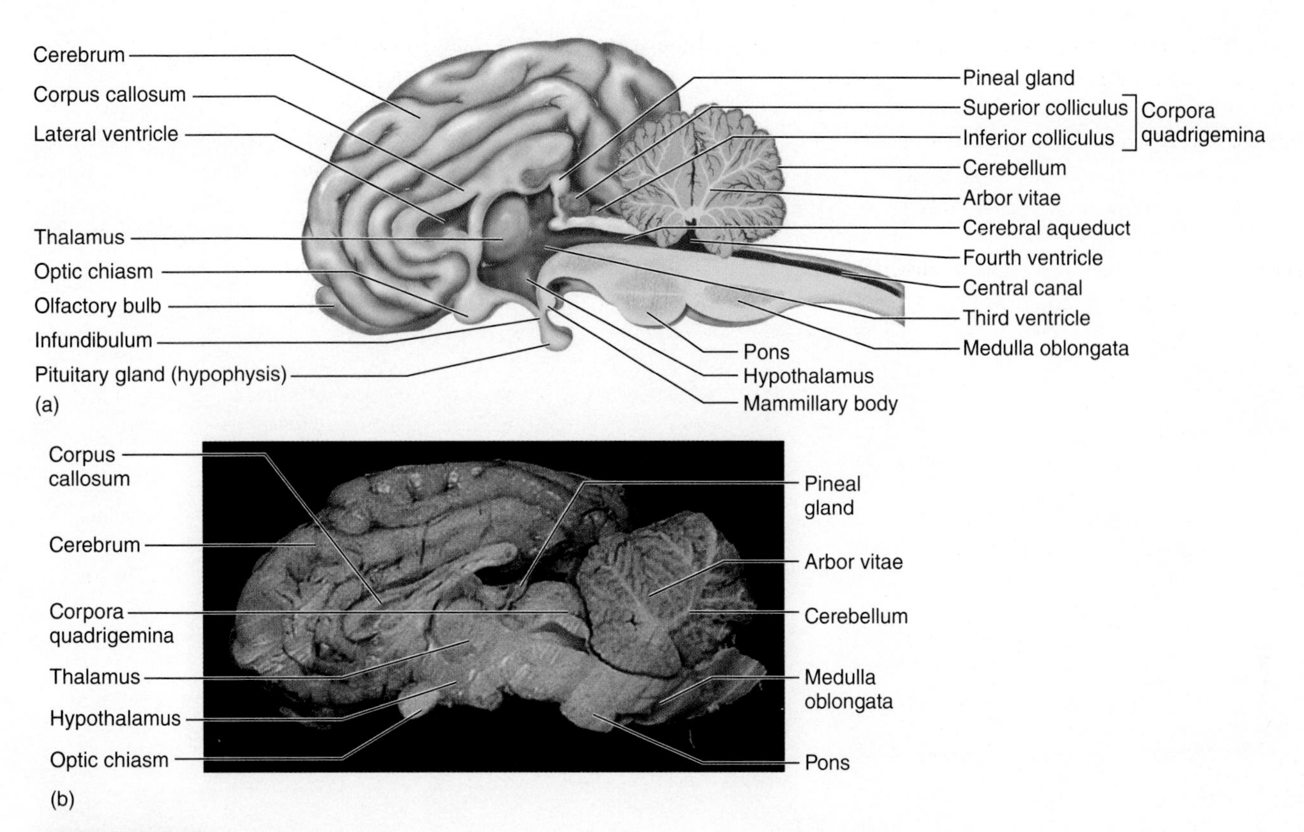

FIGURE 17.17 **Sheep Brain, Longitudinal Section** (a) Diagram; (b) photograph.

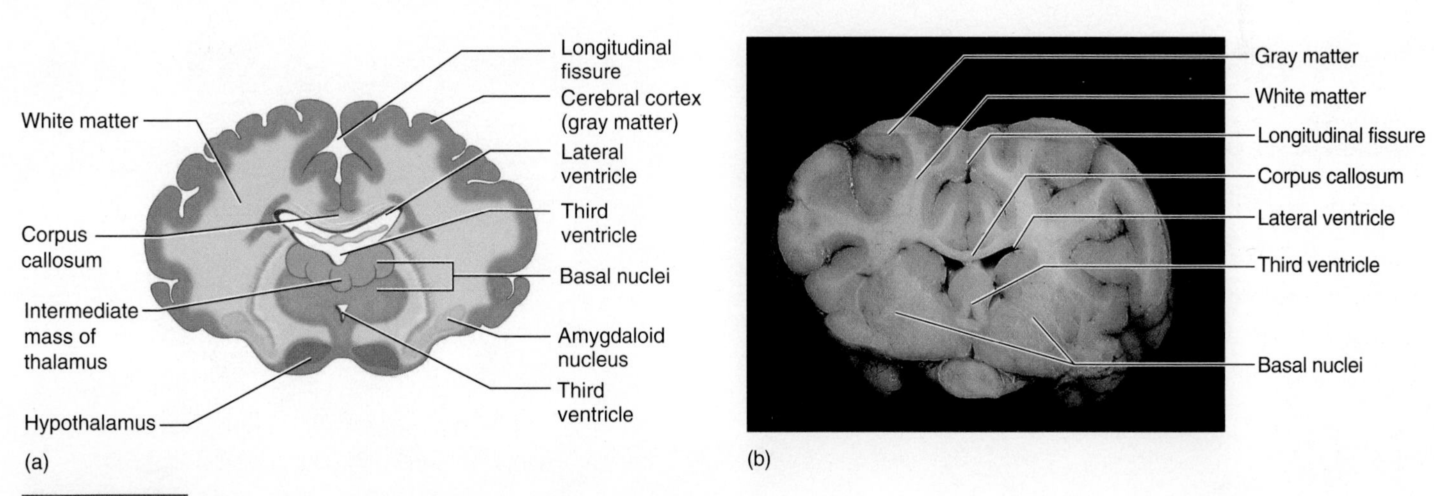

FIGURE 17.18 **Sheep Brain, Coronal Section** (a) Diagram; (b) photograph.

Exercise 17

Brain and Cranial Nerves

Name ______________________________ Date ______________

1. Which of the meninges is located next to the brain? ______________________________

2. Name the major veins that take blood from the brain. ______________________________

3. The basilar artery in the brain receives blood from what two arteries? ______________________________

4. What fluid is found in the ventricles of the brain? ______________________________

5. Into what space does fluid flow from the cerebral aqueduct? ______________________________

6. In terms of shape, compare and contrast a sulcus to a gyrus. ______________________________

7. What role do the convolutions play in the brain? ______________________________

8. What are all the lobes of the cerebrum called? ______________________________

____

9. What is the function of the precentral gyrus? ______________________________

____

10. What sense does the temporal lobe alone interpret? ______________________________

____

11. What physical depression separates the temporal lobe from the parietal lobe? ______________________________

____

12. What structure connects the cerebral hemispheres? ______________________________

____

13. Name the major regions of the midbrain. ______________________________

____

____

14. What function does the cerebellum have? ______________________________

____

15. The trigeminal nerve is larger than the trochlear nerve. How does this correlate with the function of both nerves?

____

____

16. John hit his forehead against the wall. What possible damage might he have done to the function of his brain, particularly the functions associated with the frontal lobe? ______________________________

____

____

17. If a stroke affected all the sensations interpreted by the brain concerning the face and the hands, what percentage of the postcentral gyrus would be affected? ______________________________

____

18. One convenient excuse that people often make for their inability to do something is to describe themselves as left-brain or right-brain individuals. Describe the effect that the loss of an entire cerebral hemisphere would have on specific functions, such as spatial awareness or the ability to speak. ______________________________

____

____

19. Aphasia is the loss of speech, and different types can occur. If the Broca area were affected by a stroke, would the content of the spoken word be affected or would the ability to pronounce the words be affected? ______________________________

____

____

20. Label the following illustration using the terms provided.

arbor vitae	corpus callosum	infundibulum	pineal gland
cerebral aqueduct	dura mater	mammillary body	pituitary gland
choroid plexus	fourth ventricle	medulla oblongata	pons
corpora quadrigemina	hypothalamus	optic chiasm	thalamus

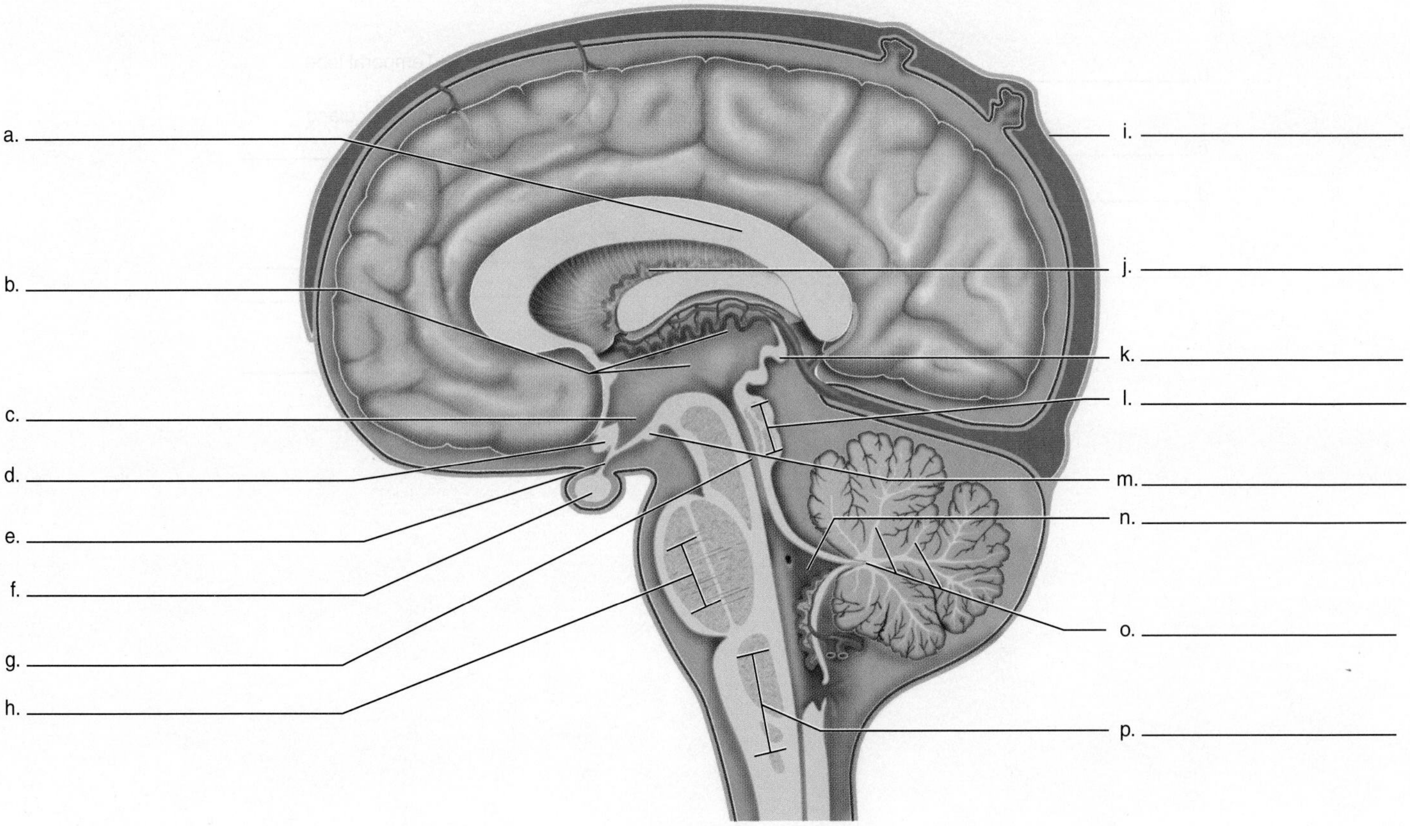

a. ______________

b. ______________

c. ______________

d. ______________

e. ______________

f. ______________

g. ______________

h. ______________

i. ______________

j. ______________

k. ______________

l. ______________

m. ______________

n. ______________

o. ______________

p. ______________

21. Where is CSF found in relation to the meninges? ______________________________

____

22. Approximately what percentage of the precentral gyrus is dedicated to the face? ______________________________

____

23. How much of the gyrus is dedicated to the hands? ______________________________

____

24. How much of the gyrus is dedicated to the trunk? ______________________________

____

25. Label the following illustration using the terms provided.

cerebellum	oculomotor nerve	pons
hypoglossal nerve	olfactory bulb	trigeminal nerve
medulla oblongata	optic nerve	vagus nerve

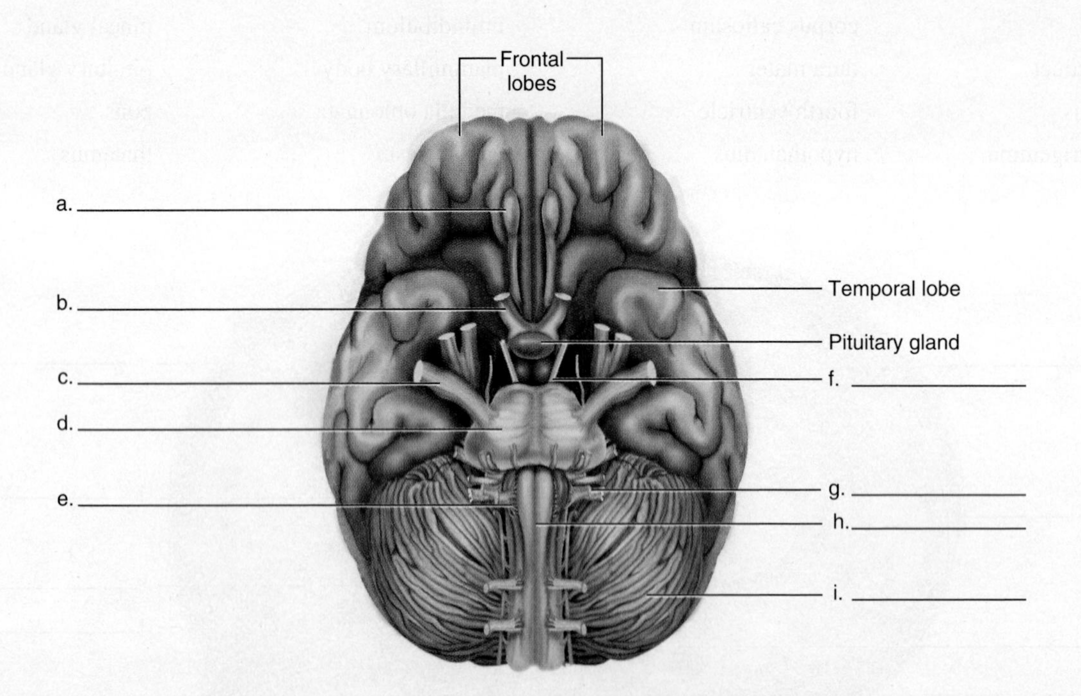

26. Describe these nerves in terms of function (sensory, motor, or both).

optic nerve ______________________________

trochlear nerve ______________________________

glossopharyngeal nerve ______________________________

hypoglossal nerve ______________________________

vagus nerve ______________________________

27. Name the cranial nerve or nerves that would innervate the following areas:

taste buds of the tongue ______________________________

internal ear ______________________________

mandible ______________________________

retina of the eye ______________________________

stomach ______________________________

lateral rectus muscle of the eye ______________________________

28. Label the following illustration using the terms provided.

arbor vitae	medulla oblongata
cerebral aqueduct	pineal gland
corpus callosum	pituitary gland
fourth ventricle	pons
hypothalamus	thalamus

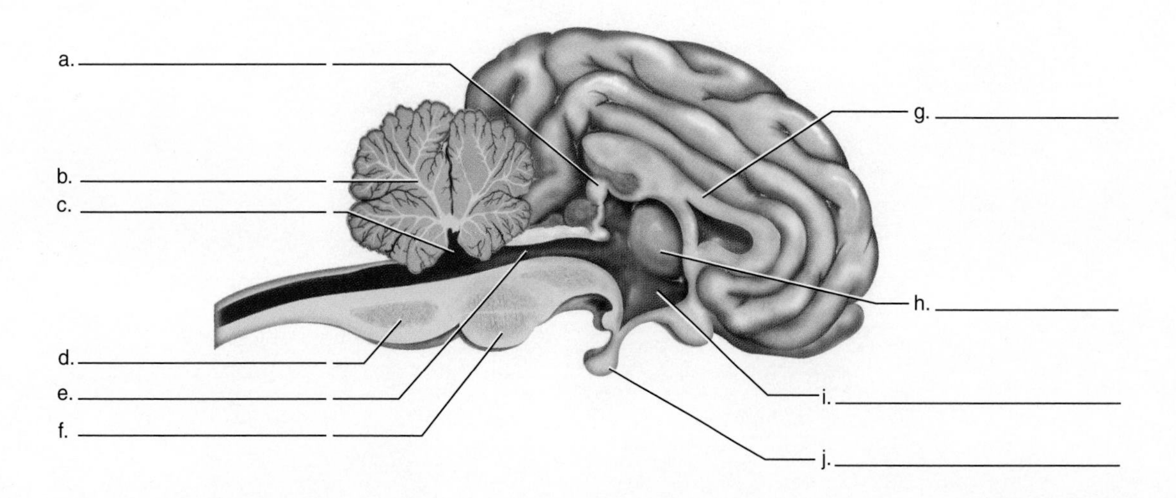

Laboratory Exercise 18

Sensory Receptors

Nervous System

Introduction

The essence of what it means to be human begins with the sense organs. Amazing sunsets in the western sky, music that touches the soul, the sense of a loved one's hand stirring the heart, and the experience of a fabulous dinner are all brought to us through the window of the sense organs. Without them we would exist in a cold, dark, soundless world. Our lives would truly be senseless.

Sensory receptors convert stimuli into neural impulses. The stimuli can come from the external environment, such as sound, or they can come from internal areas, such as hunger pangs. There are two main classes of sense—**general (somesthetic) senses** and **special senses.** General senses occur in many locations of the body and are found in places such as the skin, muscle, joints, and viscera. The senses of touch, pressure, changes in temperature, pain, blood pressure, and stretching are general senses. Special senses occur in specific locations, such as the eye, ear, tongue, and nose, and they include taste, smell, sight, hearing, and balance. These receptors are not uniformly distributed throughout the body but are absent, or few in number, in some areas, while densely clustered in other locations. This pattern of uneven distribution of sense organs is called **punctate distribution.**

For the sensory system to operate, several factors need to be present. There can be no perception without a sensation (an environmental event). The primary types of sensations are called **modalities.** Examples of modalities are light, heat, sound, pressure, and specific chemicals. These environmental modalities stimulate receptors. **Receptors** are receiving entities of the body that respond to specific stimuli. They transform the stimulus to neural signals that are transmitted by sensory nerves and neural tracts to the brain, which interprets the message. If any link in this sensory chain is broken, perception does not occur.

Receptors respond to specific modalities and can be classified according to the stimuli they receive. The human body has **photoreceptors** in the eyes, which detect light. There are **thermoreceptors,** located in the skin, which detect changes in temperature. **Osmoreceptors** monitor changes in osmolarity. **Proprioceptors** detect changes in tension, such as those in joints or in muscles. **Pain receptors,** or **nociceptors,** are naked nerve endings in the skin or viscera. **Mechanoreceptors** perceive mechanical stimuli (e.g., touch receptors or receptors that determine hearing or equilibrium in the ear). **Baroreceptors** respond to changes in pressure, such as blood pressure, and **chemoreceptors** respond to changes in the chemical environment (e.g., taste and smell).

The sense of taste, or **gustation,** is received predominantly by taste buds in the tongue, although there are also receptors in the soft palate and pharynx. The sense of taste travels through the facial nerves, glossopharyngeal nerves, and vagus nerves to the medulla oblongata. From there, some fibers travel to either the hypothalamus or amygdala where autonomic reflexes (such as swallowing) occur. Other fibers travel to the thalamus and then to higher brain centers, such as the postcentral gyrus, where the sense of taste is determined. From the postcentral gyrus, fibers take neural impulses to the orbitofrontal cortex, where sight and smell are integrated with taste. The sense of taste is influenced by a food's smell, appearance, temperature, texture, and even the mood of the individual.

Examine a model, chart, or diagram of an inferior view of the brain and locate the olfactory nerves, olfactory bulb, and olfactory tract. The sense of smell, or **olfaction,** originates in the nose when particles stimulate hair cells in the olfactory epithelium (a specialized neuroepithelium in the upper nasal cavities) and is transmitted to the brain by the olfactory nerves and tracts. Some fibers travel to the temporal lobe, where the perception of smell occurs, while others travel to the hippocampus or amygdala, where the memory of smell is stored or the emotional response of smell occurs.

The eye consists of an anterior portion, visible as we look at the face of an individual, and a posterior portion located in the orbit of the skull. Light from the external environment travels through a number of transparent structures that bend and focus the light on the retina, which is the receptive layer of the eye that converts light energy to neural impulses. These impulses travel from the eyes to the optic nerves to the occipital lobes of the brain, where they are interpreted as sight. In most people, eyesight accounts for much of their accumulated knowledge.

The ear is a complex sense organ that performs two major functions, hearing and equilibrium. Hearing is a form of **mechanoreception,** because the ear receives mechanical vibrations (sound waves) and translates them into nerve impulses. This process begins with the vibrations reaching the outer ear and ends up being interpreted as sound in the temporal lobe of the brain. Equilibrium (balance) involves receptors in the inner ear, visual cues, and **proprioception** (a form of mechanoreception involving the perception of gravity or of forces applied to a structure). There are two types of equilibrium sensed by the inner ear. These are static equilibrium and dynamic equilibrium. In **static equilibrium,** an individual is able to determine his or her nonmoving position (such as standing upright or lying down). In **dynamic,** or **kinetic, equilibrium,** motion is detected. Sudden acceleration, abrupt turning, and spinning are examples of dynamic equilibrium.

Learning Objectives

At the end of this exercise you should be able to

1. define the terms "modality" and "receptor";
2. list the major receptor types in the body;
3. explain punctate distribution of sensory receptors;
4. list the two major chemoreceptors located in the region of the head;
5. trace the sense of smell from the nose to the integrative areas of the brain;
6. identify the major structures of the mammalian eye;
7. describe the six extrinsic muscles of the eye and their effect on the movement of the eye;
8. explain how mechanical sound vibrations are translated into nerve impulses;
9. list the structures of the outer, middle, and inner ear; and
10. describe the structure of the cochlea.

Materials

Microscopes
Prepared slides of thick skin (with lamellar and tactile corpuscles)
Models and charts of the eye
Prepared slides of taste buds
Prepared slides of the eye in sagittal section
Preserved sheep or cow eyes
Dissection trays
Protective gloves
Scalpel
Animal waste disposal container
Models and charts of the ear
Prepared slides of the cochlea
Model of ear ossicles

Procedure

Touch Receptors

Examine a slide of thick skin for touch corpuscles. Two of these light touch corpuscles are **tactile corpuscles,** in the upper portion of the dermis, and **tactile discs,** located in the upper dermis and lower epidermis. These two receptors allow for the perception of very slight touch stimuli (such as a fly lightly walking over your cheek). The light touch corpuscles have **receptive fields.** In areas such as the shoulder, the receptive field is large and there are few corpuscles in a given area. In areas with fine touch discrimination, such as the fingers, there are many receptive fields in a given area. In addition, deep touch, or pressure, receptors are found in the dermis, farther away from the epidermis. **Lamellar corpuscles** sense pressure, such as when you lean against a wall or feel a vibration. Other receptors in the skin are warm receptors and cool receptors. When you are at a comfortable temperature, both of these receptors are firing. As the temperature gets warmer, more warm receptors fire and, as you get cooler, more cool receptors fire. If you increase or decrease the skin temperature beyond the perception of these receptors, pain receptors are stimulated. Pain receptors are naked nerve endings in the dermis that respond to numerous environmental stimuli. Locate the tactile corpuscles and lamellar corpuscles in a prepared slide of skin and compare them with figure 18.1.

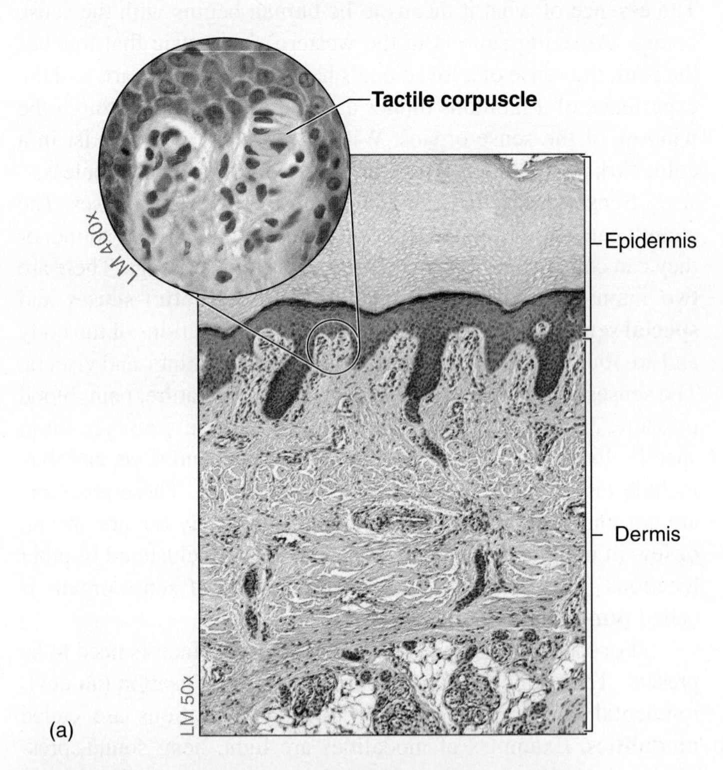

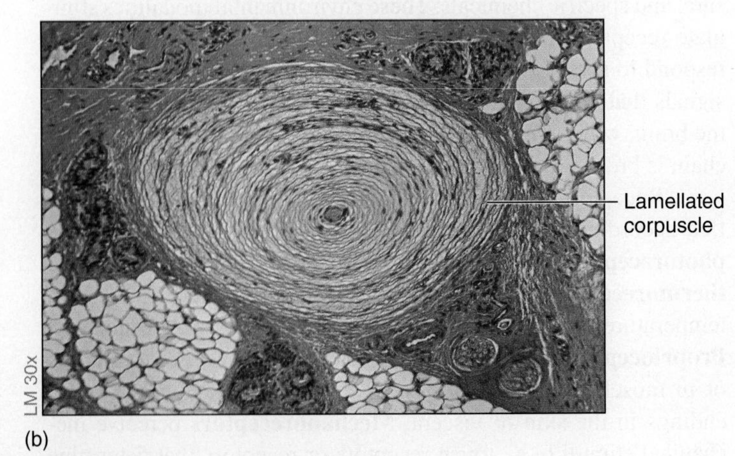

FIGURE 18.1 **Skin Receptors (100×)** (a) Tactile corpuscle; (b) lamellar corpuscle.

Examination of Taste Buds

Examine the prepared slide of taste buds and compare them with figure 18.2. Note how they are located on the sides of the vallate papillae on the tongue. Papillae are raised structures on the tongue. The taste buds appear lighter than the surrounding tissue (like microscopic onions cut in long section). Taste buds are composed of neural tissue and epithelial tissue. The sense of taste is transmitted by the facial, glossopharyngeal, and vagus nerves and is interpreted in the postcentral gyrus of the parietal lobe and other parts of the cerebral cortex. There are five primary tastes: sweet, sour, bitter, salty, and umami.

Transmission of the Sense of Olfaction to the Brain

Examine a model, chart, or diagram of a median section of the head or a model of the brain, and locate the **olfactory nerve fibers, cribriform foramina** in the **cribriform plate, olfactory bulb,** and **olfactory tract.** Compare the lab charts or models with figure 18.3.

Odor molecules touch the **olfactory mucosa** and stimulate the hair cells of the olfactory neurons. The transmission of the sense of smell occurs through the cribriform plate of the ethmoid bone to the olfactory bulb at the base of the frontal lobe of the brain. From here the sense of smell is transmitted to two regions—one in the limbic system and another in the temporal lobe of the brain.

External Features of the Eye

Examine a model of the eye and note the external features, as compared with figure 18.4. The **pupil** is located in the center of the eye and is surrounded by the colored **iris.** The eye has a **sclera,** or "white of the eye," which is composed of dense irregular connective tissue. It is covered by a membrane known as the **conjunctiva,** which continues underneath the eyelids. The sclera is a protective portion of the eye that serves as an attachment point for the muscles of the eye and helps maintain the **intraocular pressure** (the pressure inside the eye). This pressure maintains the shape of the eye and keeps the retina adhered to the back wall of the eye. Numerous blood vessels traverse the sclera, and if they become dilated, they give the eye the appearance of being "bloodshot." The sclera is continuous with the transparent **cornea** in the front of the eye.

The eyelids join at the **lateral commissure** and the **medial commissure.** There is a small piece of tissue near the medial commissure known as the **lacrimal caruncle,** and it is

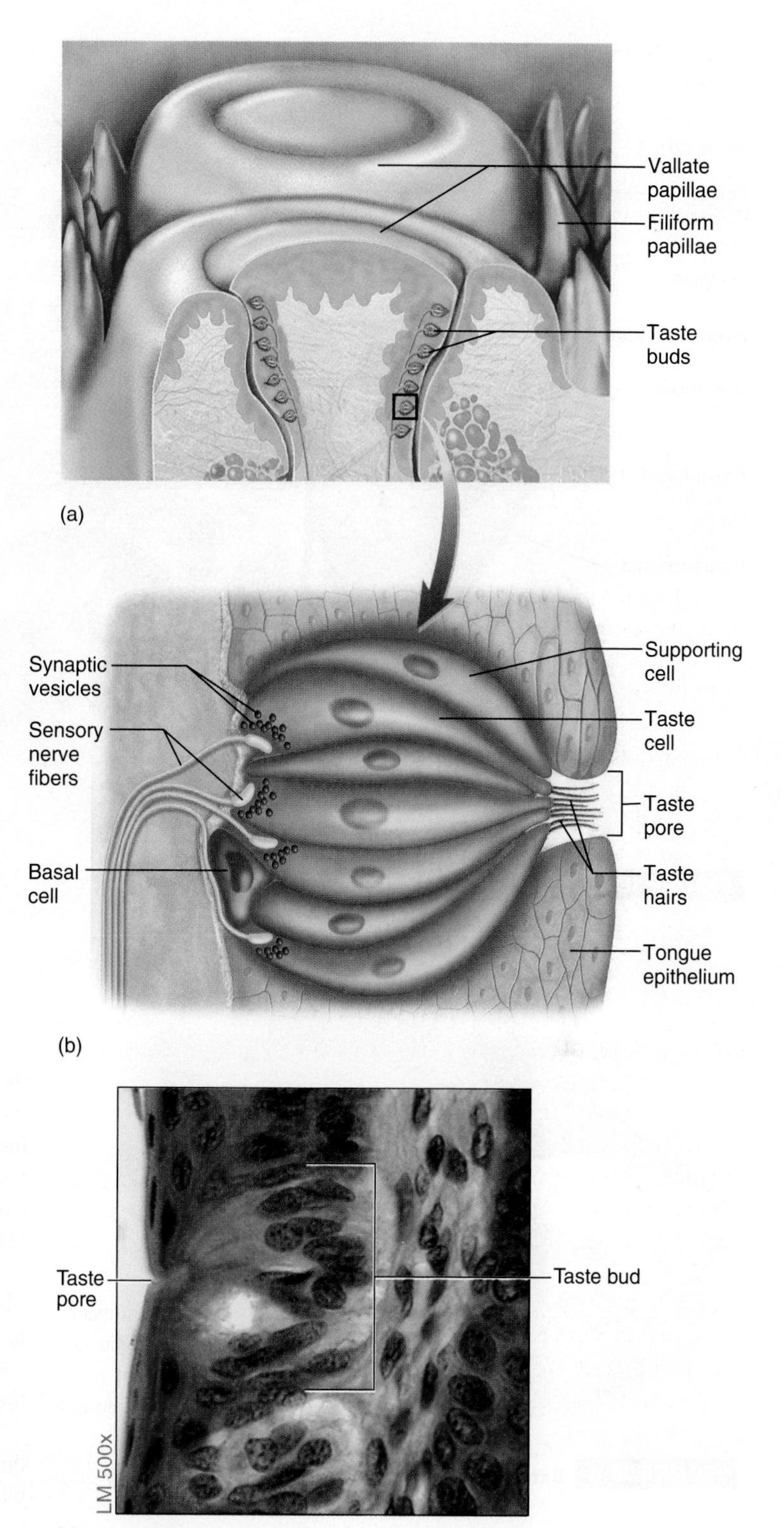

FIGURE 18.2 Taste Buds (a) Taste buds on sides of a tongue papilla; (b) details of taste buds. Photomicrograph: (c) taste buds (100×).

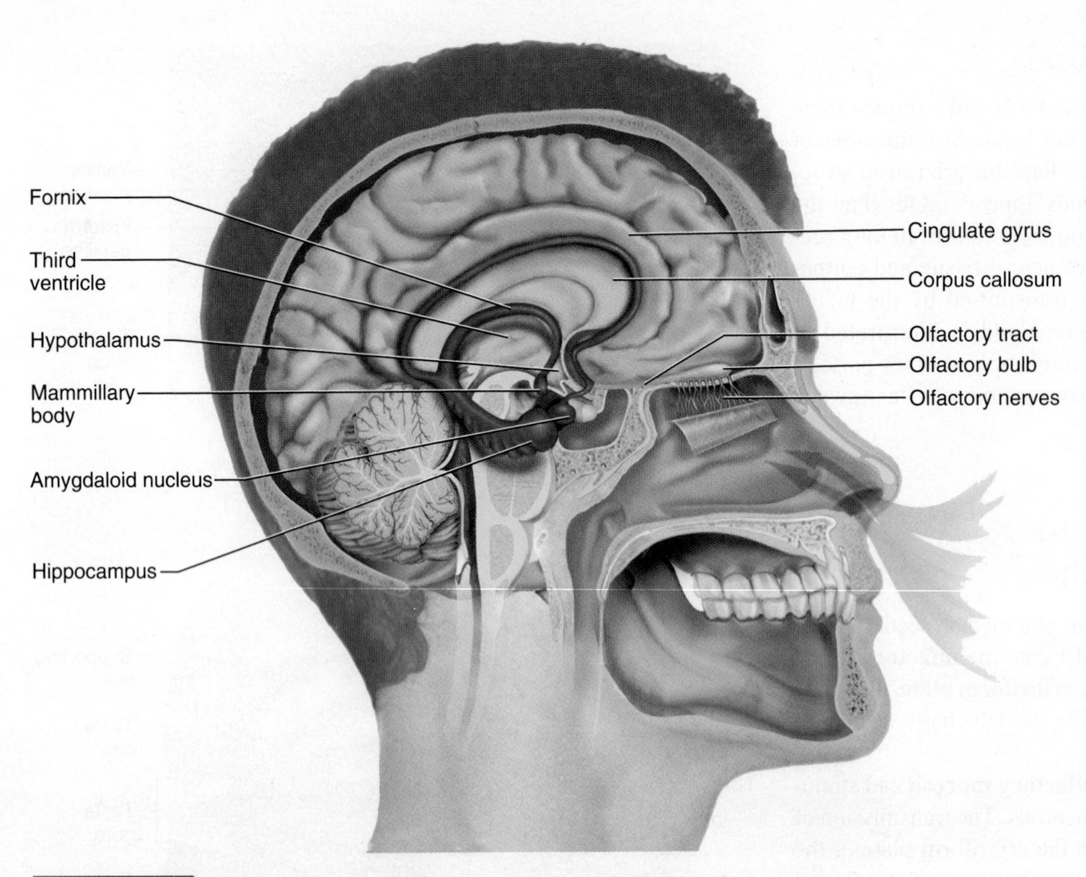

FIGURE 18.3 Olfactory Transmission

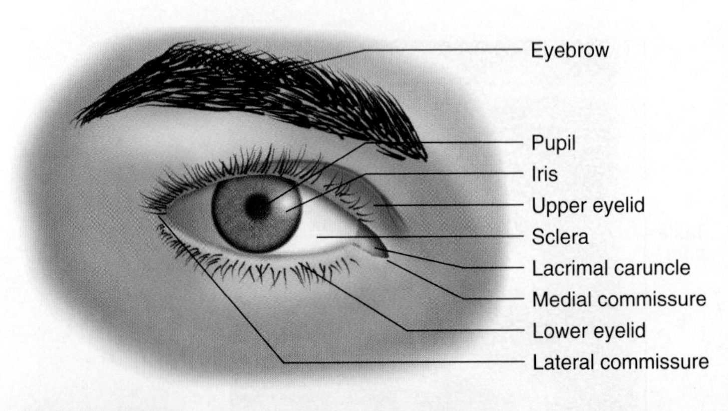

FIGURE 18.4 External Anatomy of the Eye

the site where tears drain from the eye. **Tarsal glands** secrete an oil, which reduces evaporation from the surface of the eye. Examine the **upper eyelid** and **eyelashes** and the **lower eyelid** and eyelashes, which prevent material from entering the eyes and (in the case of the eyelids) reduce visual stimulation when we sleep. Look also at the **eyebrow** located on the supraorbital ridge.

Attached to the sclera are the **extrinsic muscles** of the eye. There are six extrinsic muscles that, in coordination, move the eye in quick and precise ways. Locate these muscles on a model and compare them with figure 18.5. These muscles and their action on the eye are listed in table 18.1.

A structure important in the maintenance of the exterior of the eye is the **lacrimal apparatus.** This consists of the **lacrimal gland** located superior and lateral to the eye (fig. 18.6). Lacrimal secretions bathe and protect the eye and clean dust from its surface. The fluid drains into the lacrimal canal through the **nasolacrimal duct** into the nasal cavity.

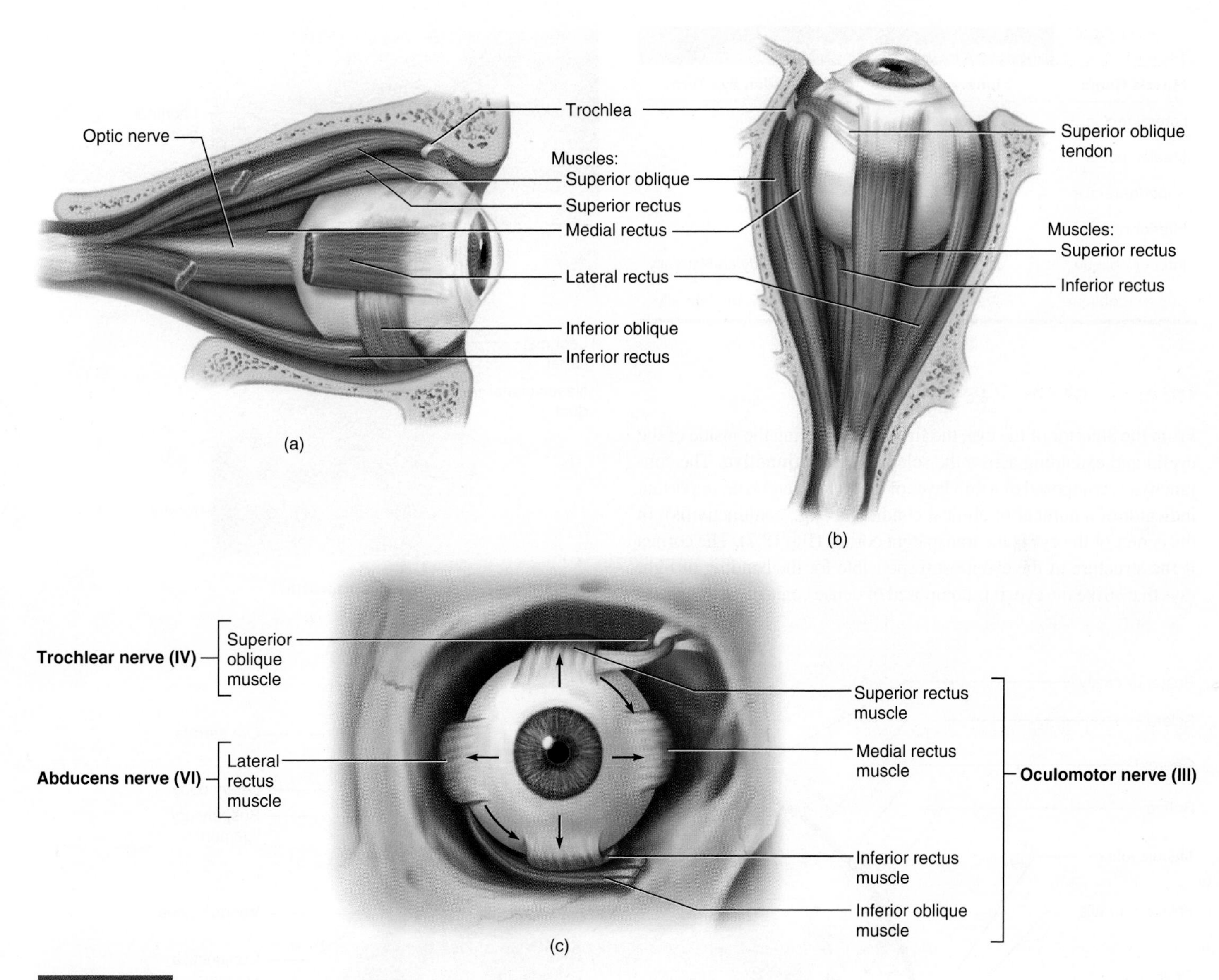

FIGURE 18.5 **Right Eye, External Features** (a) Lateral view; (b) superior view; (c) anterior view.

TABLE 18.1 Extrinsic Muscles of the Eye

Muscle Name	Innervation	Direction Eye Turns
Lateral rectus	VI (abducens)	Laterally
Medial rectus	III (oculomotor)	Medially
Superior rectus	III (oculomotor)	Superiorly
Inferior rectus	III (oculomotor)	Inferiorly
Inferior oblique	III (oculomotor)	Superiorly and laterally
Superior oblique	IV (trochlear)	Inferiorly and laterally

Interior of the Eye

From the anterior of the eye, the first layer covering the inside of the eyelid and extending across the sclera is the **conjunctiva.** The conjunctiva is composed of a thin layer of epithelium and is an important indicator of a number of clinical conditions (e.g., conjunctivitis). In the center of the eye is the transparent cornea (fig. 18.7). The cornea is the structure of the eye most responsible for the bending of light rays that strike the eye. It is composed of dense connective tissue and

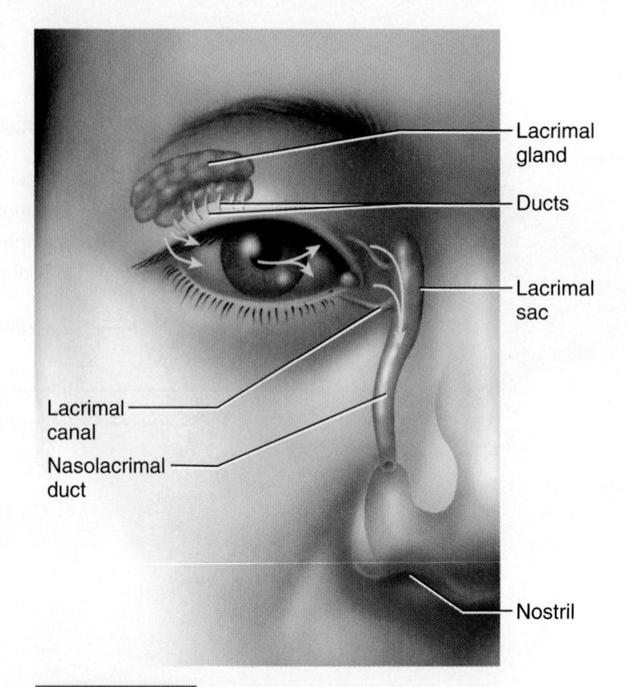

FIGURE 18.6 Lacrimal Apparatus

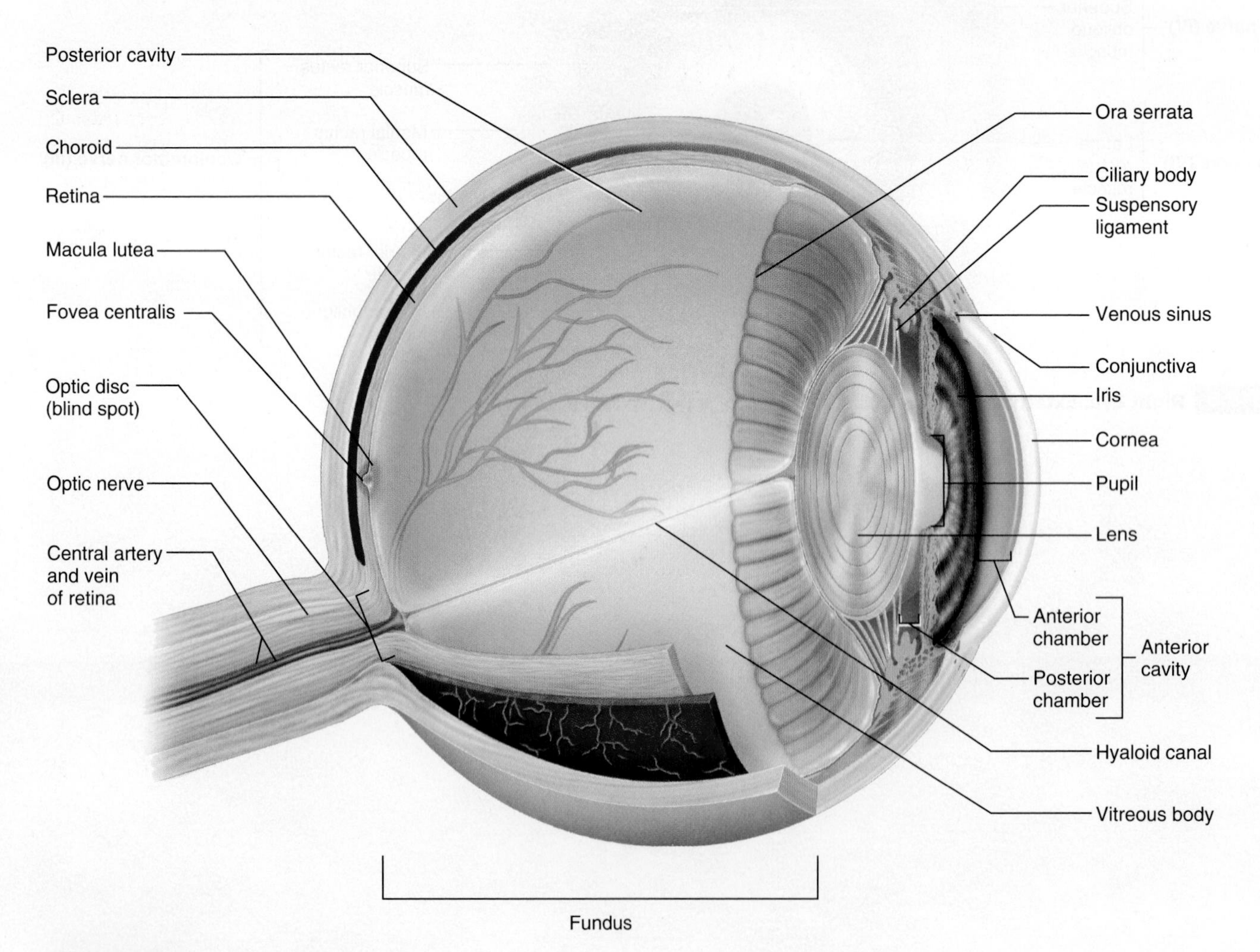

FIGURE 18.7 Sagittal Section of the Eye

is avascular. Why would the presence of blood vessels in the cornea be a visual liability? The lens adjusts the images in order to focus. It is composed of dense connective tissue and is avascular.

Directly posterior to the cornea is the **anterior cavity,** which is subdivided into the **anterior chamber,** between the cornea and the iris, and the **posterior chamber,** between the iris and the lens. The anterior cavity is filled with **aqueous humor,** which is produced by the **ciliary body.** Only a few milliliters of aqueous humor are produced each day, and this amount is absorbed by the **venous sinus (canal of Schlemm),** as illustrated in figure 18.7.

The **iris** is what gives us a particular eye color. People with blue or gray eyes are more sensitive to ultraviolet light than those with brown eyes, due to the protective pigment **melanin** in brown eyes. There are two main sets of structures in the iris. The pupillary constrictor muscle contracts in bright light, reducing the diameter of the **pupil** (the space enclosed by the iris), and the pupillary dilator contracts in dim light, increasing the diameter of the pupil. Posterior to the pupil is the lens, which is made of a crystalline protein. The **lens** is more pliable in youth and becomes less elastic as a person ages. Because of the loss of this elasticity, people in their forties usually begin to use reading glasses. The ciliary muscle in the ciliary body contracts, and the suspensory ligaments that attach to the lens loosen, decreasing the pull on the lens. The lens becomes rounder, allowing for close focusing.

Deep to the lens is the **posterior cavity,** or **vitreous chamber** (fig. 18.7). This cavity occupies most of the posterior portion, or **fundus,** of the eye. The posterior cavity is filled with **vitreous body,** a clear, jellylike fluid that maintains the shape of the eyeball. Most of the posterior cavity is bounded by three layers, or **tunics.** The outermost one, the **sclera,** and the cornea make up the **fibrous layer** of the eye. Deep to the sclera is the **choroid,** a pigmented, vascular layer. The blood vessels found in this layer nourish the eye, and the pigmentation prevents light from scattering and blurring vision. The choroid, the iris, and the ciliary body make up the **vascular layer** of the eye. The layer closest to the vitreous humor is the **retina.** This is the inner layer of the eye. These structures are seen in figure 18.7.

Humans have stereoscopic sight, which gives us depth of vision. This is due to the eyes having overlapping visual fields, seen as the green region in figure 18.8. Light strikes the retina at the back of the eye and is transmitted via the optic nerves, optic chiasm, and optic tract to the occipital lobe of the brain, where the impulses are interpreted as sight.

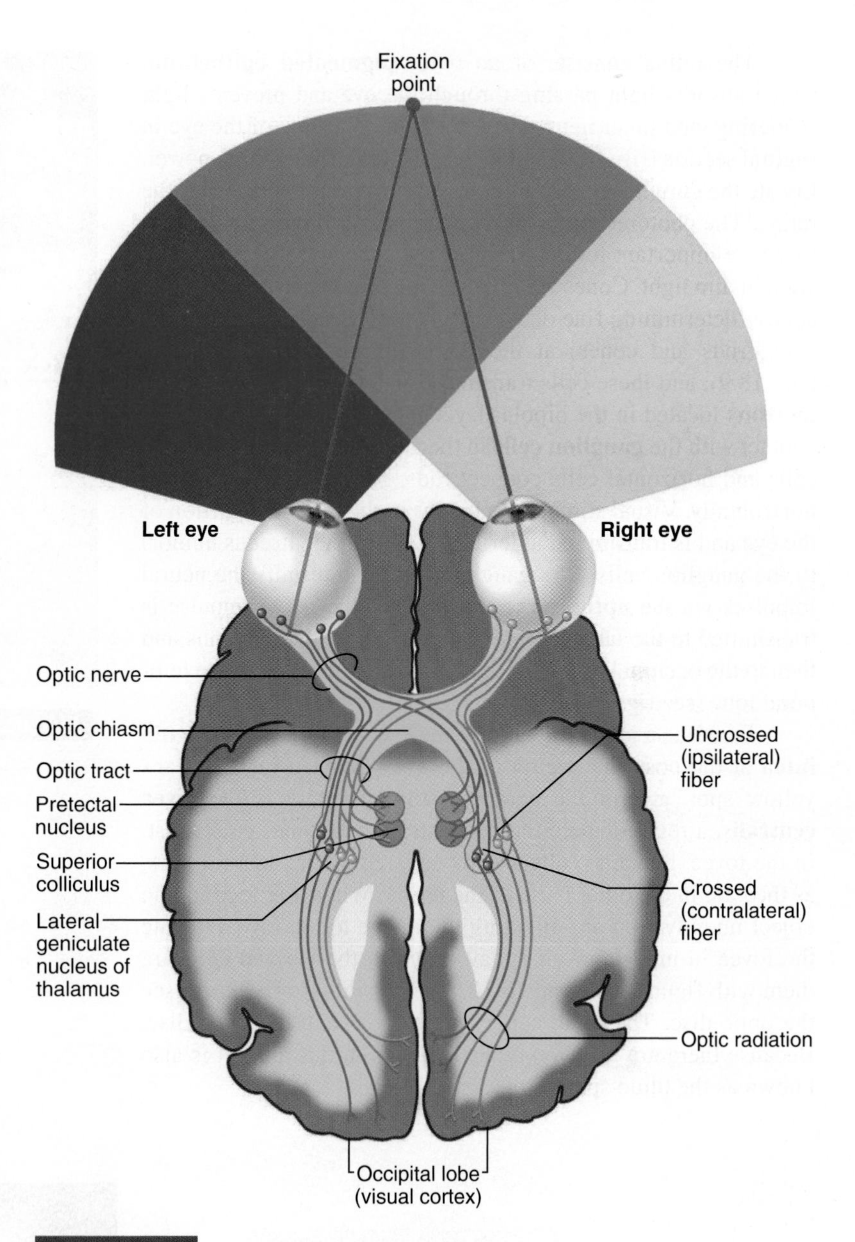

FIGURE 18.8 Visual Pathway to the Brain

The retina consists of an outer **pigmented epithelium,** which absorbs light passing through the eye and prevents light scattering, and inner **neural layer.** Examine a slide of the eye in sagittal section (fig. 18.9) and look at the retina under high power. Locate the **ganglionic, bipolar,** and **photoreceptive** layers of the retina. The photoreceptive layer is composed of **rods** and **cones.** Rods are important for determining the shape of objects and for sight in dim light. Cones are involved in color vision and in visual acuity (determining fine detail). Light strikes the photoreceptive cells (rods and cones) at the posterior portion of the retina (fig. 18.9), and these cells transmit visual signals to the **bipolar neurons** located in the bipolar layer. The bipolar neurons are in contact with the **ganglion cells** in the ganglionic layer. Amacrine cells and horizontal cells connect rods, cones, and bipolar cells horizontally. Visual stimulation begins in the posterior portion of the eye and is transmitted anteriorly (toward the vitreous humor) to the ganglion cells. The ganglionic layer transmits the neural impulses via the **optic nerve.** From there the nerve impulse is transmitted to the lateral geniculate nucleus of the thalamus and then to the occipital region of the brain and integrated in the temporal lobe (see fig. 18.8).

Examine a model or chart of the eye and locate the **macula lutea** at the posterior region of the eye. "Macula lutea" means yellow spot, and in the center of this structure is the **fovea centralis,** a region where the concentration of cones is greatest. In the fovea, the cone cells are not covered by the neural layers, as they are in the other parts of the retina. When you focus on an object intently, you are directing the image to the fovea. Locate the fovea in models or charts available in the lab and compare them with figures 18.7 and 18.10. You should also be able to see the optic disc. The optic nerve enters the eye at the optic disc. Because there are no photoreceptive cells in this area, it is also known as the blind spot.

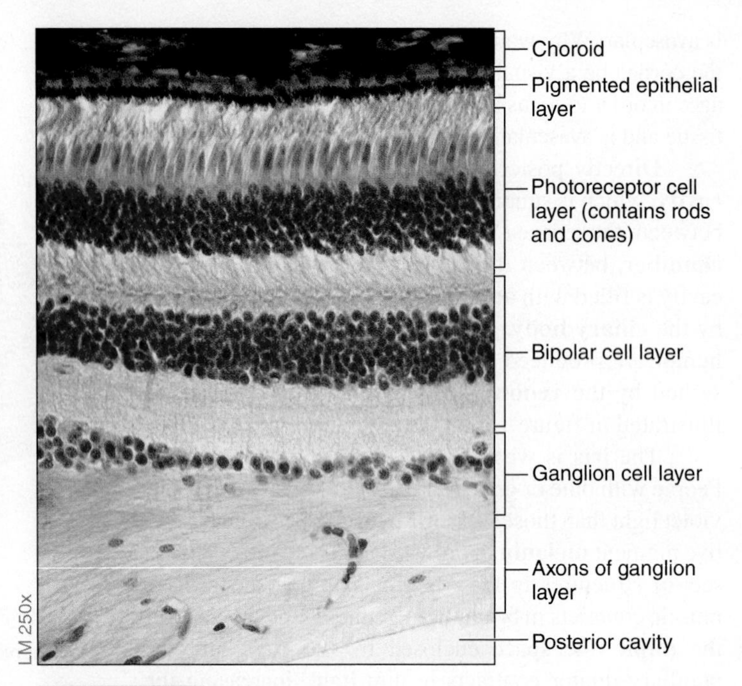

FIGURE 18.9 **Retina (250×)**

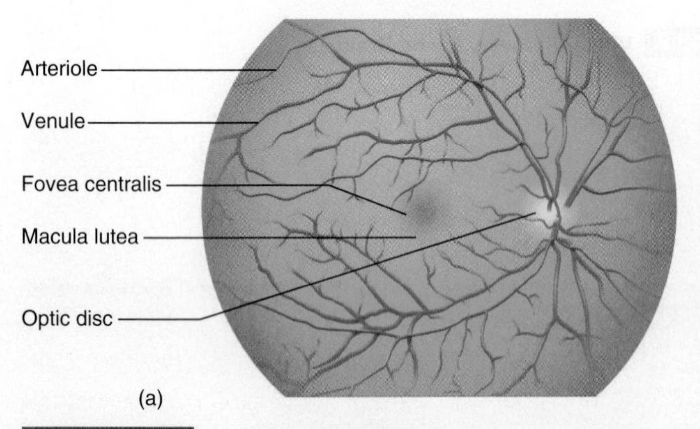

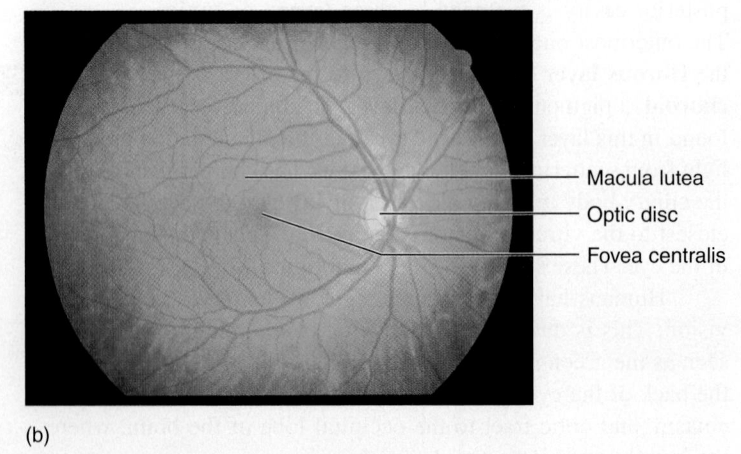

FIGURE 18.10 **Eye, Posterior View** (a) Diagram; (b) photograph.

Dissection of a Sheep or Cow Eye

Rinse a sheep or cow eye in running water and place it on a dissection tray. Obtain a scalpel, scissors, and a blunt probe. Be careful with the sharp instruments and cut *away from* the hand holding the eye. Wear protective gloves while you perform the dissection. Using a scalpel or scissors, make a coronal section of the eye behind the cornea (fig. 18.11). Do not squeeze the eye with force or thrust the blade sharply, because you may squirt yourself with vitreous humor. Find the **optic nerve.** Cut through the eye entirely and note the jellylike material in the posterior cavity. This is the **vitreous body.** Look at the posterior portion of the eye. Note the beige **retina,** which may have pulled away from the posterior surface of the eye. The retina is attached at the optic nerve and at the ora serrata. If the retina pulls away from the rest of the interior of the eye, it is called a detached retina. Posterior (or superficial) to the retina is the darkened **choroid.** The choroid in humans is very dark, but you may see an iridescent color in your specimen. This is the **tapetum lucidum,** which improves night vision in some animals. The tapetum lucidum produces the "eye shine" of nocturnal animals. Also examine the tough, white **sclera,** which envelops the choroid.

Now examine the anterior portion of the eye. Is the **lens** in place? The lens in your specimen probably will not be clear because the preserving fluid denatures the protein of the lens. Normally the lens is transparent and allows for light penetration.

The lens is held to the **ciliary body** by the **suspensory ligaments** (figs. 18.7 and 18.11). These ligaments pull on the lens and alter its shape for distant vision. Locate the ciliary body at the edge of the suspensory ligaments.

Now turn the eye over. Is there any **aqueous humor** left in the **anterior cavity**? Make an incision through the **conjunctiva** and **cornea** into the anterior cavity. Can you determine the region

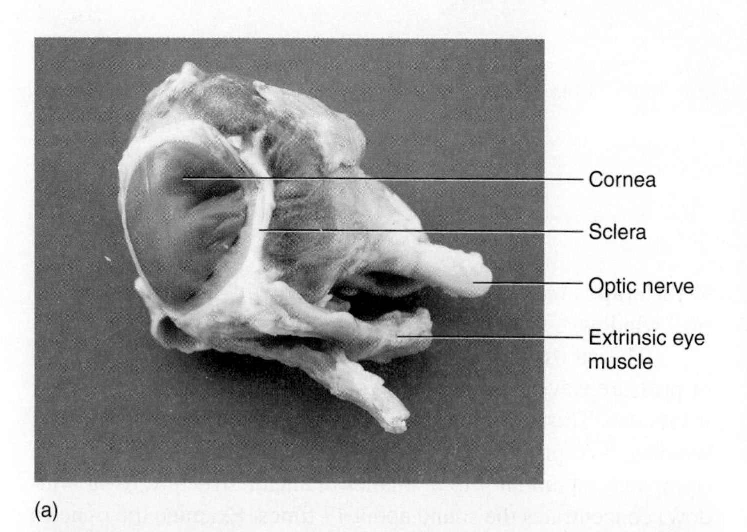

(a)

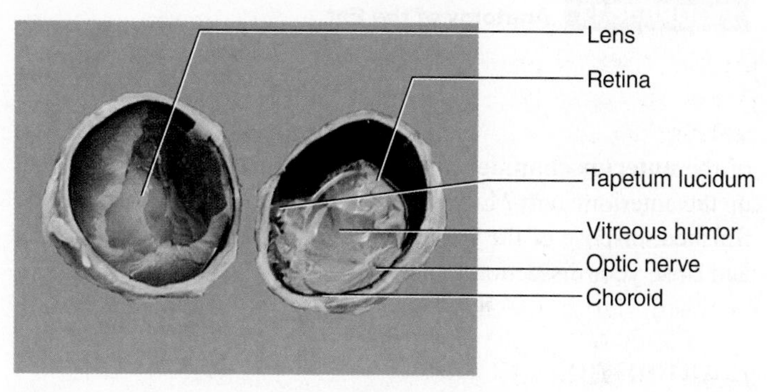

(c)

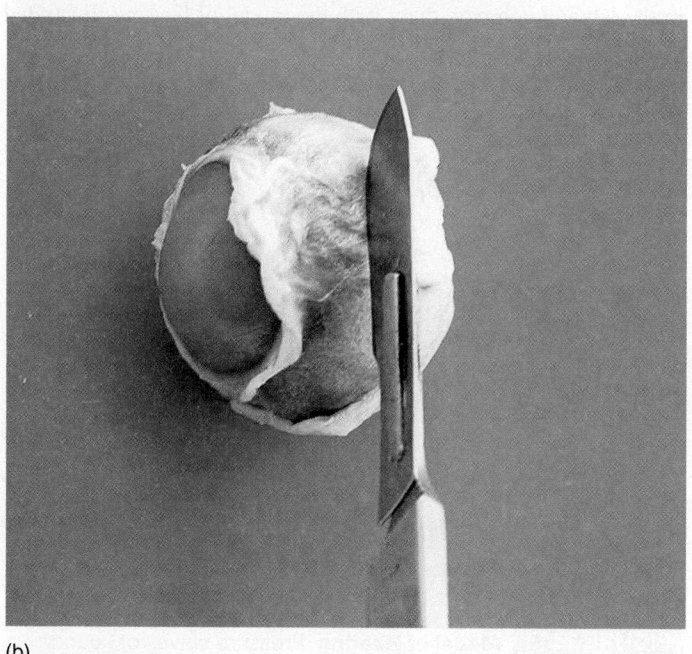

(b)

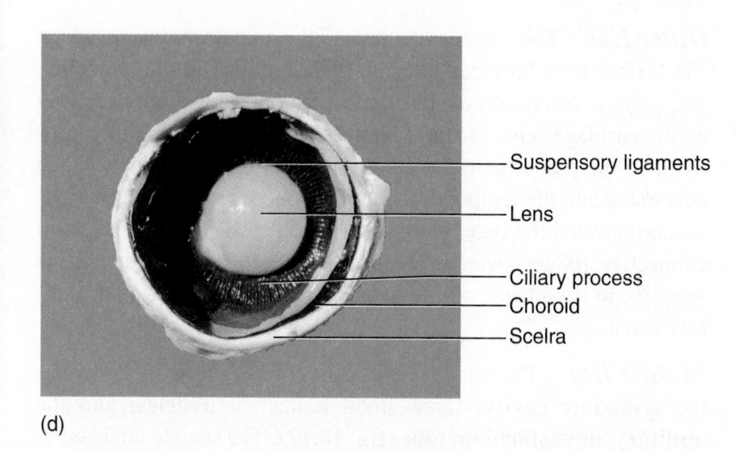

(d)

FIGURE 18.11 Dissection of a Sheep Eye (a) External features; (b) coronal section; (c) eye with vitreous humor; (d) anterior eye without vitreous humor.

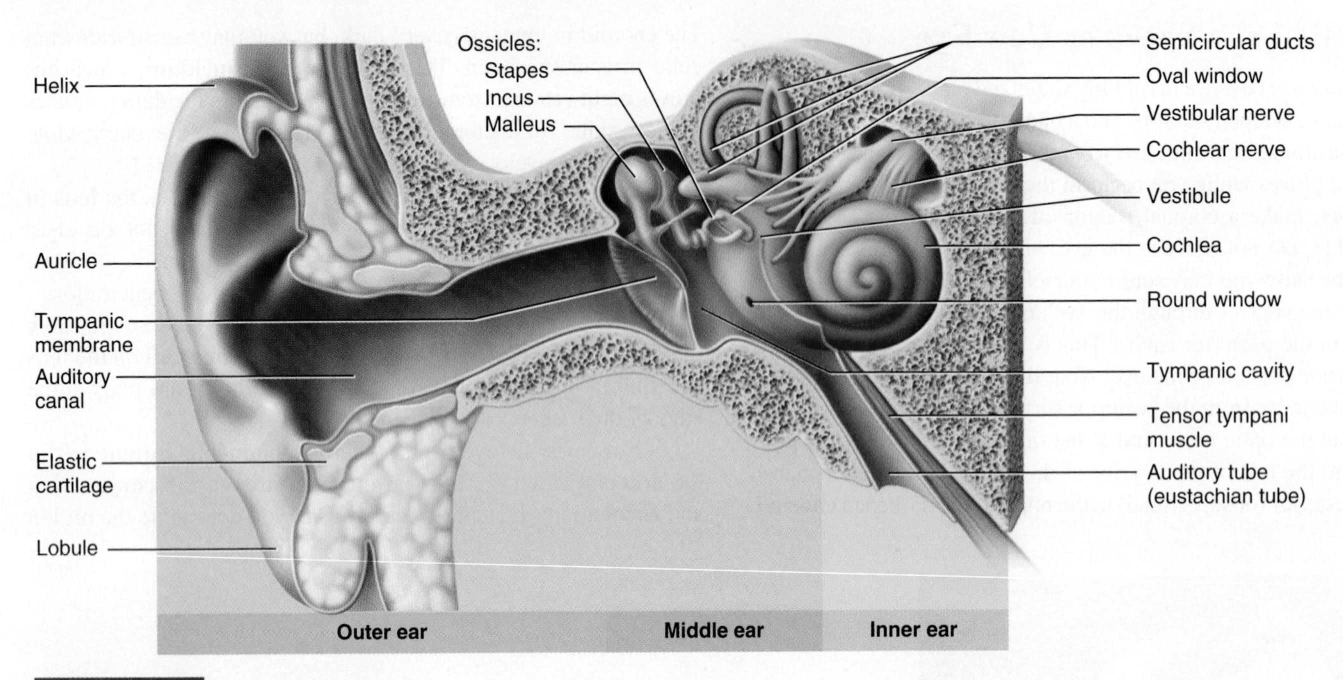

FIGURE 18.12 **Anatomy of the Ear**

of the **anterior chamber** and the **posterior chamber** that makes up the anterior cavity? Locate the **iris** and the **pupil.** When you are finished, dispose of the specimen in a designated waste container and rinse your dissection tools.

Anatomy of the Ear

The ear can be divided into three regions—the outer, middle, and inner ear. The **outer ear** consists of auditory structures superficial to the **tympanic membrane** (eardrum). The **middle ear** contains the tympanic membrane, tympanic cavity, ear ossicles, and auditory tube, and the **inner ear,** located in the petrous portion of the temporal bone, consists of the cochlea, vestibule, and semicircular ducts. Examine the charts and models in the lab and compare them with figure 18.12 as you read about the individual regions of the ear.

Outer Ear The outer ear consists of the **auricle,** or **pinna,** which can further be subdivided into the **helix** and the **lobule** (**earlobe**). The helix is composed of stratified squamous epithelium overlying elastic cartilage. This cartilage allows the ears to bend significantly. Deep to the pinna the outer ear forms the **auditory canal,** which penetrates into the temporal bone. The **tympanic membrane** is the border between the outer ear and the middle ear. It is composed of connective tissue covered by epithelial tissue. The membrane is sensitive to sound and vibrates as sound is funneled down the auditory canal.

Middle Ear The middle ear consists of a main cavity known as the **tympanic cavity;** three small bones, or **ossicles;** and the **auditory,** or **eustachian,** tube (fig. 18.12). The ossicle attached to the tympanic membrane is the **malleus** (*malleus* = hammer). The malleus is attached to the **incus** (*incus* = anvil), which is attached to the **stapes** (*stapes* = stirrup) (fig. 18.12), which is next to the oval window.

The ear ossicles transfer sound to the cochlea. Sound consists of pressure waves. As these waves strike the tympanic membrane, it vibrates. This vibration is conducted by the ossicles to the oval window. The process of moving from a large-diameter structure (tympanic membrane) to a smaller-diameter structure (oval window) concentrates the sound about 17 times. Examine the ossicles in the lab and compare them with the model of hearing illustrated in figure 18.13.

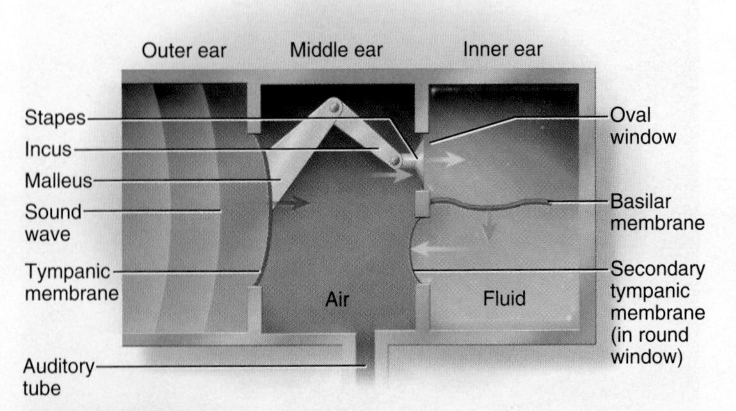

FIGURE 18.13 **Model of Hearing** Pressure waves of sound vibrate the tympanic membrane, which transfers sound to the oval window and finally to the basilar membrane, where vibration is converted to neural impulses.

The auditory tube connects the middle ear to the nasopharynx (see fig. 18.12) and provides for the equalization of pressure between the middle ear and the external environment when changes of pressure occur (such as during changes of elevation). The auditory tube can be a conduit for microorganisms that travel from the nasopharynx to the middle ear and lead to middle ear infections, particularly in young children, in whom the tube is more horizontal than in adults.

Inner Ear The inner ear is encased in two complex structures and filled with two separate fluids. The outermost structure is the **bony labyrinth** (*labyrinth* = maze). Inside the bony labyrinth is **perilymph,** a clear fluid that is external to the **membranous labyrinth.** The fluid enclosed by the membranous labyrinth is the **endolymph,** which is important in both hearing and equilibrium. The membranous labyrinth is enclosed inside the bony labyrinth like a tube within a tube.

The inner ear is composed of three separate regions—the cochlea, the vestibule, and the semicircular ducts (figs. 18.14 and 18.15).

Cochlea The **cochlea** is a structure that resembles a seashell (*cochlea* = snail; figs. 18.14 and 18.15). It is involved in hearing. As the sound waves travel down the auditory canal, they cause the

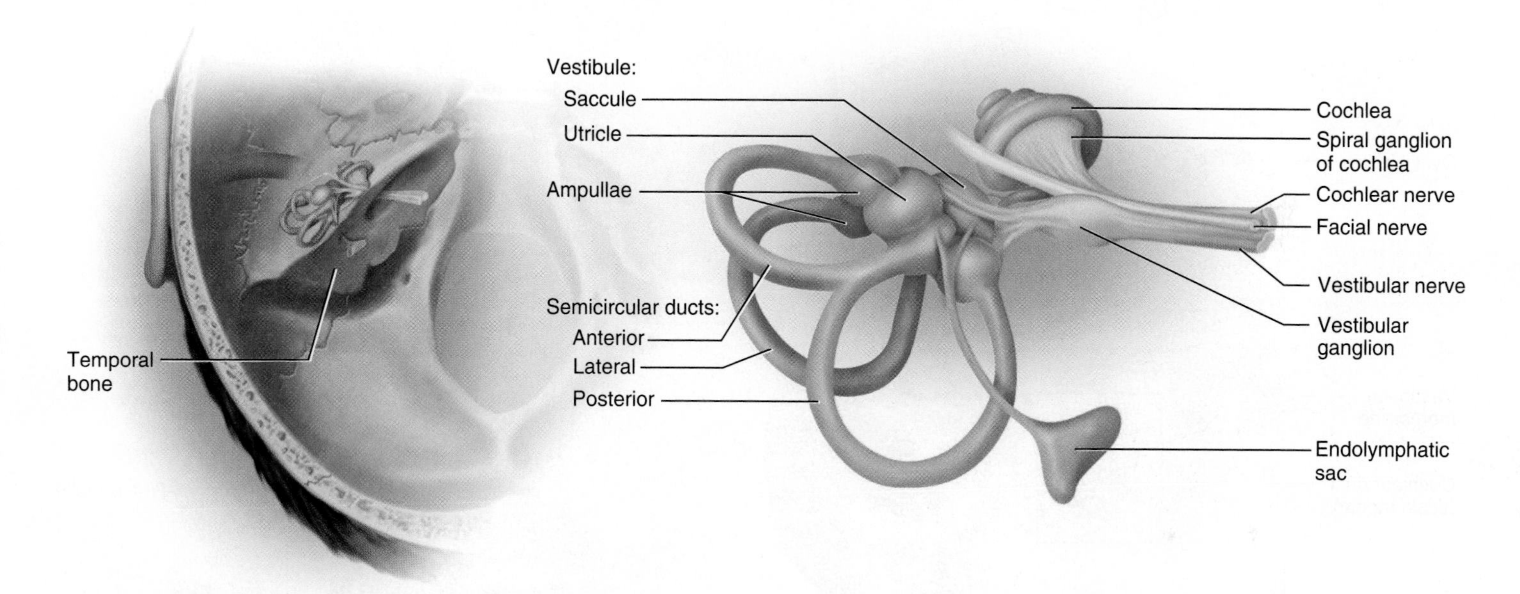

FIGURE 18.14 **Anatomy of the Inner Ear**

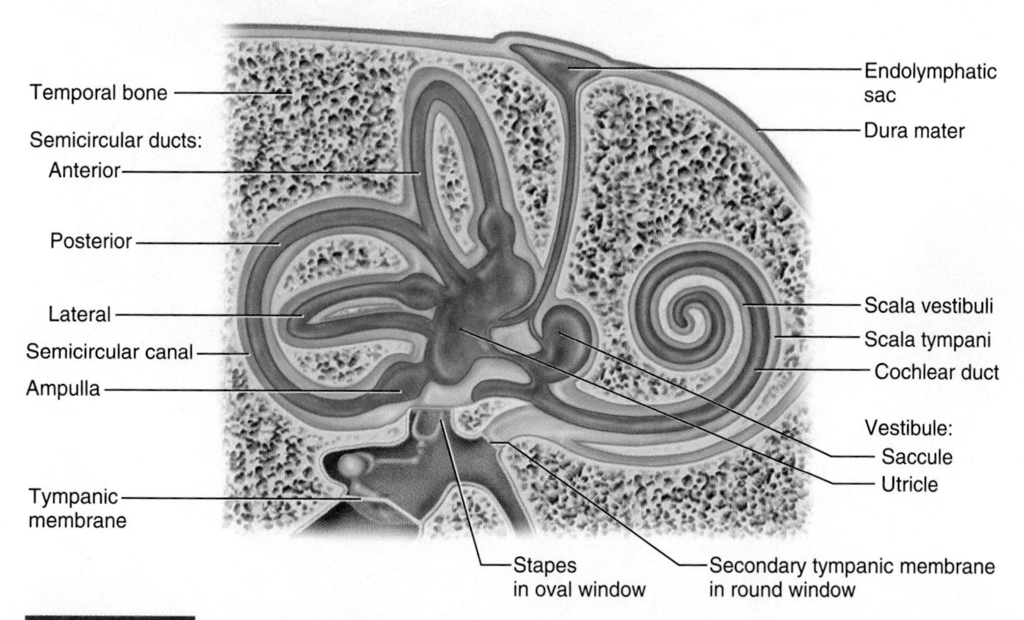

FIGURE 18.15 **Inner Ear—Membranous and Bony Labyrinth** Perilymph is found in the regions shown in green, and endolymph is found in the regions shown in blue.

tympanic membrane to vibrate. This vibration rocks the ear ossicles, which are connected to the inner ear. As the stapes vibrates, it moves back and forth in the **oval window,** causing fluid to move back and forth in the cochlea. The cochlea also has a **round window,** which allows the vibration from the ossicles to move fluid back and forth. Sound waves are measured by their amplitude (loudness) and wavelengths (pitch). These wavelengths are measured in cycles per second, also known as hertz (Hz).

In the human ear, high-frequency sounds with vibrations of up to 20,000 Hz stimulate the region of the cochlea closest to the middle ear. Low-frequency sounds with vibrations down to 20 Hz stimulate the region of the cochlea farther from the middle ear. In this way the cochlea can perceive sounds of varying wavelengths at the same time.

Microscopic Section of Cochlea If you examine a microscopic section of the cochlea in cross section, you will see a number of chambers. The chambers are clustered in threes. Find the **scala vestibuli (vestibular duct), scala media (cochlear duct),** and **scala tympani (tympanic duct)** on the microscope slide. Compare these with figures 18.16 and 18.17.

Note the **spiral organ,** or the **organ of Corti,** in the area between the **vestibular membrane** and the **basilar membrane.** The spiral organ in the slide is in cross section, but remember that it runs

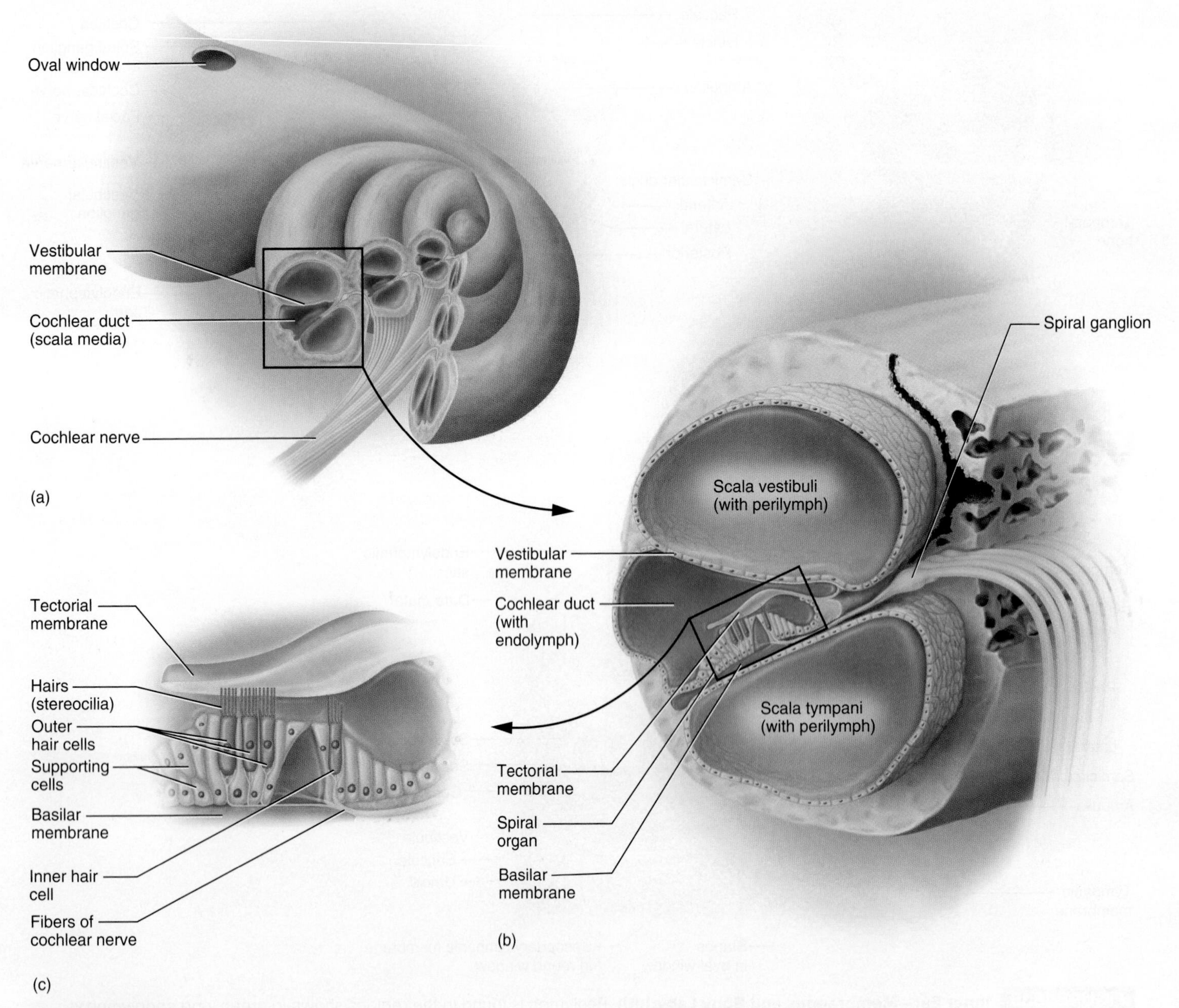

FIGURE 18.16 **Anatomy of the Cochlea** (a) Overview of cochlea; (b) details of chambers; (c) spiral organ.

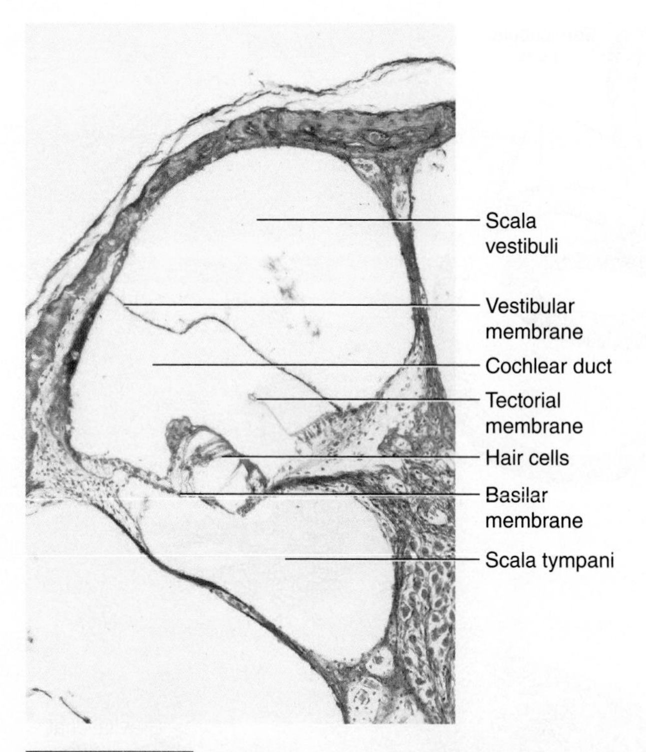

FIGURE 18.17 Photomicrograph of Cross Section of the Cochlea (100×)

the length of the cochlea. The spiral organ is sensitive to sound waves. As a particular region of the spiral organ is stimulated, the basilar membrane and the gelatinous **tectorial membrane** vibrate independently of each other. This produces movement of the stereo cilia on the hair cells. The hair cells send impulses to the cochlear branch of the **vestibulocochlear nerve.** These impulses eventually reach the **auditory cortex** of the **temporal lobe,** where they are interpreted as sound (fig. 18.18).

Vestibule Another part of the inner ear is the vestibule. The **vestibule** consists of the **utricle** and **saccule** as seen in figure 18.19. These two chambers are involved in the interpretation of static equilibrium and acceleration. The utricle and saccule have regions known as **maculae,** which consist of hair cells with stereocilia (apical modifications of cells) and a kinocilium (an apical cluster of specialized cilia) grouped together with an overlying gelatinous mass and calcium carbonate stones called **otoliths.** These can be seen in figure 18.19. As the head undergoes linear acceleration or is tipped by gravity, the otoliths cause the cilia to bend, indicating that the position of the head has changed. Static equilibrium is perceived not only from the vestibule but from visual cues as well. When the visual cues and the vestibular cues are not synchronized, then a sense of imbalance or nausea can occur.

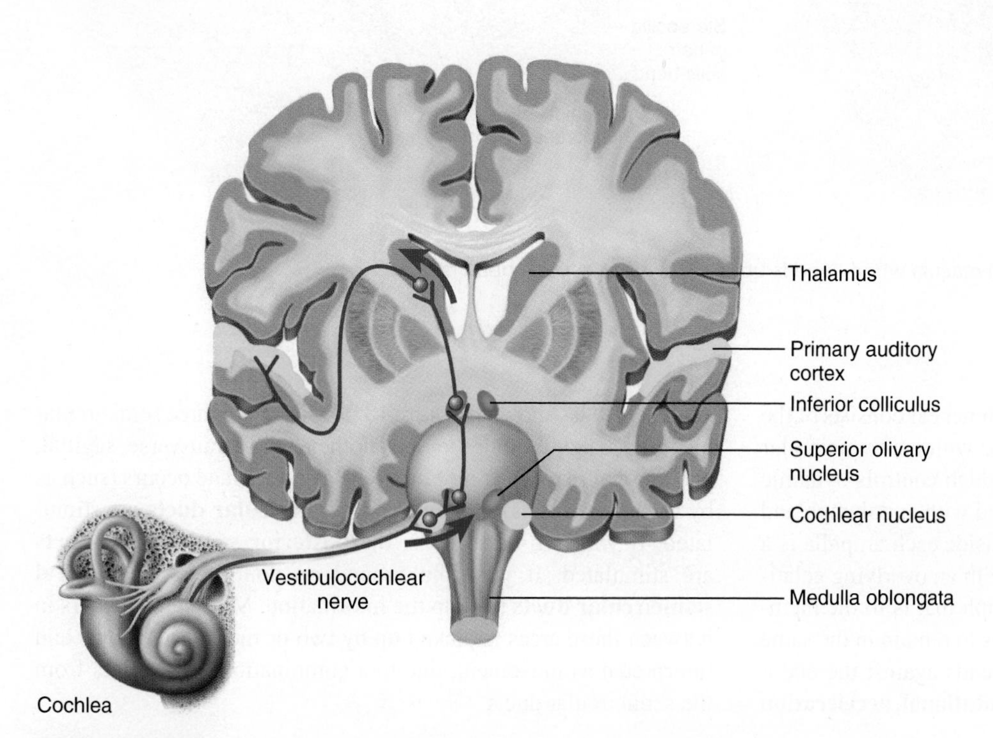

FIGURE 18.18 Interpretive Pathway of Hearing, Anterior View Sound impulses from the cochlea travel via nerves and tracts to the primary auditory cortex.

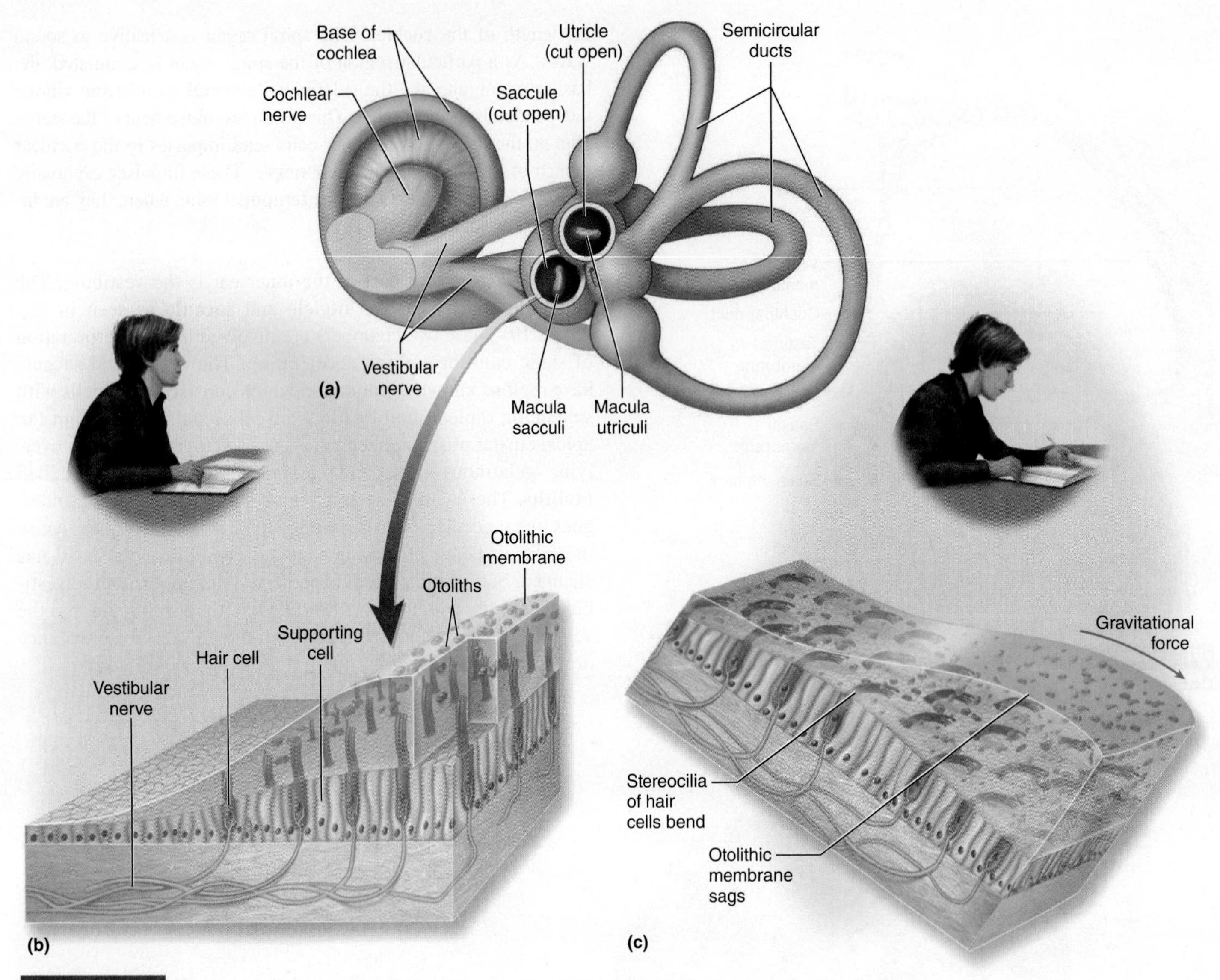

FIGURE 18.19 **Vestibule** (a) Inner ear; (b) macula when head is upright; (c) macula when head is tilted.

Semicircular Ducts The third part of the inner ear consists of the **semicircular ducts** (fig. 18.20; each located within a semicircular canal), which send impulses to the brain which controls dynamic equilibrium. Each semicircular duct is filled with endolymph and is expanded at the base into an **ampulla.** Inside each ampulla is a **crista ampullaris,** a cluster of **hair cells** with an overlying gelatinous mass called the **cupula.** The endolymph that is in the membranous labyrinth has inertia; that is, it tends to remain in the same place. As the head is turned, the cupula bends against the endolymph, the hair cells bend, and **angular,** or **rotational, acceleration** is perceived, as shown in figure 18.20. There are three semicircular ducts, each at 90° angles to one another (in the transverse, sagittal, and coronal planes). If motion in the sagittal plane occurs (such as by doing back flips), the **anterior semicircular ducts** are stimulated. If you turn cartwheels, the **posterior semicircular ducts** are stimulated. If you spin around on your heels, the **lateral semicircular ducts** pick up the information. Motion that occurs in between these areas is picked up by two or more of the ducts and interpreted as movement, due to a combination of impulses from the semicircular ducts.

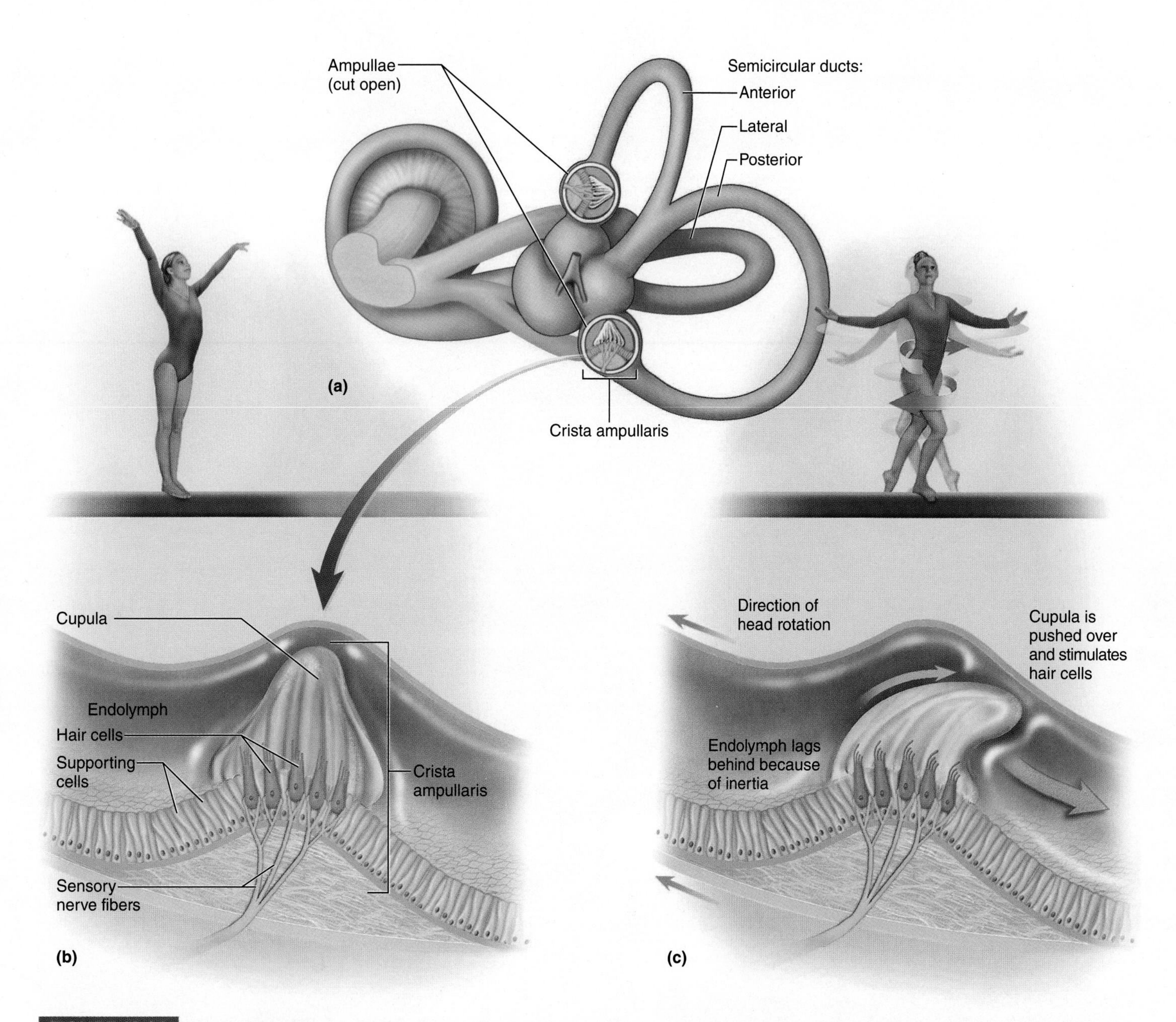

FIGURE 18.20 **Semicircular Ducts** (a) Overview of the ducts with cutaway of crista ampullaris; (b) detail of the crista ampullaris at rest; (c) movement of the crista ampullaris in motion.

Exercise 18

Sensory Receptors

Name ______________________ Date ____________

1. Distinguish among the functions of lamellar corpuscles, tactile corpuscles, and pain receptors in the skin.

2. What is punctate distribution?

3. What is a modality?

4. What kind of receptor is sensitive to the following modalities?
   a. light
   b. touch
   c. temperature
   d. sound
   e. smell

5. What kind of receptor determines the weight of an object when you pick it up?

6. What nerves transmit the sense of smell to the brain?

7. What structures are involved in taking the sense of taste from the taste buds to the brain?

8. Where are the taste buds located?

9. Fill in the following illustration using the terms provided.

anterior cavity (anterior chamber)	lens	retina
choroid	optic nerve	sclera
ciliary body	pupil	suspensory ligaments
cornea		

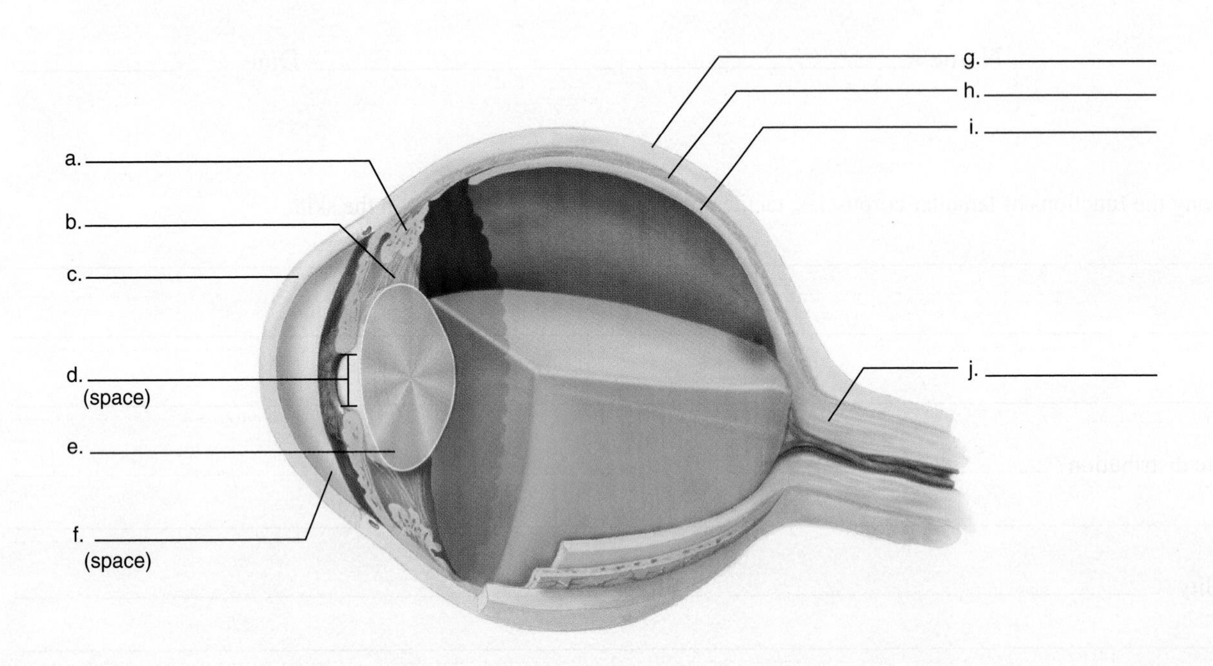

10. Since the lens is made of protein, what effect might the preserving fluid used in lab have on the structure of the lens?

____

How would this affect the clarity? ____________________

____

11. How does the vitreous body differ from the aqueous humor in terms of location and viscosity? ____________________

____

12. What layer of the eye converts visible light into nerve impulses? ____________________

____

13. What nerve takes the impulse of sight to the brain? ____________________

____

14. What is the common name for the sclera? ____________________

15. How would you define an extrinsic muscle of the eye? ____________________

16. What gland produces tears? ____________________

17. What is the name of the transparent layer of the eye in front of the anterior chamber? ____________________

18. The iris of the eye has what function? ____________________

19. What is the middle tunic of the eye called? ____________________

20. Is the lens anterior or posterior to the iris? ____________________

21. What is at the area of the eye where the blind spot is found? ____________________

22. What are the three general regions of the ear? ____________________

23. The ear performs two major sensory functions. What are they? ____________________

24. What area is found between the scala vestibuli and the scala tympani? ____________________

25. What part of the inner ear is involved in perceiving static equilibrium? ____________________

26. What tube is responsible for the equalization of pressure when you change elevation? ____________________

27. What is the name of the space that encloses the ear ossicles? ______________________________

____

28. Place the ear ossicles in sequence from the tympanic membrane to the oval window. ______________

____

____

29. Fill out the illustration using the terms provided.

auditory canal	cochlea	helix	stapes
auditory tube	earlobe	semicircular ducts	tympanic membrane

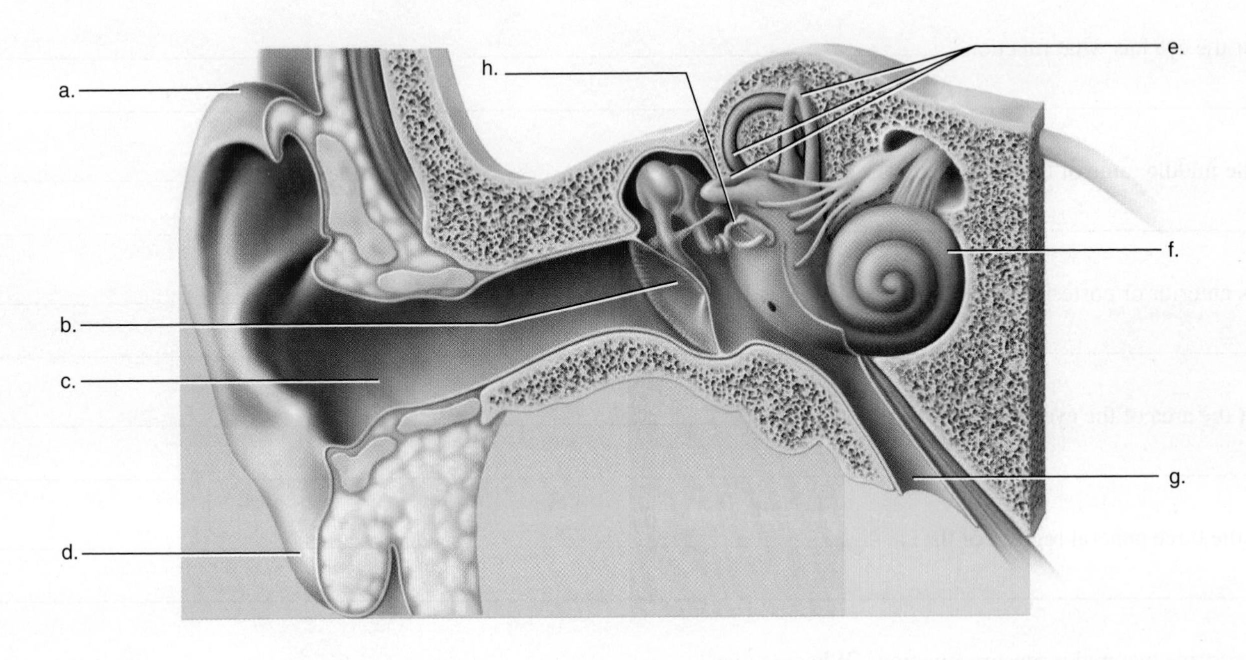

The Endocrine System

Endocrine System

Introduction

The **endocrine system** consists of glands that produce chemical messengers called **hormones** that are picked up by blood capillaries and travel throughout the body. This type of release is called a "ductless secretion," which is characteristic of the endocrine system. Hormones enter the interstitial fluid and then travel by blood vessels, which act as highways to carry them throughout the body. Areas that are receptive to hormones are called **target cells.** Many organs of the body, such as the stomach, heart, and kidney, produce hormones and thus have endocrine functions. This exercise focuses on the glands that have a major endocrine component.

Hormones can have many effects. Some of these actions are growth, changes in development (maturation), metabolism, sexual development, regulation of the sexual cycle, and homeostasis.

The glands that secrete material in ducts or tubules (such as sweat and salivary glands) or directly to a surface (ovary) are known as **exocrine glands.** Exocrine glands do not use the cardiovascular system for transport.

Learning Objectives

At the end of this exercise you should be able to

1. discuss how the secretions of the endocrine glands differ from those of the exocrine glands;
2. list the major endocrine organs of the human body;
3. identify endocrine organs in histological slides; and
4. name the hormones produced by the endocrine organs.

Materials

Models and charts of endocrine glands
Model or chart of median section of head
Microscopes
Microscope slides:
- Thyroid
- Pituitary
- Adrenal gland
- Pancreas
- Testis
- Ovary

Procedure

Anatomy of the Major Endocrine Organs

Locate the major endocrine glands in charts or models in the lab and compare them with figure 19.1, including the hypothalamus, pineal gland, pituitary gland, thyroid, parathyroids, thymus, pancreas, adrenals, and gonads (testes or ovaries). Once you have noted their location, you can proceed with a more detailed study.

Hypothalamus The **hypothalamus** has a major role in controlling endocrine functions. It is located in the inferior portion of the forebrain and is connected to the anterior pituitary by the hypophyseal portal system. Primary capillaries in the hypothalamus lead to the secondary capillaries in the anterior pituitary. The hypothalamus has neurons that extend into the posterior pituitary by the hypothalamohypophyseal tract, so the posterior pituitary can actually be considered an extension of the hypothalamus. The hypothalamus secretes both releasing hormones and inhibitory hormones that either stimulate hormone secretion from their target areas or prevent the release of hormones from these areas. Hormones such as **gonadotropic-releasing hormone (GnRH)** stimulate the pituitary to release **follicle-stimulating hormone,** and **thyrotropic-releasing hormone (TRH)** stimulates the pituitary to release thyroid-stimulating hormone (TSH). A tropic hormone is one that causes another endocrine gland to release hormones. Locate the hypothalamus on models or charts in the lab and in figures 19.1 and 19.2.

Pineal Gland The **pineal gland** develops from the diencephalon of the brain. Locate the pineal gland in a model or chart of a median section of the head and compare it with figure 19.1. The pineal gland secretes the hormone **melatonin.** The role of melatonin is not completely understood in humans. The production of melatonin from the pineal gland increases at night, so it has been named the "hormone of darkness." The level decreases during the day, and it is thought that these circadian rhythms contribute to sleep cycles. There is a condition known as seasonal affective

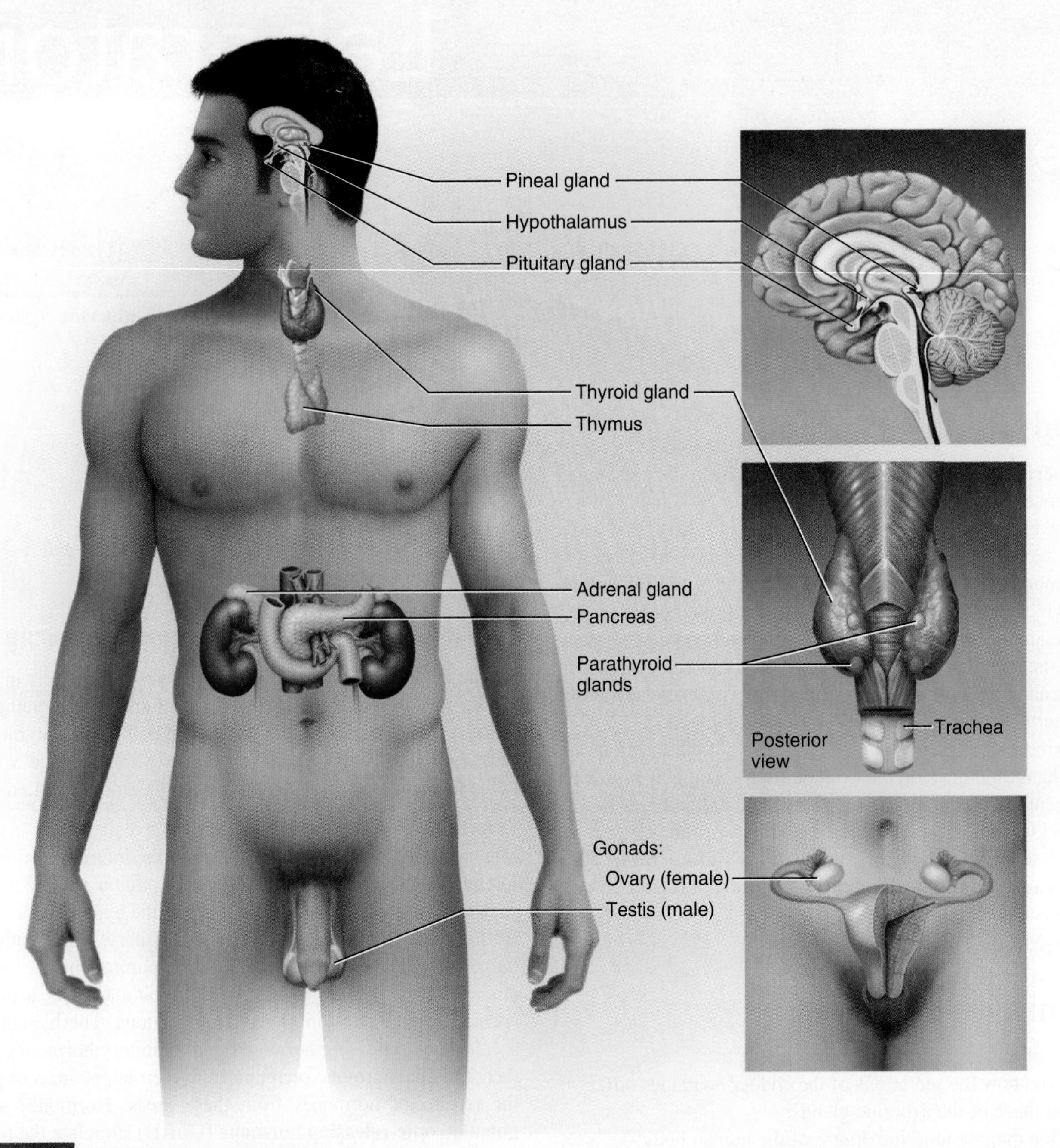

FIGURE 19.1 **Major Endocrine Glands**

disorder (SAD) that causes depression in people, typically during winter months. The increased levels of melatonin during this time of year have been implicated in SAD.

Pituitary Gland Figures 19.1 and 19.2 also illustrate another endocrine gland, called the **pituitary gland,** or **hypophysis.** The pituitary is divided into the **anterior pituitary,** or **adenohypophysis,** and a **posterior pituitary,** or **neurohypophysis.** The pituitary is suspended from the base of the brain by a stalk called the **infundibulum** and is enclosed in the sella turcica of the splenoid bone.

Adenohypophysis The adenohypophysis originates from the roof of the oral cavity during embryonic development. The result of this development can be seen if you examine a prepared slide of the pituitary gland. The cells of the anterior pituitary have the same embryonic origin, yet they have differentiated into several types of specialized cells. In the histological section, note that the adenohypophysis is composed of cuboidal cells (fig. 19.3). These cells, generally grouped into acidophil and basophil cells, produce a number of hormones that have broad effects throughout the body.

FIGURE 19.2 Pituitary Gland (a) Overview of the pituitary gland; (b) target areas of the anterior pituitary.

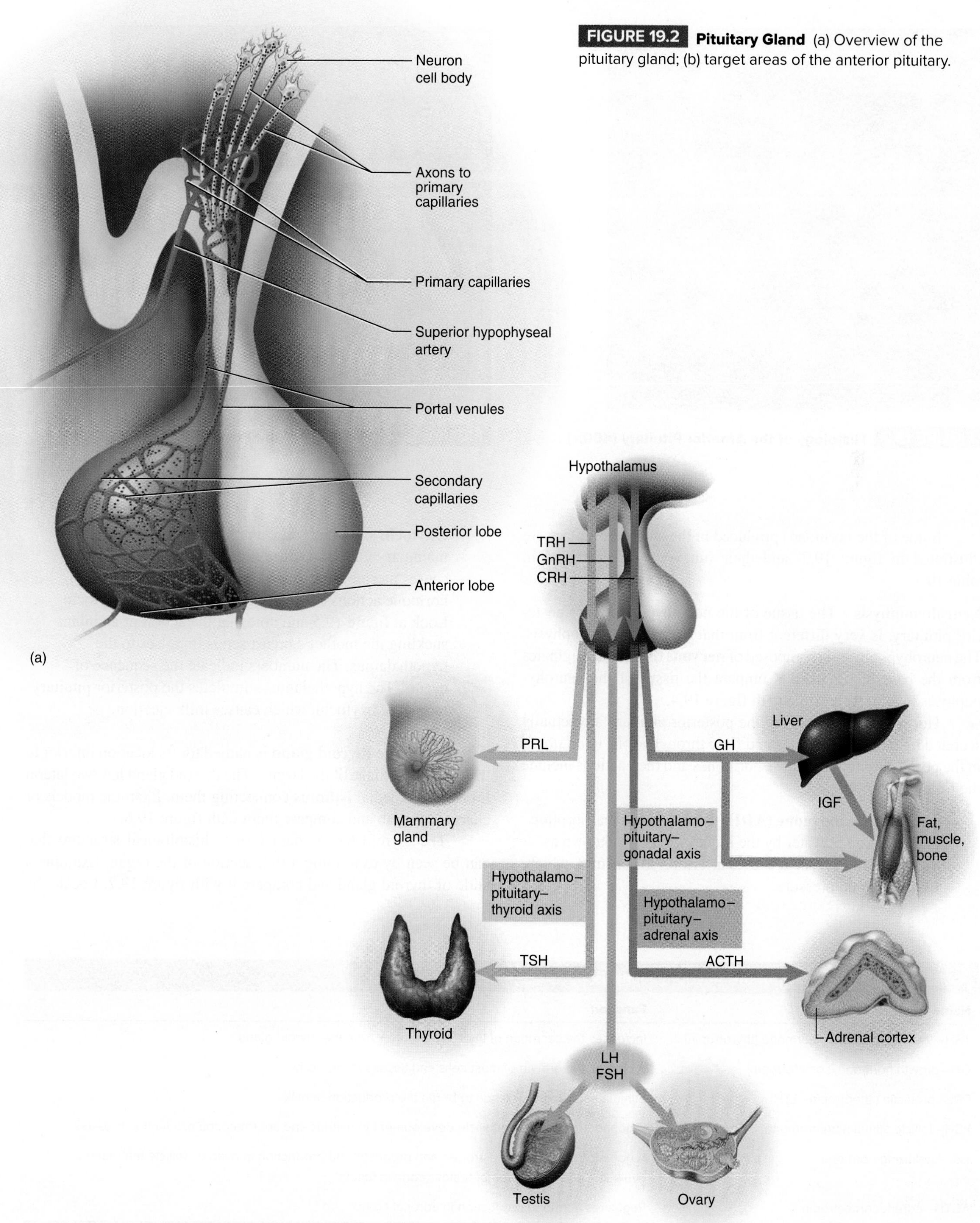

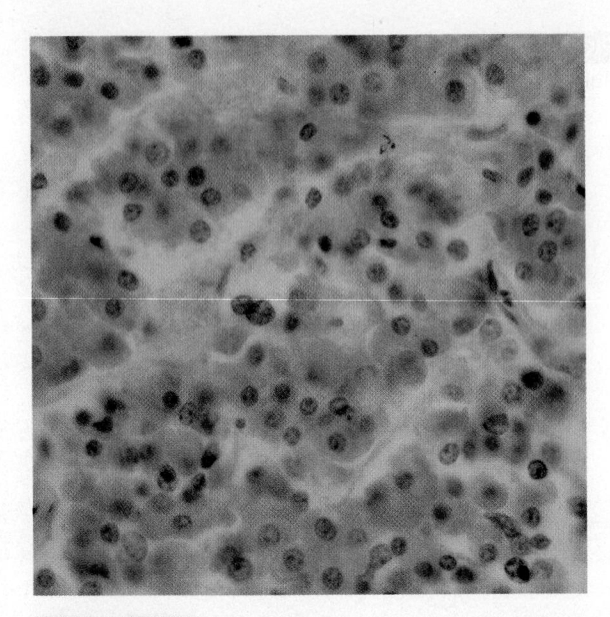

FIGURE 19.3 Histology of the Anterior Pituitary (400×)

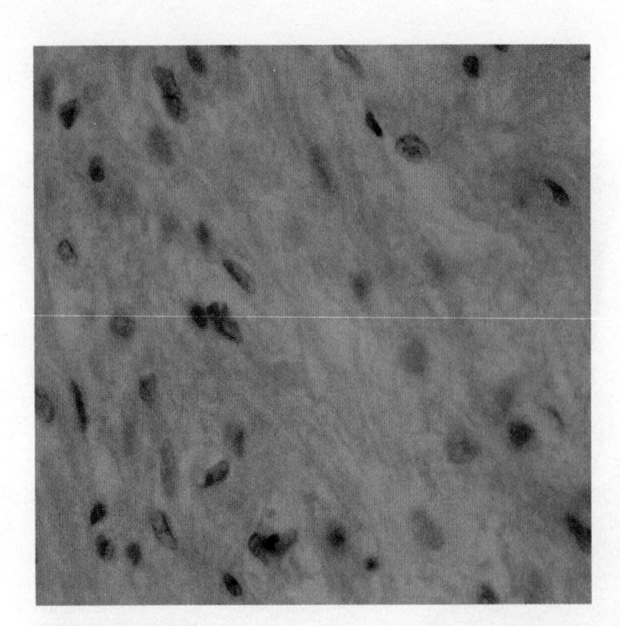

FIGURE 19.4 Histology of the Posterior Pituitary (400×)

Some of the hormones produced in the anterior pituitary are illustrated in figure 19.2, and their functions are described in table 19.1.

Neurohypophysis The tissue of the neurohypophysis, or posterior pituitary, is very different from that of the adenohypophysis. The neurohypophysis is composed of **nervous tissue** that originates from the base of the brain. Compare the tissue of the neurohypophysis in a prepared slide with figure 19.4.

Hormones released from the posterior pituitary are actually secreted by the hypothalamus and flow through axons to be *stored* in the posterior pituitary. These hormones and their actions include the following:

- **Antidiuretic hormone (ADH)** stimulates the reabsorption and retention of water by the kidneys. It is also known as **vasopressin** because it causes arterioles to constrict, which elevates blood pressure.
- **Oxytocin** stimulates the contraction of the cells of the mammary glands, resulting in the release of milk, and causes uterine contractions. External influences on hormone actions can be seen in the release of oxytocin. Look at figure 19.5 and note that the action of an infant suckling the mother's breast sends impulses to the hypothalamus. The numbers indicate the sequence of events. The hypothalamus stimulates the posterior pituitary to release oxytocin, which causes milk ejection.

Thyroid The **thyroid gland** is named for its location inferior to the thyroid cartilage of the larynx. The thyroid gland has two lateral **lobes** and a medial **isthmus** connecting them. Examine models or charts in the lab and compare them with figure 19.6.

The thyroid has a characteristic histological structure that can be seen by examining a thin section of the organ. Examine a slide of thyroid gland and compare it with figure 19.7. Locate the

TABLE 19.1 Hormones of the Adenohypophysis

Name	Function
TSH—thyroid-stimulating hormone (thyrotropin)	Increases the secretion of thyroid hormones from the thyroid gland
GH—growth hormone (somatotropin)	Promotes the growth of most cells and tissues in the body
PRL—prolactin (luteotropin—LTH)	Stimulates mammary glands to begin the production of milk
FSH—follicle-stimulating hormone	A gonadotropin causing follicle development in ovaries and spermatozoa production in testes
LH—luteinizing hormone	A gonadotropin causing estrogen and progesterone production in ovaries, follicle maturation, ovulation, and production of testosterone in testes
ACTH—adrenocorticotropin	Regulates hormone production in adrenal cortex

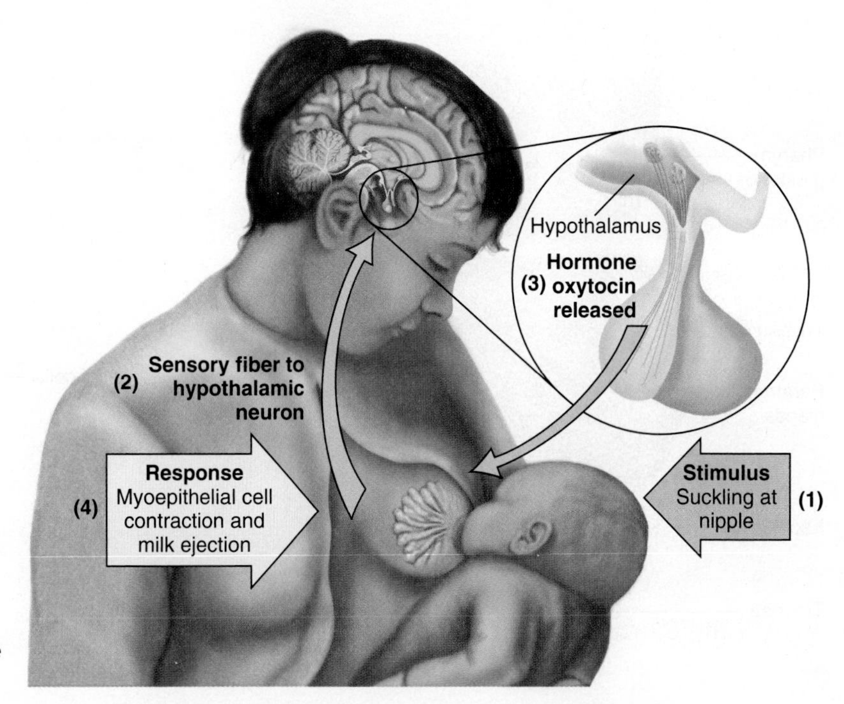

FIGURE 19.5 External Influences on Hormonal Action (1) External stimulus of suckling. (2) Sensory neurons take the impulse to the brain, where (3) the hypothalamus stimulates the posterior pituitary to release oxytocin and (4) milk is released from the breast.

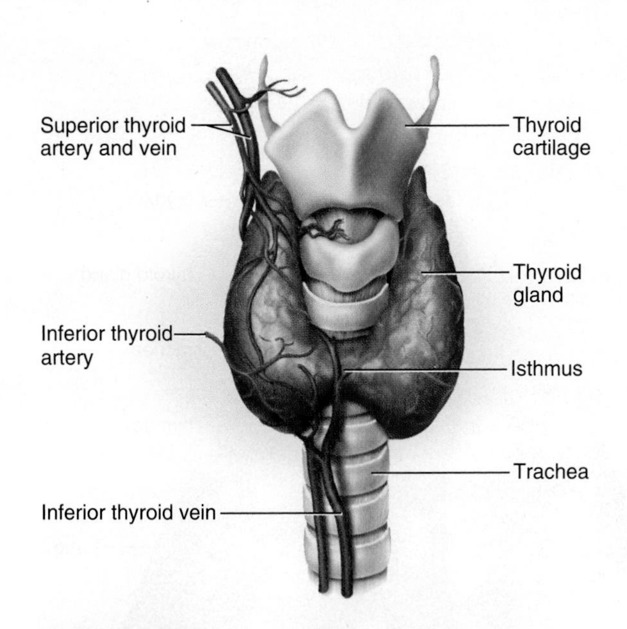

FIGURE 19.6 Anatomy of the Thyroid Gland

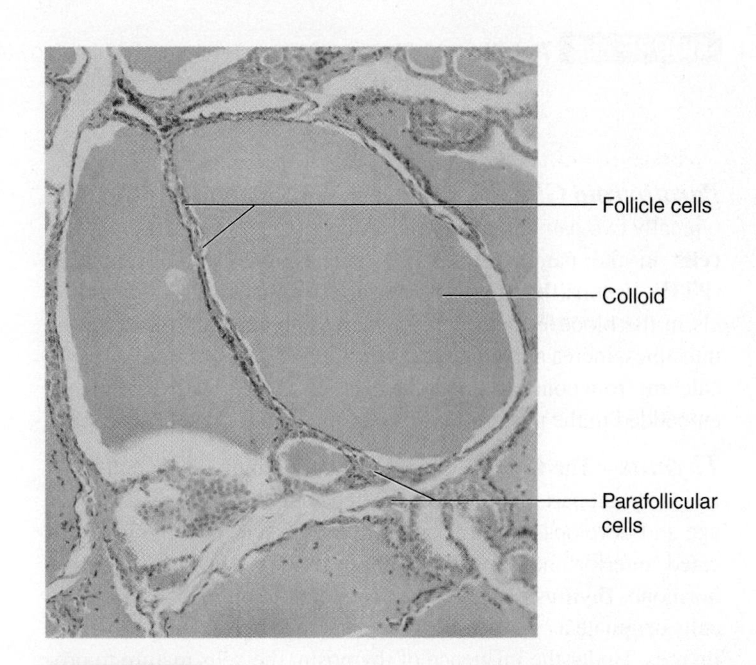

FIGURE 19.7 Histology of the Thyroid Gland (100×)

follicle cells that surround the **colloid,** a storage region for thyroid hormones. Colloid is mostly composed of thyroglobulin, which is a large-molecular-weight compound. Thyroid hormones are formed inside the larger thyroglobulin molecule. Also locate the **parafollicular cells** in the spaces between the follicles.

The thyroid gland secretes three hormones. Two of these are **T_3,** or **triiodothyronine** (a molecule that includes three iodine atoms), and **T_4,** or **thyroxine** (containing four iodine atoms per molecule). These hormones increase **basal metabolic rates** and are stored in the colloid. Another hormone of the thyroid gland is **calcitonin.** Calcitonin decreases blood calcium levels by causing excretion of calcium by the kidneys and deposition of calcium in bone by decreasing osteoclast activity and formation. The role of calcitonin is important in children and may protect individuals from hypercalcemia. Calcitonin is produced by the parafollicular, or C, cells. Calcitonin is antagonistic to parathyroid hormone (PTH) (discussed next).

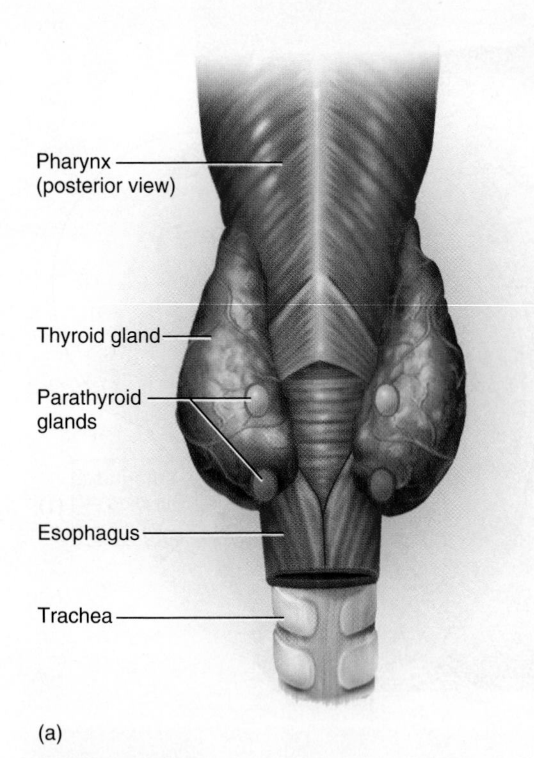

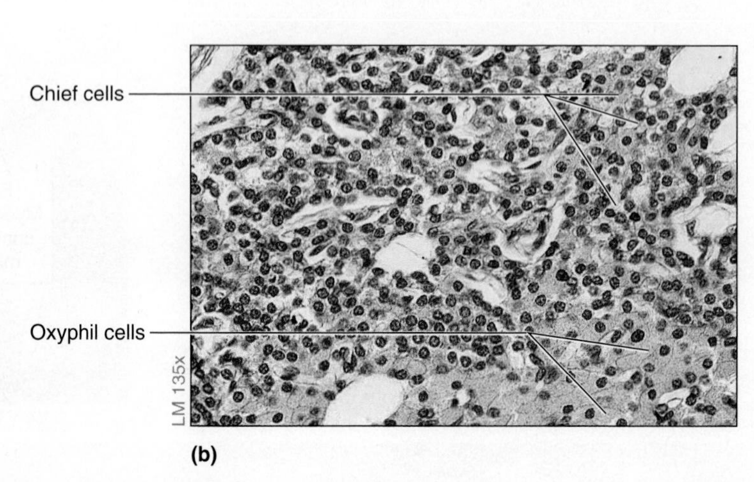

FIGURE 19.8 **Parathyroid Glands** (a) Gross anatomy; (b) histology (400×).

Parathyroid Glands In the posterior portion of the thyroid are typically two pairs of organs called the **parathyroid glands. Chief cells** in the parathyroid gland secrete **parathyroid hormone (PTH),** or **parathormone,** responsible for increasing calcium levels in the blood. This occurs by increasing calcium uptake in the intestines, increasing kidney reabsorption of calcium, and releasing calcium from bone. Examine the location of the parathyroid glands embedded in the posterior surface of the thyroid gland (fig. 19.8).

Thymus The **thymus** is active in young individuals and plays an important part in immunocompetency. It becomes smaller as we age and develops more fibrous and fatty tissue. The thymus is located anterior and superior to the heart (fig. 19.9). It secretes a hormone, **thymosin,** that causes the maturation of T cells. These cells originate as stem cells in the bone marrow and migrate to the thymus. Under the influence of thymosin, the cells mature to provide "cellular immunity" against **antigens** (foreign material, such as bacteria and viruses, which cause immune reactions). The T cells migrate from the thymus predominantly to the lymph nodes and the spleen to carry out their functions.

Pancreas The **pancreas** is a mixed gland in that it has an exocrine function and an endocrine function. Locate the pancreas in figures 19.1 and 19.10*a* and in models or charts in the lab. The exocrine function is digestive in nature, because pancreatic juice secreted through the pancreatic ducts contains both buffers and digestive enzymes. The endocrine function of the pancreas consists of the secretion of the hormones **insulin** and **glucagon,**

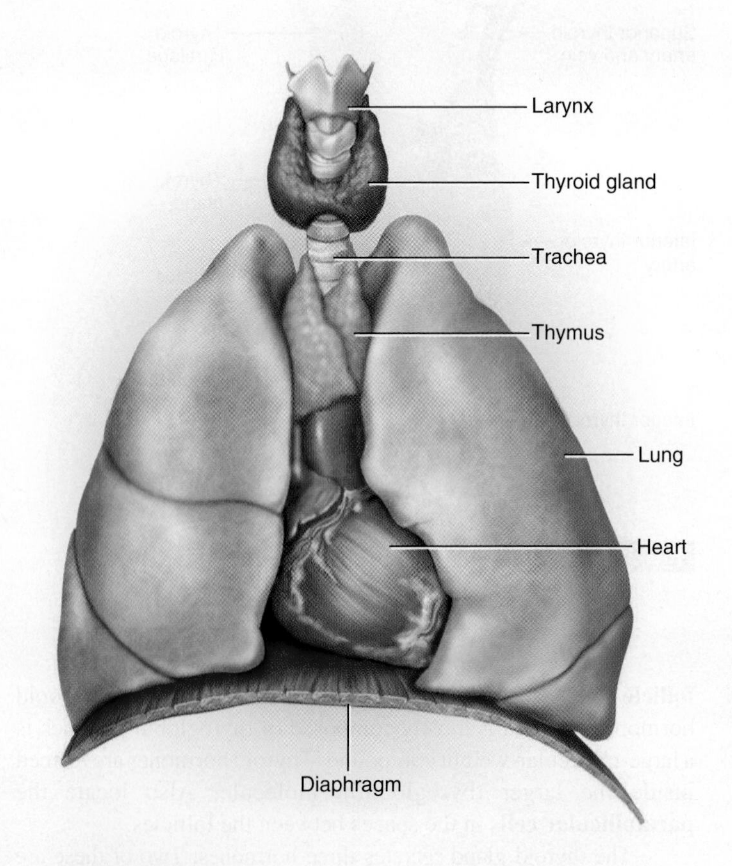

FIGURE 19.9 **Thymus**

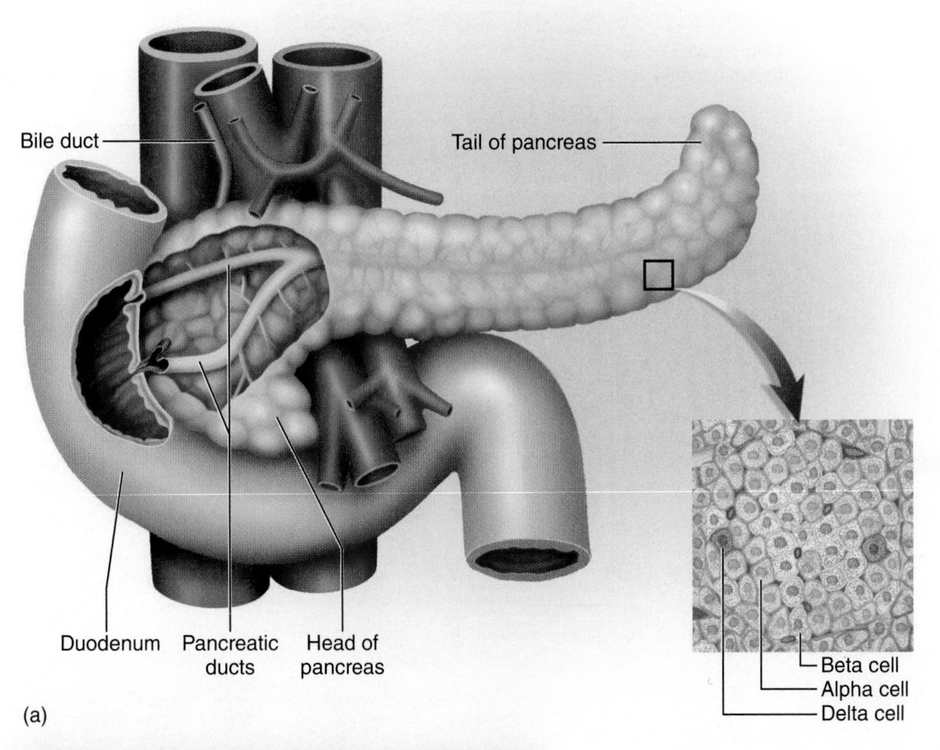

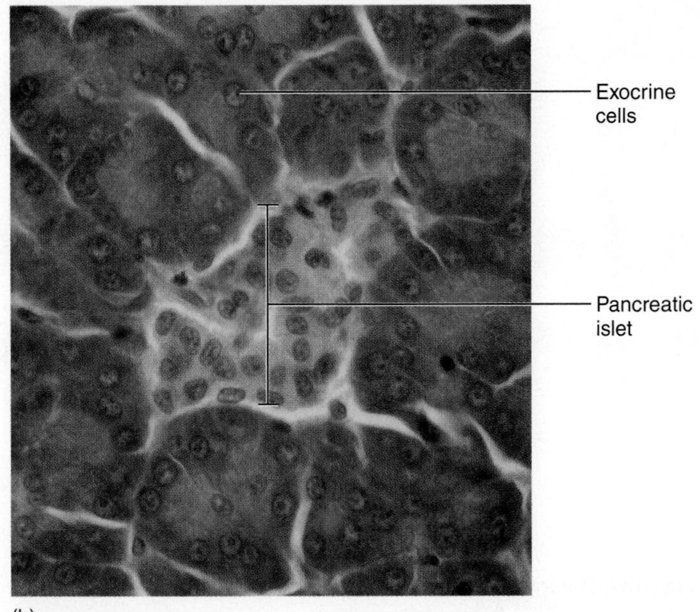

FIGURE 19.10 **Pancreas** (a) Gross anatomy; (b) histology (400×).

which regulate blood glucose levels. When blood glucose levels drop, glucagon converts glycogen (a starch storage product) to glucose. Glucagon is produced in specialized cells **(alpha [α] cells)** in clusters called **pancreatic islets (islets of Langerhans)** (fig. 19.10). Pancreatic islets also produce insulin, which lowers the blood glucose level. Insulin, which is produced in **beta (β) cells,** stimulates the conversion of glucose to glycogen. The pancreas also contains **delta cells,** which secrete somatostatin. Somatostatin is secreted after the ingestion of a meal and inhibits the secretion of insulin and glucagons from the pancreas. This may play a role in increasing efficiency in digestion. Locate the pancreatic islets on a microscope slide and compare them with those in figure 19.10.

Adrenal Glands The **adrenal glands** (*ad* = next to; *renal* = kidney) are superior to the kidneys. Each adrenal gland is composed of an outer **cortex** and an inner **medulla.** These are illustrated in figure 19.11*a*.

Examine a microscope slide of an adrenal gland and locate the cortex and the medulla. The hormones secreted from the adrenal cortex are called **corticosteroid hormones** and are controlled by ACTH. They are important in water and ion (Na^+, K^+) balance in the body. They are also important for carbohydrate, protein, and fat metabolism, as well as stress management. The cortex can be divided into three regions, all of which secrete corticoid hormones. The outermost is the **zona glomerulosa** (fig. 19.11*b*), which consists of clusters of cells that predominately secrete **mineralocorticoids** (especially **aldosterone**). Inside this layer (closer to the medulla) is the **zona fasciculata,** which consists of parallel bundles of cells that primarily secrete **glucocorticoids,** especially **cortisol (hydrocortisone).** Glucocorticoids regulate protein and fat catabolism and gluconeogenesis (production of glucose from amino acids). They are also involved in reducing

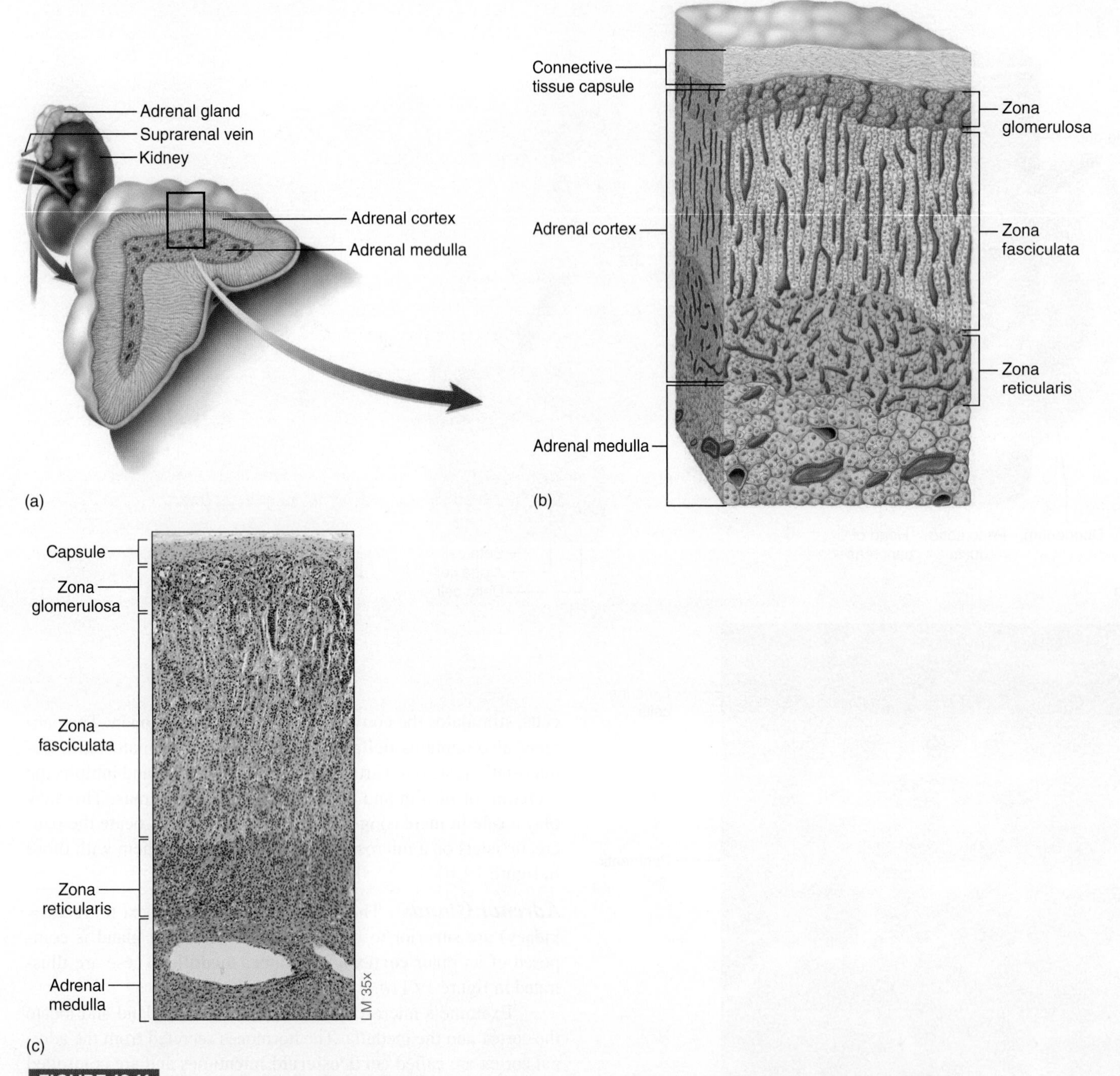

FIGURE 19.11 **Adrenal Cortex and Medulla** Diagram (a) gross anatomy; (b) histology; (c) photomicrograph (100×).

inflammation. The deepest cortical layer is the **zona reticularis,** which consists of a branched pattern of cells that produce both glucocorticoids and **sex hormones** (**androgens** and **estrogens**). The hormone **androstenedione** is a weak androgen, yet both males and females produce androgens from the adrenal glands in small amounts. Androstenedione is secreted from the adrenal glands but is converted to testosterone in other tissues. Most of the androgens in males are produced by the testes. Estrogen (estradiol) is secreted in small amounts in both sexes by the zona reticularis.

The hormones **epinephrine** and **norepinephrine** are produced in the adrenal medulla. Stimulation of the adrenal glands by the sympathetic nervous division causes the release of epinephrine and norepinephrine from the gland, which increases heart rate and prepares the body for fight-or-flight reactions.

Gonads The gonads, testes and ovaries, produce not only sex cells but also hormones, so they are considered endocrine glands. The gonads are stimulated by follicle-stimulating hormone (FSH) from the anterior pituitary, which causes the production and maturation of the sex cells (**spermatozoa** or **oocytes**). The gonads are also under the influence of luteinizing hormone (LH), which increases the level of hormone production, such as estrogen and testosterone, by the gonads.

Testes In the male the **testes** produce **testosterone,** a hormone responsible for the development of the male genitalia (penis and scrotum) during embryological and fetal development. Testosterone also aids FSH in the production of **spermatozoa,** which occurs in the seminiferous tubules. Testosterone also controls secondary sex characteristics, such as the development of facial and body hair, the expansion of the larynx (which produces a deeper voice), and the increased muscle and bone mass seen in males. The testes are mixed glands that have both an endocrine and an exocrine function. The testes are illustrated in figure 19.1. The endocrine function is testosterone production, and the exocrine function is the production of spermatozoa. **Inhibin** is a hormone produced and secreted by the testes, and it is involved in negative feedback, providing regulation for testosterone production. The exocrine function of the testis is explored in laboratory exercise 28 on the male reproductive system.

Examine a microscope slide of a testis and find the **seminiferous tubules** and **interstitial cells.** Refer to figure 19.12 for assistance. The interstitial cells are found between the tubules and produce testosterone.

Ovaries In females the **ovaries** produce **estrogen** and **progesterone** (fig. 19.1). Ovaries are also considered mixed glands that produce hormones as an endocrine function and **oocytes** (eggs) as an exocrine function. The ovaries are illustrated in figure 19.13. Female hormones are also responsible for secondary sex characteristics in women, such as the development of breasts, an additional subcutaneous adipose layer, and a higher voice. Estrogen and progesterone also influence the development of the endometrium, cause maturation of the oocytes, and regulate the menstrual cycle. "Estrogen" is a generic term for several hormones produced by the female, including **estradiol.** Inhibin is also secreted by the ovary and regulates the levels of estrogen and progesterone. During pregnancy the placenta has a major role as an endocrine gland in the secretion of estrogen and progesterone.

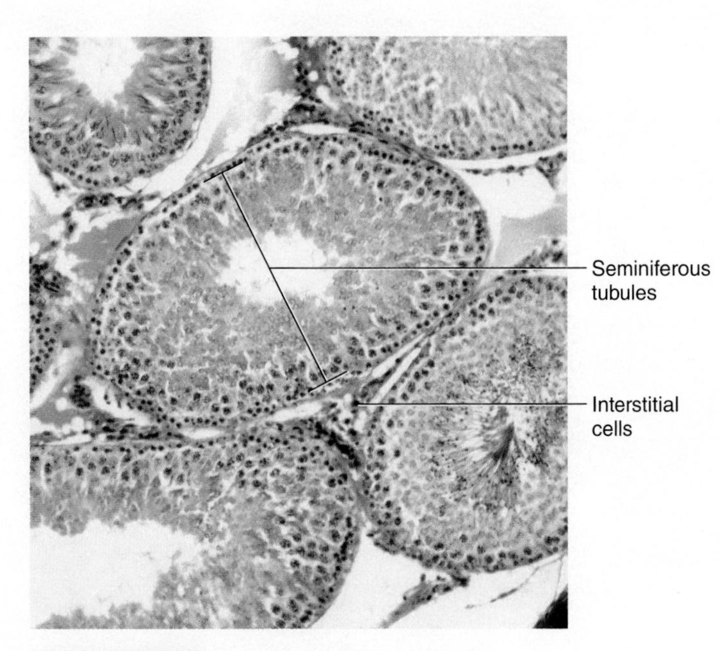

FIGURE 19.12 **Histology of the Testis (100×)**

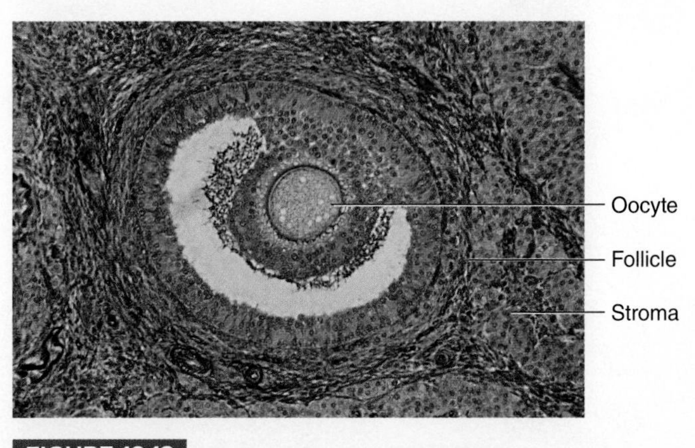

FIGURE 19.13 **Histology of the Ovary (40×)**

Exercise 19

Endocrine System

Name ______________________ Date __________

1. What name is given to cells that are receptive to hormones? ______________________

2. Melatonin is secreted by what gland? ______________________

3. In what specific part of what gland is ADH stored? ______________________

4. What is the effect of TSH and where is it produced? ______________________

5. What does glucagon do as a hormone and where is it produced? ______________________

6. Which hormones in the adrenal gland control water and electrolyte balance? ______________________

7. What is the primary gland that secretes epinephrine? ______________________

8. Where is growth hormone produced? ______________________

9. What is another name for T_3? ______________________

10. What connects the two lobes of the thyroid gland? ____________________

11. What impact does parathormone have on calcium levels in the blood? ____________________

12. Interstitial cells produce which hormone? ____________________

13. What hormone causes a *decrease* in calcium levels in the blood? ____________________

14. Label the endocrine glands indicated in the following illustration, using the terms provided.

    adrenal glands
    hypothalamus
    ovaries
    pancreas
    parathyroid glands
    pineal gland
    pituitary gland
    testes
    thymus
    thyroid gland

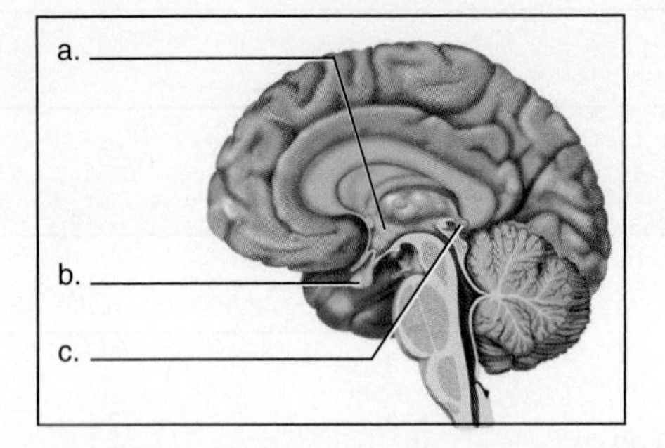

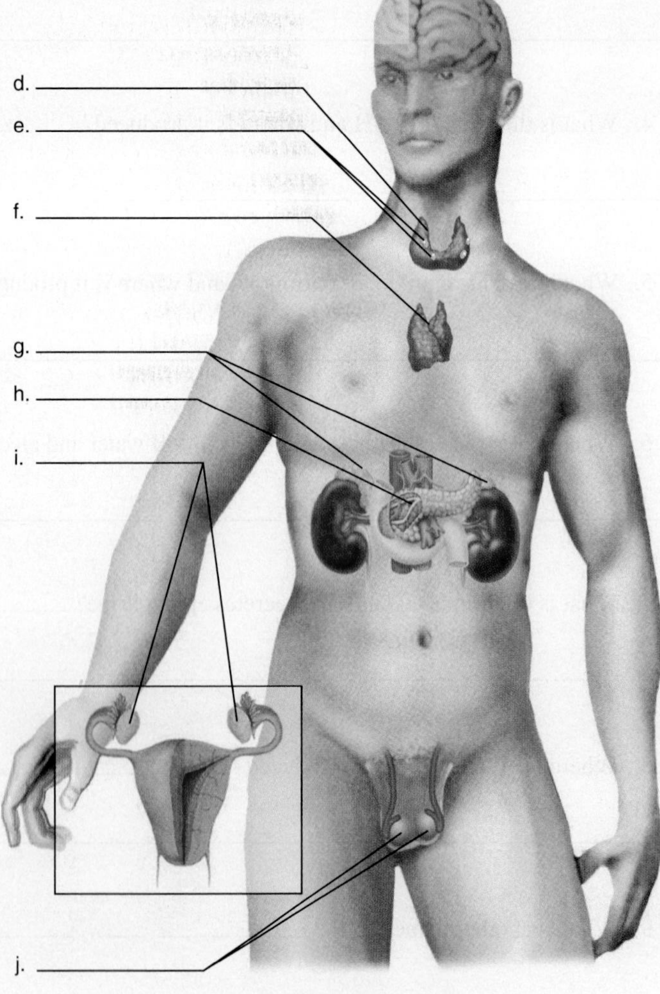

15. Identify the three layers of the adrenal cortex as illustrated and list a hormone produced by each layer.

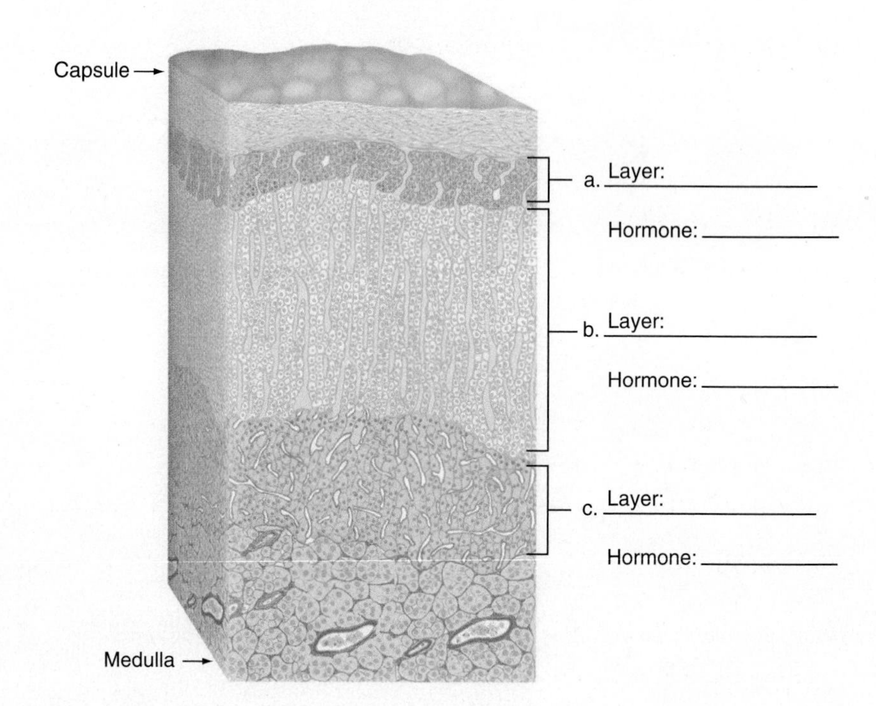

Laboratory

Exercise 20

Blood Cells

Cardiovascular System

Introduction

Blood is a connective tissue that consists of two parts, formed elements and plasma. Formed elements make up about 45% of the blood volume and can further be subdivided into red blood cells (erythrocytes), white blood cells (leukocytes), and platelets (thrombocytes). Platelets are not cells, and this is why the fraction of blood is called formed elements and not blood cells. Plasma constitutes approximately 55% of the blood volume and contains water, lipids, dissolved substances, colloidal proteins, and clotting factors.

The study of blood is important because it is the fluid medium of the cardiovascular system and because it has significant clinical implications. Changes in the numbers and types of blood cells may be used as indicators of disease.

Caution The risk of bloodborne diseases has been significantly minimized by the use of sterilized human blood and nonhuman mammal blood in this exercise. However, use the same precautions as if you were handling fresh and potentially contaminated human blood. Your instructor will determine whether to use animal blood (from nonhuman mammals), sterilized blood, or synthetic blood. In any of these cases, follow strict procedures for handling potentially pathogenic material.

1. Wear protective gloves during the procedures.
2. Do not eat or drink in the lab.
3. If you have an open wound, do not participate in this exercise, or make sure the wound is *securely* covered.
4. Your instructor may elect to have you use your own blood. Due to the potential for disease transmission, such as AIDS or hepatitis, in fresh blood samples, **make sure you keep away from other students' blood and keep them away from your blood!**
5. Place all disposable material in the biohazard bag.
6. Place all used lancets in the sharps container and place all glassware that is to be reused in a 10% bleach or other disinfectant solution.
7. After you have finished the exercise, clean and disinfect the countertops with a 10% bleach or other disinfectant solution.

Plasma is the fluid portion of blood and is about 92% water by volume. The remainder mostly consists of proteins, such as albumins, globulins, and fibrinogen. **Albumins** are produced by the liver and make up the majority of the plasma proteins. Some **globulins** are made by the plasma cells and make up the next largest amount of proteins, but most globulins are made in the liver. Globulins have an important function for immunity (as antibodies) and they aid in solute transport and blood clotting. **Fibrinogen** is a clotting protein, and this and other clotting factors are produced by the liver. Plasma also contains ions (Na^+, K^+, and Cl^-), nutrients, hormones, and wastes.

In this exercise you examine the nature of blood cells.

Learning Objectives

At the end of this exercise you should be able to

1. discuss the composition of blood plasma;
2. distinguish among the various formed elements of blood;
3. describe hematopoiesis;
4. determine the percentage of each type of leukocyte in a differential white cell count; and
5. describe the significance of an elevated level of a particular white blood cell concerning a disease state or an allergic reaction.

Materials

Prepared slides of human blood with Wright's or Giemsa stain
Compound microscopes
Lab charts or illustrations showing the various blood cell types
Protective gloves and goggles
Roll of paper towels
Vial of mammal blood or
 Sterile cotton balls
 Alcohol swabs
 Sterile, disposable lancets
 Adhesive bandages

Dropper bottle of Wright's stain
Squeeze bottle of distilled water or phosphate buffer solution
Large finger bowl or staining tray
Toothpicks
Clean microscope slides
Coverslips
Pasteur pipette and bulbs
Hand counter
Biohazard bag or container
10% bleach or other disinfectant container
Sharps container

Procedure

Examination of Blood Cells

Your lab instructor will direct you to use a prepared slide of blood; fresh, nonhuman mammal blood; or your blood. If you are going to compare both the prepared slide of blood and the one you make, then examine the prepared slide of blood first. If you are using a prepared slide of blood, you can move to the section "Microscopic Examination of Blood Cells." If you are to make a smear, be sure to read the following directions.

Preparing a Fresh Blood Smear Your lab instructor will direct you to use either fresh, nonhuman mammal blood or your own blood. If you are using provided mammal blood, wear protective gloves and withdraw a small amount of blood from the vial with a clean Pasteur pipette. Place a drop of the blood on a clean microscope slide; then proceed to step 8 in the following procedure.

Withdrawing Your Own Blood Obtain the following materials: a clean, sterile lancet; an alcohol swab; an adhesive bandage; protective gloves; sterile cotton balls; two clean microscope slides; and a paper towel.

1. Arrange the materials on the paper towel in front of you on the countertop.
2. Warm your hands with water or by rubbing them to increase blood flow to the fingers.
3. Select a finger on your nondominant hand (the one you don't use for writing). Clean the end of the donor finger with the alcohol swab, and let the hand from which you will withdraw blood hang by your side for a few moments to collect blood in the fingertips. You will be puncturing the pad of your fingertip (where the fingerprints are located lateral to the center of the finger).
4. Peel back the covering of the lancet and hold on to the blunt end as you withdraw it from the package. Do not touch the sharp end or lay the lancet on the table before puncturing your finger.
5. Prick your finger quickly, and wipe away the first drop of blood that forms with a sterile cotton ball. Throw the lancet in the sharps container. Never reuse the lancet or set it down on the table.
6. Place the second drop of blood that forms on the microscope slide about 2 cm away from one of the ends of the slide (fig. 20.1), and place another cotton ball on your finger.
7. Put an adhesive bandage on your finger.
8. Whether you are using prepared blood or your own blood, use another clean slide to spread the blood by touching the drop with the edge of the slide and *push the blood* across the slide (fig. 20.1). This should produce a smooth, thin smear of blood. Let the blood smear dry completely. Place the slide used to spread the blood in the 10% bleach or other disinfectant solution.
9. After the slide is dry, place it in a large finger bowl elevated on toothpicks or on a staining tray.
10. Cover the blood smear with several drops of Wright's stain from a dropper bottle. Let the stain remain on the slide for 1 to 2 minutes.
11. After this time add water or a prepared phosphate buffer solution to the slide. You can rock the slide gently with gloved hands or blow on it to stir the stain and water. A metallic green material should come to the surface of the slide. Let the slide remain covered with stain and water for 3 to 8 minutes.

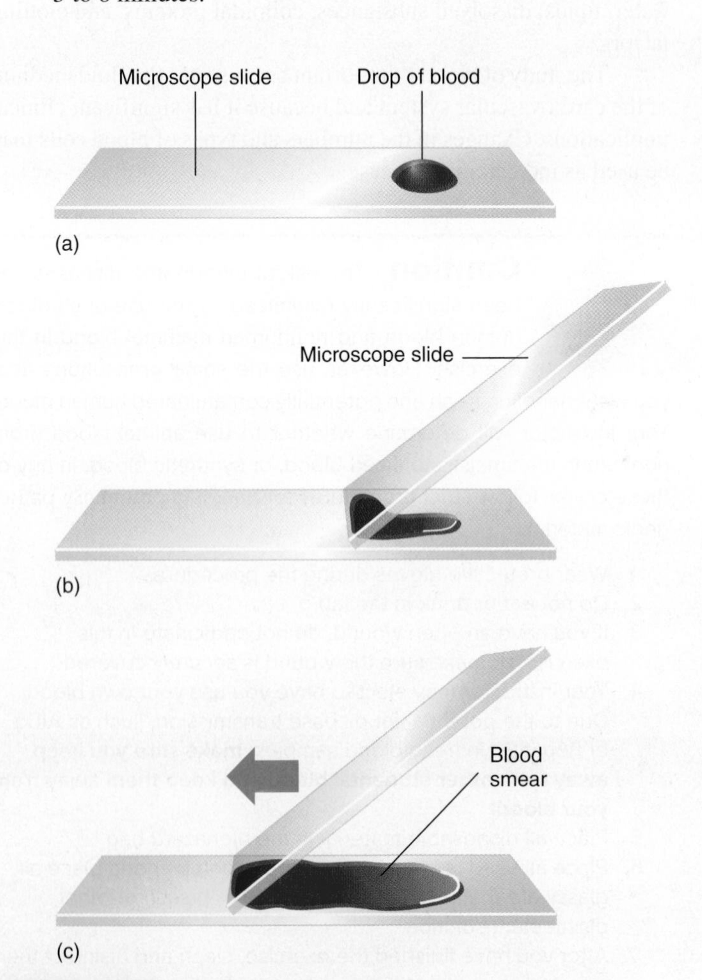

FIGURE 20.1 Making a Blood Smear (a) Placing blood on slide; (b) touching glass slide to front of blood drop; (c) spreading blood across slide.

12. Wash the slide gently with distilled water until the material is light pink and then stand it on edge to dry. You may also stain blood using an alternate stain (such as Giemsa stain), following your lab instructor's procedure.
13. Place all blood-contaminated disposable material in the biohazard container.
14. Once the slide is completely dry, it can be examined under the high-power or oil immersion lens of your microscope.

Microscopic Examination of Blood Cells

Overview As you examine the blood slide, you should refer to figures 20.2–20.4. You will have to look at many blood cells in order to see all the cells that are included in this exercise.

Erythrocytes Examine a slide of blood stained with either Wright's or Giemsa stain. **Erythrocytes** are the most common cells you will find on the slide. There are about 5 million erythrocytes per microliter. They do not have a nucleus but appear as pink, biconcave discs (like doughnuts with the holes partially filled in). Erythrocytes contain the pigment **hemoglobin,** which carries oxygen. Blood formation is called **hematopoiesis** or **hemopoiesis.** The specific production of erythrocytes is called **erythropoiesis.** Erythrocytes are about 7.5 micrometers in diameter, on average, and have a life span of about 120 days, after which time they are broken down by the spleen or liver. The iron portion of the hemoglobin is used by the bone marrow to make more hemoglobin, the proteins are broken down to amino acids and reabsorbed for general use by the body, and the heme is metabolized to bilirubin in the liver and secreted as part of the bile. Compare what you see under the microscope with figure 20.2.

Platelets There are about 130,000 to 360,000 platelets per microliter of blood. Platelets, or thrombocytes, are involved in clotting and consist of small fragments of megakaryocytes. Examine the slide for small, purple fragments that may be single or clustered, and compare them with the platelets in figure 20.2.

Leukocytes There are far fewer **leukocytes,** or **white blood cells,** in the blood than erythrocytes. The number of leukocytes in a healthy adult is about 5000 to 10,000 cells per microliter of blood. Leukocytes are formed in bone marrow and lymphoid tissue. The life span of a white blood cell varies from a few hours to several months. Many are capable of ameboid movement as they squeeze between cells (a process termed *diapedesis*), and engulf foreign particles, or cellular debris.

Leukocytes can be divided into two groups based on the presence or absence of granules in their cytoplasm. These two groups are the **granular** and **agranular leukocytes.** Examine the slide under high power or oil immersion and identify the different leukocytes as discussed in the following sections.

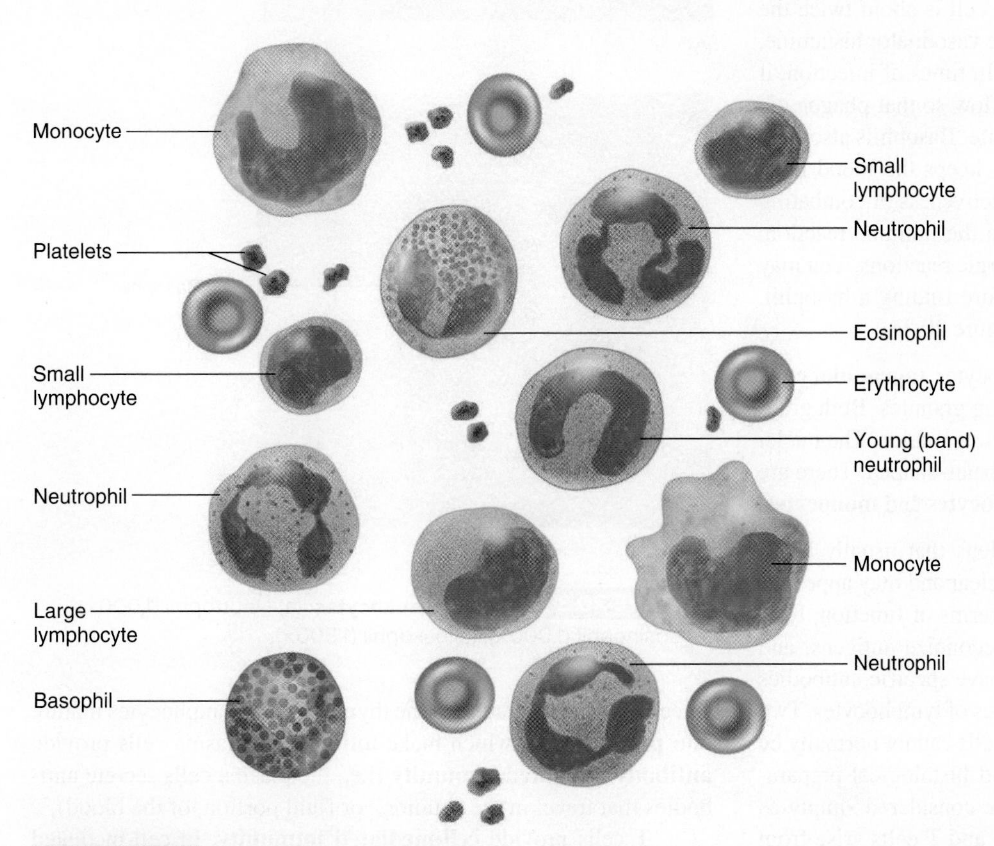

FIGURE 20.2 Formed Elements of Blood

Granular Leukocytes Granular leukocytes **(granulocytes)** are so named because they have granules in their cytoplasm. They are also known as **polymorphonuclear (PMN) leukocytes** due to the variable shape of their nuclei (which are lobed, not round). The three types of granular leukocytes are **neutrophils, eosinophils,** and **basophils.**

Neutrophils are the most common of all leukocytes. Neutrophils typically live for about 10 to 12 hours and have ameboid capabilities. They move by diapedesis between blood capillary cells and in the interstitial areas between cells of the body. Neutrophils move toward infection sites, squeeze between adjacent endothelial cells that make up blood capillaries, move into the interstitial spaces of tissues, and destroy foreign material. They are the major phagocytic white blood cell. The granules of neutrophils absorb very little stain so their granules are light pink to light purple, but neutrophils can be distinguished from other leukocytes by their three- to five-lobed nucleus. They are about one and a half times the size of erythrocytes. Examine your slide for neutrophils and compare them with figure 20.3*a*.

Eosinophils typically have a two-lobed nucleus with pink-orange granules in the cytoplasm. The term "eosinophil" actually means eosin-loving (eosin is a pink-orange stain and is picked up by the granules). They are about twice the size of erythrocytes. Eosinophils combat infections caused by multicellular pathogens, such as parasitic worms, and they are involved in allergic reactions. Compare figure 20.3*b* with your slide as you locate the eosinophils.

Basophils are rare. The granules stain very dark (blue-purple), and sometimes the nucleus is obscured because of the dark-staining granules. The nucleus is S-shaped and the cell is about twice the size of an erythrocyte. Basophils contain the vasodilator histamine, which increases blood flow to the tissues. In times of infection, it is beneficial to have an increase in blood flow so that phagocytic cells or antibodies can reach the infection site. Basophils also contain heparin, which is an anticoagulant. It keeps the blood from clotting too quickly. Clots decrease the effectiveness of combating infection as they can seal off the site from the immune reaction. They are involved in inflammatory and allergic reactions. You may have to look at 200 to 300 leukocytes before finding a basophil. Compare the basophil in your slide with figure 20.3*c*.

Agranular Leukocytes Agranular leukocytes **(agranulocytes)** are so named because they lack dark-staining granules. Both granulocytes and agranulocytes have nonspecific granules. The nuclei are not lobed but may be dented or kidney bean–shaped. There are two types of agranular leukocytes—**lymphocytes** and **monocytes.**

Lymphocytes have a large, unlobed nucleus that usually has a flattened or dented area. The cytoplasm is clear and may appear as a blue halo around the purple nucleus. In terms of function, lymphocytes do not need prior exposure to recognize antigens, and they are remarkable in that separate cells have specific antibodies for specific antigens. There are multiple types of lymphocytes. Two of them are the **B cells** and **T cells.** These cells cannot normally be distinguished from one another in standard histological preparations (for example, Wright's stain) and are considered simply as lymphocytes in this exercise. Both B cells and T cells arise from fetal bone marrow. B cells probably mature in the fetal liver and spleen and T cells mature in the thymus. B cell lymphocytes mature into **plasma cells,** which make **antibodies.** Plasma cells provide **antibody-mediated immunity** (i.e., the plasma cells secrete antibodies that travel in the "humor," or fluid portion, of the blood).

T cells provide **cell-mediated immunity.** In cell-mediated immunity, the cells themselves (not antibodies in the blood plasma)

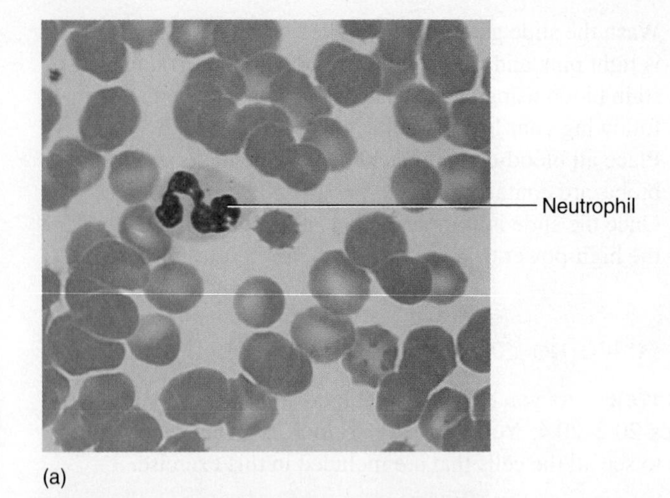

(a)

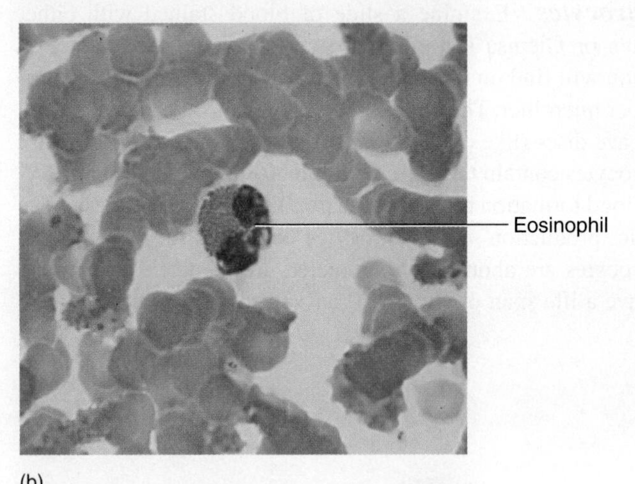

(b)

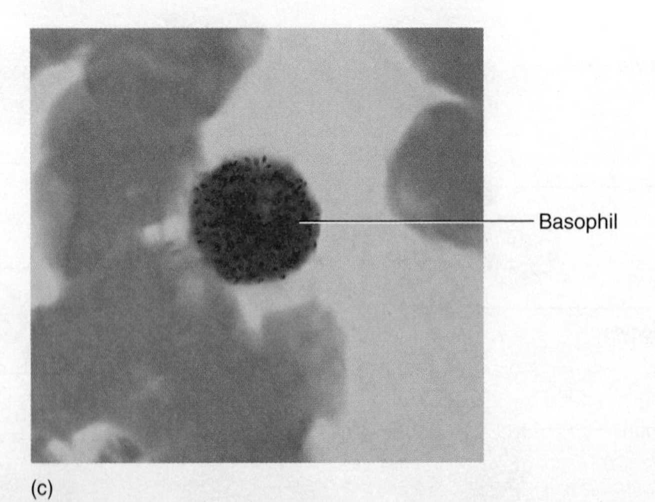

(c)

FIGURE 20.3 **Granular Leukocytes** (a) Neutrophil (1,000×); (b) eosinophil (1,000×); (c) basophil (1,500×).

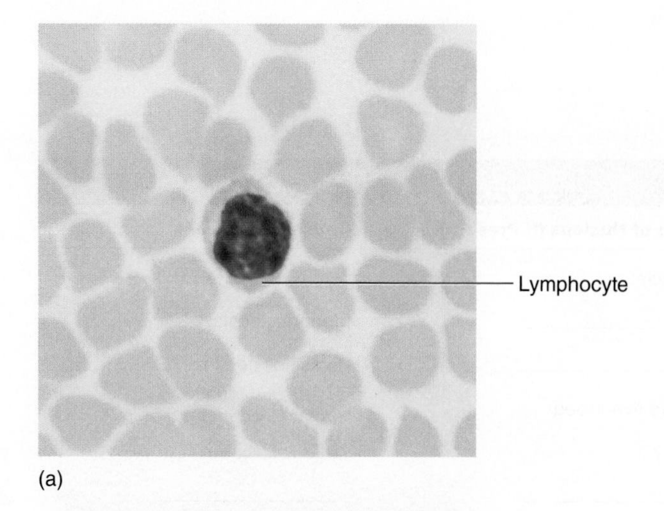

(a)

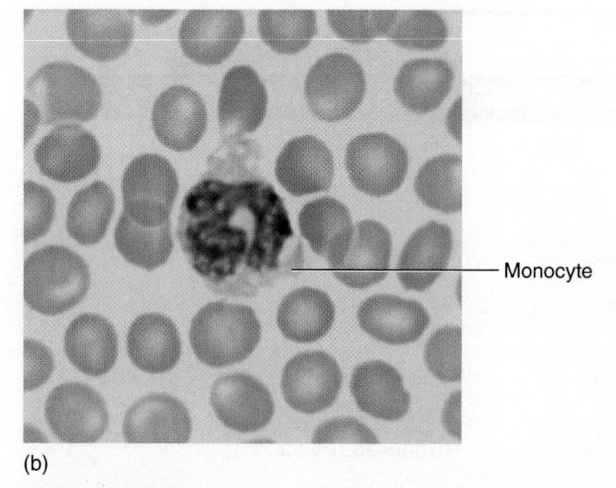

(b)

FIGURE 20.4 Agranular Leukocytes (1,000×) (a) Lymphocyte; (b) monocyte.

move close to and destroy some types of bacteria or virus-infected cells. T cells also attack tumors and transplanted tissues. Most of the T cells are found in the lymph nodes, thymus, and spleen. They enter the bloodstream via the lymphatic tissue. **Natural killer (NK) cells** are lymphocytes that attack bacteria, transplanted cells, and cancer cells. Compare your slide with figure 20.4*a* for lymphocytes.

Monocytes are very large and often have a kidney– or horseshoe-shaped nucleus. They are about three times the size of erythrocytes and are activated by T cells. Monocytes are important in that they can turn into **macrophages** which are major phagocytic cells, and they are important in presenting foreign antigens to T lymphocytes. Monocytes move into tissues from the blood and become macrophages. Locate the large cells on your slide and compare them with figure 20.4*b*.

Differential White Blood Cell Count

Rapid and inexpensive diagnosis of disease is a goal of modern health care. Narrowing the field of potential disease states is frequently done with a differential leukocyte count. For example, if a patient comes to the hospital with a fever, it could be due to many things, including viral infection, reactions to medications, metabolic disorders, and vascular disease. By taking a small sample of blood and examining the percentages of leukocytes, many diseases can be eliminated from the diagnosis. If, for example, there is an elevated lymphocyte count, it may mean that there is a severe viral infection in the body. It could also mean that there is an autoimmune disease or cancer in the lymph system, but it helps to direct the health-care provider to look for answers in certain areas and not in others. In this part of the exercise, you examine a blood smear and count leukocytes to determine their percentages. Once you have identified the various leukocytes on the blood slide, conduct a differential white blood cell count.

1. One lab partner should look into the microscope and methodically call out the names of the different types of leukocytes seen.
2. The other lab partner should record how many of each type are found and keep track of the overall number with the use of a hand counter until 100 cells are counted.
3. Scan the slide in a systematic way, so that you don't count any cell twice. One method is illustrated in figure 20.5. Tally your results.

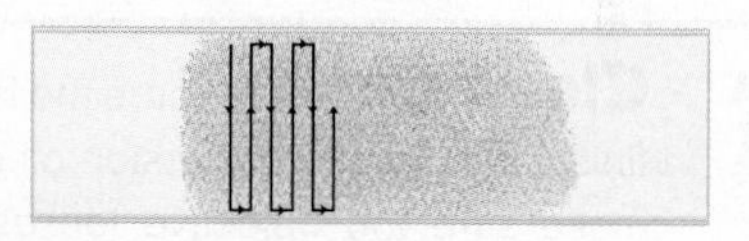

FIGURE 20.5 Counting Leukocytes

Neutrophils: ______________________

Eosinophils: ______________________

Basophils: ______________________

Lymphocytes: ______________________

Monocytes: ______________________

Neutrophils represent about 60% to 70% of all the leukocytes. They increase in number in appendicitis or acute bacterial infections.

Eosinophils represent about 2% to 4% of all leukocytes. Eosinophils increase in number during allergic reactions and parasitic infections (e.g., trichinosis).

Basophils represent about 0.5% of all leukocytes. They increase in number during allergies and radiation.

Lymphocytes make up 25% to 33% of the leukocytes. These cells increase in times of viral infection, such as infectious mononucleosis, and antibody–antigen reactions.

Monocytes make up about 3% to 8% of the leukocytes. They increase in times of chronic infections, such as tuberculosis.

Leukopenia is a decrease in the number of circulating leukocytes in the blood. Common causes are radiation therapy, chemotherapy, and some medications. **Leukemia** is an increase in the number of leukocytes. Even though the leukocytes are abundant, these cells are often immature or deformed and they do not function normally.

Fill in chart 1 with what you know about the formed elements. Read all the information given before filling in the chart.

CHART 1 Characteristics of Formed Elements

Formed Element	Granules (If Present)	Shape of Nucleus (If Present)	Cause for Increase
Erythrocyte	No granules	No nucleus	________
________	Not obvious	________	Mononucleosis
________	Orange-staining	________	Parasitic infections
________	________	Two- to five-lobed	________
________	Not obvious	Kidney	________
Basophil		________	________

Review the general properties of blood in table 20.1 and table 20.2.

Clean Up Make sure the lab is clean after you finish. If you used immersion oil on the microscope, make sure the objective lenses are wiped clean. (Use clean lens paper only!) Place any slide with fresh blood on it or any material contaminated with bodily fluid in the bleach or other disinfectant solution. Place all gloves or contaminated paper towels in the biohazard container. All sharps material (broken slides or coverslips, lancets, etc.) should be placed in the sharps container. Clean the counters with a towel and a 10% bleach or other disinfectant solution.

TABLE 20.1 General Properties of Blood

Volume	Female: 4–5 L Male: 5–6 L
pH	7.35–7.45
Mean salinity	0.9%
Hematocrit	Female: 37% to 48% Male: 45% to 52%
Hemoglobin	Female: 12–16 g/dL Male: 13–18 g/dL
Platelet count	130,000–360,000/uL
Total WBC* count	5000–10,000/uL

*White blood cell.

TABLE 20.2 Summary of Formed Elements in Blood

	Size	Number	Characteristics
Erythrocytes	7.5 μm	5 million/μL	Live 120 days on average No nucleus
Platelets	2–4 μm	130,000–360,000/μL	Cell fragments
Leukocytes		7,000/μL	Nucleus present
Granular leukocytes			
Neutrophils	9–12 μm	60–70% of leukocytes	Three- to five-lobed nucleus Granules indistinct
Eosinophils	10–14 μm	2–4% of leukocytes	Two-lobed nucleus, Orange granules
Basophils	8–10 μm	< 0.5% of leukocytes	S-shaped nucleus Large, dark granules
Agranular leukocytes			
Lymphocytes	5–17 μm	25–33% of leukocytes	Nucleus appearing dented Thin rim of cytoplasm in some
Monocytes	12–15 μm	3–8% of leukocytes	Large, kidney-shaped nucleus

Review

Exercise 20

Blood Cells

Name ______________________ Date __________

1. What are the three main types of formed elements? ______________________

2. What is the most common plasma protein? ______________________

3. What is another name for a platelet? ______________________

4. Which is the most common blood cell? ______________________

5. What is another name for a leukocyte? ______________________

6. What leukocyte is most numerous in a normal blood smear?

7. How many erythrocytes are normally found per microliter of blood?

8. What is an average number of leukocytes found per microliter of blood?

9. B cells and T cells belong to what class of agranular leukocytes?

10. How does a differential leukocyte count aid in medical diagnosis?

11. In counting 100 leukocytes, you are accurately able to distinguish 15 basophils. Is this a normal number for the white blood cell count, and what possible health implications can you draw from this?

12. What is the function of the thrombocytes of the blood? ______________________________

____

____

____

13. Formed elements constitute what percentage of the total blood volume? ______________________________

____

____

____

14. Label the formed elements in the following illustration.

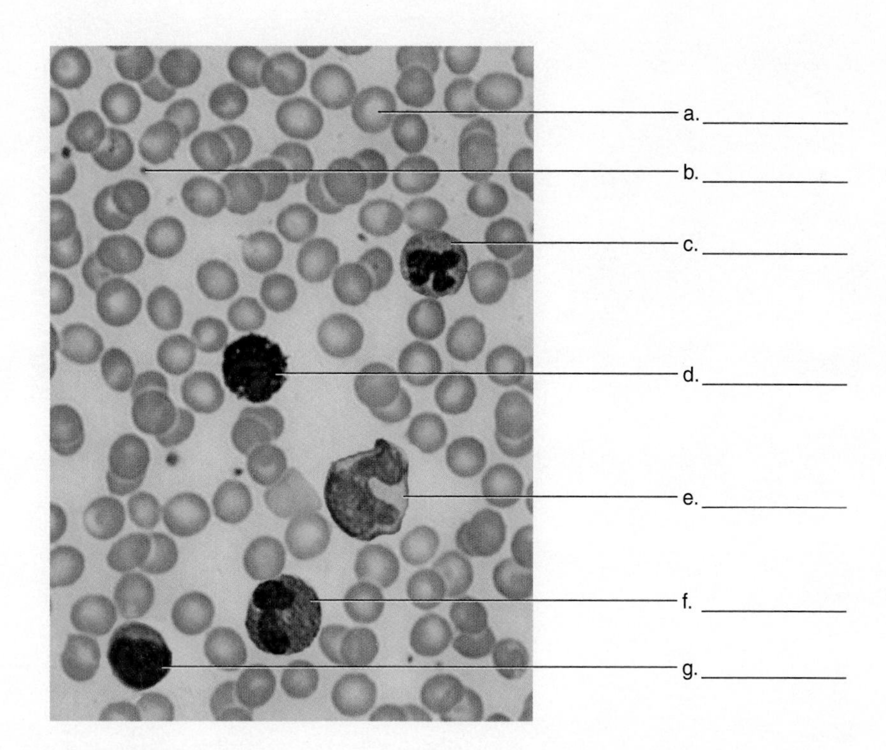

Laboratory

Exercise 21

The Heart

Cardiovascular System

Introduction

The heart is a muscular pump, which provides the force for circulating blood throughout the body. Blood has two major circulation routes. One is the circulation from the heart to the pulmonary trunk, to the lungs, and back to the heart. This is the **pulmonary circulation.** The other is from the heart to the aorta and throughout the rest of the body (head, upper limbs, trunk, lower limbs) and back to the heart. This is the **systemic circulation.** Blood travels from the heart in vessels called **arteries,** and blood returns to the heart through veins. The blood vessels and the heart together constitute the **cardiovascular system.**

In this exercise you examine models of the heart and preserved sheep and human hearts, if available. As you look at the preserved material, try to see how the structure of the heart relates to its function.

Learning Objectives

At the end of this exercise you should be able to

1. list the three layers of the heart wall;
2. describe the position of the heart in the thoracic cavity;
3. describe the significant surface features of the heart;
4. describe the internal anatomy of the heart;
5. find and name the anatomical features on models of the heart and in the sheep heart;
6. describe the blood flow through the heart and the function of the internal parts of the heart;
7. discuss the functioning of the heart valves and their role in circulating blood through the heart; and
8. distinguish the systemic circulation from the pulmonary circulation.

Materials

Models and charts of the heart
Preserved sheep hearts
Preserved human hearts (if available)
Blunt probes (Mall probes)
Dissection pans
Scalpel
Sharps container
Protective gloves
Waste container
Microscopes
Prepared slides of cardiac muscle

Procedure

Heart Location and Membranes

The heart is located deep in the thorax between the lungs in a region known as the mediastinum. The **mediastinum** contains the heart, the membranes surrounding the heart (the pericardia), and other structures, such as the esophagus and descending aorta. The mediastinum is located between the sternum, lungs, and thoracic vertebrae and is illustrated in figure 21.1.

If you were to open the chest cavity, the first structure you would see is the **parietal pericardium.** This tough, outer connective tissue sheath encloses the heart and has two layers: a tough, outer connective tissue sheath called the **fibrous layer** and an inner layer called the **serous layer.** Deep to the parietal pericardium is the **pericardial cavity,** which contains a small amount of **serous pericardial fluid.** This fluid reduces the friction between the outer surface of the heart and the parietal pericardium. The heart wall itself has an outer layer known as the **epicardium,** or **visceral pericardium.** Locate these pericardial layers in figure 21.1.

Heart Wall

The heart wall is composed of three major layers. The outermost layer is the epicardium, or visceral pericardium, composed of epithelial and connective tissue. The middle layer, the **myocardium,** is the thickest of the three layers. It is mostly made of cardiac muscle. You may wish to review the slides of involuntary cardiac muscle and note the intercalated discs, branching fibers, and fine

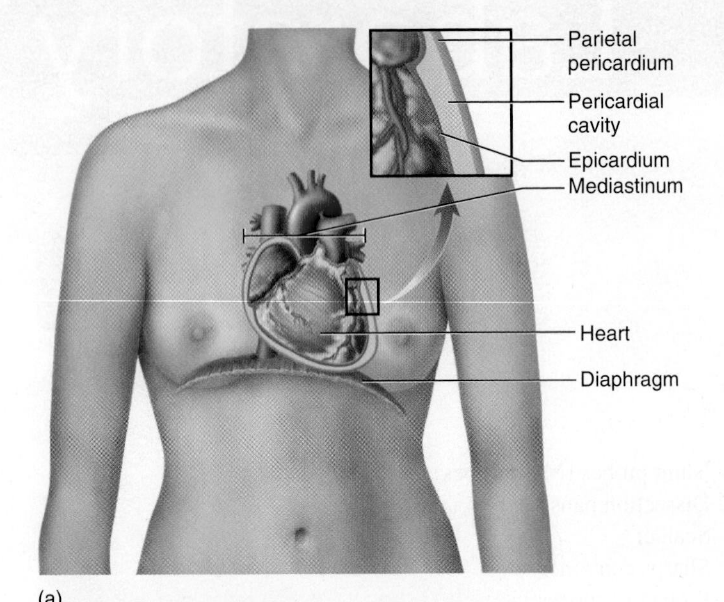

(a)

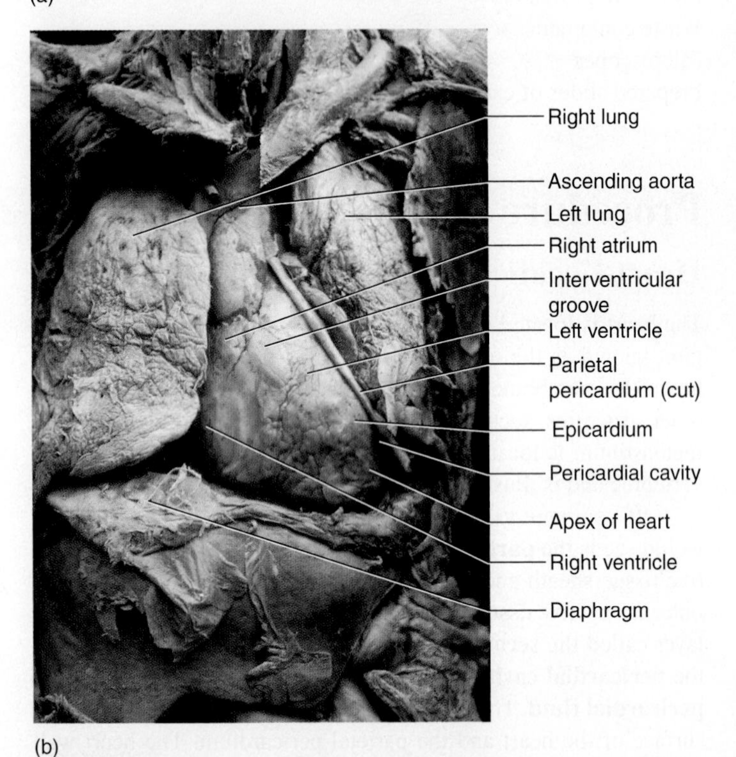

(b)

FIGURE 21.1 Heart in Thoracic Cavity and Heart Coverings (a) Coronal section; (b) photograph of cadaver.

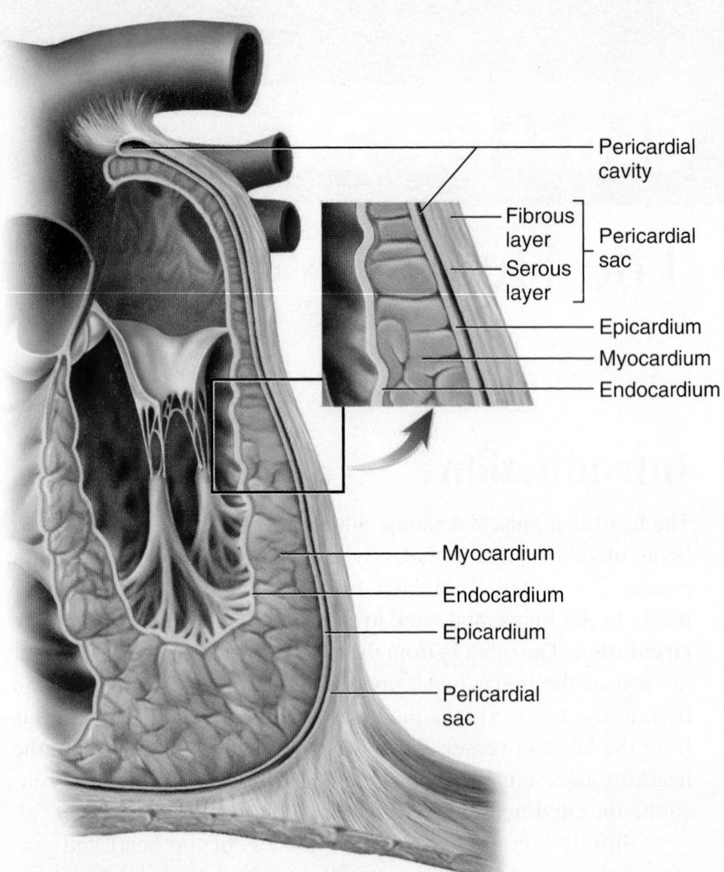

FIGURE 21.2 Pericardial Layers and Heart Wall

striations of cardiac tissue (as described in laboratory exercise 4). The cardiac muscle is arranged spirally around the heart, and this arrangement provides a more efficient wringing motion to the heart. The inner layer of the heart wall is known as the **endocardium,** which is a serous membrane consisting of **endothelium** (simple squamous epithelium) and connective tissue. These layers are seen in figure 21.2.

It is best to examine heart models before dissecting a sheep heart unless your instructor directs you to do otherwise. Heart models are color-coordinated and labeled to make the structures easier to locate. In traditional anatomical models and charts, the pulmonary trunk and pulmonary arteries are colored blue to indicate that they carry deoxygenated blood.

Examination of the Heart Model

Overview The heart has four chambers, two superior atria (AY-tree-uh) and two inferior ventricles. Blood with low oxygen levels enters the heart in the right atrium (fig. 21.3) and flows into the right ventricle. Once blood is in the right ventricle, the contraction of the ventricular wall sends blood to the lungs. The blood is oxygenated in the lungs and returns to the heart by entering the left atrium. Blood moves from the left atrium to the left ventricle and is then pumped from there to the rest of the body.

Interior of the Heart Examine a model of the interior of the heart, and locate the right and left ventricles. Note that the wall of the **right ventricle** is thinner than the wall of the left ventricle. This is so because blood from the right ventricle is pumped a short distance to the lungs while blood in the left ventricle is pumped more extensively through the body. The ventricles are separated by the **interventricular septum,** which forms a wall between the two ventricular chambers. Compare the model with figure 21.3.

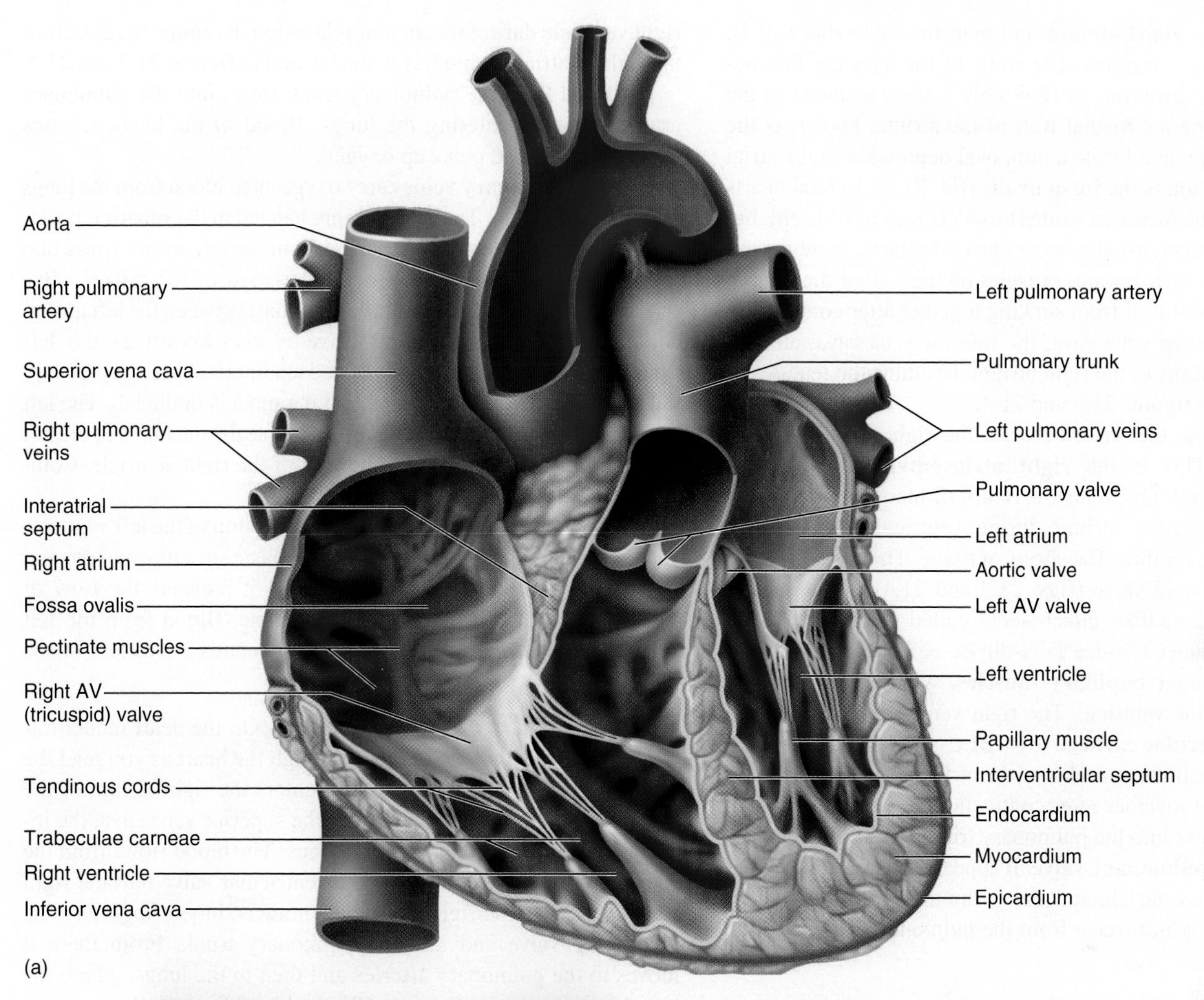

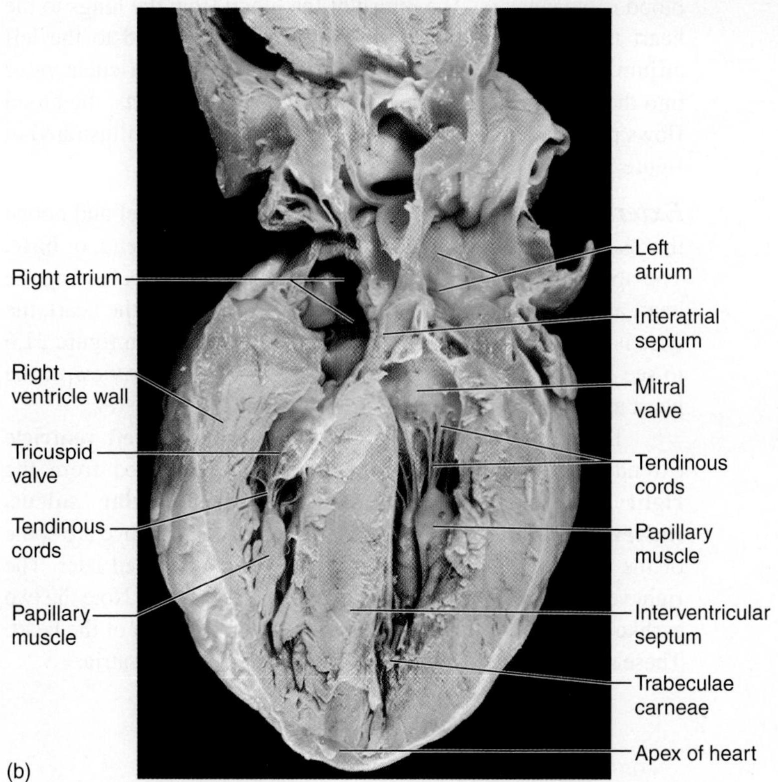

FIGURE 21.3 **Heart, Frontal Section** (a) Diagram; (b) photograph.

Examine the **right atrium** and note how thin the wall is, compared with the ventricles. The walls of the atria are thin because blood in the atria has to flow only a short distance to the ventricles. Examine the medial wall of the atrium, known as the **interatrial septum,** and locate a thin, oval depression in the atrial wall. This depression is the **fossa ovalis** (fig. 21.4). In fetal hearts this is the site of the **foramen ovale** (for-AYE-men o-VAL-eh), but closure of the foramen usually occurs just after birth. Note the extensive **pectinate** (PEK-tin-ate) **muscles** on the wall of the atrium. These keep the atrial wall from sticking together after contraction. Blood in the superior vena cava, the inferior vena cava, and the coronary sinus returns to the right atrium. Examine the features of the right atrium in figures 21.3 and 21.4.

Now examine the valve between the right atrium and the right ventricle. This is the **right atrioventricular valve,** or **tricuspid valve,** and it prevents the return of blood from the right ventricle into the right atrium during ventricular contraction. Examine the valve for three flat sheets of tissue. These are the three **cusps** of the tricuspid valve (figs. 21.3 and 21.4). The tricuspid valve has thin, threadlike attachments called **tendinous cords** (**chordae tendineae;** COR-dee TEN-din-ee-ee). These tough cords are attached to larger **papillary muscles,** which are extensions from the wall of the ventricle. The right ventricle wall has small struts called **trabeculae carneae** (trah-BEC-you-lee CAR-nee-ee), which, like the pectinate muscles of the atria, keep the ventricular wall from sticking together after contraction. The blood from the right ventricle flows into the pulmonary trunk toward the lungs.

Locate the **pulmonary valve.** It appears as three small cusps between the right ventricle and the pulmonary trunk and keeps blood from flowing in reverse from the pulmonary trunk into the right ventricle during ventricular relaxation. Examine the details of the right ventricle in models in the lab and in figures 21.3 and 21.4.

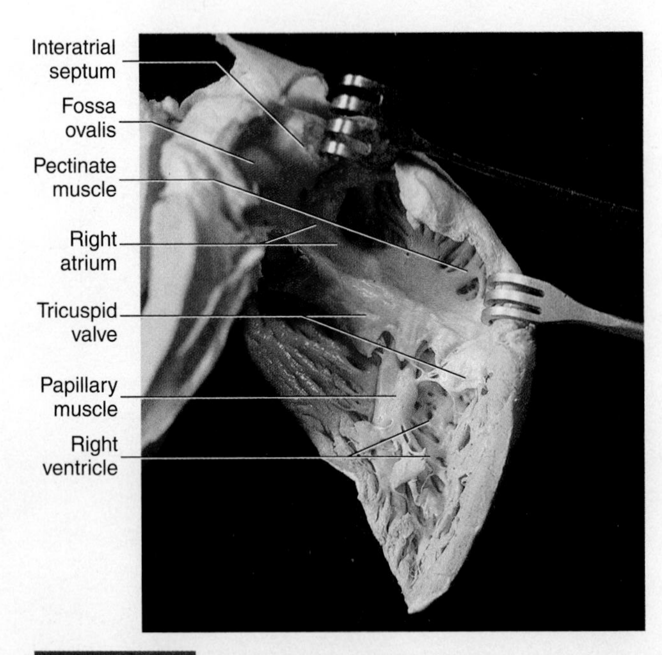

FIGURE 21.4 **Details of the Right Atrium**

Blood from the pulmonary trunk flows into the pulmonary arteries prior to entering the lungs. Blood in the lungs releases carbon dioxide and picks up oxygen.

The **pulmonary veins** carry oxygenated blood from the lungs into the **left atrium.** These vessels are located in the superior, posterior portion of the left atrium. Blood from the left atrium flows into the left ventricle. Locate the two large cusps of the **mitral valve** (named for the shape of a bishop's miter hat) between the left atrium and left ventricle. The mitral valve is also known as the **left atrioventricular valve.** It has attached tendinous cords and papillary muscles, which you should locate in the models in the lab. The left ventricle has trabeculae carneae as well. Note the thickness of the left ventricle wall, compared with the wall of the right ventricle. Compare the left side of the heart to figure 21.3.

The **aortic valve** is located at the junction of the left ventricle and the ascending aorta. It has the same basic structure and general function as the pulmonary valve in that it prevents the flow of blood from the aorta into the left ventricle. Blood from the left ventricle moves into the aorta and subsequently to the rest of the body.

Blood Flow Through the Heart On the heart model follow the pathway that blood takes through the heart as you read the remainder of this paragraph. Blood enters the right atrium of the heart from three sources. These are the superior vena cava, the inferior vena cava, and the coronary sinus. The blood flows from the right atrium through the right atrioventricular valve into the right ventricle. When the right ventricle contracts, blood flows past the pulmonary valve and into the pulmonary trunk. From there it moves to the pulmonary arteries and then to the lungs, where the blood is oxygenated. The return of the blood from the lungs to the heart is by the pulmonary veins, which carry blood to the left atrium. Blood then flows through the left atrioventricular valve into the left ventricle. When the left ventricle contracts, the blood flows past the aortic valve and into the aorta. This is illustrated in figure 21.5.

Exterior of the Heart Examine the heart model and notice that the heart has a pointed end, or **apex,** and a blunt end, or **base.** The apex of the heart is inferior, and the great vessels leaving the heart are located at the base (therefore, in the case of the heart, the base is superior to the apex). Compare the model with figure 21.6 to see how the **aorta** curves to the left in an anterior view of the heart and is posterior to the **pulmonary trunk.**

Locate the anterior features of the heart. The **left ventricle** extends to the apex of the heart and is delineated from the **right ventricle** by the **anterior interventricular sulcus, (interventricular groove).** In this interventricular groove are some of the **coronary arteries** and **cardiac veins,** discussed later. The right ventricle occupies less area than the left ventricle. Note the two earlike flaps that occur on the anterior, superior region of the heart. These structures are the **auricles,** which are part of the atria.

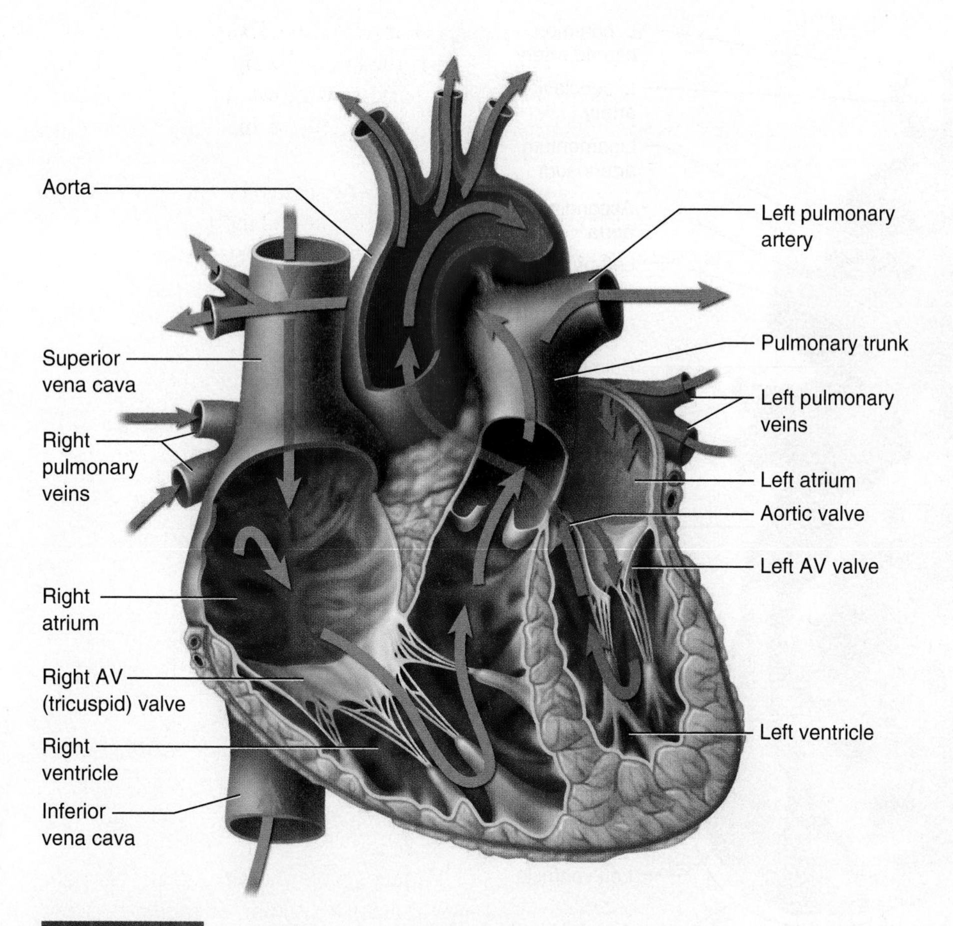

FIGURE 21.5 **Blood Flow Through the Heart** Purple arrows indicate deoxygenated blood; orange arrows, oxygenated blood.

If you examine the heart from the posterior side, you will see the atria more clearly. At the junction of the **right atrium** and the right ventricle is the **atrioventricular sulcus,** or **groove.** The **coronary sinus,** a large venous chamber that carries blood from the cardiac veins to the right atrium, is located in this sulcus. Locate the **superior vena cava** and the **inferior vena cava,** two vessels that also return blood to the right atrium. Locate the **pulmonary veins,** which carry blood from the lungs to the left atrium. Compare the heart model with figure 21.7.

The major vessels of the heart are illustrated in figures 21.6 to 21.9. Locate the **pulmonary trunk, pulmonary arteries, ligamentum arteriosum** (what remains of the fetal blood vessel that shunts blood between the pulmonary trunk and aortic arch), **ascending aorta, pulmonary veins, superior vena cava, inferior vena cava, coronary arteries,** and **cardiac veins**.

The heart tissue is nourished by coronary arteries. The **left coronary artery** arises from the **ascending aorta** and then branches into the **anterior interventricular artery** (left anterior descending artery) and the **circumflex branch.** The **right coronary artery** also arises from the ascending aorta and branches to form the **posterior interventricular branch** and the **right marginal branch.** These major arteries of the heart supply blood to the myocardium. On the return flow from the heart muscle, the **great cardiac vein** follows the depression of the interventricular groove and the atrioventricular groove to the **coronary sinus.** On the posterior, right side of the heart, the **middle cardiac vein** leads to the coronary sinus, which empties into the right atrium. Locate these vessels on the external surface of the model of the heart and compare them with figure 21.9.

Dissection of the Sheep Heart

Caution Be careful when handling preserved materials. Ask your instructor for the proper procedure for working with preserving fluid and for handling and disposal of the specimen. Do not dispose of animal material in the sinks. Place it in an appropriate waste container.

The sheep heart is similar to the human heart and usually is readily available as a dissection specimen. Dissection of anatomical material is valuable in that you can examine structures that are represented more accurately in preserved material than in models. Also, the preserved material has greater flexibility and is

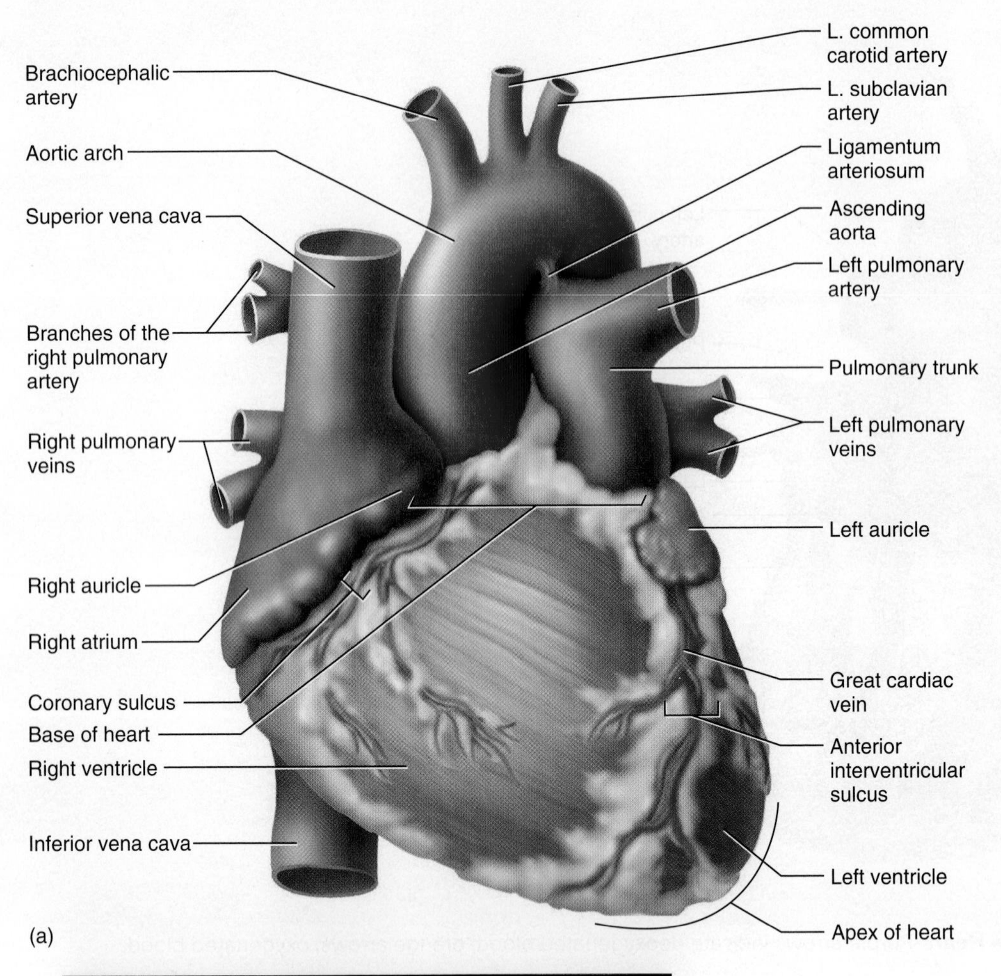

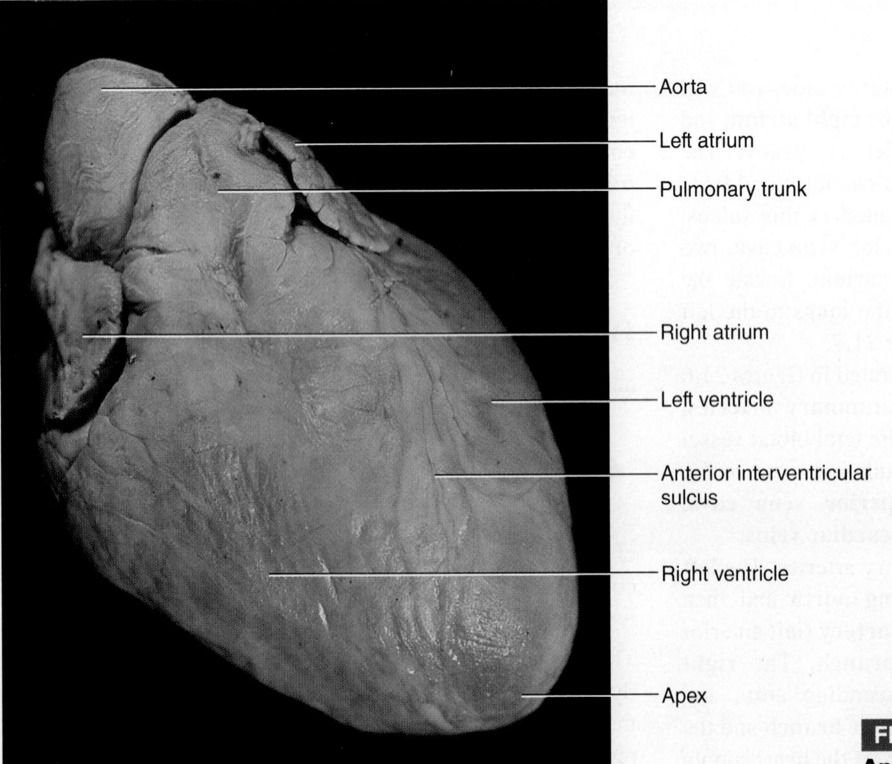

FIGURE 21.6 Surface Anatomy of the Heart, Anterior View (a) Diagram; (b) photograph.

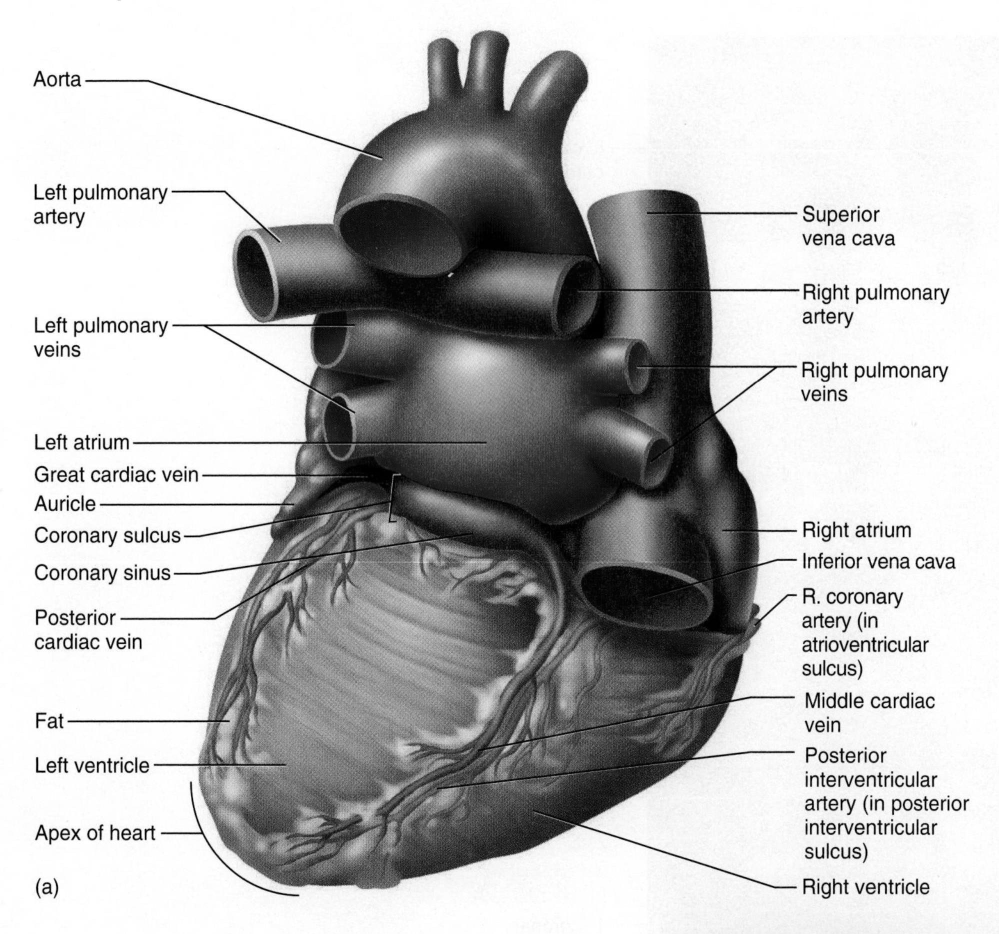

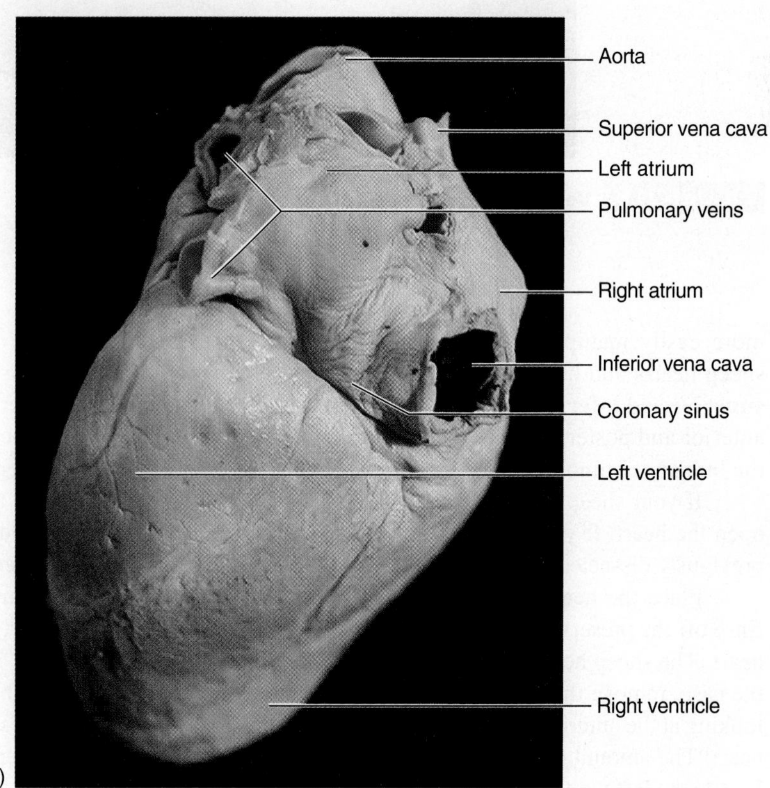

FIGURE 21.7 Surface Anatomy of the Heart, Posterior View (a) Diagram; (b) photograph.

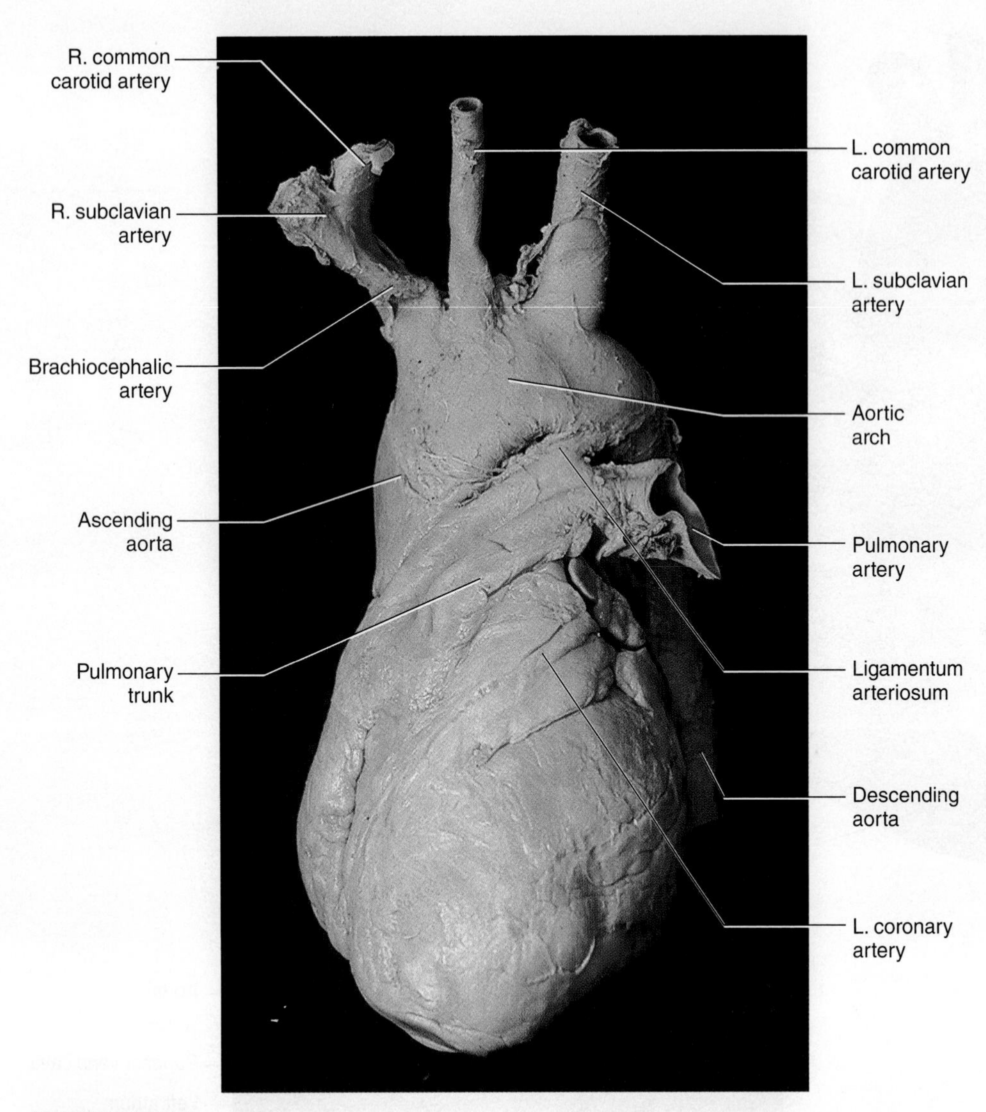

FIGURE 21.8 **Vessels of the Heart, Anterior View**

more easily manipulated. There are some differences between sheep hearts and human hearts, especially in the position of the **superior** and **inferior venae cavae**. In sheep, these are called the anterior and posterior venae cavae, but they are referred to using the human terminology.

If your sheep heart has not been dissected, you will need to open the heart. If your sheep or other mammalian heart has been previously dissected, you can skip the next paragraph.

Place the heart under running water for a few moments to rinse off the preserving fluid. Examine the external features of the heart. The sheep heart may still be in the pericardial sac. If this is the case, remove the sac before proceeding. Determine if you are looking at the anterior or posterior surface. Note the fat layer on the heart. The amount of fat on the human or sheep heart is variable. Locate the **left ventricle,** the **right ventricle,** the **interventricular sulcus,** the **right atrium,** and the **left atrium.** Note the **auricles** that extend from the anterior surface of the atria. Carefully remove the adipose tissue from the major vessels of the heart.

Using a sharp scalpel, make an incision along the right side of the heart (lateral side) from the apex of the heart to the lateral side of the right atrium. If you are unsure about how to proceed during any part of the dissection, ask your instructor for directions. Make another long cut from the lateral side of the left atrium through the lateral side of the left ventricle. You will have made a frontal section of the heart if you cut through the **interventricular septum.** Once you have opened the heart, compare the structures of the sheep heart with the human heart in figure 21.3.

Locate the vessels of the heart by inserting a blunt metal probe into the vessels and determining which chamber the vessel goes to or comes from. Place the heart in anatomical position and insert the probe into the large, anterior vessel that exits toward the specimen's left side. The blunt end of the probe should enter into the **right ventricle.** Be careful and do not tear the heart valves. The vessel you have placed the probe into is the **pulmonary trunk.** The pulmonary trunk may still have the **pulmonary arteries** attached.

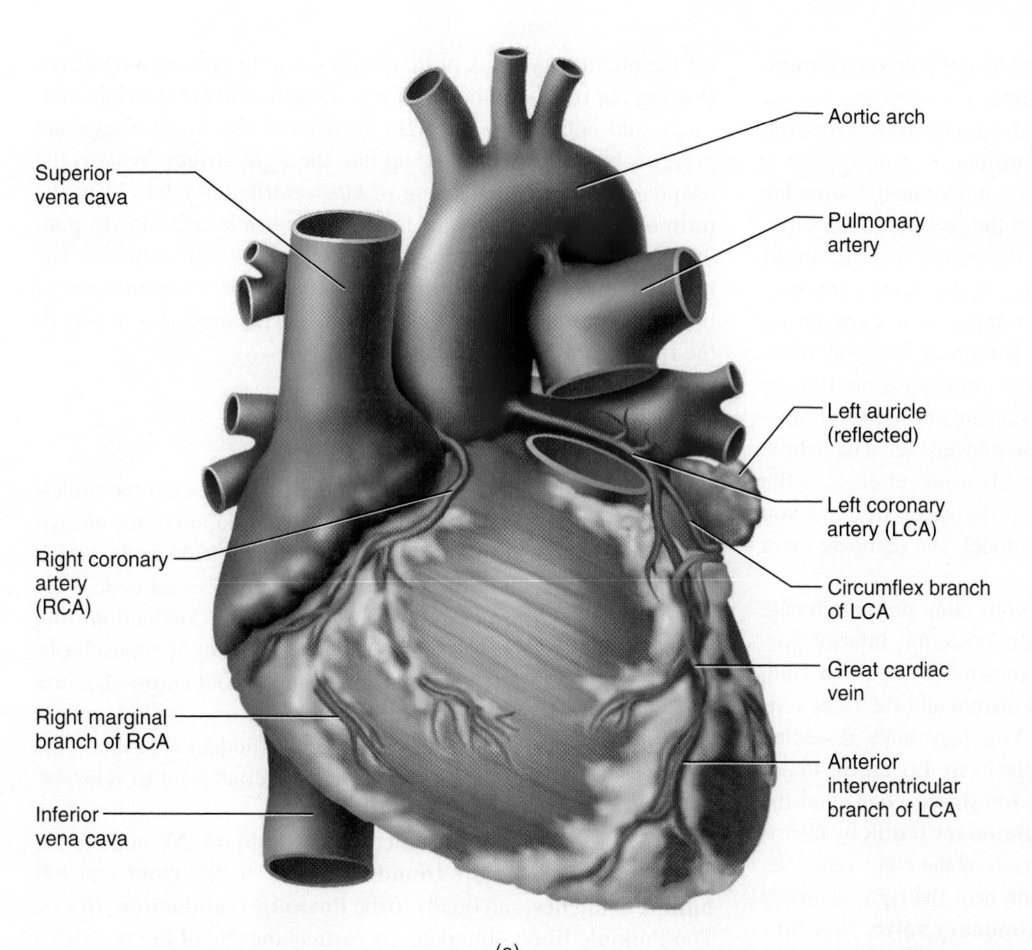

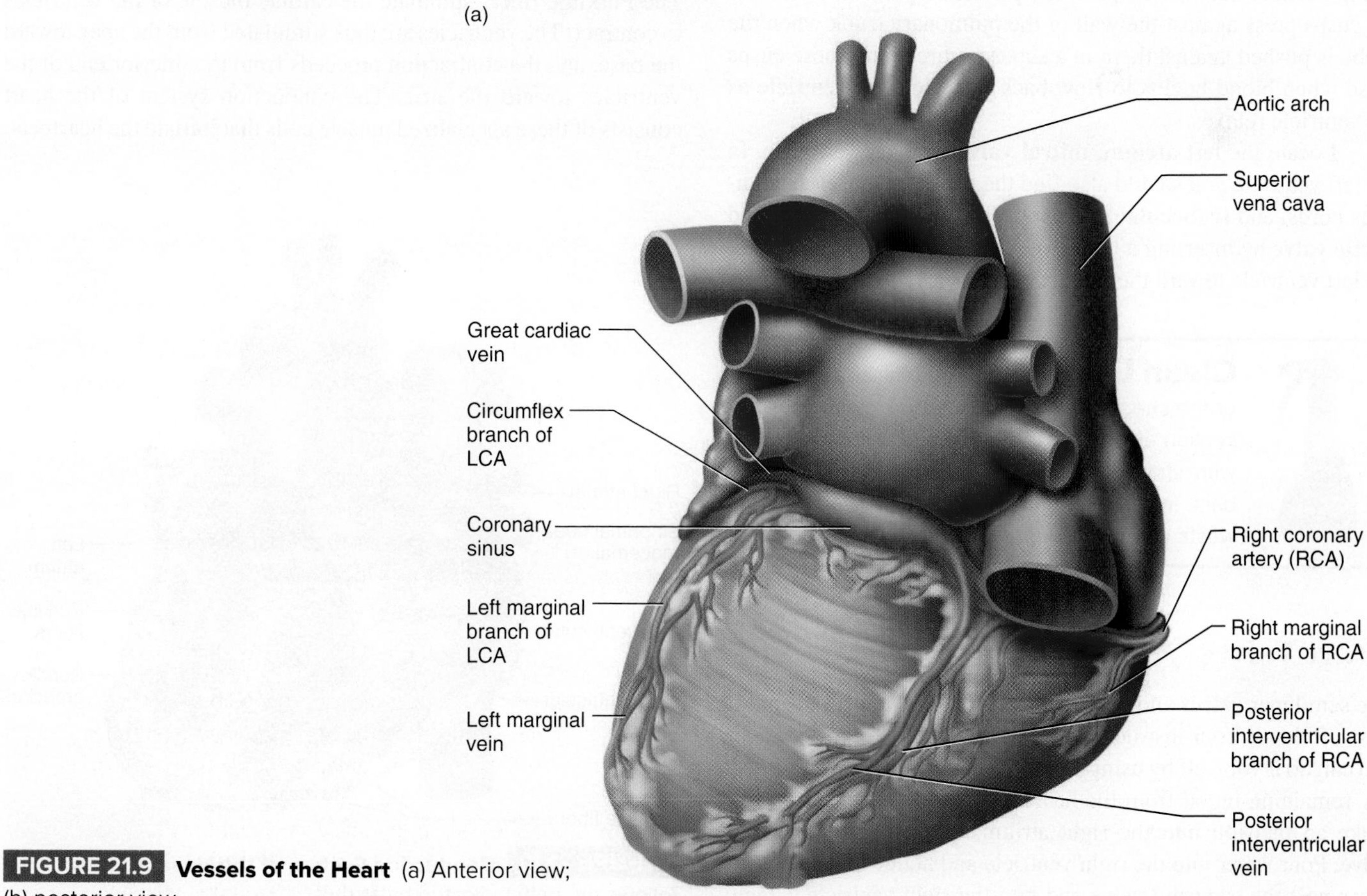

FIGURE 21.9 **Vessels of the Heart** (a) Anterior view; (b) posterior view.

Locate the large vessel directly posterior to the pulmonary trunk (see fig. 21.6). This is the **ascending aorta.** If the vessels are cut farther away from the heart, you can see the **aortic arch.** Insert the probe into this vessel and into the left ventricle.

Turn the heart to the posterior surface and locate the **superior vena cava** and **inferior vena cava.** Insert the probe into the superior and inferior vena cavae, pushing the probe into the **right atrium.** If you find only one large opening in the atrium, you may have cut through either the superior or inferior vena cava during your initial dissection. The probe can be felt through the wall more easily here than in a ventricle because the atrial walls are thinner than those of the ventricles. On the left side either the **pulmonary veins** appear as four separate veins, or you may just see a large hole on each side of the left atrium if the vessels were cut close to the atrial wall. Locate the same structures in the sheep heart as you found on the posterior side of the heart model, and compare them with figure 21.7.

Cut into the right atrium and use your blunt probe to locate the opening of the **coronary sinus** in the posterior, inferior portion of the atrium. It is small and somewhat difficult to find. Examine the opening between the right atrium and the right ventricle to locate the **tricuspid valve.** You may have dissected through one of the cusps as you opened the heart. Locate the major features of the right ventricle. Find the **tendinous cords** and the **papillary muscles.** You can find the **pulmonary trunk** by inserting a blunt probe into the superior portion of the right ventricle. Make an incision in the pulmonary trunk near the right ventricle to expose the three thin cusps of the **pulmonary valve.** Note how the cusps press against the wall of the pulmonary trunk when the probe is pushed against them in a superior direction. These cusps close when blood begins to flow back into the right ventricle as the ventricle relaxes.

Locate the **left atrium, mitral valve,** and **left ventricle.** In the left ventricle you should also find the papillary muscles, tendinous cords, and **trabeculae carneae**. You can find the **aorta** and **aortic valve** by inserting a blunt probe toward the superior end of the left ventricle toward the middle of the heart.

Clean Up When you have finished your study of the sheep heart, make sure you clean your dissection equipment with soap and water. Be careful with sharp blades. Place the sheep heart either back in the preserving fluid or in the appropriate waste container as directed by your instructor.

Heart Valves

The semilunar valves and atrioventricular valves prevent the backflow of blood. Your instructor may demonstrate the procedure, or you can do it yourself by using a fresh or thawed sheep heart. Flush any remaining blood from the heart before you locate the valves. Make an incision into the right atrium, exposing the **tricuspid valve.** Pour water into the right ventricle and notice how the water flows past the tricuspid valve and into the right ventricle. Clamp off the pulmonary trunk or tie it with string to prevent blood from flowing out of the pulmonary trunk. *Gently* squeeze the right ventricle, and notice how the right atrioventricular valve closes and prevents blood from backing up into the right atrium. What is the adaptive value for the closing of atrioventricular valves? Cut the **pulmonary trunk** close to the right ventricle and slowly pour water into it as if you were trying to fill the right ventricle. The **pulmonary valve** should fill with water and close the entrance of the right ventricle, preventing backflow. This normally occurs as the right ventricle begins diastole.

Conduction in the Heart

Locate the structures of the heart's conduction system on models or charts in the lab and in figure 21.10. The initiation of the electrical impulse in the heart begins at the **sinoatrial (SA) node,** which is commonly known as the **pacemaker**. The sinoatrial node is located in the superior portion of the right atrium. Conduction from the sinoatrial node travels across the atria, causing the muscles of the atria to contract. The impulse that spreads out across the atria reaches the **atrioventricular (AV) node.** The impulse has a slight delay (0.1 second) in the node before being conducted farther. This delay allows the atrial cardiac muscle to contract prior to ventricular firing.

The electrical impulse then travels from the AV node to the **atrioventricular bundle (bundle of His),** to the **right** and **left bundle branches,** and finally to the **Purkinje (conduction) fibers.** The Purkinje fibers stimulate the cardiac muscle of the ventricles to contract. The ventricles are thus stimulated from the apex toward the base, and the contraction proceeds from the inferior end of the ventricles toward the atria. The conduction system of the heart consists of these specialized muscle cells that initiate the heartbeat.

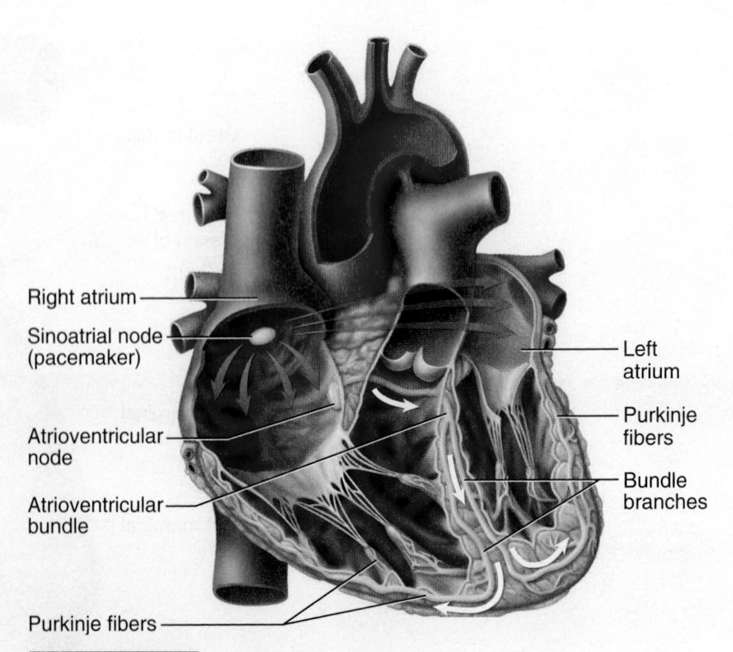

FIGURE 21.10 Conduction System of the Heart Conduction follows the path indicated by arrows.

Exercise 21

The Heart

Name ______________________ Date __________

1. The heart is located between the lungs in an area known as the: ______________________

2. What arrangement of cardiac muscle fibers makes the heart efficient? ______________________

3. What is the innermost layer of the heart wall called? ______________________

4. Which cell type makes up most of the myocardium? ______________________

5. Is the apex of the heart superior or inferior to the rest of the heart? ______________________

6. What is the name of the depression on the anterior surface of the heart that is between the two ventricles?

7. Are auricles extensions of the atria or the ventricles? ______________________

8. What separates the left atrium from the right atrium? ______________________

9. What is the name of the thin spot between the atria? ______________________________

____

10. The mitral valve is located between what two chambers of the heart? ______________________________

____

11. Name the structure between the left atrioventricular valve and the papillary muscle. ______________________________

____

12. What adaptation do you see with the walls of the left ventricle being thicker than those of the right ventricle? ______________________________

____

____

13. What is the function of the aortic valve? ______________________________

____

____

14. What is another name for the tricuspid valve? ______________________________

____

15. What three vessels take blood to the right atrium? ______________________________

____

____

16. The great cardiac vein and the middle cardiac vein lead to what vessel? ______________________________

____

17. What blood vessels nourish the heart tissue? ______________________________

____

____

18. Label the following illustration using the terms provided.

aorta	left ventricle (wall)
apex	mitral valve
interatrial septum	right atrium
interventricular septum	right ventricle (wall)
left atrium	tendinous cords

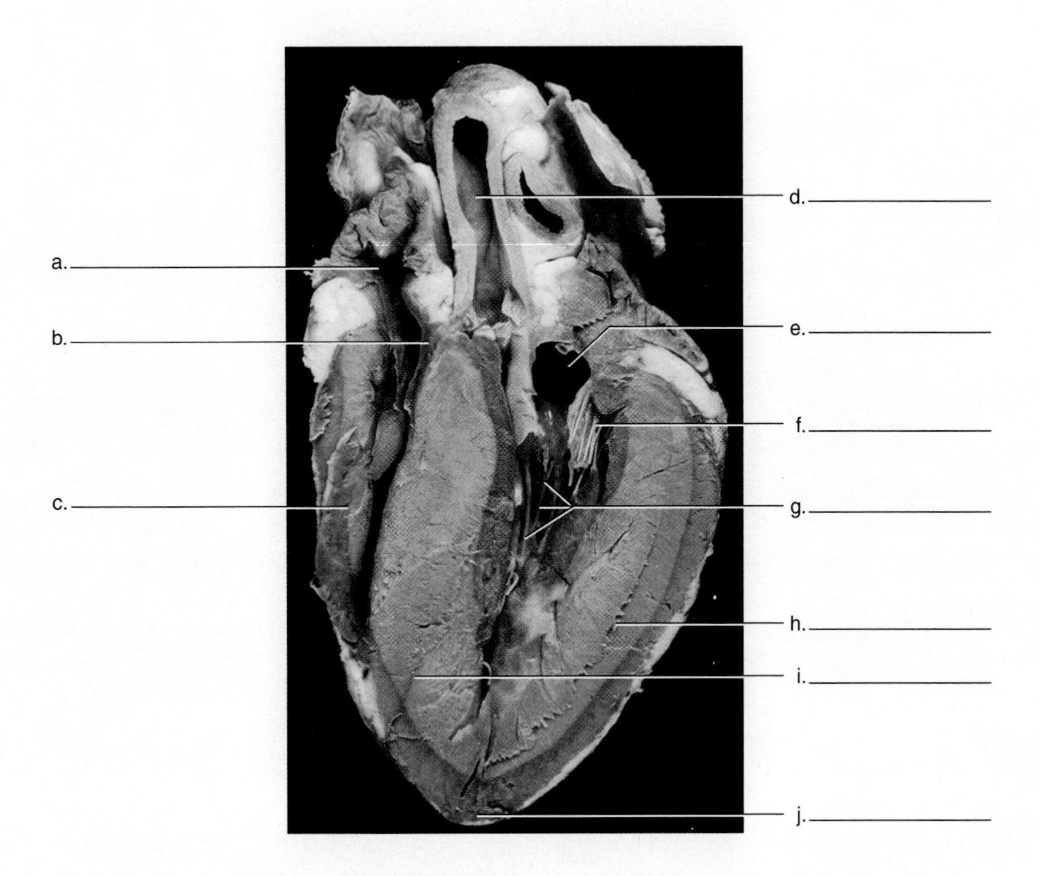

19. The sinoatrial node has a common name. What is it? ____________________

____

20. After the AV node depolarizes, what structures conduct the impulse to the myocardium of the ventricles? ____________________

____

____

Laboratory

Exercise 22

Introduction to Blood Vessels and Blood Vessels 1: Blood Vessels of the Axial Region

Cardiovascular System

Introduction

The blood vessels are the conduits of the cardiovascular system that (1) carry oxygen, nutrients, and other materials to the cells and (2) remove wastes from the tissue fluid near the cells. These vascular conduits consist of numerous vessels, including arteries, arterioles, capillaries, venules, and veins. **Arteries** are defined as blood vessels that carry blood *away from* the heart. Most arteries carry oxygenated blood, but there are a few exceptions, such as the pulmonary artery which takes deoxygenated blood to the lungs. Because of the higher blood pressure found in them, arteries have thicker walls than veins. Arteries are frequently named for the region of the body they pass through, such as the *brachial* artery which is found in the region of the arm, or the *femoral* artery located in the thigh region. Arteries are also named for the organs they serve, such as the *renal* artery that takes blood to the kidney or the *splenic* artery.

Arteries become progressively smaller and branch into **arterioles**. Arterioles become even smaller and become **capillaries** where nutrients, water, and oxygen are provided to the cells of the body. The capillaries pick up carbon dioxide and waste from the interstitial fluid surrounding the cells. On the return flow, blood returns via **venules** to **veins** and back to the heart under relatively low pressure. Veins are defined as blood vessels that carry blood toward the heart. Veins resemble tributaries of rivers; smaller veins flow into larger veins just as small creeks flow into rivers. Veins can be superficial (close to the skin) or deep. The deep veins of the body frequently travel alongside the major arteries and take on the arterial names (e.g., the femoral vein, axillary vein, and subclavian vein), while superficial veins have names specific to themselves (e.g., the great saphenous vein). Veins have thinner walls than arteries, and superficial veins contain valves for a one-way flow of blood to the heart. In this exercise you learn the basic structure of blood vessels and study the arteries and veins of the axial region.

Learning Objectives

At the end of this exercise you should be able to

1. draw a cross section of the wall of a generalized blood vessel, showing the three layers;
2. compare the structure of the wall between arteries, veins, and capillaries;
3. differentiate between conducting arteries and distributing arteries;
4. describe the sequence of major arteries that branch from the aortic arch;
5. describe the sequence of major arteries that originate from the abdominal aorta or its derivatives;
6. identify major axial arteries or veins in models or a cadaver;
7. describe the major organs that receive blood from the axial arteries;
8. trace the blood flow from a selected organ back to the heart;
9. distinguish a portal system from normal venous return flow;
10. describe the major digestive organs that supply blood to the hepatic portal system; and
11. name the vessels that take blood to, or receive blood from, a particular axial artery or vein.

Materials

Microscopes
Prepared slides of arteries and veins in cross section
Models of the blood vessels of the body
Charts and illustrations of the arterial and venous system
Cadaver (if available)
Prepared slide of arteriosclerosis

Procedure

Overview of Blood Vessels

Obtain a prepared microscope slide of a cross section of artery and vein. Both arteries and veins have walls that consist of three layers. The outer layer is known as the **tunica externa (adventia)** and consists of a connective tissue sheath. The middle layer, or **tunica media,** is composed of smooth muscle in both arteries and veins. The tunica media is thicker in arteries, and there may be pronounced elastic tissue in the wall of the tunica media in arteries.

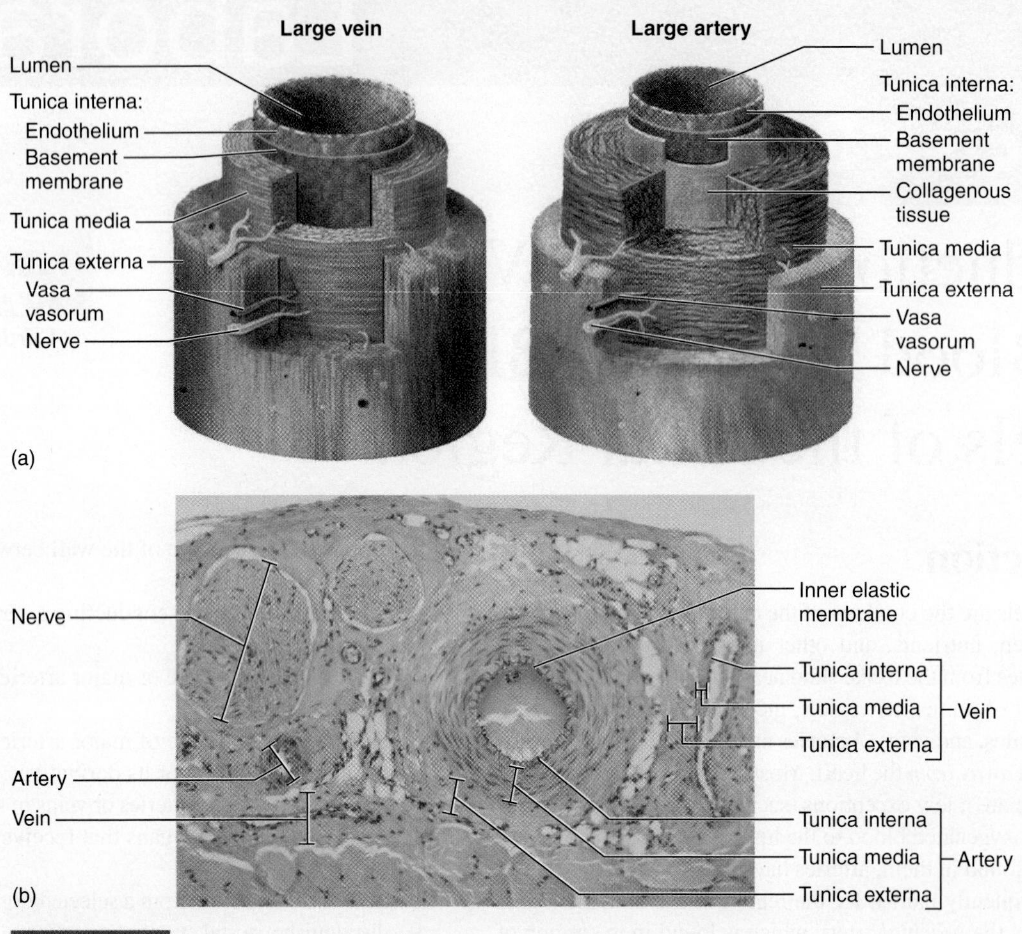

FIGURE 22.1 **Artery and Vein Cross Section** (a) Diagram; (b) photomicrograph of an artery, a vein, and a nerve (40×).

The wall of a blood vessel requires oxygen, which is supplied by blood vessels that feed the tissue from the outside. Blood vessels that nourish other blood vessels are known as the **vasa vasorum.** The innermost layer (near the blood) is the **tunica interna (tunica intima)** and consists of a thin layer of connective tissue and a thin layer of simple squamous epithelium, known as **endothelium.** Endothelium is the layer closest to the blood. In arteries there is an **inner elastic lamina,** a thin layer of elastic tissue. These layers are seen in figure 22.1. Examine a prepared slide of an artery and vein and locate these layers.

Conducting (elastic) arteries are larger arteries close to the heart. Their appearance is made distinctive by the presence of significant amounts of elastic tissue in the tunica media. **Distributing (muscular) arteries** are found farther away from the heart and have more smooth muscle in the tunica media. There is a gradual transition between these two types of arteries.

A feature of veins not found in arteries is valves. You can easily distinguish veins from arteries in a prepared slide where the vessels are cut in cross section. Veins have a large lumen (though the vein is frequently collapsed in prepared sections) and a thin tunica media relative to its overall size. You may also find circular structures in these slides that are solid, not hollow. These are nerves. Compare your observations of your slide with figure 22.1.

Major Arteries and Veins of the Body

Examine the models, charts, and illustrations in the lab, and locate the major arteries of the body. Compare figure 22.2 with the following list of some of the major arteries of the body. Place a check mark next to the name of the artery when you locate it.

_____ Internal carotid artery

_____ Subclavian artery

_____ Axillary artery

_____ Brachial artery

_____ Radial artery

_____ Ulnar artery

_____ Descending aorta

_____ Common iliac artery

_____ Internal iliac artery

_____ External iliac artery

_____ Femoral artery

_____ Anterior tibial artery

Examine the models, charts, and illustrations in the lab and locate the major veins of the body. An overview of the major veins of the body is shown in figure 22.3. Place a check mark next to the

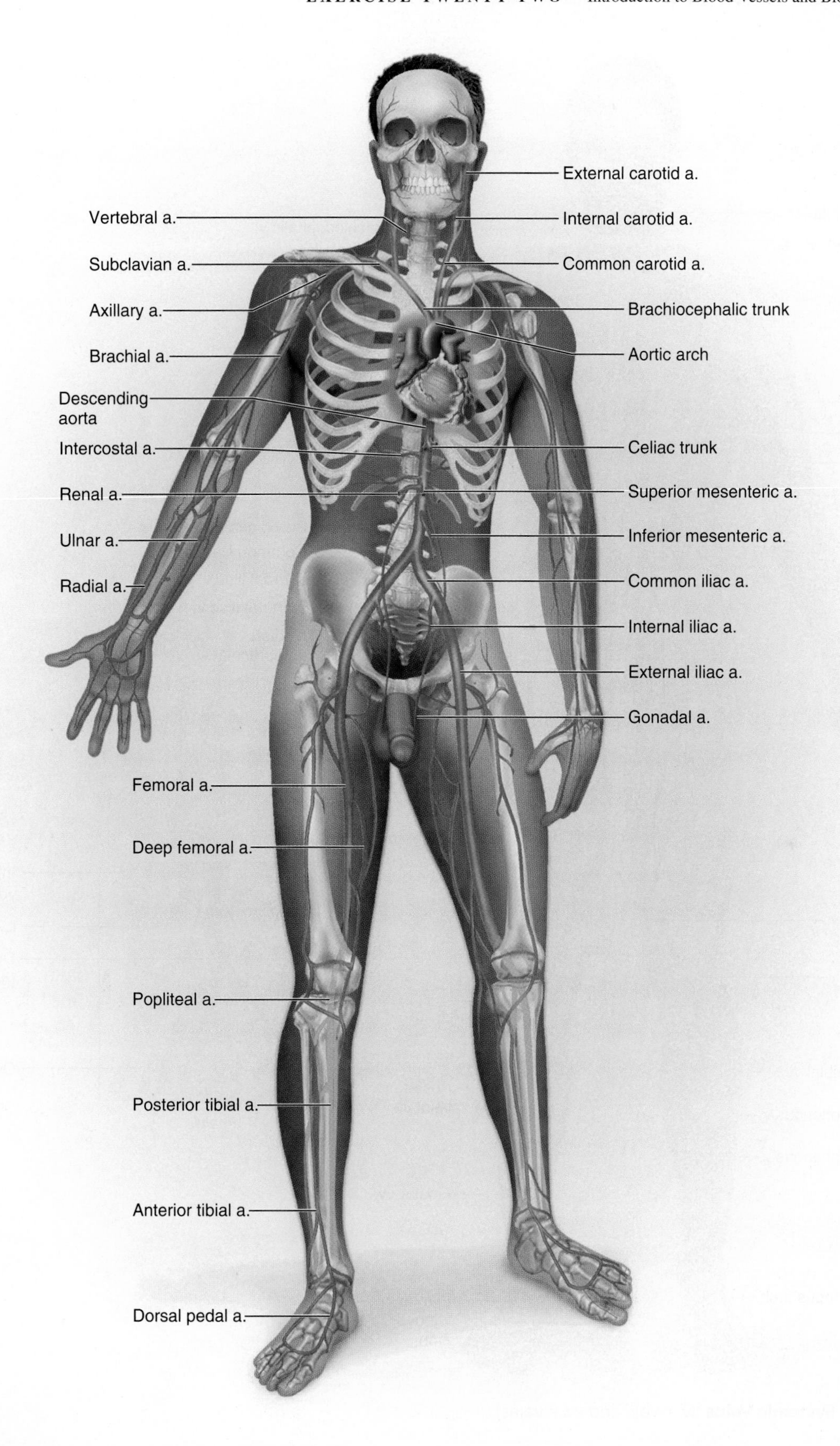

FIGURE 22.2 **Major Arteries of the Body** (a. = artery)

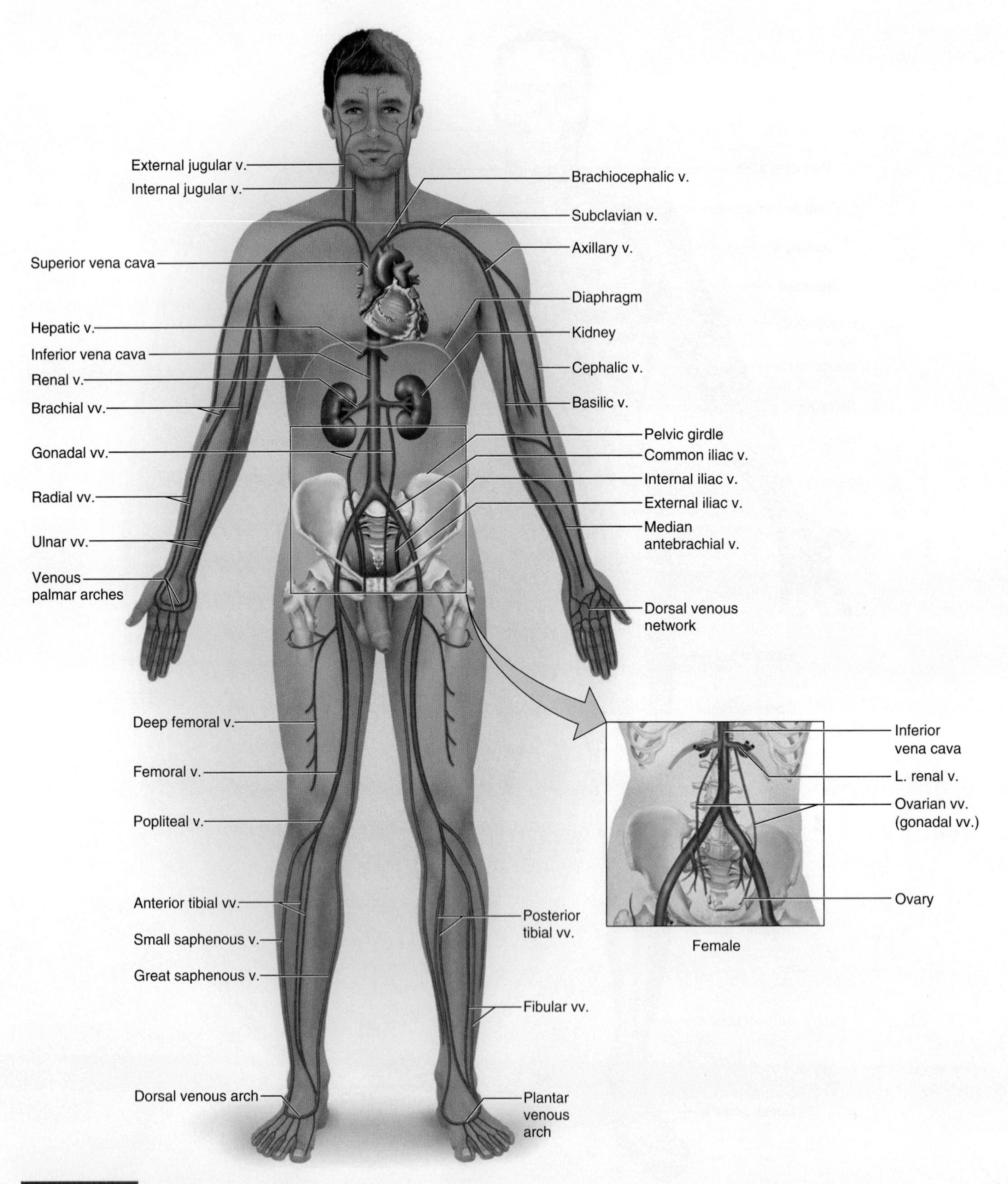

FIGURE 22.3 **Major Systemic Veins** (v. = vein and vv. = veins)

name of the vein in the following list when you locate it on figure 22.3 and in charts or models in the lab.

_____ Internal jugular vein

_____ External jugular vein

_____ Brachiocephalic vein

_____ Superior vena cava

_____ Axillary vein

_____ Cephalic vein

_____ Basilic vein

_____ Inferior vena cava

_____ Common iliac vein

_____ External iliac vein

_____ Internal iliac vein

_____ Femoral vein

_____ Great saphenous vein

_____ Anterior tibial veins

Pulmonary Circuit

Examine models or charts of blood flow to and from the lungs in the lab and compare them to figure 22.4. Locate the vessels in this circuit as you read the following description. The pulmonary circulation involves the heart pumping blood to the lungs for oxygenation, removing carbon dioxide, and returning blood to the heart.

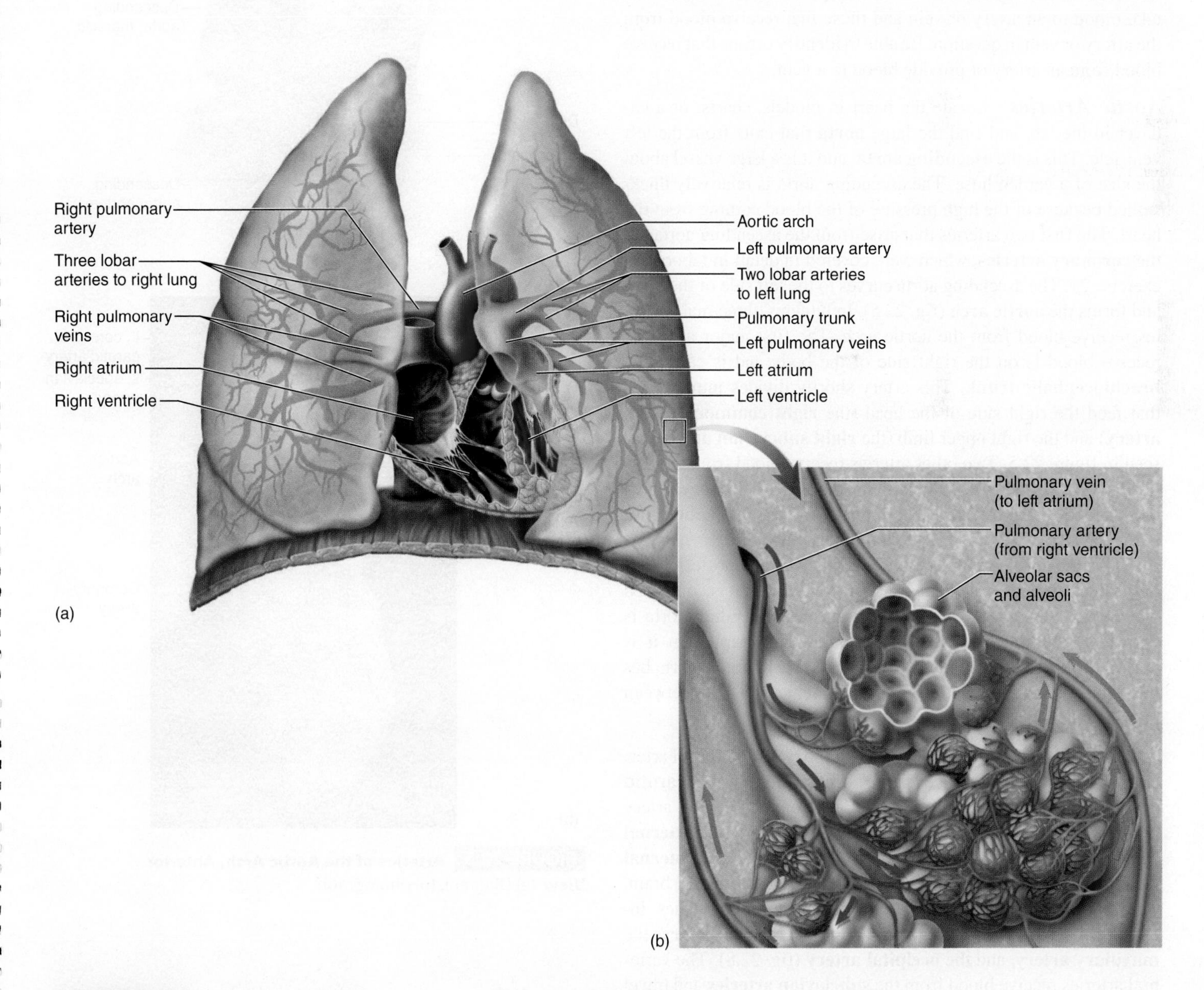

FIGURE 22.4 **Pulmonary Circulation** (a) Overview; (b) details in the lung.

Blood in the **right ventricle** exits the heart by the **pulmonary trunk.** The blood enters the left and right **pulmonary arteries.** These arteries are unusual in that they carry deoxygenated blood. Most arteries carry oxygenated blood. Anatomical models usually color these vessels blue to indicate that they carry deoxygenated blood. From the pulmonary arteries, blood enters the **lobar arteries** of the lungs and eventually flows to the capillaries, where the blood is oxygenated and carbon dioxide is released. The return flow from the lungs eventually enters the **pulmonary veins.** There are two pulmonary veins from each lung, and these enter the heart at the **left atrium.**

Systemic Vessels of the Axial Region

For the remainder of the lab exercise you should find the arteries and veins of the regions that are described. Name the vessels that take blood to an artery or vein and those that receive blood from the artery or vein in question. Be able to identify organs that receive blood from an artery or provide blood to a vein.

Aortic Arteries Locate the heart in models, charts, or a cadaver in the lab, and find the large **aorta** that exits from the left ventricle. This is the **ascending aorta,** and it is a large vessel about the size of a garden hose. The ascending aorta is relatively thick-walled because of the high pressure of the blood coming from the heart. The first two arteries that arise from the ascending aorta are the **coronary arteries,** which were covered in detail in laboratory exercise 21. The ascending aorta curves to the left side of the body and forms the **aortic arch** (fig. 22.5). In humans, three main arteries receive blood from the aortic arch. The first major artery to receive blood is on the right side of the body and is called the **brachiocephalic trunk.** This artery shortly divides into arteries that feed the right side of the head (the **right common carotid artery**) and the right upper limb (the **right subclavian artery**), as seen in figure 22.5. Two other arteries receive blood from the aortic arch. On the left side is the **left common carotid artery,** which takes blood to the left side of the head, and the **left subclavian artery,** which takes blood to the left upper limb.

As the aortic arch turns inferiorly behind the posterior part of the heart, it becomes the **descending aorta,** which is composed of two segments. Above the diaphragm the descending aorta is known as the **thoracic aorta,** and below the diaphragm it is known as the **abdominal aorta** (fig. 22.6). The thoracic aorta has numerous branches, called **intercostal arteries,** that run between the ribs (fig. 22.7).

Arteries of the Head and Neck The two main sets of arteries that travel through the neck to the head are the **common carotid arteries** and the **vertebral arteries.** Each common carotid artery branches just below the angle of the mandible to form the **external carotid artery,** which takes blood to the face, and the **internal carotid artery,** which passes through the carotid canal to the brain. The external carotid artery on each side has several branches, including the **facial artery,** the **superficial temporal artery,** the **maxillary artery,** and the **occipital artery** (fig. 22.8). The vertebral arteries receive blood from the **subclavian arteries** and travel through the transverse foramina of the cervical vertebrae and then

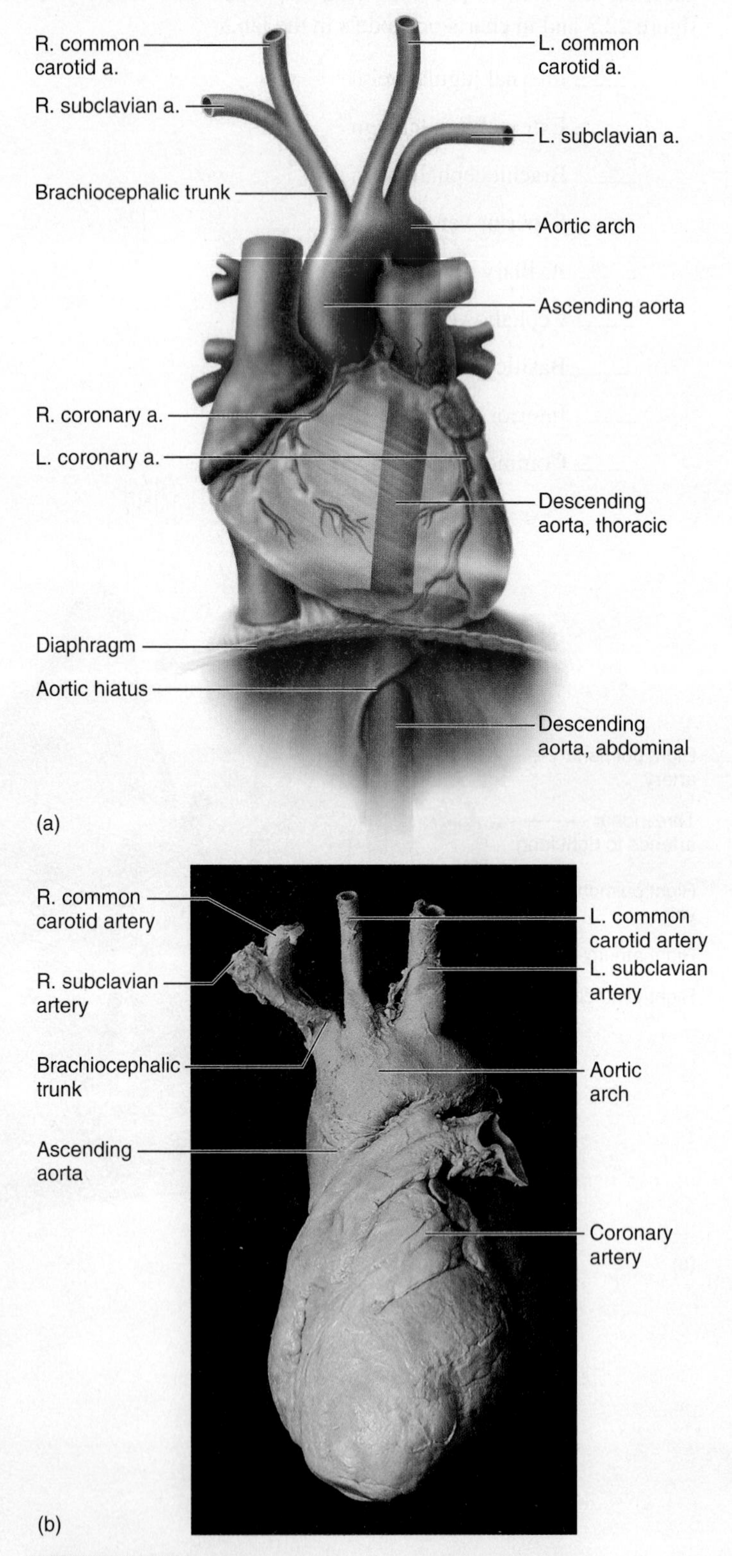

FIGURE 22.5 Arteries of the Aortic Arch, Anterior View (a) Diagram; (b) photograph.

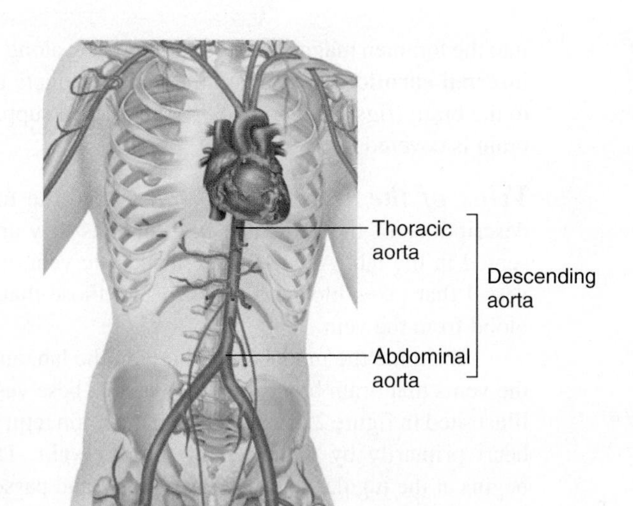

FIGURE 22.6 **Thoracic and Abdominal Aorta**

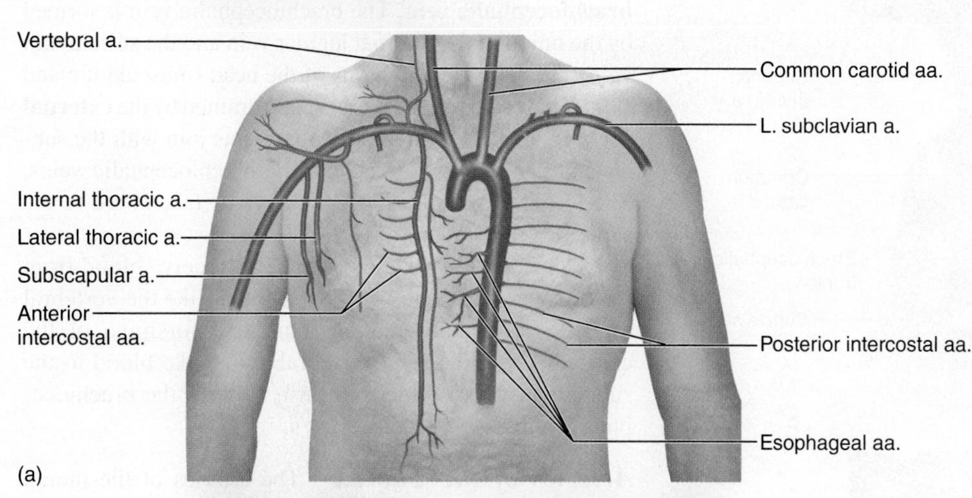

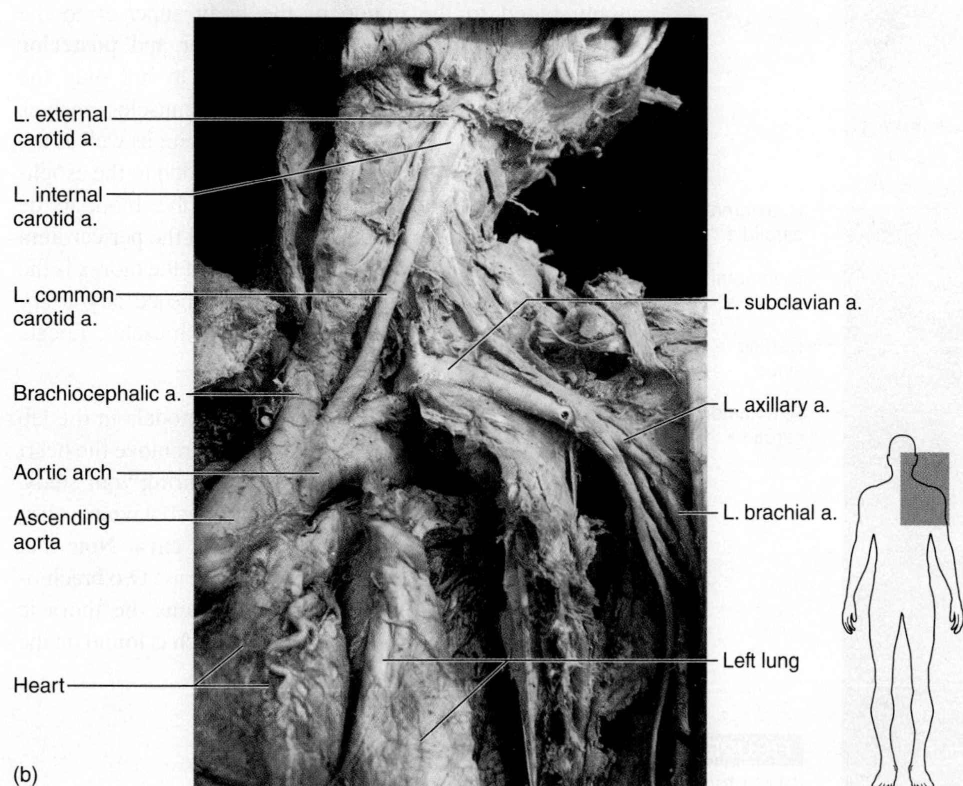

FIGURE 22.7 **Branches of the Subclavian Artery** (a) Diagram; (b) photograph of cadaver.

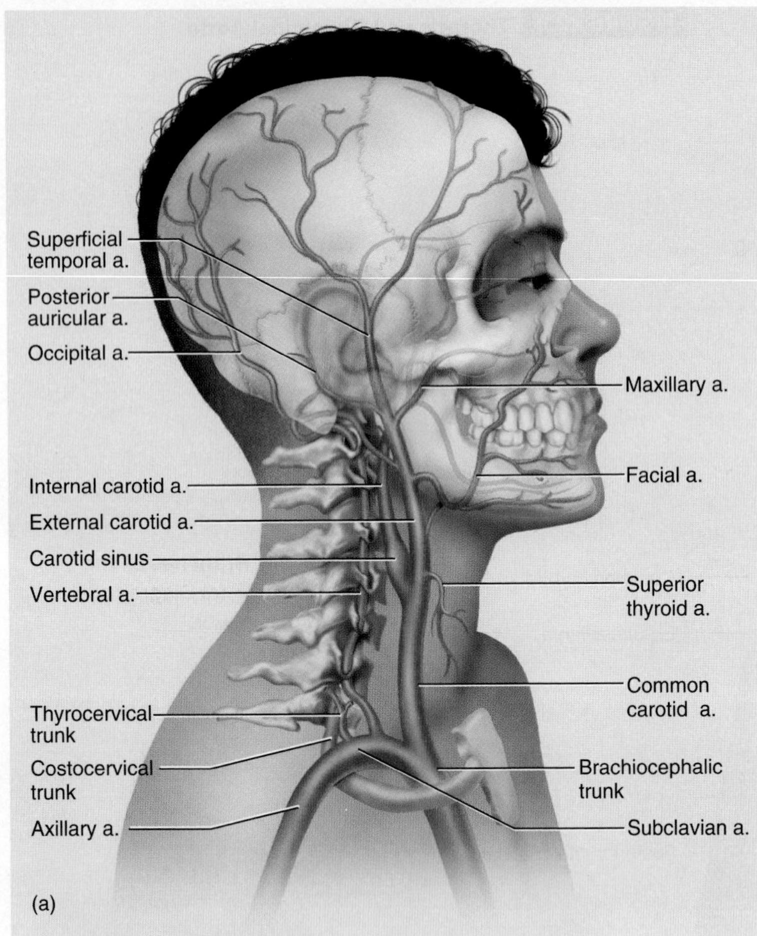

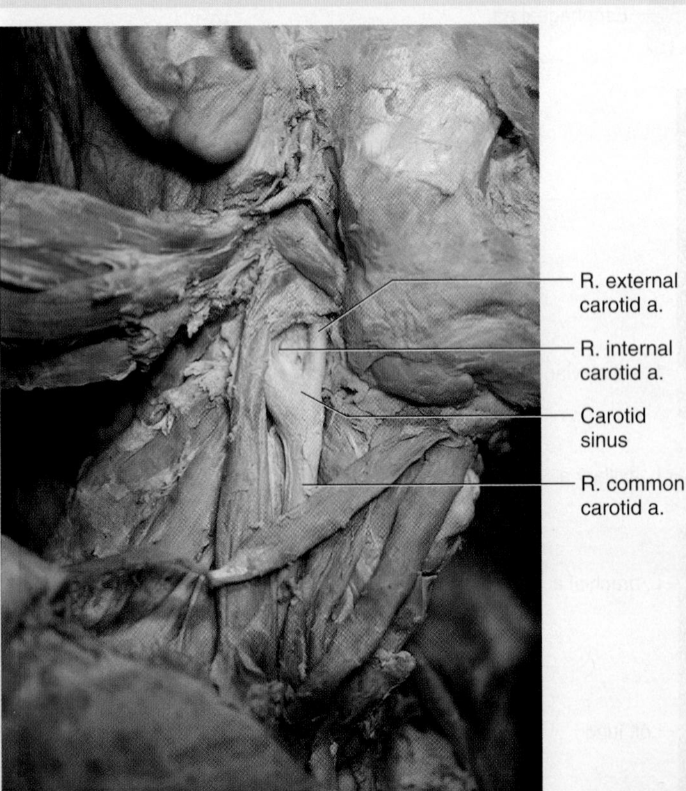

FIGURE 22.8 Arteries of the Head (a) Diagram; (b) photograph of a cadaver.

into the foramen magnum of the skull. These, along with the **internal carotid arteries,** are the main suppliers of blood to the brain (figs. 22.2 and 22.8). The arterial supply to the brain is covered in Exercise 17.

Veins of the Head and Neck Read the following descriptions of the veins and find them as they are represented in the lab. As you locate a specific vein, name the vessel that takes blood to the vein and those that receive blood from the vein.

Examine the models and charts in the lab, and locate the veins that drain blood from the head. These vessels are illustrated in figure 22.9. Blood from the brain returns to the heart primarily by the **internal jugular vein.** This vein begins at the jugular foramen of the skull and passes along the lateral aspect of the neck as it moves toward the **brachiocephalic vein.** The brachiocephalic vein is formed by the union of the internal jugular vein and the **subclavian vein.** The superficial regions of the head (musculature and skin of the scalp and face) are mostly drained by the **external jugular vein.** The external jugular veins join with the subclavian veins prior to reaching the brachiocephalic veins. The left and right brachiocephalic veins take blood to the superior vena cava.

The **vertebral veins** (fig. 22.9) receive blood from the muscles and bones of the neck and, like the vertebral arteries, travel through the transverse foramina of the cervical vertebrae. The vertebral veins take blood to the subclavian veins, which, in turn, flow to the brachiocephalic veins.

Arteries of the Thorax The arteries of the thorax supply blood to the region of the body superior to the diaphragm. Among these are the **anterior** and **posterior intercostal arteries** which supply blood to not only the intercostal muscles but also some thoracic muscles, such as the pectoralis muscles, and the serratus anterior as well as the vertebrae. The **esophageal arteries** send blood to the esophagus, and the **internal thoracic artery** supplies blood to the breast tissue, the anterior thoracic wall, and the pericardium and diaphragm. Another significant artery of the thorax is the **subscapular artery,** which branches from the axillary artery and takes blood to the scapula and nearby muscles. Locate these muscles in the lab and in figure 22.7.

Veins of the Thorax If you look at models in the lab for the veins of the thorax, you will have to remove the heart and lungs from the models at some time during your study. The primary veins of the thorax are the subclavian veins, brachiocephalic veins, and **superior vena cava.** Note that there is one brachiocephalic artery but there are two brachiocephalic veins. The primary vein that drains the thoracic organs is the **azygos** (AZ-ih-goss) **vein** which is found on the

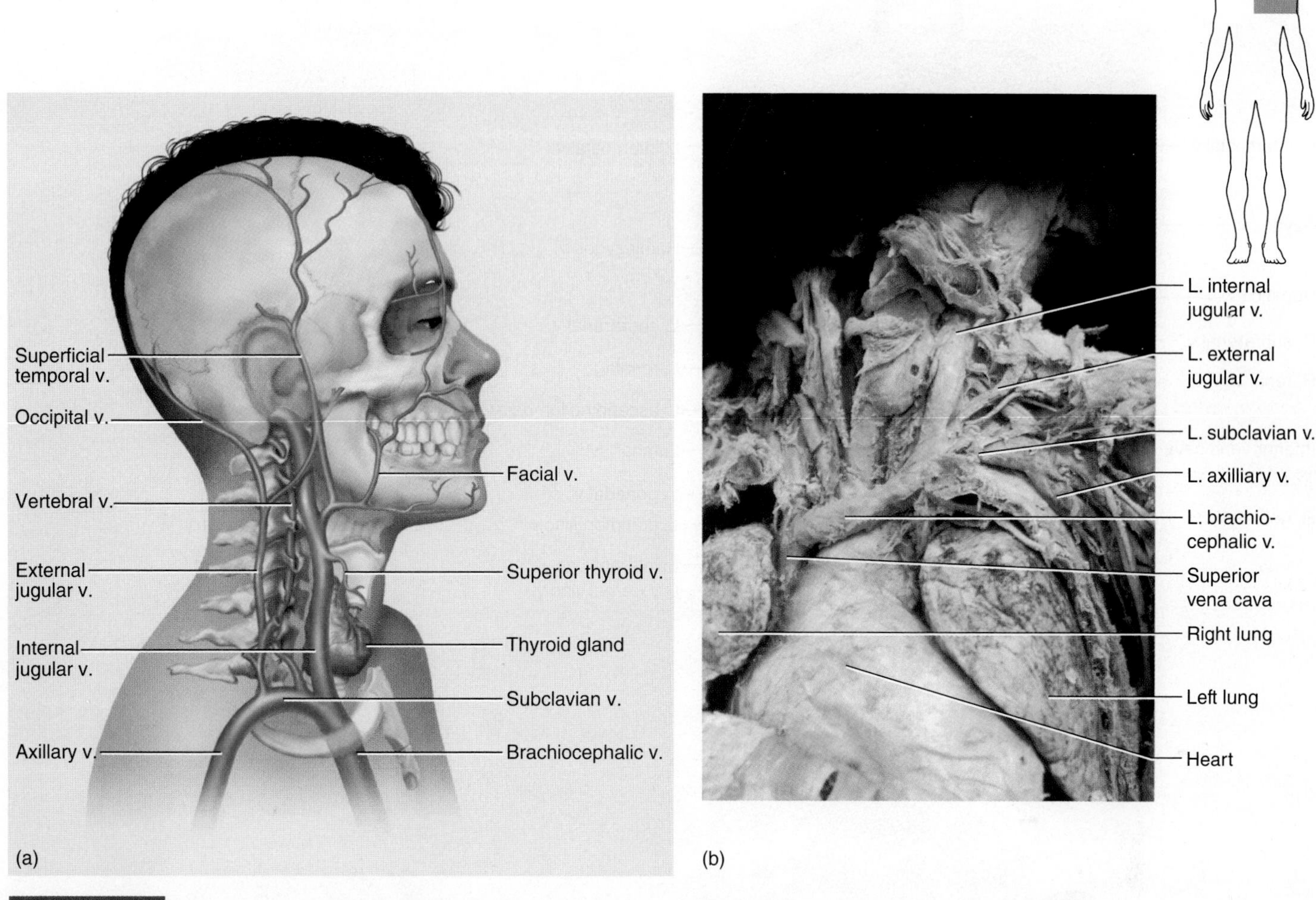

FIGURE 22.9 **Veins of the Head and Neck** (a) Diagram of right side; (b) photograph of left side of cadaver.

right, posterior thoracic wall. The **hemiazygos vein** is smaller and is found on the left, posterior thoracic wall. It drains into the azygos vein which takes blood to the superior vena cava. Examine figure 22.10 for these veins.

Arteries of the Abdomen and Pelvis The **abdominal aorta** is the portion of the **descending aorta** inferior to the diaphragm. It passes through a hole in the diaphragm known as the **aortic hiatus.** The first major branch of the abdominal aorta is the **celiac trunk** (also known as the **celiac artery**). The celiac trunk splits into three separate arteries: the **splenic artery,** taking blood to the spleen, pancreas, and part of the stomach; the **left gastric artery,** taking blood to the stomach and esophagus; and the **common hepatic artery,** taking blood to the liver, stomach, duodenum, and pancreas. Locate the celiac trunk and the branches of the celiac in figure 22.11.

Just below the celiac trunk is the **superior mesenteric artery.** This vessel takes blood from the abdominal aorta and continues through the mesentery until it reaches the small intestine and proximal portions of the large intestine, including the cecum, the ascending colon, and part of the transverse colon. The major branches of the superior mesenteric artery are the **ileal** and **jejunal arteries,** the **ileocolic artery,** and the right and middle **colic arteries.** These are illustrated in figure 22.12.

The next vessels to branch from the aorta are the paired **suprarenal arteries,** which take blood to the adrenal glands. Inferior to the suprarenal arteries are the **left** and **right renal arteries.** These arteries take blood to the kidneys. Locate these vessels in the lab and in figure 22.13.

The **gonadal arteries** branch inferior to the renal arteries and descend to either the testes or the ovaries. The **inferior mesenteric artery** is the next vessel to branch from the aorta, and it takes blood to the lower portion of the large intestine, including part of the transverse colon, the descending colon, the sigmoid colon, and the rectum. These can be located in figures 22.12. and 22.13.

The abdominal aorta terminates by dividing into the two **common iliac arteries.** The common iliac arteries take blood to the internal and **external iliac arteries.** The **internal iliac artery** takes blood to the pelvic region, including branches that feed the rectum, pelvic floor, external genitalia, groin muscles, hip muscles, uterus, ovary, and vagina. These arteries are illustrated in figure 22.13.

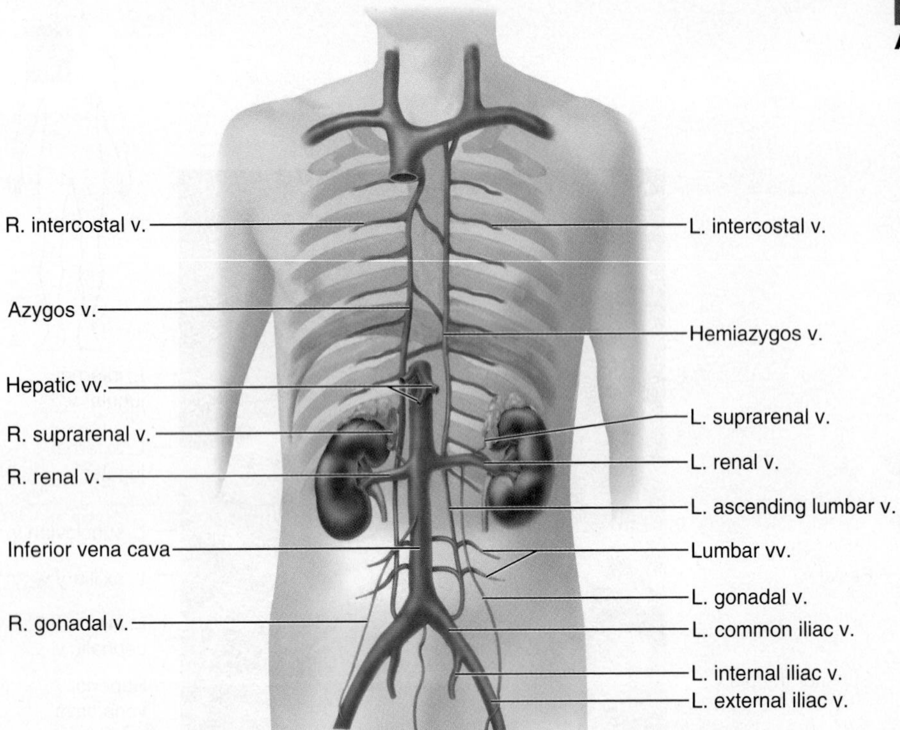

FIGURE 22.10 Veins of the Thorax, Abdomen, and Pelvis

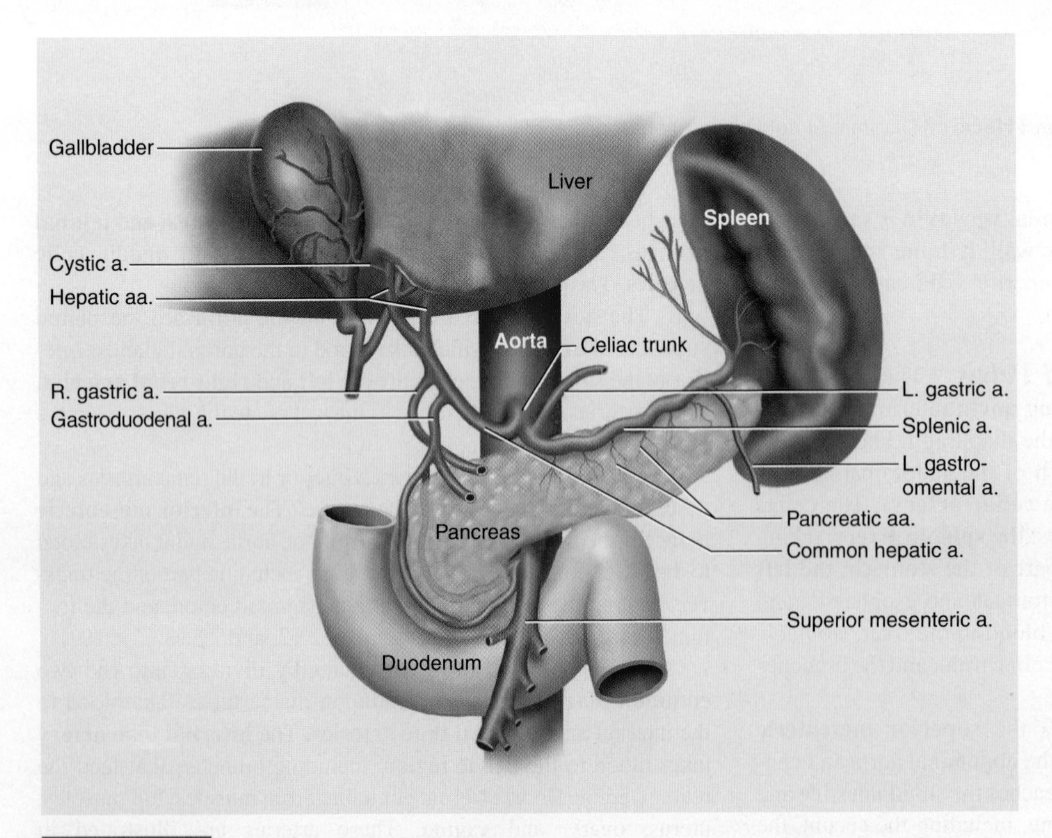

FIGURE 22.11 Celiac Trunk and Its Branches

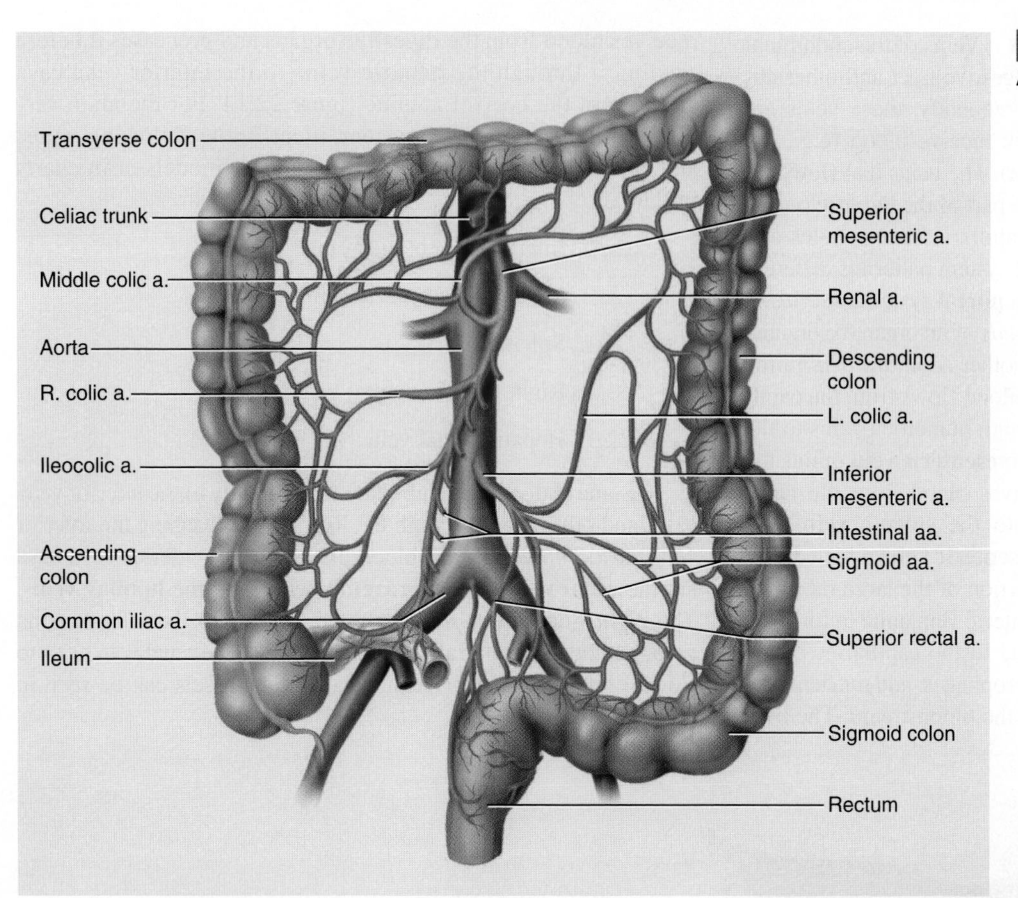

FIGURE 22.12 Middle Abdominal Arteries

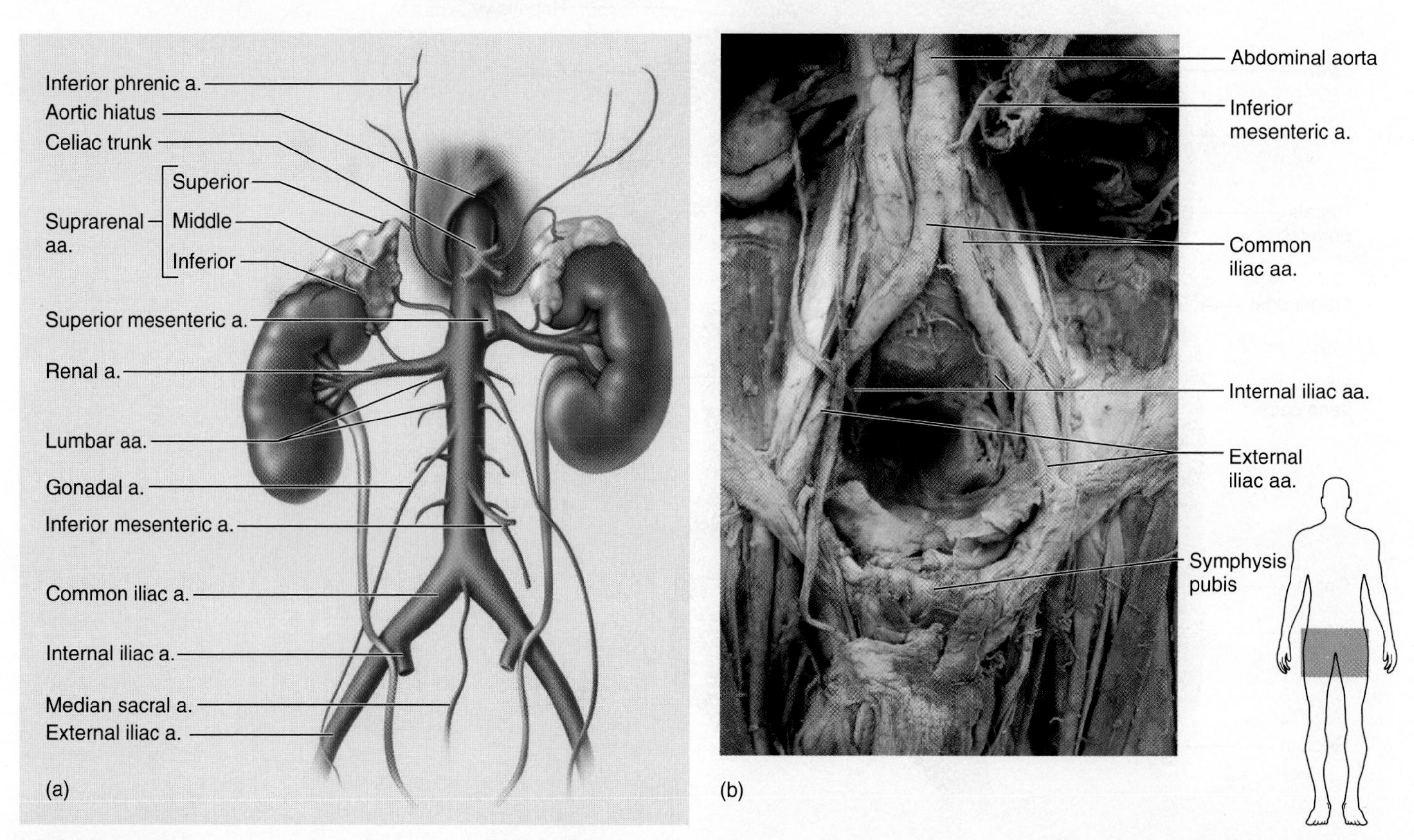

FIGURE 22.13 Abdominal Arteries (a) Diagram of abdominal arteries; (b) photograph of lower abdominal and pelvic arteries in human cadaver.

Veins of the Abdomen and Pelvis Veins of the abdominal region drain the major organs of the digestive tract and other abdominal organs, such as the spleen. Frequently, these veins are named for the organs from which they receive blood (e.g., the splenic vein takes blood from the spleen). The veins that flow into the liver before returning to the heart are part of the **hepatic portal system.** Most veins take blood from capillaries and venules and return the blood to the heart. The portal system pattern is different from the normal venous blood flow. In a **portal system,** a series of vessels takes blood from the *capillary beds* of an organ (or organs) through a series of veins and then to another *capillary bed.* In the case of the **hepatic portal system,** the blood flows from the capillary beds of the abdominal organs through numerous veins to the capillary bed of the liver. The **inferior mesenteric vein** drains the distal part of the large intestine, receives blood from the **right gastro-omental vein,** and empties into the **splenic vein.** The splenic vein joins with the **superior mesenteric vein,** which drains the small intestine and the proximal portion of the large intestine. The splenic vein and the superior mesenteric vein unite to form the **hepatic portal vein,** which takes blood to the capillaries of the liver, where the blood is cleaned by macrophages, and nutrients are either stored in the liver or passed into the bloodstream. The liver receives blood from the digestive organs and processes it before sending it through the **hepatic veins** to the inferior vena cava and toward the heart. Examine figure 22.14. For the main vessels of the hepatic portal system, identify the following veins, and check them off when you find them on models or in charts in the lab.

_____ Inferior mesenteric vein

_____ Superior mesenteric vein

_____ Splenic vein

_____ Right gastro-omental vein

_____ Hepatic portal vein

Some abdominal veins take blood directly to the inferior vena cava, and others pass through the liver before reaching the inferior vena cava. Those that take blood directly to the inferior vena cava are the **renal veins,** the **suprarenal veins,** and the **lumbar veins.** The **right gonadal vein** takes blood directly to the inferior vena cava, but the **left gonadal vein** takes blood to the renal vein prior to flowing into the inferior vena cava. These vessels can be seen in figure 22.15.

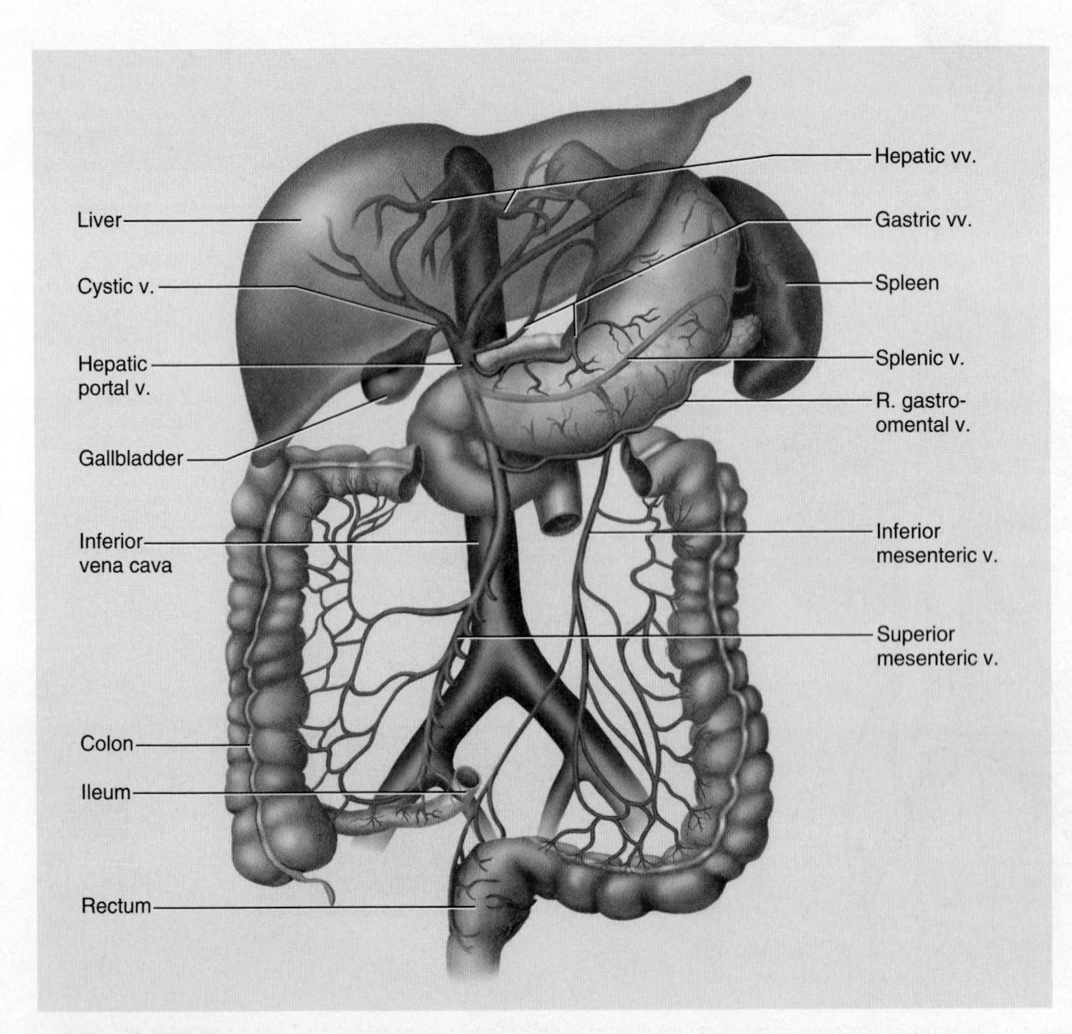

FIGURE 22.14 **Hepatic Portal System**

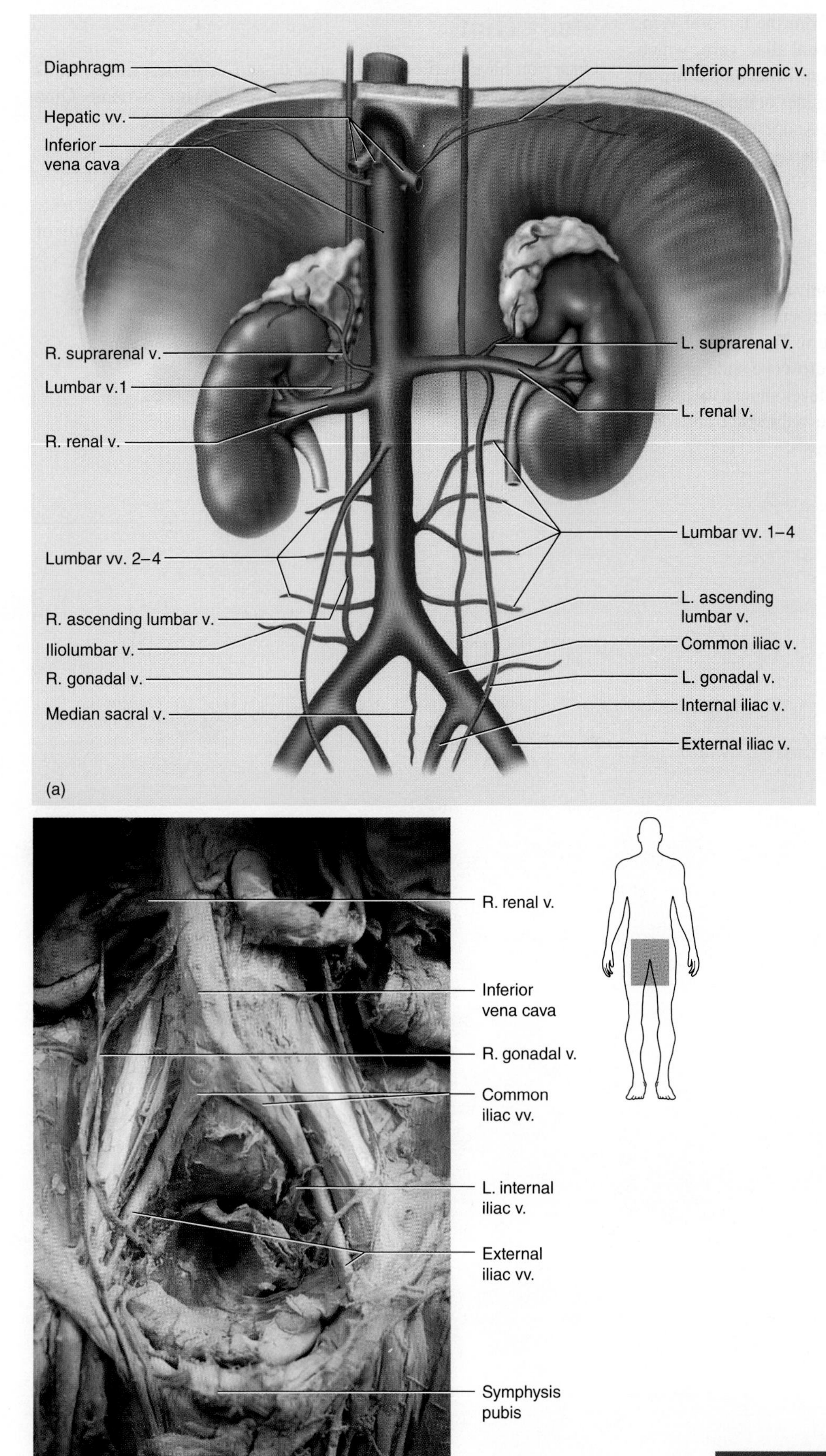

FIGURE 22.15 Veins of the Pelvis (a) Diagram; (b) photograph of cadaver.

The **external iliac vein** takes blood from the femoral vein. The external iliac vein joins with the **internal iliac vein,** which drains the region of the pelvis and their union forms the **common iliac vein.** The common iliac veins from both sides of the body unite and form the **inferior vena cava,** which travels superiorly along the right side of the vertebrae, taking blood to the right atrium of the heart. Locate these veins and compare them with figure 22.15.

Arteriosclerosis

Arteriosclerosis is a condition known commonly as hardening of the arteries. Cholesterol, fatty acids, and other molecules can be deposited in the arterial wall, deep to the endothelium. This can produce a hard region known as **plaque.** Examine a microscope slide or an illustration of arteriosclerosis, and note the development of plaque under the endothelial layer. Draw what you see in the space provided.

Drawing of an artery with arteriosclerosis:

Study Hint

Once you have studied all the arteries and veins described in this exercise, select an artery or vein for your lab partner to name. Quiz each other on the charts or models available in the lab.

Cat Anatomy

If you are using cats, turn to section 6, "Blood Vessel Anatomy of the Cat," on page 425 of this lab manual.

Review

Exercise 22

Introduction to Blood Vessels and Blood Vessels 1: Blood Vessels of the Axial Region

Name ______________________________ Date ______________

1. Label the following illustration with the major arteries of the body. First try to complete the illustration, and then review the material in this exercise to determine your accuracy.

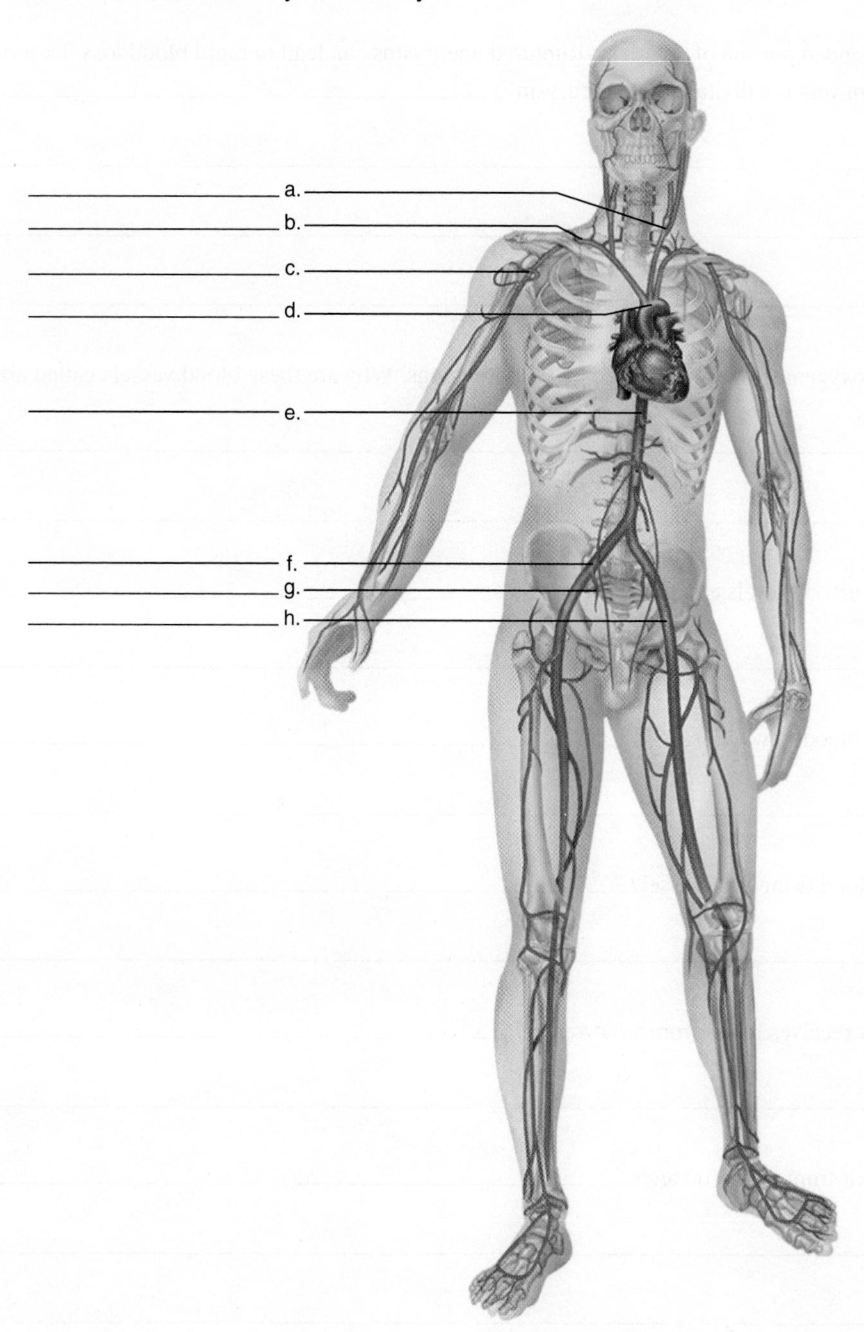

2. Which veins (superficial/deep) have names that do not correlate with arteries? ______

3. What is the name of the outermost layer of a blood vessel? ______

4. What kinds of blood vessels have valves? ______

5. An aneurysm is a weakened, expanded portion of an artery. Ruptured aneurysms can lead to rapid blood loss. Describe the significance of an aortic aneurysm versus a digital artery aneurysm. ______

6. The pulmonary arteries carry deoxygenated blood from the heart to the lungs. Why are these blood vessels called arteries?

7. Blood from the common carotid artery travels to what two vessels next? ______

8. The internal carotid artery takes blood to what organ? ______

9. The descending aorta receives blood from what vessel? ______

10. The right common carotid artery receives blood from what vessel? ______

11. Name three blood vessels that exit from the aortic arch. ______

12. Name the section of the descending aorta inferior to the diaphragm. ______

______

13. Blood from the celiac artery flows into three different blood vessels. What are these vessels? ______

______

______

14. Blood from the superior mesenteric artery feeds which major abdominal organs? ______

______

15. What vessels take blood to the kidneys? ______

______

16. The ovaries or testes receive blood from which arteries? ______

______

17. Blood in the inferior mesenteric artery travels to what organs? ______

______

18. Where does blood in the external iliac artery come from? ______

______

19. The internal jugular vein takes blood from what area? ______

______

20. What veins pass through the transverse vertebral foramina? ______

______

21. The external jugular veins drain what area? ______

______

22. What is the functional nature of a portal system, and how is it different from normal venous return flow?

______

______

______

23. What major vessels take blood to the hepatic portal vein?

24. What is arteriosclerosis?

25. In what part of the arterial wall does cholesterol plaque develop?

Blood Vessels 2: Blood Vessels of the Appendicular Region

Cardiovascular System

Introduction

This exercise is a continuation of the study of blood vessels with an examination of the blood vessels of the upper and lower limbs as well as a study of fetal circulation. Many of the blood vessels of the appendicular region are named for their location, such as the *brachial* artery or the *femoral* vein. Superficial veins generally have names unique to them such as the *basilic vein* and *great saphenous vein.* In this exercise you locate the arteries and veins of the limbs and determine where the blood in that vessel comes from and where it goes.

Learning Objectives

At the end of this exercise you should be able to

1. list the arteries and veins of the upper limbs;
2. list the arteries and veins of the lower limbs; and
3. name the vessels that take blood to or from a particular artery or vein.

Materials

Models of the blood vessels of the body
Charts and illustrations of arterial and venous systems
Cadaver (if available)

Procedure

Arteries of the Upper Limbs If you review the position of the brachiocephalic artery, you will remember that it is a short tube that takes blood from the aortic arch and then branches into the **right subclavian artery** and the **right common carotid artery.** The right subclavian artery has a branch that takes blood to the **right axillary artery.** The **left subclavian artery** comes from the aortic arch and has a branch that takes blood to the **left axillary artery** (fig. 23.1).

The axillary artery on each side of the body takes blood to the **brachial artery.** The brachial artery has an obvious pulse that can be palpated by finding a groove on the medial, distal region of the arm between the biceps brachii and brachialis muscles. This location is the site for the placement of the diaphragm of a stethoscope during blood pressure measurement. The brachial artery supplies blood to the muscles of the arm and the humerus. The brachial artery bifurcates (splits in two) to form the **radial artery** on the lateral side of the antebrachium and the **ulnar artery** on the medial side. These two arteries supply blood to the forearm and hand. The radial artery can be palpated just lateral to the flexor carpi radialis muscle at the wrist, which is a common site for the clinical measurement of the pulse. The radial artery and ulnar artery become united in the palm as a **superficial palmar arch** and a **deep palmar arch.** They join, or anastomose, and send out the small **digital arteries** that supply blood to the fingers (fig. 23.1).

Veins of the Upper Limbs Examine the models and charts in the lab and locate the veins of the upper limbs. The fingers are drained by the small **digital veins,** which lead to the **venous palmar arches.** The major superficial veins of each upper limb are the **basilic vein,** on the anterior, medial side of the forearm and arm, and the **cephalic vein,** on the anterior, lateral side of the forearm and arm. The two vessels have many anastomosing branches (cross-connections) between them. One of the significant anastomosing veins is the **median cubital vein,** which crosses the anterior cubital fossa and is a common site for the withdrawal of blood. Locate these superficial veins in figure 23.2. The deep veins of the forearm are the **radial veins** and the **ulnar veins,** which can be found traveling near the arteries of the same name. The deep veins of the arm are the **brachial veins,** next to the brachial artery in the proximal portion of the arm. The brachial veins are formed by the union of the radial and ulnar veins and merge superiorly with the **basilic vein** to form the short **axillary vein.** The axillary vein connects with the cephalic vein to form the **subclavian vein.** The blood of the upper limbs is carried to the heart by the left and right subclavian veins, which flow to the **brachiocephalic veins,** then to the **superior vena cava,** and finally to the right atrium of the heart. Locate the deep veins of the upper limbs in figure 23.2.

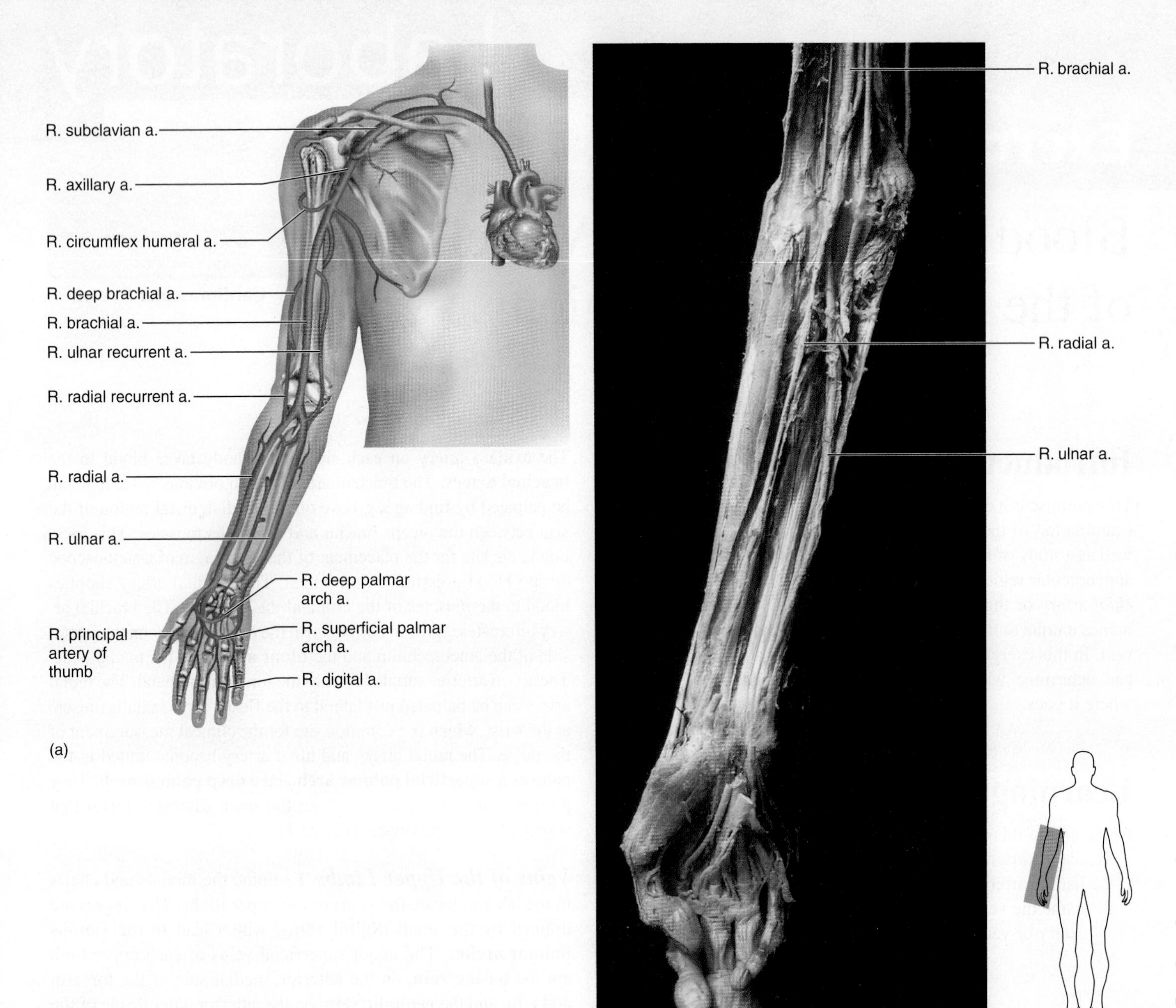

FIGURE 23.1 **Arteries of the Upper Limb** (a) Diagram; (b) photograph of a cadaver.

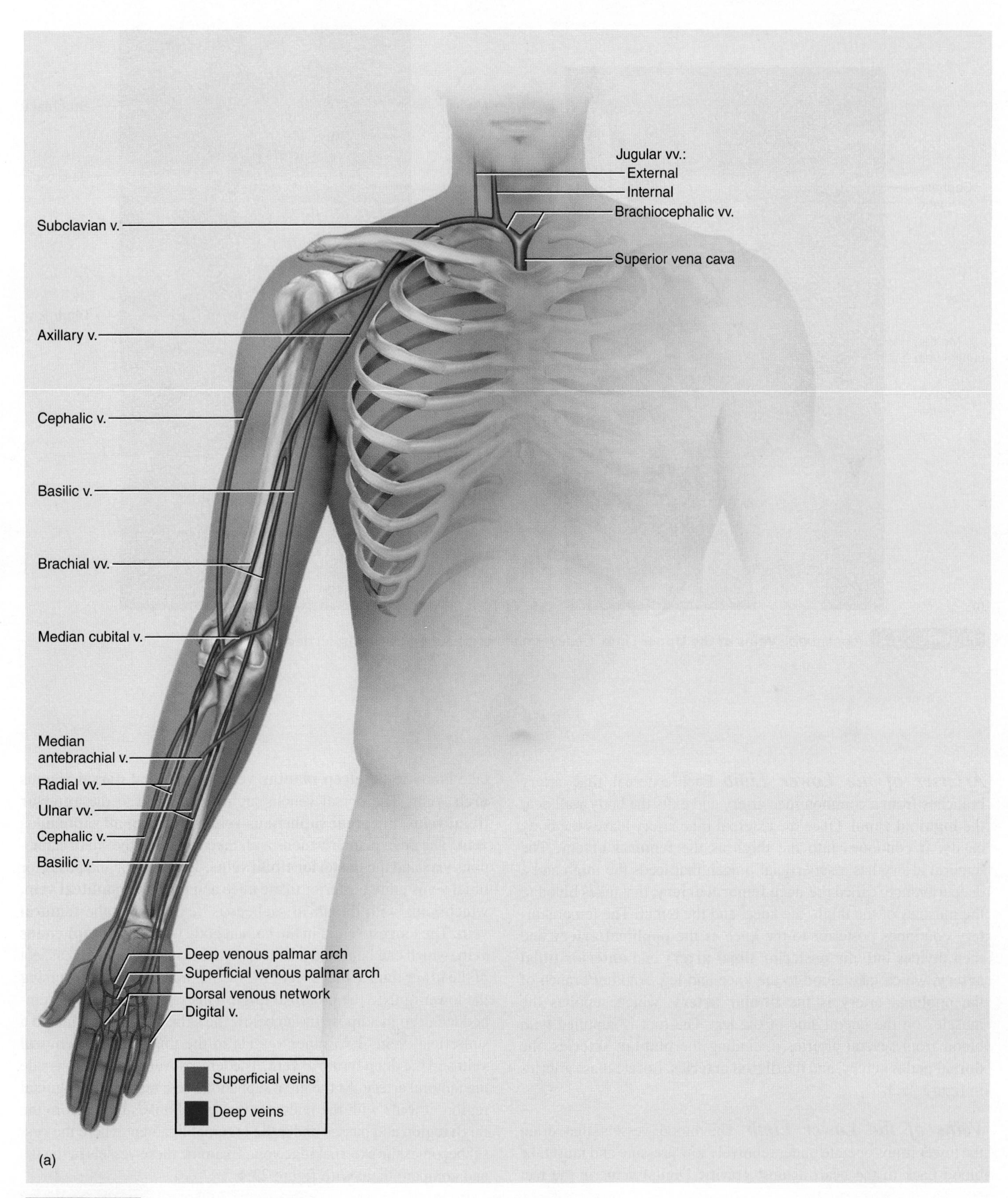

FIGURE 23.2 **Veins of the Upper Limb** (a) Diagram of superficial and deep veins.

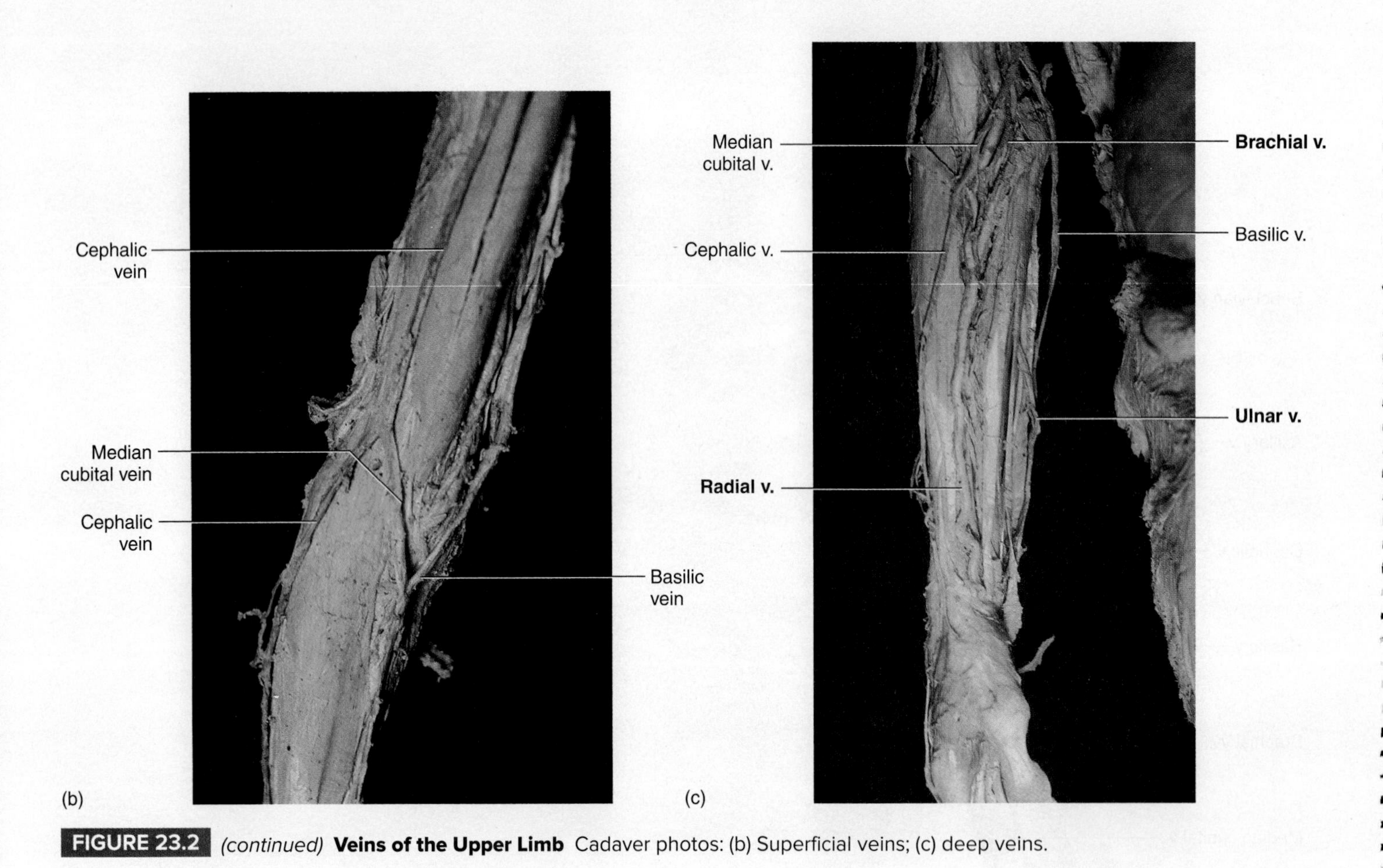

FIGURE 23.2 *(continued)* **Veins of the Upper Limb** Cadaver photos: (b) Superficial veins; (c) deep veins.

Arteries of the Lower Limb Each external iliac artery branches from a common iliac artery and exits the body wall near the **inguinal canal.** Once the external iliac artery leaves the body cavity, it continues into the thigh as the **femoral artery.** The femoral artery has a superficial branch that feeds the thigh and a deeper branch, called the **deep femoral artery,** that takes blood to the muscles of the thigh, the knee, and the femur. The femoral artery continues posterior to the knee as the **popliteal artery** and then divides into the **posterior tibial artery** and **anterior tibial artery,** which take blood to the knee and leg. Another branch of the popliteal artery is the **fibular artery,** which supplies the muscles on the lateral side of the leg. The foot is supplied with blood from several arteries, including the **plantar arteries,** the **dorsal pedal artery,** and the **digital arteries.** Locate these arteries in figure 23.3.

Veins of the Lower Limb The blood vessels that drain the lower limbs operate under relatively low pressure and must take blood back to the heart against gravity. Digital veins in the feet take blood to the **deep plantar venous arch** and **dorsal venous arch** veins. The dorsal venous arch takes blood to the **anterior tibial veins,** the **great saphenous vein,** and the **small saphenous vein.** The deep plantar venous arch takes blood to the small saphenous vein and the **posterior tibial veins.** The anterior and posterior tibial veins unite posterior to the knee and form the **popliteal vein,** which joins with the small saphenous vein to form the **femoral vein.** The longest vessel in the human body is the **great saphenous vein,** which can be found just deep to the skin on the medial aspect of the lower limb at the level of the medial malleolus and traversing the lower limb to the proximal thigh. This vessel frequently is embedded deep in adipose tissue below the skin, yet it is considered a superficial vein. Two other vessels in the thigh are the **femoral vein** and the **deep femoral vein.** The femoral vein travels alongside the femoral artery. As the great saphenous vein reaches the inguinal region, it joins with the femoral vein, which takes blood from the thigh region and passes under the inguinal ligament where the vessel becomes the external iliac vein. Examine these vessels in the lab and compare them with figure 23.4.

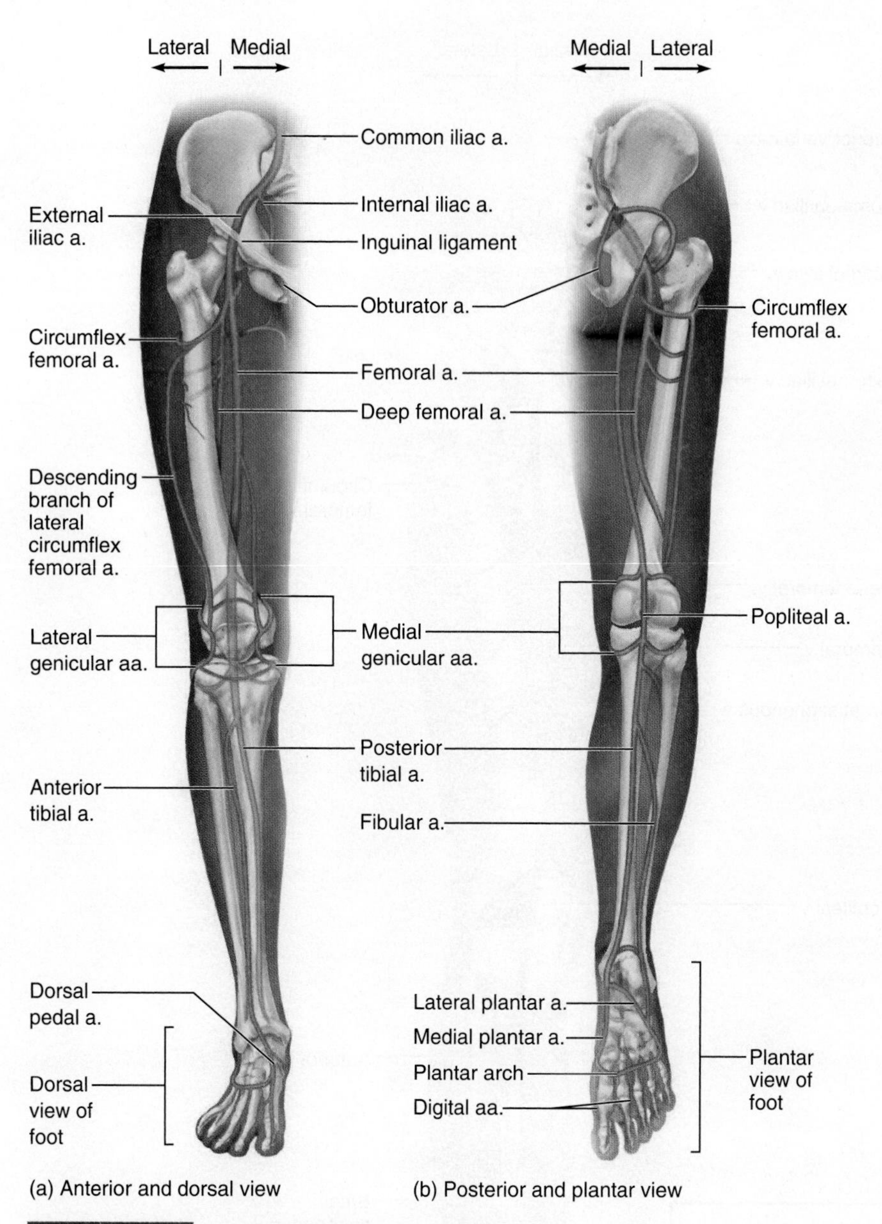

FIGURE 23.3 **Arteries of the Lower Limb**

Fetal Circulation

The pathway of fetal blood is somewhat different from that of the adult in that the lungs are nonfunctional (in terms of blood oxygenation) in the fetus. Oxygen and nutrients move from the maternal side of the placenta to the fetal bloodstream, while carbon dioxide and metabolic wastes move from the fetal bloodstream to the placenta. Examine figure 23.5 and note how the numbers in the description are placed in the diagram. From the placenta **1** the blood flows through the **umbilical vein 2,** which is located in the **umbilical cord.** The blood from the umbilical vein travels through the **ductus venosus 3,** which serves as a shunt to the inferior vena cava **4** of the fetus. The maternal blood from the umbilical vein, which is relatively high in oxygen and nutrients, mixes with the deoxygenated, nutrient-poor fetal blood from the inferior vena cava, and thus the fetus receives a mixture of blood.

The blood from the inferior vena cava travels to the right atrium of the heart. While in the right atrium, the blood travels to the right ventricle (which pumps blood to the lungs) and through a hole in the right atrium called the **foramen ovale 5.** Since the lungs do not oxygenate blood in the fetus, the foramen ovale serves as a bypass route, taking most of the blood away from the lungs and to the chambers of the heart that will pump blood to the body. Blood in the right ventricle is pumped to the pulmonary trunk, where another shunt vessel, the **ductus arteriosus 6,** carries blood to the aortic arch, bypassing the lungs. The lungs do receive some blood, but it is for the nourishment of the lung tissue, not for gas exchange. Blood from the heart exits the left ventricle and passes through the aorta to the systemic arteries. Blood travels down the internal iliac arteries **7** to the pelvic organs and lower limbs, and some moves into the **umbilical arteries 8,** which carry blood to the placenta. At birth, the pressure changes in the newborn's lungs and heart cause the closing of a flap of tissue over the foramen ovale, leaving a thin spot in the **interatrial septum** known as the **fossa ovalis.** Lack of closure of this foramen can lead to a condition known as "blue baby."

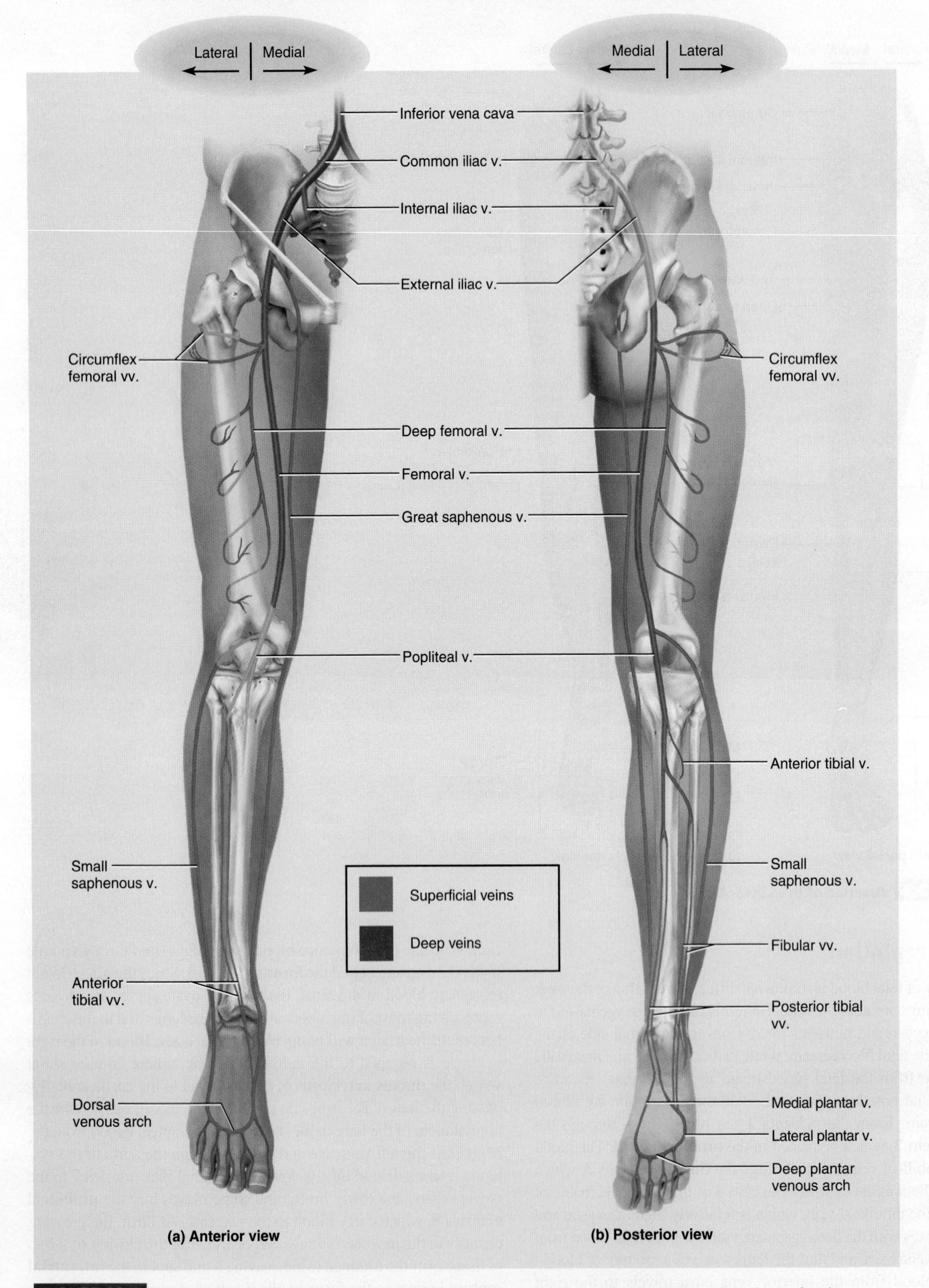

FIGURE 23.4 **Veins of the Lower Limb** (a) Diagram of entire lower limb; (b) photograph of upper right thigh of cadaver.

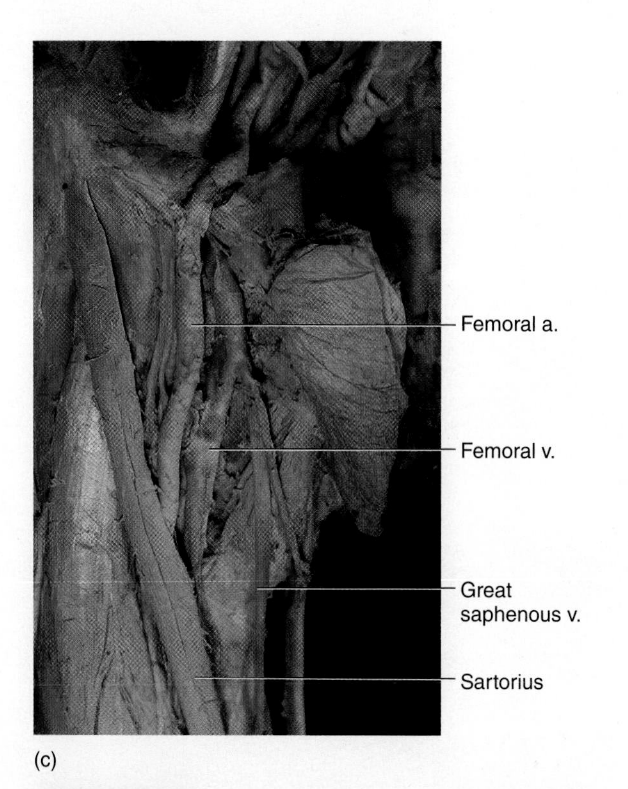

(c)

FIGURE 23.4 *(continued)* **Veins of the Lower Limb**
(a) Diagram of entire lower Limb; (b) photograph of upper right thigh of cadaver.

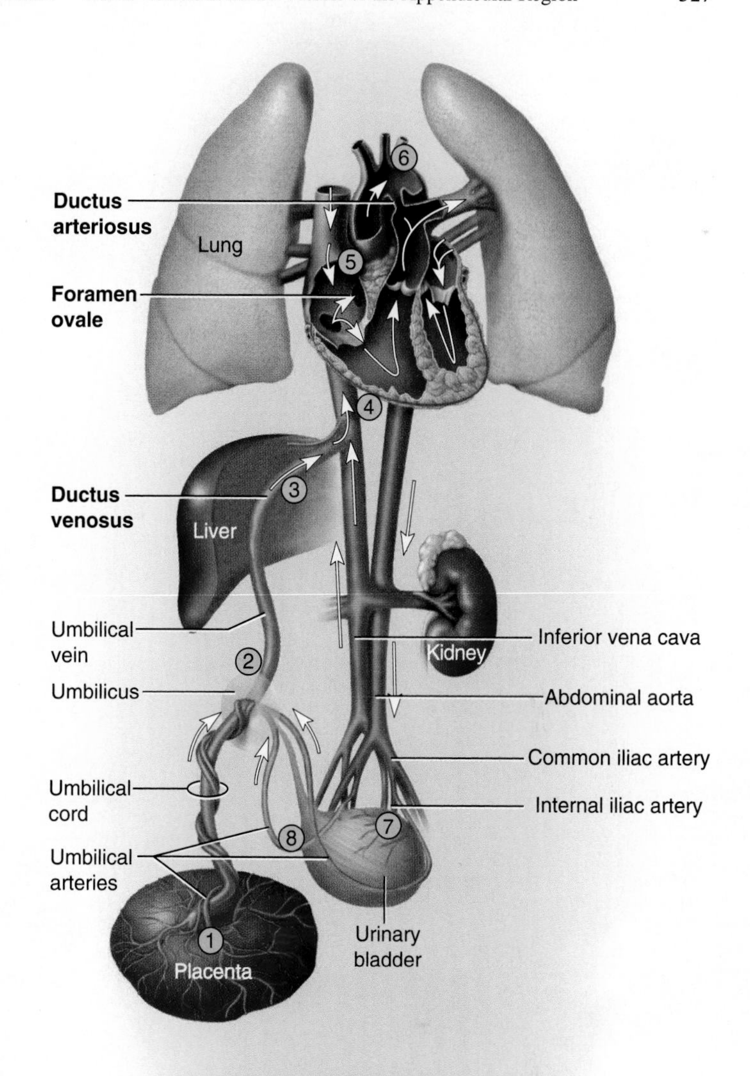

FIGURE 23.5 **Fetal Circulation in the Human**

Cat Anatomy

If you are using cats, turn to section 6, "Blood Vessel Anatomy of the Cat," on page 425 of this lab manual.

Review

Exercise 23

Blood Vessels 2: Blood Vessels of the Appendicular Region

Name ______________________________ Date ______________

1. Where does blood in the right subclavian artery come from? ______________________________

2. Blood from the left subclavian artery flows into what vessel as it moves toward the left arm? ______________________________

3. Blood in the radial artery comes from what blood vessel? ______________________________

4. Is the radial vein a superficial or deep vein? ______________________________

5. Where is the median cubital vein found? ____________________

6. Where is the cephalic vein located? ____________________

7. Blood from the right axillary vein travels next to what vessel? ____________________

8. What vessel receives blood from the ulnar vein? ____________________

9. What artery takes blood directly to the femoral artery? ____________________

10. Blood from the popliteal artery comes directly from what artery? ____________________

11. What vessels take blood to the left femoral vein? ______________________

____

____

____

12. In what region of the body is the great saphenous vein? ______________________

____

____

____

13. Where does blood flow after it leaves the femoral vein? ______________________

____

____

____

14. In the fetal heart, what is the name of the shunt between the pulmonary trunk and the aortic arch?

____

____

____

15. Name the opening between the atria in the fetal heart. ______________________

____

____

____

16. Label the following illustration, using the terms provided.
    anterior tibial artery
    axillary artery
    brachial artery
    dorsal pedal artery
    femoral artery
    radial artery
    ulnar artery

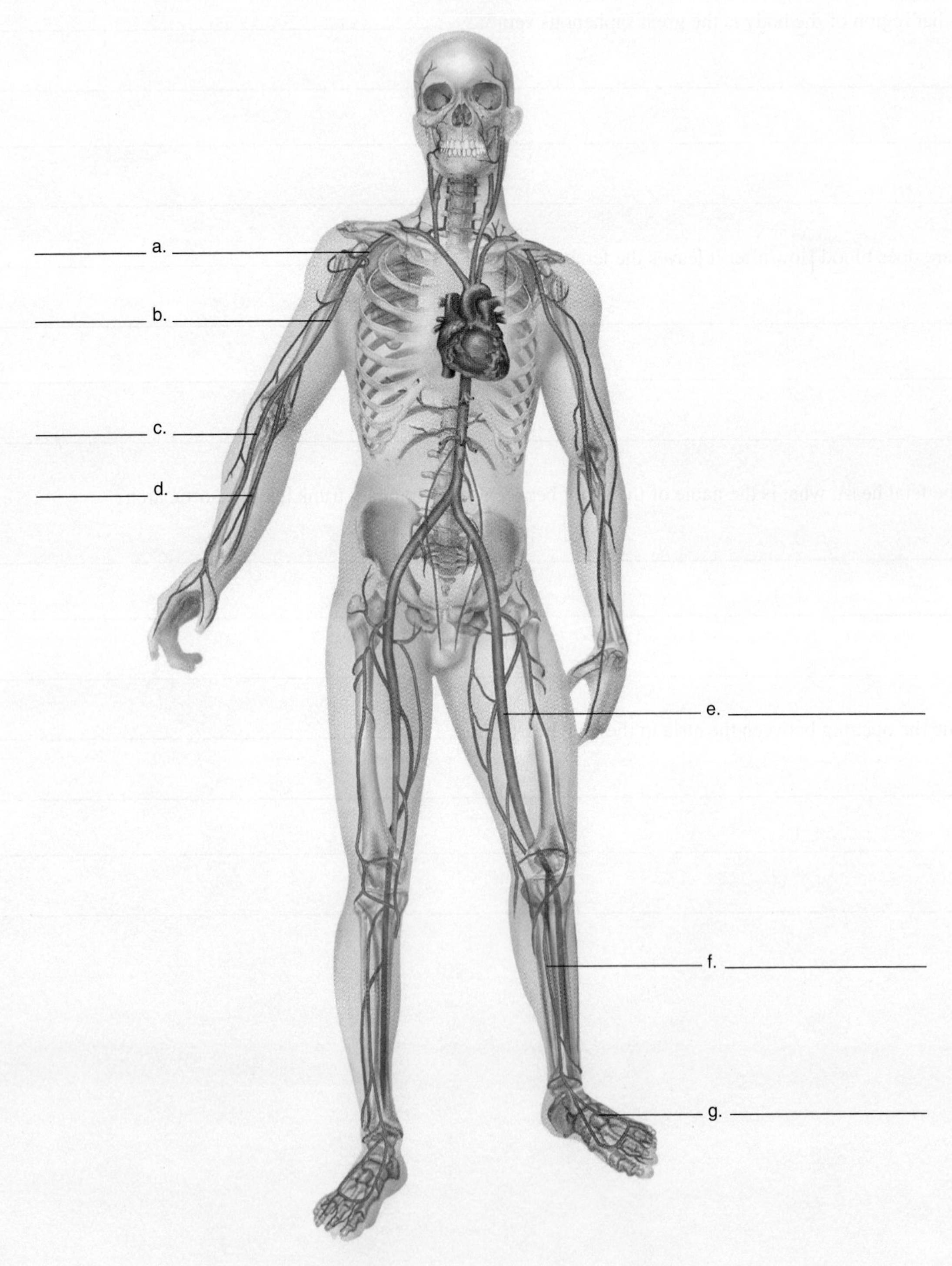

17. Label the following illustration, using the terms provided.
    - basilic vein
    - brachial vein
    - cephalic vein
    - femoral vein
    - great saphenous vein
    - median cubital vein

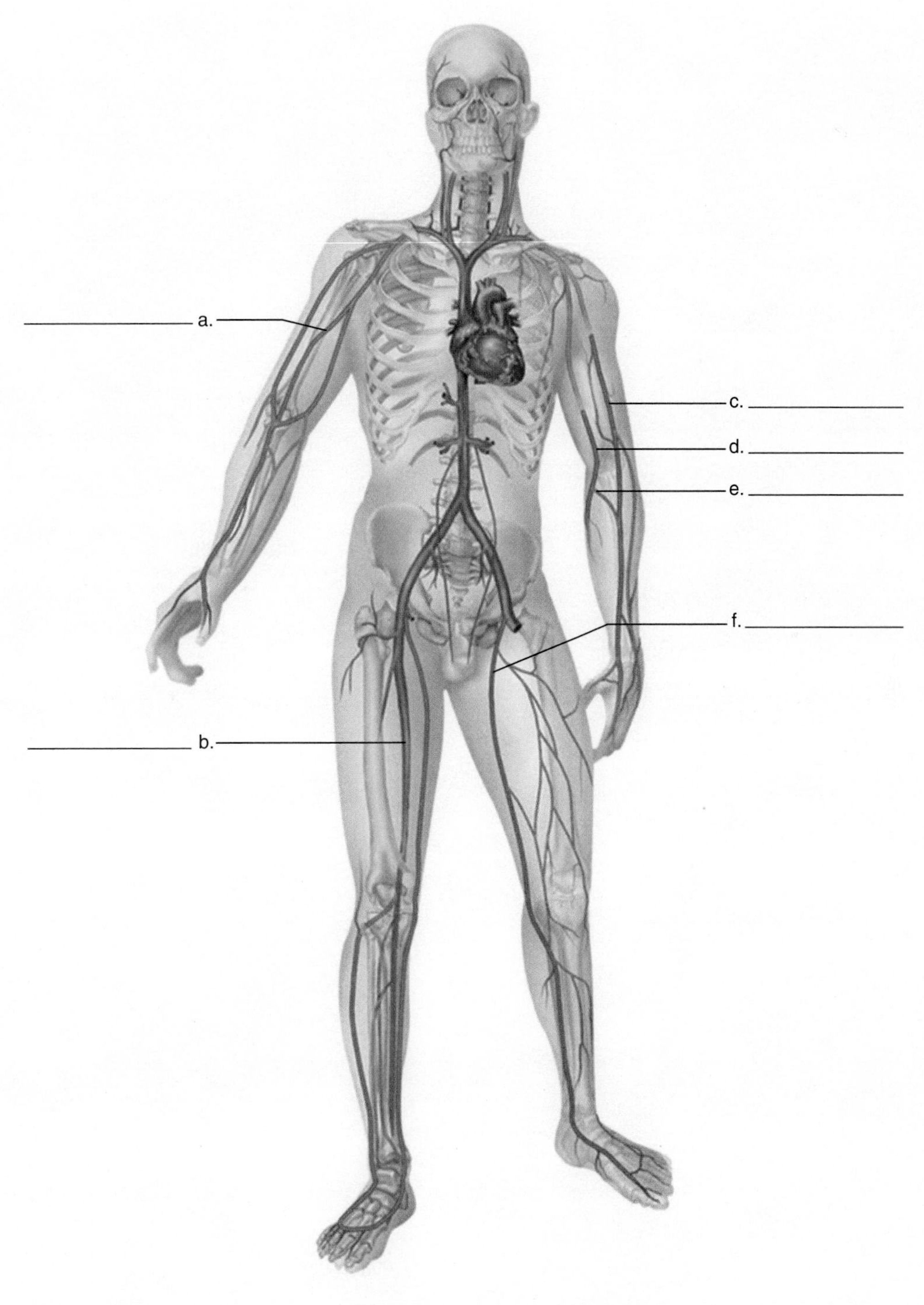

Laboratory

Exercise 24

The Lymphatic System

Lymphatic System

Introduction

Blood cells and plasma protein typically stay in the blood vessels, yet fluid from the plasma leaks from the capillaries and bathes the cells of the body. This fluid flows between the cells of the body and is known as **interstitial fluid.** It provides nutrients to the cells and receives dissolved wastes from the cells along with cellular debris. Most of this fluid flows back into the capillaries, but some is picked up by **lymphatic capillaries,** which return the fluid, now known as **lymph,** to the **collecting vessels.** Collecting vessels take the lymph back to the venous system. In this way the remaining interstitial fluid is returned to the cardiovascular system. **Lymphatic vessels** refer to any type of lymph vessel (lymph capillary, collecting duct, etc.).

In addition to returning fluid to the cardiovascular system, the lymph system protects the body and is instrumental in the absorption of lipids from the digestive system. Lipids in the small intestine are converted to **chylomicrons** (phospholipids and other molecules) in the cells that line the digestive tract. These chylomicrons are conducted into the lymphatic system, which then transports the material through lymphatic vessels to the cardiovascular system.

As the fluid flows through the lymphatic vessels, lymph nodes phagocytize the cellular debris and foreign material (bacteria, viruses) that may have entered the lymphatic system. Lymph nodes also activate B cells and T cells, providing immunity. The spleen recycles old blood cells; the tonsils provide immunity, as does the lymphatic tissue; and the red bone marrow produces some lymphocytes.

Learning Objectives

At the end of this exercise you should be able to

1. identify the valves in lymph vessels;
2. locate the major histological features of a lymph node;
3. demonstrate on a model the location of the tonsils and other lymph organs;
4. describe the function of the spleen and thymus; and
5. identify selected lymph vessels.

Materials

Charts, diagrams, and models of the lymph system
Microscopes
Microscope slides of lymphatic vessels with valves
Torso models
Cadaver (if available)

Procedure

Lymphatic System

The lymphatic system is difficult to study in preserved specimens because it collapses at death. The lymphatic system is composed of **lymphatic capillaries, collecting vessels, lymph nodes, lymph organs,** and **lymphatic (lymphoid) tissue.** An overview of the system is illustrated in figure 24.1. Examine charts and models in the lab and compare them with this figure.

Lymph originates as interstitial fluid that bathes the cells of the body. This fluid flows from blood capillaries and carries oxygen, nutrients, and other dissolved materials to the cells. Some of the interstitial fluid does not return to the blood capillary bed but enters the lymphatic capillaries by way of small, valvelike slits in the lymphatic capillary wall. This process is illustrated in figure 24.2.

Lymphatic Vessels Once the lymph is in the lymphatic capillaries, it travels through the **collecting vessels.** As with the capillaries, there is a one-way flow that occurs in the collecting vessels due to the presence of valves. Examine a prepared slide of a collecting vessel and note the presence of the **valve.** Valves prevent the backflow of lymph away from the vascular system into which they drain. Compare your slide with figure 24.3.

Use figure 24.4 to examine the drainage pattern of lymph in the body. The **lymphatic vessels** lead to regions of the body where lymph nodes are found. Nodes are clustered in the groin (inguinal region), axilla, antecubital fossa, popliteal region, neck, thorax, and abdomen. The **thoracic duct** (a large lymphatic vessel) drains most of the body, taking lymph to the left subclavian vein, where the fluid is returned to the cardiovascular system. Identify the

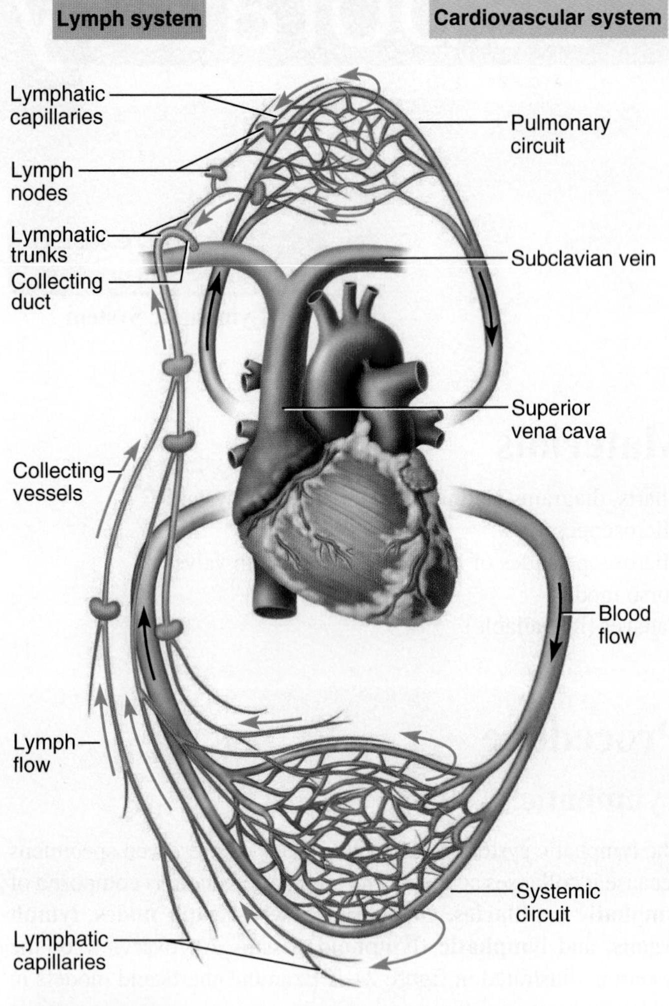

FIGURE 24.1 **Flow of Interstitial Fluid and Lymph**

Capillary bed
Tissue fluid
Tissue cell
Lymphatic capillary
Arteriole
Venule

(a)

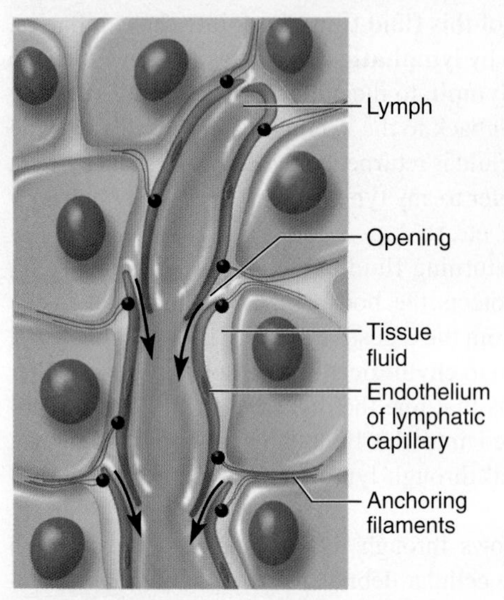

(b)

FIGURE 24.2 **Lymphatic Capillary** (a) Overview of cardiovascular and lymphatic capillaries; (b) detail of lymphatic capillaries with valves.

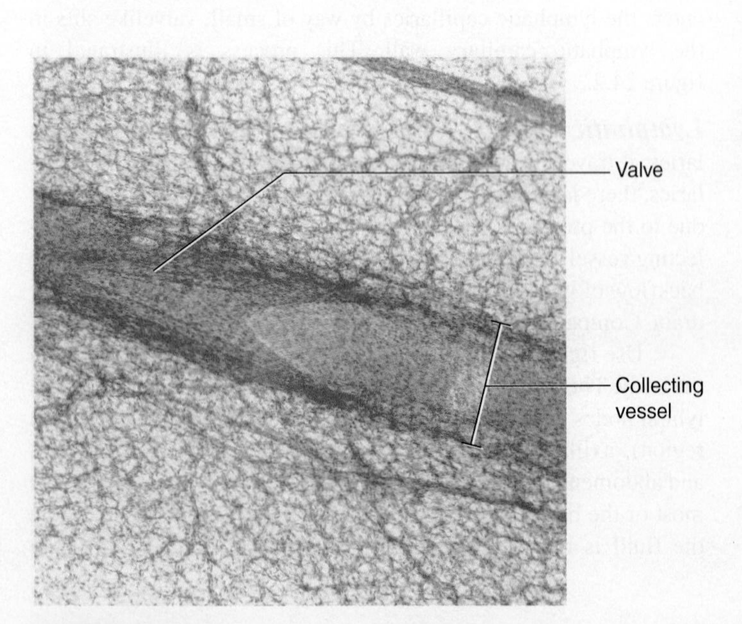

FIGURE 24.3 **Collecting Vessel with Valve**

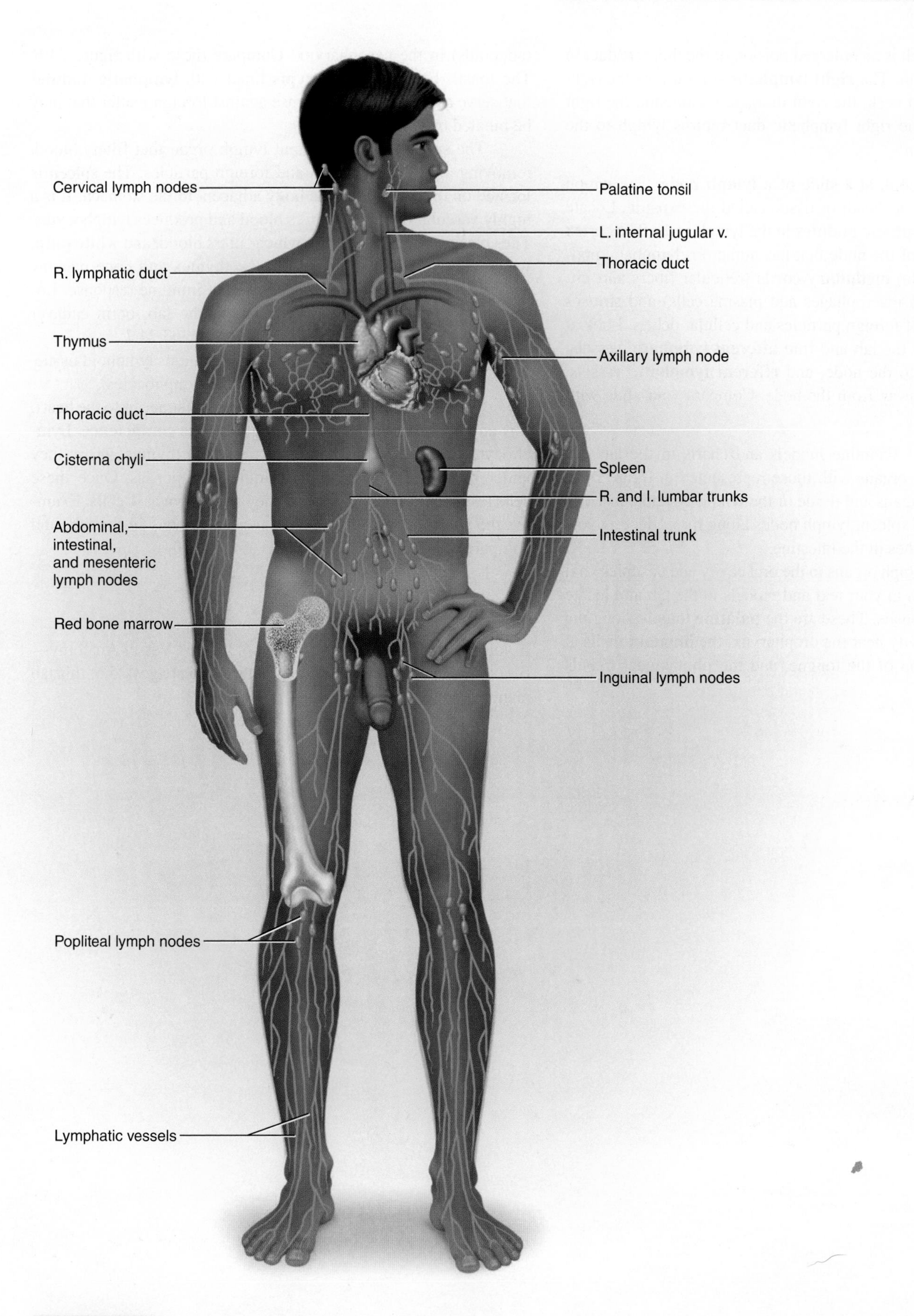

FIGURE 24.4 **Lymphatic Organs and Vessels of the Body**

cisterna chyli, which is an enlarged portion of the thoracic duct in the abdominal region. The **right lymphatic duct** drains the right side of the head and neck, the right thoracic region, and the right upper extremity. The right lymphatic duct returns lymph to the right subclavian vein.

Lymph Node Look at a slide of a **lymph node.** The lymph node is enclosed by a sheath of tissue called the **capsule.** Locate the dark purple **lymphatic nodules** in the lymph node. The cortex is the outer region of the node that has numerous lymphatic nodules. In the **medulla, medullary cords** (reticular fibers and immune cells, such as macrophages and plasma cells) and **sinuses** cleanse the lymph of foreign particles and cellular debris. Look at models or charts in the lab and find **afferent lymphatic vessels,** which take lymph to the node, and **efferent lymphatic vessels,** which take lymph away from the node. Compare your slide with figure 24.5.

Lymph Organs Examine models and charts in the lab and compare the lymph organs with those represented in figure 24.4. The major lymph organs and tissue of the lymph system consist of the tonsils, thymus, spleen, lymph nodes along the collecting vessels, and Peyer patches in the intestine.

Tonsils are lymph organs in the oral cavity and nasopharynx. Look at illustrations in your text and models in the lab and locate the three pairs of tonsils. These are the **palatine tonsils** along the sides of the oral cavity near the oropharynx, the **lingual tonsils** at the posterior portion of the tongue, and the **pharyngeal tonsils** (adenoids) in the nasopharynx. Compare these with figure 24.6. The tonsils have **tonsillar crypts** lined with **lymphatic nodules** that serve as a first line of defense against foreign matter that may be inhaled or swallowed.

The **spleen** is an important lymph organ that filters blood, removing aging erythrocytes and foreign particles. The spleen is located on the left side of the body adjacent to the stomach. It is a highly vascular organ that filters blood and produces lymphocytes. The spleen contains **red pulp,** which filters blood, and **white pulp,** which contains lymphocytes from the thymus and bone marrow and releases more of these cells during an immune response. Locate the spleen in models and charts in the lab, or in cadaver specimens, and compare it with figures 24.4 and 24.7.

The appendix is another organ that contains lymphoid tissue. It destroys intestinal bacteria and produces lymphocytes.

The **thymus** is superficial to the vessels superior to the heart. It is an important site in determining **immune competence.** Lymphocytes travel from the bone marrow to the thymus, where they undergo maturation essential to immune responses. Once these cells become immunocompetent, they are known as **T cells.** Examine the charts and models, or cadaver specimens in the lab, and compare them with figure 24.8.

Cat Anatomy

If you are using cats, turn to section 6, "Blood Vessel Anatomy—Blood Vessels and Lymph System," starting on page 425 of this lab manual.

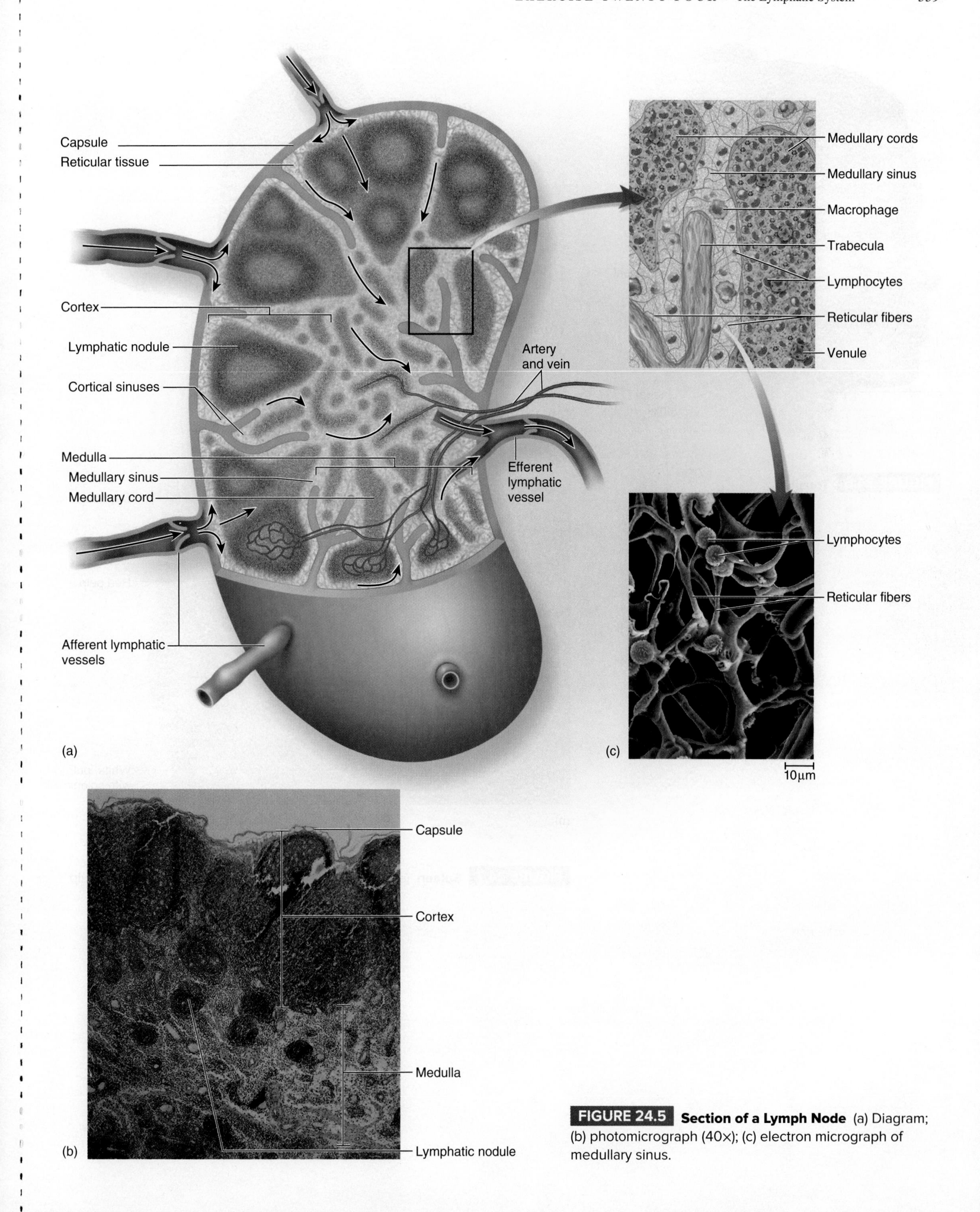

FIGURE 24.5 **Section of a Lymph Node** (a) Diagram; (b) photomicrograph (40×); (c) electron micrograph of medullary sinus.

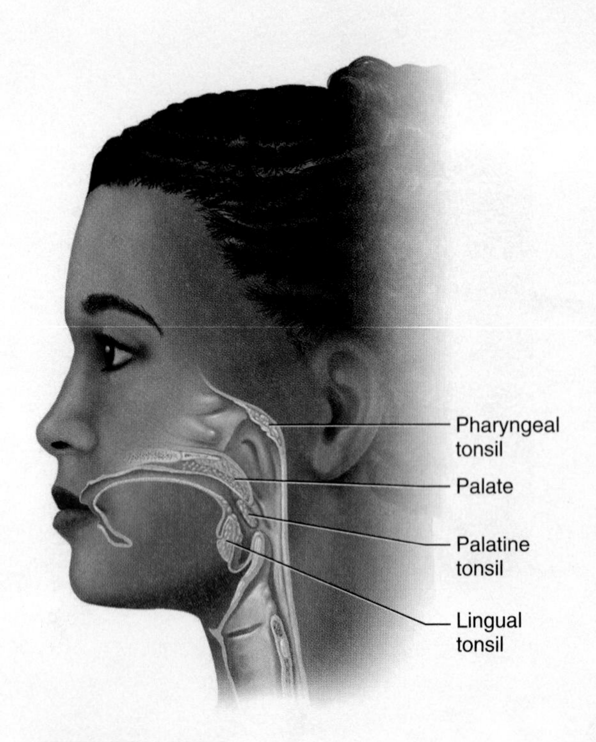

FIGURE 24.6 **Tonsils**

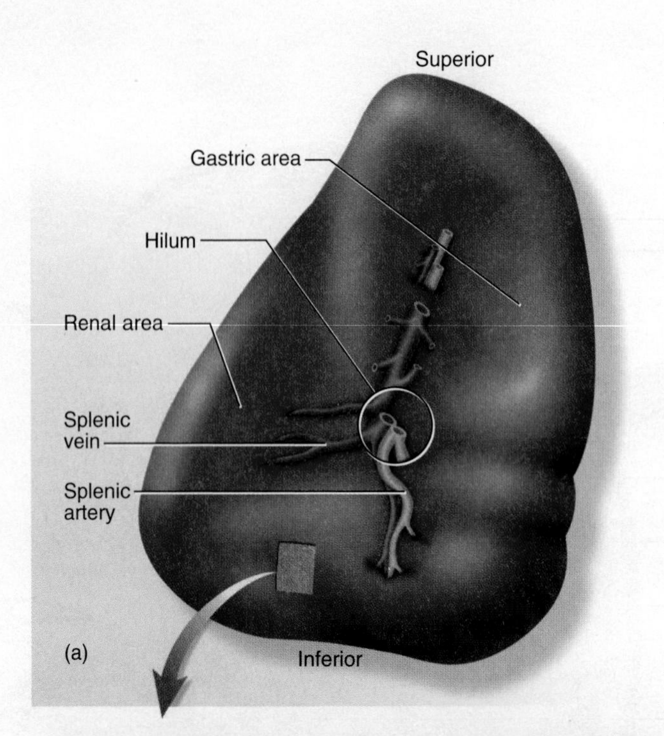

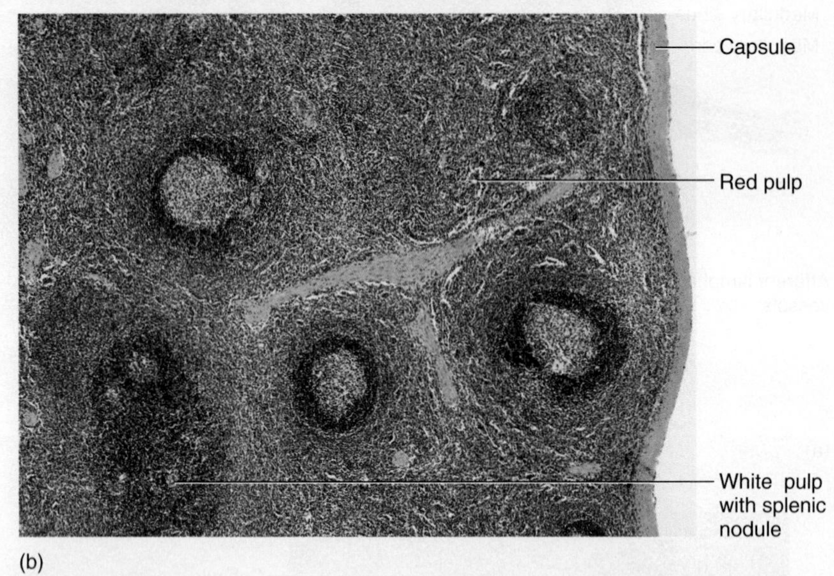

FIGURE 24.7 **Spleen** (a) Medial surface; (b) histology with red and white pulp.

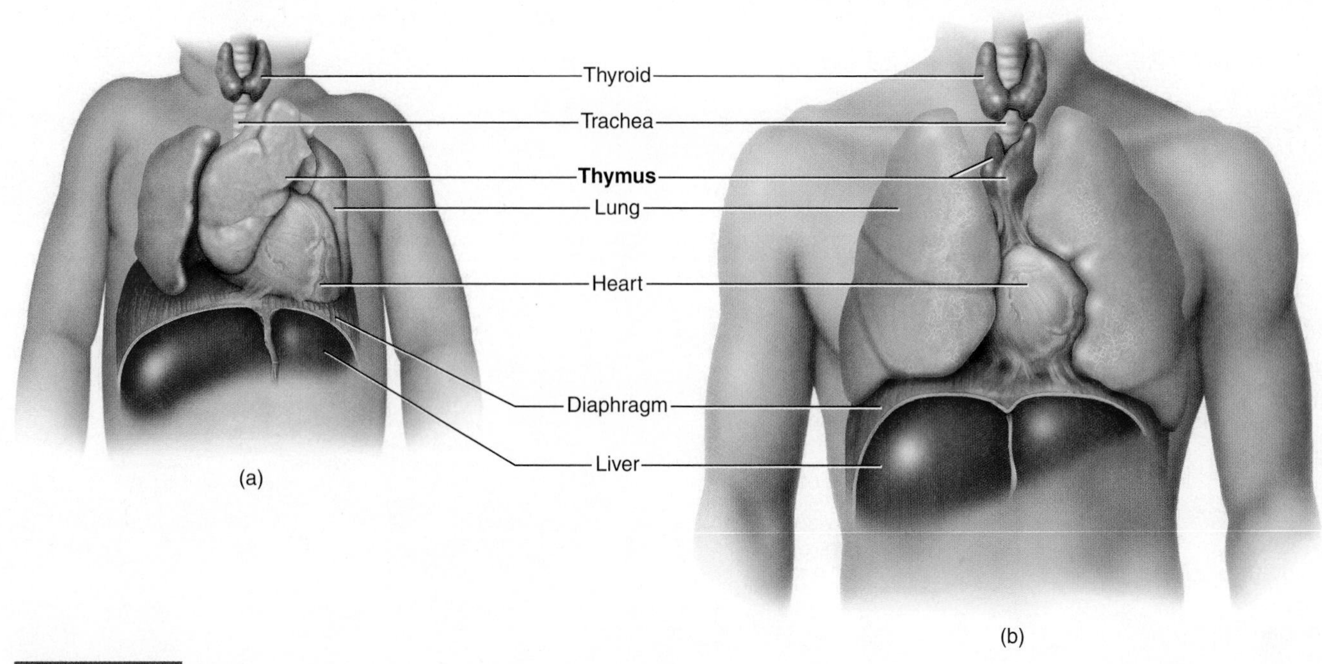

FIGURE 24.8 **Thymus** (a) Young individual; (b) mature adult.

The Lymphatic System

Name ______________________________ Date ____________

1. What kind of vessels carry lymph from the lymph capillaries to the veins? ______________________________

____

2. Once interstitial fluid enters the lymphatic vessels, what is it called? ______________________________

____

3. What are the names of the inner region and outer region of a lymph node? ______________________________

   Describe their function. ______________________________

____

____

4. What kind of vessel takes lymph to a lymph node? ______________________________

____

5. The adenoids are enlarged ______________________________ tonsils.

6. Which tonsils are found on the sides of the oral cavity? ______________________________

____

7. What tonsils are located at the back of the tongue? ______________________________

____

8. Blood is filtered by which lymph organ in the adult? ______________________________

____

9. What part of the spleen is involved in producing lymphocytes? ______________________________

____

10. In the analysis of breast cancer, lymph nodes of the axillary region are removed and a biopsy is performed. The removal of the nodes is done to determine if cancer has spread from the breast to other regions of the body. What effect would the removal of lymph nodes have on the drainage of the pectoral region? ______________________________

____

____

11. Label the illustration of the lymph system, using the terms provided.

cistern chyli	right lymphatic duct	thoracic duct
collecting vessels	spleen	thymus
lymph nodes		

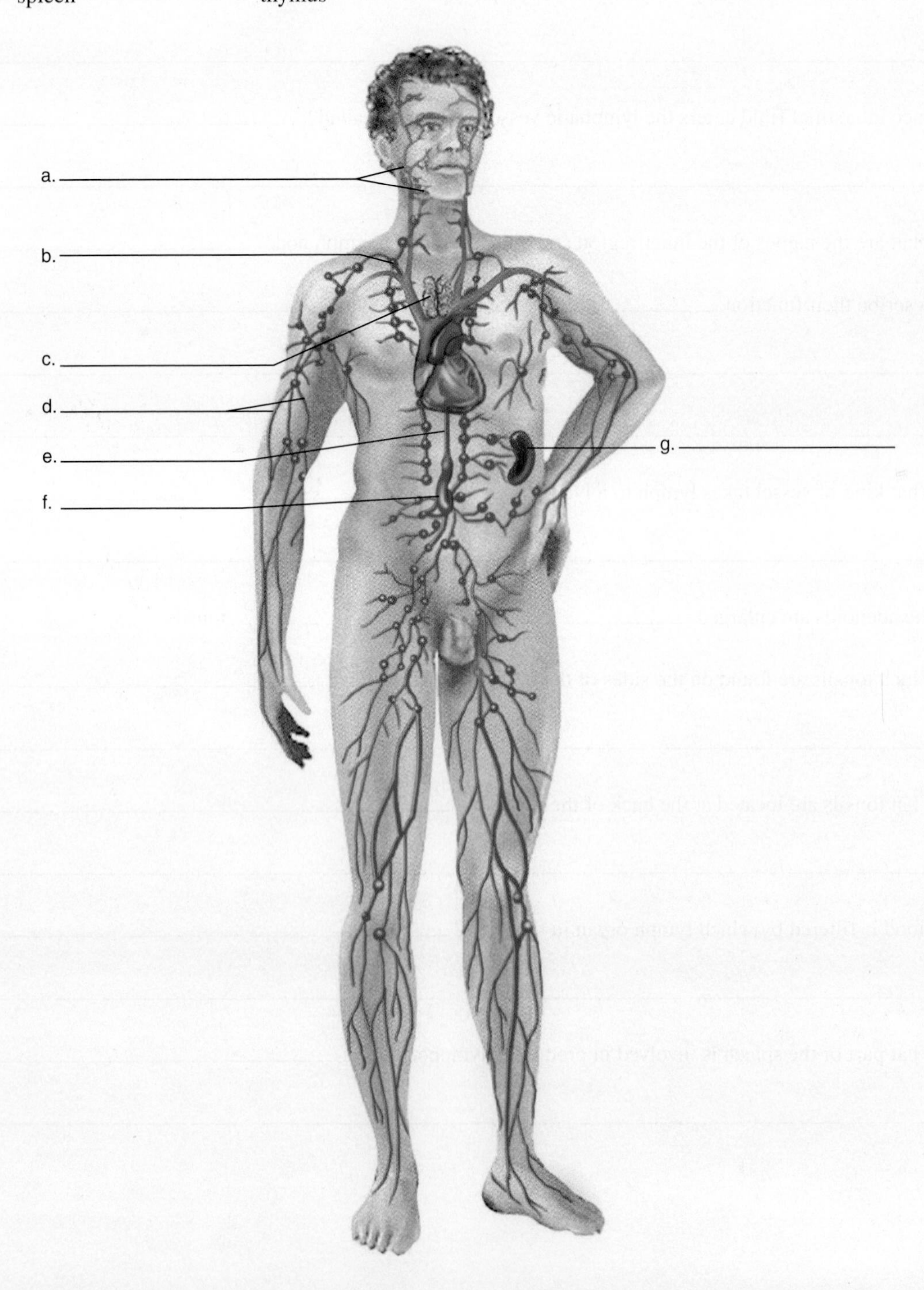

Laboratory

Exercise 25

The Respiratory System

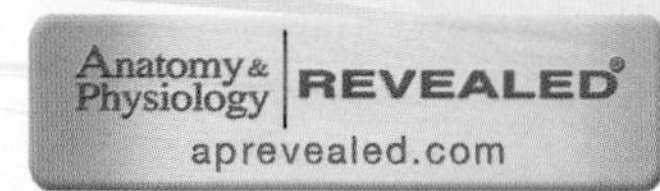

Respiratory System

Introduction

The respiratory system exchanges oxygen and carbon dioxide with the atmosphere. Cells use oxygen as the terminal electron acceptor, and carbon dioxide is a waste product of cellular respiration. Atmospheric oxygen moves into the lungs and diffuses into the cardiovascular system. It subsequently reaches the individual cells of the body while the metabolic waste product, carbon dioxide, is released from the intercellular environment and travels via the blood to the lungs, where it is released by exhalation. Too much carbon dioxide in the blood increases the hydrogen ion concentration, producing acidosis of the blood. Too few hydrogen ions in the blood cause alkalosis. Acidosis or alkalosis disrupts normal metabolic processes in the blood. The respiratory system is therefore integrated with the other body systems. The **conducting division** involves the airflow, while the **respiratory division** is the part of the system where gas exchange occurs. As you study the anatomy of the system, be aware of the role that other systems play in respiration, such as the cardiovascular system, which is the only system that transports oxygen and carbon dioxide, and the muscular system, which increases demand for oxygen during times of exertion.

Learning Objectives

At the end of this exercise you should be able to

1. list the organs and significant structures of the respiratory system;
2. define the role of the respiratory system in terms of the overall function of the body;
3. explain the physical reason for the tremendous surface area of the lungs;
4. identify the cartilages of the larynx; and
5. distinguish among a bronchus, bronchiole, and respiratory bronchiole.

Materials

Lung models or detailed torso model, including a median section of head
Model of larynx
Microscopes
Prepared microscope slides of lung tissue
Prepared microscope slides of "smoker's lung"
Prepared microscope slide of trachea
Charts and illustrations of the respiratory system

Procedure

Look at the charts and models of the respiratory system and compare them with figure 25.1. Locate the following structures and place a check mark next to the terms when you locate them:

_____ Nose
_____ Anterior naris
_____ Nasal cavity
_____ Pharynx
_____ Larynx
_____ Trachea
_____ Pleural cavity
_____ Bronchus
_____ Lungs, left and right

Overview of the Respiratory System

Nose and Nasal Cartilages Examine a median section of a model or chart of the head, and look for the **nose, nasal cartilages, nares (nostrils),** and **nasal septum.** The nasal cavity is divided by the nasal septum into the two nasal fossae. The nasal septum is composed of the perpendicular plate of the ethmoid bone, the vomer, and the septal cartilage. Examine these features in figures 25.1 and 25.2.

The entrance of the nares is protected by guard hairs. The region of the nose, just posterior to the nares, is the **nasal vestibule,** which is lined with stratified squamous epithelium. Behind the vestibule is the **nasal cavity,** which is lined with a **mucous membrane** that moistens air entering the respiratory system. The membrane consists of **respiratory epithelium,** composed of **pseudostratified ciliated columnar epithelium** with **goblet cells.** This mucous membrane overlies a superficial venous plexus that warms the air.

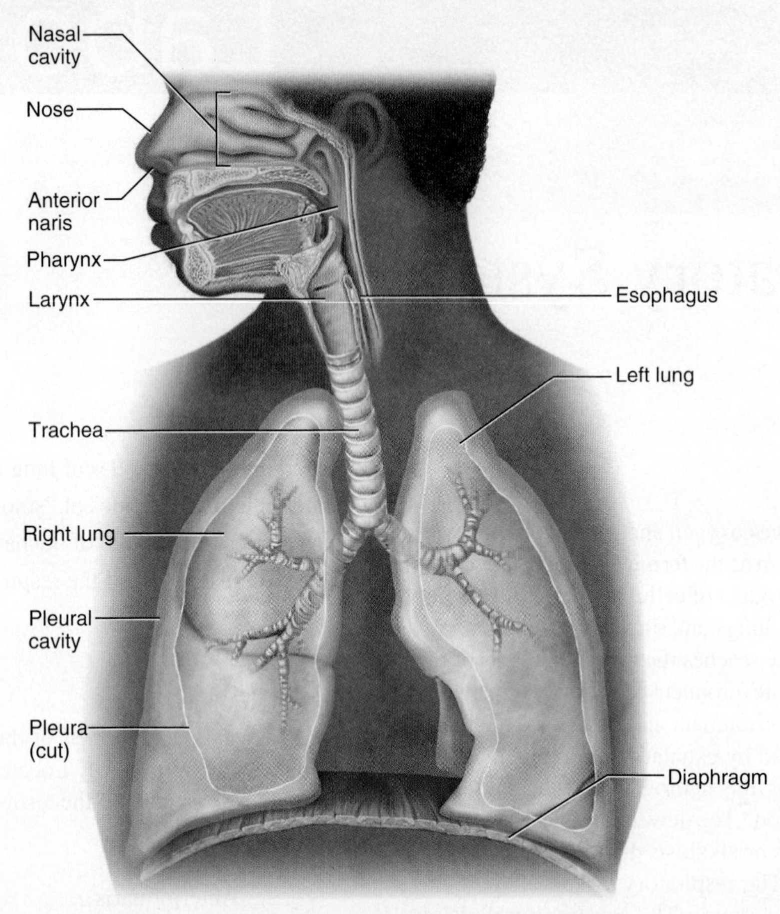

FIGURE 25.1 Overview of the Anatomy of the Respiratory System

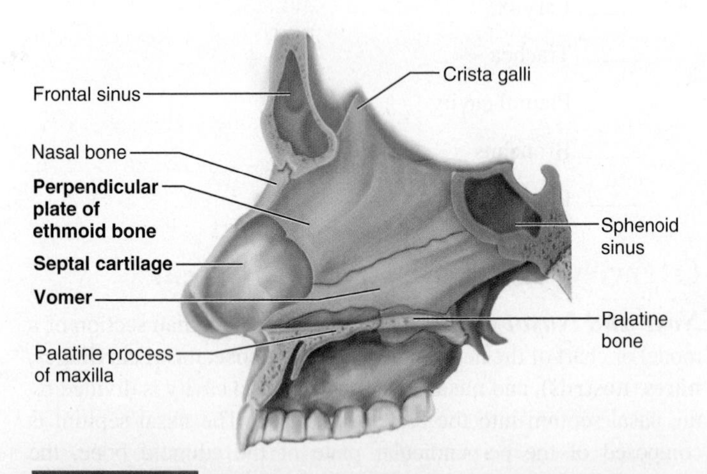

FIGURE 25.2 Structures That Make Up the Nasal Septum

The lateral walls of the cavity have three protrusions that push into the nasal cavity. These are the **nasal conchae** or **turbinates** (TUR-bih-nates). They cause the inhaled air to swirl in the nasal cavity and come into contact with the mucous membrane. This moistens the air. Locate the **superior, middle,** and **inferior conchae** in the nasal cavity. The nasal cavity terminates where two openings, the **posterior nasal apertures,** or **choanae** (co-AH-nee), lead to the **pharynx.** Locate these structures on the materials in the lab and compare them with figure 25.3.

Pharynx The pharynx is divided into three regions based on location. The uppermost area is the **nasopharynx** (NAZE-oh-FAIR-inks), which is directly posterior to the nasal cavity. The nasopharynx has two openings on the lateral walls, which are the openings of the **auditory,** or **eustachian, tubes.** These equalize pressure in the middle ear. Inferior to this is the **oropharynx.** The oropharynx is a common passageway for food, liquid, and air. The **uvula** is a small, pendulous structure that partially separates the **oral cavity** from the oropharynx. The uvula flips upward during swallowing, reducing the chance of fluids entering the nasopharynx. The most inferior portion of the pharynx is the **laryngopharynx** (la-RING-go-FAIR-inks), located posterior to the larynx. Find the regions of the pharynx in figure 25.3.

Larynx The **larynx** (LAIR-inks) is commonly known as the "voice box" because it is an important organ for sound production in humans. It controls the pitch of the voice, while the shape of the oral cavity and the placement and size of the paranasal sinuses are responsible for the sonority, or sound quality, of the voice. The larynx is at the level of the fourth through sixth cervical vertebrae and consists of a number of cartilages. The most prominent

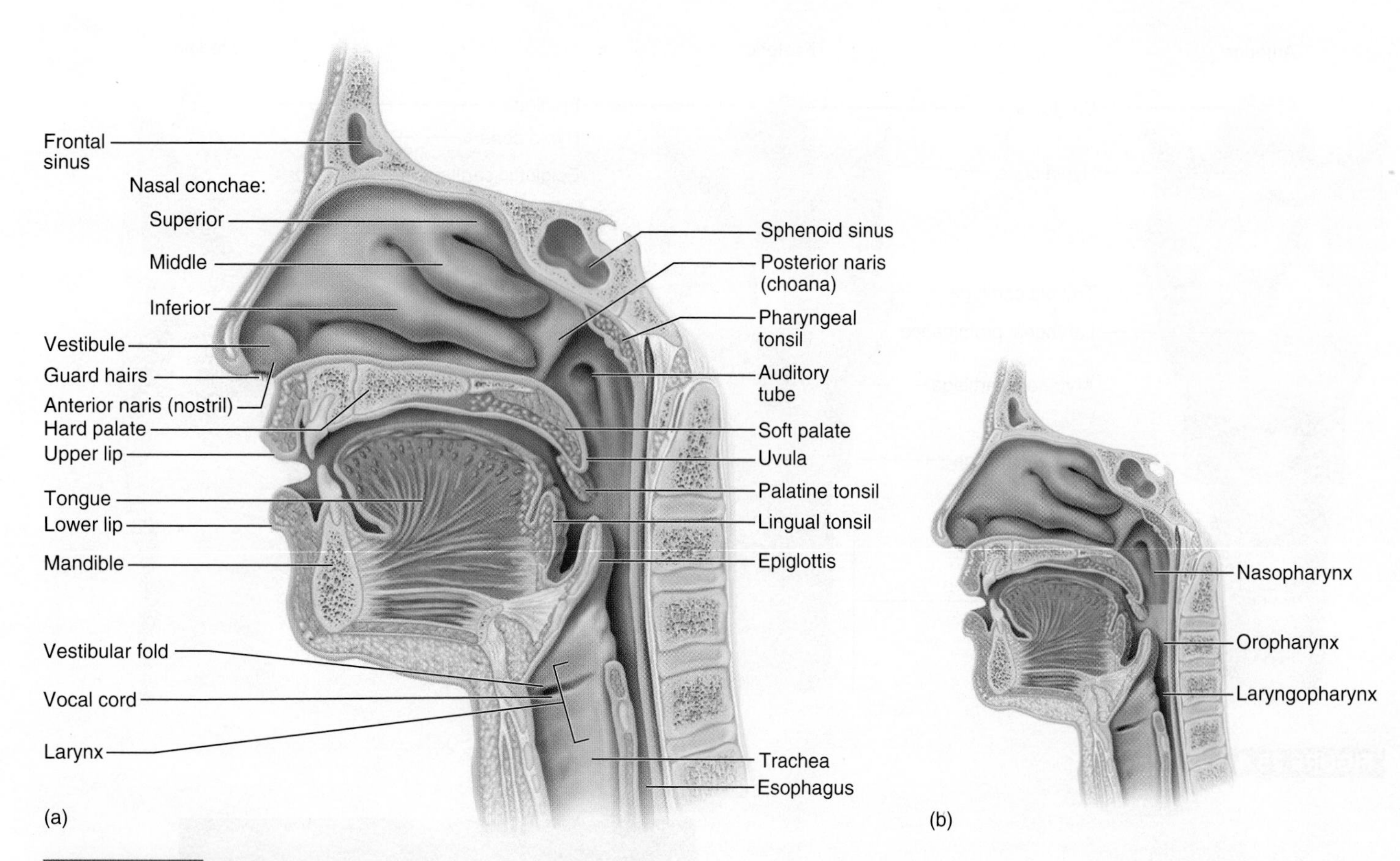

FIGURE 25.3 **Upper Respiratory Tract** (a) Median section of the head with nasal septum removed; (b) three regions of the pharynx.

cartilage in the larynx is the **thyroid cartilage,** which is a shield-shaped structure made of hyaline cartilage.

The thyroid cartilage is more prominent in males because it increases in size under the influence of testosterone. Its common name is "Adam's apple." Inferior to the thyroid cartilage is the **cricoid** (CRY-coyd) **cartilage.** The cricoid cartilage is also composed of hyaline cartilage, and it is a thin band of cartilage when seen from the anterior aspect but increases in size at its posterior surface. Superior to the cricoid cartilage in the posterior wall of the larynx are the paired **arytenoid** (AR-ih-TEE-noyd) **cartilages.** These cartilages attach to the posterior end of the **true vocal folds (vocal cords).** Movement of the arytenoid cartilages pulls on the true vocal folds, causing them to stretch, thus increasing the pitch of the voice. This occurs by the contraction of **intrinsic muscles** attached to the arytenoid cartilages from the back, while the vocal cords are held stationary by the thyroid cartilage in the front. Superior to the true vocal folds are the **vestibular folds (false vocal cords)** (fig. 25.4).

At the very posterior, superior edge of the larynx are the **corniculate** and **cuneiform** (cue-NEE-ih-form) **cartilages.** These are also made of hyaline cartilage. The most superior structure of the larynx is the **epiglottis,** which is composed of elastic cartilage and mucous membrane. During swallowing, the epiglottis protects the opening of the larynx, which is known as the **glottis.** In the swallowing reflex, muscles pull the epiglottis down over the glottis. This is not a perfect system, as anyone knows who has started swallowing a liquid and responded by laughing at a joke. Inhalation at the beginning of the laugh causes fluid to move into the larynx and trachea, irritating the respiratory lining. This causes another reflex called the **cough reflex,** which propels the liquid out of the respiratory system. Locate the structures of the larynx on models or charts in the lab and in figure 25.4.

Trachea and Bronchi The trachea (TRAY-kee-uh) is commonly known as the "windpipe" because it conducts air from the larynx to the lungs. The **trachea** is a straight tube whose lumen is kept open by C-shaped **tracheal cartilages.** Examine these cartilages by running your fingers gently down the outside of your throat. Palpate the cartilage rings below the larynx. The tracheal cartilages are composed of hyaline cartilage. At the most inferior portion of the trachea is a keel-shaped cartilage, the **carina** (ca-RY-na; *carina* = keel). The carina divides the incoming air into the bronchi. Locate the features of the trachea in figures 25.5 and 25.6. The trachea is also lined with respiratory epithelium. Obtain a prepared slide of the trachea and find the tracheal cartilage, respiratory epithelium, and **posterior tracheal membrane** (which includes the trachealis muscle—absent in some slide preparations) (fig. 25.6). The membrane allows the esophagus to expand into the trachea during swallowing.

The trachea splits into two tubes, which enter the lungs. These tubes are the **main bronchi** (BRONK-eye). Each lung receives air from a main bronchus, which contains hyaline cartilage and is lined with respiratory epithelium. The main bronchi of the lung divide into the **lobar bronchi,** and these further divide to form

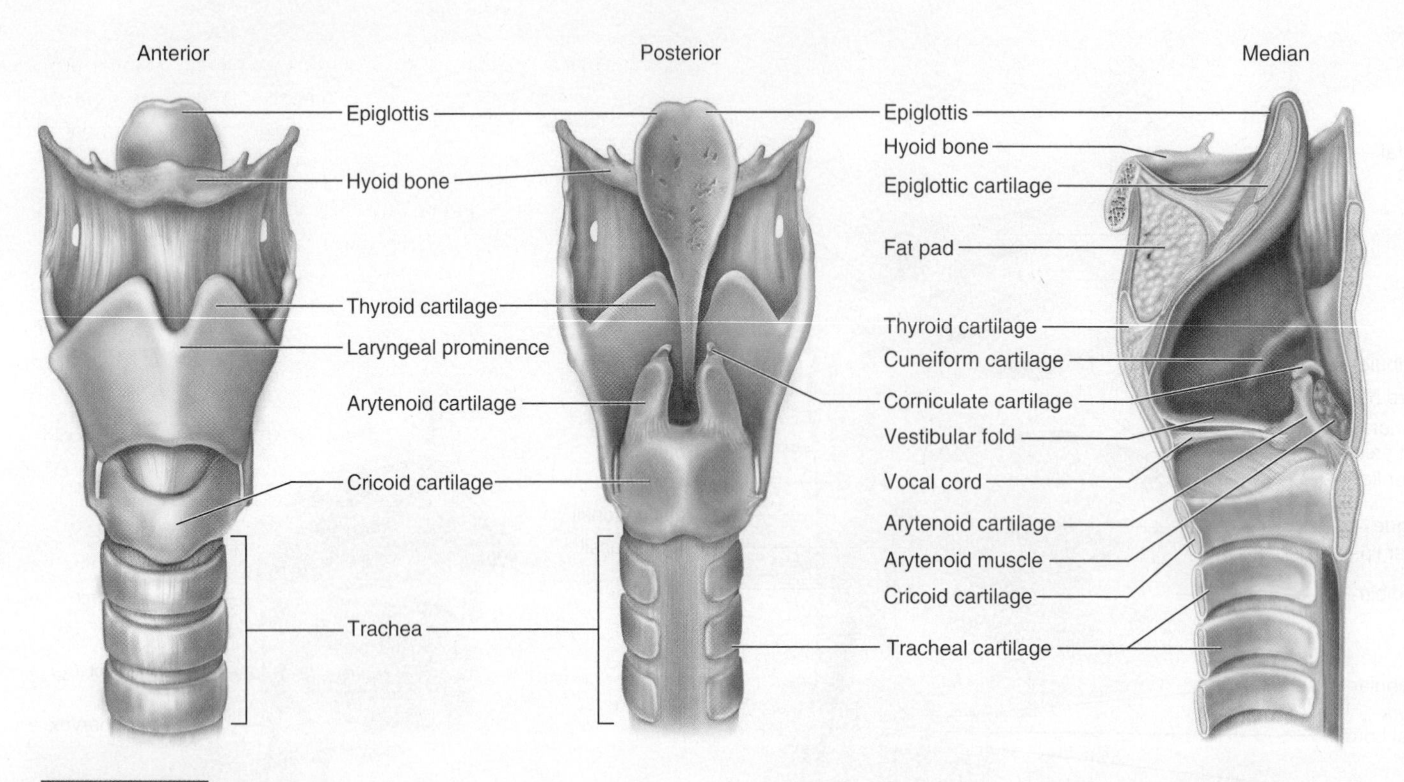

FIGURE 25.4 **Larynx**

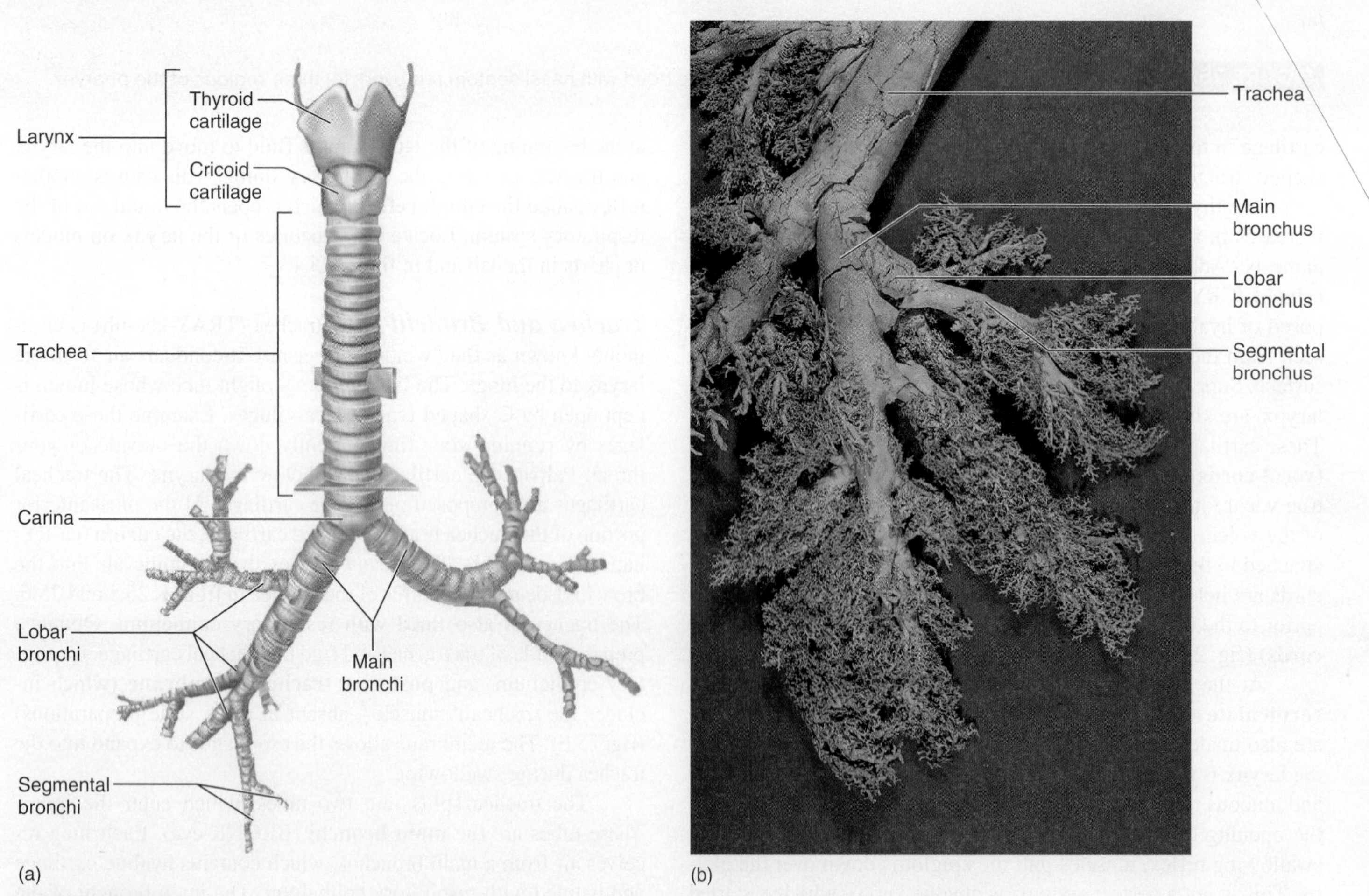

FIGURE 25.5 **Larynx, Trachea, and Bronchial Tree** (a) Diagram anterior view; (b) cast of bronchial tree.

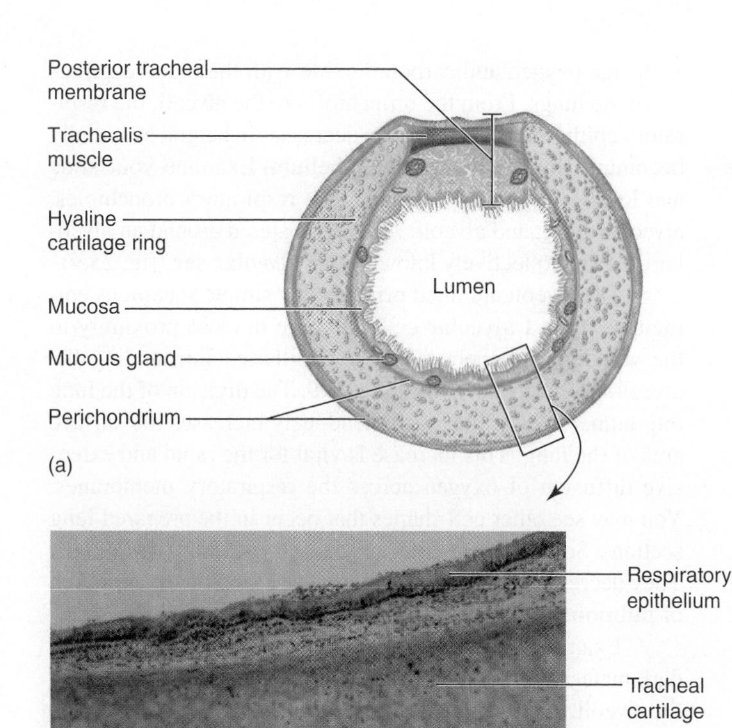

FIGURE 25.6 **Trachea, Cross Section** (a) Diagram; (b) photomicrograph (100×).

segmental bronchi. The extensive branching of the bronchi produces what is known as the bronchial tree (fig. 25.5). Each segmental bronchus takes air to its specific bronchopulmonary segment.

Lungs There are two **lungs** in humans; the right lung has **three lobes,** known as the **superior, middle,** and **inferior lobes,** separated by a superior **horizontal fissure** and an inferior **oblique fissure.** The left lung has **two lobes** with an **oblique fissure** and the indentation left by the heart known as the **cardiac impression.** The lobes of the left lung are the **superior** and **inferior lobes.** Look at models or charts in the lab and identify the major features, as shown in figure 25.7.

The lungs are enveloped by the **pleural cavities** on each side of the mediastinum. The **parietal pleura** is the outer membrane on the chest cavity wall, and the membrane that is adhered to the surface of the lungs is the **visceral pleura.** The space between the membranes is known as the pleural cavity. These are shown in figure 25.8.

Histology of the Lung The bronchi continue to divide until they become **bronchioles,** small respiratory tubules with smooth muscle in their walls. They differ from bronchi in that they have no cartilage. Obtain a prepared slide of lung and scan first under low power and then under higher powers. The bronchioles in the lung further divide into **respiratory bronchioles,** which are so named for small structures called alveoli attached to their walls. The respiratory bronchioles lead to passageways known as **alveolar ducts,** which branch into alveoli. **Alveoli** are air sacs in the lung that

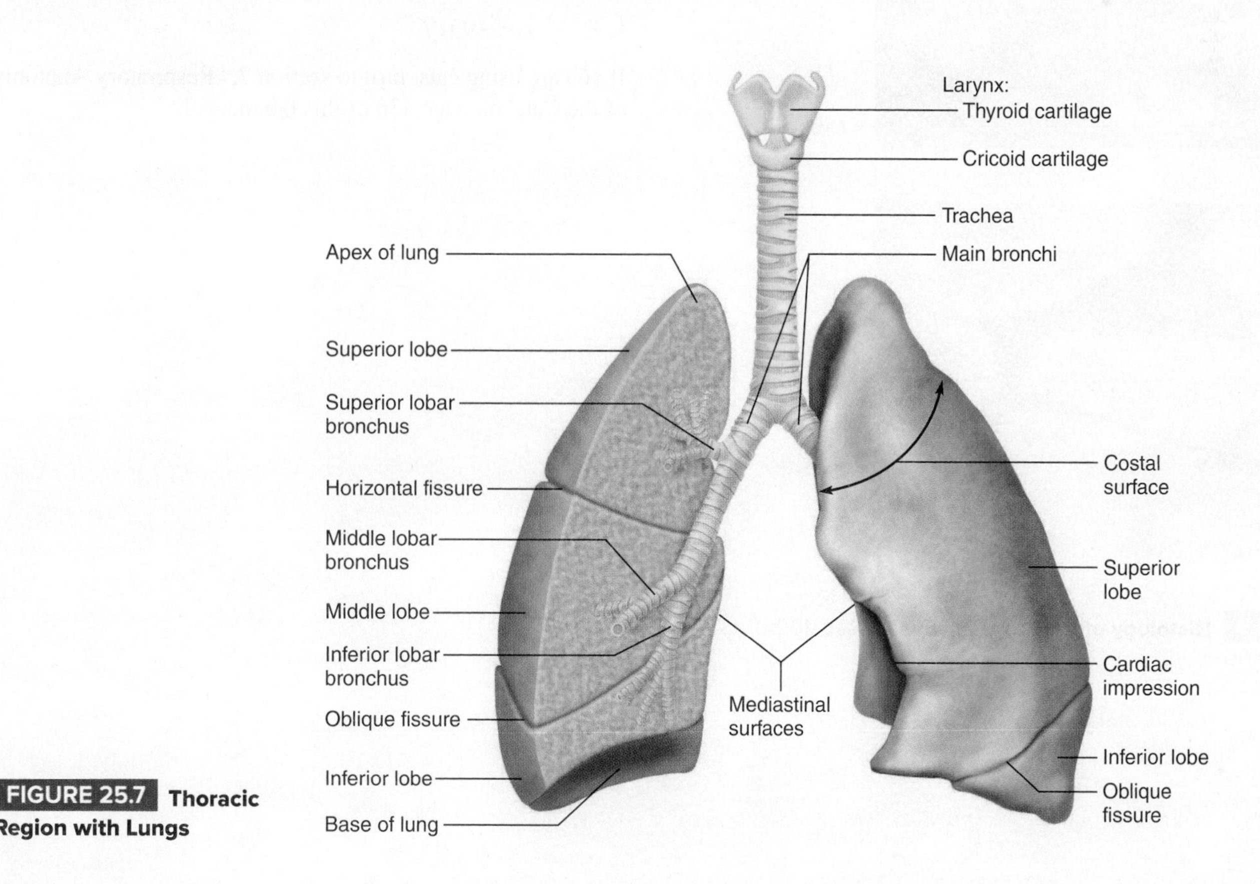

FIGURE 25.7 **Thoracic Region with Lungs**

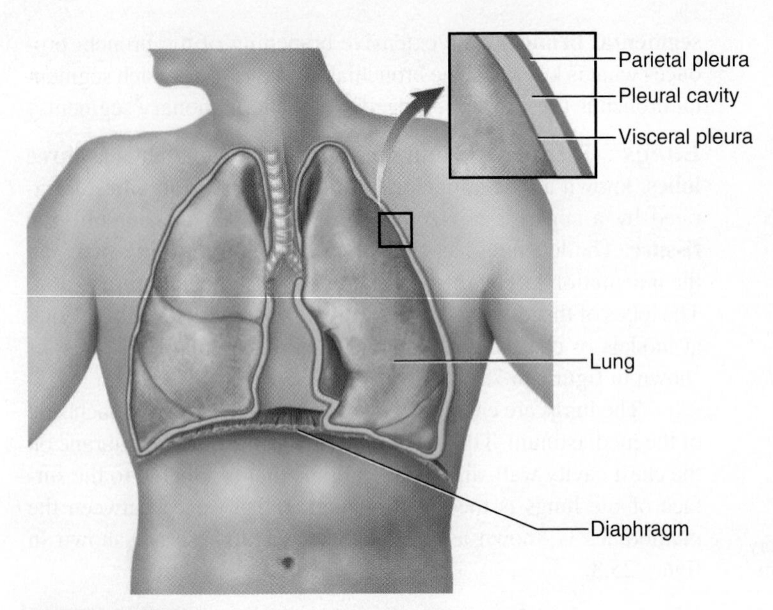

FIGURE 25.8 **Pleural Membranes and Cavity**

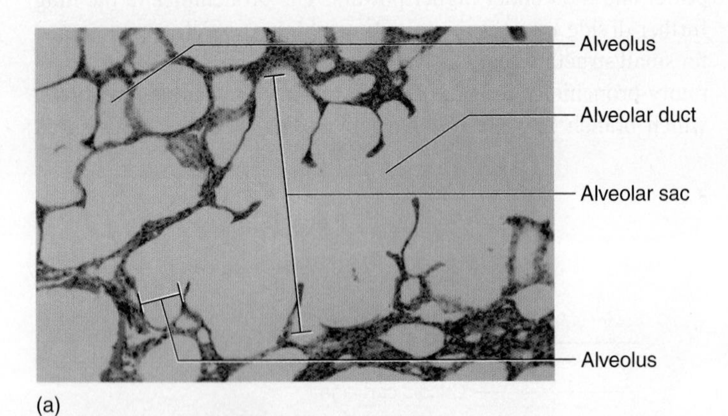

(a)

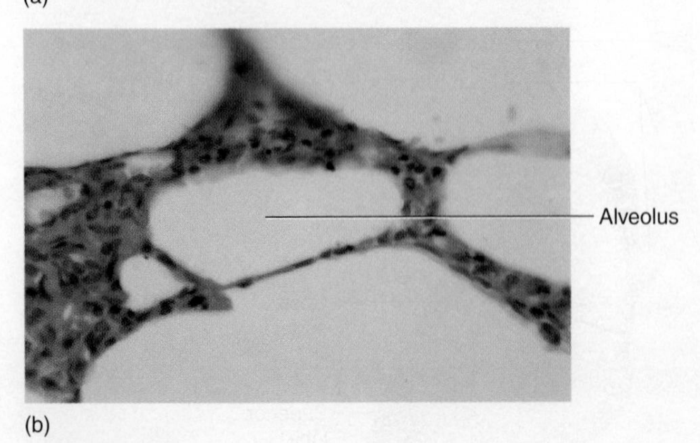

(b)

FIGURE 25.9 **Histology of the Lung** (a) Alveolar sac (100×); (b) alveolus (400×).

exchange oxygen and carbon dioxide with the blood capillaries of the lungs. From the bronchioles to the alveoli, the respiratory epithelium progressively decreases in height, eventually becoming simple squamous epithelium. Examine your slide and locate the bronchi, bronchioles, respiratory bronchioles, alveolar ducts, and alveoli. Alveoli clustered around an alveolar duct are collectively known as an **alveolar sac** (fig. 25.9).

The alveoli are lined primarily by simple squamous epithelium **(type I alveolar cells)** and are in close proximity to the vascular endothelium of the capillaries surrounding the alveoli. These are seen in figure 25.9. The division of the lung into numerous small sacs tremendously increases the surface area of the lung. This increase is vital for the rapid and extensive diffusion of oxygen across the respiratory membranes. You may see other cell shapes that occur in the prepared lung sections. Some of these cells are **great (type II) alveolar cells.** They decrease the surface tension of the lung by the secretion of **pulmonary surfactant.**

Examine a prepared slide of smoker's lung. Note the dark material in the lung tissue and the general destruction of the alveoli. Breakdown of the alveoli leads to a disease known as **emphysema.**

Pathway of Air Air moves from the nares to the nasal vestibule and into the nasal cavity. From there the warmed and moistened air passes through the pharynx and into the larynx, trachea, and bronchial tree. The air moves into the alveolar ducts and finally to the alveolus. Exhalation is essentially the reverse process.

Cat Anatomy

If you are using cats, turn to section 7, "Respiratory Anatomy of the Cat," on page 436 of this lab manual.

The Respiratory System

Name ______________________ Date __________

1. What is the common name for the nares? ______________________

2. The apex of the nose is composed of cartilage. What functional adaptation does cartilage have over bone in making up the external framework of the nose? ______________________

3. The nasal cavity is divided in two by what structure? ______________________

4. Name the three structures that make up the nasal septum. ______________________

5. What is the function of respiratory epithelium and the venous plexuses in the nasal cavity? ______________________

6. From the nasal cavity, what is the name of the opening into the nasopharynx? ______________________

7. What is the name of the space posterior to the oral cavity and superior to the laryngopharynx? ______________________

8. What is the name of the structure that prevents fluid in the oral cavity from entering the nasopharynx during swallowing?

____

9. What is the name of the large cartilage of the anterior larynx? ____________________

10. What is the structure that covers the glottis, protecting the larynx from fluid during swallowing? ____________________

11. Which lung has just two lobes? ____________________

12. What membrane attaches directly to the external surface of the lungs? ____________________

13. The trachea branches into two tubes that go to the lungs. What are these tubes called? ____________________

14. Where is the bronchial tree found? ____________________

15. What small structure in the lung is the site of exchange of oxygen with the blood capillaries? ____________________

16. The surface area of the lungs in humans is about 70 square meters. How can this be if the lungs are located in the small space of the thoracic cavity? ____________________

What role do alveoli play in the nature of the surface area? ____________________

17. Emphysema is a destruction of the alveoli of the lungs. What effect does this have on the surface area of the lungs?
____

18. Fill in the following illustration of the human respiratory system using the terms provided.

cricoid cartilage	main bronchus	nasopharynx
epiglottis	middle lobe	superior lobe
inferior lobe	nasal cavity	trachea

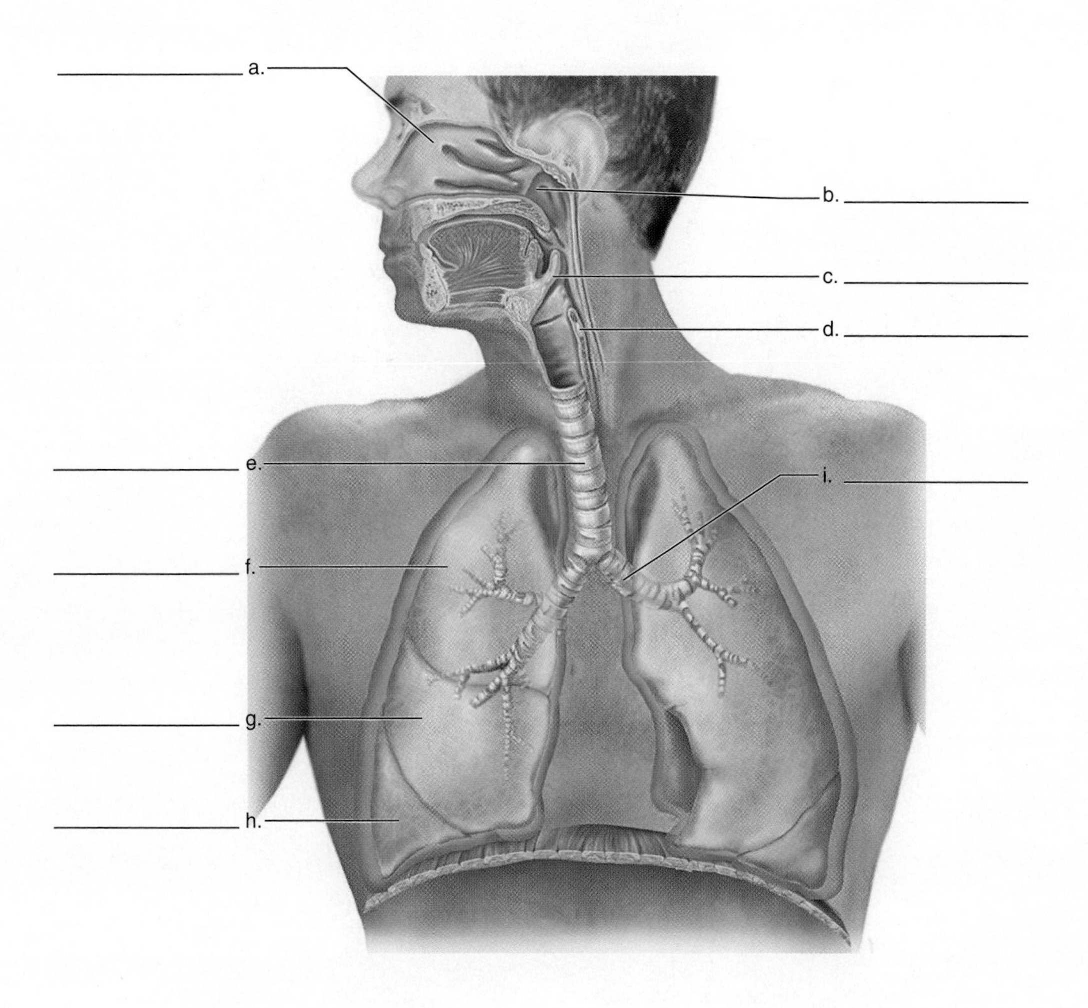

Laboratory

Exercise 26

The Digestive System

Digestive System

Introduction

The digestive system can be divided into two major parts—the **digestive tract** and the **accessory organs.** The digestive tract is a long tube that runs from the mouth to the anus; it comes into contact with food or the breakdown products of digestion. The digestive tract is also known as the **alimentary canal.** The portion of the alimentary canal from the stomach to the anus is the **gastrointestinal (GI) tract.** Some of the organs of the digestive tract are the esophagus, stomach, small intestine, large intestine, and anus. The accessory organs are important in that they secrete many substances necessary for digestion, yet these organs do not come into direct contact with ingested material. Examples of accessory organs are the salivary glands, liver, gallbladder, and pancreas.

The functions of the digestive system are many and include the ingestion of food, the physical breakdown of food, the chemical breakdown of food, nutrient storage, nutrient and water absorption, and the elimination of indigestible material. In this exercise you examine the anatomy of the digestive system, correlating the structure of the digestive organs with their functions.

Learning Objectives

At the end of this exercise you should be able to

1. list, in sequence, the major organs of the digestive tract;
2. describe the basic function of the accessory digestive organs;
3. note the specific anatomical features of each major digestive organ;
4. describe the layers of the wall of the digestive tract;
5. describe the major functions of the stomach, small intestine, and large intestine; and
6. distinguish among different regions of the alimentary canal by their histology.

Materials

Models, charts, or illustrations of the digestive system
Hand mirror
Skull, human teeth, or cast of teeth
Cadaver (if available)
Microscopes
Microscope slides:
- Esophagus
- Stomach
- Small intestine
- Large intestine
- Liver

Procedure

Overview of the Digestive System

Begin this exercise by examining a torso model or charts in the lab and compare them with figure 26.1. Locate the major digestive organs and, when you find them, place a check mark next to the following terms.

_____ Mouth

_____ Pharynx

_____ Esophagus

_____ Stomach

_____ Small intestine

_____ Large intestine (colon)

_____ Liver

_____ Pancreas

_____ Anus

_____ Gallbladder

_____ Salivary glands (parotid, submandibular, and sublingual glands)

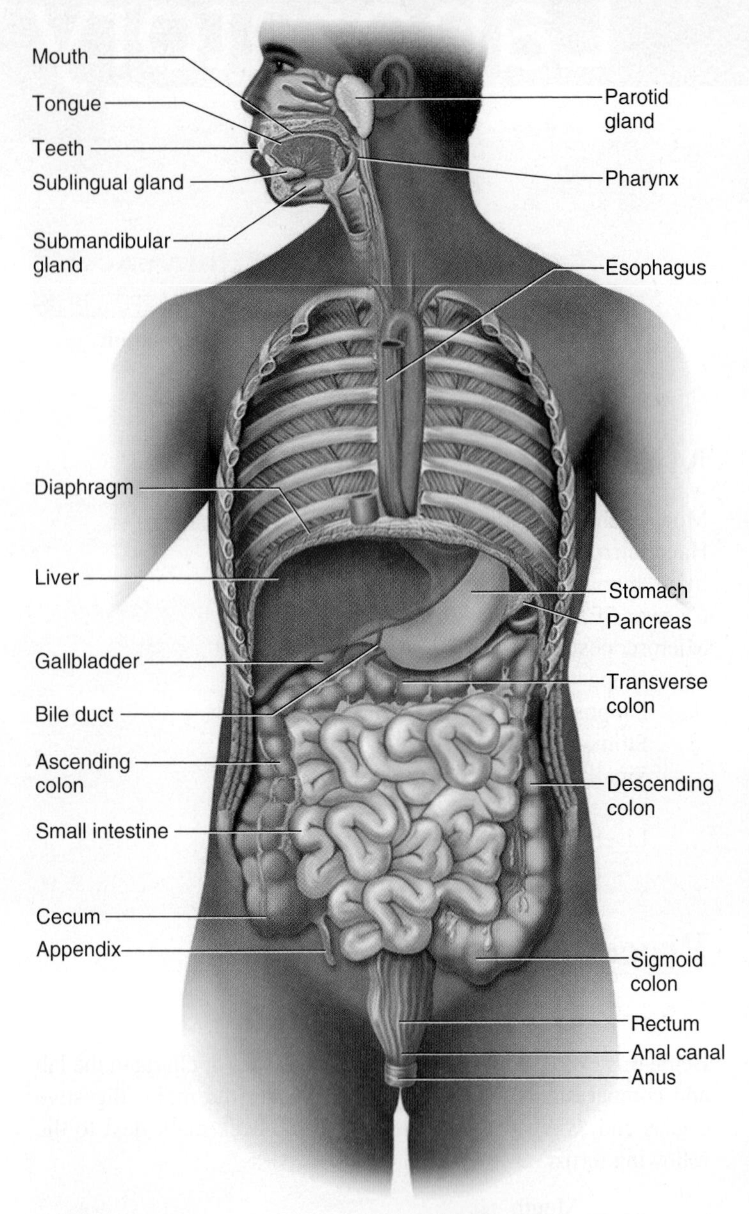

FIGURE 26.1 **Overview of the Digestive System**

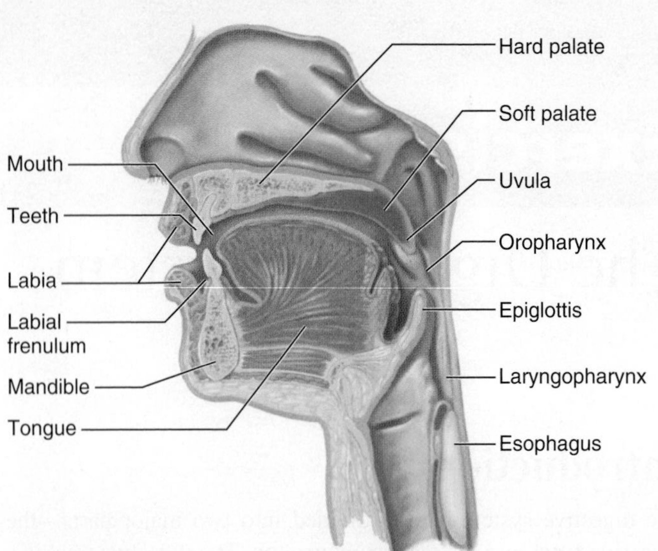

FIGURE 26.2 **Mouth, Median Section**

Organs of the Digestive Tract

Begin your study of the digestive tract with the **mouth (oral cavity).** Examine a median section of the head, as represented in figure 26.2, and locate the major anatomical features.

At the beginning of the digestive tract is the mouth, which begins as an opening surrounded by lips, or **labia.** The **labial frenulum** is a membranous structure that keeps the lip adhered to the gums, or **gingivae.** The mouth is a space anterior to the **oropharynx,** and medial to the cheeks. The hard and soft palates form the roof of the mouth, and the mylohyoid muscle is the inferior border. The **hard palate** is composed of the palatine bones and the palatine processes of the maxillae. The **soft palate** is composed of connective tissue and a mucous membrane. At the posterior portion of the mouth is the **uvula** (small grape), a structure that is suspended from the posterior edge of the soft palate. The uvula helps prevent food or liquid from moving into the nasal cavity during swallowing. The oral cavity is mostly lined with nonkeratinized stratified squamous epithelium, which protects the underlying tissue from abrasion.

In a median section of the head, the tongue appears as a broad, fan-shaped muscular structure. The **tongue** is made of many skeletal muscles, glands, epithelial tissue, and the lingual tonsils. The tongue is important in speech, taste, moving food toward the teeth for chewing, and pushing food to the oropharynx for swallowing. The tongue is held down to the floor of the mouth by a thin mucous membrane called the **lingual frenulum.** On the superior surface of the tongue are four types of **papillae,** or raised areas. These are the **foliate, fungiform, filiform,** and **circumvallate papillae.** Papillae increase the frictional surface of the tongue. The tongue also has taste receptors located in **taste buds.** Taste buds are found on the tongue along the sides of the circumvallate papillae and at the apex of the fungiform papillae. The sense of taste is covered in laboratory exercise 18. Using a mirror, examine your tongue and locate the papillae. The mouth is important in digestion for the physical breakdown of food. This process is driven by powerful muscles called the **muscles of mastication.** The **masseter** and the **temporalis muscles** are involved in the closing of the jaws, and the **pterygoid muscles** are important in the sideways grinding action of the molar and premolar teeth.

Structure of Teeth Examine models of teeth, dental casts, or real teeth on display in the lab and compare them with figure 26.3. A tooth consists of a **crown,** a **neck,** and a **root.** The crown is the exposed part of the tooth; the neck, a constricted portion of the tooth, is normally at the surface of the gingivae; and the root is embedded in the jaw. Examine a model or an illustration of a longitudinal section of a tooth and find the outer **enamel,** an extremely hard material. Inside this layer is the **dentin,** which is made of

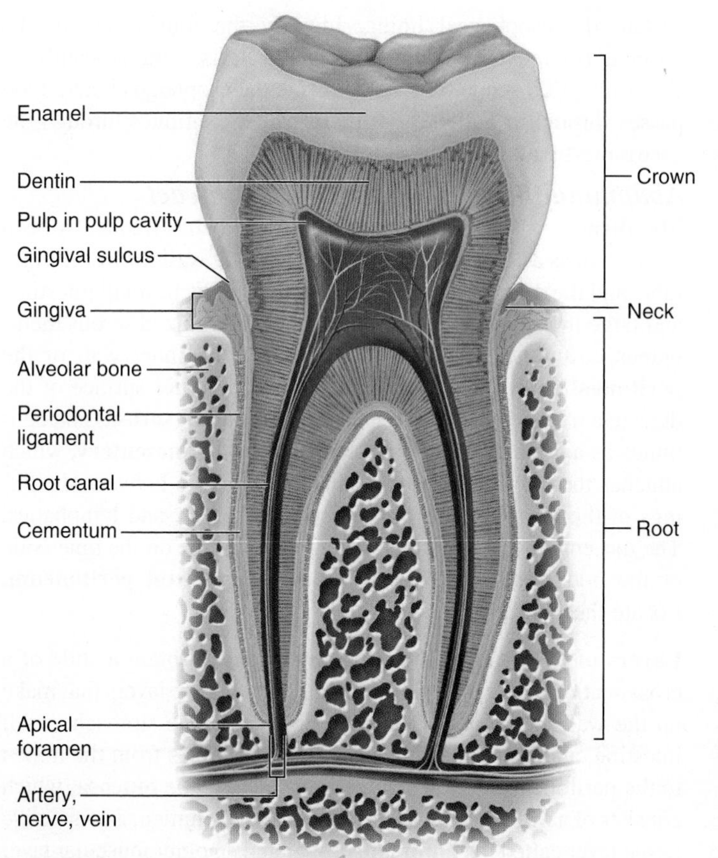

FIGURE 26.3 **Tooth, Longitudinal Section**

bonelike material. The innermost portion of the tooth is the **pulp cavity,** which leads to the **root canal,** a passageway for nerves and blood vessels into the tooth. The nerves and blood vessels enter the tooth through the **apical foramen** at the tip of the root of the tooth. The teeth are set in depressions in the mandible or maxilla called **alveoli** and are anchored to the bone by periodontal ligaments.

There are four types of teeth in the adult mouth:

- **Incisors** are flat, bladelike front teeth that nip food. There are eight incisors in the adult mouth.
- **Canines,** or **cuspids,** are pointed teeth just lateral to the incisors that shear food. There are four cuspids in the adult, and they can be identified as the teeth that have just one cusp, or point.
- **Premolars,** or **bicuspids,** are lateral to the cuspids and grind food. There are eight premolars in the adult mouth, and they can be identified by their two cusps.
- **Molars** are the most posterior. Like the premolars, these teeth grind food. There are 12 molars in the adult mouth (including the third molars, or **wisdom teeth**). Molars typically have three to five cusps.

Examine a human skull or dental cast and identify the characteristics of the four types of teeth (fig. 26.4). Humans have two sets of teeth: the **primary,** or **deciduous, teeth** (milk teeth) appear

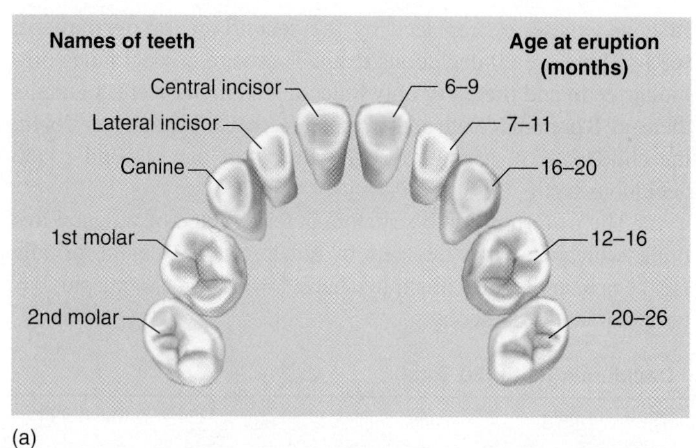

(a)

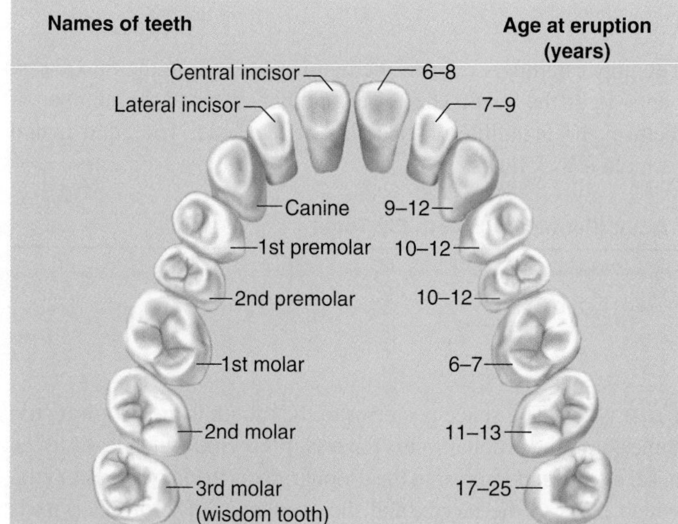

(b)

(c)

FIGURE 26.4 **Dentition** (a) Upper teeth of a child. Notice the absence of premolar teeth. (b) Upper teeth of an adult. (c) Replacement of deciduous teeth with permanent teeth.

first, and these are replaced by the **secondary,** or **permanent, teeth.** There are 20 deciduous teeth. There are no deciduous premolar teeth, and there are only 8 deciduous molar teeth. In adults there are 8 premolar teeth and 12 molar teeth. Compare figure 26.4*a*, the child pattern, to figure 26.4*b*, the adult pattern, and *c*, the deciduous teeth.

The pattern of tooth structure is represented by a dental formula, which describes the teeth by quadrants. The dental formula for the primary teeth is illustrated here: I = incisor, C = cuspid, P = premolar, and M = molar.

Deciduous Teeth (20 Total)

Dental formula	1	C	P	M	One side (quadrant)
	2	1	0	2	Top
	2	1	0	2	Bottom

The upper numbers refer to the number of teeth in the maxilla on one side of the head. The lower numbers refer to the number of teeth in the mandible on one side of the head. The adult dental formula is as follows:

Adult (Permanent) Teeth (32 Total)

1	C	P	M
2	1	2	3
2	1	2	3

Pharynx The space posterior to the mouth is the **oropharynx.** Superior to the oropharynx is the **nasopharynx,** which leads to the nasal cavity, and inferior to the oropharynx is the **laryngopharynx,** which leads to the larynx and the esophagus. The oropharynx is lined with nonkeratinized stratified squamous epithelium and is a common passageway for food, liquids, and air. Muscles around the wall of the oropharynx and laryngopharynx are the **pharyngeal constrictor muscles** and are involved in swallowing. Food is moved by the tongue to the region of the pharynx where it is propelled into the **esophagus.** Locate the oropharynx and the esophagus in figures 26.1 and 26.2.

Esophagus The esophagus conducts food from the laryngopharynx, through the diaphragm, and into the stomach. The esophagus has an inner layer of stratified squamous epithelium near the lumen (the space where ingested material travels throughout the alimentary canal) and an outer connective tissue layer called the **adventitia.** Unlike the trachea, which is an open tube, the esophagus is a closed tube that begins approximately at the level of the sixth cervical vertebra. As a lump of food, or **bolus,** enters the esophagus, skeletal muscle begins to move it toward the stomach. The middle portion of the esophagus is composed of both skeletal and smooth muscle, while the lower portion of the esophagus is made of smooth muscle. In the lower region of the esophagus the smooth muscle contracts, moving the bolus by a process known as **peristalsis.** Locate the esophagus in charts and models in the lab and compare it with figure 26.1. The lower portion of the esophagus has the **lower esophageal sphincter,** which prevents the backflow of stomach acids. Esophageal reflex (heartburn) occurs if the stomach contents pass through the lower esophageal sphincter and irritate the esophageal lining. Identify the four layers of the esophagus—mucosa, submucosa, muscularis, and adventitia—under the microscope. The space inside the esophagus where food passes through is called the **lumen,** which continues through the gastrointestinal tract.

Abdominal Portions of the Digestive Tract

Membranes The inner structure of the body has frequently been referred to as a "tube within a tube." The body wall forms the outer tube, and the digestive tract, including the stomach, small intestine, and large intestine, forms the inner tube. Specialized serous membranes cover the various organs and line the inner wall of the **peritoneal cavity.** The membrane lining the outer surface of the digestive tract is called the **visceral peritoneum (serosa)** and continues as a double-folded membrane called the **mesentery,** which attaches the tract to the back of the body wall. In between the linings of the mesentery are arteries, veins, nerves, and lymphatics. The mesentery is continuous with the membrane on the inner side of the body wall, where it is called the **parietal peritoneum.** Locate these three membranes in figure 26.5.

Layers of the Wall of the Digestive Tract Obtain a slide of a cross section of the digestive tract and look for the layers that make up the wall of the digestive tract. In general, the stomach, small intestine, and large intestine have the same layers from the lumen to the peritoneal cavity. The innermost layer is the **mucosa,** which consists of a **mucous membrane** closest to the lumen; a connective tissue layer called the **lamina propria;** and an outer muscular layer called the **muscularis mucosae.**

The next layer is the **submucosa,** which is mostly composed of **connective tissue** and contains numerous blood vessels. The next layer is the **muscularis externa,** which is typically made of two or three layers of smooth muscle. The muscularis externa propels material through the digestive tract and mixes ingested material with digestive juices. The outermost layer is called the **serosa,** or **visceral peritoneum,** and this layer is closest to the peritoneal cavity. Examine a microscope slide of the digestive tract (small intestine) and locate the layers as represented in figure 26.6.

Stomach Examine a model of the stomach or charts in the lab and compare these with figure 26.7. The **stomach** is located on the superior middle and left portion of the abdomen and receives its contents from the esophagus. The food that enters the stomach is stored and mixed with enzymes and hydrochloric acid (HCl) to form a soupy material called **chyme.** The stomach can have a pH as low as 1 or 2. Chyme remains in the stomach as the acids denature proteins and enzymes reduce proteins to shorter fragments. The acid of the stomach also has an antibacterial action, as most microbes do not grow well in conditions of low pH.

Locate the upper portion of the stomach called the **cardiac region,** or **cardia.** The region of the stomach that extends superiorly as a domed section is called the **fundus,** or **fundic region.** The main part of the stomach is called the **body,** and the terminal portion of the stomach, closest to the small intestine, is called the **pyloric region.** The pyloric region has an expanded area called the **antrum** and a narrowed region called the **pyloric canal** that leads to the duodenum.

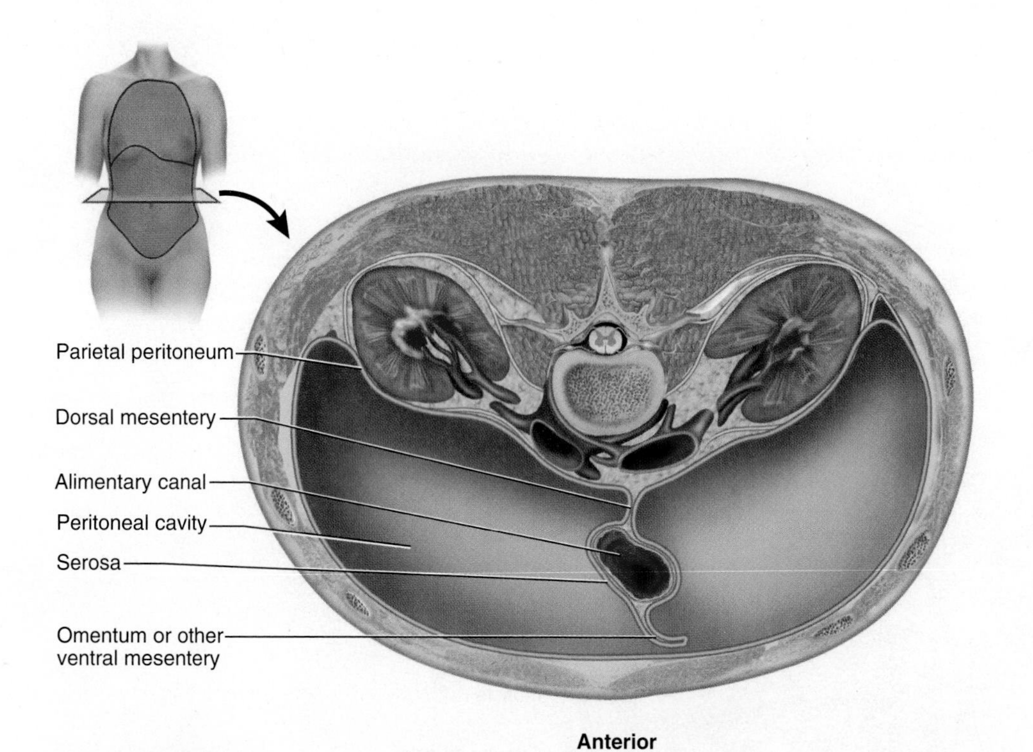

FIGURE 26.5 **Membranes of the Digestive Tract (Idealized Drawing)**

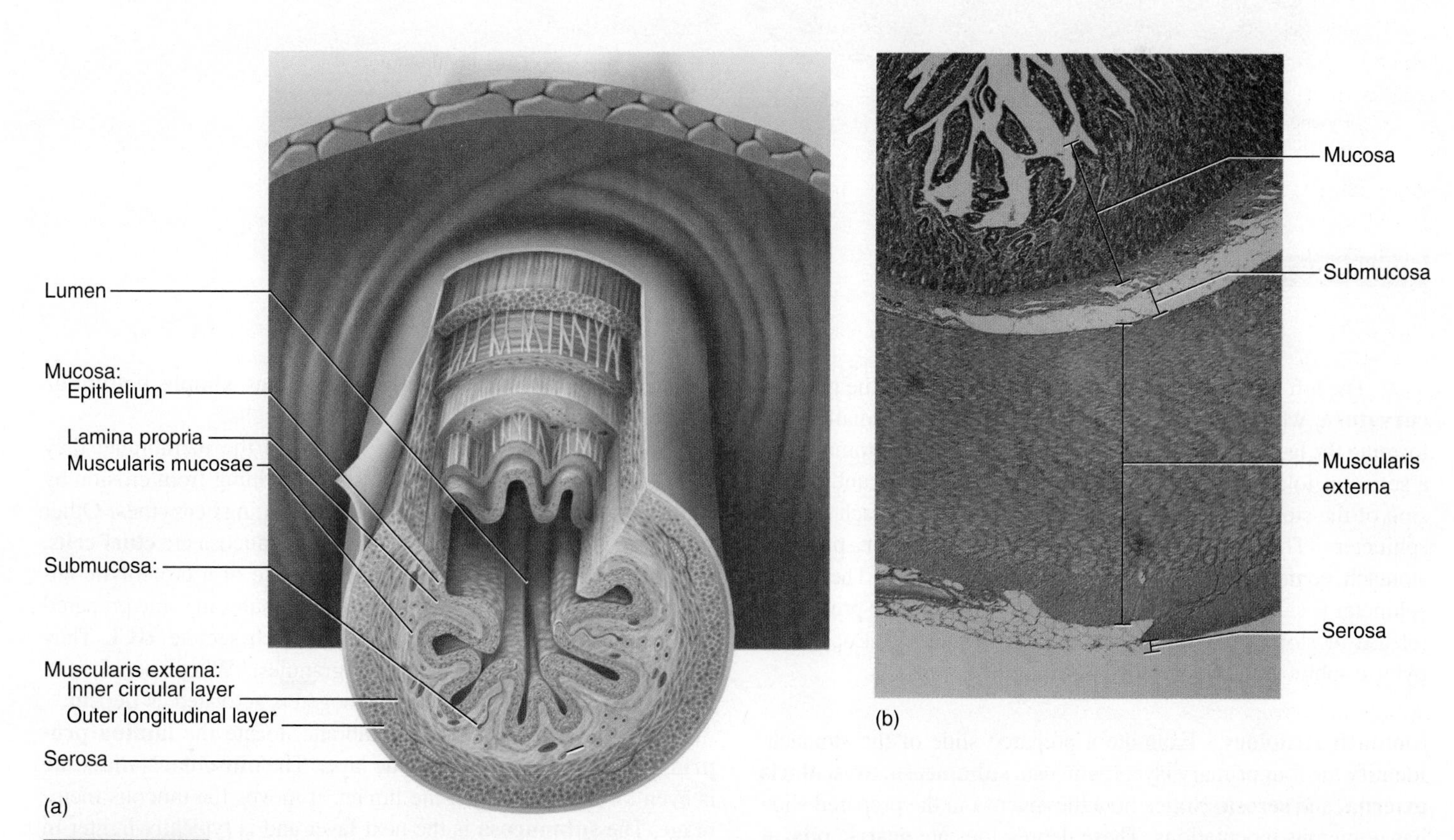

FIGURE 26.6 **Digestive Tract, Cross Section** (a) Diagram; (b) photomicrograph (100×).

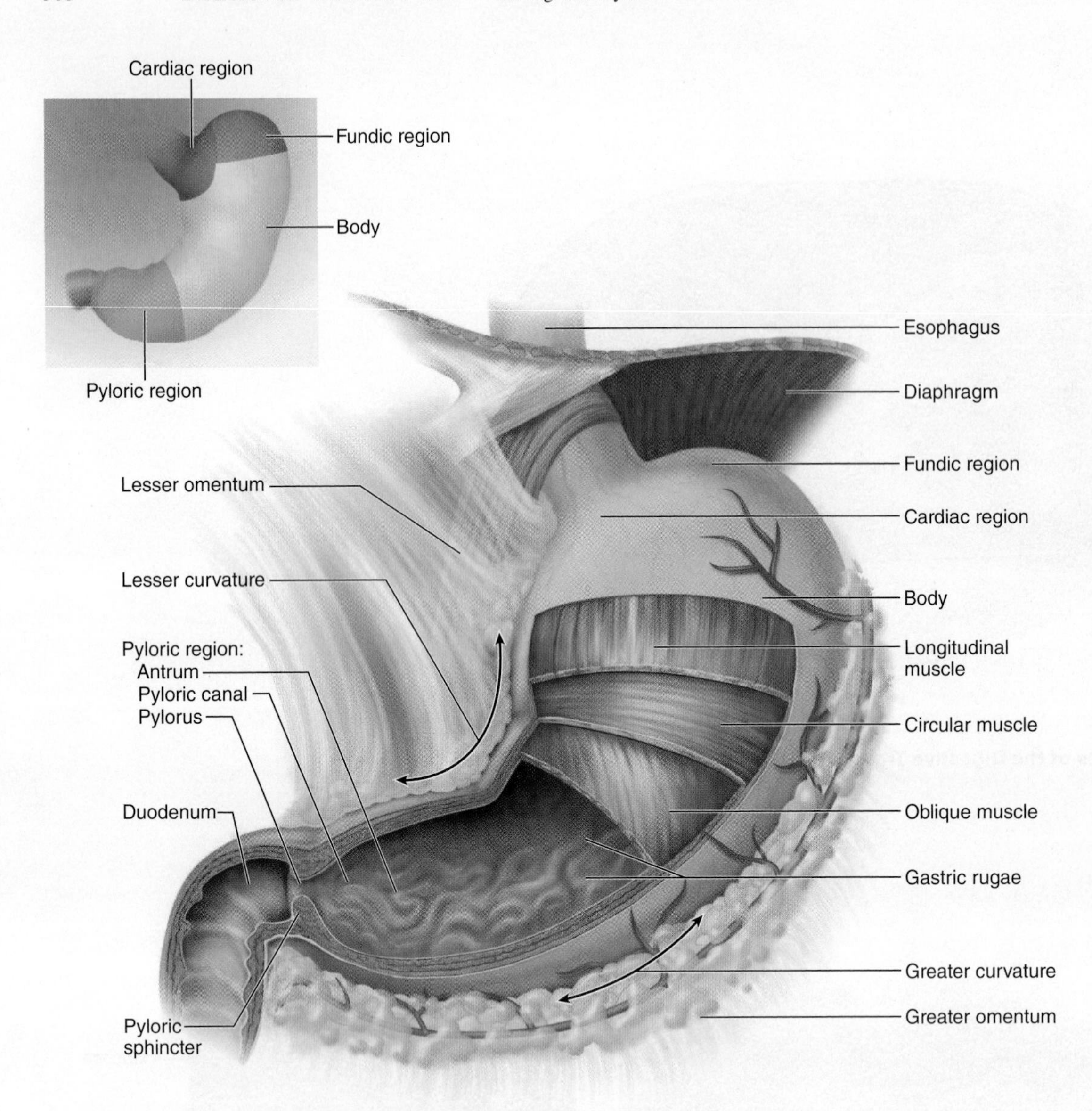

FIGURE 26.7 **Gross Anatomy of the Stomach**

The left side of the stomach is arched and forms the **greater curvature,** while the right side of the stomach is a smaller arc, forming the **lesser curvature.** The inner surface of the stomach has a series of folds called **rugae,** which allow for significant expansion of the stomach. The contents are held in the stomach by two sphincters. The upper one, the **esophageal sphincter,** prevents stomach contents from moving into the esophagus. The lower sphincter is called the **pyloric sphincter;** it prevents the premature release of stomach contents into the small intestine. Locate the pyloric sphincter at the terminal portion of the stomach.

Stomach Histology Examine a prepared slide of the stomach. Identify the four primary layers: **mucosa, submucosa, muscularis externa,** and **serosa.** Notice how the mucosa in the prepared slide has numerous indentations. These depressions are **gastric pits,** at the bottom of which are the **gastric glands** in the inner lining of the stomach. The **mucous membrane** contains **simple columnar epithelium,** along with several specialized cells.

Surface mucous cells are located in the membrane; they secrete mucus, which protects the stomach lining from erosion by stomach acid and proteolytic (protein-digesting) enzymes. Other specialized cells that you might find in the mucosa are **chief cells,** which secrete **pepsinogen** (the inactive state of a proteolytic enzyme). Chief cells contain blue-staining granules in some prepared slides. Other cells are **parietal cells,** which secrete **HCl.** They typically contain orange-staining granules. When pepsinogen comes into contact with HCl, pepsinogen is activated as **pepsin.**

Deeper to the mucous membrane, locate the **lamina propria,** which is a connective tissue layer. The **muscularis mucosae** is even farther away from the lumen; it moves the mucous membrane. The **submucosa** is the next layer and is typically lighter in color in prepared slides.

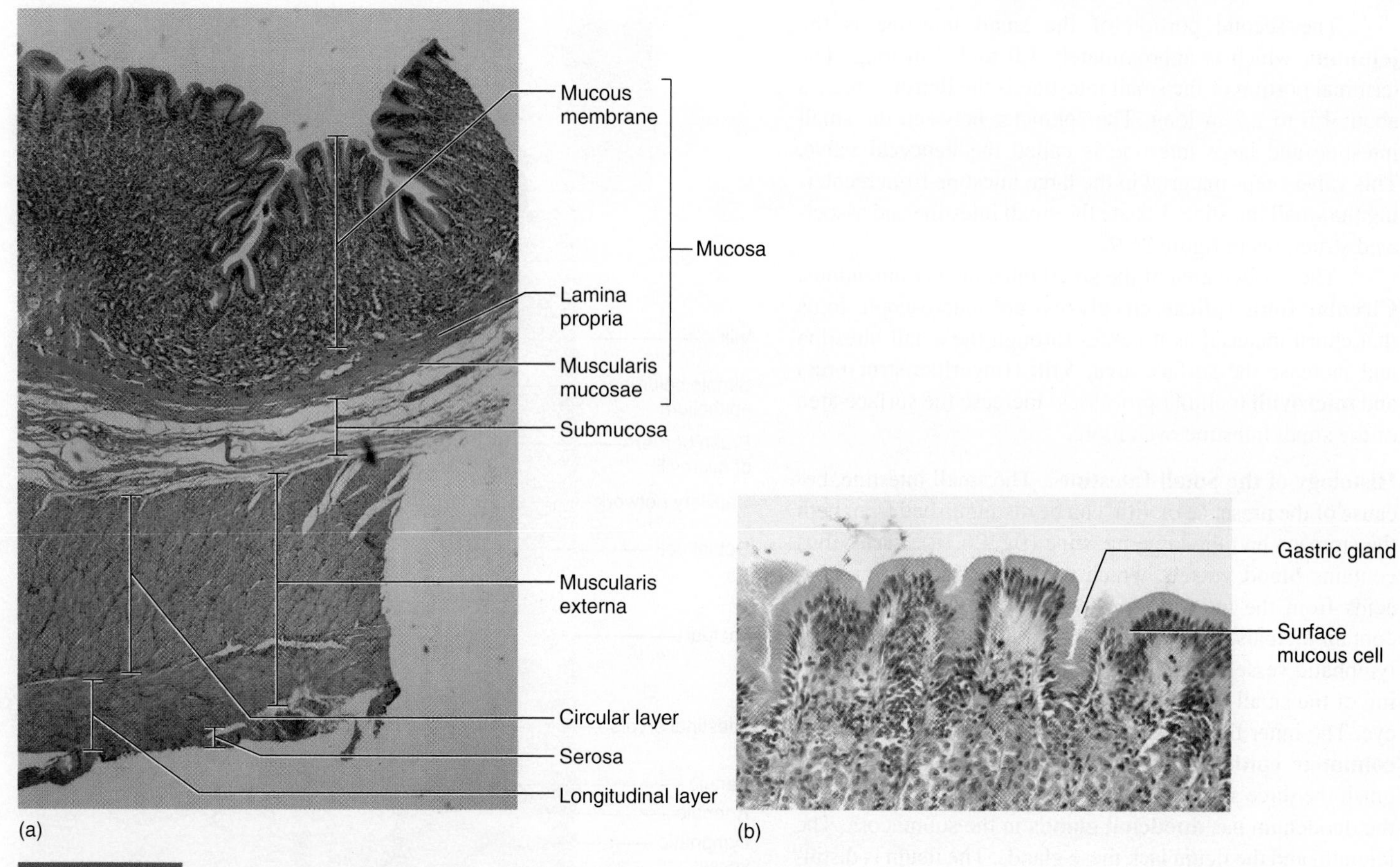

FIGURE 26.8 Histology of the Stomach (a) Overview (40×); (b) mucosa (100×).

The next layer of the stomach is the **muscularis externa.** The muscularis externa consists of an inner **oblique layer,** a middle **circular layer,** and an outer **longitudinal layer.** The muscularis externa moves chyme from the stomach through the pyloric sphincter and into the small intestine. It also mixes the chyme.

The outermost layer is the **serosa,** and it is composed of a thin layer of connective tissue and **simple squamous epithelium.** Locate these structures and compare them with figure 26.8.

Small Intestine The **small intestine** is approximately 4 to 5 m long in living humans yet can be longer in cadaveric specimens due to the stretching of the smooth muscle. Movement through the small intestine occurs by peristalsis, which is smooth muscle contraction. It is called the small intestine because it is small in diameter. The small intestine is typically 3 to 4 cm in diameter when empty. The primary function of the small intestine is nutrient absorption.

Locate the three major regions of the small intestine on a model or chart and compare them with figure 26.9. The first part of the small intestine is the **duodenum,** which forms a C-shaped curve attached to the pyloric region of the stomach. The duodenum is approximately 25 cm long. It receives fluid from both the **pancreas** and the **gallbladder.** The junction between the duodenum and the jejunum is known as the **duodenojejunal flexure.**

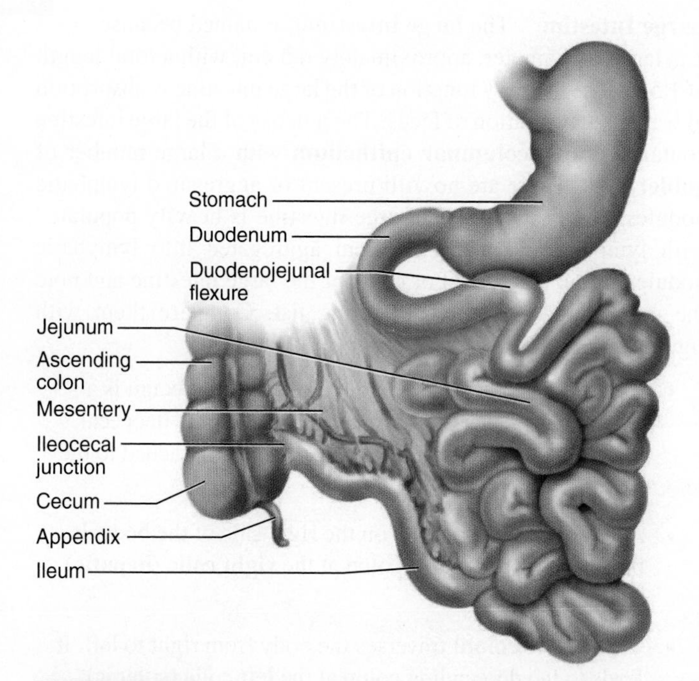

FIGURE 26.9 Gross Anatomy of the Small Intestine

The second portion of the small intestine is the **jejunum,** which is approximately 1.0 to 1.7 m long. The terminal portion of the small intestine is the **ileum,** which is about 1.6 to 2.7 m long. The sphincter between the small intestine and large intestine is called the **ileocecal valve.** This valve keeps material in the large intestine from reentering the small intestine. Locate the small intestine and associated structures in figure 26.9.

The surface area of the small intestine is tremendous. **Circular folds (plicae circulares)** are macroscopic folds that churn material as it passes through the small intestine and increase the surface area. **Villi** (fingerlike structures) and **microvilli** (cellular processes) increase the surface area of the small intestine even more.

Histology of the Small Intestine The small intestine, because of the presence of villi, can be distinguished from both the stomach and the large intestine (fig. 26.10). Each villus contains **blood vessels,** which transport sugars and amino acids from the intestine to the liver. In addition, the villi contain **lacteals,** which transport lipids as chylomicrons via lymphatic vessels to the bloodstream. The villi give the lining of the small intestine a velvety appearance to the naked eye. The inner lining of the small intestine contains **simple columnar epithelium** with **goblet cells.** You may distinguish the three sections of the small intestine by noting that the duodenum has **duodenal glands** in the submucosa. The jejunum and the ileum lack these glands. The ileum is distinguished by the presence of **aggregated lymphatic nodules,** or **Peyer patches,** present in the submucosa. These lymphatic nodules produce lymphocytes, which protect the body from the bacterial flora in the lumen of the small intestine. Compare the prepared slides of the small intestine with figure 26.11.

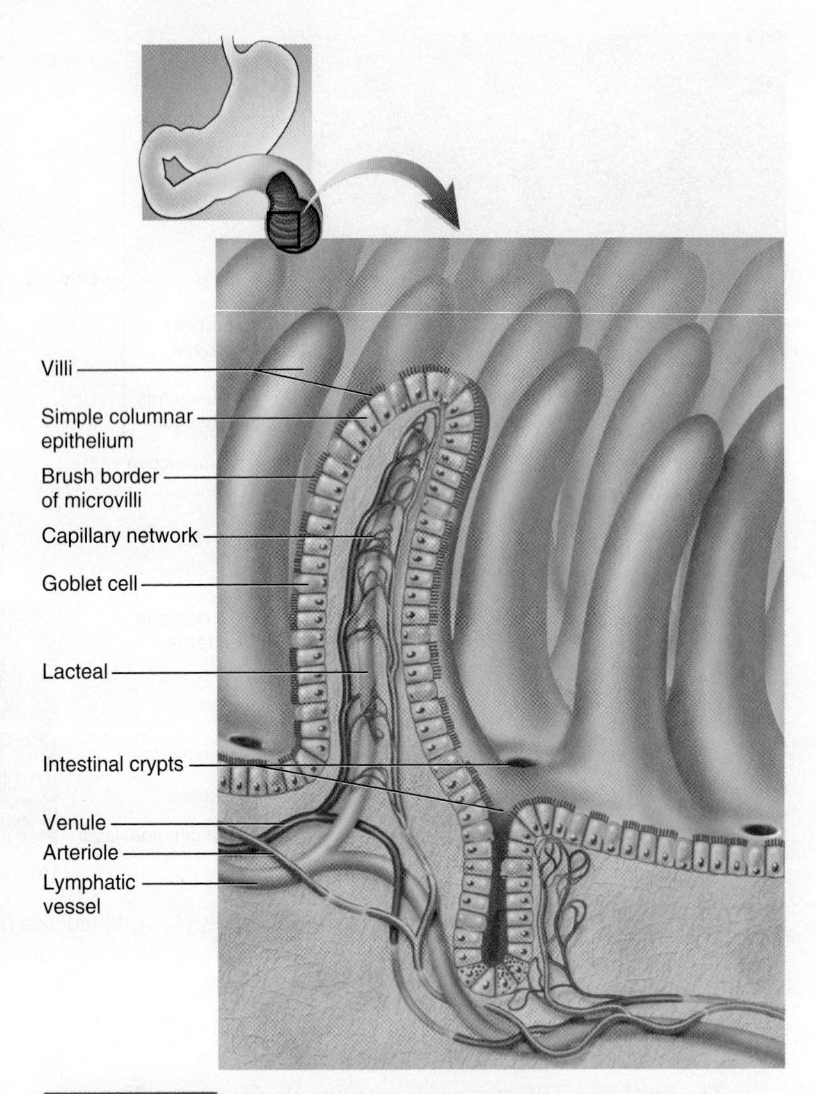

FIGURE 26.10 Villi of the Small Intestine

Large Intestine The **large intestine,** so named because it is large in diameter, approximately 6.5 cm, with a total length of 1.5 m. The primary function of the large intestine is absorption of water and formation of feces. The mucosa of the large intestine contains **simple columnar epithelium** with a large number of **goblet cells.** There are no villi present or aggregated lymphatic nodules, yet the wall of the large intestine is heavily populated with lymphocytes, many of them aggregated into lymphatic nodules. Look at a model or chart of the large intestine and note the major regions in the following list. Compare them with figure 26.12.

- **Cecum:** first part of the large intestine. The cecum is a pouch that joins with the small intestine at the ileocecal junction. The appendix (described later) is attached to the cecum.
- **Ascending colon:** found on the right side of the body. It becomes the transverse colon at the **right colic (hepatic) flexure.**
- **Transverse colon:** traverses the body from right to left. It leads to the descending colon at the **left colic (splenic) flexure.**
- **Descending colon:** passes inferiorly on the left side of the body and joins with the sigmoid colon
- **Sigmoid colon:** S-shaped segment of the large intestine in the pelvic region
- **Rectum:** straight section of colon in the pelvic cavity. The rectum has superficial veins in its wall called **hemorrhoidal veins.** When these enlarge, they produce hemorrhoids.

The large intestine has some unique structures. The longitudinal layer of the muscularis externa of the large intestine is thickened along the length of the colon into three bands called **taeniae coli.** These muscles contract and form pouches called **haustra** (singular, *haustrum*). Another unique feature of the large intestine is fat lobules along the outer wall called **omental appendages.** Locate these structures in figure 26.12.

Fecal material passes through the large intestine by peristalsis and is stored in the rectum and sigmoid colon. Defecation occurs as mass peristalsis causes a bowel movement.

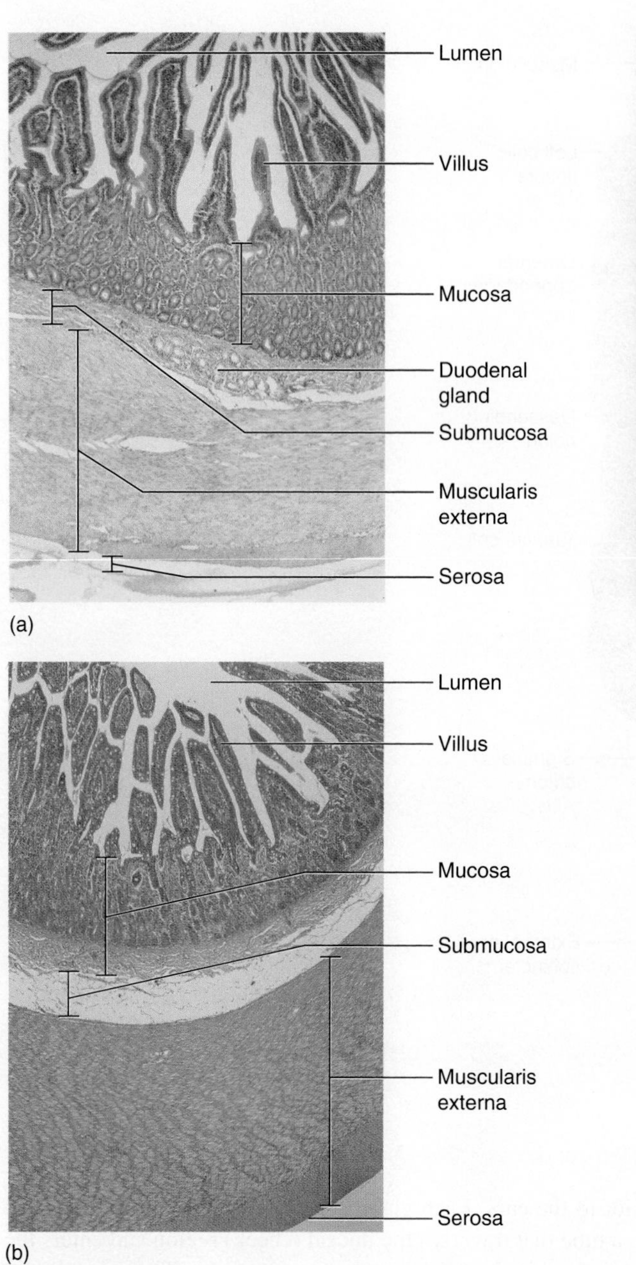

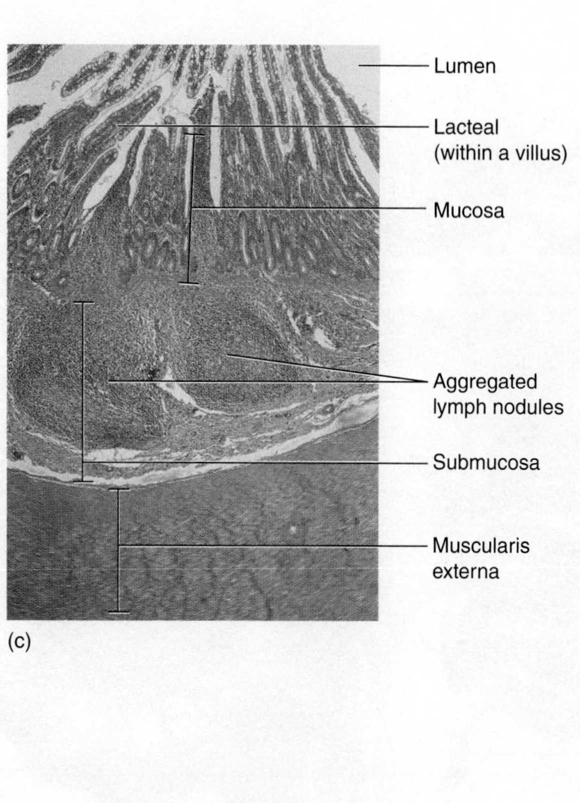

FIGURE 26.11 Three Sections of Small Intestine (40×) (a) Duodenum; (b) jejunum; (c) ileum.

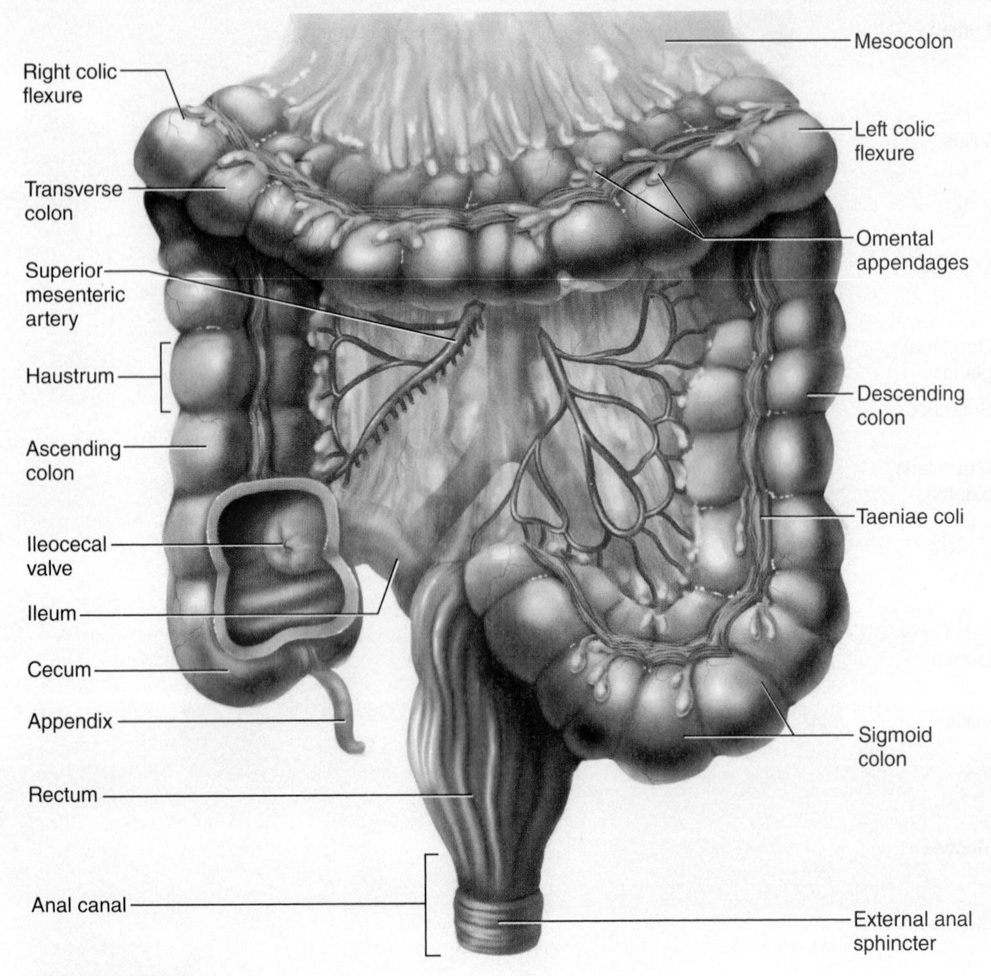

FIGURE 26.12 **Gross Anatomy of the Large Intestine and Anal Canal**

Histology of the Large Intestine Examine a prepared slide of the large intestine and compare it with figure 26.13. Locate the mucosa, submucosa, muscularis externa, and serosa in your slide. The large intestine is distinguished from the small intestine by the absence of villi, and it is distinguished from the stomach by the presence of large numbers of goblet cells. Examine your slide for these characteristics.

Anal Canal The **anal canal** is not part of the large intestine but is a short tube lined with stratified squamous epithelium which resists abrasion during defecation. The anal canal leads to an external opening, the **anus.** Locate the anal canal in figure 26.12.

Accessory Structures

Salivary Glands The **salivary glands,** located in the head, secrete **saliva** into the mouth. Saliva is a watery secretion that contains **mucus,** a protein lubricant, **salivary amylase,** a starch-digesting enzyme, and lingual lipase, a lipid-digesting enzyme. The average adult secretes about 1.5 liters of saliva per day.

There are three pairs of salivary glands. The first of these, the **parotid glands,** are located superficial to the masseter muscles just anterior to the ears. Each gland secretes saliva through a **parotid duct,** a tube that traverses the buccal (cheek) region and enters the mouth just posterior to the upper second molar. The second pair, the **submandibular glands,** are located just medial to the mandible on each side of the face. The submandibular glands secrete saliva into the mouth by a single duct on each side of the oral cavity inferior to the tongue. The third pair, the **sublingual glands,** are located inferior to the tongue and these glands open into the mouth by several ducts. Locate these salivary glands on a model of the head and in figure 26.14.

Appendix The **appendix** is about the size of the little finger and is attached to the cecum (at the region of the ileocecal valve). The appendix has lymph nodules, but its function is not apparent. Locate the appendix on a torso model or chart in the lab and compare it with figures 26.12 and 26.15.

Omenta The **lesser omentum** is an extension of the peritoneum that forms a double fold of tissue between the stomach and the liver. The **greater omentum** is a section of peritoneum that attaches to the inferior margin of the stomach and drapes

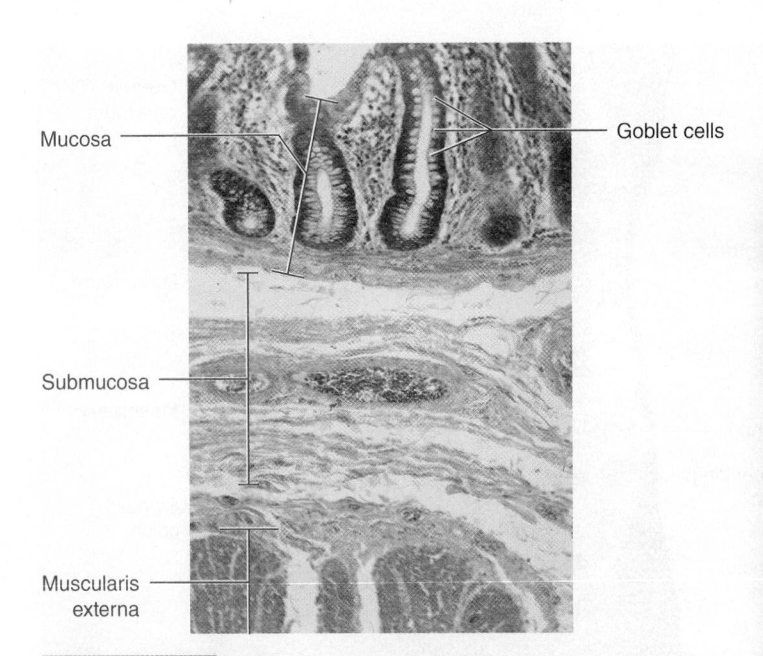

FIGURE 26.13 Histology of the Large Intestine (100×)

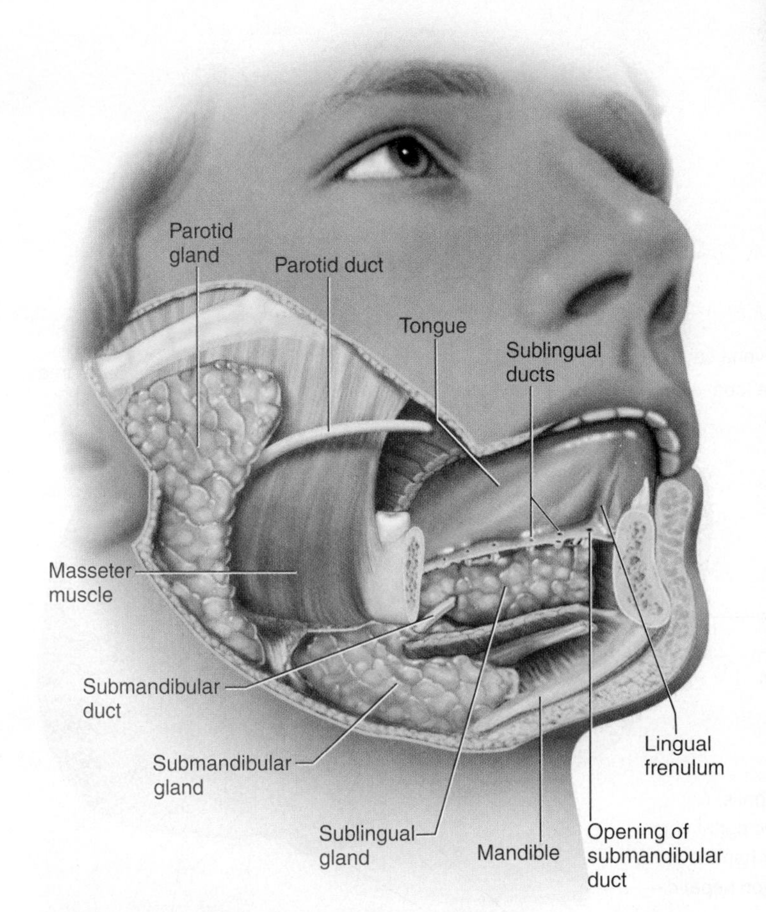

FIGURE 26.14 Salivary Glands

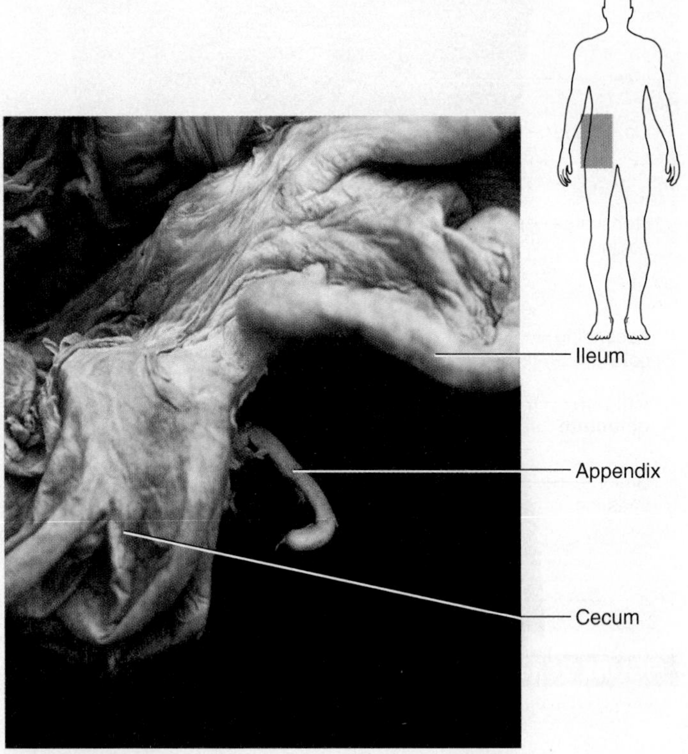

FIGURE 26.15 Appendix

over the intestines like a fatty curtain. In the large intestine the membrane that attaches the colon to the posterior body wall is called the **mesocolon.** Locate the lesser and greater omenta in figure 26.16.

Liver The **liver** is a complex organ with numerous functions, one of which is digestive (the secretion of bile) but most of which are not. The liver processes digestive material from the vessels returning blood from the intestines and has a role in either moving material into the bloodstream or storing it in the liver tissue. The liver produces blood plasma proteins, detoxifies harmful chemicals that have been produced by the body or introduced into the body, and produces bile.

Examine a model or chart of the liver and note its relatively large size. The liver is located mainly on the right side of the body and is divided into four lobes—the **right, left, quadrate,** and **caudate lobes.** Only two lobes of the liver can be seen from the anterior side, and these are the large right lobe and the smaller left lobe. These two lobes are separated by a slip of mesentery called the **falciform ligament,** which suspends the liver from the diaphragm. In an inferior view of the liver, all four lobes can be seen. The quadrate lobe is located in the middle portion of the liver, adjacent to the gallbladder and anterior to the caudate lobe. Locate the **gallbladder,** which is on the inferior aspect of the liver, and compare the gallbladder and liver structures with figure 26.17.

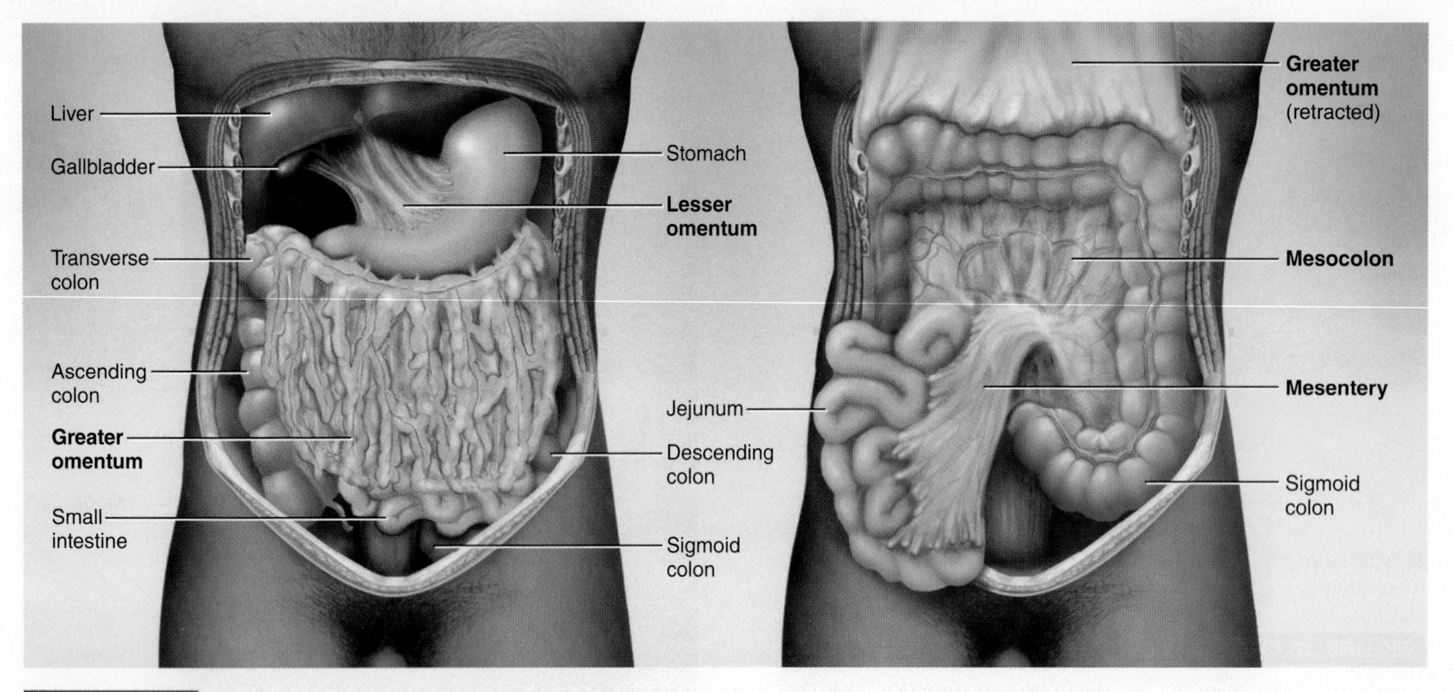

FIGURE 26.16 **Omenta, Mesentery, and Mesocolon**

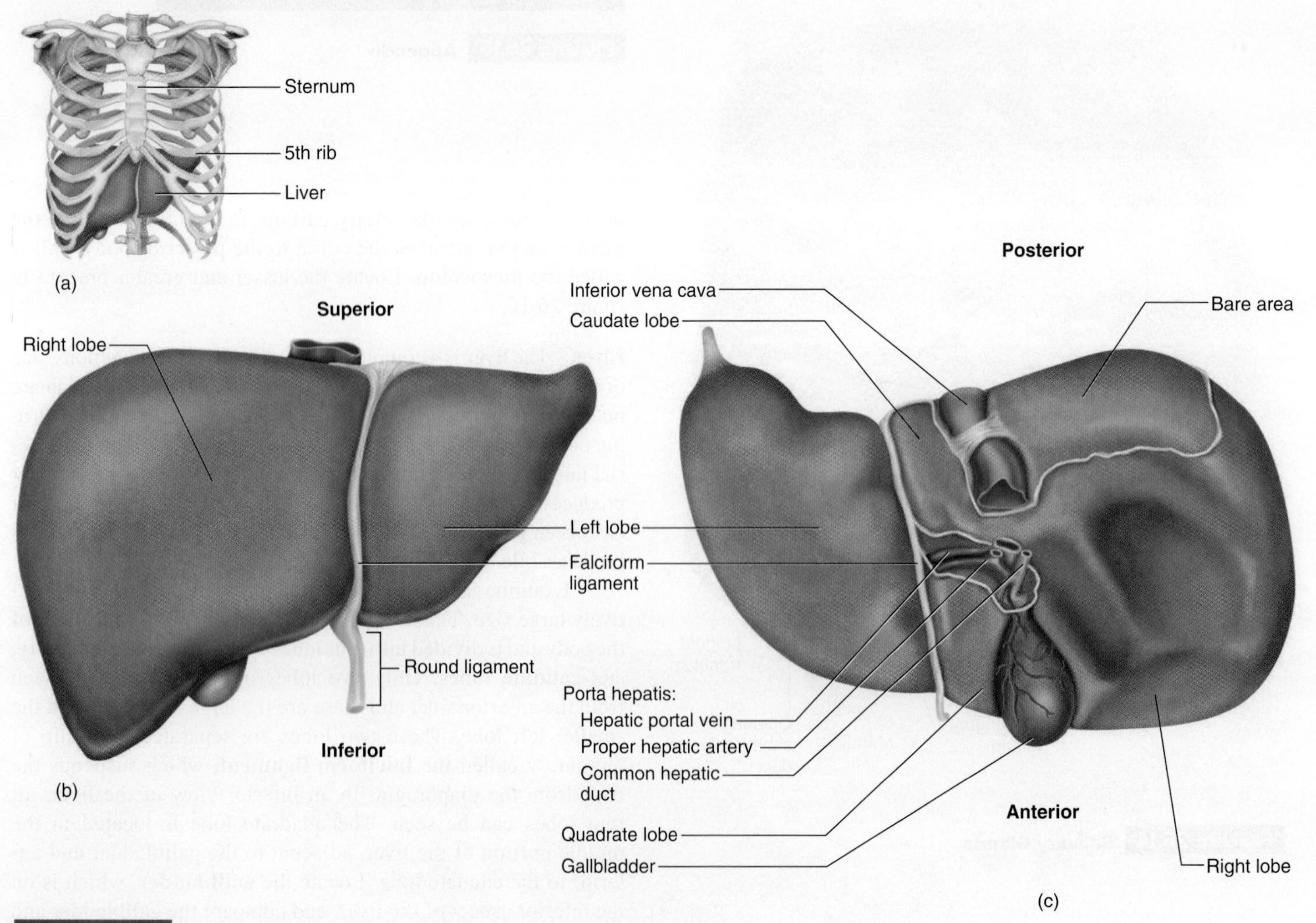

FIGURE 26.17 **Liver** (a) Overview in body; (b) anterior view; (c) inferior view.

Liver Histology Examine a prepared slide of liver tissue. Note the hexagonal structures in the specimen. These are the **liver lobules.** Each lobule has a blood vessel in the middle called the **central vein.** Vessels that carry blood to the central vein are the **hepatic sinusoids,** and these are lined with a double row of cells called **hepatocytes.** Hepatocytes carry out the various functions of the liver. The tissue of the liver is extremely vascular and functions as a sponge. Oxygenated blood from the hepatic artery and deoxygenated blood from the hepatic portal vein mix in the liver. **Hepatic macrophages** are cells that occur in the hepatic sinusoids and function as phagocytic cells. Locate the lobules, central vein, hepatic sinusoids, bile ductule, and hepatocytes in a prepared slide, using figure 26.18 to aid you.

Gallbladder The **gallbladder** releases bile, which emulsifies lipids, into the duodenum. The lipids break into smaller droplets, which increase the surface area for digestion. The gallbladder is located just inferior to the liver. The liver is the site of bile production. Bile is secreted from hepatocytes into **bile canaliculi,** which take the bile to **bile ductules.** The canaliculi flow between the plates of the liver lobules, and the bile ductules run perpendicular to the bile canaliculi. The bile eventually flows from the bile ductules into the **left** and **right hepatic ducts.** These join to form the **common hepatic duct.** A short tube, the **cystic duct,** takes bile to and from the gallbladder, where it is concentrated. As the stomach begins to empty its contents into the duodenum, the gallbladder constricts and bile flows from the gallbladder back into the cystic duct and into the **bile duct,** which empties into the duodenum. Sometimes the bile duct empties into the duodenum as a singular tube, but it usually joins the pancreatic duct to form the **hepatopancreatic ampulla.** Locate these structures on a model and compare them with figure 26.19.

Pancreas The **pancreas** is located inferior to the stomach on the left side of the body. It has both endocrine and exocrine functions. The hormonal function of the pancreas is covered in laboratory exercise 19. The pancreas consists of a **tail** near the spleen, an elongated **body,** and a rounded **head** near the duodenum. Digestive enzymes and buffers that neutralize stomach acids pass from the tissue of the pancreas into the **pancreatic duct** and then into the duodenum. Locate the pancreatic structures, as represented in figure 26.19.

Cat Anatomy

If you are using cats, turn to section 8, "Digestive Anatomy of the Cat" on page 438 of this lab manual.

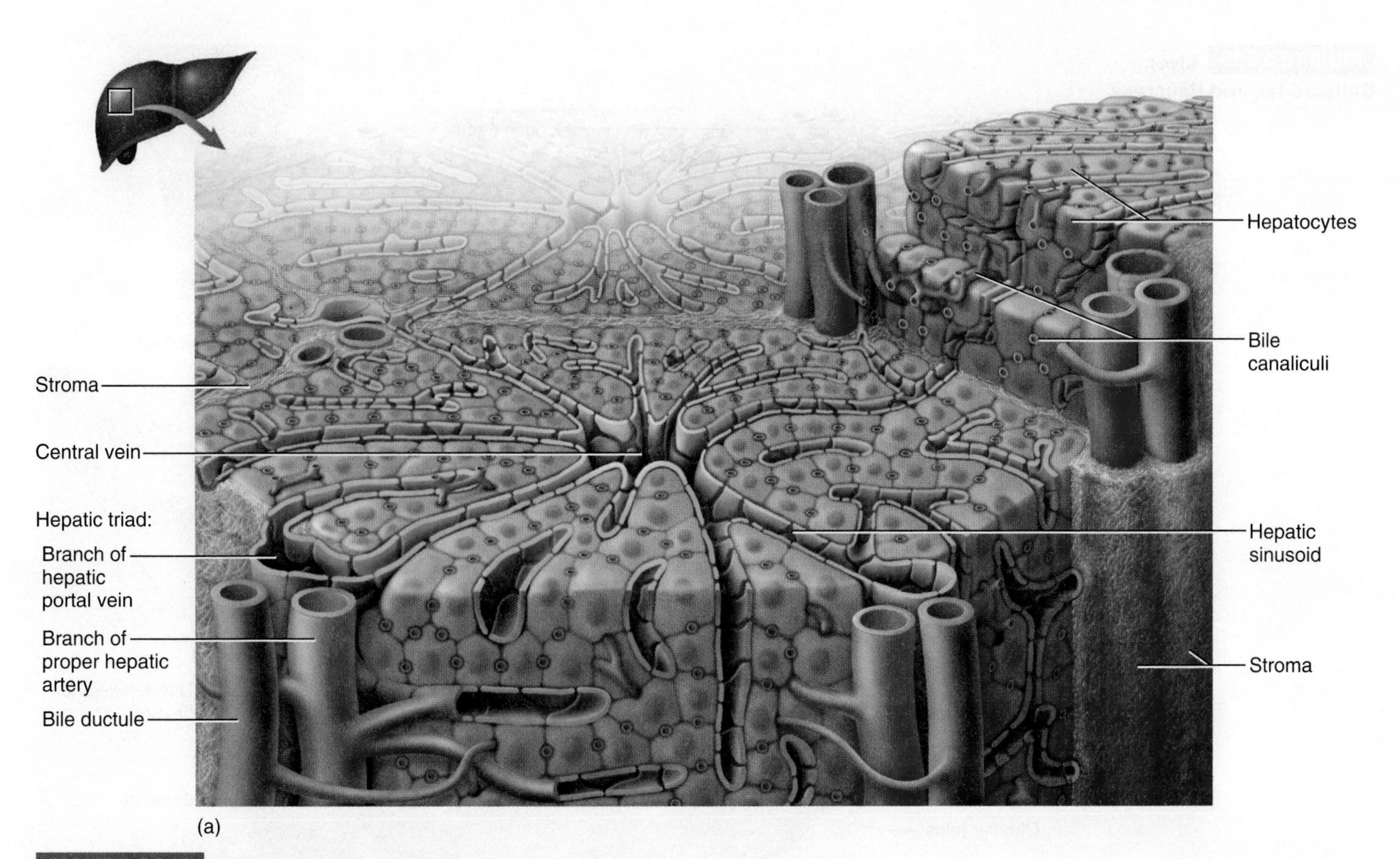

FIGURE 26.18 **Histology of the Liver** (a) Diagram.

FIGURE 26.18 (*continued*) **Histology of the Liver** (b) Photomicrograph of liver lobule (100×).

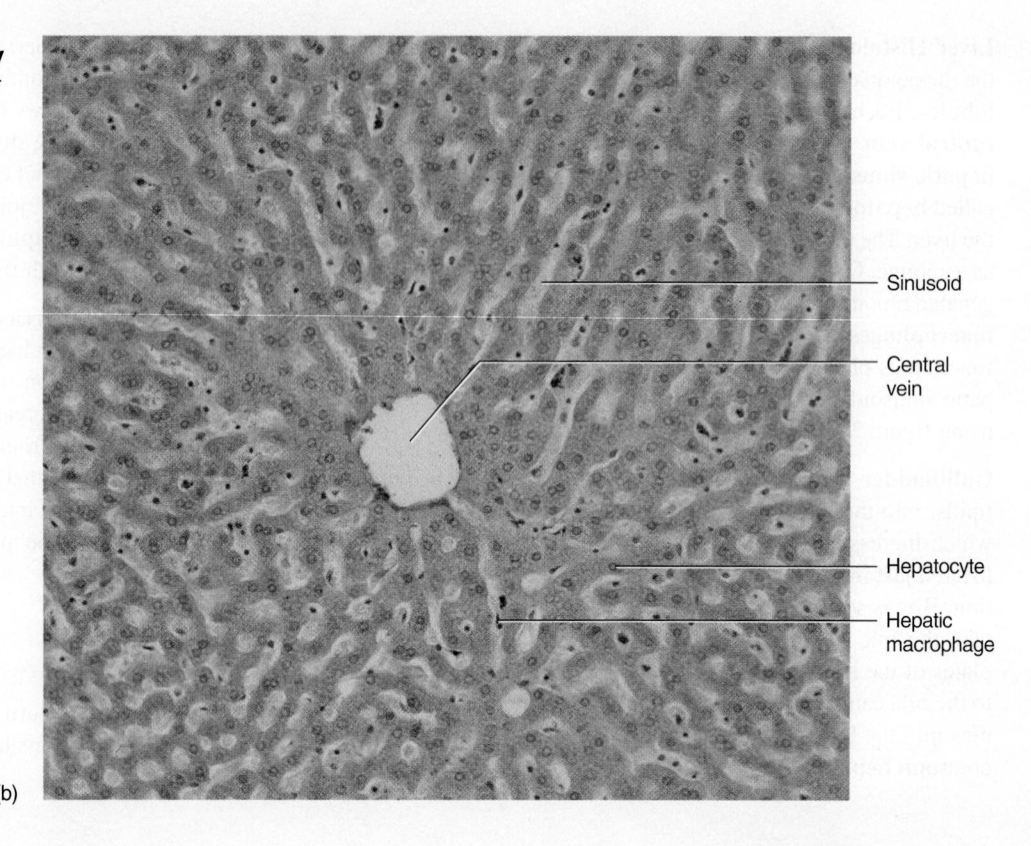

FIGURE 26.19 Liver, Gallbladder, and Pancreas

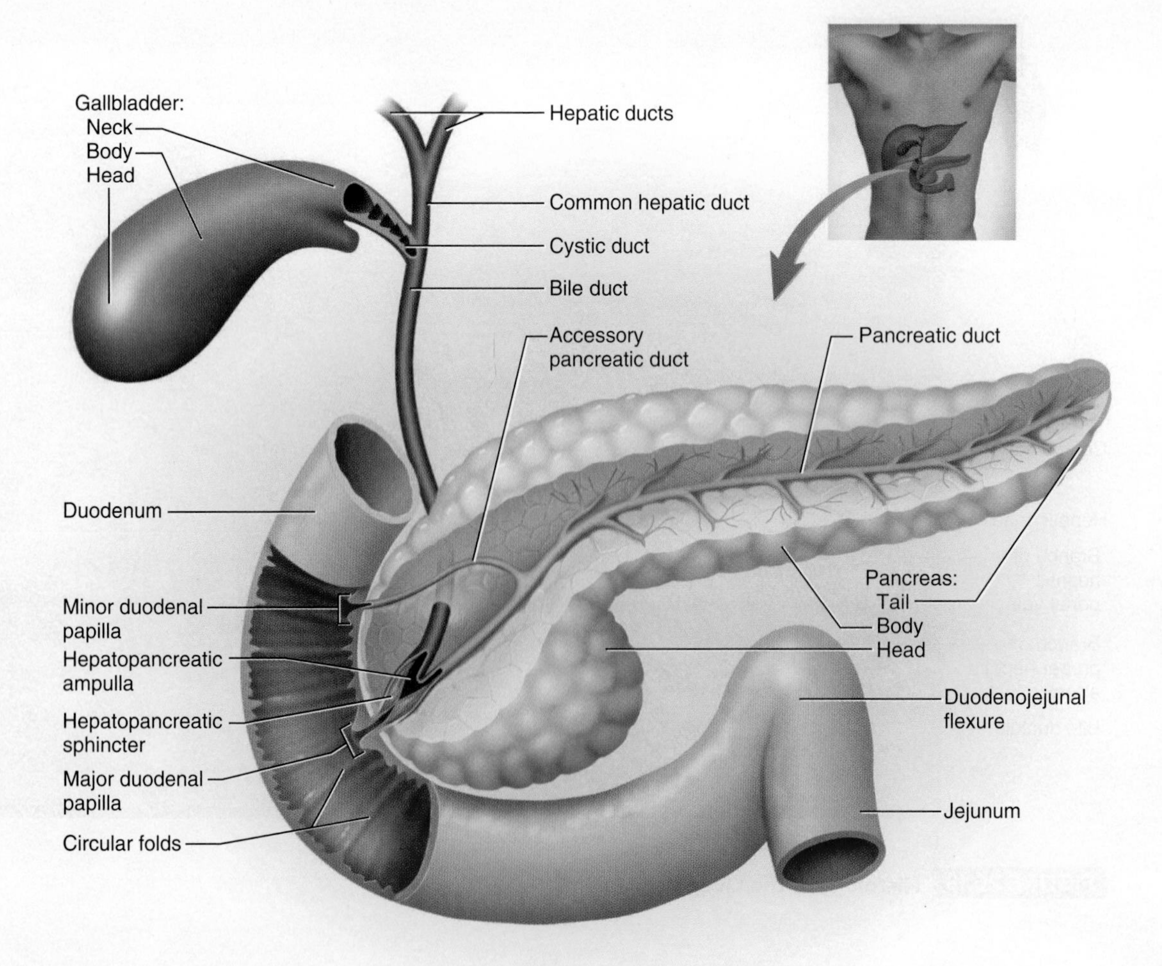

Exercise 26

The Digestive System

Name ______________________________ Date ______________

1. The pancreas belongs to what part of the digestive system (alimentary canal/accessory structure)? ______________

____

2. The descending colon belongs to what part of the digestive system (alimentary canal/accessory structure)? ______________

____

3. Name the layer lining the outer surface of the stomach and small intestine. ______________

____

4. What is partially digested food in the stomach called? ______________

____

5. What is the name of the portion of the stomach closest to the small intestine? ______________

____

6. What cell type constitutes the inner lining of the mucosa (near the lumen) of the small intestine? ______________

____

7. How long is the large intestine? ______________

____

8. What is the middle length of the small intestine called? ______________

____

9. In what specific structure do you find lacteals in the digestive tract? ______________

____

10. What membrane holds the tongue to the floor of the oral cavity? ______________

____

11. What region of the tooth is found above the neck? ______________________________

12. What is the layer of a tooth superficial to the dentin? ______________________________

13. Which adult teeth are directly posterior to the canine teeth? ______________________________

14. What kind of muscle lines the small intestine? ______________________________

15. What is the name for a segment, or pouch, in the large intestine? ______________________________

16. Name the salivary gland directly anterior to the ear. ______________________________

17. The lesser omentum is found between what two organs? ______________________________

18. Where is bile stored? ______________________________

19. Stomach acidity is approximately pH 2. How does this affect bacterial growth in the digestive tract? ______________________________

20. Trace the flow of bile from the liver to the duodenum, listing all the structures that come into contact sequentially with the bile on its journey. ______________________________

21. How does the large intestine differ from the small intestine in terms of length? ______________________________

22. How does the large intestine differ from the small intestine in terms of diameter? ______________________________

23. Name two functions of the pancreas. ______________________________

____

24. Describe the pathway that food takes from the mouth to the anus, listing each structure that food comes into contact with.

____

____

25. Label the following illustration using the terms provided.

appendix	liver	sigmoid colon
ascending colon	mouth	small intestine
duodenum	parotid gland	stomach
esophagus	rectum	tongue

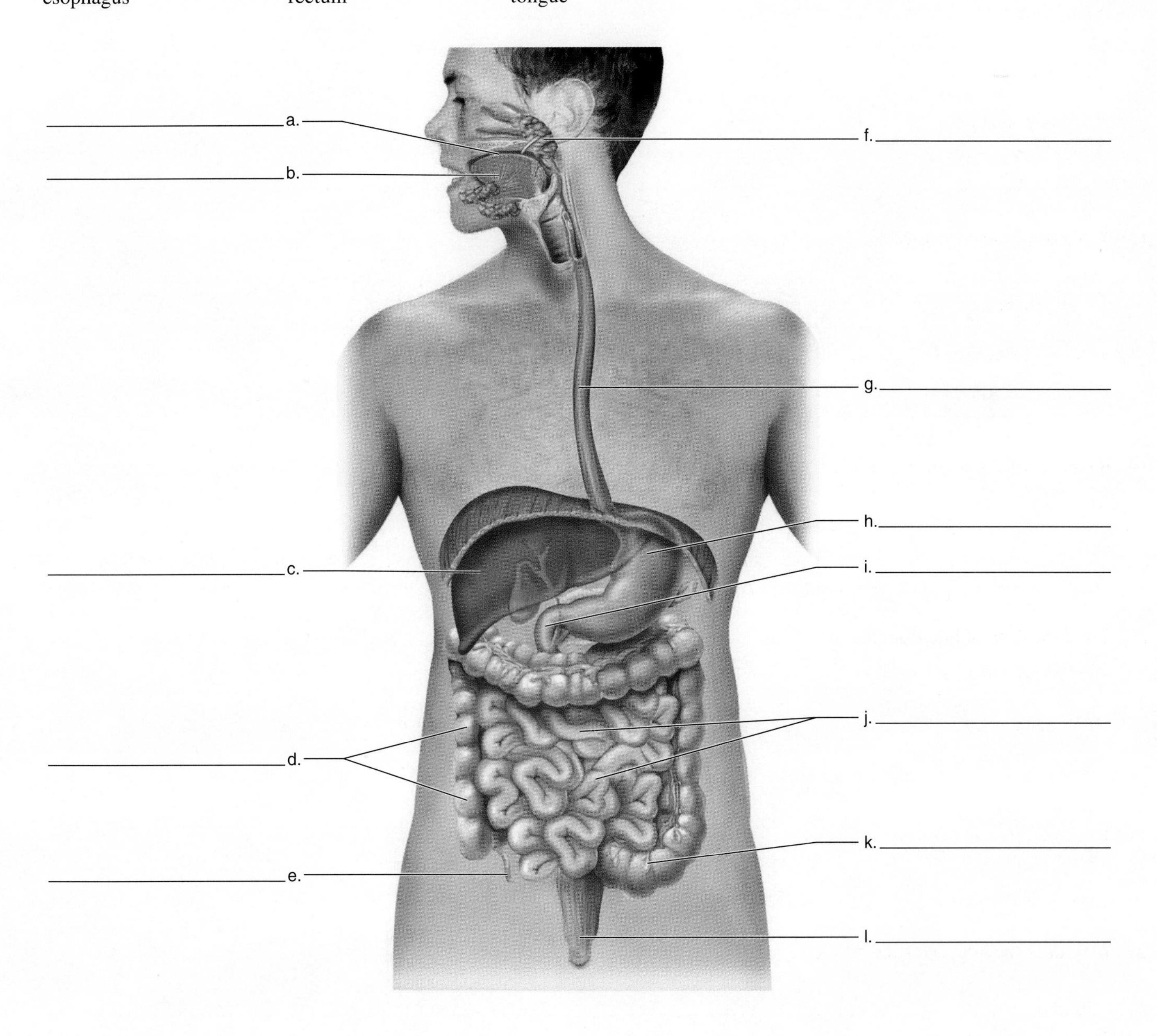

Laboratory

Exercise 27

The Urinary System

Urinary System

Introduction

The organs of the urinary system consist of two kidneys, two ureters, a single urinary bladder, and a single urethra. The urinary system filters dissolved material from the blood, regulates ions (electrolytes) and fluid volume, concentrates and stores waste products, and reabsorbs metabolically important substances, returning them to the circulatory system. **Filtration** occurs when one or more substances pass through a selectively permeable membrane while others do not. Filtration in the kidney involves both metabolic waste products (urea) and material beneficial to the body. Not all filtered material is desirable to have **excreted** from the body. Glucose and other materials, such as sodium and potassium ions, are **reabsorbed** from the kidney back into the circulatory system. The kidney **secretes** urea; some drugs; hydrogen and hydroxyl ions; and hormones, such as erythropoietin and the enzyme renin. Finally the kidneys excrete metabolic wastes, hydrogen ions, toxins, water, and salts.

In this exercise you examine the gross and microscopic anatomy of the urinary system, study the major organs as represented in humans, and dissect a mammalian kidney, if available.

Learning Objectives

At the end of this exercise you should be able to

1. identify the major organs of the urinary system;
2. describe the blood flow through the kidney;
3. describe the flow of filtrate through the kidney;
4. name the major parts of the nephron;
5. trace the flow of urine from the kidney to the exterior of the body;
6. distinguish among the parts of the nephron in histological sections; and
7. compare male and female urinary anatomy.

Materials

Models and charts of the urinary system
Models and illustrations of the kidney and nephron
Microscopes
Microscope slides of kidney and bladder
Samples of renal calculi (if available)
Preserved specimens of sheep or other mammalian kidney
Dissection trays and materials
Scalpels
Forceps
Blunt probes
Protective (barrier) gloves
Waste container

Procedure

You can begin the study of the urinary system by locating its principal organs. Look at figure 27.1 and compare it with material available in the lab. Find the **kidneys,** the **ureters,** the **urinary bladder,** and the **urethra.**

Kidneys

The kidneys are **retroperitoneal** (posterior to the parietal peritoneum) and are embedded in **renal fat pads (perirenal fat).** These adipose pads cushion the kidneys, which are found mostly below the protection of the rib cage. The kidneys are located adjacent to the vertebral column approximately at the level of T12 to L3. The right kidney is slightly more inferior than the left. This is illustrated in figure 27.1.

1. Examine a model of the kidney and compare it with figure 27.2.
2. Locate the outer **renal capsule,** a tough connective tissue layer, the outer **cortex,** and the inner **medulla** of the kidney. The kidney has a depression on the medial side where the **renal artery** enters the kidney and the **renal vein** and the **ureter** exit the kidney. This depression is called the **hilum.**
3. Examine a coronal section of the kidney and locate the **renal pyramids,** found inside the renal medulla. Each renal pyramid ends in a blunt point called the **renal papilla** (fig. 27.2). Urine drips from many papillae toward the middle of the kidney. The renal pyramids are separated by **renal columns,** which are extensions of the cortex.

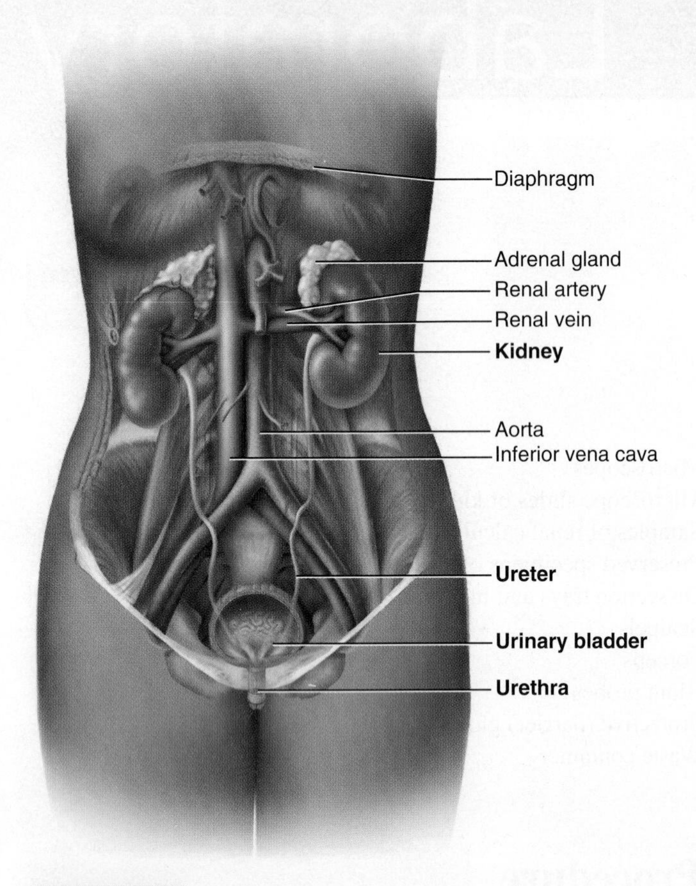

FIGURE 27.1 **Major Organs of the Urinary System**

The urine drips into the minor calyces (singular, *calyx*), which enclose the renal papillae. Minor calyces act somewhat as funnels that collect fluid and lead to the major calyces. These, in turn, conduct urine into the large **renal pelvis.** The renal pelvis is found in a space known as the **renal sinus.** The renal pelvis is like a glove in a coat pocket. The pocket is the renal sinus, and the membranous glove that occupies the space is the renal pelvis.

4. Locate these structures in figure 27.2. The renal pelvis is connected to the ureter at the medial side of the kidney.

Blood Flow Through the Kidney There are two fluid flows in the kidney. One is the flow of the filtrate and the other is the flow of blood. The kidney filters material from the blood and returns important material such as water, glucose, and sodium to the blood. It is not a perfect system as some urea is also returned to the cardiovascular system. There is an arterial system that takes blood to the cortex of the kidney and several capillary beds, which is somewhat unusual compared to other organs. The venous system returns blood to the inferior vena cava. The blood enters the kidney by way of the **renal artery.** The first branches are the **segmental arteries,** which take blood from the renal artery. The segmental arteries are found inside the renal sinus. The kidney is divided into lobes, and there are **interlobar arteries,** which take blood from the segmental arteries and pass through the renal columns. The interlobar arteries are relatively large and make a sharp bend, becoming the

FIGURE 27.2 **Gross Anatomy of the Kidney, Entire and Coronal Section**
(a) Diagrams of a frontal section of the kidney and details of renal pyramid.

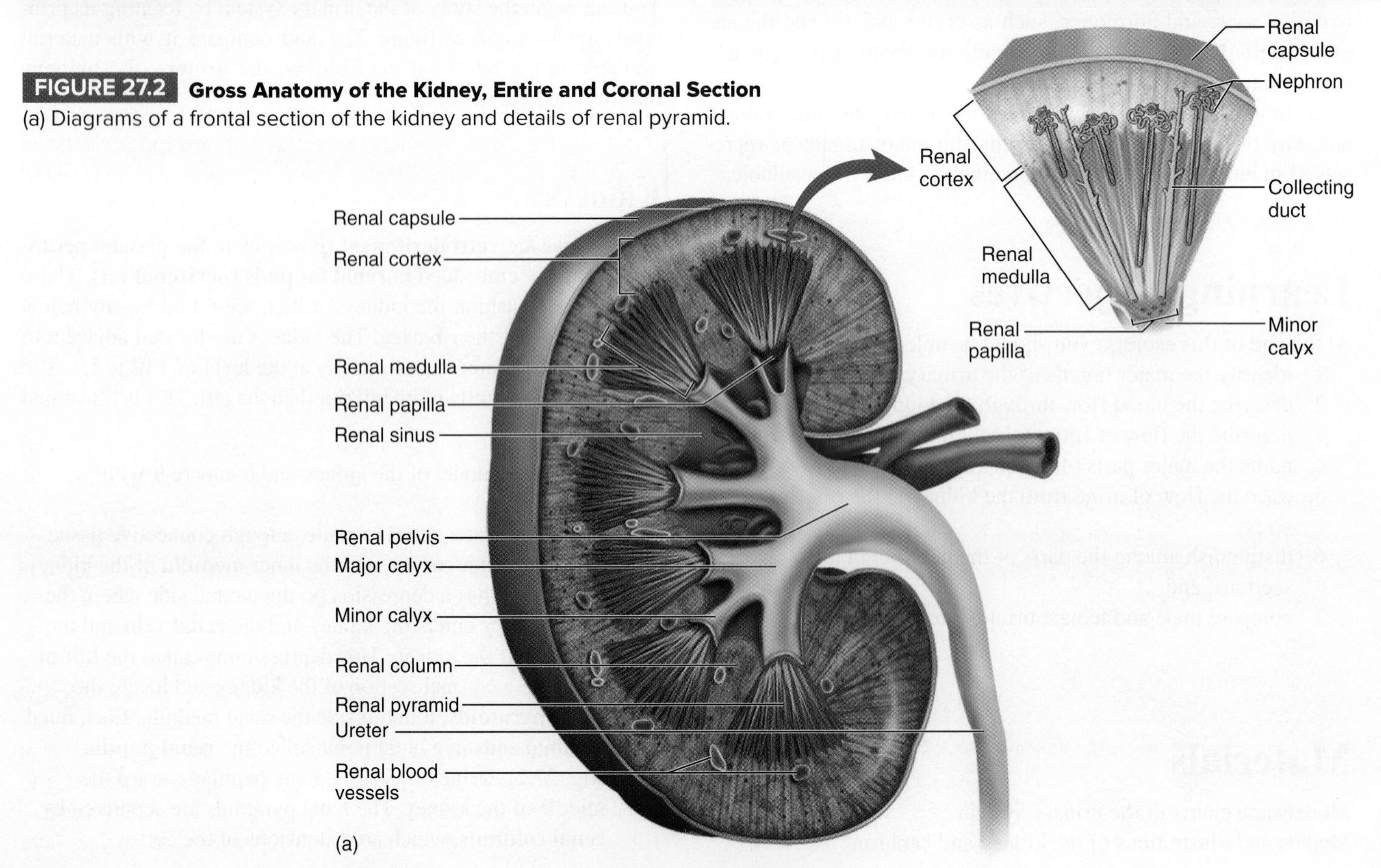

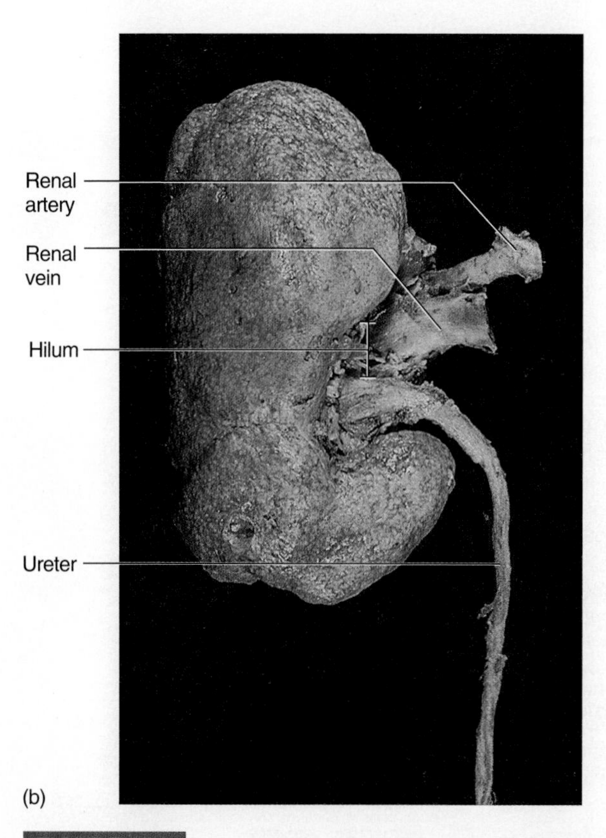

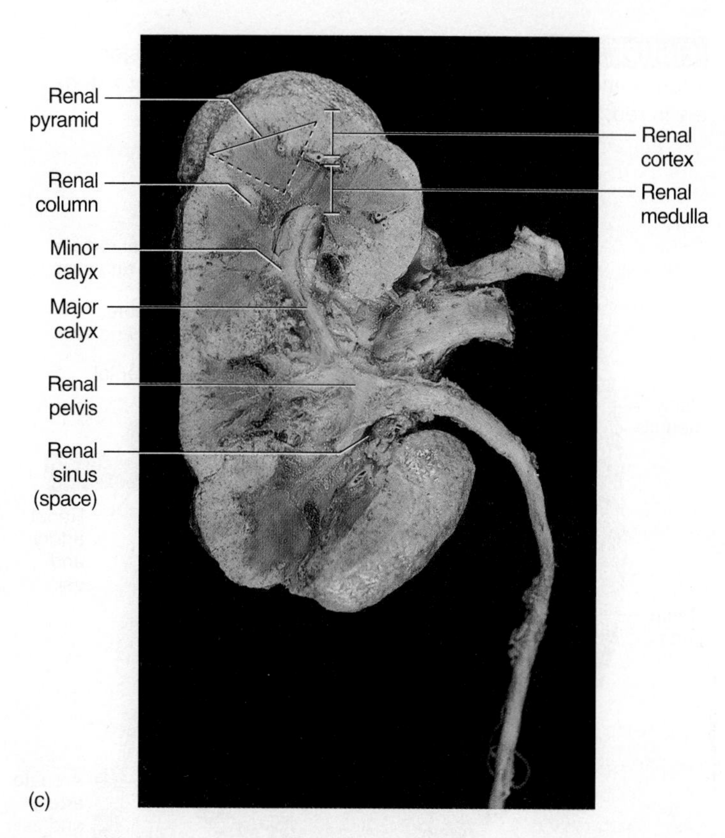

FIGURE 27.2 (*Continued*) **Gross Anatomy of the Kidney, Entire and Frontal Section** Photographs: (b) entire kidney; (c) coronal section.

arcuate arteries. The arcuate arteries form arcs between the cortex and the medulla, and they obtain their name from these arcs. Branching from the arcuate arteries are the **cortical radiate arteries** that move into the cortex of the kidney.

These blood vessels can be seen in figure 27.3. From one of the cortical radiate arteries, blood flows into an **afferent arteriole,** into the **glomerulus** (a tuft of capillaries), and to the **efferent arteriole.** The blood then enters the **peritubular capillaries,** where reabsorption and secretion take place. In regions of the cortex near the medulla other vessels branch from the efferent arteriole. These are the **vasa recta.** They represent only a small number of capillaries in the kidney, but they are important in producing a concentrated urine by reabsorption and secretion. The vasa recta are found in association with special nephrons called **juxtamedullary nephrons.**

Thus, there are three capillary beds in the kidney, the glomeruli (singular, *glomerulus*), the peritubular capillaries, and the vasa recta. The blood flow in the kidney forms a **portal system,** which is defined as a group of blood vessels in which blood flows from one capillary bed (the glomerulus) to another capillary bed (the peritubular capillaries or vasa recta) with an arteriole or venule between them prior to returning to the heart. Blood returns via the **cortical radiate veins,** the **arcuate veins,** the **interlobar veins** to the **renal vein,** and finally to the inferior vena cava. The separation between the cortex and the medulla occurs at the level of the arcuate veins (fig. 27.3).

Microanatomy of the Kidney Before you examine the sections of kidney under the microscope, become familiar with the structure of the **nephron.** Look at figure 27.4 and locate the **renal corpuscle, proximal convoluted tubule, nephron loop (loop of Henle),** and **distal convoluted tubule.** These four structures make up the nephron.

Blood travels to the nephron via the afferent arteriole. When the blood reaches the glomerulus, in the **glomerular capsule,** the plasma is filtered by blood pressure forcing fluid across the capillary membranes. Blood pressure is the driving force for renal filtration, and this pressure is opposed by hydrostatic pressure in the glomerular capsule space outside of the glomerulus and the osmotic pressure that occurs due to the proteins left in the blood plasma. This filtered fluid, present in the nephron, is called **filtrate.** The glomerulus and the glomerular capsule are known as the **renal corpuscle.**

As the filtrate flows through the nephron, water, glucose, and many electrolytes are returned to the blood. The urea is concentrated as it passes through the entire nephron and into the **collecting duct,** a tube that receives the end product of the nephrons.

The amount of reabsorption in the kidney varies with the material that flows through it. In normal conditions, glucose is completely reabsorbed into the blood. About 99% of the water in the filtrate is reabsorbed. Only 50% of the urea is excreted into the urine. The remainder of the urea returns to the blood and is filtered again. Urea is concentrated in the collecting duct, and

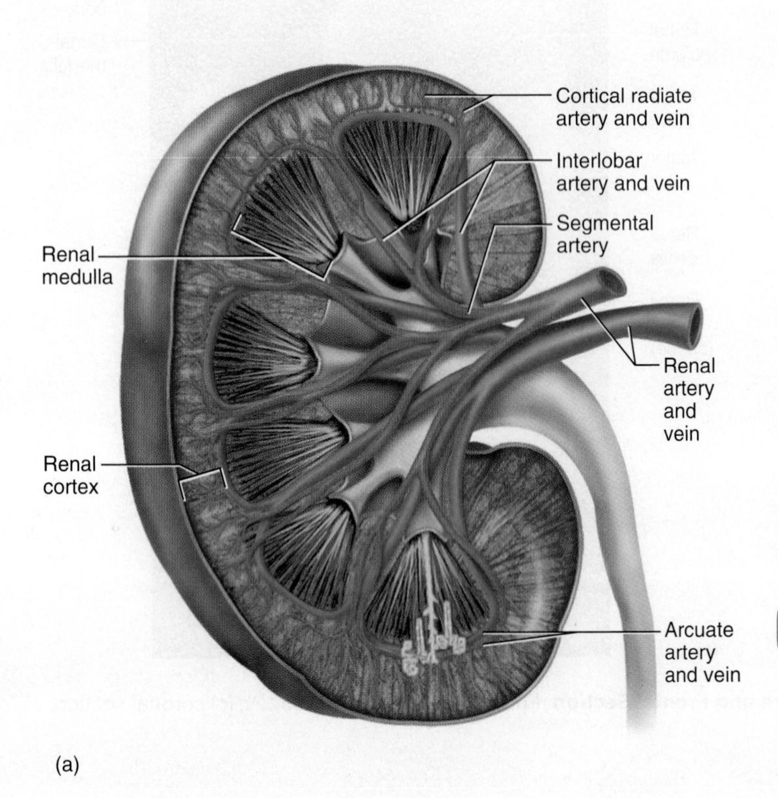

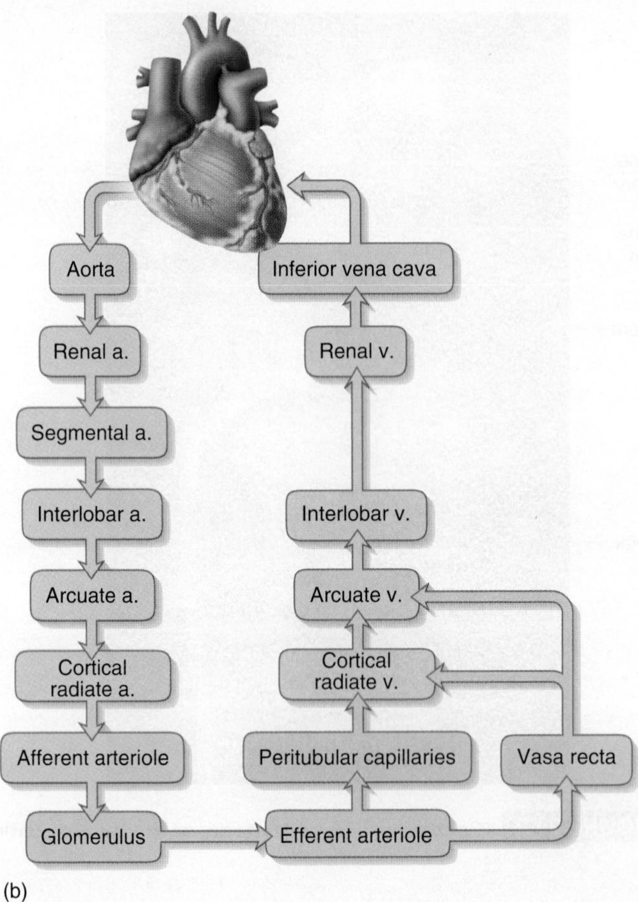

FIGURE 27.3 Blood Flow Through the Kidney (a) Major arteries and veins of the kidney; (b) diagram of blood flow. Arteries are in red; veins are in blue.

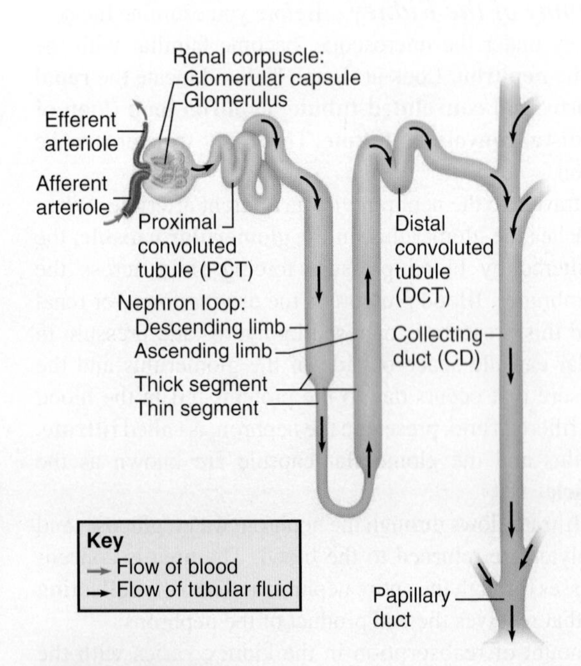

FIGURE 27.4 Nephron

some of it diffuses into the medulla, increasing the osmolarity in the medulla. This increases the flow of water out of the nephron, thus concentrating the urine. There are two types of nephrons in the kidney. The majority of the nephrons are found in the cortex of the kidney and are thus called **cortical nephrons.** Those found in the medulla are called juxtamedullary nephrons and are far fewer in number, some extending to near the tip of the renal pyramids.

In summary, urine is produced from filtered blood. Some of the liquid portion of blood flows from the glomerulus into the nephron. Materials valuable to the body, such as glucose and other solutes, are reabsorbed by the nephron and returned to the cardiovascular system by the peritubular capillaries. The main metabolic by-product, urea, is removed from the kidney and passes as urine from the collecting ducts to the minor calyces. The volume of urine and some of the constituents found in urine are controlled by hormones such as aldosterone and antidiuretic hormone (ADH). Aldosterone increases the reabsorption of sodium and reduces water volume. Antidiuretic hormone causes the distal convoluted tubules to reabsorb water, also decreasing the urine output.

Procedure for Microscopic Examination of the Kidney

1. Examine a kidney slide under low power. You should see the cortex of the kidney, which has a number of round structures scattered throughout. These are the glomeruli. The medullary region of the kidney slide has open, parallel spaces called **collecting ducts.** Compare the slide to figure 27.5.
2. Examine the slide under higher magnification and locate the glomerulus and the glomerular capsule (fig. 27.5). The capsule is composed of simple squamous epithelium and specialized cells called **podocytes.**
3. Examine the outer edge of the capsule around the glomerulus. If you move the slide around in the cortex, you should find the proximal convoluted tubules with the **brush border** or **microvilli** on the inner edge of the tubule. The inner surface of the tubule appears fuzzy. The microvilli increase the surface area for reabsorption in the proximal convoluted tubule.

   The distal convoluted tubules do not have brush borders; therefore, the inner surface of a tubule does not appear fuzzy. The cells of the distal convoluted tubules generally have darker nuclei and cytoplasm that is relatively clear when compared with the cells of the proximal convoluted tubules (fig. 27.5).
4. Examine the medulla of the kidney under high magnification and locate the thin-walled nephron loop and the larger-diameter collecting ducts (fig. 27.5).

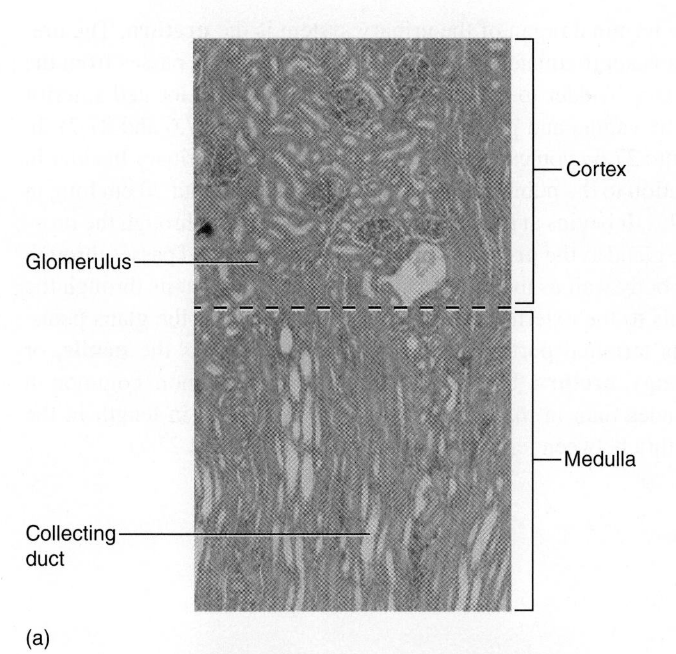

(a)

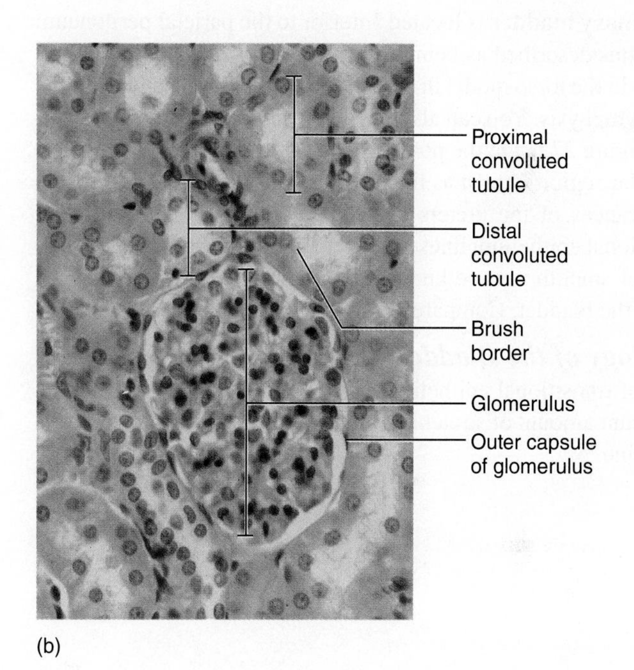

(b)

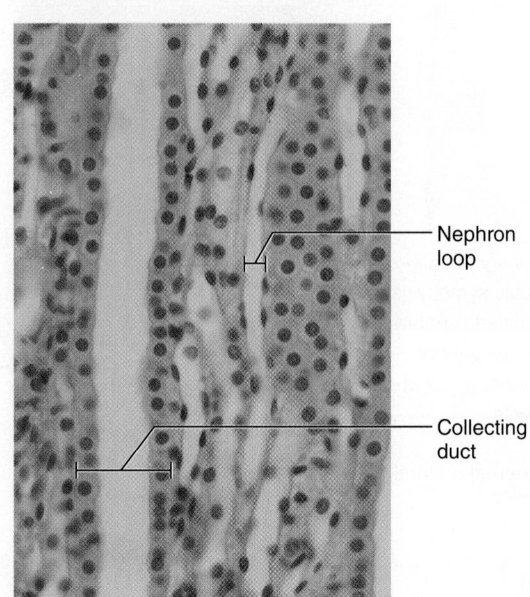

(c)

FIGURE 27.5 **Photomicrographs of Kidney** (a) Overview (40×); (b) cortex (400×); (c) medulla (400×).

Renal Calculi Examine renal calculi (kidney stones) if they are available in the lab.

Ureters

The **ureters** are long, thin tubes that conduct urine from the kidneys to the urinary bladder. The ureters have **transitional epithelium** as an inner lining and smooth muscle in their wall. Urine is moved from the kidney by **peristalsis** to the urinary bladder. Examine the models in the lab and compare them with figure 27.1.

Urinary Bladder

The **urinary bladder** is located anterior to the parietal peritoneum and is thus described as being **anteperitoneal.** Locate the urinary bladder in the torso model in the lab. It is found just posterior to the pubic symphysis. You can also see the position of the urinary bladder in figure 27.6. On the posterior wall of the urinary bladder is a triangular region known as the trigone. The **trigone** is defined by the entrances of the ureters and the inferior exit of the urethra. Transitional epithelium lines the inner surface of the bladder, while layers of smooth muscle known as the **detrusor** are found in the wall of the bladder. Compare models in the lab to figure 27.7.

Histology of the Bladder The urinary bladder has an inner lining of transitional epithelium. This epithelium can withstand a significant amount of stretching (distension) when the bladder fills with urine.

1. Examine a slide of transitional epithelium under the microscope and compare it with figure 27.8. Transitional epithelium is also discussed and illustrated in laboratory exercise 4.
2. Look at the inner surface of the prepared section for the epithelial layer. The cells are shaped somewhat like teardrops. Transitional epithelium can be distinguished from stratified squamous epithelium in that the cells of transitional epithelium from an empty bladder do not flatten at the surface of the tissue. Below the transitional epithelium is an underlying layer known as the **lamina propria.**
3. Examine the smooth muscle layers of the urinary bladder that make up the detrusor muscle.

Urethra

The terminal organ of the urinary system is the **urethra.** The urethra is approximately 3 to 4 cm long in females. It passes from the urinary bladder to the **external urethral orifice,** located anterior to the vagina and posterior to the clitoris (figs. 27.6 and 27.7). In figure 27.6*a* you can also see the position of the urinary bladder in relation to the pubic symphysis. The urethra is about 20 cm long in males. It begins at the urinary bladder and passes through the prostate gland as the **prostatic urethra.** It continues and passes through the body wall as the **membranous urethra,** then exits through the penis to the external urethral orifice at the tip of the glans penis. This terminal portion of the urethra is known as the **penile,** or **spongy, urethra.** Urinary bladder infections are more common in females than in males because of the difference in length in the urethra between females and males (figs. 27.6 and 27.7).

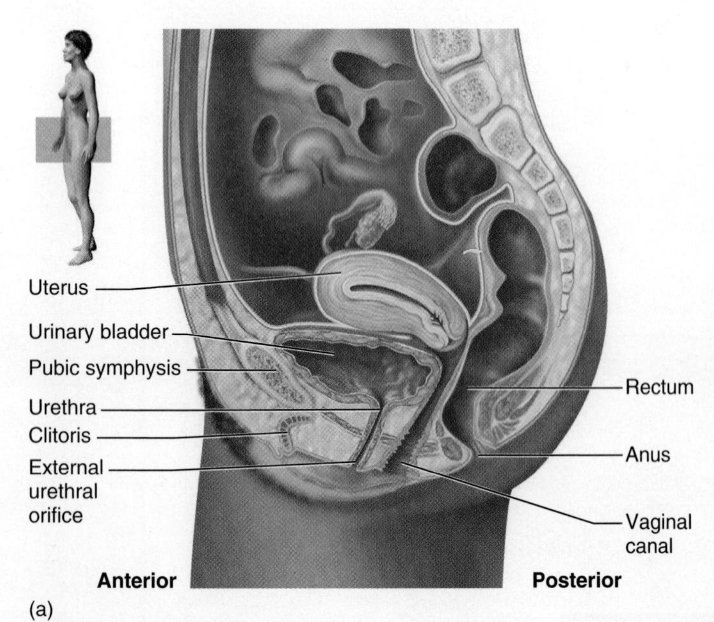

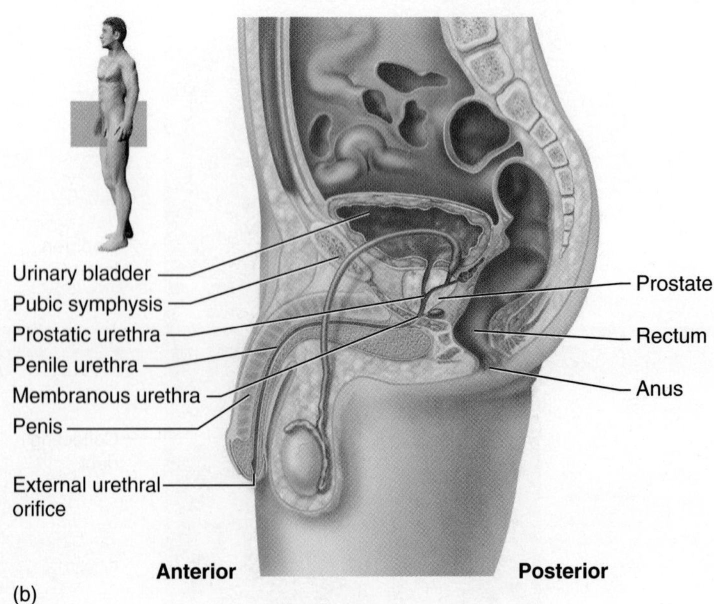

FIGURE 27.6 Median Section of the Female and Male Pelves (a) Female pelvis; (b) male pelvis.

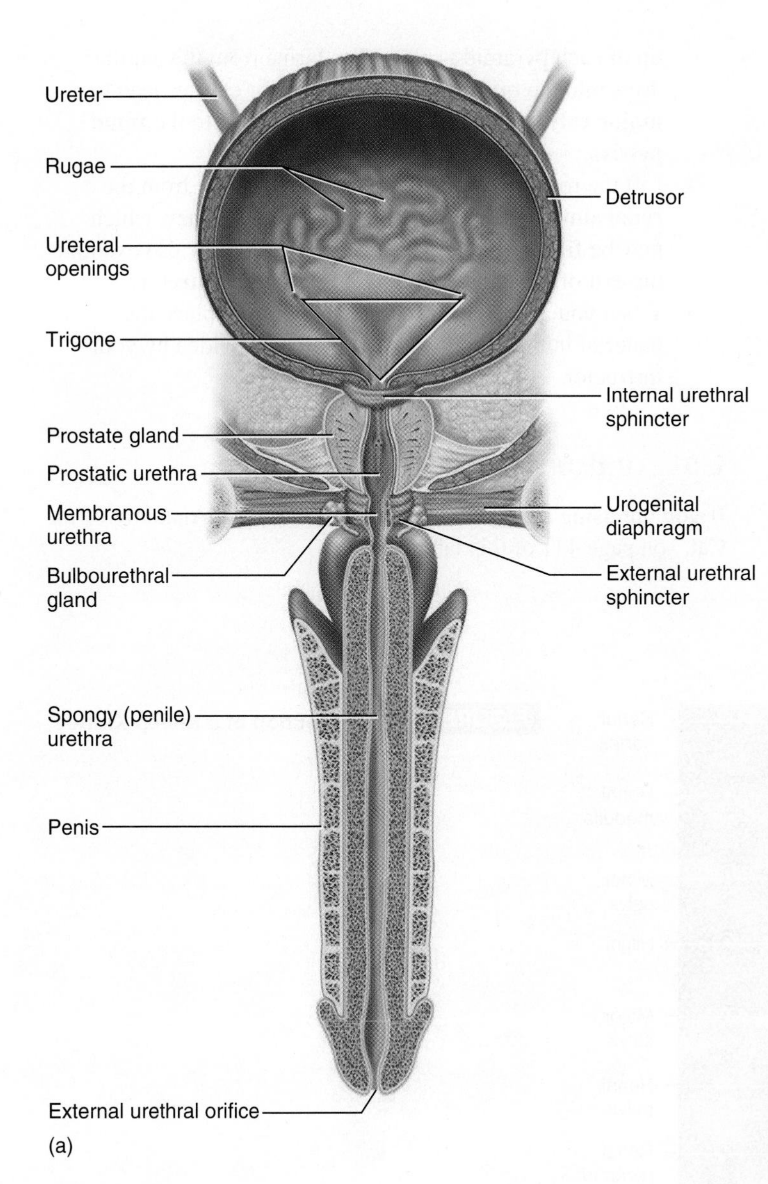

FIGURE 27.7 **Bladder and Urethra, Frontal Sections** (a) Male; (b) female.

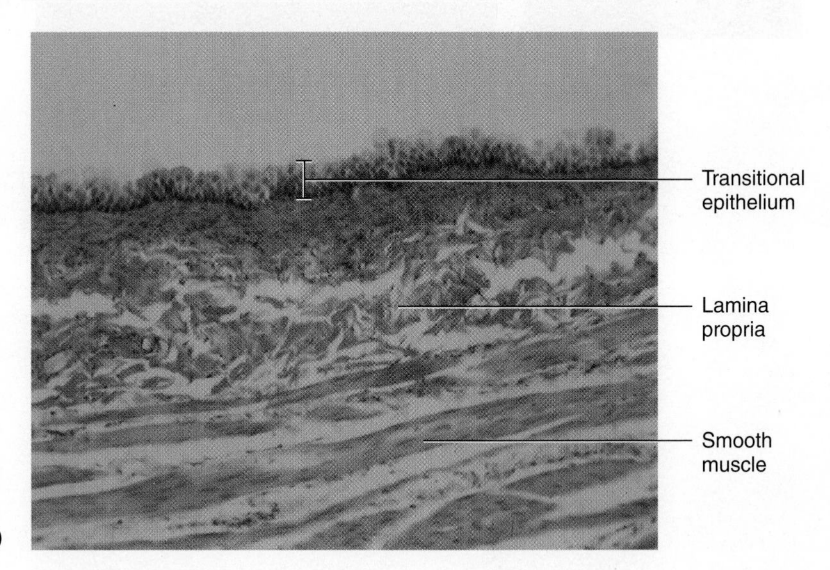

FIGURE 27.8 **Histology of the Bladder (100×)**

Dissection of the Sheep Kidney

1. Place a sheep kidney on a dissection tray and bring the kidney and dissection equipment to your table. Examine the outer **capsule** of the kidney. You may see some tubes coming from an indentation in the kidney. The indentation is the **hilum,** and the tubes are the **renal artery, renal vein,** and **ureter.** The renal artery is smaller in diameter and has a thicker wall than the renal vein. The ureter has an expanded portion near the hilum.
2. Make an incision in the sheep kidney a little off center in the frontal plane (fig. 27.9). This off-center section allows you to better see the interior structures of the kidney.
3. Locate the **renal cortex,** which is the outer layer of the kidney. You should also locate the **renal medulla,** which has triangular regions known as the **renal pyramids.** At the tip of each pyramid is a **papilla.** Urine from the papillae drips into the **minor calyx.** Many minor calyces lead to a **major calyx.** The **major calyces** take fluid to the **renal pelvis.**
4. Lift the renal pelvis somewhat to pull it away from the **renal sinus.** The sinus is the space in the kidney, which may be filled with adipose tissue. You should also examine the exit of the renal pelvis as it becomes the **ureter.**
5. When you are finished with the dissection, place the material in the proper waste container provided by your instructor.

Cat Anatomy

If you are using cats, turn to section 9, "Urinary Anatomy of the Cat," on page 441 of this lab manual.

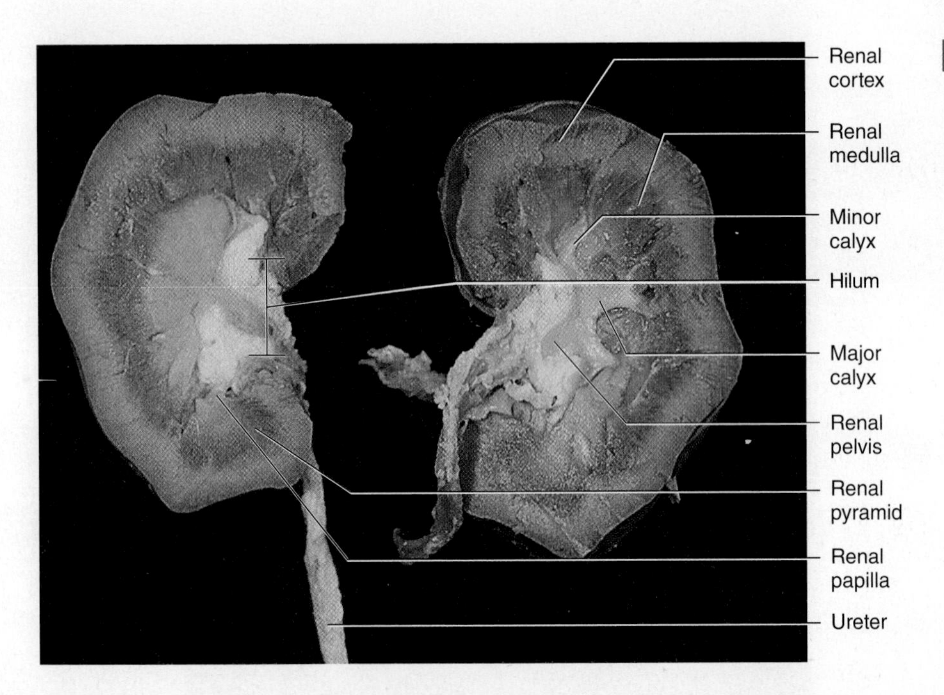

FIGURE 27.9 Dissection of a Sheep Kidney

Exercise 27

The Urinary System

Name ______________________ Date __________

1. Match the descriptions in the left column with the terms in the right column.

____ outermost part of the kidney	a. renal vein
____ storage organ of the urinary system	b. major calyx
____ takes blood from the kidney	c. minor calyx
____ separates the renal cortex from the medulla	d. urinary bladder
____ receives urine from the renal papilla	e. renal capsule
____ leads directly to the renal pelvis	f. arcuate arteries

2. Describe the kidneys with regard to their position to the parietal peritoneum. ______________________

3. What takes urine directly to the urinary bladder? ______________________

4. What anatomical feature in females is responsible for the higher level of urinary tract infections in females? ______________________

5. What is the functional value of the increased surface area caused by the microvilli of the proximal convoluted tubule? ______________________

6. On the posterior bladder there is a triangular region. What is it called? ______

7. What is the name of the cluster of capillaries in the kidney where filtration occurs? ______

8. Distal convoluted tubules flow directly into what structures? ______

9. What is a renal papilla? ______

10. Blood in the glomerulus flows to what arteriole? ______

11. Which shows the greatest anatomical difference between the sexes: ureters, urinary bladder, or urethra? ______

12. Trace the path of filtrate and urine from the glomerulus to the external urethral opening. ______

13. What histological feature distinguishes a proximal convoluted tubule from a distal convoluted tubule? ______

14. Name the parts of the nephron. ______

15. What type of tissue lines the bladder? ______

16. Fill out the following illustration, using the terms provided.

major calyx	renal capsule	renal pyramid
minor calyx	renal cortex	renal vein
renal artery	renal pelvis	ureter

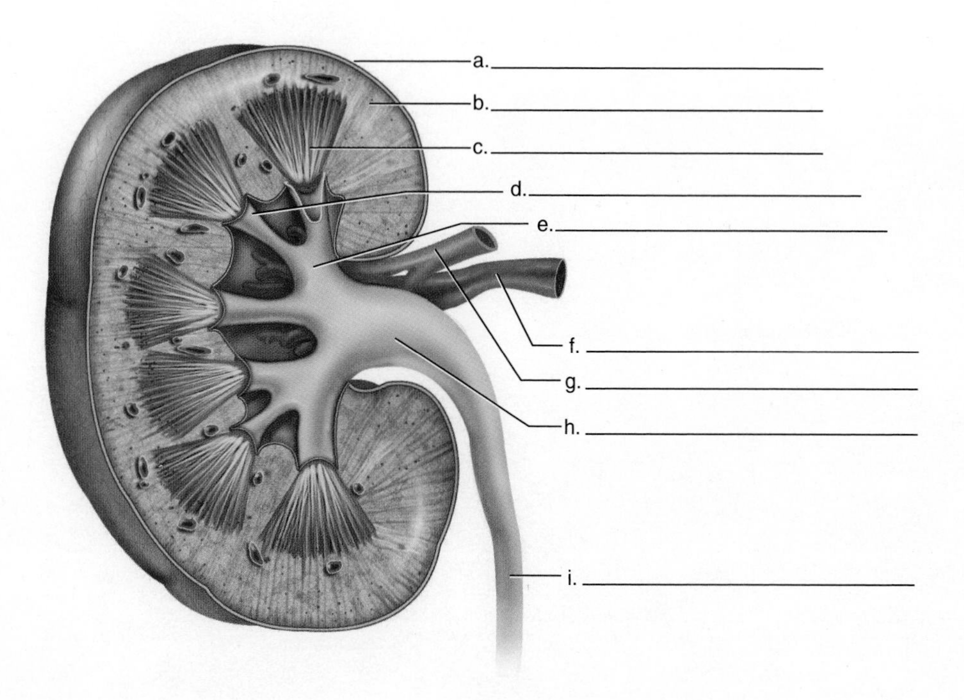

Laboratory

Exercise 28

The Male Reproductive System

Reproductive System

Introduction

The male reproductive system produces male **gametes (spermatozoa),** transports the gametes to the female reproductive tract, and secretes the male reproductive hormone, **testosterone.** The **gonad,** or gamete-producing structure, of the male reproductive system is the *testis* (plural, *testes*). The testes are considered mixed glands in that they have both an exocrine and an endocrine function. The exocrine nature of the gland is the secretion of gametes through ducts or tubules. The endocrine function involves the ductless secretion of testosterone. In this exercise you examine the gross anatomy of the male reproductive system and the histology of the system.

Learning Objectives

At the end of this exercise you should be able to

1. identify the gamete-producing organ of the male reproductive system;
2. describe the anatomy of the major structures of the male reproductive system;
3. describe the formation of spermatozoa in the testis;
4. list the pathway that spermatozoa follow from production to expulsion;
5. describe the anatomy of the spermatic cord;
6. list the four components of semen and where they are produced; and
7. name the three cylinders of erectile tissue in the penis.

Materials

Charts, models, and illustrations of the male reproductive system
Microscopes
Prepared slides of a cross section of testis and sperm smear

Procedure

Overview of the Gross Anatomy of the Male Reproductive System

Examine the models and charts of the male reproductive system available in the lab and locate the listed structures in figure 28.1. When you find the structures, place a check mark next to the name.

_____ Testis
_____ Epididymis
_____ Scrotum (scrotal sac)
_____ Ductus deferens
_____ Seminal vesicle
_____ Prostate gland
_____ Bulbourethral gland
_____ Penis
_____ Urethra

Testes

The **testes** are paired organs wrapped in a tough connective tissue sheath called the **tunica albuginea** (fig. 28.1) and covered with an extension of the peritoneum called the **tunica vaginalis.** They lie outside of the body cavity, where the temperature is somewhat cooler and are surrounded by the **scrotum (scrotal sac),** which envelops the testes. The testes are the site of **spermatozoa (sperm)** production, and this process must occur at about 35°C. The scrotum is lined with a layer of muscle called the **dartos muscle.** It is composed of smooth muscle that contracts when the testes are cold, thus bringing them closer to the body. When the environment around the testes is warm, the dartos muscles relax and the testes descend from the body, thus becoming cooler. The **cremaster** are muscles that attach to the testes. They contract, which brings the testes up, warming them, and relax to lower the testes. Examine charts and models of the testes and compare them with figures 28.1 and 28.2.

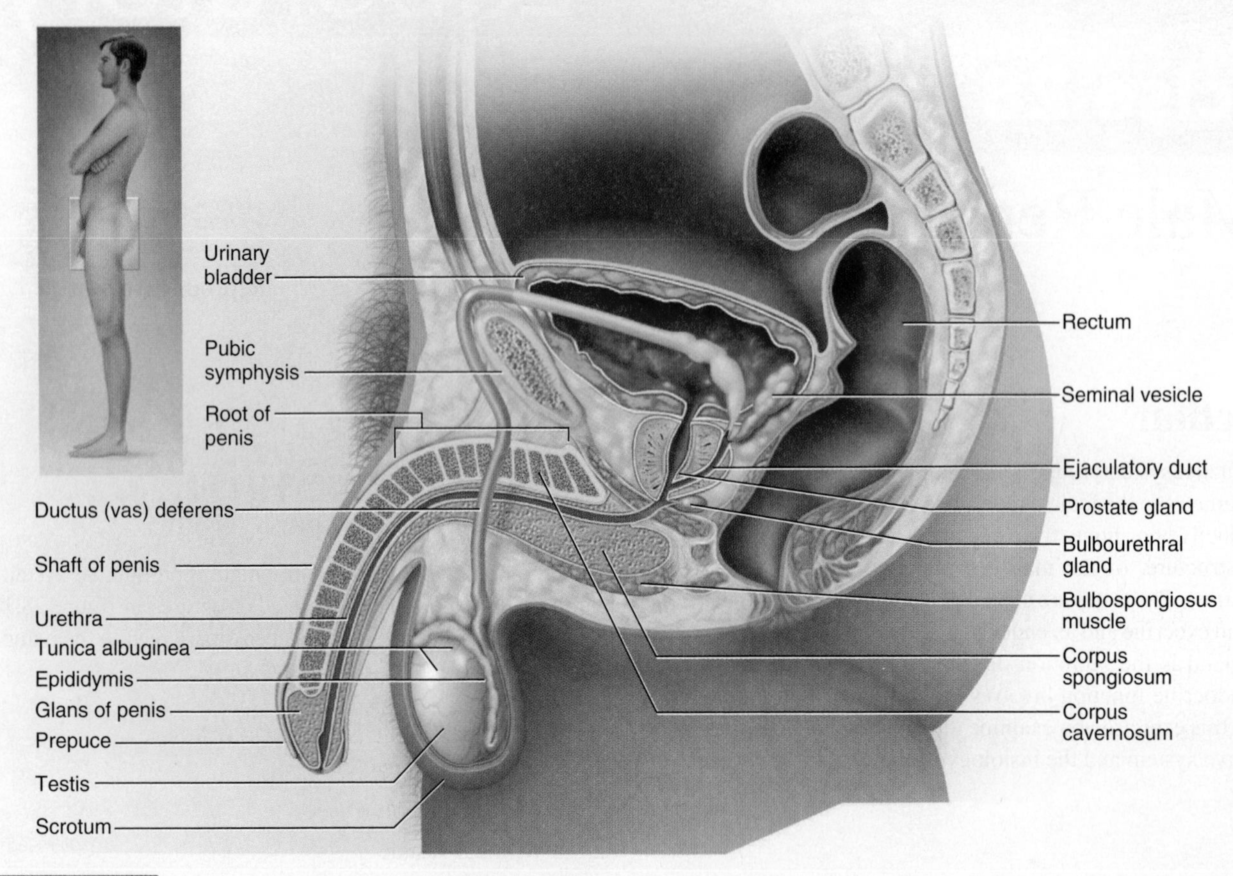

FIGURE 28.1 **Male Reproductive System, Median View**

Histology of the Testis

1. Examine the testis under low power in a prepared slide. Numerous tubules are seen in cross section. These are the **seminiferous tubules.** The gametes, or spermatozoa, are produced in seminiferous tubules in the testis (figs. 28.3 and 28.4).
2. Find the triangular clusters of cells in between the tubules. These are called **interstitial cells.** They produce the male sex hormone, **testosterone.**
3. Examine the seminiferous tubules under high magnification. You should see the outer row of cells called the **spermatogonia.** These cells reproduce by mitosis to produce **primary spermatocytes. Sustentacular (Sertoli) cells** nourish, support, and move the sperm cells during their development.

   The primary spermatocytes undergo meiosis, or reduction division, to eventually produce the sex cells (spermatozoa). The primary spermatocytes divide to form **secondary spermatocytes,** which are found closer to the lumen. The secondary spermatocytes become **spermatids.** Spermatids lose their remaining cytoplasm and mature into **spermatozoa.**
4. Examine a prepared slide of testis and compare it with figures 28.3 and 28.4. Locate the spermatogonia, primary and secondary spermatocytes, spermatids, and spermatozoa.

   After you have examined your prepared slide of a section of testis, draw what you see in the following space. Include the interstitial cells, seminiferous tubules, spermatogonia, spermatocytes, spermatids, and spermatozoa. You may need to look at several sections to see all these items. Your drawing may be a composite of many tubules.

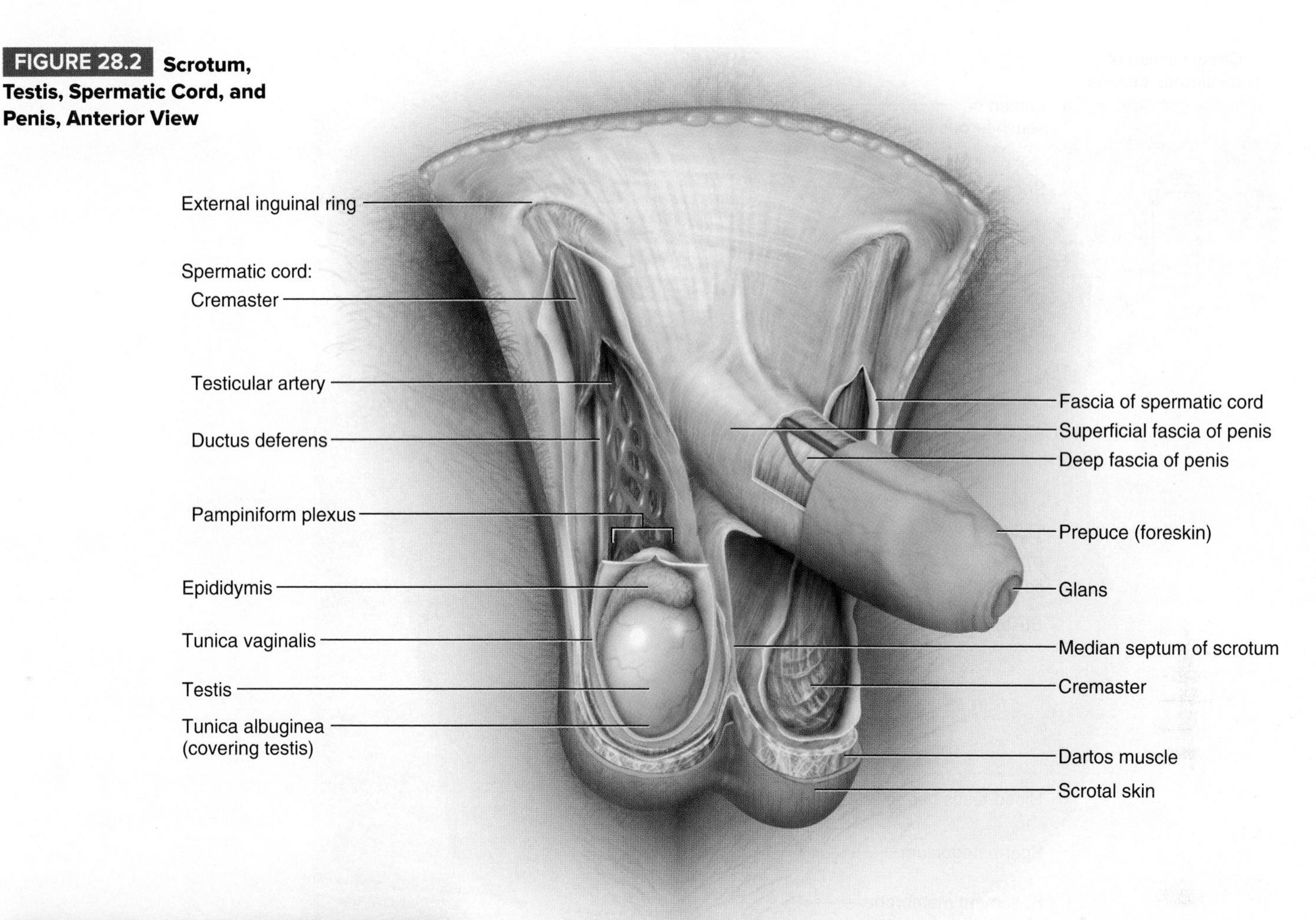

FIGURE 28.2 Scrotum, Testis, Spermatic Cord, and Penis, Anterior View

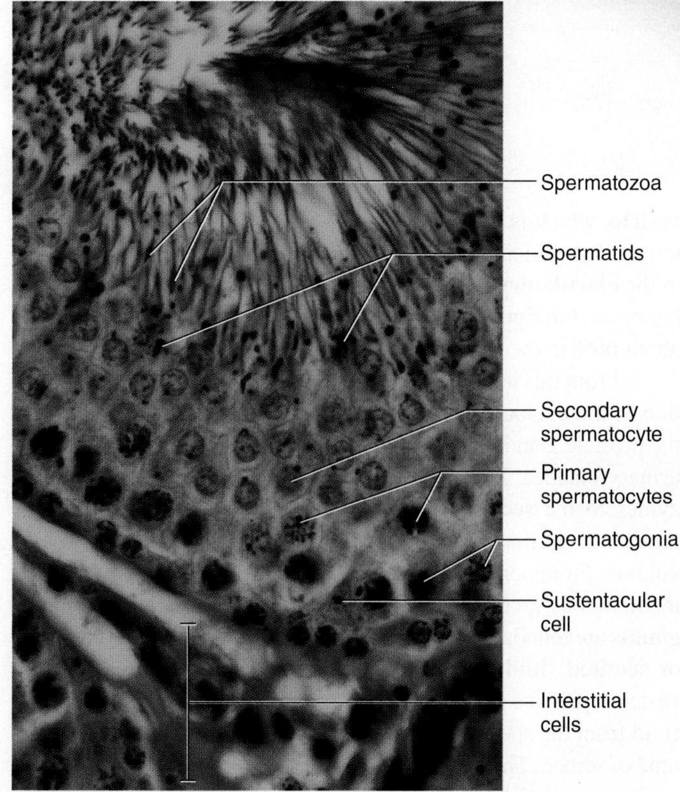

FIGURE 28.3 Histology of the Testis (400×)

Spermatozoa The structure of an individual spermatozoa (sperm) consists of a **head** and a **tail.** The head contains the genetic information (DNA) as well as a cap known as the **acrosome.** The acrosome contains digestive enzymes, which digest the exterior covering of the female gamete. The midpiece of the tail of the sperm contains **mitochondria** that provide ATP to move the spermatozoa. The remainder of the tail of the sperm is a flagellum that propels the sperm forward. Examine a prepared slide of the sperm and compare it with figure 28.5. Your instructor may want you to look at this slide with the oil immersion lens. After you focus the slide on high power, swing the high-power lens away from the slide, add a drop of immersion oil to the slide, and use *only* the fine focus knob to examine the slide. Make sure to wipe the slide clean of immersion oil when you are done.

Epididymis Spermatozoa from each testis travel from tubules in the testis via **efferent ductules** to the **rete testis,** which is a network of tubules. Sperm travel from the rete testis into the **epididymis,** where they are stored and mature. Each epididymis has a rounded **head,** an elongated **body,** and a tapering **tail** that leads to the **ductus deferens** (plural, *ductuli deferentes*). Sperm maturation, or capacitation, occurs in the epididymis. If spermatozoa are removed from the testis proper, they are not capable of fertilizing the female oocyte (egg). Spermatozoa move slowly

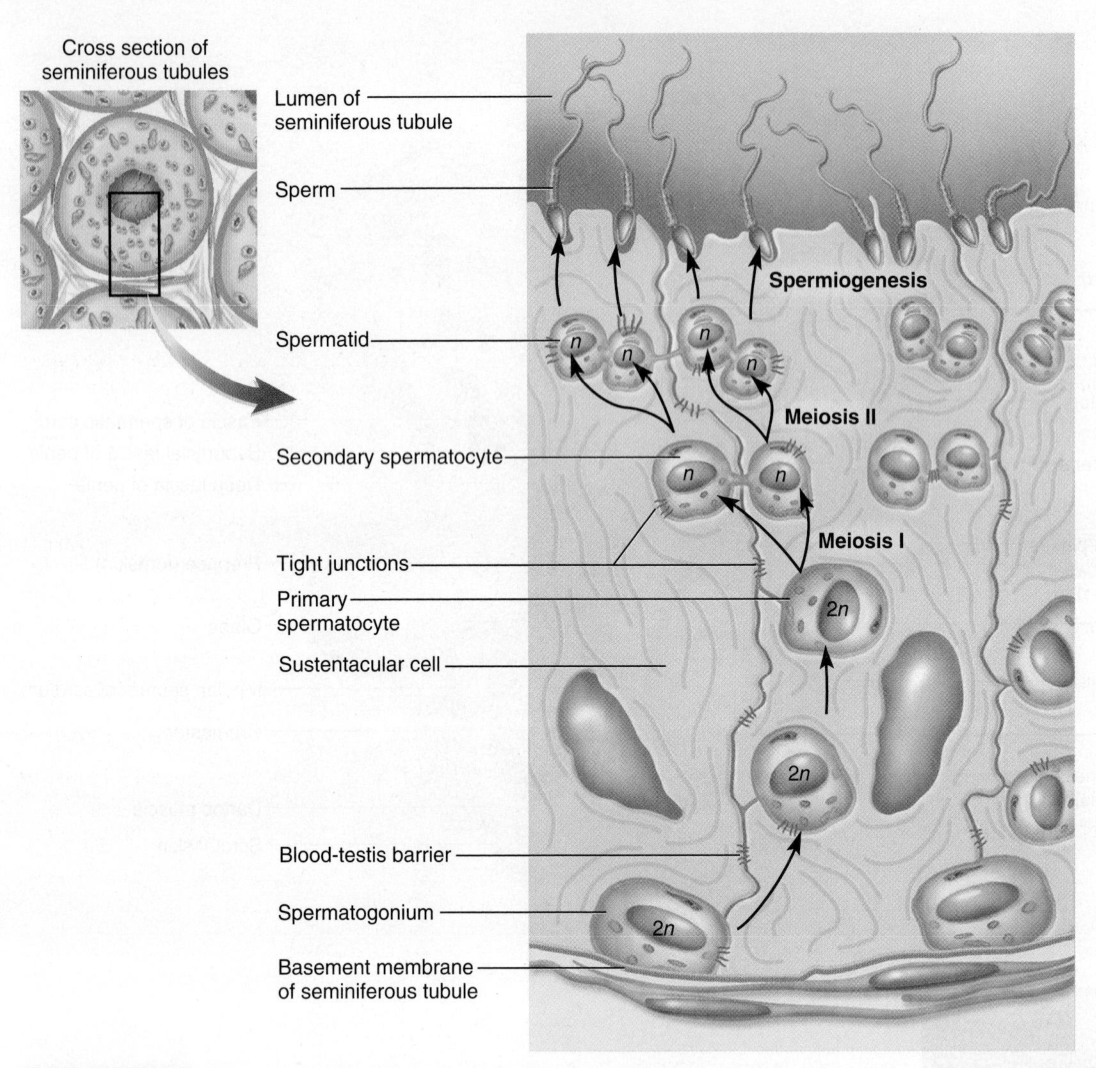

FIGURE 28.4 **Spermatogenesis in the Seminiferous Tubule** Diploid cells (2*n*) reduce their chromosome number by half to haploid cells (*n*) in this process.

through coiled tubules of the epididymis, which lead to the ductus deferens. Examine a model or chart of the longitudinal section of a testis and epididymis, and locate the structures by comparing them with figure 28.6.

Spermatic Cord and Accessory Glands Spermatozoa travel from the epididymis into the ductus deferens. The ductus deferens is enclosed in the **spermatic cord,** a complex structure consisting of the ductus deferens, the **testicular artery** and **vein,** the **testicular nerves,** and the **cremaster.** The cremaster is a cluster of skeletal muscle fibers. The spermatic cord is longer on the left side than on the right; therefore, the left testis is lower than the right. Locate the structures of the spermatic cord in figures 28.2 and 28.6.

The spermatic cord ends at the **inguinal canal** and attaches to the body wall. The ductus deferens arches around the posterior aspect of the urinary bladder (fig. 28.1). You can trace the course of the ductus deferens until each reaches near the inferior portion of the bladder. The ductuli deferentes enlarge somewhat here to form the **ampullae.** Each ductus deferens joins with a **seminal vesicle,** which is a gland that adds fluid to the spermatozoa. The union of the ductuli deferentes and the two seminal vesicles leads to the **ejaculatory ducts.** Locate the seminal vesicle and the ejaculatory duct in figure 28.7. The fluid from the seminal vesicles adds about 60% to the final volume of semen.

From this location the ejaculatory ducts, near the inferior portion of the bladder, join with the prostatic urethra, passing through the prostate gland. The **prostate gland** is located just inferior to the urinary bladder. The prostate gland adds a buffering fluid and enzymes to the secretions of the testes and seminal vesicles. The prostate fluid makes up slightly less than 30% of the final semen volume. From here the sperm cells and prostatic fluid flow into the membranous urethra. Here the paired **bulbourethral (Cowper's) glands** are found, which add a lubricant to the seminal fluid. **Semen** or **seminal fluid** consists of secretions from the testes, seminal vesicles, prostate gland, and bulbourethral glands. Spermatozoa and fluid from the spermatic ducts make up about 10% of the total volume of semen. The semen passes out of the body cavity through the **spongy,** or **penile, urethra** of the penis. Locate the portions of the urethra and the accessory glands in figure 28.7.

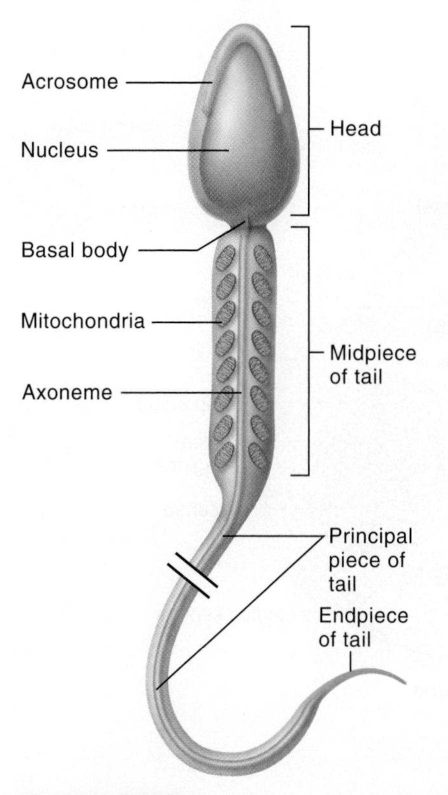

FIGURE 28.5 **Isolated Sperm** Tail is much longer than shown.

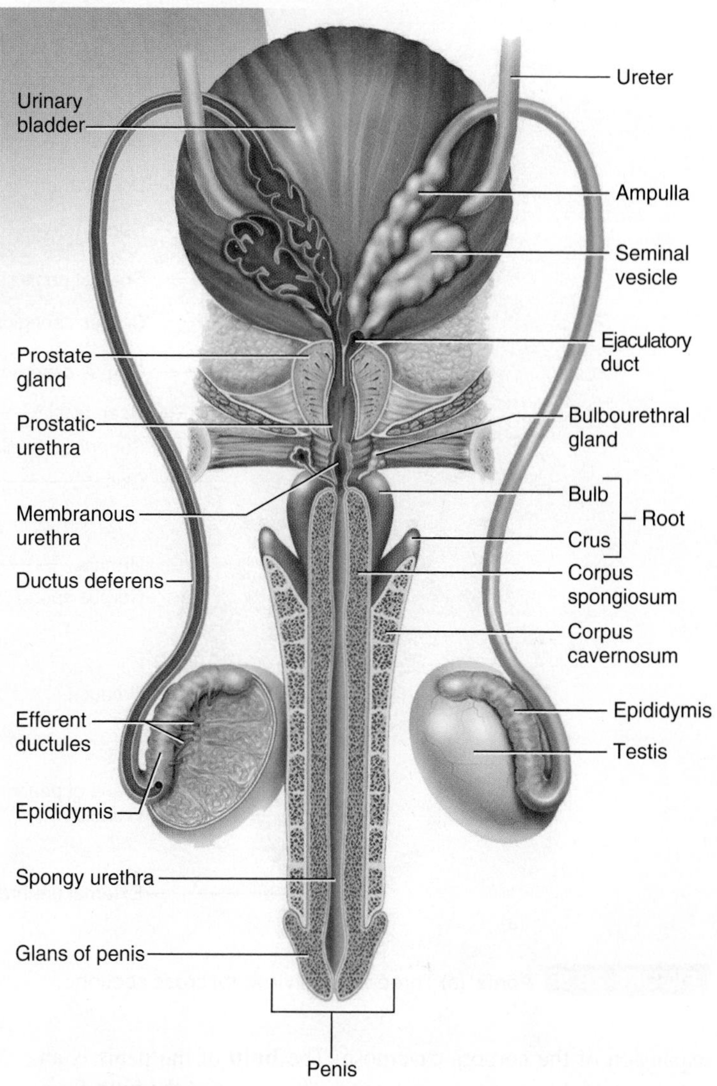

FIGURE 28.7 **Urinary Bladder with Seminal Vesicles, Posterior View**

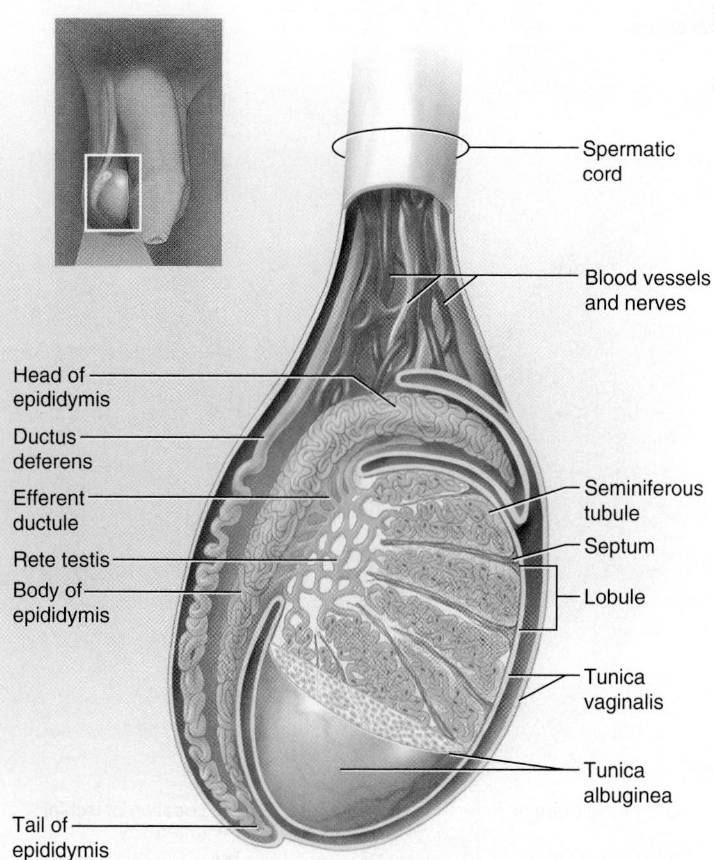

FIGURE 28.6 **Testis and Epididymis, Lateral View**

External Genitalia

Penis The **penis** consists of an elongated **shaft** and a distally expanded **glans.** The glans is covered with the **prepuce,** or **foreskin,** which is removed in some males by a procedure called **circumcision.** At the inferior portion of the glans is a fold of tissue called the **frenulum** that connects the margin of the glans to the prepuce. It is a region richly supplied with nerve endings. The glans is a distal, expanded region that stimulates the vagina of the female. The erect penis is, on average, about 16 cm in length. The various structures of the penis can be seen in figures 28.1, 28.2, and 28.8.

The penis contains three cylinders of **erectile tissue.** The **corpus spongiosum** is the cylinder of erectile tissue that contains the **spongy (penile) urethra.** The two **corpora cavernosa** (singular, *corpus cavernosum*) are located dorsal to the corpus spongiosum. Examine a model or chart of a cross section of penis and compare it with figure 28.8. The proximal parts of the cylinders of erectile tissue are anchored to the body. Locate the **crus,** which is an

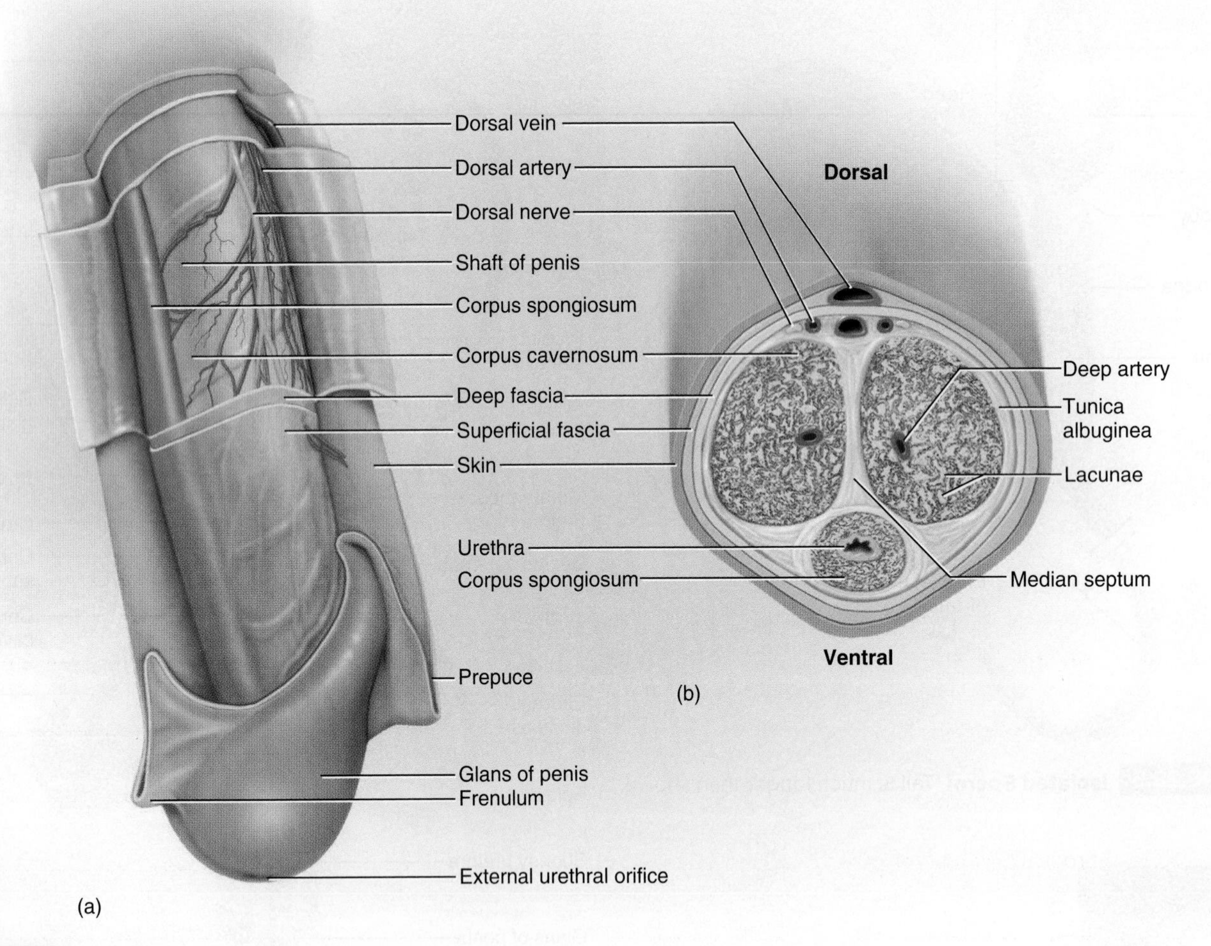

FIGURE 28.8 **Penis** (a) Three-fourths view; (b) cross section.

expansion of the corpora cavernosa. The **bulb** of the penis is an extension of the corpus spongiosum. The crus and the bulb form the **root** of the penis. The corpus spongiosum expands distally to form the glans. Note the **dorsal arteries** and **deep arteries** of the penis. These take blood to the penis. Locate the **dorsal vein** of the penis. When the arteries of the penis dilate, the erectile tissues engorge with blood and the penis becomes erect. The erection subsides as the arteries constrict, decreasing blood flow into the penis. Examine a model of the penis and find the features listed in figures 28.7 and 28.8.

Perineum The floor of the pelvis as seen from the outside is referred to as the **perineum.** It is a diamond-shaped structure defined by the pubic symphysis at the anterior point, the lateral points being the ischial tuberosities, and the posterior point being the coccyx. It can be divided into a posterior **anal triangle** and an anterior **urogenital triangle.** The anal triangle surrounds the anus, and the urogenital triangle encloses the penis and scrotum (fig. 28.9).

Cat Anatomy

If you are using cats, turn to section 10, "Reproductive Anatomy of the Cat," on page 442 of this lab manual.

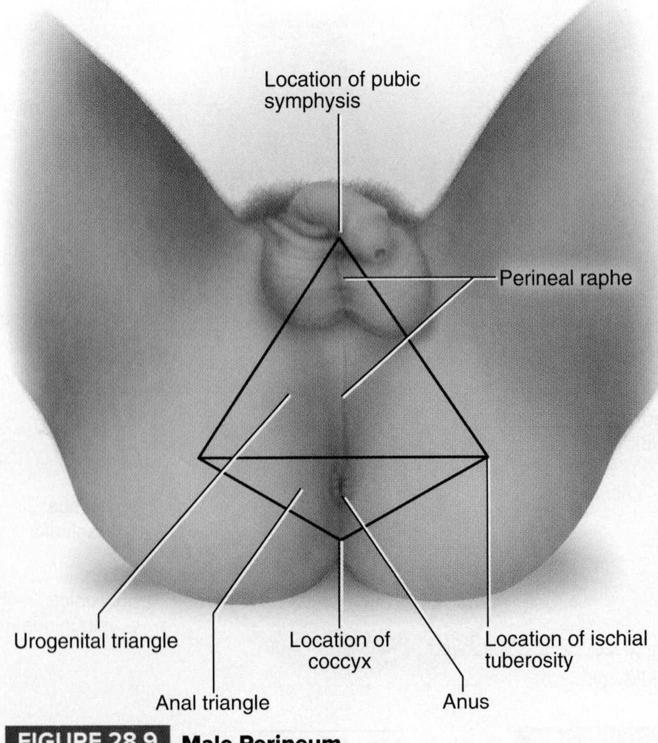

FIGURE 28.9 **Male Perineum**

Exercise 28

The Male Reproductive System

Name ______________________________ Date ______________

1. What is the gonad in the male reproductive system? ______________________________

2. The testes are considered mixed glands because they have both an endocrine and an exocrine function. Describe the endocrine and exocrine products that come from the testes. ______________________________

3. Proper sperm production must occur at what temperature? ______________________________

4. Male sterility can result from excessively high temperatures around the testes. What mechanism occurs in the scrotum to counteract the effects of high temperature? ______________________________

5. Trace the movement of sperm from the testes to where they exit the penis. List all the vessels they come into contact with as they move through the system. ______________________________

6. What is the name of the structure in the testis where spermatozoa are produced? ______________________________

7. What are the cells that initiate spermatozoa production? ______________________________

8. Where is the cremaster found? ______________________________

____

9. List all the structures involved in producing semen. ______________________________

____

____

____

10. How do spermatozoa differ from semen? ______________________________

____

11. A vasectomy is the cutting and tying of the two ductuli deferentes at the level of the spermatic cords. Review the percentage of spermatozoa that composes semen, and determine what effect a vasectomy has on semen volume.

____

____

12. Which one of the seminal fluid glands is not a paired gland? ______________________________

____

13. Where is the glans of the penis located? ______________________________

____

14. What is the cylinder of erectile tissue inferior to the corpora cavernosa? ______________________________

____

15. Match the structure in the left column with the function in the right column.

Structure	**Function**
_____ Testis	a. secretes lubricant
_____ Corpus cavernosum	b. secretes buffering solution and enzymes
_____ Epididymis	c. produces spermatozoa
_____ Prostate gland	d. erectile tissue in the penis
_____ Bulbourethral gland	e. place for maturation of spermatozoa

16. Name the three sections of the urethra and where they occur. ______________________________

____

____

17. Label the following illustration using the terms provided.

bulb of penis	epididymis	scrotum
bulbourethral gland	glans penis	seminal vesicle
corpus cavernosum	prepuce	testis
ductus deferens	prostate gland	urinary bladder

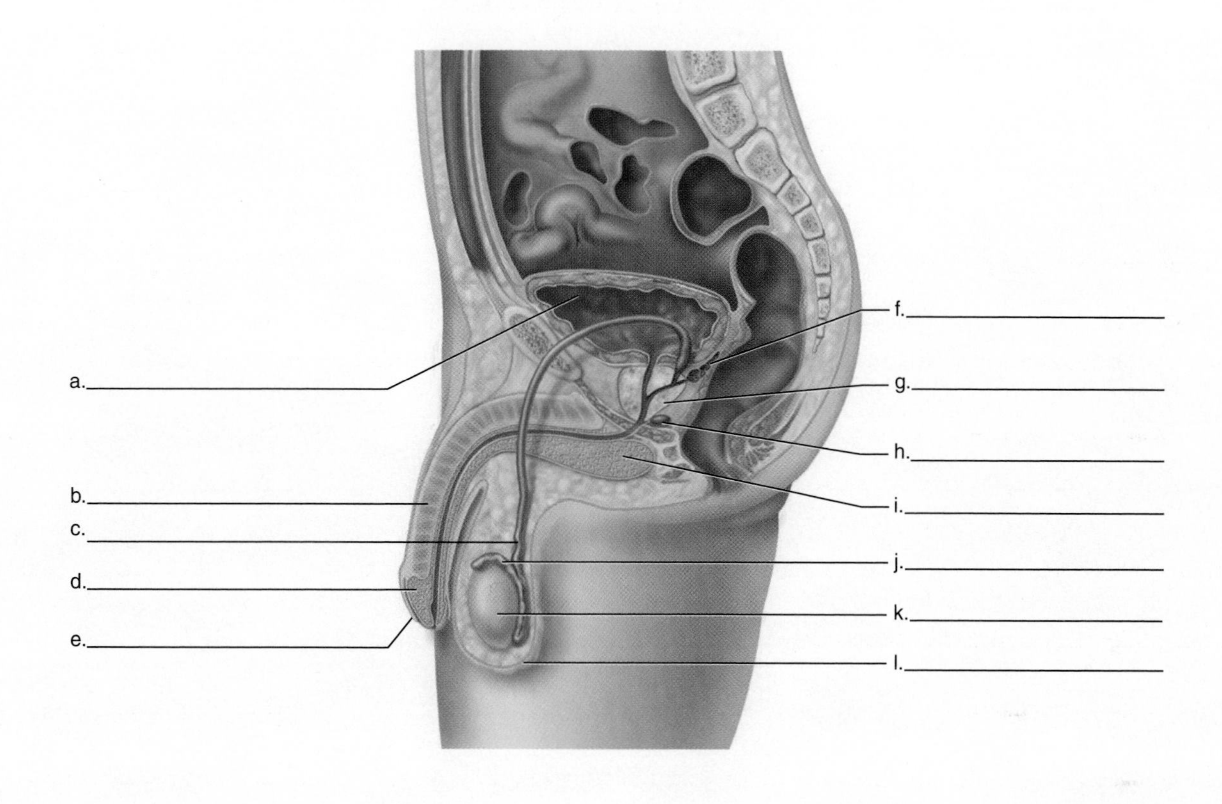

Laboratory

Exercise 29

The Female Reproductive System and Development

Reproductive System

Introduction

The female reproductive system is functionally more complex than the male reproductive system. In the male, the reproductive system produces gametes and delivers them to the female reproductive system. The female reproductive system not only produces gametes and receives the gametes from the male, but also provides space and nutrients for the developing **conceptus.** Finally, the female reproductive system delivers the child into the outer environment. In this exercise you learn about the structure of the female reproductive system and development.

Learning Objectives

At the end of this exercise you should be able to

1. identify the gamete-producing organ of the female reproductive system;
2. trace the pathway of a gamete from the ovary to the site of implantation;
3. list the structures of the vulva;
4. describe the function of each structure in the female reproductive system;
5. name the layers of the uterus from superficial to deep; and
6. describe the early development of humans.

Materials

Charts, models, and illustrations of the female reproductive system
Microscopes
Prepared slides of ovary and uterus

Procedure

Overview of the Gross Anatomy of the Female Reproductive System

Examine a model or chart of the female reproductive system and locate the following major reproductive organs there and in figure 29.1. When you find the structure, check it off in the list.

_____ Ovary

_____ Uterine tube

_____ Uterus

_____ Vagina

_____ Clitoris

_____ Labia minora (singular, *labium minus*)

_____ Labia majora (singular, *labium majus*)

Ovaries

The ovaries are the gamete-producing organs of the female reproductive system. The ovaries are a mixed gland having both an endocrine and an exocrine function. The exocrine function is the production of oocytes, and the endocrine function is the production of the female sex hormones—estrogen (estradiol) and progesterone.

Each **ovary** is an ovoid organ approximately 3 to 4 cm long (fig. 29.1). The ovaries produce **oocytes,** shed from the outer surface of the ovary during **ovulation.** From here the oocytes (which may mature into ova, if fertilized) move into the **uterine (fallopian) tube.** The ovaries are not directly attached to the uterine tube, and the oocytes must move from the surface of the ovary into the uterine tube.

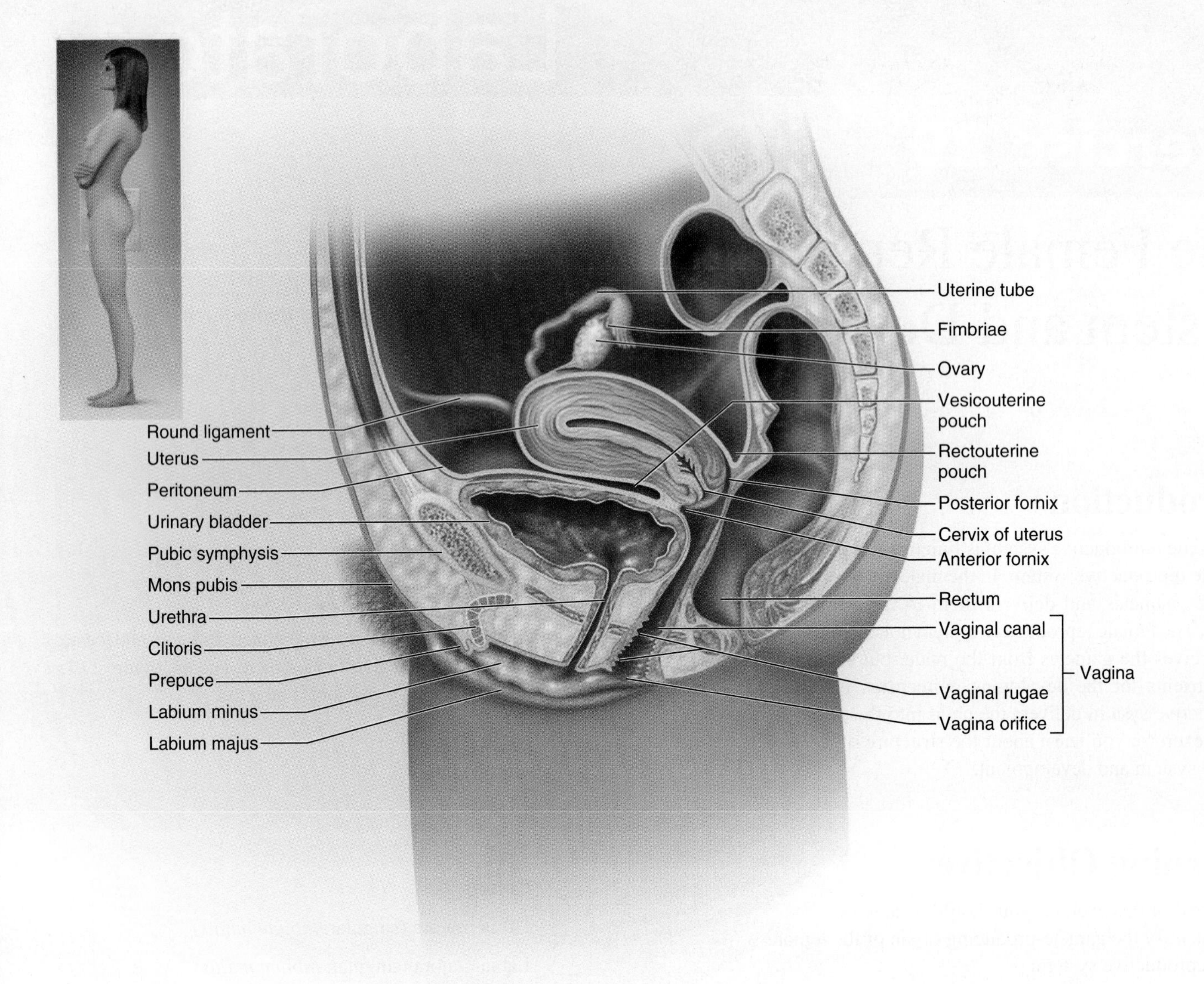

FIGURE 29.1 **Female Reproductive System, Median View**

Histology of the Ovary Examine a prepared slide of the ovary (cat or human) under the microscope on low power. Locate the **stroma,** which is a vascular, fibrous tissue in the middle of the ovary. The stroma is divided into a superficial **cortex** and a deep **medulla.** Look for circular structures in the ovary. These are the **ovarian follicles,** which include an oocyte surrounded by follicular cells. The smallest of the follicles are the **primordial follicles.** These contain primary oocytes. Locate the primordial follicles in your slide and compare these with the follicles in figure 29.2.

You should also locate the enlarging **primary** and **secondary follicles.** Some of the follicles may contain **oocytes**. All follicles contain oocytes, but you may not see them if the sectioning plane did not pass through the oocyte. Primary follicles and secondary follicles contain **secondary oocytes.** As a follicle develops fluid, it becomes a **tertiary follicle.** The largest follicles in the ovary are the **graafian follicles,** or **mature ovarian follicles,** and one may be present in your slide. During ovulation in humans, usually one secondary oocyte is shed from the ovary. In cats, many oocytes may be shed. Follicles produce estrogen, which contributes to the endocrine function of the ovary.

Examine your slide and compare it with figures 29.2 and 29.3. Draw what you see in your slide in the following space.

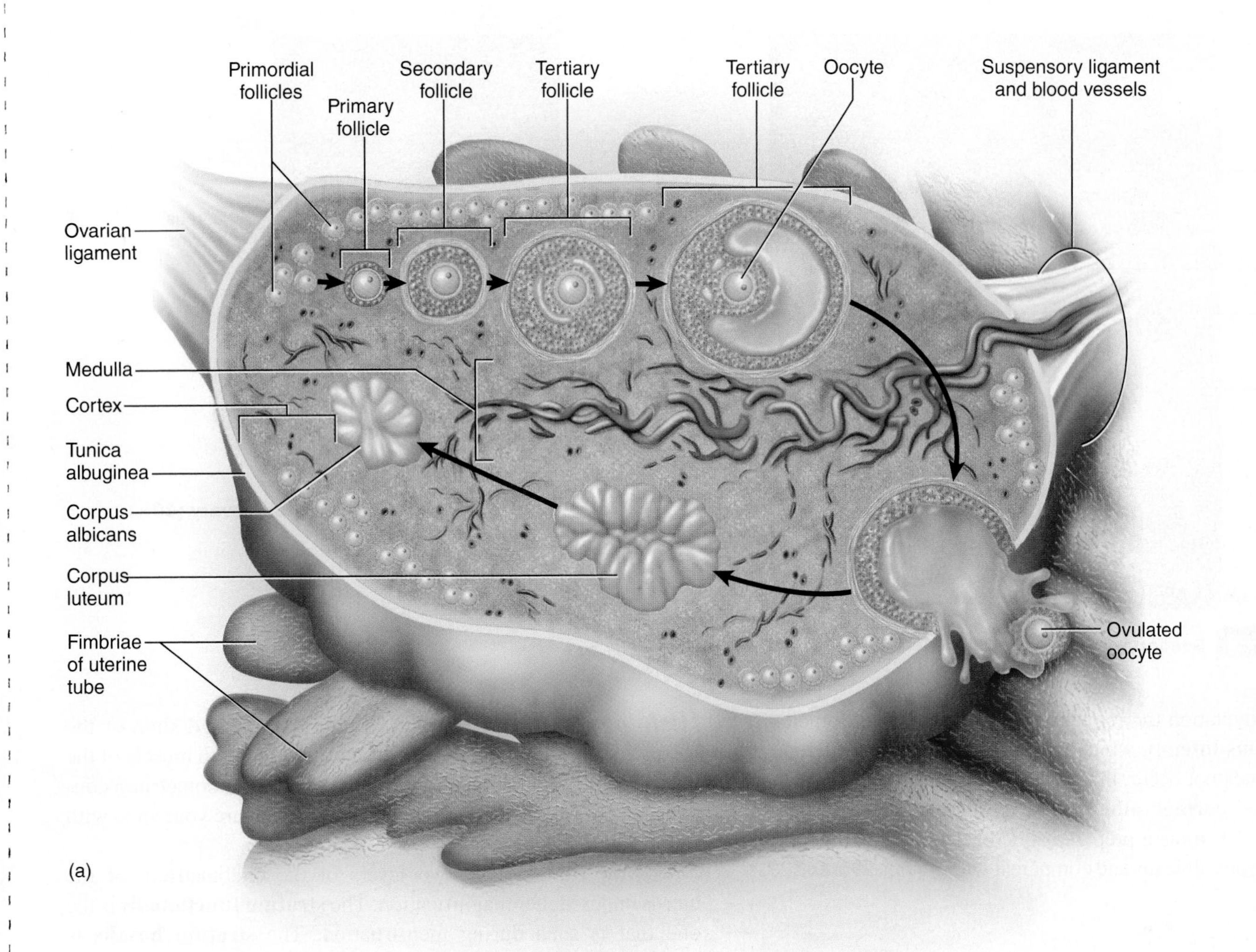

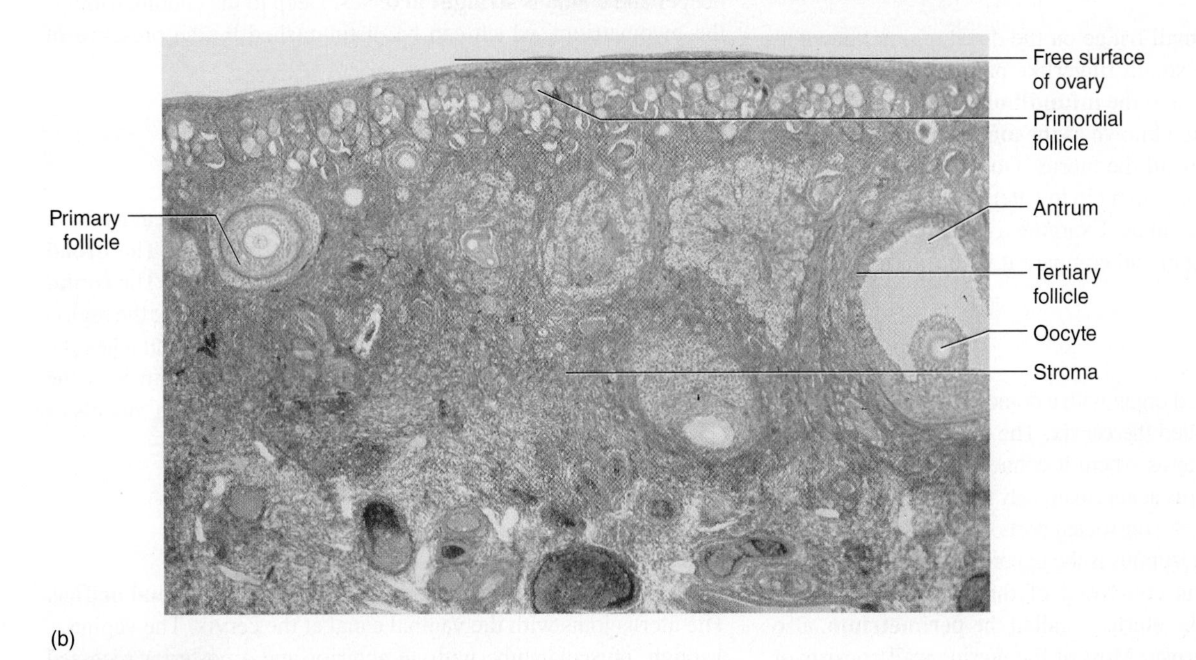

FIGURE 29.2 Histology of the Ovary (a) Diagram (arrows indicate a time line of development from primordial follicles to the corpus albicans); (b) photomicrograph (40×).

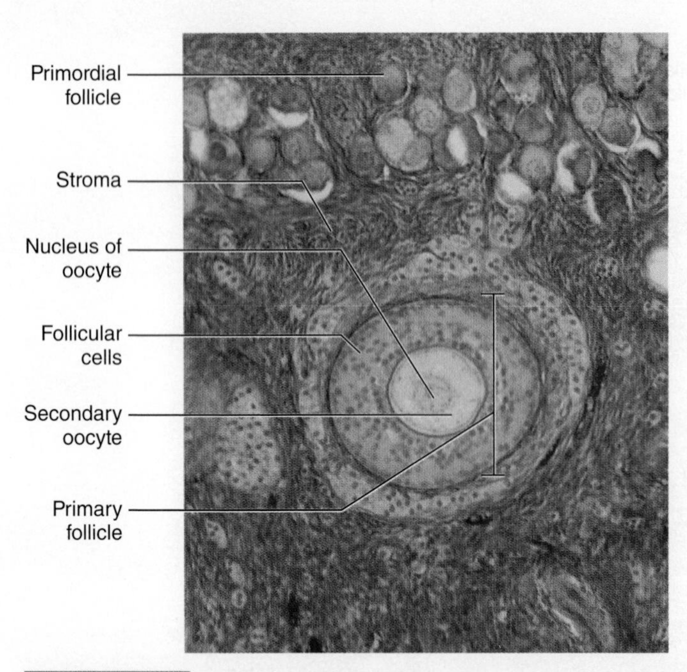

FIGURE 29.3 **Secondary Oocyte in Follicle (100×)**

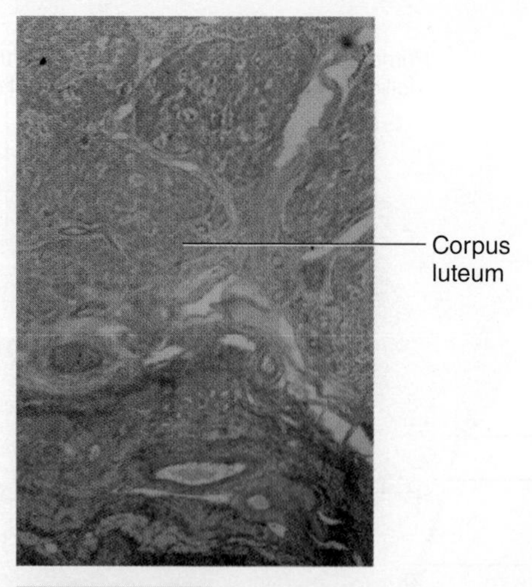

FIGURE 29.4 **Postovulatory Ovary (40×)**

After ovulation the remains of a mature ovarian follicle become a **corpus luteum,** which primarily secretes progesterone. If pregnancy does not occur, the corpus luteum decreases in size and becomes the **corpus albicans.** This progression is seen in figure 29.2*a*. Examine a prepared slide of the ovary with a corpus luteum or corpus albicans and compare it with figures 29.2 and 29.4.

Uterine Tubes

The uterine tube has a small fringe on the distal region known as the **fimbriae.** These are small, fingerlike projections attached to an expanded region known as the **infundibulum.** The uterine tube also has an enlarged region known as the **ampulla** and a narrower portion, the **isthmus,** toward the uterus. Oocytes move down the uterine tubes by ciliary movement in the uterine tube. Fertilization takes place in the uterine tubes. Examine a model or chart of the female reproductive system and compare it with figure 29.5.

Uterus

The **uterus** is a pear-shaped organ with a domed **fundus,** a **body,** and a circular, inferior end called the **cervix**. The cervix is a cylindrical, terminal portion of the uterus where it connects to the vagina. The uterine tubes enter the uterus at approximately the junction of the fundus with the uterine body. A constricted portion of the inferior uterus is called the **isthmus.** The isthmus is the upper third of the cervix.

The uterine wall is composed of three layers. The outer (superficial) surface of the uterus is called the **perimetrium,** also known as the **uterine serosa.** Most of the uterine wall consists of the **myometrium,** a thick layer of smooth muscle, and the innermost (deepest) layer of the uterus is the **endometrium,** which is the mucosa of the uterus. Examine the models or charts in the lab and locate the structures in figure 29.5.

Histology of the Uterus Examine a prepared slide of the uterus and locate the outer **perimetrium,** the smooth muscle of the myometrium, and the inner **endometrium.** The endometrium contains **spiral arteries** and **uterine glands.** Compare your slide with figures 29.5 and 29.6.

Now examine the two layers of the endometrium of the uterus under higher magnification. The **stratum functionalis** is the one that is shed during menstruation. The **stratum basalis** is deeper and contains **straight arteries.** Deep to the endometrium is the myometrium, which can be distinguished by the presence of smooth muscle.

Ligaments

The uterus and ovaries are suspended in the pelvic cavity by a number of connective tissue sheaths called ligaments. The **broad ligament** anchors the uterus to the anterior pelvic wall. The **round ligament** attaches the uterus to the anterior pelvic wall at the region of the inguinal canal. The **ovarian ligament** directly attaches the ovary to the uterus, and the **suspensory ligament** attaches the ovaries to the lumbar region. Locate these structures in models or charts in the lab and in figure 29.5.

Vagina

The **vagina** consists of the **vaginal canal** and the **vaginal orifice.** The uterus joins with the vaginal canal at the cervix. The vagina is a tough, muscular tube with an anterior and a posterior recessed region around the cervix known as the **fornix.** The vaginal canal is about 8 to 10 cm long, although it can stretch considerably during intercourse and delivery. It is located between the urethra, on the anterior side, and the rectum, which is posterior. The outer layer of

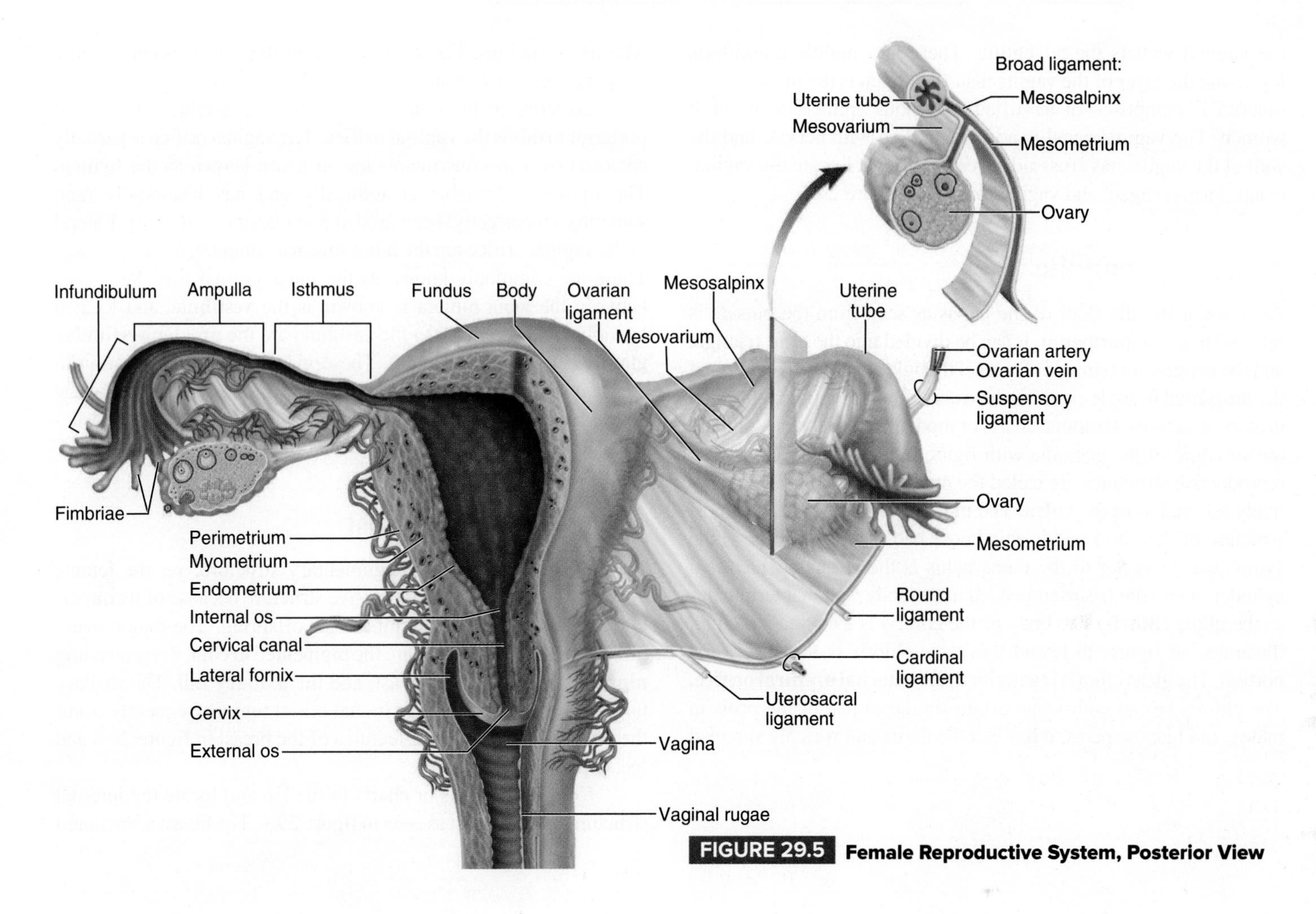

FIGURE 29.5 **Female Reproductive System, Posterior View**

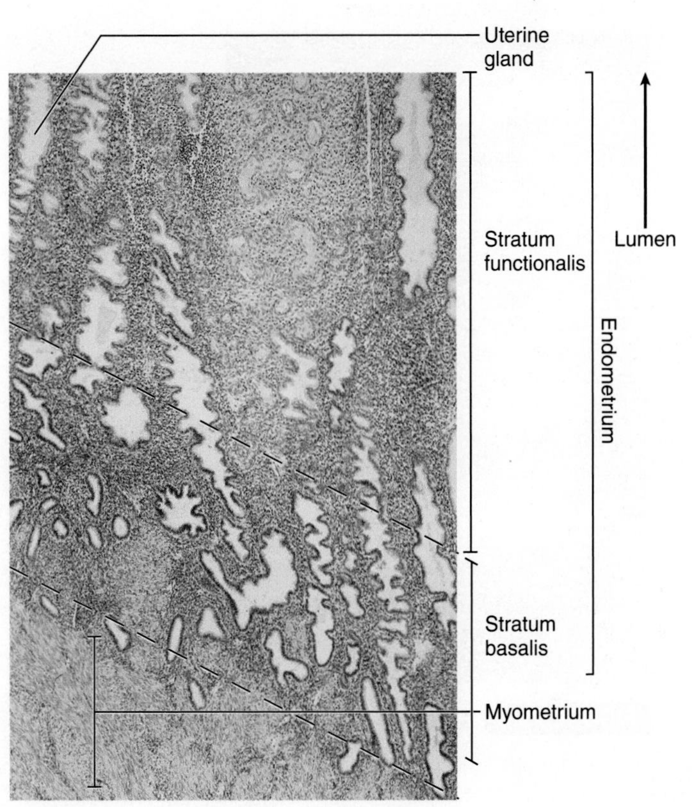

FIGURE 29.6 **Histology of the Uterus (40×)**

the vaginal wall is the adventitia. There is a middle muscularis layer, and the layer of the vagina near the lumen is the mucosa. The mucosa is composed of stratified squamous epithelium in adult women. The vaginal canal is poorly supplied with nerves, and the wall of the vagina has cross ridges called **rugae.** Locate the vaginal canal, fornix, rugae, and vaginal orifice in figure 29.1.

External Genitalia

As in the male, the floor of the pelvis as seen from the outside is referred to as the **perineum.** It can be divided into the **anal triangle** and the **urogenital triangle.** The anal triangle contains the **anus,** and the urogenital triangle contains the external female reproductive and urinary structures. Examine charts or models in the lab and compare the structure of the genitalia with figure 29.7. The external female reproductive structures are called the outer genitalia and are collectively referred to as the **vulva.** The **mons pubis** is the anterior-most structure of the vulva and is an adipose pad that overlies the pubic symphysis. Posterior to the mons pubis is the **clitoris,** which is a cylinder of erectile tissue embedded in the body wall that terminates as the **glans clitoris.** The body of the clitoris is a curved structure, illustrated in figure 29.1, and the glans clitoris is the superficial portion. The glans clitoris is anterior to the **external urethral orifice.** The clitoris has an embryonic origin similar to that of the penis in males, and like the penis, it has erectile tissue and is richly supplied with nerve endings. The anterior edge of the clitoris is enclosed by the **prepuce,** which is an extension of the labia minora.

Posterior to the clitoris is the external urethral orifice and posterior to this is the **vaginal orifice.** The vaginal orifice is partially enclosed by a mucous membrane structure known as the **hymen.** The hymen is variable anatomically and has historically (and sometimes incorrectly) been used as an indicator of virginity. Lateral to the vaginal orifice are the **labia minora** (singular, *labium minus*). These are commonly known as the inner vaginal lips. The space between the labia minora is known as the **vestibule,** and located laterally and posteriorly to the vestibule are the **greater vestibular glands (Bartholin's glands).** These glands provide lubrication to the vagina during intercourse. Lateral to the labia minora are the paired **labia majora** (singular, *labium majus*). Locate these structures on models or charts in the lab and in figure 29.7.

Anatomy of the Breast

The human breast is an integumentary structure, yet the female breast is discussed as a reproductive structure because of its importance as a source of nourishment for offspring. The major structures of the external breast are the pigmented **areola,** the protruding **nipple,** the **body** of the breast, and the **axillary tail.** The axillary tail is of clinical importance in that breast tumors frequently occur there. Examine the surface features of the breast in figure 29.8 and locate the structures listed.

Compare models or charts in the lab and locate the internal structures of the breast as seen in figure 29.9. The breast is anchored

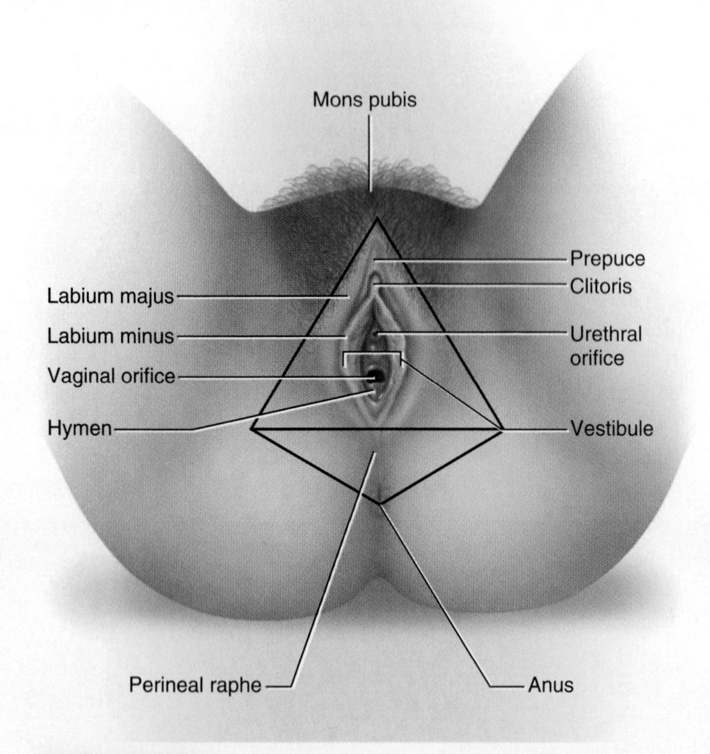

FIGURE 29.7 **Female Perineum and External Genitalia**

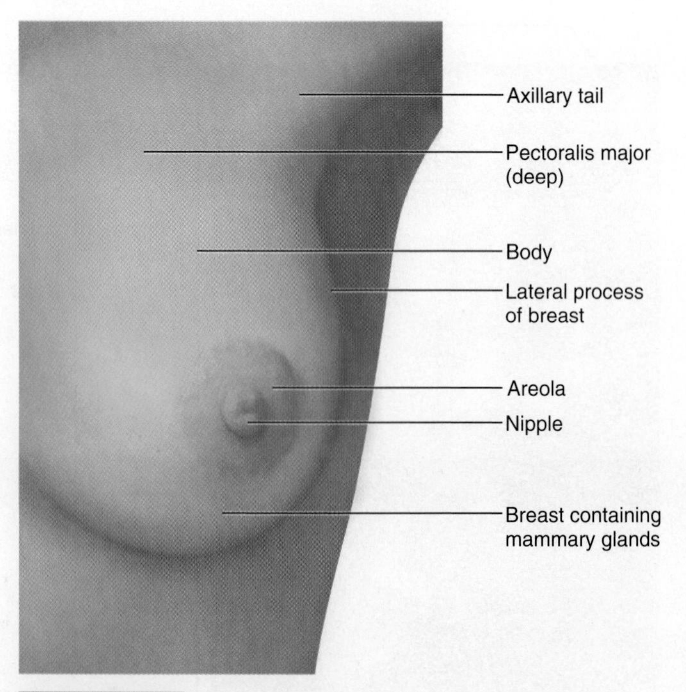

FIGURE 29.8 **Surface Features of the Female Breast**

to the pectoralis major muscle and dermis of the skin with **suspensory ligaments.** Much of the breast is composed of **adipose tissue,** and embedded in the adipose tissue are the **mammary glands.** The mammary glands are responsible for the production of milk in lactating females. The glands are clustered in **lobes,** and there are about 15 to 20 lobes in each breast. In lactation the mammary glands increase in size and lead to **lactiferous ducts,** which subsequently lead to **lactiferous sinuses (ampullae).** These sinuses exit via the nipple. Humans have several sinuses leading to each nipple. The mammary glands in females begin to undergo changes prior to puberty and become functional glands after delivery of a child. Note the features of the breast in figure 29.9.

Stages of Development

Early Development The union of a sperm and an egg in a process known as **fertilization** initiates a remarkable phenomenon of growth and differentiation from the single-celled **zygote** to the adult human. The sperm and egg pronuclei fuse, and the genetic information from the mother and father forms the genes of the **conceptus.** Fertilization usually occurs in the uterine tube, and the zygote divides into 2 cells, then 4, 8, 16, and so on until a solid cluster of cells, called a **morula,** is formed. The morula is so named because it looks like a mulberry (*Morus* = genus name for mulberry). The morula continues to divide until it becomes a hollow ball of cells known as the **blastocyst.** The covering of cells on the outside of the blastocyst is called the **trophoblast,** and the cluster of cells on the inside is known as the **embryoblast,** or **inner mass.** Implantation occurs if the blastocyst attaches to the uterine wall. Review these stages on models and charts in the lab and in figure 29.10. The placenta is important during development, as it produces hormones that maintain pregnancy.

Embryonic Tissues The conceptus continues to progress and three embryonic tissues form. These tissues are the **ectoderm,** the **mesoderm,** and the **endoderm.** The ectoderm gives rise to the outer layer of skin and the nervous tissue, the mesoderm gives rise to bones and muscles, and the endoderm gives rise to many internal organs, such as digestive and respiratory organs. The derivations from the embryonic tissues are outlined in table 29.1. The early development of these layers is illustrated in figure 29.11.

Cat Anatomy

If you are using cats, turn to section 10, "Reproductive Anatomy of the Cat," "Female Cat," on page 443 of this lab manual.

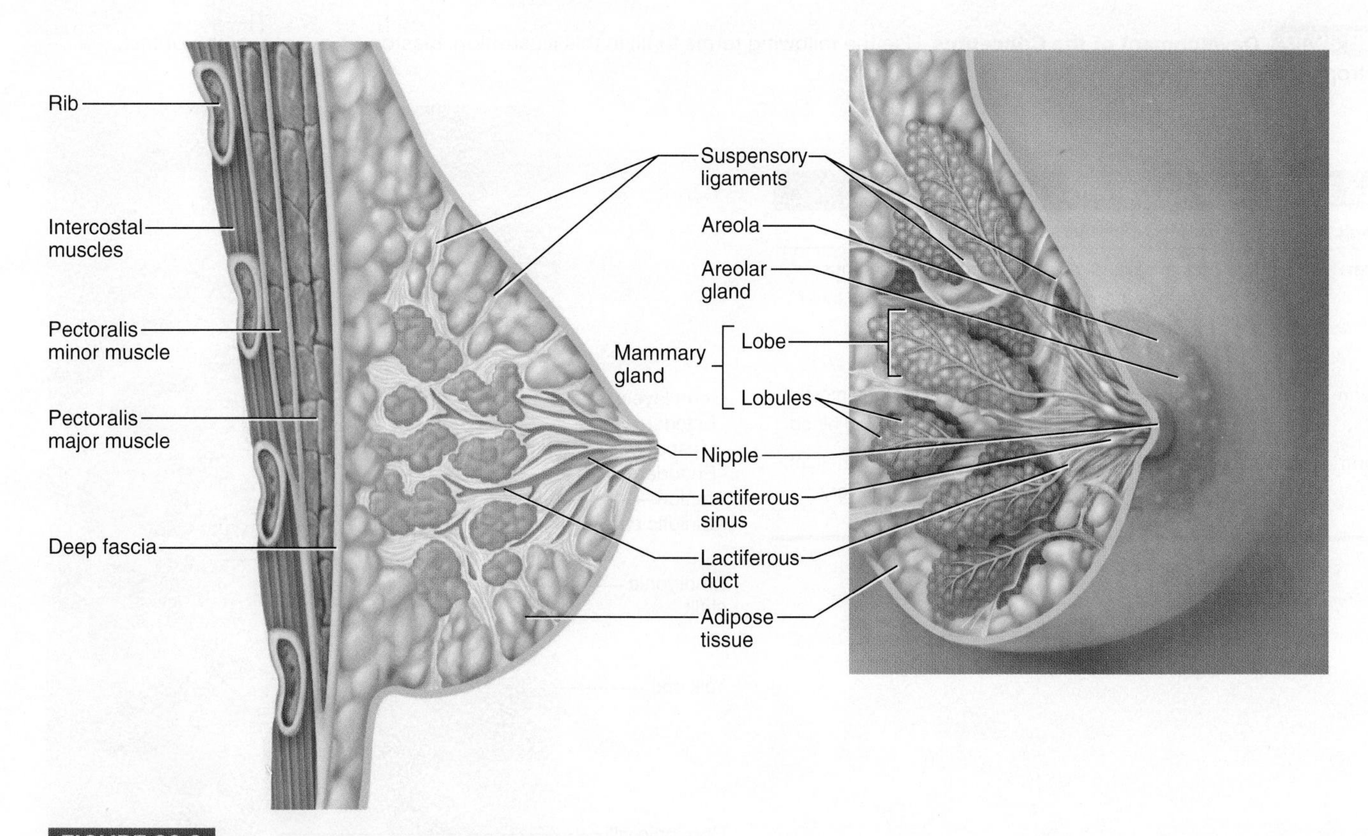

FIGURE 29.9 **Interior of the Female Breast**

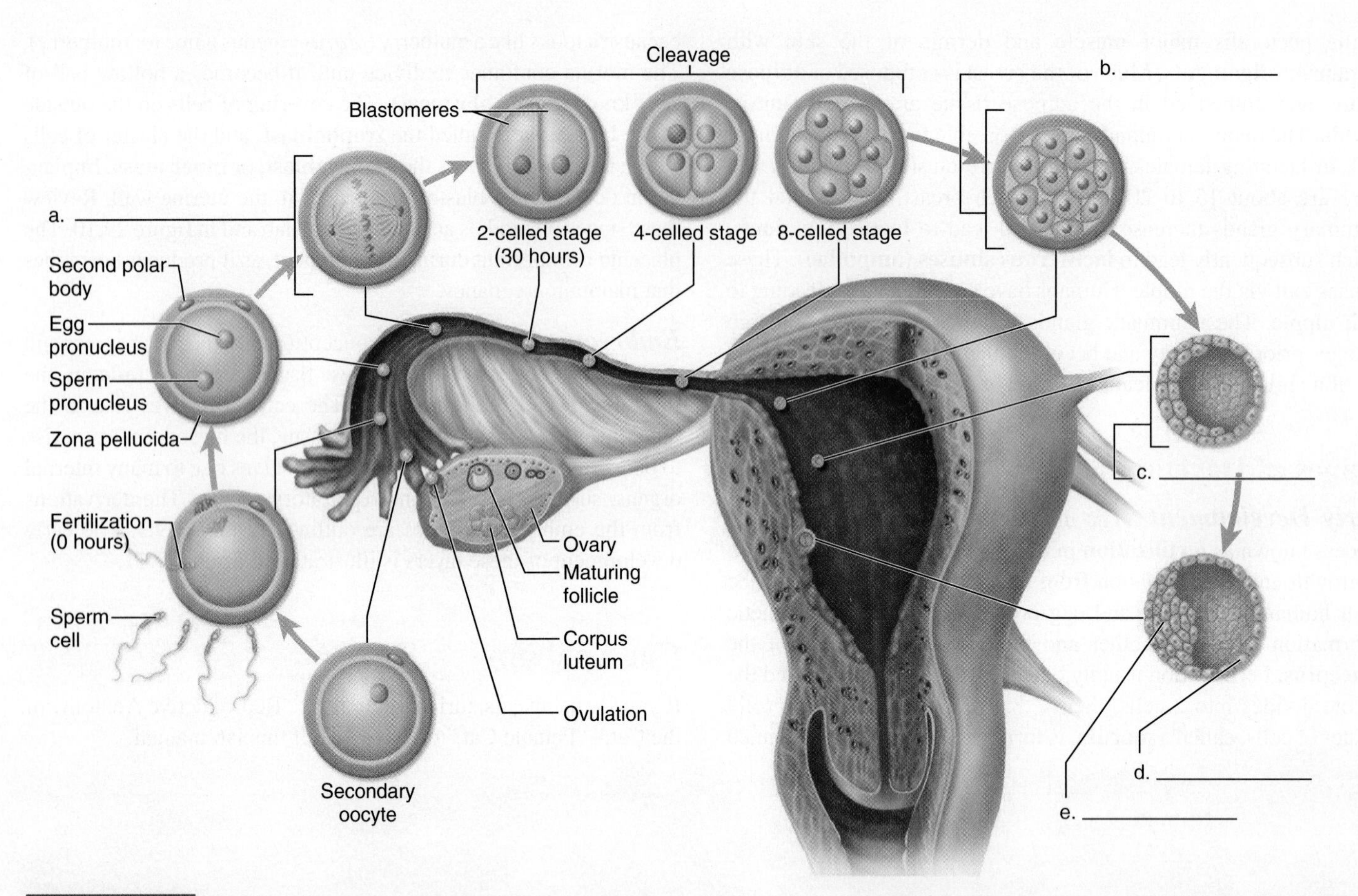

FIGURE 29.10 Development of the Conceptus Use the following terms to fill in this illustration: blastocyst, zygote, embryoblast, morula, trophoblast.

TABLE 29.1 Derivations of Embryonic Tissues

Embryonic Tissues	Adult Derivatives
Ectoderm	Epidermis and most of its derivatives, nervous system, outer surface of the eye, outer and inner ear, and epithelia of the mouth, nose, and anus
Mesoderm	Bone; bone marrow; skeletal, cardiac, and most of the smooth muscle; dermis of the skin; and blood
Endoderm	Most of the digestive and respiratory epithelium, most of the digestive glands (except salivary glands), and urinary bladder epithelium

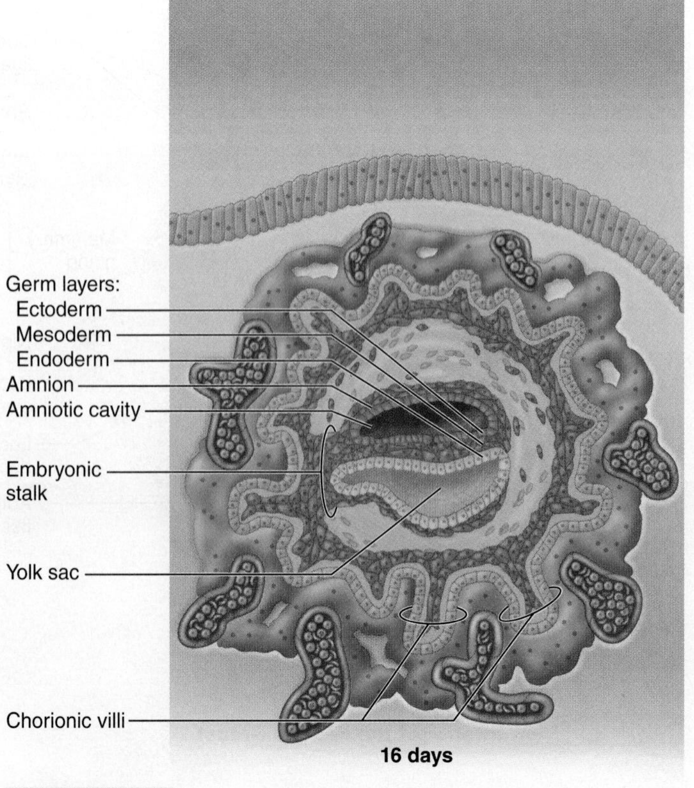

FIGURE 29.11 Embryonic Development

Review

Exercise 29

The Female Reproductive System and Development

Name ______________________________ Date ______________

1. What are the gonads in the female reproductive system? ______________________________

____

2. Where is the fornix in the female reproductive system? ______________________________

____

3. What is the background substance of the ovary called? ______________________________

____

4. What is the inner layer of the uterus called? ______________________________

____

5. The ovaries attach to the uterus by what structure? ______________________________

____

6. Which is more anterior, the urethral opening or the clitoris? ______________________________

____

7. What is the name for the expulsion of the oocyte from the ovary? ______________________________

____

8. What is the layer of the endometrium closest to the myometrium called? ______

9. Ectopic pregnancies are those that occur outside of the endometrial layer of the uterus. Explain how pregnancies can occur in the uterine tube (thus, a tubal pregnancy) or in the abdominopelvic cavity. ______

10. Trace the pathway of milk from the mammary glands to expulsion. ______

11. What is the name of the part of the breast nearest to the shoulder? ______

12. What are the milk-producing glands of the breast called? ______

13. A zygote is formed from the fusion of what two cells? ______

14. Embryonic tissue consists of three layers. What are these called? ______

15. What is a morula? ______

16. Label the following illustration using the terms provided.

cervix	fundus	urinary bladder
clitoris	labium minus	vaginal canal
fornix	rugae	vaginal orifice

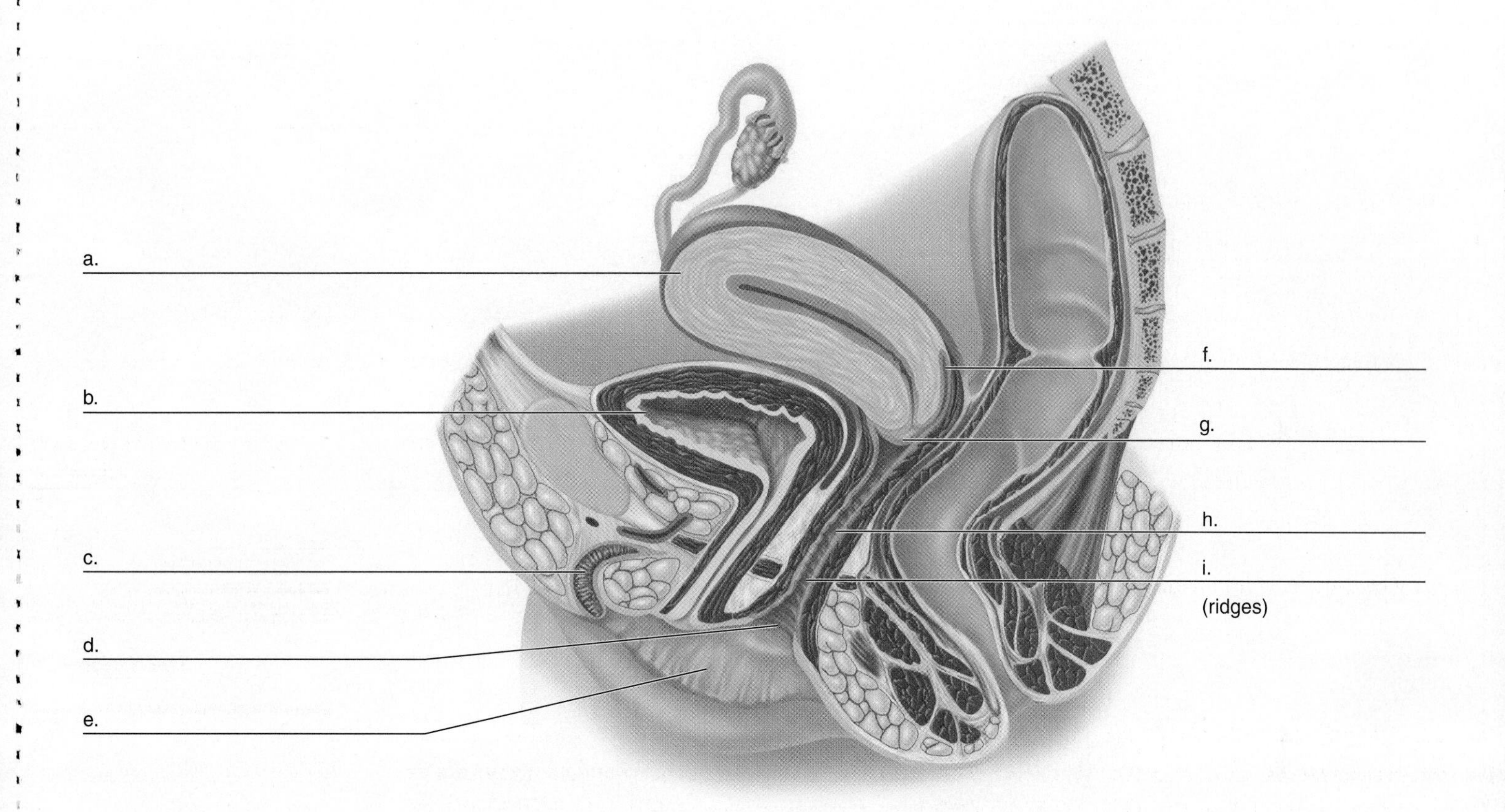

Cat Anatomy

The cat is a reasonable study specimen for understanding human anatomy. The fact that cats and humans have many of the same bones, muscles, and organs reflects the evolutionary mammalian pattern. There are some noticeable differences between cats and humans, though. As cats are quadrupeds, they have a skeletal system and muscles adapted to walking on all fours.

Because this is the study of human anatomy, human terms will be used to describe anatomical features of the cat as a general rule in these lab exercises. For example, the *anterior vena cava* in cats is homologous to the *superior vena cava* in humans. There are some structures in the cat, such as the *epitrochlearis* muscle, that are not found in the human. These structures will retain the cat name.

The exercises in the main portion of the lab manual direct you to the appropriate cat anatomy section. Sections and topics are listed with their page numbers in the following table.

Materials for Cat Dissection

Each time you dissect your cat, you should have the appropriate area and tools for dissection. Remember that many people are sensitive to the dissection of animals, so please be aware of this concern as you do your work. A list of materials for dissection follows.

- Cats
- Large plastic bag to store cat
- String to keep bag closed
- Tag to identify specimen
- Paper towels or clean cloth (an old T-shirt works well) to keep cat moist during storage
- Dissection trays—large enough to hold cat and keep materials in place
- Scalpel and new blades
- Protective gloves
- Blunt (Mall) probe
- Pins to hold parts down or for identification
- Forceps
- Sharp scissors
- Animal waste disposal container

Additional Materials to Have on Hand During Dissection Labs

First aid kit in lab or prep area

Sharps container for used scalpel blades

Dissection Concerns

Wear an apron or an old overshirt, protective gloves, and safety glasses when dissecting. Dissection instruments are sharp, and care must be taken when using scalpels. Do not cut *down* into the specimen but, rather, lift structures gently and try to make incisions so the scalpel blade cuts *laterally*. Cut *away* from yourself and your lab partners.

If you do cut yourself, notify the instructor immediately! Wash the cut with antimicrobial soap, cover the wound with a sterile bandage, and seek medical advice to reduce the chance of infection.

The laboratory should be well ventilated, but if your eyes burn and you develop a headache, get some fresh air for a moment. If you dissect without having your face directly over the specimen, it may help. Occasionally, students have allergic reactions to formaldehyde. This usually consists of a feeling of restriction of breath either during or after the lab. If you have this experience, notify your instructor.

At the end of the exercise, wash your dissection equipment with soap and water, taking extra precaution with the scalpel blade. Place all used blades or sharp material in the **sharps container** in the lab. Remove all the excess animal material on the dissection trays and put it in the appropriate animal waste container. *Do not dump any animal material in the lab sinks!* Wash the dissection trays and place them in the appropriate area to dry.

Cat Care

Keep your cat in a plastic bag. It is recommended that you retain the skin of the cat as a protective wrapping when you are finished with the day's dissection or that you take a small towel (or an old T-shirt) and wrap the specimen in the cloth, soaking it in a cat wetting solution before you place it back in the bag. Usually, one lab period is required to skin the cat and remove the superficial fascia over the muscles, to prepare the specimen for further study.

Cleanup

When you are done with your dissection, carefully place your cat back in the plastic bag. Place all waste material in the appropriate animal waste disposal container, not down the sinks in the lab.

Section 1: Skeletal Anatomy of the Cat

The cat skeleton differs from the human skeleton in several ways. Humans are bipedal, and the vertebral column reflects this. In humans, the lower vertebrae are larger than the upper vertebrae. The vertebrae in cats are more evenly matched in size because cats are quadrupeds and the weight on the vertebral column is distributed more evenly. Humans have a coccyx, while cats have tail vertebrae that are fully functional.

Other skeletal differences can be seen in that humans walk on the entire inferior surface of the foot, while cats walk on their toes. Examine a cat skeleton in the lab and note the general features of the skeleton (fig. S1.1).

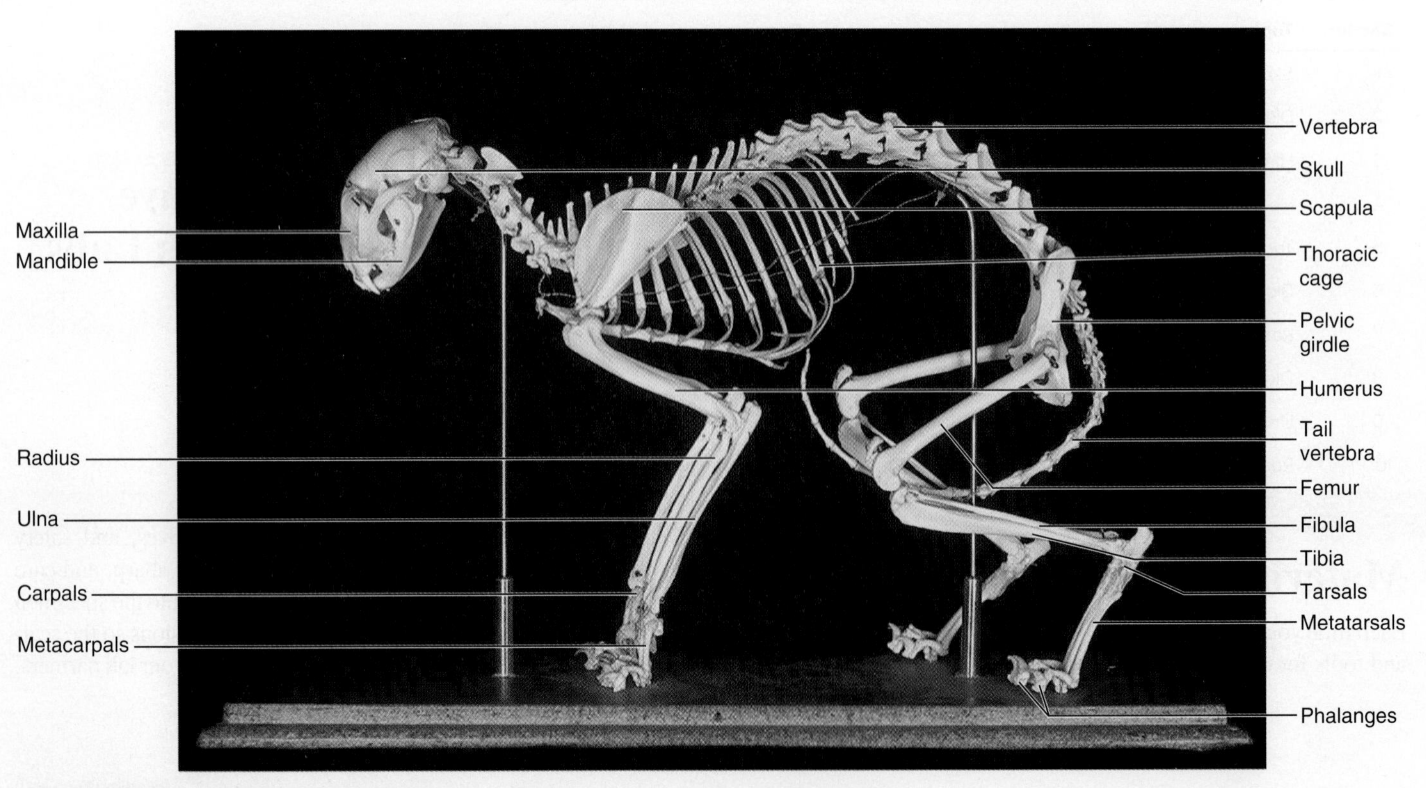

FIGURE S1.1 Cat Skeleton, Lateral View

Section 2: Dissection, Overview, and Forelimb Muscles of the Cat

The objective in using a cat for the exercise on musculature is to provide a study specimen for dissection and application to the human system. Differences occur between cat musculature and human musculature, but you should focus on similar structures in order to gain an appreciation of human musculature. Read all the material in the exercise prior to beginning the dissection. Dissection is a skill that requires separation of the overlying structures from those underneath, while keeping intact as much material as possible.

External Features

Take a dissection tray and a cat specimen to your table. Remove the cat from the plastic bag and place it on the tray. The plastic bag may contain excess fluid; this should be disposed of properly as directed by your instructor. Place the cat on its back and determine whether it is male or female. Both sexes have multiple **teats** (nipples), yet the males have a **scrotum** near the base of the tail with a **prepuce** (penile foreskin) ventral to the scrotum. Females have a **urogenital opening** anterior to the anus without a scrotum and prepuce. Once you have identified the sex of your specimen, compare yours with others in the class, so that you can identify the sexes externally. If you have difficulty determining the sex, ask your instructor for help.

Examine the cat and notice the **vibrissae** (vib-RISS-ee), or whiskers, in the facial region. Other variations from humans are the presence of **claws** and **friction pads** on the extremities and a **tail.** Review the planes of the body as illustrated in laboratory exercise 1 before beginning the dissection. The terms used in that exercise will be of great importance in the dissection procedures.

Removal of the Skin

Begin the dissection by lifting the skin in the pectoral region with a forceps and making a small cut with a scalpel or sharp scissors in the midline. Work a blunt probe gently into the cut to free the skin from the underlying fascia and muscle somewhat. Be careful—the muscles are close to the skin. Insert your scalpel, blade side up, and make small incisions in the skin, cutting away from the underlying muscle, as illustrated in figure S2.1.

Make a cut that runs up the midline of the sternal region to the neck. Likewise, cut posteriorly to an area anterior to the genital region. Leave the genitals intact and carefully make an incision that runs perpendicular to your first cut. Likewise, make a perpendicular incision along the upper thoracic region (fig. S2.2). Continue your caudal incision along the medial aspect of the thigh. Watch

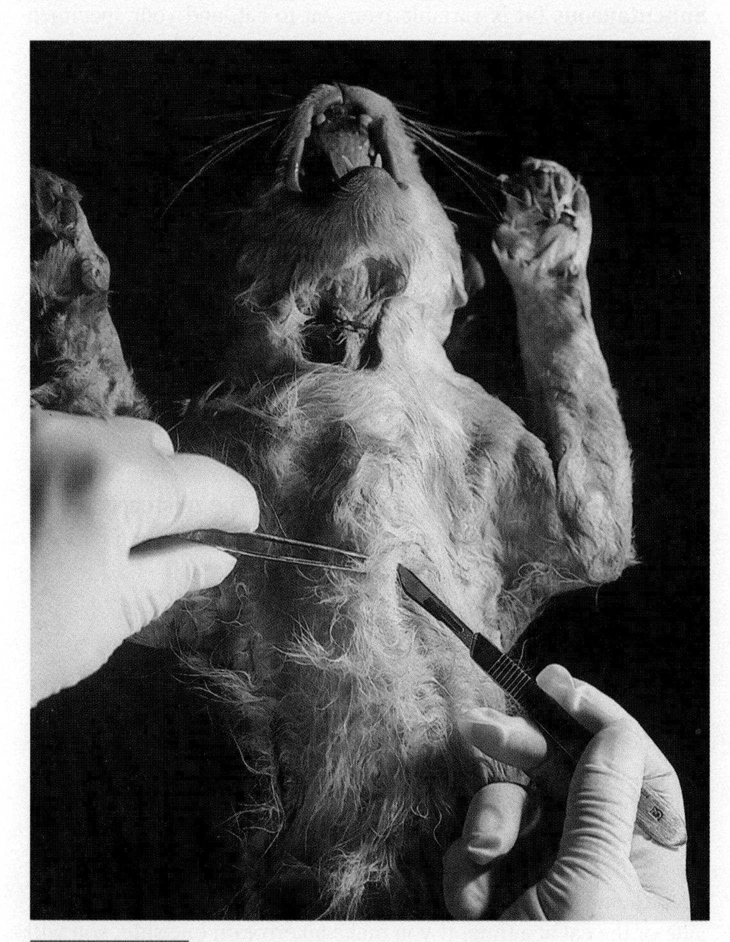

FIGURE S2.1 Lateral Cutting with a Scalpel

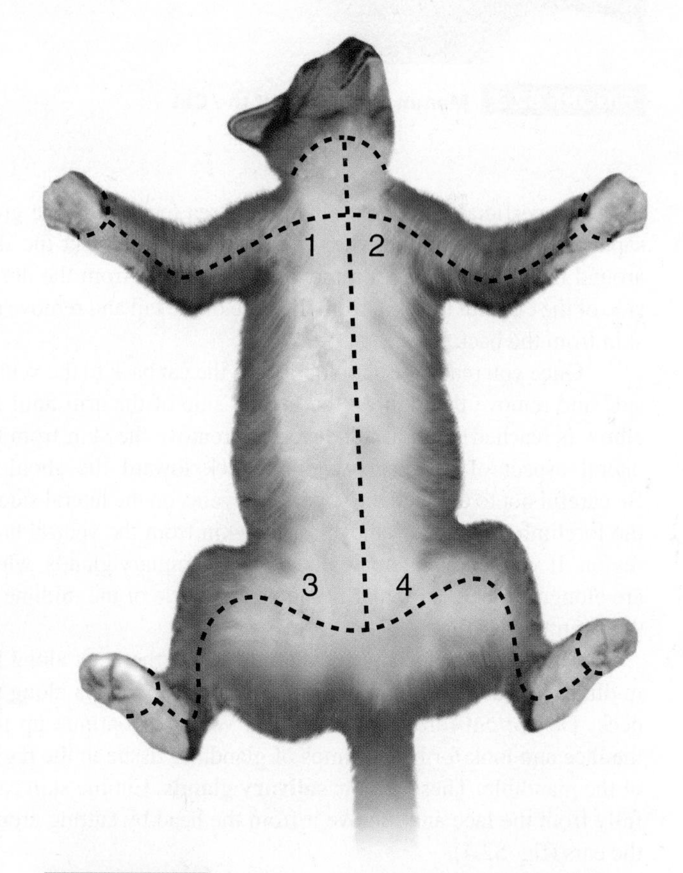

FIGURE S2.2 Removal of the Skin of the Cat After cutting down the midline, make incisions into the limbs (follow the numbers) and gently remove the skin from underlying structures.

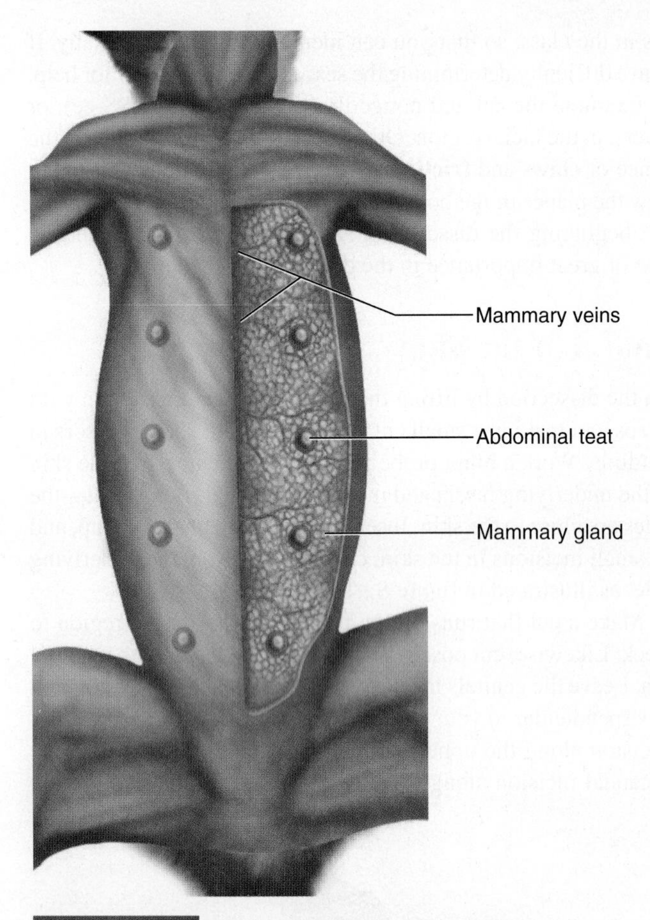

FIGURE S2.3 **Mammary Glands of the Cat**

out for superficial blood vessels in this area (especially the great saphenous vein) and stop when the knee is reached. Cut the skin around the knee and begin removing it as a layer from the dorsal side of the cat. Cut the skin from the base of the tail and remove the skin from the back.

Once you reach the shoulders, turn the cat back to the ventral side and remove the skin on the medial side of the arm until the elbow is reached. Stop at this level and remove the skin from the lateral aspect of the arm, working back toward the shoulder. Be careful not to damage the superficial veins on the lateral side of the forelimb. Continue removal of the skin from the ventral body region. If your cat is a female, locate the mammary glands, which are elongated, beige, lobular tissue on each side of the midline on the ventral side (fig. S2.3).

Make an incision on the ventral side of the neck along the midline. Be careful—there are numerous blood vessels along the neck. Do *not* cut through these blood vessels. Continue up into the face and look for beige lumps of glandular tissue in the region of the mandible. These are the **salivary glands.** Cut the skin carefully from the face and remove it from the head by cutting around the ears (fig. S2.4).

You should be able to remove all the skin from the cat at this time. Examine the undersurface of the skin and note the superficial muscles that cause the skin to move. These are the **cutaneous** (que-TAY-nee-us) **maximus** and the **platysma** (plah-TISS-mah).

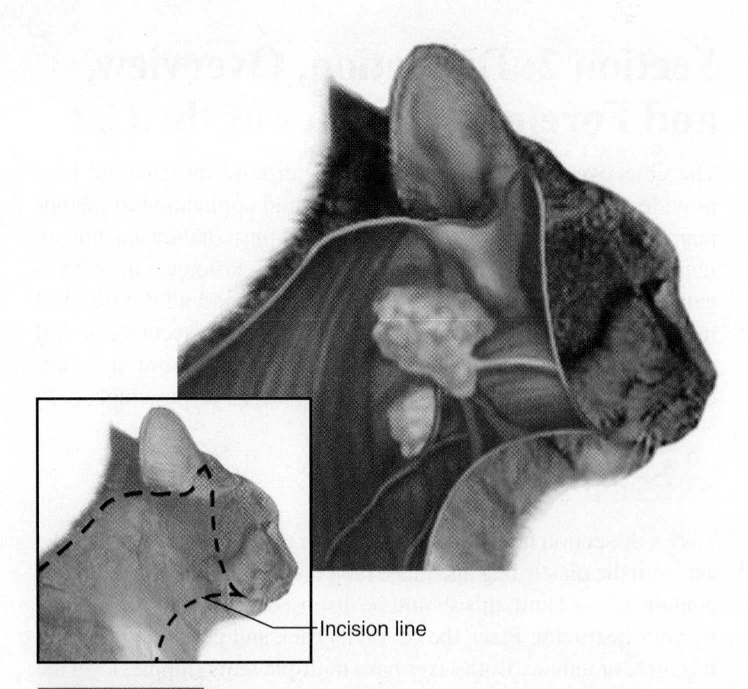

FIGURE S2.4 **Removal of Skin from the Head of a Cat**

Return to the cat and remove as much fat as you can. **Subcutaneous fat** is variable from cat to cat, and your specimen may have little or may have significant amounts. The muscle also is covered by fascia, a connective tissue wrapping. Remove the fascia from the muscle, so that the fiber direction is apparent.

Once you have removed the skin and the superficial fascia, identify the major muscles of the cat. Look for the large latissimus dorsi muscle of the back and the external abdominal oblique muscle. Find the deltoids and triceps brachii muscles of the shoulder region, as well as the gluteus and biceps femoris muscles of the hip and thigh region. Compare your cat with figure S2.5.

Individual Muscles of the Cat

Begin the **dissection** of the muscles by understanding that the term "dissect" means to separate. When you isolate one muscle from another, locate the tendons of that muscle. The **tendon** is the attachment point of the muscle to a bone. Broad, flat tendons are known as **aponeuroses** (AH-poh-nyeu-ROH-seez).

Cat Dissection

If you did not already remove the skin from the forelimb of the cat, you should do so now. This can be accomplished by making a longitudinal incision along the length of the forelimb. Be very careful not to cut the tendons, blood vessels, or nerves. Pull the skin off as if you were removing a pair of knee socks. As you get to the tips of the digits, cut the skin from the digits. Pay particular attention and keep the tendons intact. It is a good procedure to dissect only one side of the cat at a time. If you make an error on one side, you will have the other side to dissect.

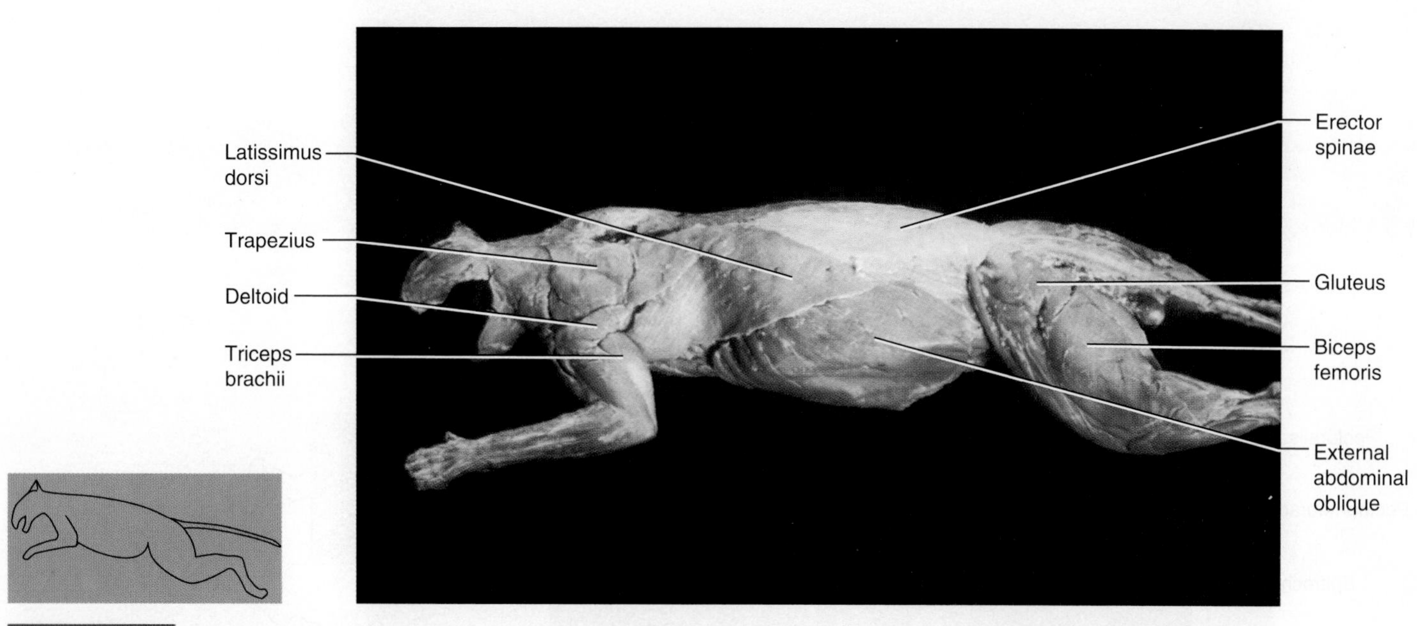

FIGURE S2.5 Major Muscles of the Cat

As you dissect the cat, you can leave many of the muscles of the forelimb intact. Deeper muscles can generally be seen by moving the more superficial muscles off to the side.

Upon locating a muscle, tug gently on it to locate its attachment points. Pulling too hard may rip the muscle. The outer wrapping of the muscle is known as the **fascia,** and it should be removed to find the fiber direction of the muscle. The main part of the muscle is known as the **belly.** You may need to cut a muscle to locate deeper muscles. This is done by **transecting** the muscle, which is to cut the muscle into two sections perpendicular to the fiber direction. Lift the muscle with a probe to transect it. Do not cut underlying muscles. After transecting the muscle, you may want to **reflect** it, or pull it toward its attachment site. When separating two muscles, you may find a cottonlike material in between them. This is loose connective tissue that forms part of the fascia.

Thoracic Muscles

Pectoral Muscles of the Cat There are four major muscles in a superficial view of the pectoral region of the cat. These are the pectoantebrachialis, pectoralis major, pectoralis minor, and xiphihumeralis. The **pectoantebrachialis** has no corresponding muscle in the human. It attaches on the sternum and on the forelimb. Transect and reflect the pectoantebrachialis to see the pectoralis major. The **pectoralis major** attaches to the sternum and then laterally to the upper humerus. The **pectoralis minor** is a large muscle in cats; it attaches to the upper humerus. The **xiphihumeralis** (ZIE-fee-HEU-mur-AL-is) is another cat muscle that has no corresponding human muscle, and it attaches to the sternum and attaches laterally to the proximal humerus, along with the pectoralis major and minor. Locate these muscles on your cat and in figure S2.6.

In humans, there is a single trapezius on each side of the body. In cats, each trapezius consists of three muscles. Examine the cat from the dorsal side and locate the **clavotrapezius,** the **acromiotrapezius,** and the **spinotrapezius.** All these muscles attach to the vertebral column, with the clavotrapezius also attaching to the occipital bone. The clavotrapezius attaches to the clavicle, the acromiotrapezius on the acromion of the scapula, and the spinotrapezius on the spine of the scapula. Compare these muscles with figure S2.7.

The deltoid muscle in cats also consists of three muscles, unlike the single deltoid in humans. The deltoid muscles are named for their bony attachments. The **clavodeltoid** (clavobrachialis) attaches to the clavicle, the **acromiodeltoid** on the acromion, and the **spinodeltoid** on the spine of the scapula. Distal attachments of these muscles are on the arm or forelimb. Locate these muscles on the cat and compare them with figure S2.7.

Two other muscles of the region are the **latissimus dorsi** and the **levator scapulae ventralis.** These two muscles are similar to those in the human. Locate these muscles on the cat and compare them with figure S2.7.

Scapular Muscles That Act on the Humerus The deep muscles of the scapula can be seen by reflecting the overlying muscles. The **supraspinatus, infraspinatus, subscapularis, teres major,** and **teres minor** are roughly equivalent to the same muscles in the human. Locate the supraspinatus, infraspinatus, and teres major on the cat and find them in figure S2.8.

Forelimb Muscles The muscles that have an action on the forelimb of the cat typically either flex the forelimb or extend the forelimb as their primary actions. The **epitrochlearis** is a muscle that does not have a corresponding muscle in humans (fig. S2.9). The epitrochlearis is on the medial side of the humerus and extends the forelimb. The **biceps brachii** is also a medial muscle, and it flexes the forelimb. The **triceps brachii, anconeus, brachioradialis,** and **brachialis** are lateral or posterior muscles. The triceps brachii and the anconeus extend the forelimb, while the brachialis flexes the forelimb. Find the lateral muscles, using figures S2.9 and S2.10 as a guide.

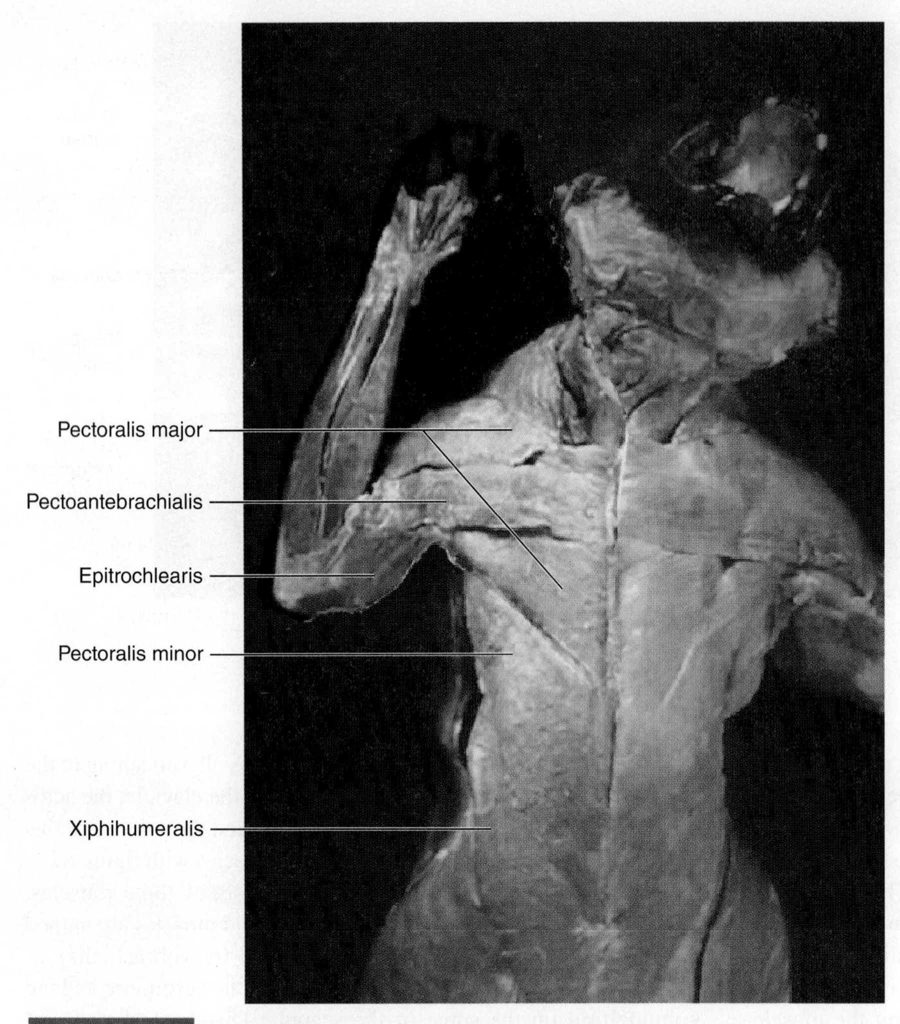

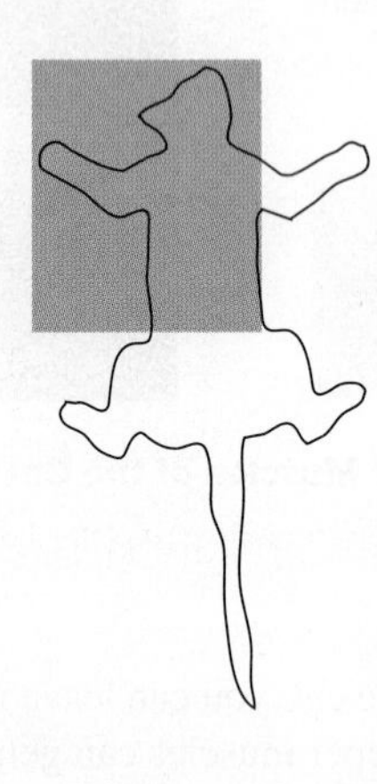

FIGURE S2.6 Muscles of the Pectoral Region of the Cat

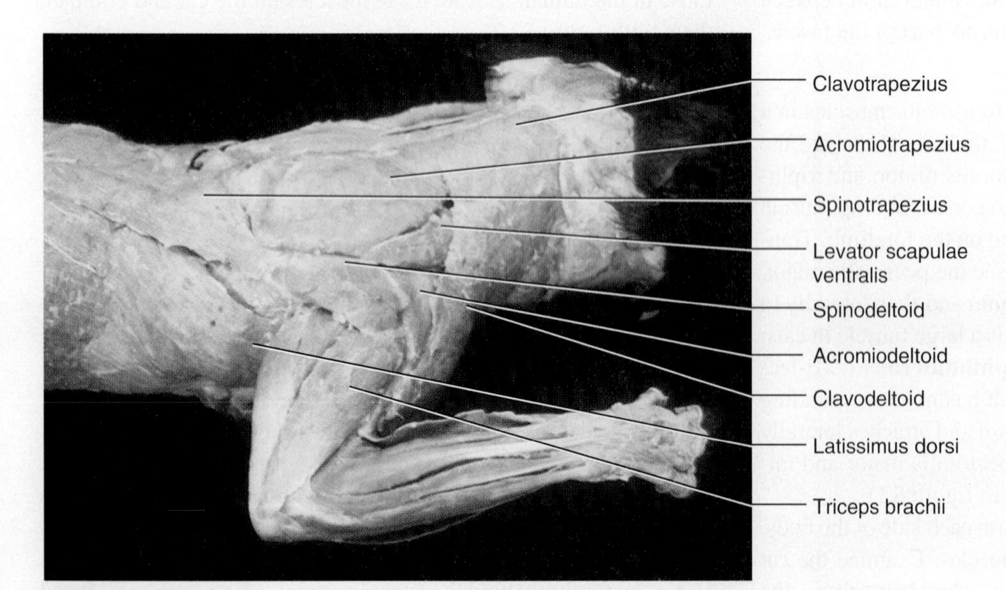

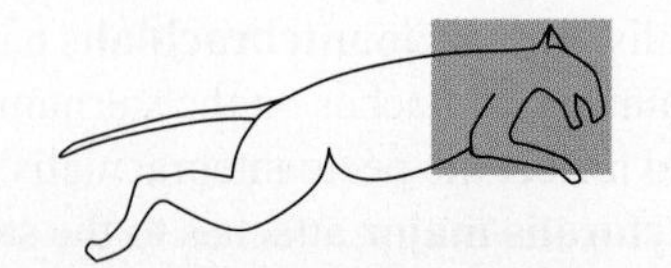

FIGURE S2.7 Superficial Muscles of the Shoulder of the Cat

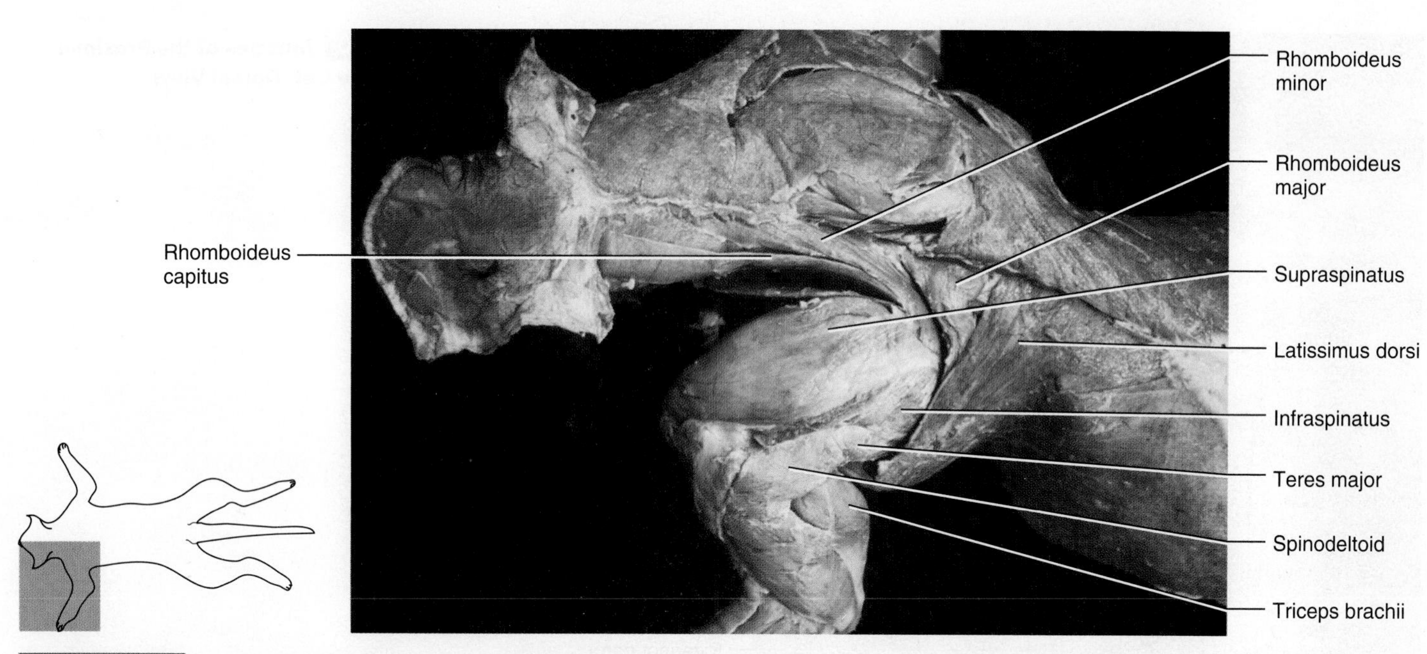

FIGURE S2.8 Deep Muscles of the Scapula of the Cat

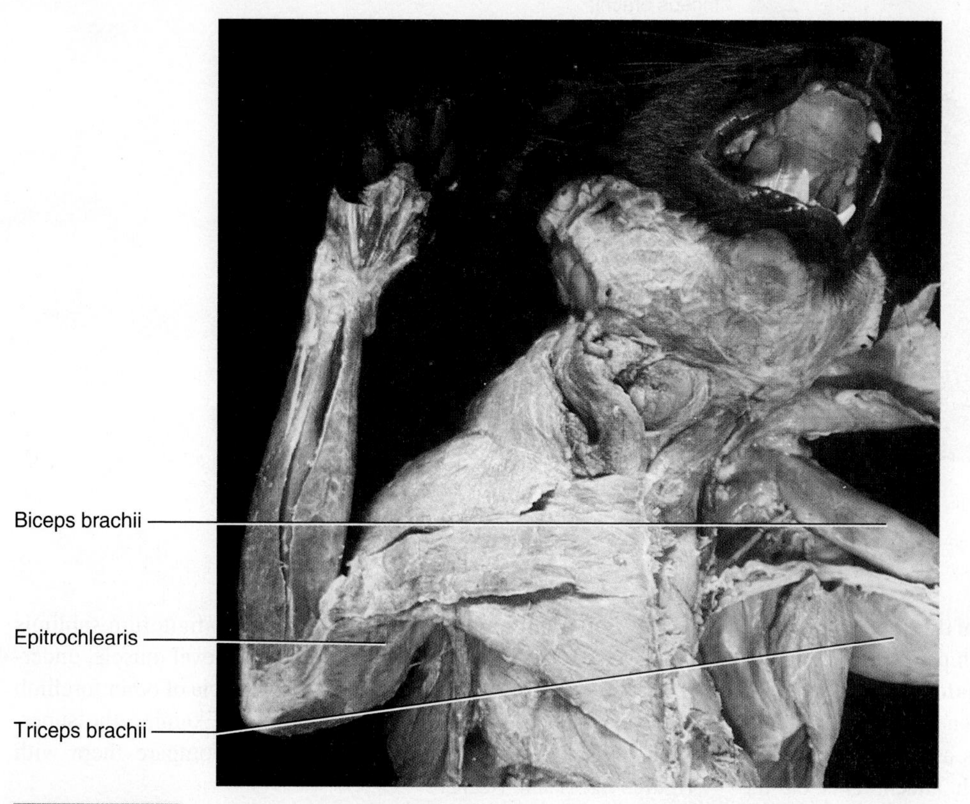

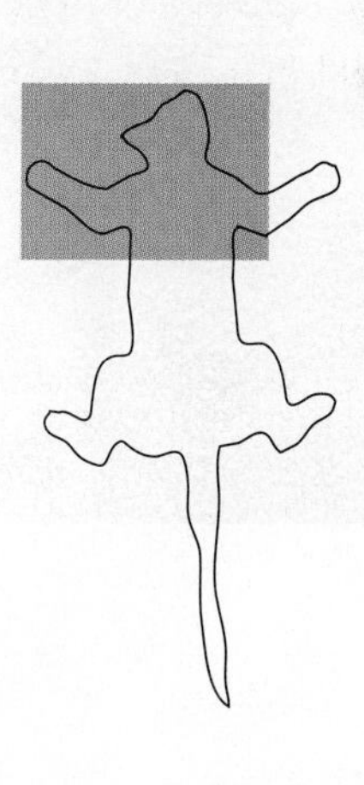

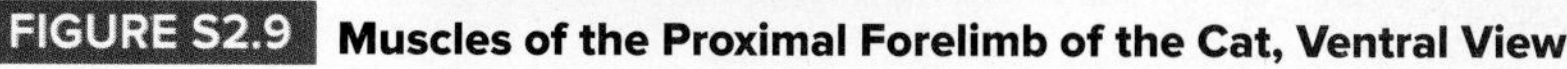

FIGURE S2.9 Muscles of the Proximal Forelimb of the Cat, Ventral View

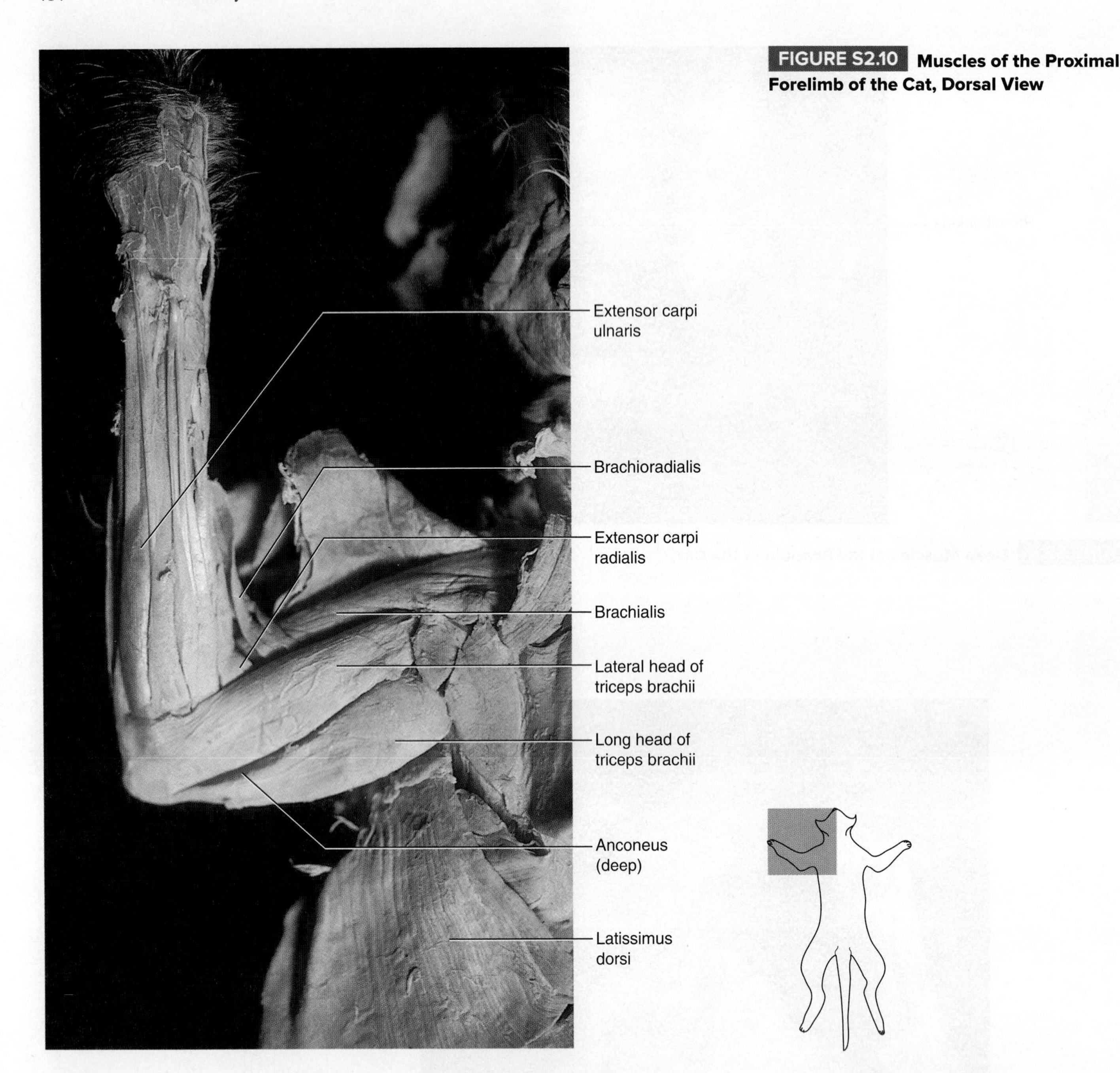

FIGURE S2.10 **Muscles of the Proximal Forelimb of the Cat, Dorsal View**

Superficial Muscles on the Medial Aspect of the Forelimb Most of the muscles of the forelimb of the cat run parallel to the radius and ulna. An exception to this is the **pronator teres,** which is a small slip of muscle that runs obliquely down the forelimb. It pronates the forearm. The **palmaris longus** is a broad, flat muscle, superficially located on the forelimb with attachments on the digits. In humans, the palmaris longus terminates in the fascia of the palm.

The **flexor carpi radialis** is a thin muscle attaching to the second and third metacarpals. It is named for its action, its attachment, and its location. The **flexor carpi ulnaris** attaches to the humerus and ulna and connects to medial metacarpals and carpals. The **flexor digitorum superficialis** (or flexor digitorum sublimis in cats) is located in the forelimb as a middle-level muscle, underneath the palmaris longus. It attaches to the fascia of other forelimb muscles, as opposed to coming from bone. Examine the superficial muscles of the medial forelimb and compare them with figures S2.11 and S2.12.

Deep Muscles on the Medial Aspect of the Forelimb Underneath the upper layer of muscles are numerous muscles with varied actions. The **supinator** is a deep muscle that runs diagonally from the lateral epicondyle of the humerus to the proximal radius. It is the deepest of the proximal forelimb muscles.

FIGURE S2.11 Superficial Muscles of the Right Medial Forelimb of the Cat

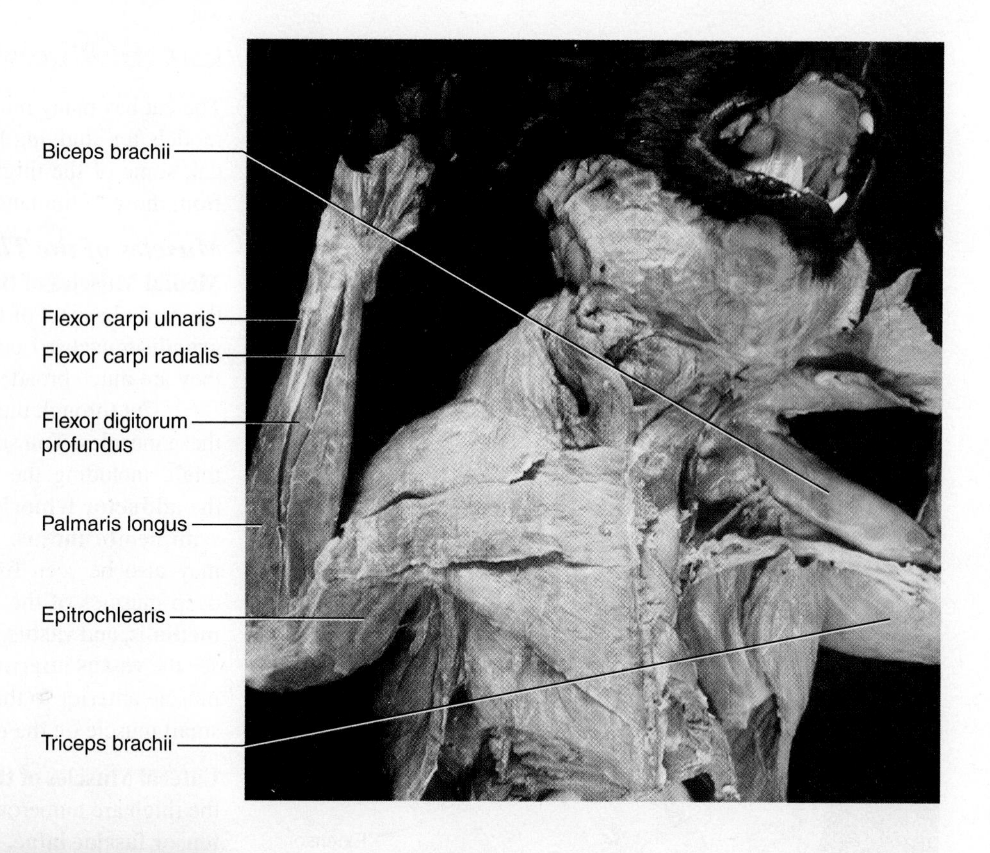

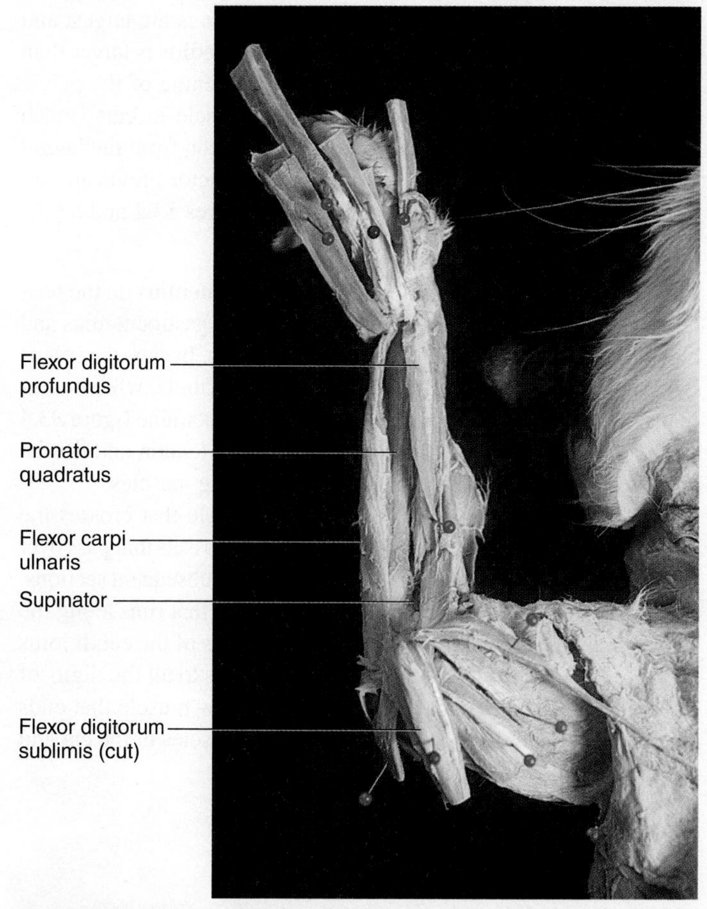

FIGURE S2.12 Deep Muscles of the Right Medial Forelimb of the Cat

The **flexor digitorum profundus** is an extensive muscle that terminates on the first through fifth digits. It replaces the flexor pollicis longus for the thumb flexion, since this muscle is absent in cats. The **pronator quadratus** is a square muscle located between the radius and ulna deep to the flexor digitorum profundus. Find the deep muscles in the cat and compare them with figure S2.12.

Muscles on the Lateral Aspect of the Forelimb The lateral muscles of the forelimb are the extensor group; the **extensor carpi radialis longus** muscle is deep to the brachioradialis and attaches to the second metacarpal. The **extensor carpi radialis brevis** is underneath the extensor carpi radialis longus and terminates on the third metacarpal.

Locate the **extensor carpi ulnaris,** which is next to the extensor digitorum lateralis, attaching to the fifth metacarpal. The **extensor digitorum communis** is on the lateral aspect of the forelimb. It is a broad muscle terminating by tendons on the second through fifth digits. Locate these muscles on the cat and in figure S2.13.

The **extensor digitorum lateralis** is specific to the cat and attaches with the tendons of the extensor digitorum communis to the digits. The **extensor pollicis brevis** is a well-developed muscle in the cat, while the abductor pollicis longus is absent in cats.

After you have examined these muscles on the cat in the lab, make sure that you put your cat away in a plastic bag and clean up your lab station. Put all waste animal material in the appropriate container (not the trash can or down the sink) and clean off your desk. After you have spent the time dissecting your cat, you will probably realize why it is important not to set food down on the lab tables.

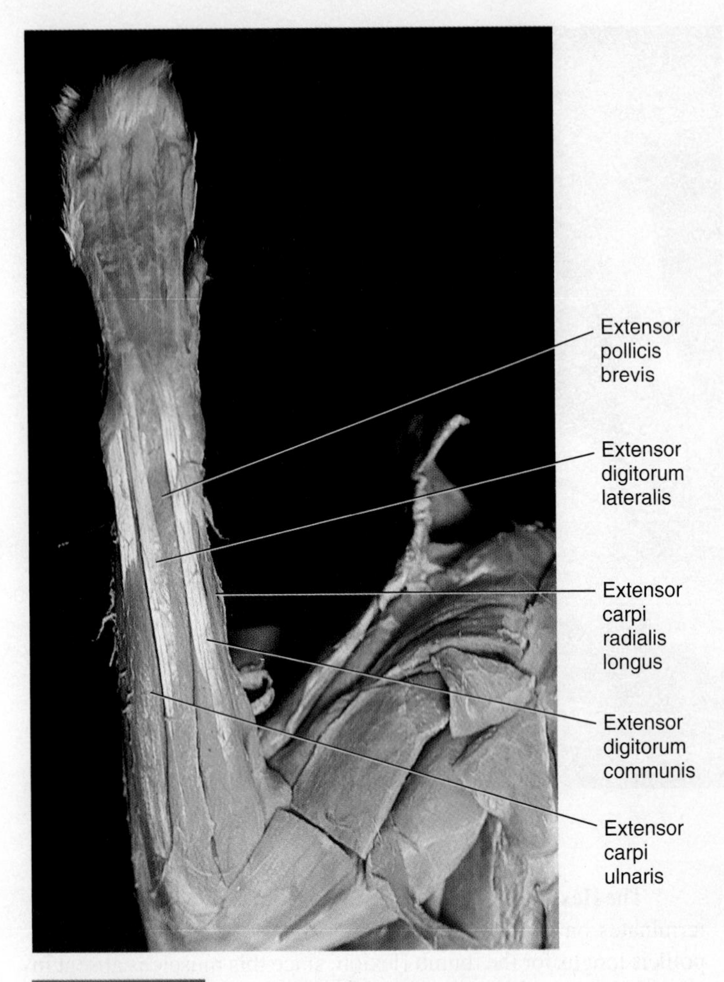

FIGURE S2.13 **Lateral Muscles of the Left Forelimb of the Cat**

Section 3: Hindlimb Muscles of the Cat

If you have not done so already, remove the skin from the lower portion of the hind limb of the cat. Be careful cutting the skin from the distal portions of the leg, so that you do not cut through the tendons of the foot. Be careful in the dissection of the cat muscles to *leave the blood vessels and nerves intact.* These structures will be observed in later sections. Pay particular attention to the **great saphenous vein,** which runs under the skin and should be preserved. Remove any excess fat and fascia from the muscles as you dissect the material. If you need to cut a muscle, bisect it perpendicular to the fiber direction, so that both attachment points can be seen.

Cat Hindlimb Musculature

The cat has many muscles of the hip and thigh that serve as good models for studying human muscles. Because cats are quadrupedal, some of the thigh muscles are different in size or placement from those of humans.

Muscles of the Thigh

Medial Muscles of the Thigh of the Cat Two major muscles on the medial aspect of the thigh in the cat are the **sartorius** and the **gracilis muscles.** Locate these muscles in figure S3.1 and note that they are much broader in the cat than in the human.

Cut through the sartorius and gracilis and reflect the ends of these muscles. You should be able to see the deeper muscles of the thigh, including the **tensor fasciae latae,** the **vastus medialis,** the **adductor femoris** (a large muscle specific to the cat), and the **semimembranosus.** The distal portion of the **semitendinosus** may also be seen from this view. Examine figure S3.1 for the deep muscles of the thigh and locate the **rectus femoris, vastus medialis,** and **vastus lateralis.** Bisect the rectus femoris in order to see the **vastus intermedius** muscle. The **adductor longus** is a thin muscle anterior to the **adductor femoris,** and it can be seen as a small muscle on the medial aspect of the thigh.

Lateral Muscles of the Thigh of the Cat On the lateral aspect of the thigh are numerous muscles, including the **biceps femoris,** the **tensor fasciae latae,** the **gluteus** muscles, and a muscle specific to the cat, the **caudofemoralis.** The biceps femoris is the largest and most lateral muscle of the thigh. The **gluteus medius** is larger than the **gluteus maximus** in cats due to the lengthening of the pelvic girdle. The **semimembranosus** is a large muscle in cats (much larger than the semitendinosus), and it can be seen from the lateral side of the cat. The adductor magnus and adductor brevis are not found in the cat. Examine these muscles in figures S3.2 and S3.3.

Muscles of the Leg

Posterior Muscles Examine the large **gastrocnemius** on the posterior aspect of the leg. The **soleus** is deep to the gastrocnemius and attaches with the gastrocnemius on the calcaneus. In cats, the soleus has only one point of proximal attachment, the fibula, while in humans the soleus attaches to the tibia and fibula. Examine figure S3.4 with your dissection. Cut through the calcaneal tendon and lift the gastrocnemius and soleus to study the underlying muscles.

The **popliteus** is a small, triangular muscle that crosses the knee joint. Do not damage the nerves and blood vessels that pass over the popliteus because you will study them in the subsequent sections.

The **flexor digitorum longus** is a muscle that runs along the medial side of the hind limb and flexes the digits of the cat. It joins with the **flexor hallucis longus,** which connects to all the digits of the hind limb. The **tibialis posterior** is a narrow muscle that ends on the tarsal bones of the foot. Locate these muscles in the cat and compare them with figure S3.5.

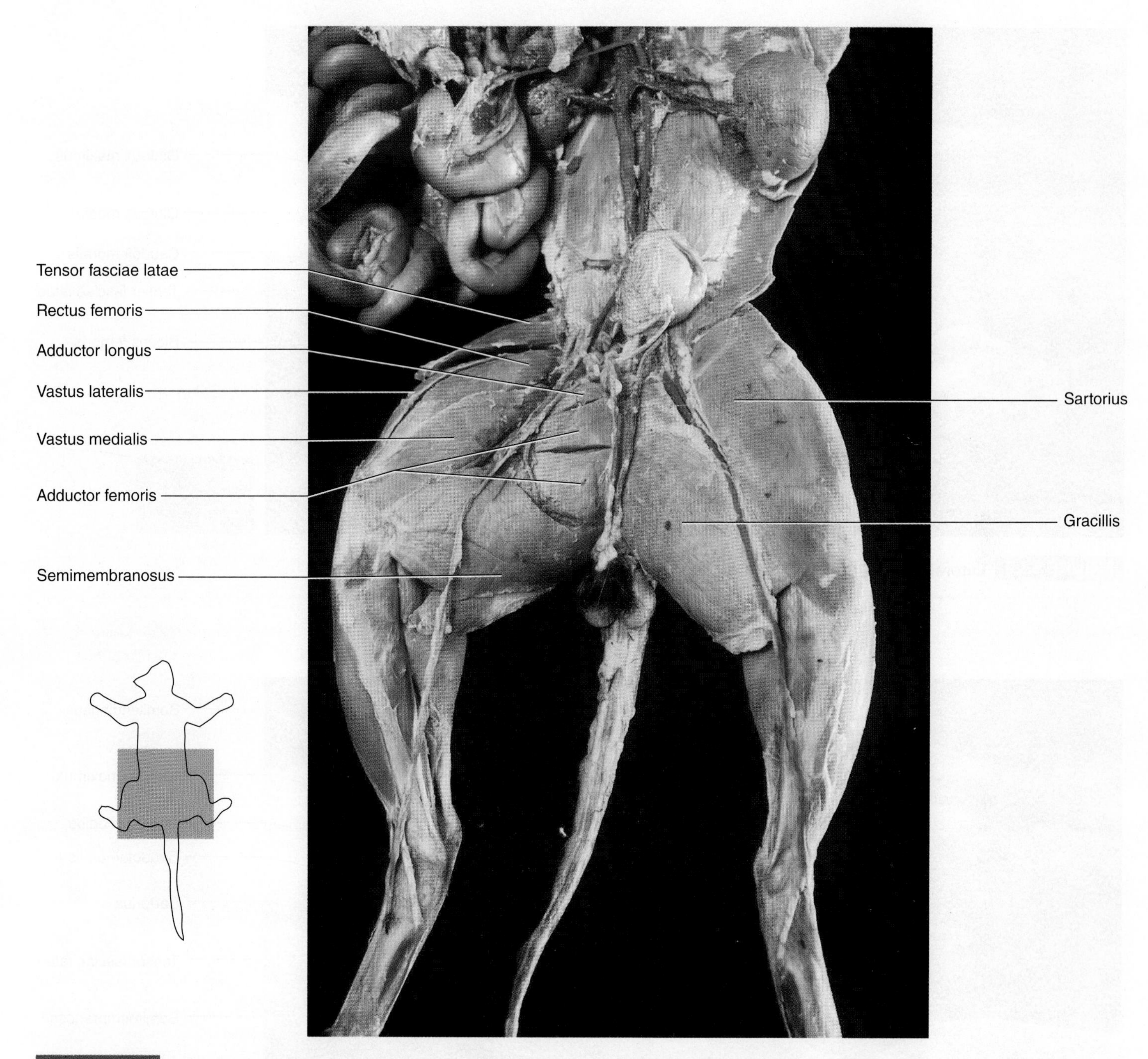

FIGURE S3.1 **Thigh Muscles of the Cat** Superficial muscles (left side of cat); deep muscles (right side of cat).

Anterior Muscles The **tibialis anterior** is a large muscle of the leg and ends on the dorsum of the foot. The **extensor digitorum longus** starts on the femur in cats and terminates on the distal phalanges in all the digits in the cat. The extensor hallucis longus is not found in cats. These muscles can be seen in figure S3.6. Examine these muscles and isolate them in your dissection.

The **fibularis longus** extends along the length of the fibula with the **fibularis brevis.** The fibularis longus attaches to the base of the metatarsals, while the fibularis brevis goes to the fifth metatarsal. The **fibularis tertius** connects to the tendon of the extensor digitorum muscle. These muscles can be seen in figure S3.7.

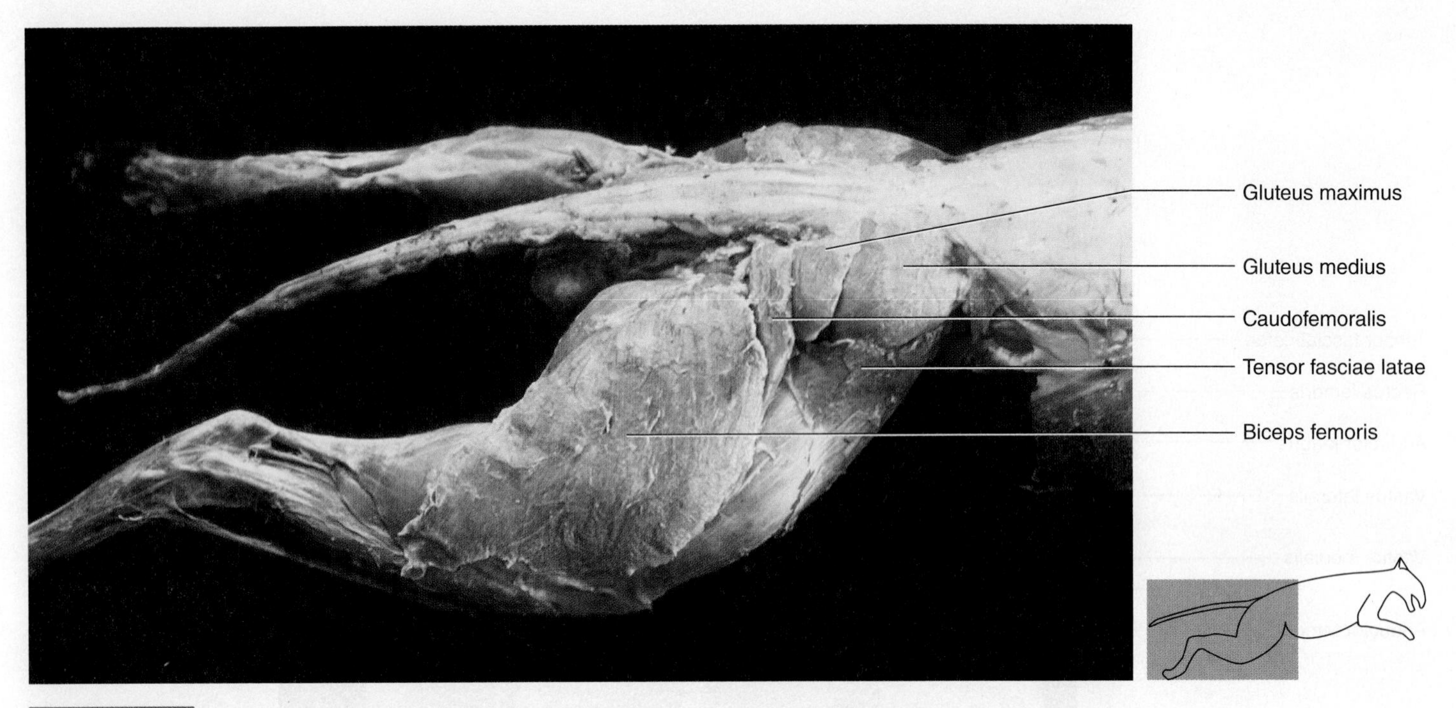

FIGURE S3.2 **Lateral Thigh Muscles of the Cat** Superficial muscles of the right side.

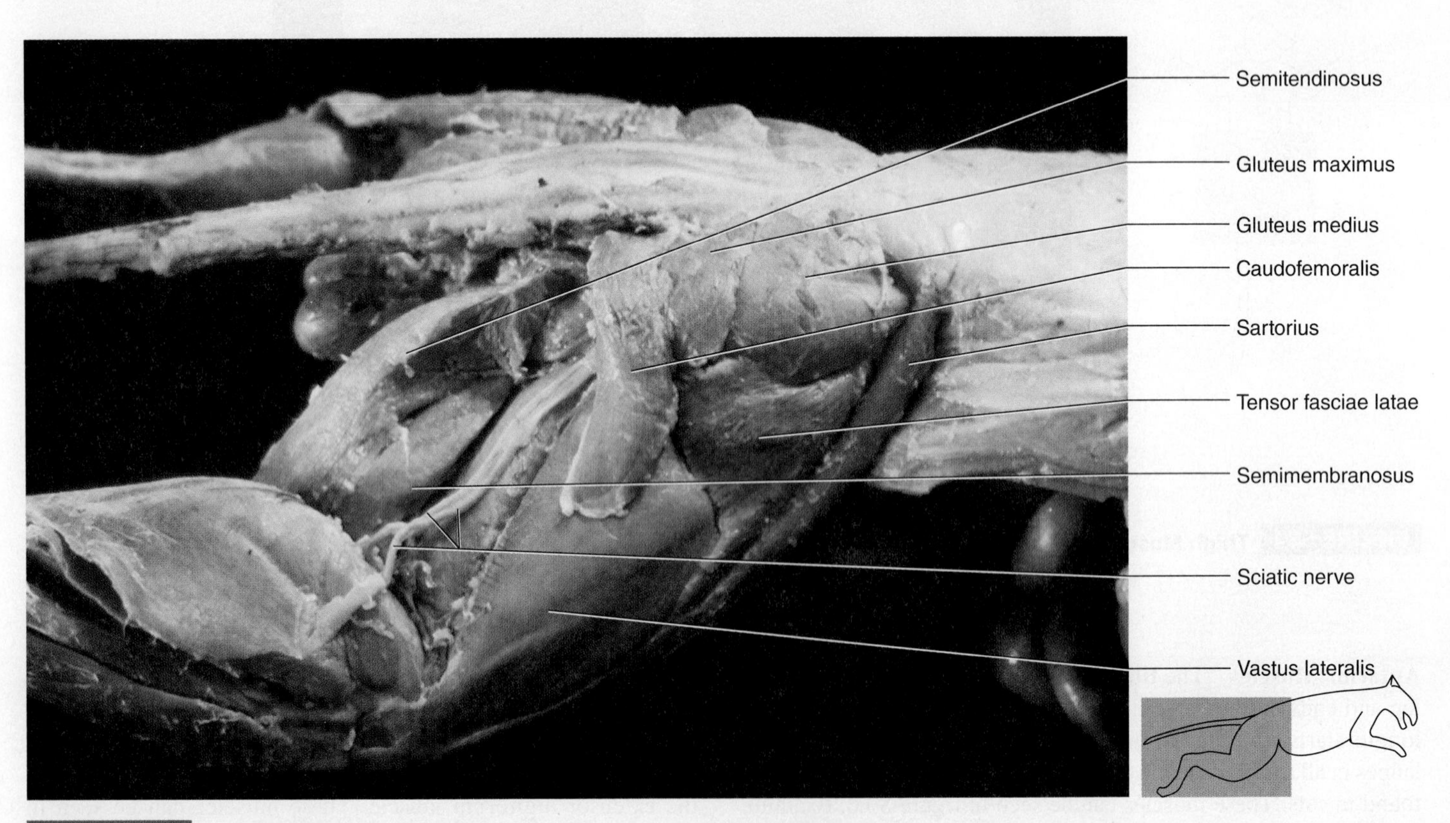

FIGURE S3.3 **Lateral Thigh Muscles of the Cat** Deep muscles of the right side.

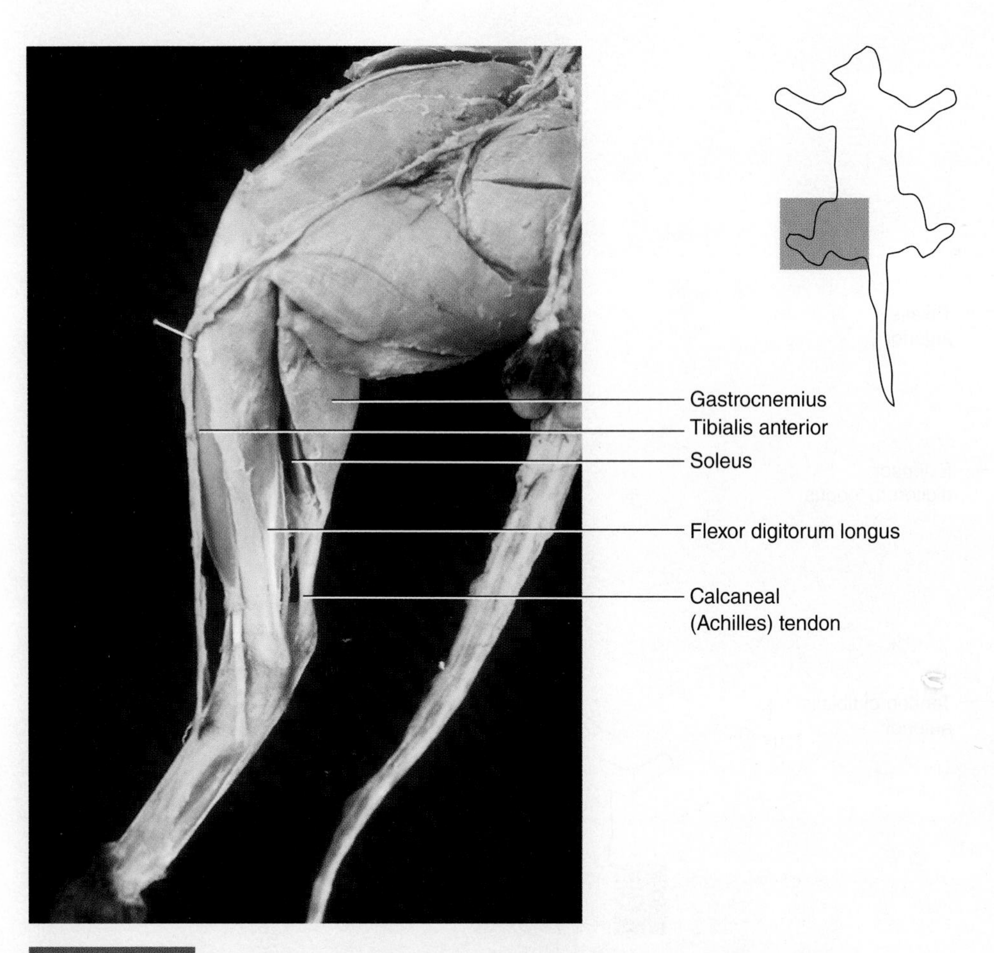

FIGURE S3.4 **Superficial Muscles of the Right Leg of the Cat, Medial View**

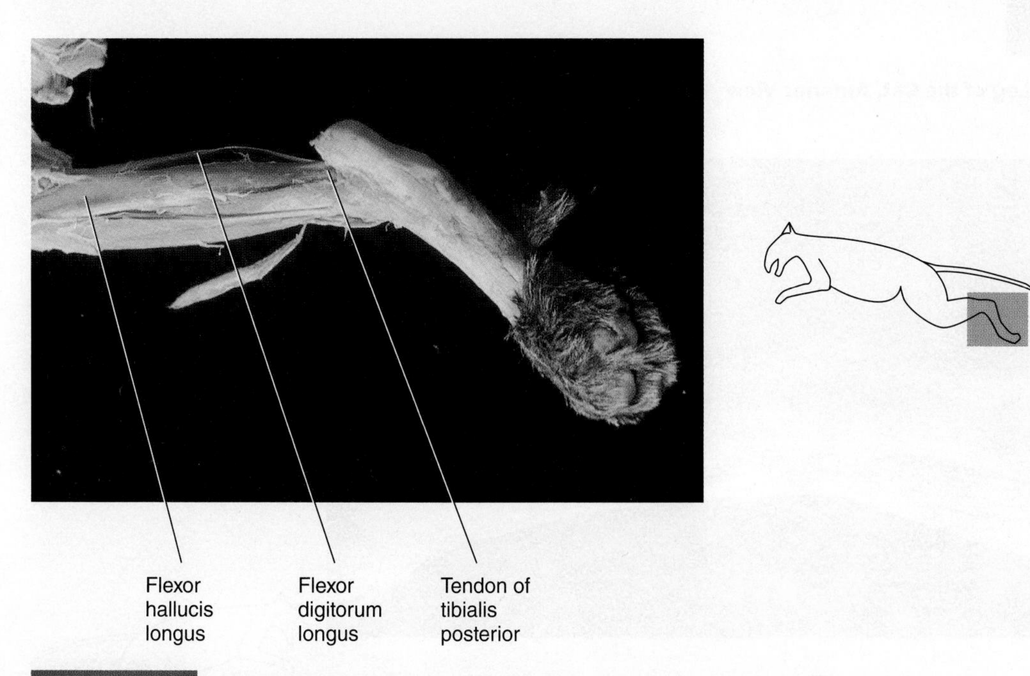

FIGURE S3.5 **Deep Muscles of the Left Leg of the Cat, Lateral View**

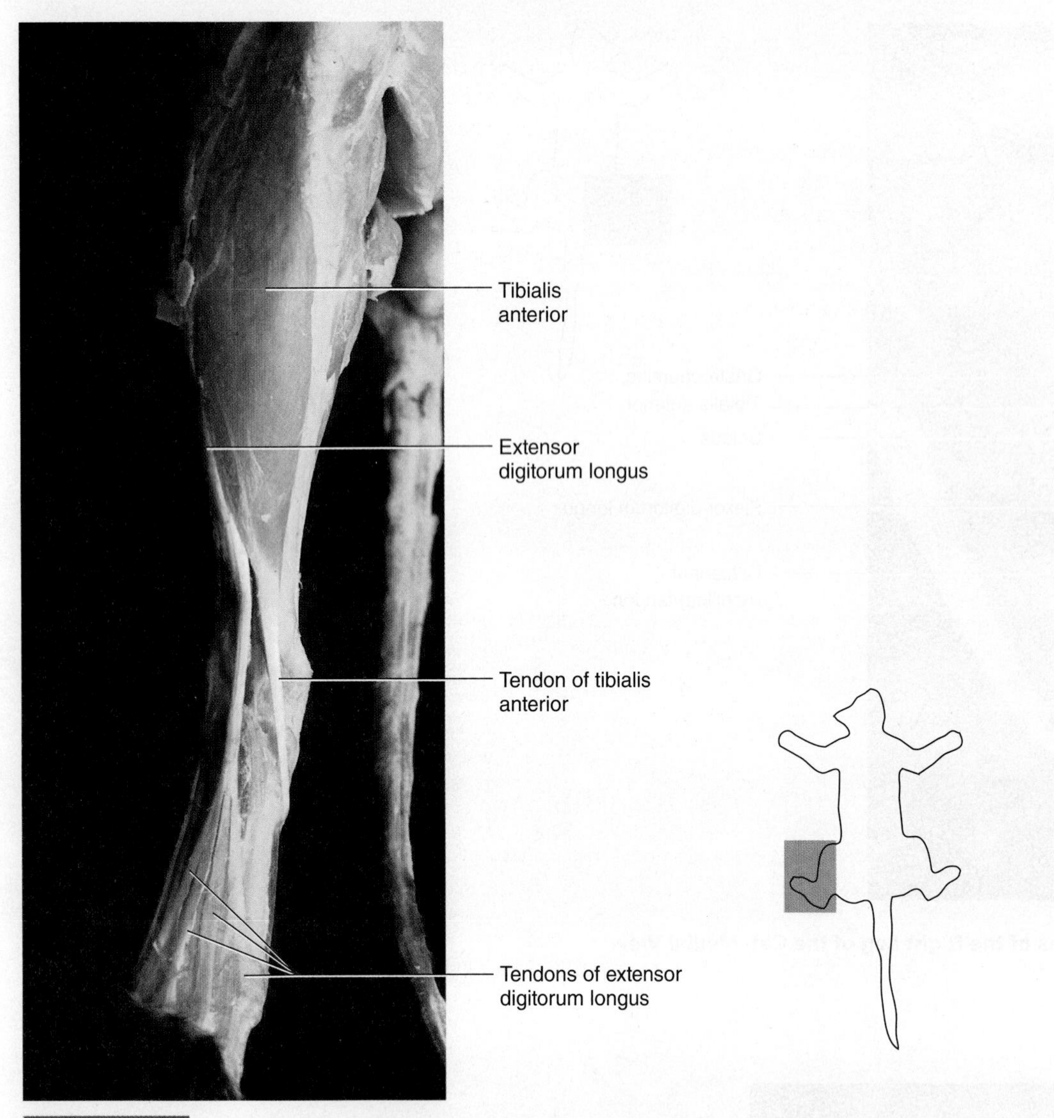

FIGURE S3.6 **Muscles of the Right Leg of the Cat, Anterior View**

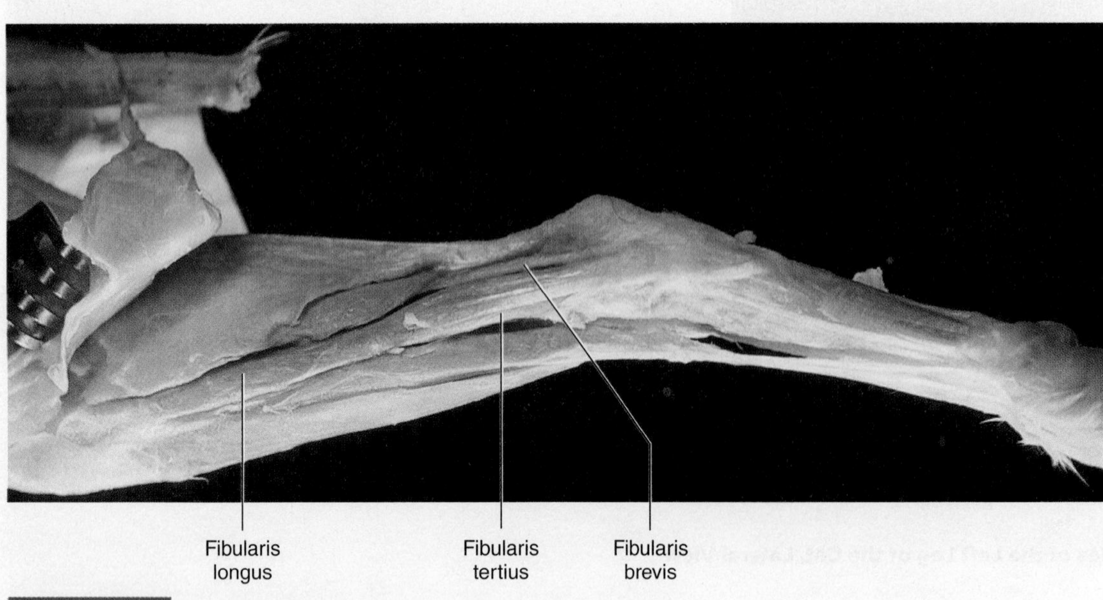

FIGURE S3.7 **Muscles of the Left Leg of the Cat, Lateral View**

Section 4: Head and Neck Muscles of the Cat

In dissecting the muscles of the head and neck, take care not to cut through the salivary glands, or the veins and arteries. You will study these structures in later sections. Dissect only one side of the head, leaving the structures on the other intact for study of the digestive system. A dorsal neck muscle of the cat is the **levator scapulae,** which can be found deep to a shoulder muscle known as the trapezius. You will need to cut a part of the trapezius, the clavotrapezius, in order to see the levator scapulae. In cats, there is an additional muscle called the **levator scapulae ventralis,** which connects to the scapular spine. Examine figure S4.1 for the levator scapulae muscles. The remainder of the muscles you will study in this section are seen from a ventral aspect. The **platysma** in the cat was removed during the skinning process, and it will not be seen unless you kept the skin with the cat. The **scalenes** can be dissected by reflecting the pectoralis minor muscle. Notice how the scalenes are composed of separate slips of muscle that run from the ribs to the neck. In humans, the muscle is more lateral than in cats. Compare your specimen with figure S4.1.

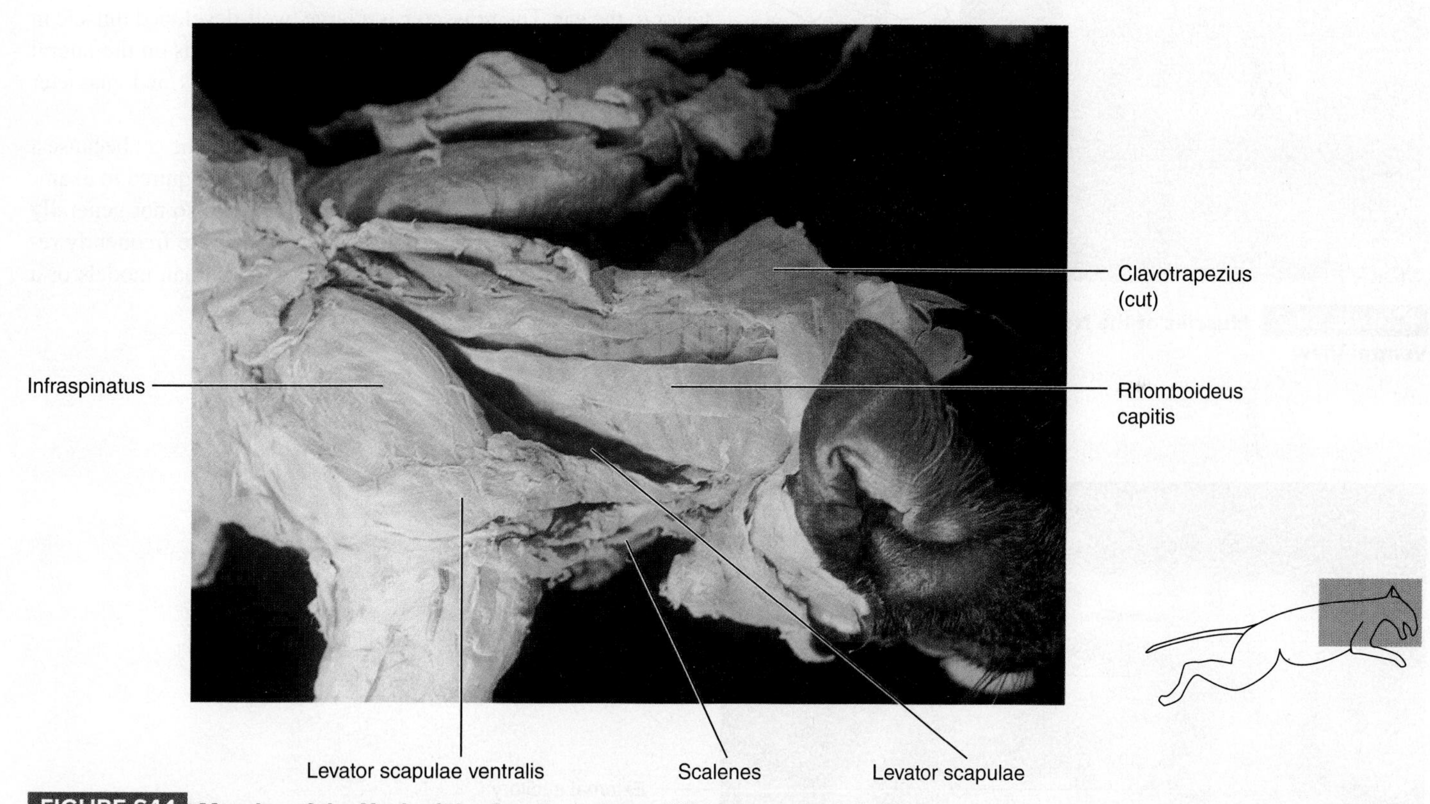

FIGURE S4.1 Muscles of the Neck of the Cat, Dorsolateral View

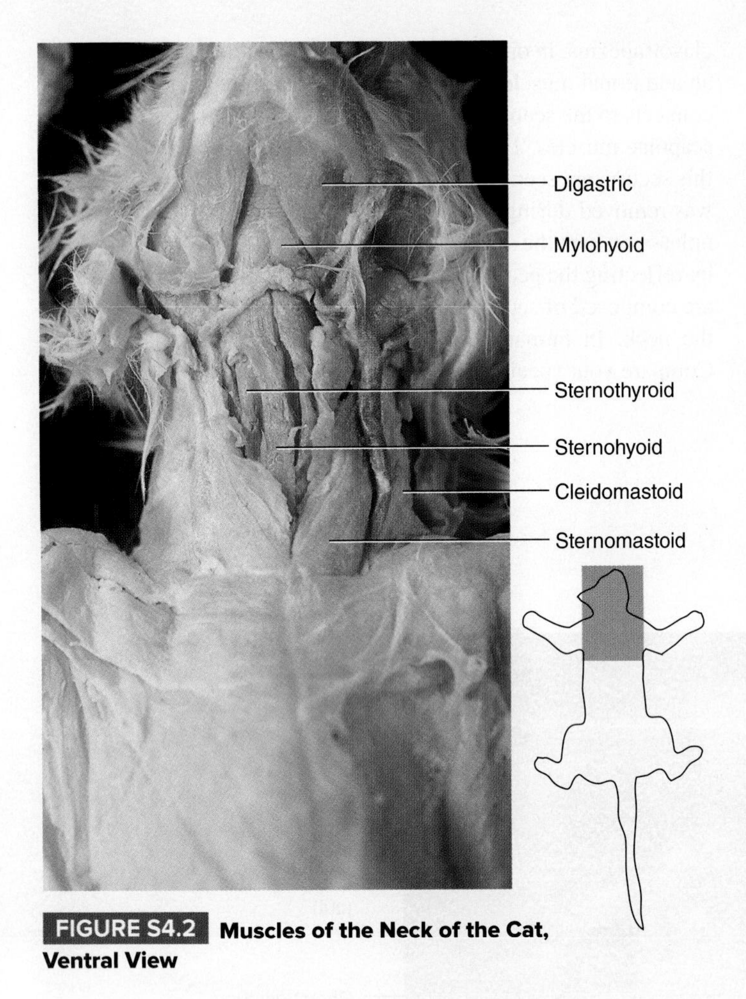

FIGURE S4.2 **Muscles of the Neck of the Cat, Ventral View**

In the cat, the **sternocleidomastoid** consists of two muscles, the **sternomastoid** and the **cleidomastoid.** The sternomastoid extends from the sternum to the mastoid process of the skull, and the cleidomastoid runs from the clavicle to the mastoid process. Underneath the sternomastoid is the most medial muscle of the neck group, the **sternohyoid.** The **sternothyroid** is deeper and more lateral than the sternohyoid. These muscles can be seen in figure S4.2.

The **digastric** muscle runs parallel to the lower edge of the mandible and underneath the submandibular gland. Deep to the digastric is the **mylohyoid,** a broad muscle that runs perpendicular to the direction of the digastric. Compare your dissection with figure S4.2.

The **occipitalis** is a posterior head muscle in the cat that attaches to the **galea aponeurotica.** The **frontalis** attaches anteriorly to the galea. Lateral to these muscles are the temporalis and masseter muscles. The **temporalis** is located more dorsally than the masseter and can be dissected by removing the skin and fascia anterior to the ear. The **masseter** is a large, well-developed muscle in the cat that attaches to the zygomatic arch and ends on the lateral surface of the mandible. Examine the temporalis and masseter muscles in figure S4.3.

The **pterygoids** are usually not dissected in the cat because a cut through the ramus of the mandible would be required to examine them. The muscles of facial expression are also not generally studied in the cat. These muscles are small and are frequently removed with the skin. Study these muscles on human models or a cadaver (if available).

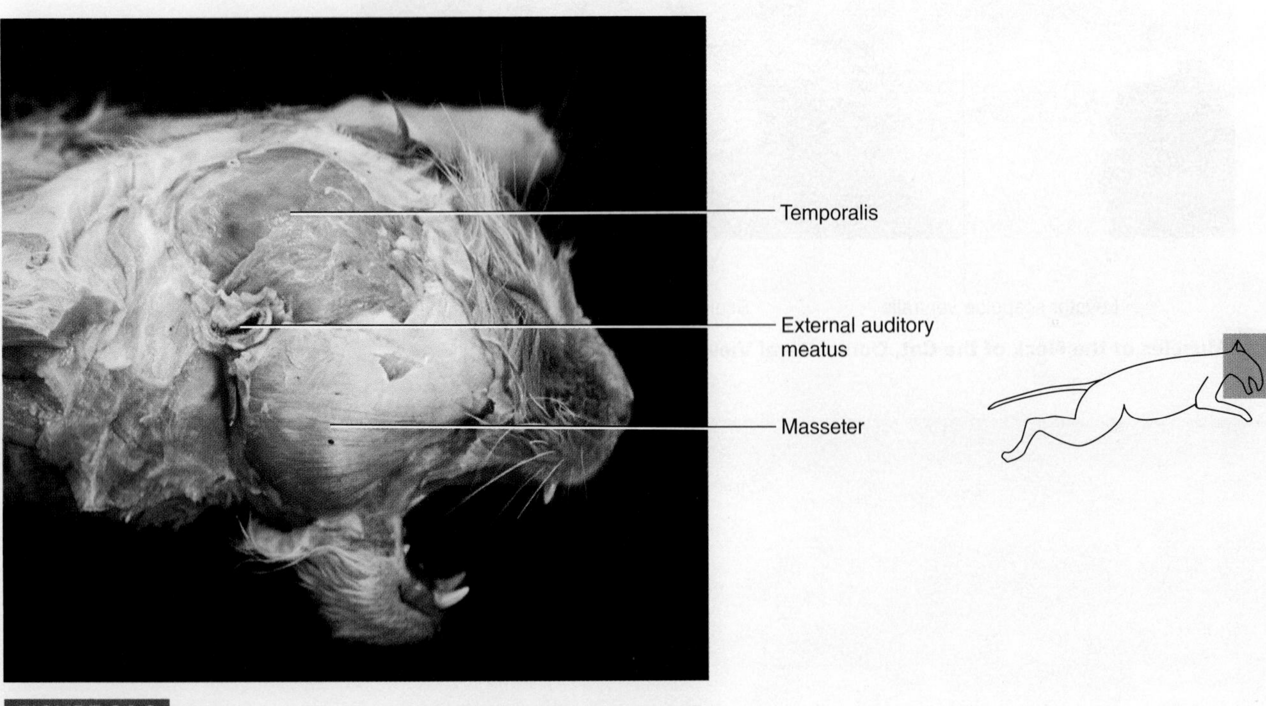

FIGURE S4.3 **Muscles of the Head of the Cat, Lateral View**

Section 5: Torso Muscles of the Cat

Abdominal Muscles

Place the cat on its back and examine the abdominal muscles. The abdominal muscles in the cat are similar to those in the human in that the **external abdominal oblique** is a broad superficial muscle on the anterior abdomen. Carefully cut through the external abdominal oblique to reveal the **internal abdominal oblique,** as illustrated in figure S5.1. Deep to this is the **transverse abdominal,** and it can be seen by carefully dissecting the internal abdominal oblique. If you cut too deeply, you will enter the abdominal cavity, so be careful in this part of the dissection. The **rectus abdominis** is a muscle that runs from the pubic region to the sternum.

Move to the thoracic region and examine the muscle of the lateral thorax dorsal to the xiphihumeralis. This is the **serratus anterior** (actually, serratus ventralis in the cat), and the scalloped edges of the muscle should be visible. Carefully separate this muscle from the others and follow where it connects to the scapula. To see the intercostal muscles, bisect the superficial chest muscles. If you have not done so already, cut through the middle of the belly of the pectoral muscles, exposing the ribs of the cat. Carefully remove the outer layer of fascia from the muscle between the ribs and locate the **external intercostal** muscle. Cutting part of this muscle away should expose the **internal intercostal** muscle. Note how the fibers run perpendicular to one another. (Do not look for the diaphragm at this time. It will be visible when you examine the lungs in section 7, "Respiratory Anatomy of the Cat.") Examine these thoracic muscles in figure S5.2.

Deep Muscles of the Back

Place the cat so that you can examine the dorsal surface. Dissect the trapezius carefully to see the **rhomboideus** muscles. These muscles attach to the vertebral column and run to the scapula. Examine the muscles of the neck and head to see the **splenius** and the **semispinalis** muscles. These are located in figure S5.3. Bisect the latissimus dorsi and the posterior portion of the external abdominal oblique to see the **erector spinae** muscles. The relative position of the cat erector spinae can be seen in figure S5.4. Compare this figure with your dissection. Locate the **iliocostalis, longissimus,** and **spinalis** of the erector spinae along with the **multifidus** in the cat.

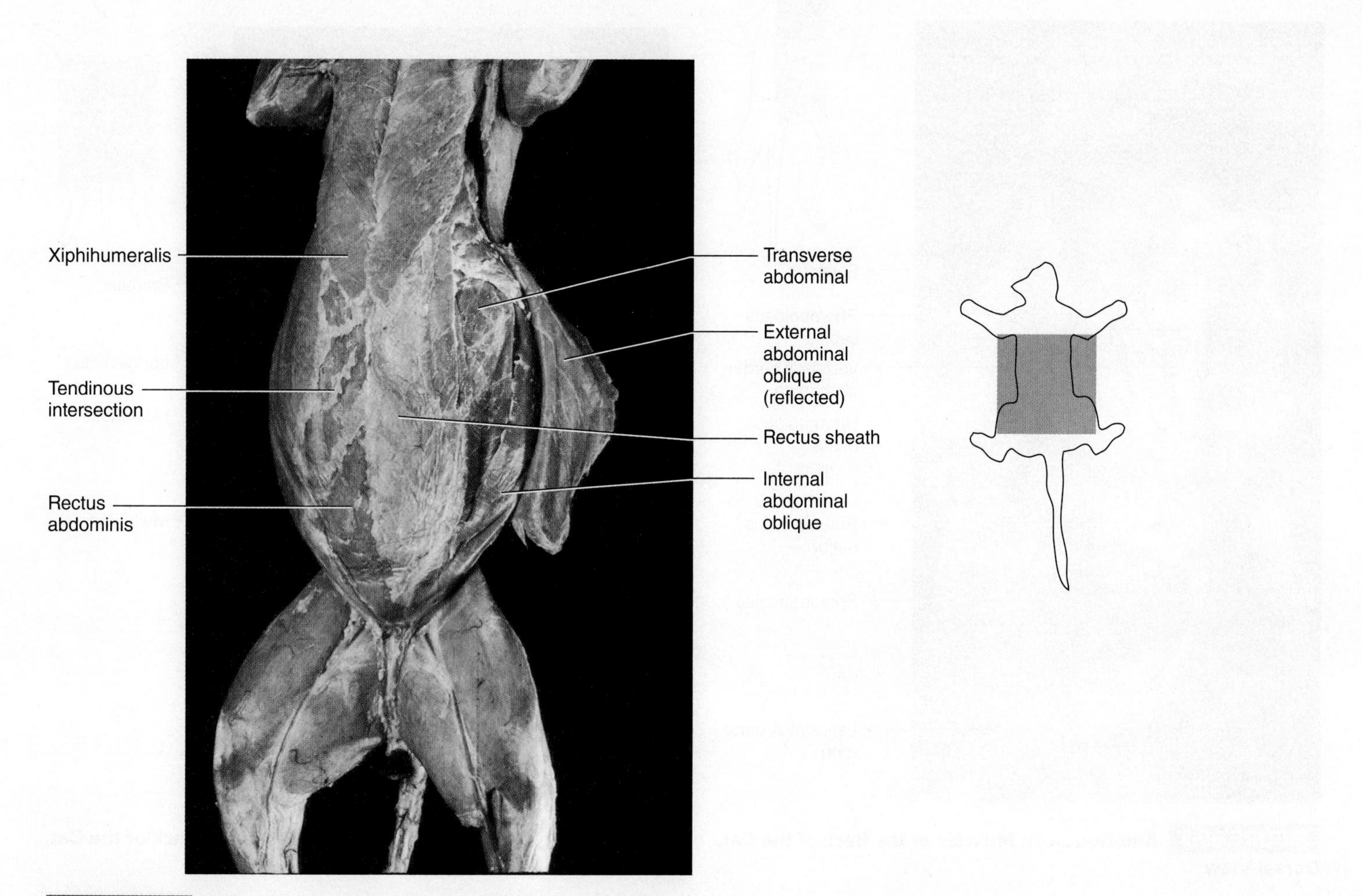

FIGURE S5.1 **Muscles of the Abdomen of the Cat, Ventral View**

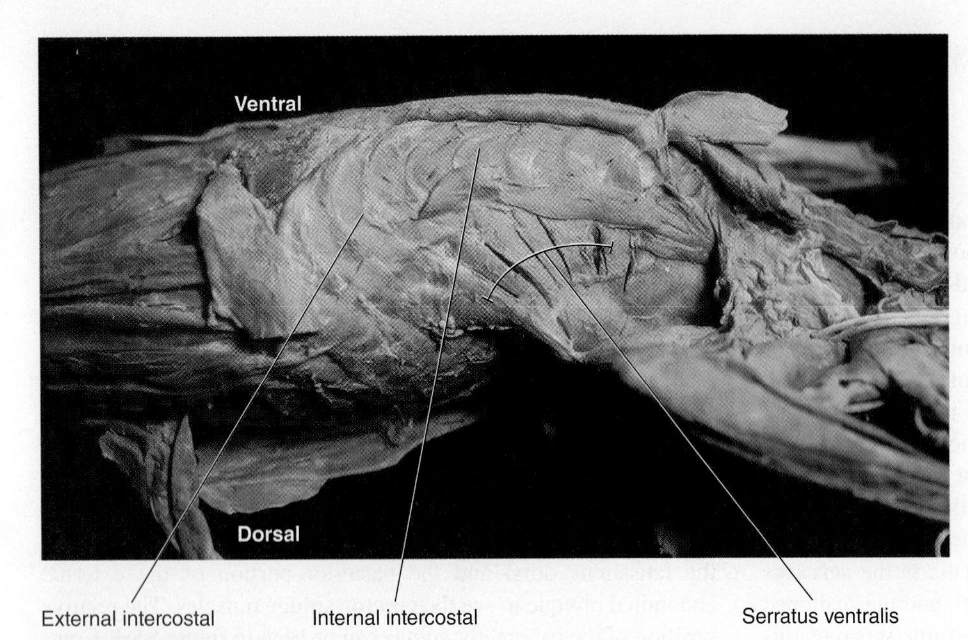

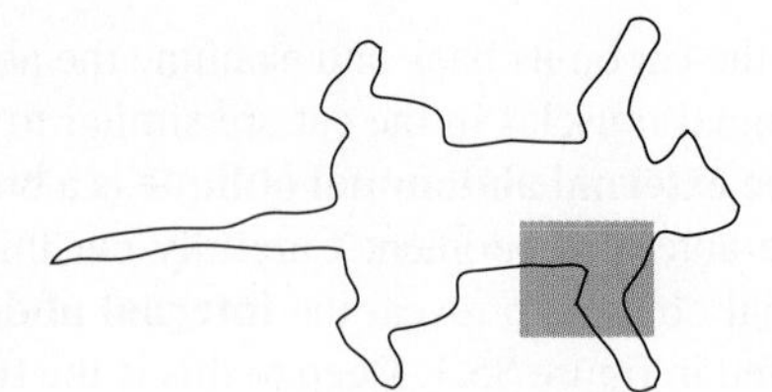

FIGURE S5.2 **Muscles of the Thorax of the Cat, Lateral View**

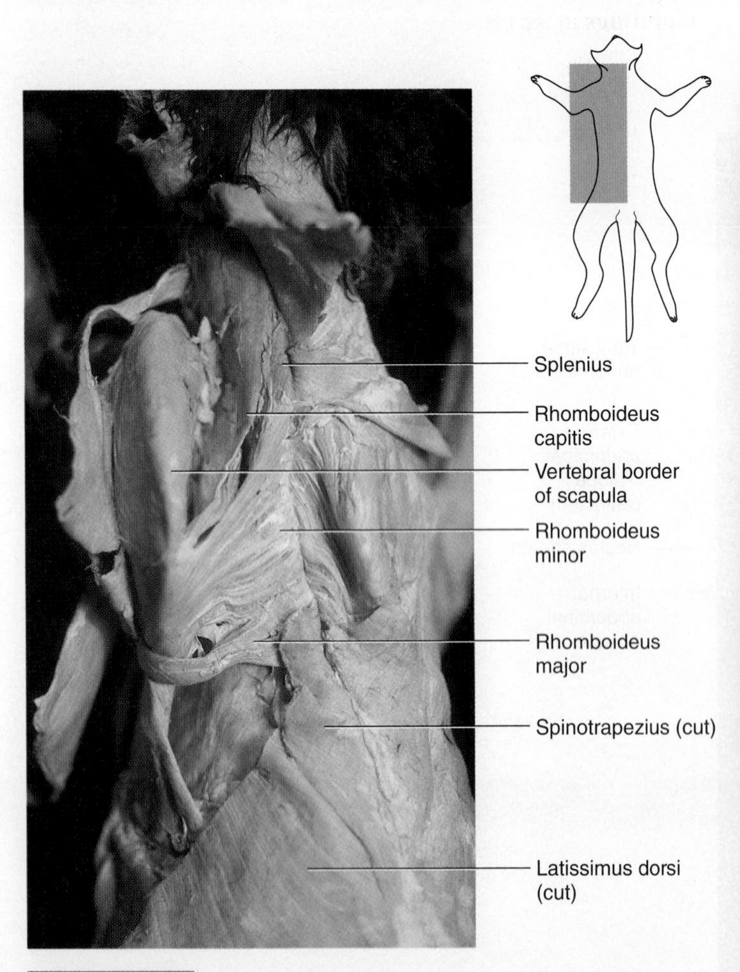

FIGURE S5.3 **Anterior, Deep Muscles of the Back of the Cat, Dorsal View**

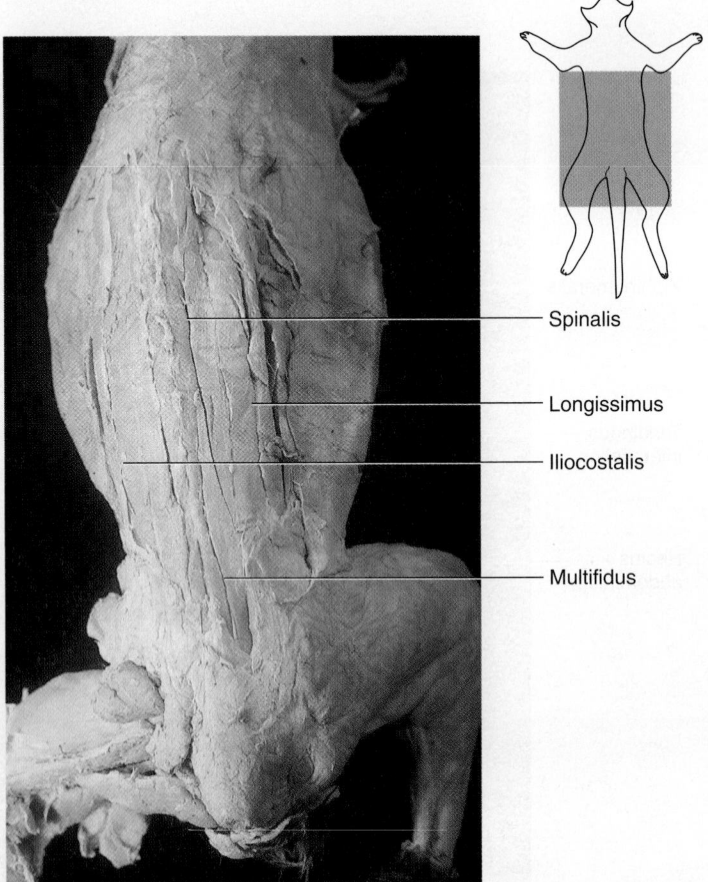

FIGURE S5.4 **Posterior, Deep Muscles of the Back of the Cat, Dorsal View**

Section 6: Blood Vessel Anatomy of the Cat—Blood Vessels and Lymph System

Open the thoracic and abdominal regions of the cat if you have not done so already. Be careful as you make the first incision into the body cavity of the cat. Cut with a sharp scissors or, if using a scalpel, use the blade facing to one side or, even better, use a lifting motion and cut upward through the tissue. Do not use a downward sawing motion as this may damage the internal organs of the cat.

Make an incision into the body wall of the cat in the belly, slightly lateral to midline. Cut into the muscle and then grasp the tissue with a forceps and lift the body wall away from the viscera. Continue cutting anteriorly through the belly. Stop at the level of the diaphragm, and then continue cutting a little off center as you cut through the costal cartilages. Use a scalpel or scissors to cut through the thoracic region.

Now make a transverse incision in the lower region of the belly, completing an inverted T-shaped incision (fig. S6.1). Lift the tissue near the ribs and expose the diaphragm. Gently and carefully snip or cut the diaphragm away from the ventral region by cutting as close to the ribs as possible.

Make the third incision by cutting transversely across the upper part of the chest. Be careful not to cut the blood vessels of the neck that were exposed during the original skinning process. Open the body cavity by pulling the two flaps laterally, exposing the thoracic and abdominal cavities. You may want to cut through the ribs at their most lateral point to facilitate the opening of the thoracic cavity.

Carefully expose the thoracic region of the cat and locate the two lungs and the heart. Note that the heart is enclosed in the pericardial sac. Look for the fatty material, called the **greater omentum**, that covers the **intestines** in the abdominal region. As you dissect the arteries of the cat, pay careful attention to the veins and nerves that travel along with the arteries. *Do not dissect other organs.* These systems (e.g., respiratory, digestive, and urogenital systems) will be studied later.

As you open the abdominal cavity, you will see the space that contains the internal organs, known as the **peritoneal,** or **body, cavity.** The lining on the inside of the body wall is the **parietal peritoneum,** and the lining that wraps around the outside of the intestines is the **visceral peritoneum.**

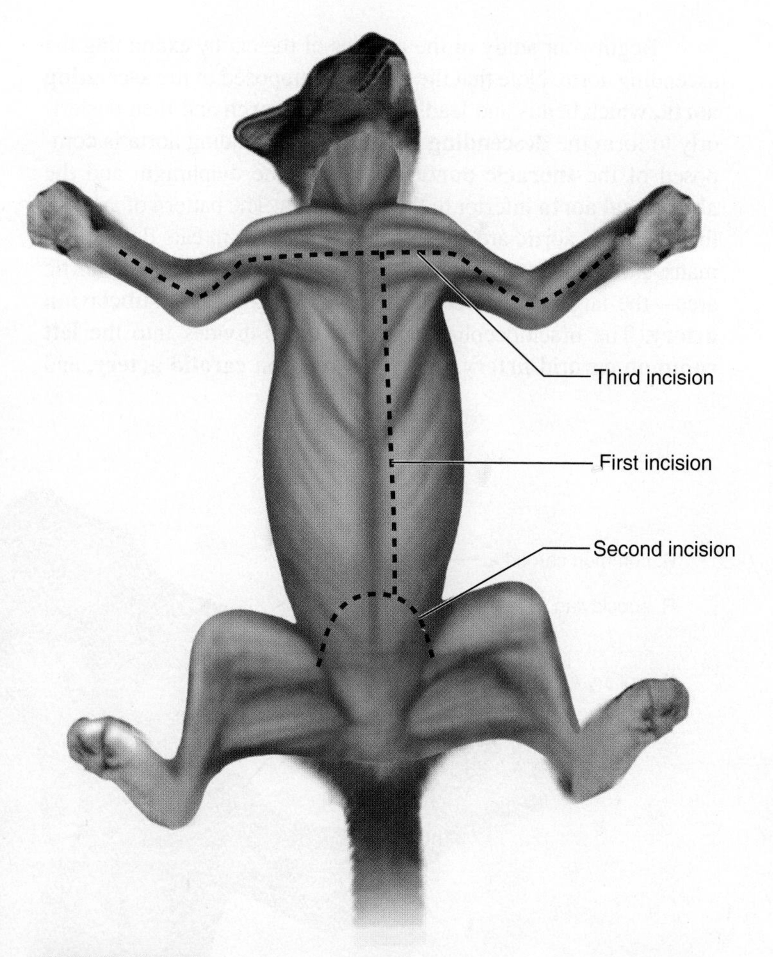

FIGURE S6.1 Incisions to Open the Thoracic and Abdominal Cavities

Arteries of the Cat

The blood vessels in the cat have been injected with colored latex. If the cat is doubly injected, the **arteries are red** and the **veins are blue.** If the cat has been triply injected, the hepatic portal vein is typically injected with yellow latex. Be careful locating the arteries in the cat. You can tease away some of the connective tissue with a dissection needle to see the vessel more clearly. Note the difference in the aortic arch arteries in the cat, compared with those of the human. Look for the **coronary arteries,** which take blood from the ascending aorta to the heart tissue.

Begin your study of the arteries of the cat by examining the ascending aorta. Note that the **aorta** is composed of the **ascending aorta,** which bends and leads to the **aortic arch** and then posteriorly to form the **descending aorta.** The descending aorta is composed of the **thoracic aorta** anterior to the diaphragm and the **abdominal aorta** inferior to the diaphragm. The pattern of arteries that leave the aortic arch is somewhat different in cats than in humans. In the cat, typically only two large vessels leave the aortic arch—the large **brachiocephalic artery** and the **left subclavian artery.** The brachiocephalic artery further divides into the **left common carotid artery,** the **right common carotid artery,** and the **right subclavian artery.** Examine figure S6.2 for the pattern in the cat. How is the pattern in the cat different from that in the human?

Several arteries arise from the **subclavian artery.** The **vertebral artery** takes blood to the head, and the **subscapular artery** takes blood to the pectoral girdle muscles, along with the **ventral thoracic artery** and **long thoracic artery.** The ventral thoracic and long thoracic arteries supply the latissimus dorsi and pectoralis muscles with blood. The **left** and **right subclavian arteries** become the **left** and **right axillary arteries,** which in turn take blood to the brachial arteries. The **brachial artery** takes

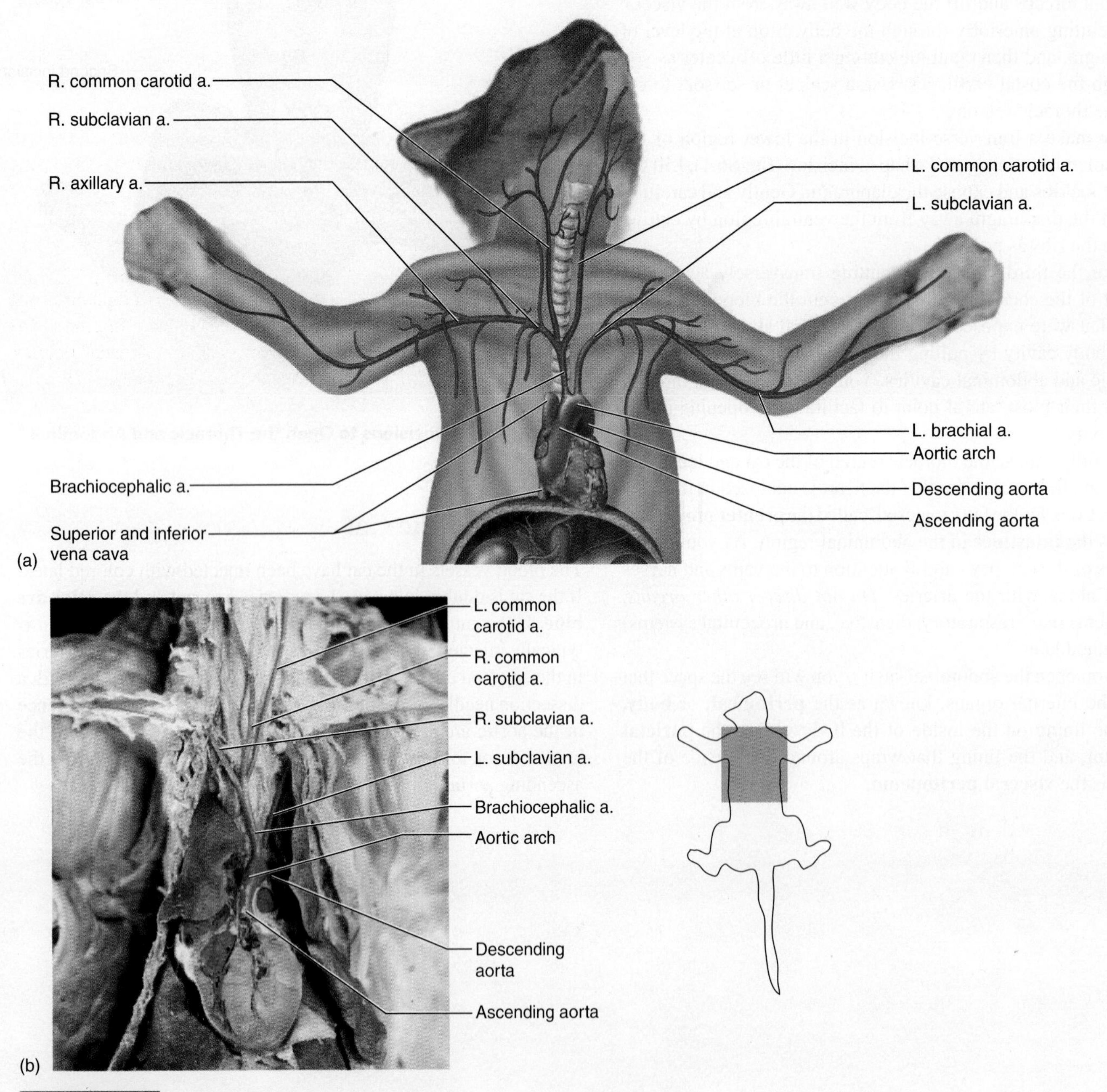

FIGURE S6.2 Aortic Arch Arteries of the Cat (a) Diagram; (b) photograph.

blood to the **radial** and **ulnar arteries.** Additional arteries that branch from the subclavian artery are the **internal mammary artery,** which takes blood to the ventral body wall, and the **thyrocervical artery,** which takes blood to the region of the shoulder and neck. The **costocervical artery** reaches the deep muscles of the back and neck (fig. S6.3).

The **common carotid artery** takes blood along the ventral surface of the neck to feed the small **internal carotid artery,** which supplies the brain. The common carotid artery also empties into the **external carotid artery,** which takes blood to the external portions of the head. The **occipital artery** receives blood from the common carotid artery and supplies the muscles of the neck.

The thoracic aorta not only takes blood to the abdominal aorta but also supplies the rib cage with blood from the **intercostal arteries. Esophageal arteries** also branch from the thoracic aorta, supplying the esophagus with blood. The thoracic aorta continues as the **abdominal aorta,** as it passes through the diaphragm.

Make sure you do not cut the blood vessels away from the organs that will be studied later. Note the major arteries of the caudal region of the cat in figure S6.4. Refer to this figure and the photographs for the remainder of this section.

Abdominal Arteries

The determination of the abdominal arteries is best done by looking for where these arteries leave the aorta and where they enter into the organs they supply. Locate the major organs of the abdominal region, such as the **stomach, spleen, liver,** and **small** and **large intestines.** Examine the specimen from the left side, gently lifting the stomach, spleen, and intestines toward the ventral right side as you look for the abdominal arteries. The **descending aorta** is located on the left side of the body, while the **inferior vena cava** is located on the right side. Locate the short **celiac artery,** which quickly divides into three major vessels—the **splenic artery,** the **left gastric artery,** and the **hepatic artery.** The splenic artery takes blood to the spleen, the hepatic artery reaches the liver, and the left gastric artery supplies the stomach, as seen in figure S6.4.

The **superior mesenteric artery** is the next major vessel to take blood from the abdominal aorta. The superior mesenteric artery travels to the small intestine and the proximal portion of the large intestine. It branches into the **jejunal arteries,** the **ileal arteries,** and the **colic arteries.** Compare your dissection with figures S6.4 and S6.5.

The paired **adrenolumbar arteries** branch on each side of the abdominal aorta and take blood to the adrenal glands, parts of the body wall, and the diaphragm. Just caudal to the adrenolumbar arteries are the paired **renal arteries,** which take blood to the kidneys. The **gonadal arteries** are the next set of paired arteries, and these take blood to the testes in male cats, in which case they are called the **testicular arteries.** If the gonadal arteries occur in females, they are called the **ovarian arteries** because they take blood to the ovaries (fig. S6.4).

A single **inferior mesenteric artery** takes blood from the abdominal aorta to the terminal portion of the large intestine. The inferior mesenteric artery can be found caudal to gonadal arteries. Locate the lower abdominal arteries in your specimen and compare them with figures S6.4, S6.5, and S6.6.

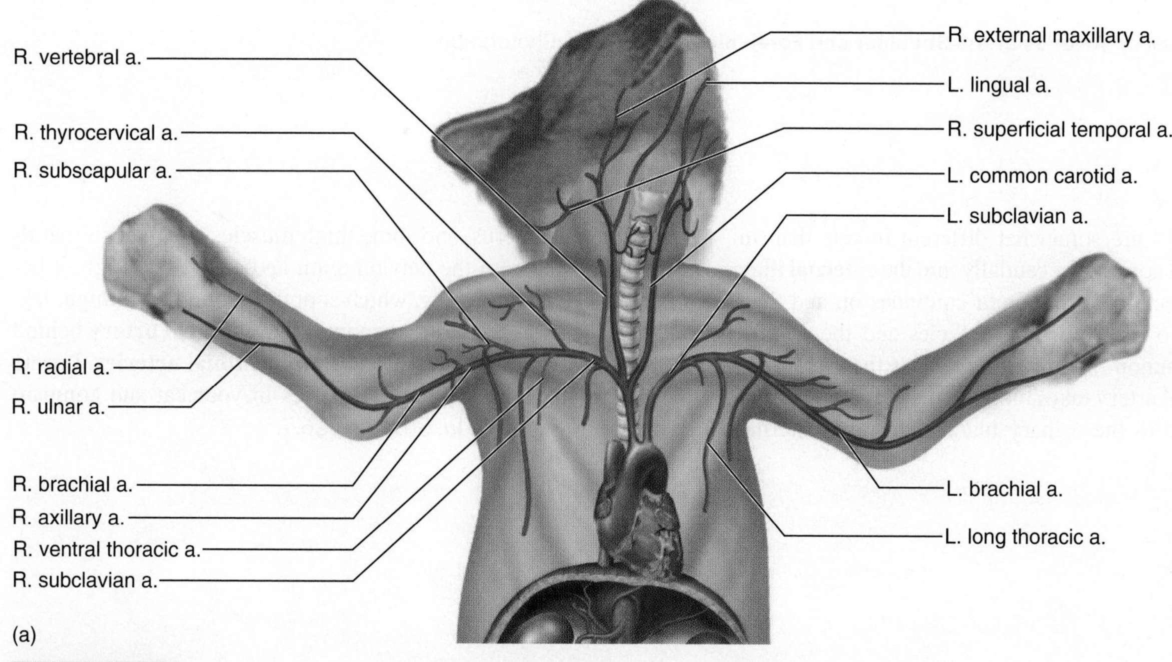

FIGURE S6.3 Arteries of the Shoulder and Forelimb of the Cat (a) Diagram.

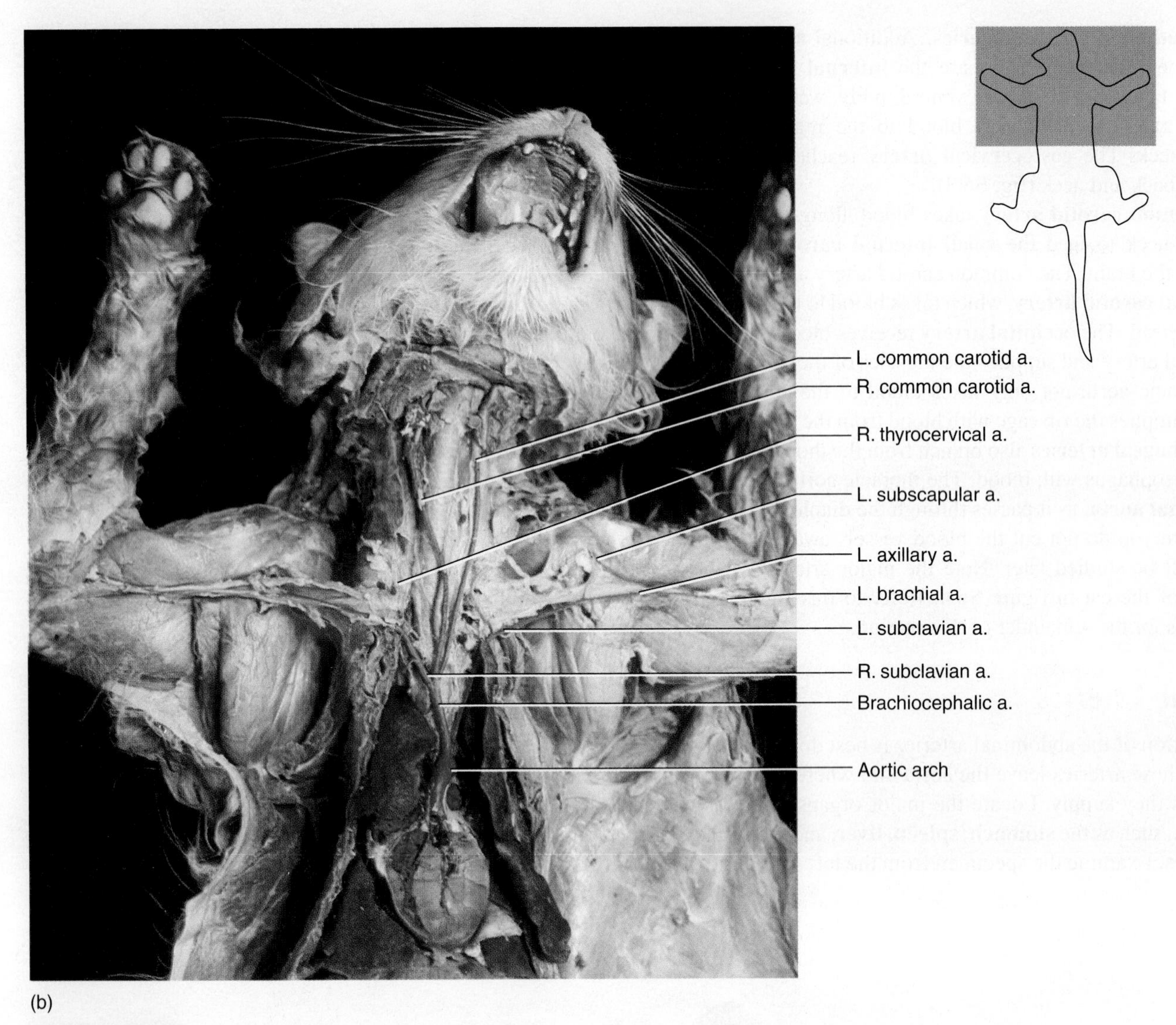

FIGURE S6.3 *(Continued)* **Arteries of the Shoulder and Forelimb of the Cat** (b) Photograph.

The pelvic arteries are somewhat different in cats than in humans. The abdominal aorta splits caudally into the **external iliac arteries,** and a short section of the **aorta** continues on and then divides to form the two **internal iliac arteries** and the **caudal artery.** There is no common iliac artery in cats as there is in humans. In cats, the caudal artery takes blood to the tail. The internal iliac artery takes blood to the urinary bladder, rectum, external genital organs, uterus, and some thigh muscles. As the external iliac artery exits from the pelvic region and enters the thigh, it becomes the **femoral artery,** which supplies blood to the thigh, leg, and foot. The femoral artery becomes the **popliteal artery** behind the knee, which further divides to form the **tibial arteries.** Locate the pelvic and lower extremity arteries in your cat and compare them with figures S6.4*a*, S6.6, and S6.7.

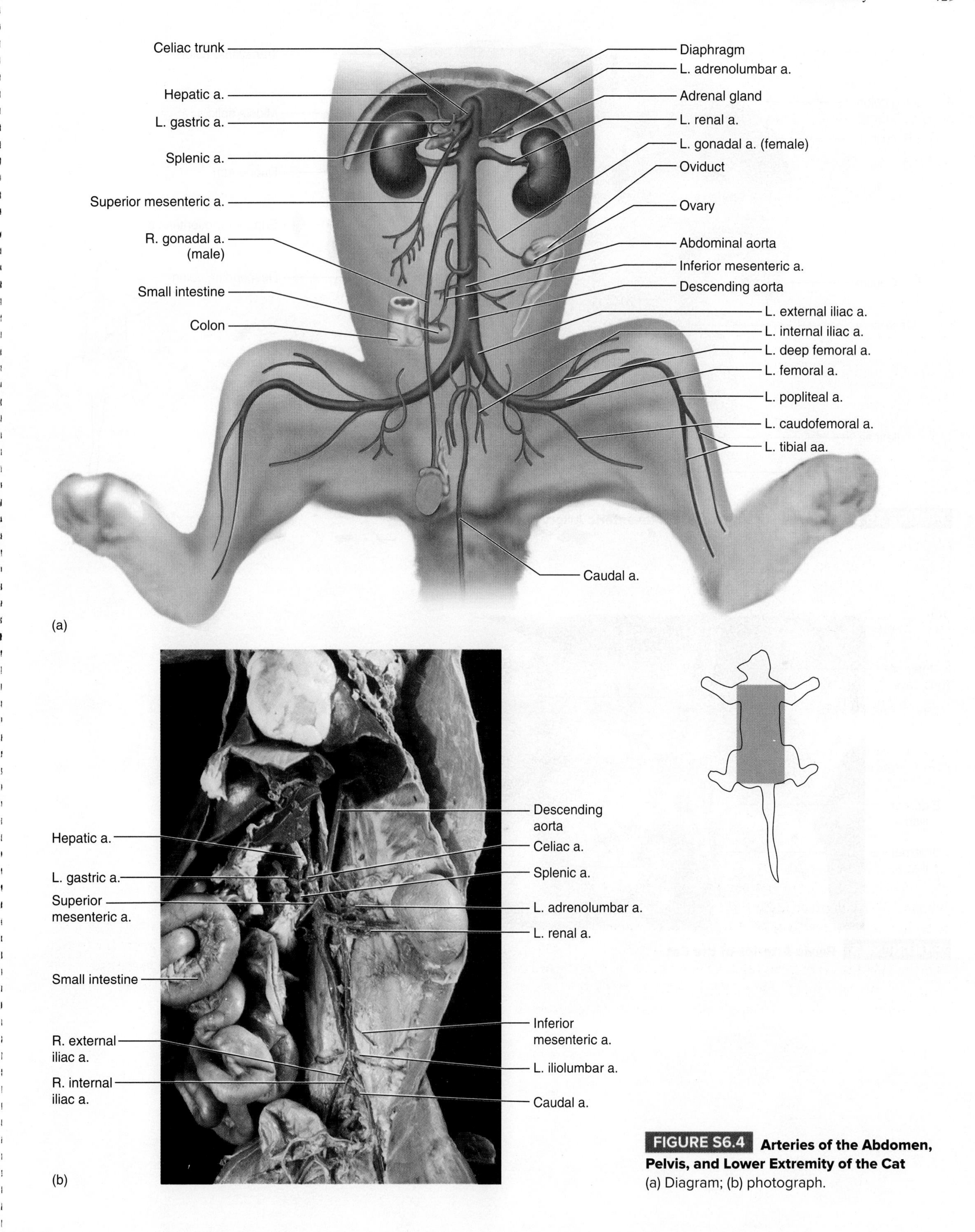

FIGURE S6.4 Arteries of the Abdomen, Pelvis, and Lower Extremity of the Cat
(a) Diagram; (b) photograph.

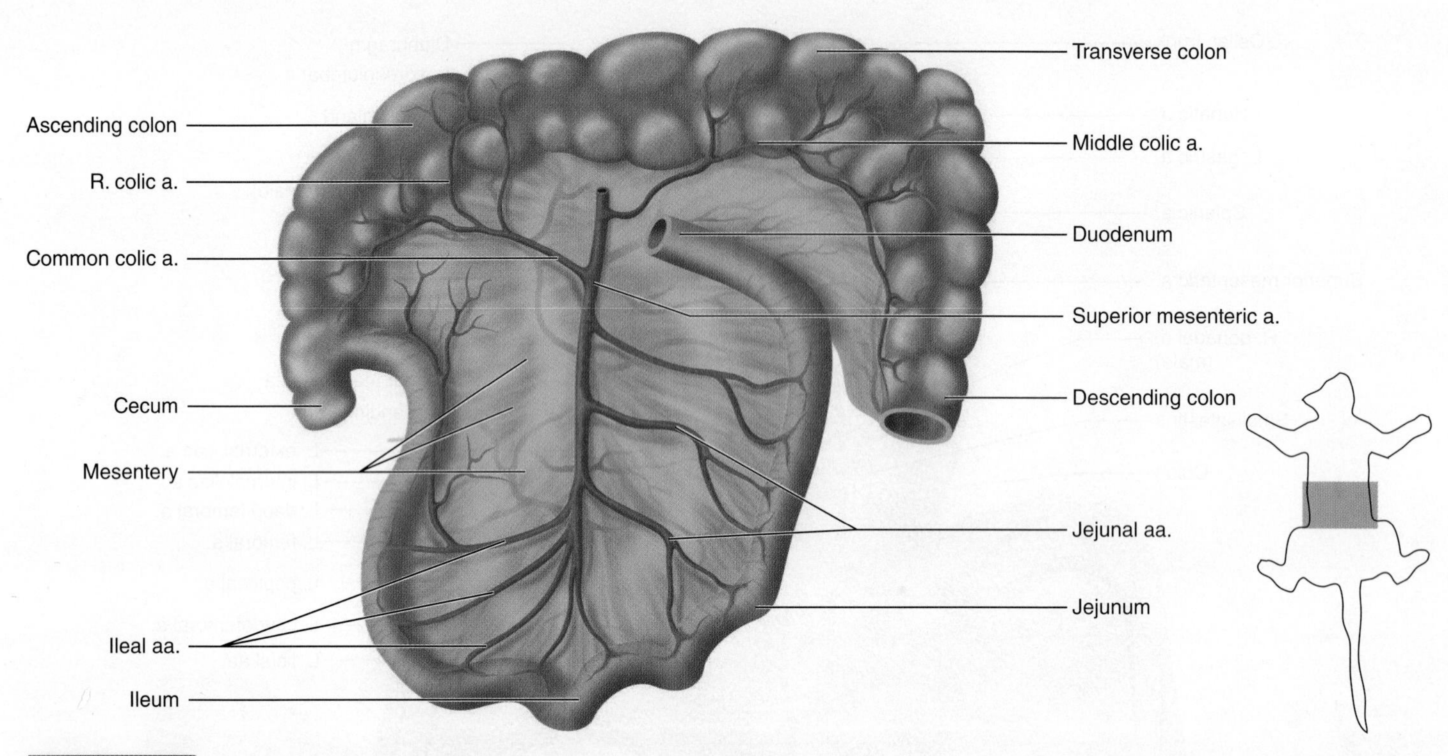

FIGURE S6.5 **Branches of the Superior Mesenteric Artery of the Cat**

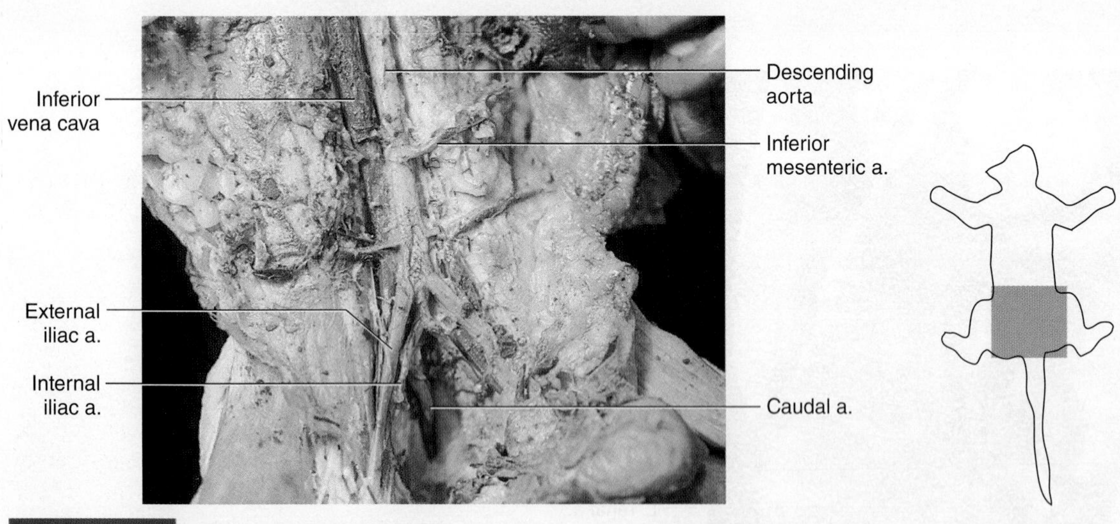

FIGURE S6.6 **Pelvic Arteries of the Cat**

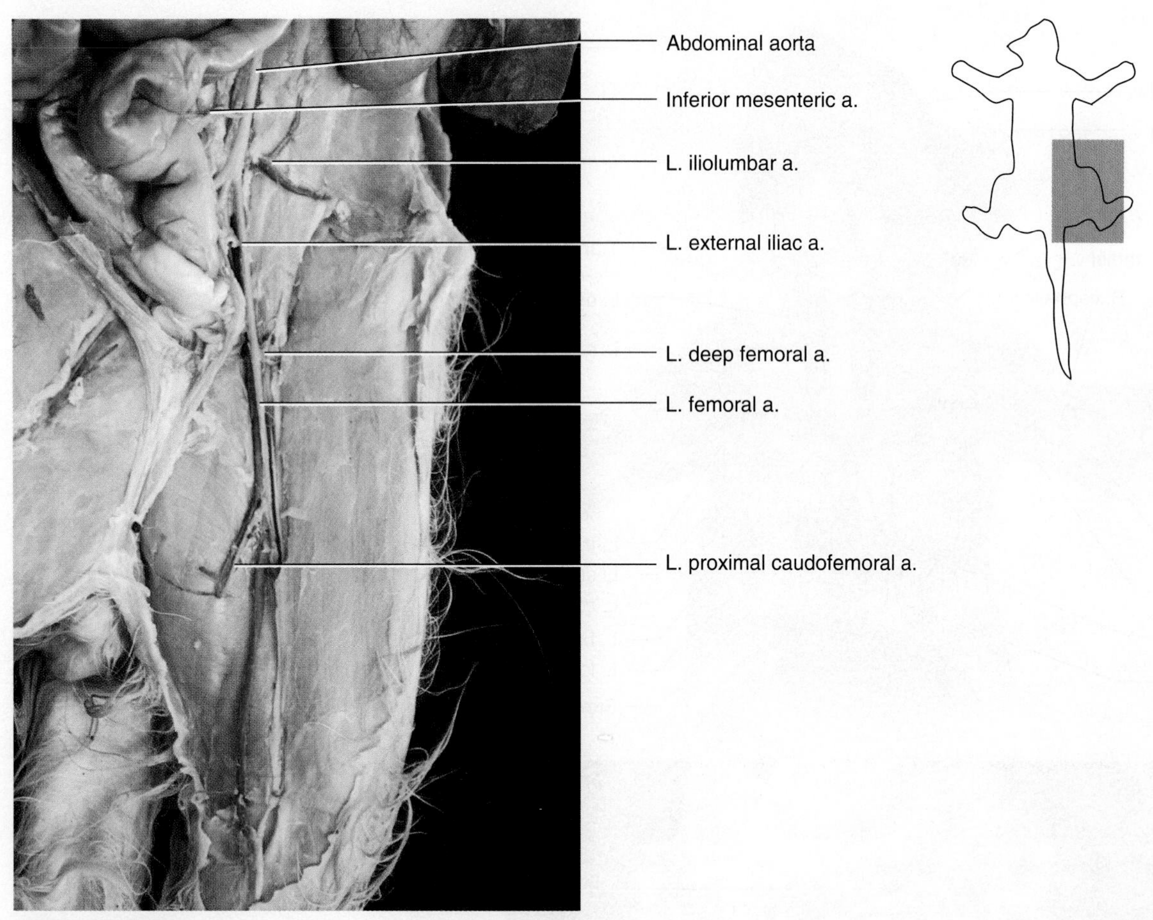

FIGURE S6.7 **Arteries of the Left Hindlimb of the Cat**

Veins of the Cat

As noted earlier, the blood vessels in the cat have been injected with colored latex. If the cat has been doubly injected, the arteries are red and the veins are blue. If the cat has been triply injected, the hepatic portal vein is also injected, typically with yellow latex.

Veins of the Head and Neck The veins in this region of the cat have a few variations from those in the human. In the cat, the **external jugular vein** is larger than the **internal jugular vein.** The reverse is true in humans. What anatomical difference between the cat and human might explain the difference in the volume of blood carried by these two vessels?

The **transverse jugular vein,** which is present in cats but absent in humans, connects the two external jugular veins. The jugular veins, along with the **costocervical veins** and the subclavian veins, unite to form the brachiocephalic veins (fig. S6.8). The **vertebral veins** take blood from the neck of the cat and empty into the subclavian veins.

Thoracic Veins Anterior to the diaphragm are the veins of the thorax. The internal mammary vein (sternal vein) may be difficult to locate, since it is frequently cut while opening up the thoracic cavity. The **internal mammary vein** receives blood from the ventral body wall. Another vein of the thoracic region is the **azygos vein.** This vessel is found on the right side of the body and takes blood from the **esophageal, intercostal,** and **bronchial veins.** Compare the veins in your cat with figure S6.8.

Abdominal Veins Be sure to identify the major organs in the digestive tract before proceeding with the study of the veins of this region. Pay particular attention to the stomach, liver, spleen, kidneys, adrenal glands, small intestine, and large intestine. The inferior vena cava is a large vessel that travels up the right side of the body in the cat (figure S6.9). If necessary, gently move some of the digestive organs to the side to find the inferior vena cava. Examine where the common iliac veins join to form the inferior vena cava. The caudal vein joins the inferior vena cava at this point. This vein is absent in humans. The gonadal veins receive blood from the testes or ovaries. The right gonadal vein takes blood to the inferior vena cava, and the left gonadal vein takes blood to the left renal vein, which subsequently leads to the inferior vena cava. The paired renal veins receive blood from each kidney and empty into the inferior vena cava. The iliolumbar veins take blood from the body wall and return it via the inferior vena cava. Other vessels that drain the body wall are the adrenolumbar veins, which receive blood from the adrenal glands in addition to the body wall. Compare the veins in your cat with figure S6.10.

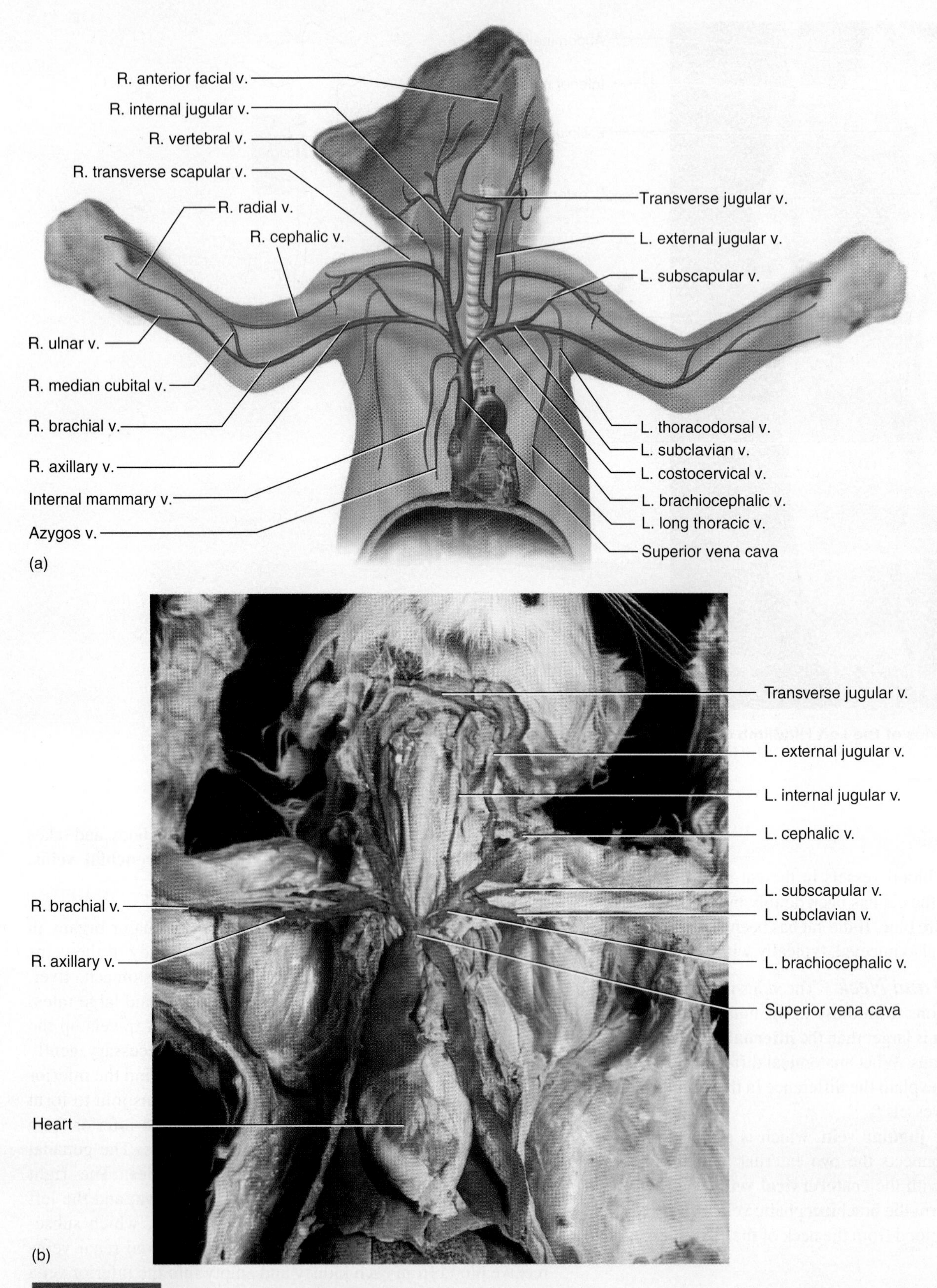

FIGURE S6.8 Anterior Veins of the Cat (a) Diagram; (b) photograph.

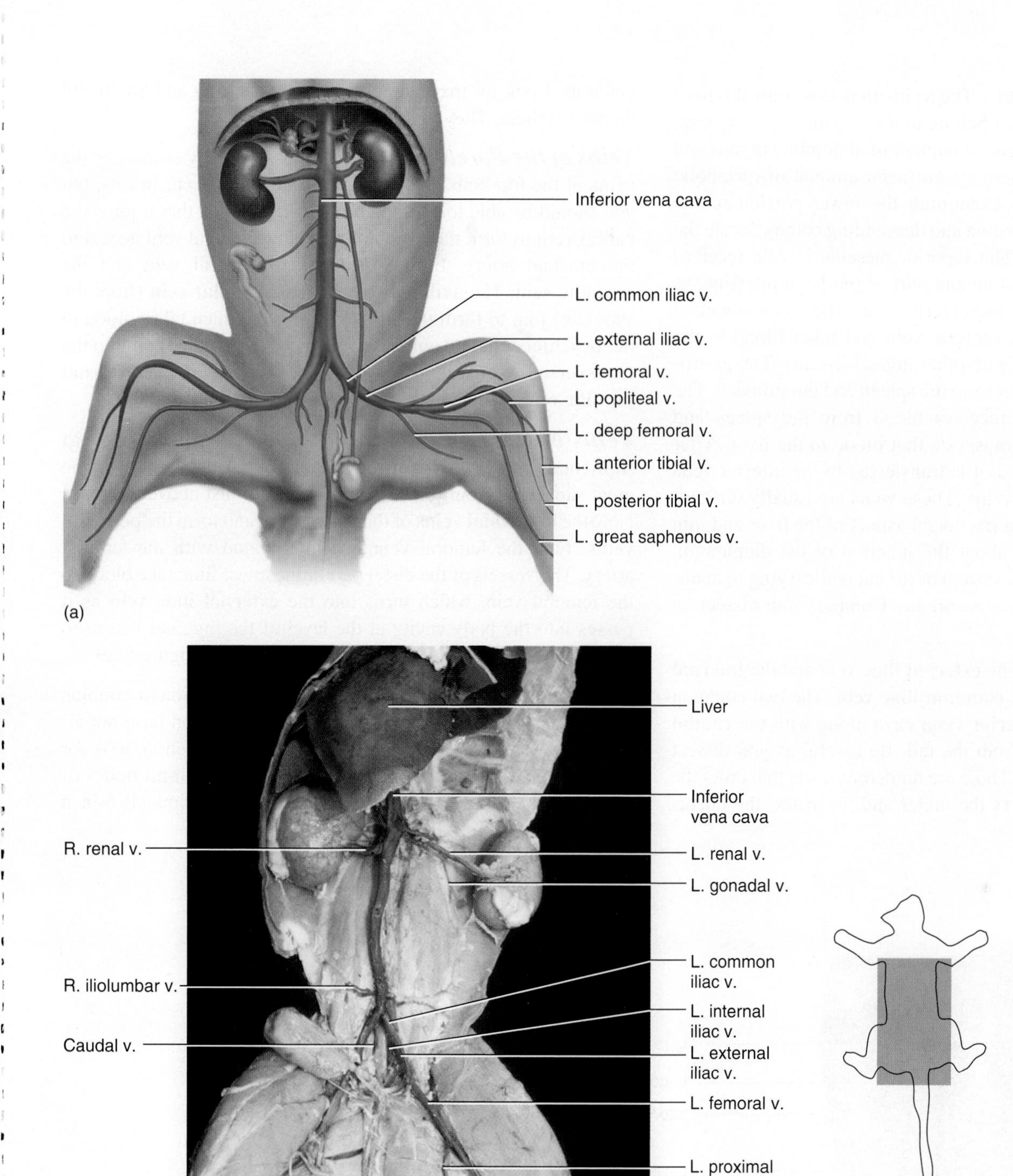

FIGURE S6.9 **Veins of the Hindlimb of the Cat** (a) Diagram; (b) photograph.

Hepatic Portal System The veins that flow into the liver before returning to the heart belong to the hepatic portal system. These vessels take blood from a number of abdominal organs and transfer it to the liver, where a significant amount of metabolic processing occurs. While examining the lower portion of the digestive tract (transverse colon and descending colon), locate the inferior mesenteric vein. The superior mesenteric vein receives blood from the small intestine and part of the large intestine, as well as from the inferior mesenteric vein. The gastro-omental vein joins the superior mesenteric vein and takes blood to the hepatic portal vein, along with other digestive veins. The gastro-omental vein receives blood from the spleen and the stomach. The hepatic portal vein thus receives blood from the spleen and the digestive organs and transports that blood to the liver. After the liver receives the blood, it is transferred to the inferior vena cava by the short hepatic veins. These veins are usually difficult to dissect, since they are at the dorsal aspect of the liver and join the inferior vena cava at about the junction of the diaphragm. Examine the hepatic portal system of the cat while trying to maintain the integrity of the digestive organs. Compare your dissection with figure S6.11.

Veins of the Pelvis The **external iliac vein** and the **internal iliac vein** join to form the **common iliac vein.** The two common iliac veins lead to the **inferior vena cava** along with the **caudal vein,** which takes blood from the tail. Be careful as you dissect these structures in the cat. There are numerous ducts that cross the external iliac veins, such as the ureter and, in males, the ductus deferens. Look for these structures in figures S6.9 and S6.10 and do not cut them. They will be studied later.

Veins of the Forelimb Begin the dissection by examining the veins of the forelimb. The basilic vein does not occur in cats, but you should be able to find the **ulnar vein** and see that it joins the **radial vein** to form the **brachial vein.** The brachial vein is next to the brachial artery. Locate the **median cubital vein** and the **cephalic vein.** The **axillary vein** and **subscapular vein** (from the shoulder) join to form the **subclavian vein,** which takes blood to the **brachiocephalic vein.** The cephalic vein continues on into the **transverse scapular vein,** which takes blood to the **external jugular vein.** Locate these veins in figure S6.8.

Veins of the Lower Limb Locate the superficial **great saphenous vein** in the cat on the medial side of the lower limb. The great saphenous vein joins the **femoral vein** just above the knee. Note the long **tibial veins** of the leg that join and form the **popliteal veins.** Find the femoral vein, which is found with the femoral artery. The vessels of the distal part of the lower limb take blood to the femoral vein, which turns into the **external iliac vein** as it passes into the body cavity at the level of the inguinal ligament. Find these veins in the cat and compare them with figure S6.9.

Lymphatic System Your instructor may want you to examine the lymphatic system in the cat in this section if you have not already done so. If you do examine the cat in this section, look for small lumps of tissue in the axilla. These are the **lymph nodes** of the region. Note the large **spleen,** which is an approximately 6-inch

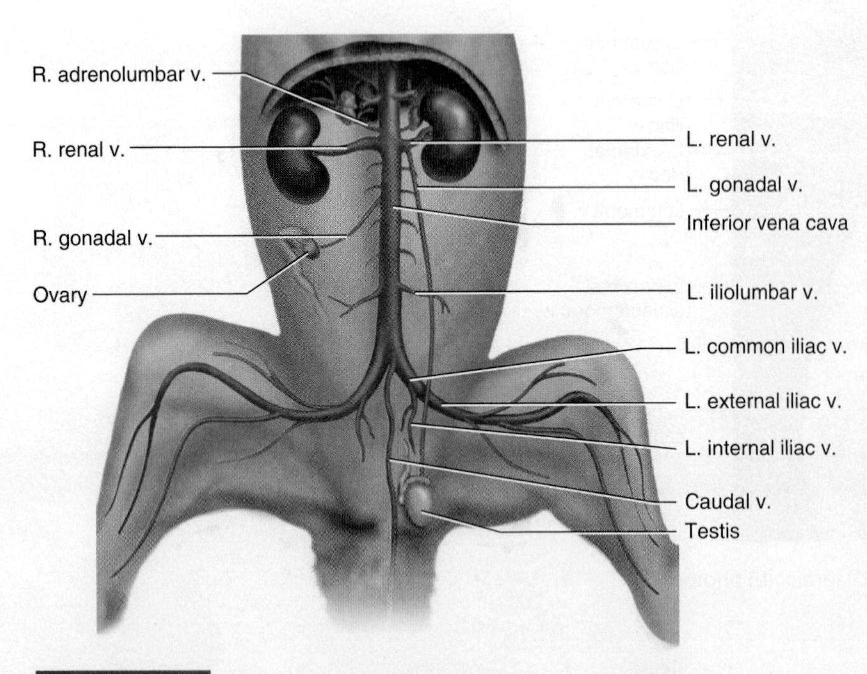

FIGURE S6.10 **Veins of the Pelvis and Abdomen of the Cat**

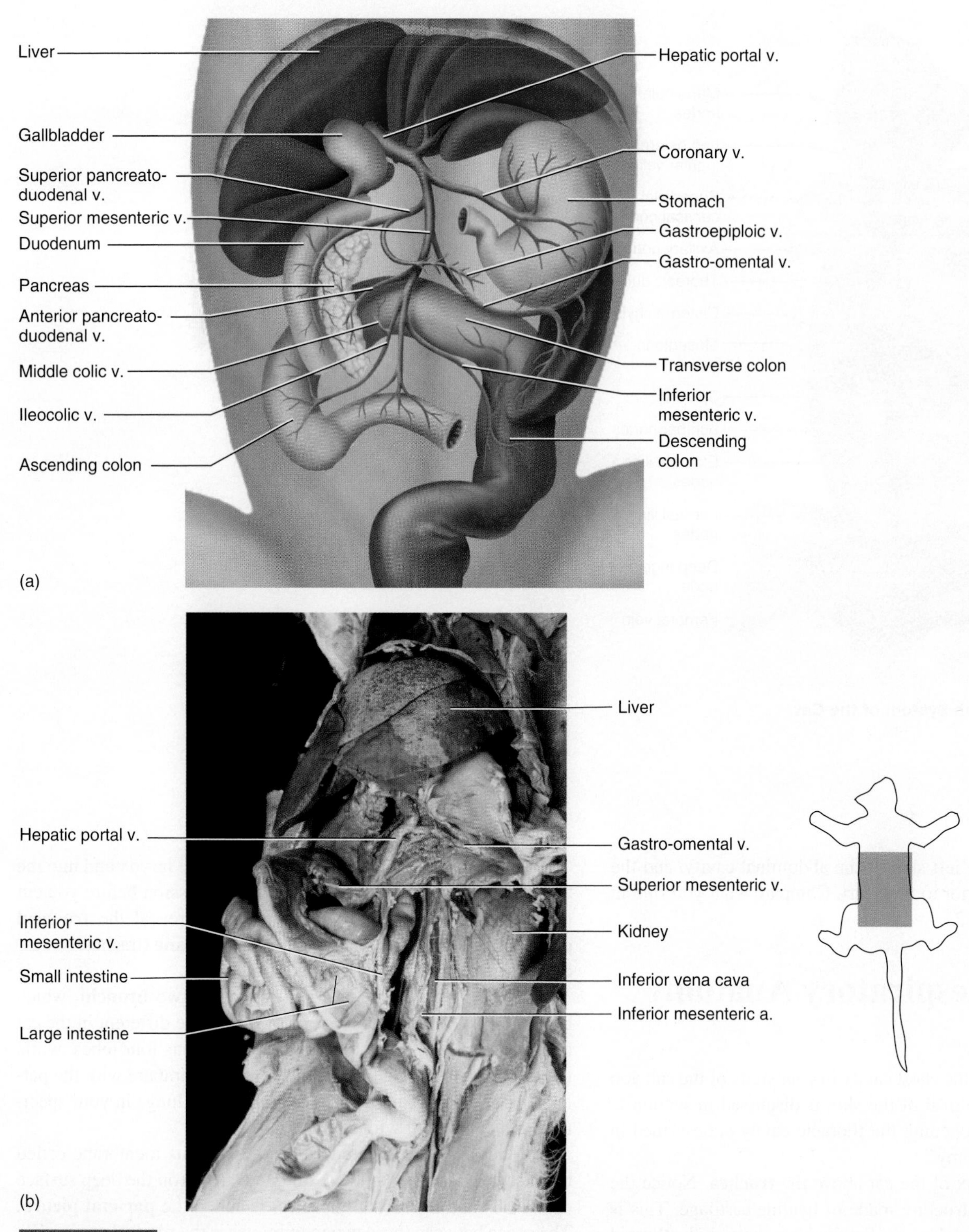

FIGURE S6.11 **Hepatic Portal System of the Cat** (a) Diagram; (b) photograph.

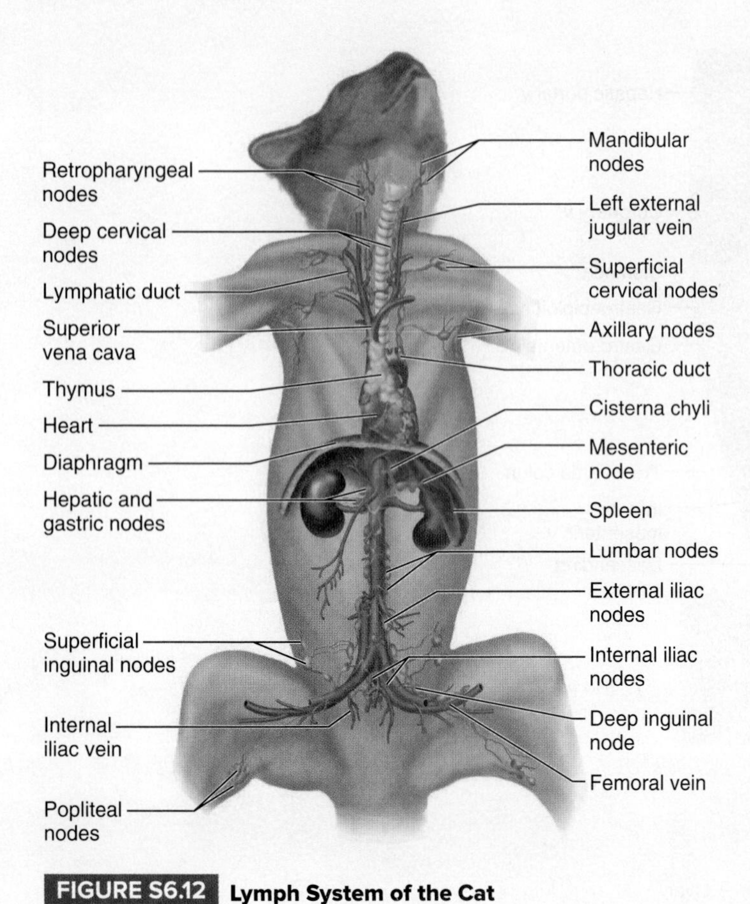

FIGURE S6.12 **Lymph System of the Cat**

elongated organ on the left side of the abdominal cavity, and the **thymus,** which is anterior to the heart. Compare what you find in the cat with figure S6.12.

Section 7: Respiratory Anatomy of the Cat

If you have not opened the chest cavity in your study of the cat, you should do so now. Removal of the skin is discussed in section 2, and the procedure for opening the thoracic cavity is described in section 6 of "Cat Anatomy."

Locate the **larynx** of the cat above the **trachea.** Notice the broad, wedge-shaped structure made of hyaline cartilage. This is the **thyroid cartilage.** Make a median incision through the thyroid cartilage and continue carefully cutting until you have cut completely through the larynx. To examine the larynx more completely, continue the median cut partway down through the trachea. Open the larynx and find the **epiglottis** of the cat. Notice how the elastic cartilage of the epiglottis is lighter in color than that of the thyroid cartilage. Find the **cricoid cartilage,** the **arytenoid cartilages,** and the **true** and **false vocal folds** (fig. S7.1).

Examine the trachea as it passes from the larynx and into the thoracic cavity. Ask your instructor for permission before you cut the trachea in cross section, which should reveal the **tracheal cartilages** and the **posterior tracheal membrane** (trachealis muscle). The trachea is ventral to the esophagus.

Notice how the trachea splits into the two **bronchi,** which then enter the lungs. The lobes of the lungs are different in the cat than in the human. The right lung in the cat has four lobes, while the left lung has three lobes. How does this compare with the pattern in humans? Examine the structure of the lungs in your specimen and compare it with figure S7.2.

The lungs are covered with a thin serous membrane called the **visceral pleura,** while the outer covering (on the deep surface of the ribs and intercostal muscles) is called the **parietal pleura.** The space between these two membranes is the **pleural cavity.** Cut into one of the lungs of the cat and examine the lung tissue. Note how the lungs look like a very fine mesh sponge. The alveoli of the lungs are microscopic.

At the inferior portion of the thoracic cavity is the diaphragm. As it contracts, the pressure in the thoracic cavity decreases and air fills the lungs. Examine the diaphragm in your specimen and compare it with figure S7.2.

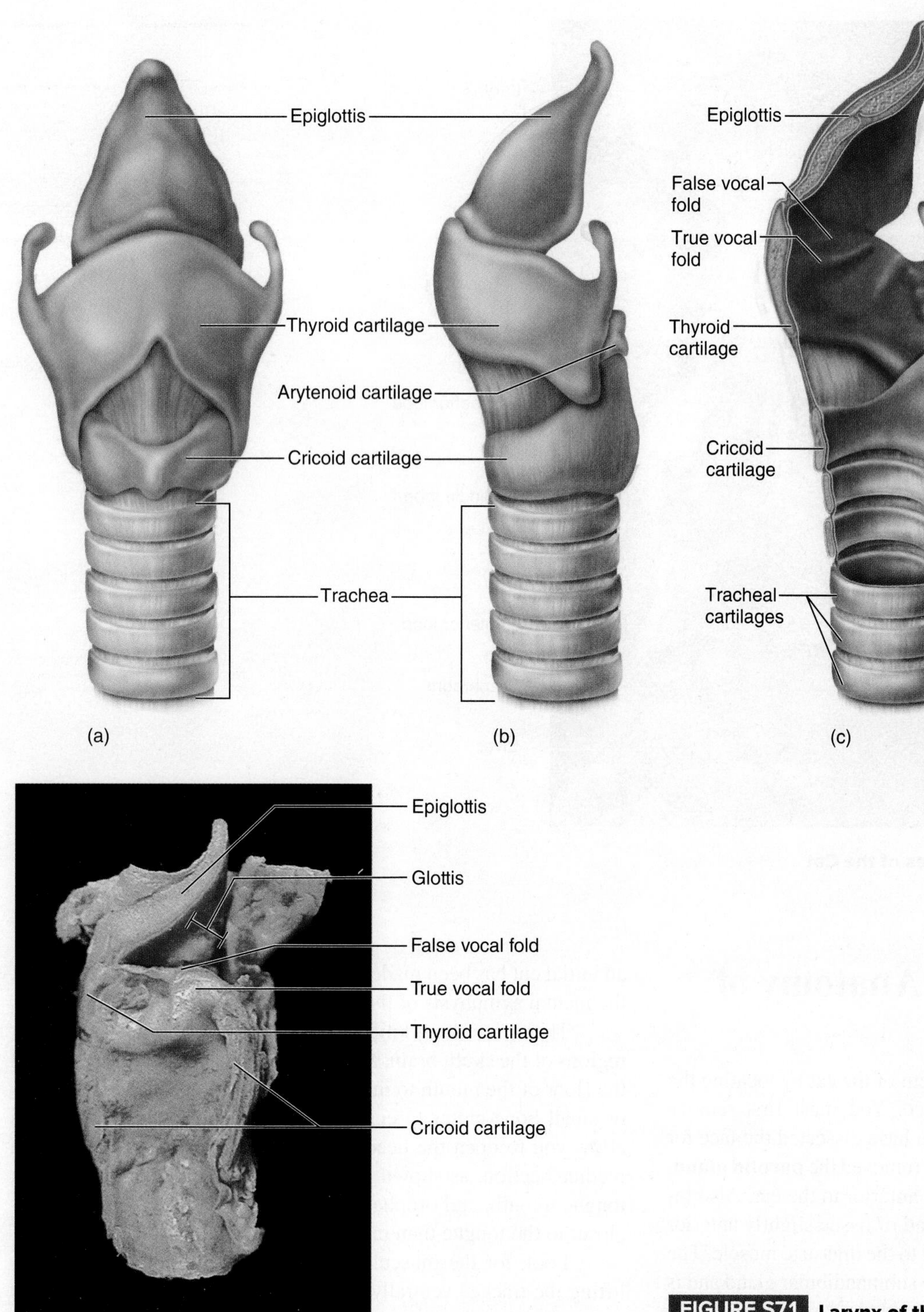

FIGURE S7.1 **Larynx of the Cat** Diagram (a) anterior view; (b) left lateral view; (c) median view. Photograph (d) median view.

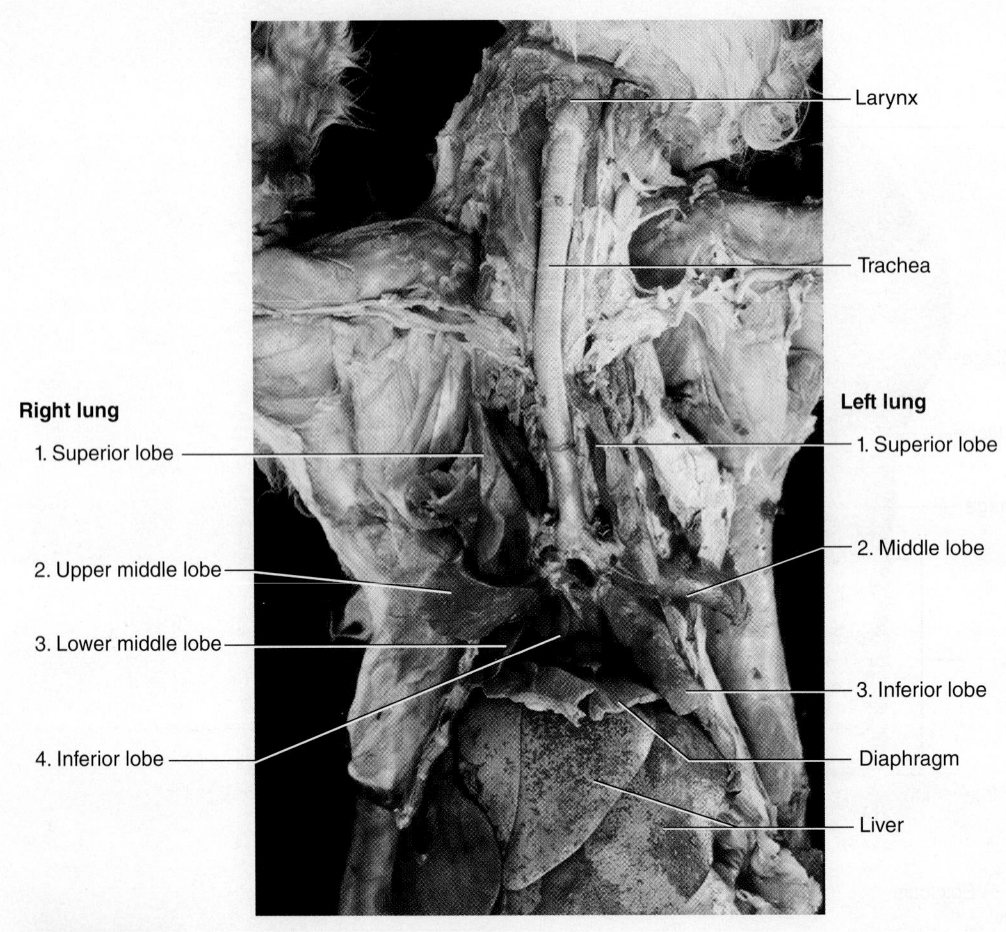

FIGURE S7.2 **Respiratory Structures of the Cat**

Section 8: Digestive Anatomy of the Cat

Begin your study of the digestive system of the cat by locating the large **salivary glands** around the face. You must first remove the skin from in front of the ear. If you have dissected the face for the musculature, you may have already removed the **parotid gland,** which is a spongy, cream-colored pad anterior to the ear. Also locate the **submandibular gland** as a pad of tissue slightly anterior to the angle of the mandible and lateral to the digastric muscle. The **sublingual gland** is just anterior to the submandibular gland and is an elongated gland that parallels the mandible. Use figure S8.1 to help locate these glands.

Examine the **tongue** of the cat and note the numerous papillae on the dorsal surface. These are **filiform papillae** and have both a digestive and a grooming function. You can lift up the tongue and examine the **lingual frenulum** on the ventral surface.

Dissection of the head of the cat should be done only if your instructor gives you permission to do so. To dissect the head, first use a scalpel and make a cut in the median plane from the forehead of the cat to the region of the occipital bone. Cut through any overlying muscle. Then use a small saw to gently cut through the cranial region of the skull, being careful to stay in the median plane. Once an initial cut has been made through the dorsal side of the head, cut the mental symphysis of the cat, thus separating the mandible.

Then use a long knife (or scalpel) to cut through the softer regions of the skull, brain, and tongue. Use a scalpel to cut through the floor of the mouth to the hyoid bone. It is best to use a scissors or small bone cutter to cut through the hyoid bone. This should allow you to open the head and examine the structures seen in a median section, as shown in figure S8.2. Locate the hard palate, tongue, mouth, and oropharynx. Note how the larynx in the cat is closer to the tongue than in humans.

Look for the muscular tube of the **esophagus** by carefully lifting the trachea ventrally away from the neck. Insert your blunt dissection probe into the mouth and gently wiggle it into the esophagus. The tongue and esophagus are represented in figure S8.2.

Abdominal Organs

To get a better view of the abdominal organs, it is best to cut the **diaphragm** away from the body wall. Carefully cut the lateral edges of the diaphragm from the ribs to reveal a fatty drape of material covering the intestines. This is the **greater omentum.** Just posterior to the diaphragm, note the dark brown multilobed **liver.** In the middle of the liver is the green **gallbladder.** If you lift the

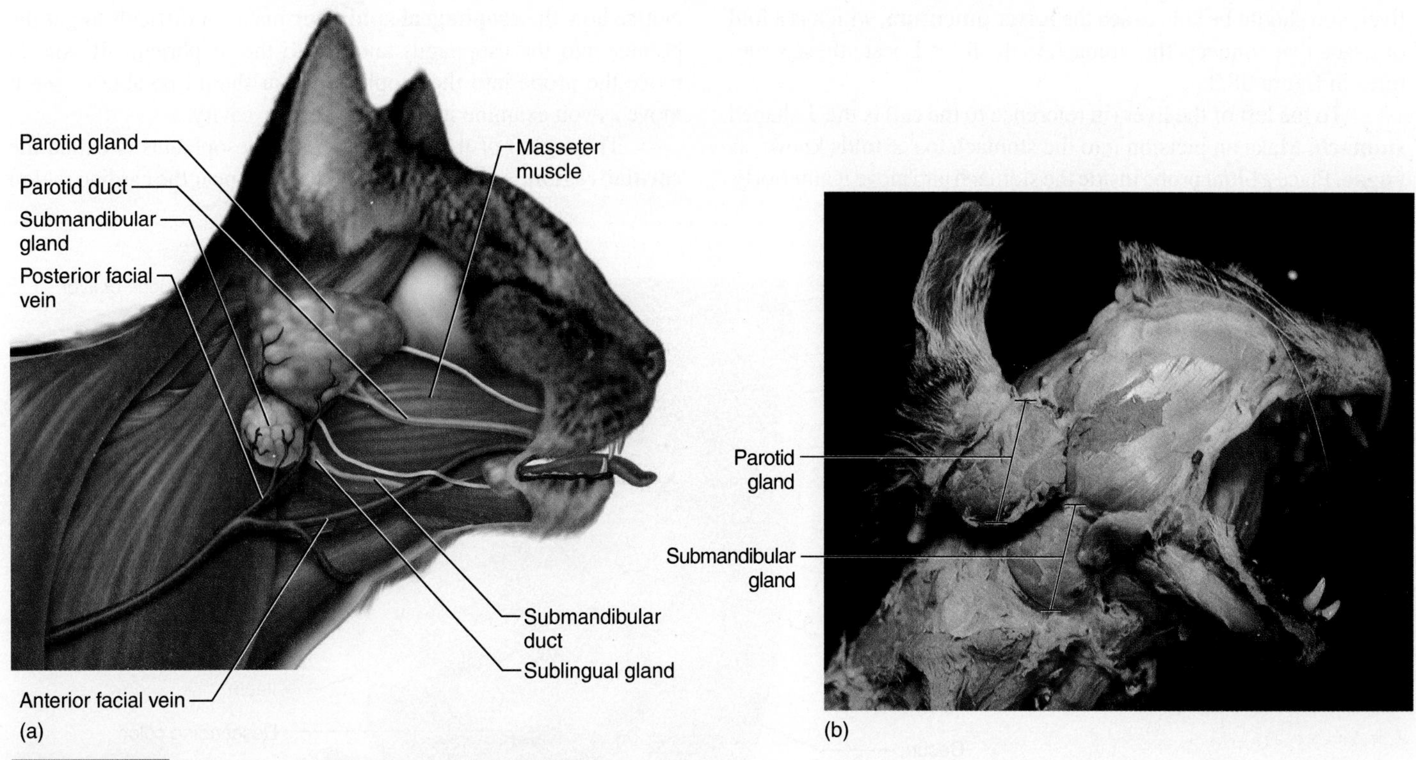

FIGURE S8.1 **Salivary Glands of the Cat** (a) Diagram; (b) photograph.

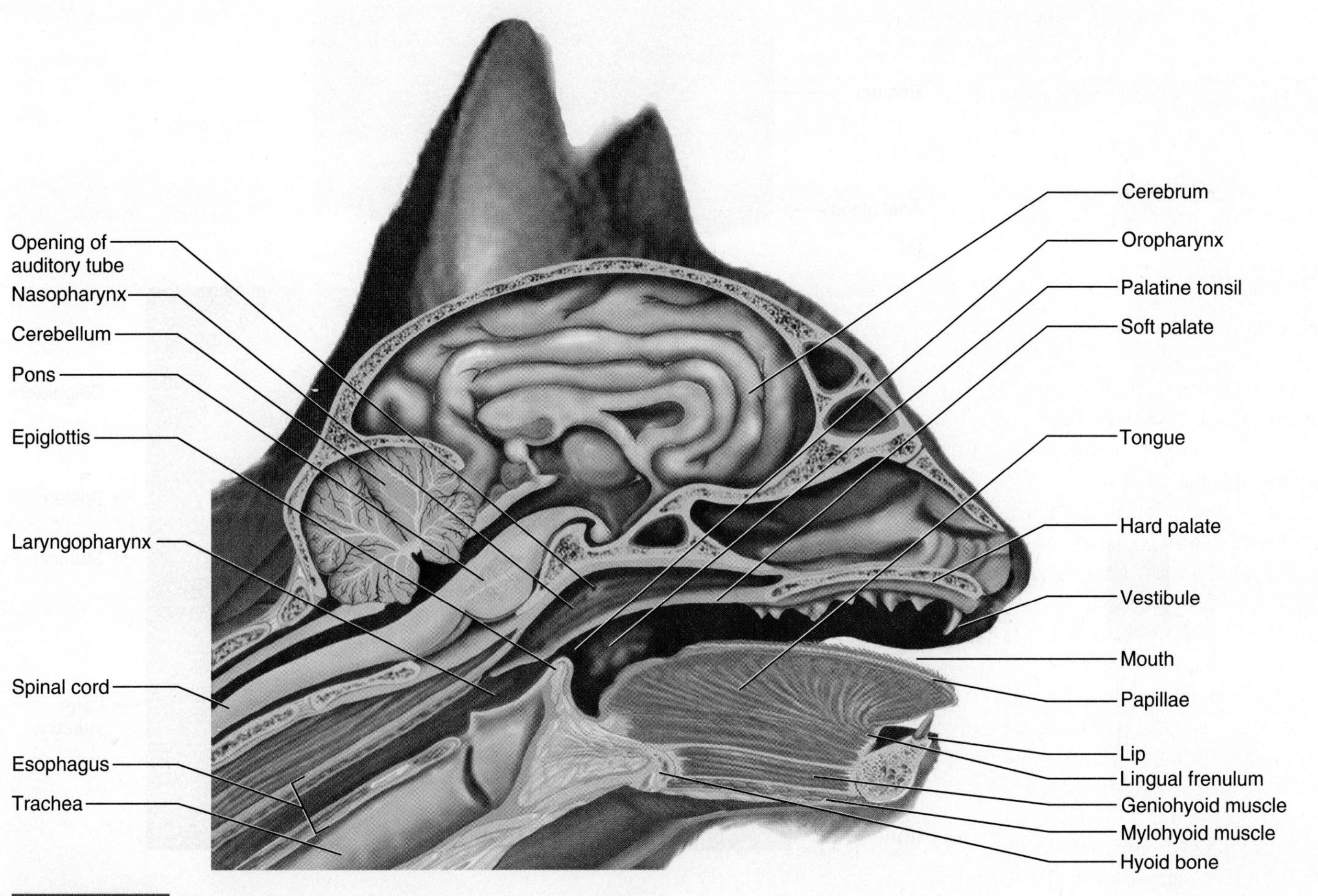

FIGURE S8.2 **Median Section of the Head and Neck of a Cat**

liver, you should be able to see the **lesser omentum,** which is a fold of tissue that connects the stomach to the liver. Locate these structures in figure S8.3.

To the left of the liver (in reference to the cat) is the J-shaped **stomach.** Make an incision into the stomach to see folds known as **rugae.** Place a blunt probe inside the stomach and move it anteriorly. Notice how the **esophageal sphincter** makes it difficult to put the pointer into the esophagus anterior to the diaphragm. If you do move the probe into the esophagus, you should be able to see it move as you examine it from the thoracic cavity.

The region of the stomach near the esophagus is called the **cardiac region.** A small part of the stomach near the cardiac region

FIGURE S8.3 Abdominal Organs of the Cat (a) Diagram; (b) photograph.

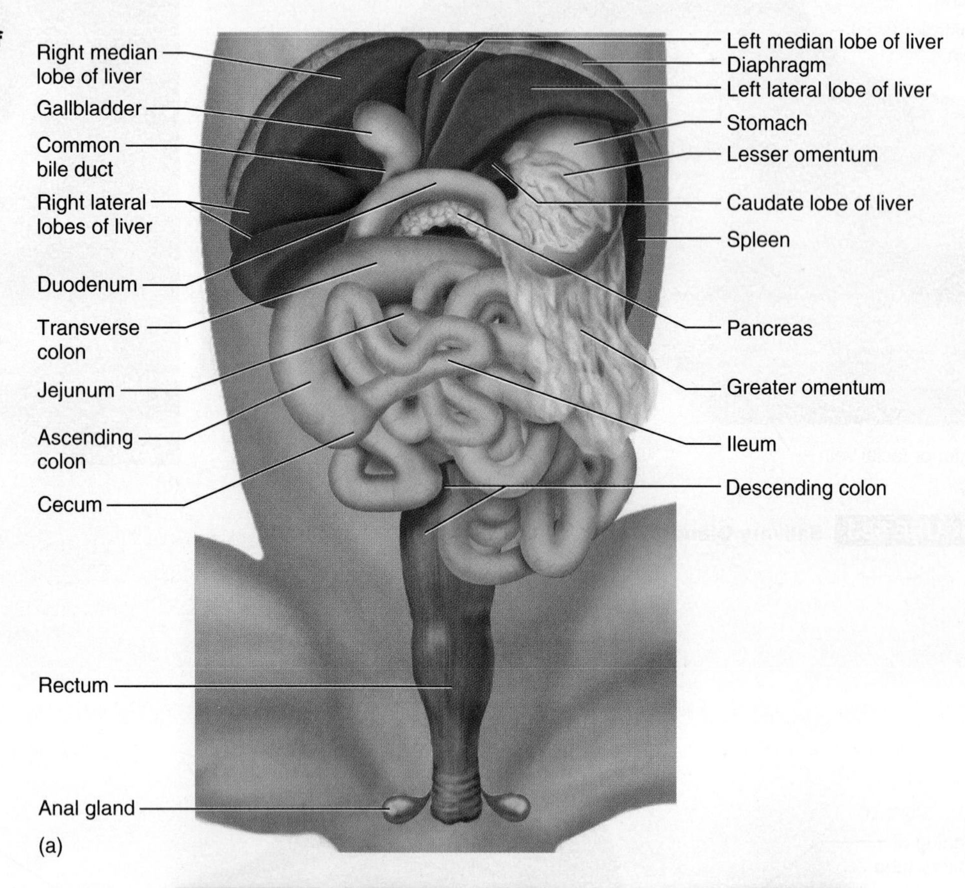

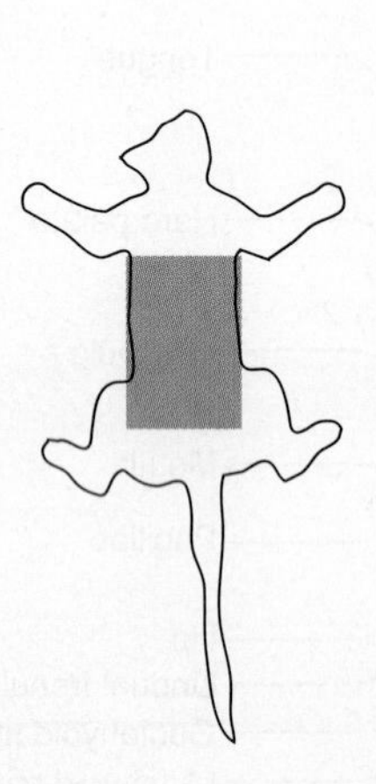

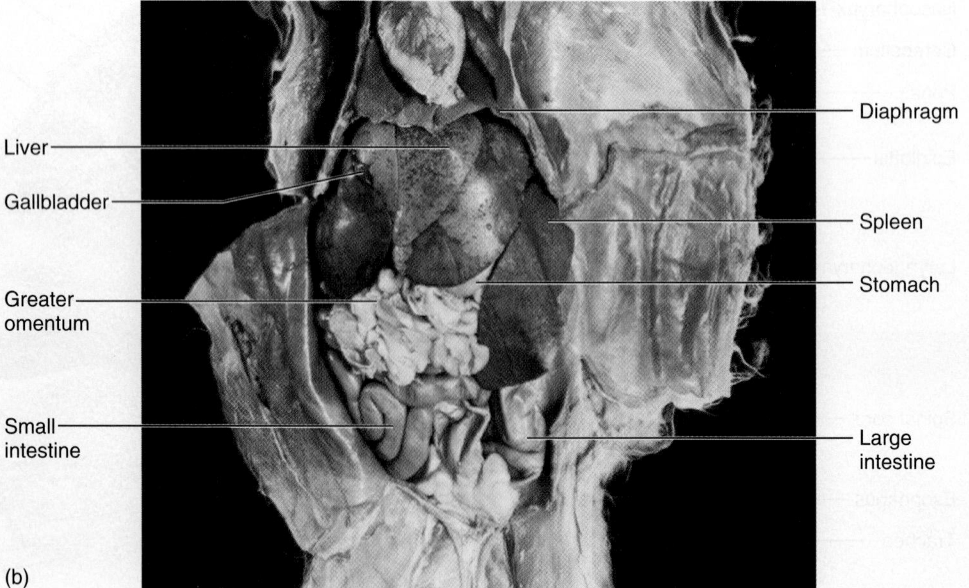

forms an anterior dome or pouch. This is the **fundus** of the stomach. The stomach is curved, with the left side of the stomach forming the **greater curvature** and the right side forming the **lesser curvature.** The main region of the stomach is the **body,** and the narrowed section of the stomach, near the small intestine, is the **pyloric region.** Locate the parts of the stomach on your cat.

Cut lengthwise into the pyloric region of the stomach and then through the duodenum to locate a tight sphincter muscle between the stomach and the duodenum. This is the **pyloric sphincter.** As you move into the duodenum, you may have to scrape some of the chyme away from the wall of the small intestine to be able to see the fuzzy texture of the intestinal wall. This texture is due to the presence of **villi.** As in the human, the small intestine is composed of three regions, a proximal **duodenum,** the middle **jejunum,** and a terminal **ileum.** Compare the stomach and small intestine with figure S8.3.

If you elevate the greater omentum, you should be able to see the **pancreas,** along the caudal side of the stomach. The pancreas appears granular and brown. The **tail** of the pancreas is near the **spleen,** which is a brown, elongated organ on the left side of the body.

The small intestine is an elongated, coiled tube about the diameter of a wooden pencil. Note the **mesentery,** which holds the small intestine to the dorsal body wall near the vertebrae. The small intestine is extensive in the cat and rapidly expands into the large intestine. The appendix is absent in the cat.

The **large intestine** in the cat is a fairly short tube with a diameter slightly larger than your thumb. The first part of the large intestine is a pouch called the **cecum.** The remainder of the large intestine can be further divided into the **ascending, transverse,** and **descending colon** and the **rectum.** Compare these with figure S8.3. Examine the **parietal peritoneum** along the inner surface of the body wall and the **visceral peritoneum** that envelops the intestines.

Section 9: Urinary Anatomy of the Cat

The kidneys are on the dorsal body wall of the cat and are located dorsal to the parietal peritoneum. Examine the kidneys and the structures that lead to and from the hilum of the kidney. Locate the renal veins that take blood from the kidney. The veins are larger in diameter than the renal arteries, and they are attached to the inferior vena cava of the cat that runs along the ventral, right side of the vertebral column. Find the renal arteries that take blood from the aorta to the kidneys. The aorta lies to the left of the inferior vena cava. Locate the ureters that run posteriorly from the kidney to the urinary bladder. Do not dissect the urethra at this time. You can locate the urethra during the dissection of the reproductive structures in section 10. Compare your dissection with figure S9.1.

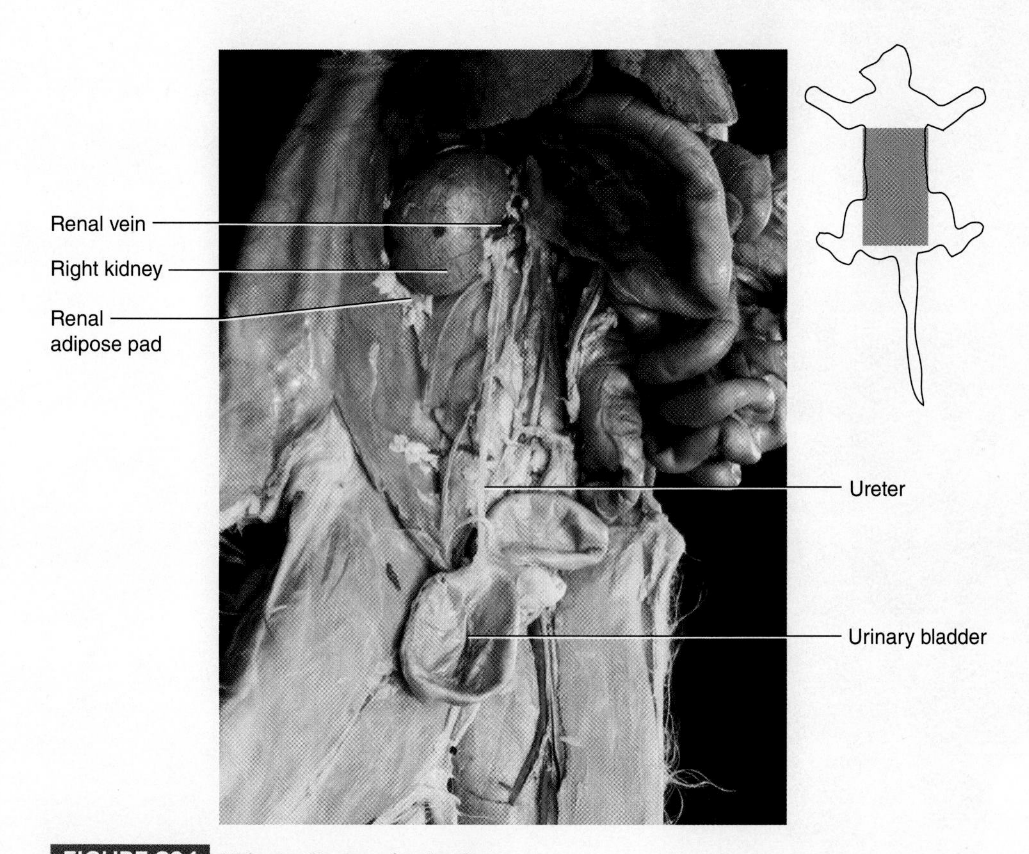

FIGURE S9.1 Urinary System in the Cat

Section 10: Reproductive Anatomy of the Cat

Check to see whether your cat is male or female. The penis may be retracted in your specimen, so look for the opening of the penile urethra and the scrotum. Team up with a lab partner or group that has a cat of a different sex than your specimen, so that you can learn both male and female reproductive systems.

Male Cat

Once you are sure you have a male cat, locate the **scrotum** and paired **testes.** Make an incision on the lateral side of the scrotum and locate the testis inside the scrotal sac. If your cat was neutered, you will not be able to locate the testis. If the testes are present, cut through the connective tissue of the scrotum (the **tunica vaginalis**) and observe both the testis and the **epididymis.** The testis is covered by a tough connective tissue membrane called the **tunica albuginea.** Use figure S10.1 as a guide. Spermatozoa move from the testis and into the epididymis, where the spermatozoa mature.

Once you have located the epididymis, proceed in an anterior direction and trace the thin **ductus deferens** from the epididymis into the **spermatic cord.** The spermatic cord traverses the body wall on the exterior and enters the body of the cat at the **inguinal canal** (an opening in the inguinal ligament). Gently insert a blunt probe into the inguinal canal to locate the ductus deferens as it passes into the peritoneal cavity. Notice how the ductus deferens arches around the ureter on the dorsal side of the urinary bladder. You may have cut one of the ductuli deferentes in an earlier exercise, so if you cannot find it on one side, look for it on the other side.

You may want to look for the accessory organs of the male reproductive system, but this takes some significant dissection. *Check with your instructor before cutting through the pelvis of your cat.* If your instructor directs you to do so, begin by cutting through the musculature of the cat at the level of the **pubic symphysis.** Make a median incision through the groin muscles, and carefully cut through the cartilage of the pubic symphysis. You should now be able to open the pelvic cavity and locate the single **prostate gland** and the paired **bulbourethral glands** (fig. S10.2). Much of the anatomy of the cat is similar to that of

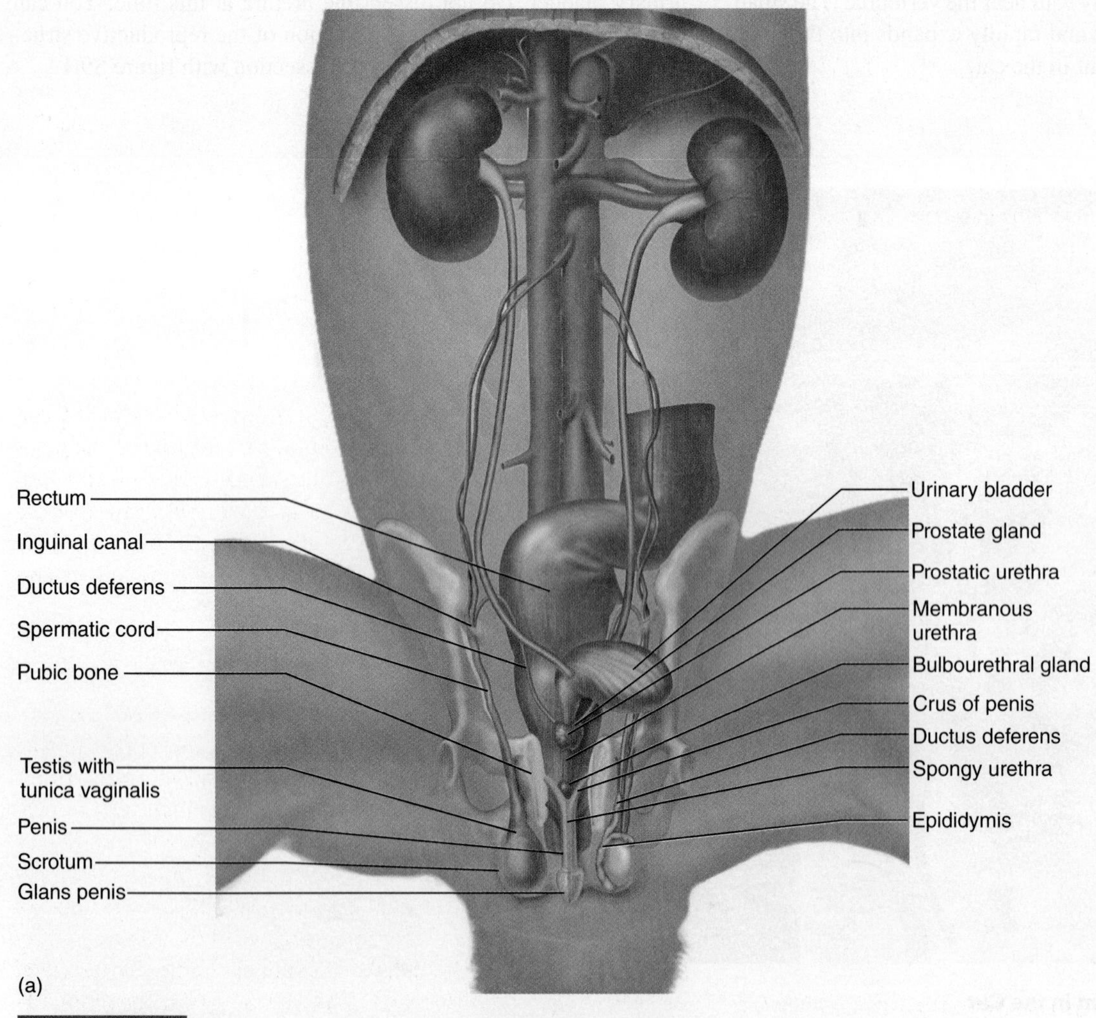

FIGURE S10.1 Male Reproductive Organs of the Cat (a) Diagram.

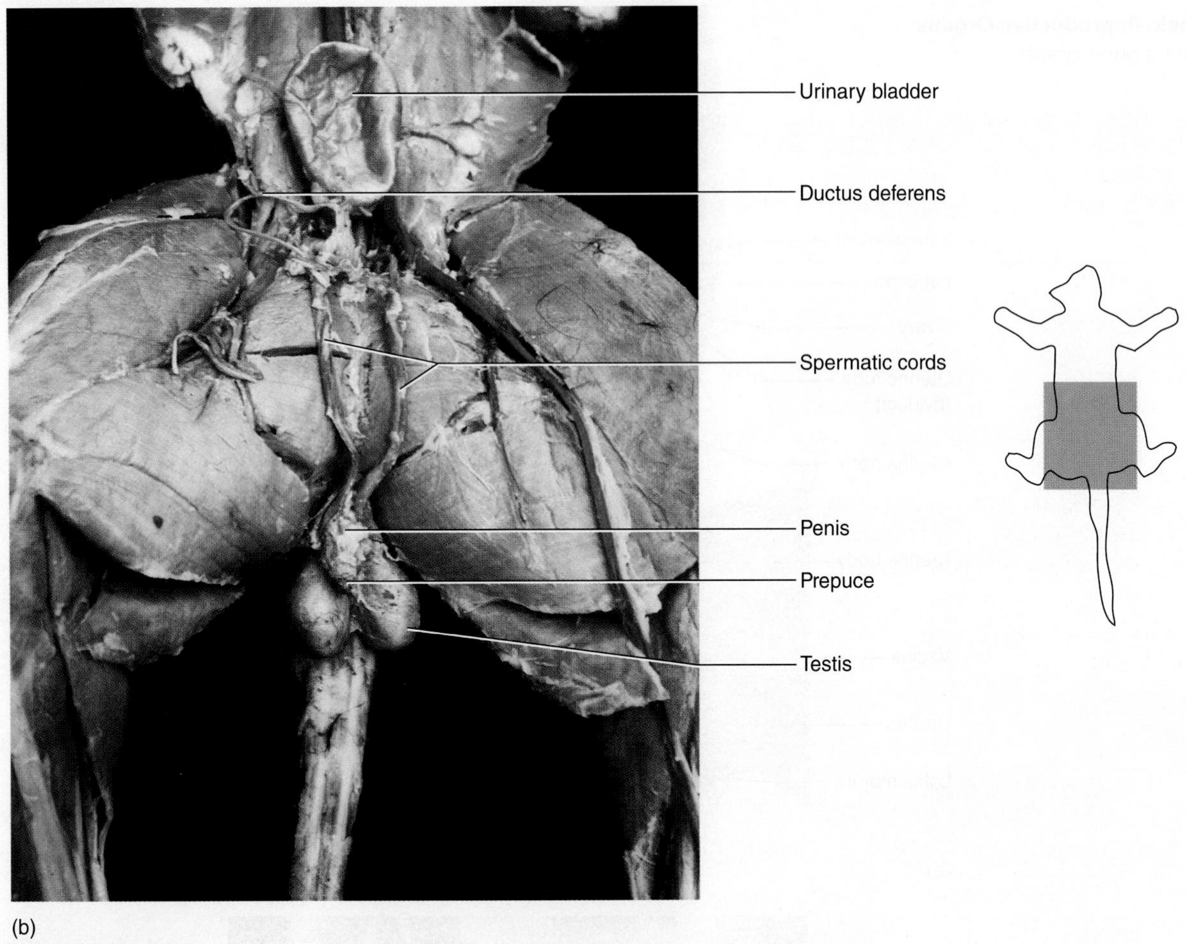

FIGURE S10.1 *(Continued)* **Male Reproductive Organs of the Cat** (b) Photograph.

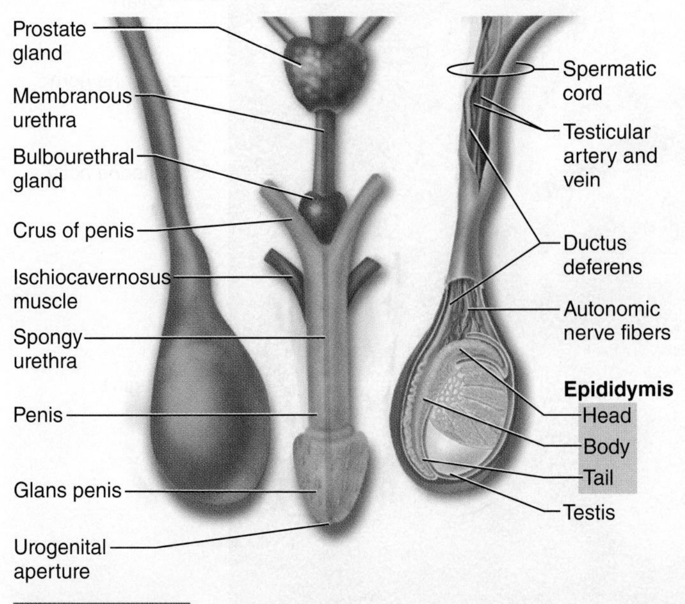

FIGURE S10.2 **Testis, Epididymis, Spermatic Cord, and Penis of the Cat**

the human except that there are no seminal vesicles in the cat. Trace the ductus deferens from the posterior surface of the bladder through the prostate gland and the penis. Either make a longitudinal section through the penis to trace the urethra in the erectile tissue or make a cross section of the penis to see the three cylinders of erectile tissue—the **corpus spongiosum** and the **two corpora cavernosa.**

Female Cat

You probably removed the multiple mammary glands of the female cat during the removal of the skin and study of the muscles. Examine the external genitalia for the **urogenital orifice.** *Check with your instructor before cutting through the pelvis of your cat.* If your instructor directs you to do so, begin by cutting through the musculature of the cat at the level of the **pubic symphysis.** Continue to cut in the median plane through the cartilage of the pubic symphysis and then cranially through the abdominal muscles as described in the dissection of the male cat. This should expose the reproductive organs (fig. S10.3).

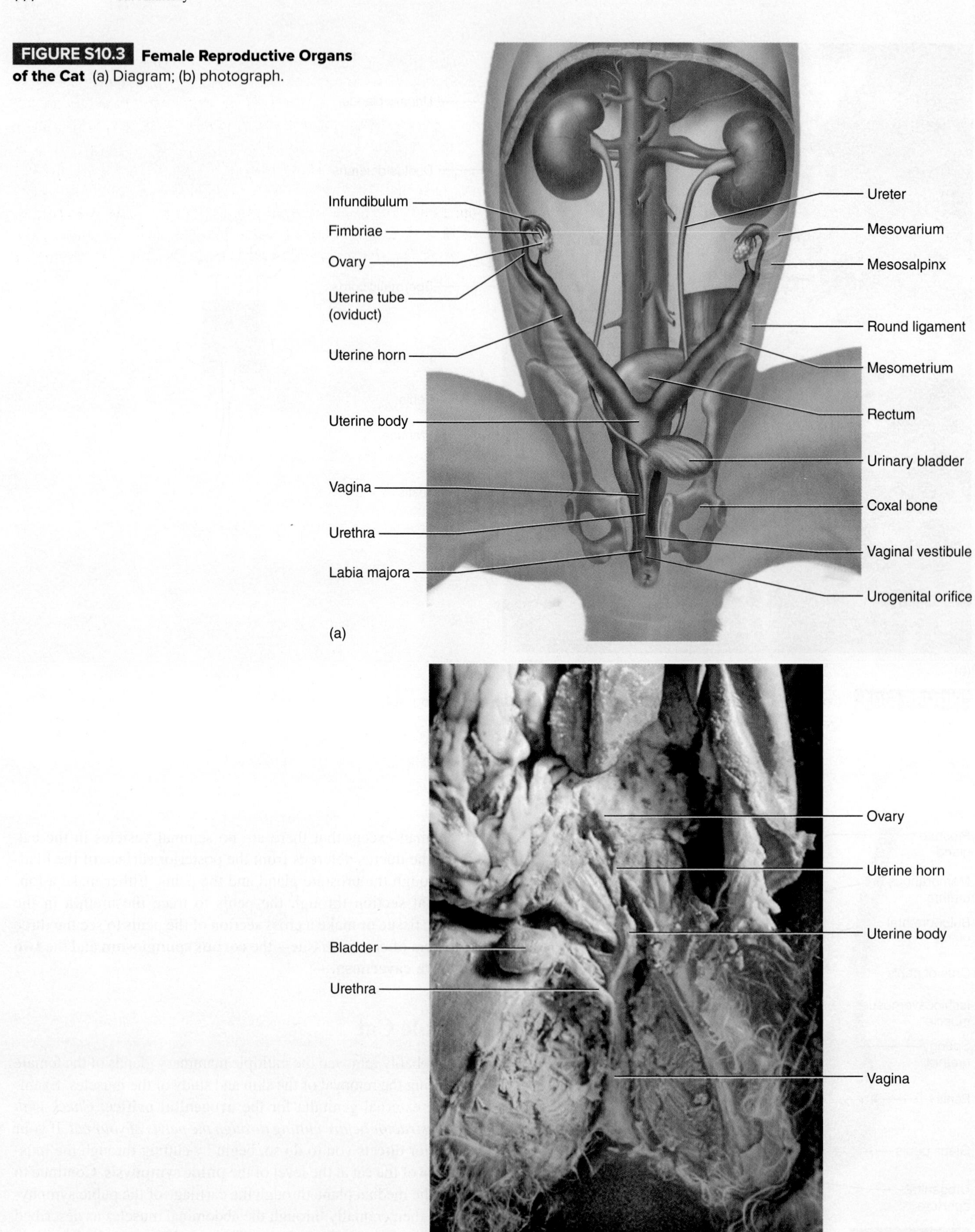

FIGURE S10.3 Female Reproductive Organs of the Cat (a) Diagram; (b) photograph.

Unlike the human female, the cat has a **horned (bipartite) uterus.** The uterus of a cat has an appearance of a Y—with the upper two branches being the **uterine horns** and the stem of the Y the **body** of the uterus. Humans normally have single births from a pregnancy, while the expanded uterus in cats facilitates multiple births. If your cat is pregnant, the uterus will be greatly enlarged, and you may find many fetuses inside.

The **ovaries** in cats are caudal to the kidneys and are relatively small organs. Examine the paired ovaries and the short **uterine tubes (oviducts)** in the cat. If you have difficulty locating them, trace the uterus toward the uterine horns and locate the ovaries. The uterine tubes in the human female are proportionally longer than those in the cat. The opening of the uterine tube near the ovary is called the **ostium,** and it receives the oocytes during ovulation.

At the termination of the uterus is the **cervix,** which leads to the **vaginal canal.** The vagina in cats is different than in humans in that the **urethral opening** is internally enclosed in the vaginal canal. This region where the vagina and the urethral opening are located is called the **vaginal vestibule.** Thus, the opening to the external environment is a common urinary and reproductive outlet called the **urogenital orifice.** Locate these structures in figure S10.3.

Appendix

Preparation of Materials

The following solutions are designed for a laboratory class of 24 students. The preparations are listed alphabetically and the solutions are the approximate volume for the needs of a class of 24. The number of the laboratory exercise or cat anatomy section follows the solution description for cross-referencing.

Bleach Solution (10%)

Mix 100 mL household bleach (sodium hypochlorite) with 900 mL tap water. (Laboratory Exercise 20)

Cat Wetting Solution

Numerous formulations are available for keeping preserved specimens moist. Some commercial preparations are available that reduce the exposure of students to formalin or phenol. You may not need any wetting solution at all if the cats are kept in a plastic bag that is securely closed. You can make a wetting solution by putting 75 mL formalin, 100 mL glycerol, and 825 mL distilled water in a 1-liter squeeze bottle. Another mixture consists of equal parts Lysol™ and water. (Cat Anatomy sections 2–10)

Phosphate Buffer Solution

Add 3.3 g potassium phosphate (monobasic) and 1.3 g sodium phosphate (dibasic) to 500 mL water. Place in squeeze bottles. (Laboratory Exercise 20)

Wright's Stain

Wright's stain is available as a commercially prepared solution from a number of biological supply houses. (Laboratory Exercise 20)

Photo Credits

Unless otherwise credited: Courtesy of Eric Wise.

Front Matter
Page viii(center right): © Francis Leroy, Biocosmos/Science Source.

Exercise 1
Figure 1.1(bottom right): © Mediscan Agency/Alamy Stock Photo; Figure 1.4(bottom left): © Life on white/Alamy Stock Photo RF; Figure 1.6: © McGraw-Hill Education/Joe DeGrandis, photographer.

Exercise 4
Figure 4.2(b): © McGraw-Hill Education; Figure 4.6(b): © Steve Gschmeissner/Science Source.

Exercise 5
Figure 5.4: © McGraw-Hill Education/Dennis Strete, photographer.

Exercise 10
Figures 10.16(a)-10.21(b): © McGraw-Hill Education/ Timothy L. Vacula, photographer.

Exercise 12
Figure 12.2(c): © McGraw-Hill Education/Christine Eckel, photographer; Figure 12.4(b): © McGraw-Hill Education/ Rebecca Gray, photographer/Don Kincaid, dissections.

Exercise 13
Figure 13.2(c): © McGraw-Hill Education/Rebecca Gray, photographer/Don Kincaid, dissections; Figure 13.9(c): © McGraw-Hill Education/ Rebecca Gray, photographer/Don Kincaid, dissections.

Exercise 14
Figure 14.1(b): © McGraw-Hill Education/Rebecca Gray, photographer/Don Kincaid, dissections; Figure 14.2(d): © McGraw-Hill Education/ Christine Eckel, photographer; Figure 14.7(b): © McGraw-Hill Education/Christine Eckel, photographer; Figure 14.8(b): © McGraw-Hill Education/ Christine Eckel, photographer.

Exercise 16
Figure 16.3: © Ed Reschke/ Photolibrary/Getty Images.

Exercise 18
Figure 18.1(a): © McGraw-Hill Education/Al Telser, photographer; Figure 18.1(b): © Carolina Biological Supply Company/ Phototake; Figure 18.2(c): © McGraw-Hill Education/ Al Telser, photographer; Figure 18.9: © McGraw-Hill Education/Al Telser, photographer; Figure18.10(b): © Steve Allen/ Getty Images RF.

Exercise 19
Figure 19.8(b): © Victor P. Eroschenko RF; Figure 19.11(c): © McGraw-Hill Education/ Al Telser, photographer; Figure 19.13 © Ed Reschke/ Photolibrary/Getty Images.

Exercise 24
Figure 24.5(c): © Francis Leroy, Biocosmos/Science Source; Figure 24.7(b): © McGraw-Hill Education/Al Telser, photographer.

Index

Note: Page references followed by the letters *f* and *t* indicate figures and tables, respectively. Unless otherwise indicated, all structures indexed are found in the human body.

A

B

D

E

F

G

H

I

J

K

L

N

O

P

S

T

U

V

W

X

Y

Z